J. Casillas, O. Cordón, F. Herrera, L. Magdalena (Eds.)

Interpretability Issues in Fuzzy Modeling

Springer-Verlag Berlin Heidelberg GmbH

Studies in Fuzziness and Soft Computing, Volume 128
http://www.springer.de/cgi-bin/search_book.pl?series=2941

Editor-in-chief
Prof. Janusz Kacprzyk
Systems Research Institute
Polish Academy of Sciences
ul. Newelska 6
01-447 Warsaw
Poland
E-mail: kacprzyk@ibspan.waw.pl

Further volumes of this series can be found on our homepage

Vol. 109. R.J. Duro, J. Santos and M. Graña (Eds.)
Biologically Inspired Robot Behavior Engineering, 2003
ISBN 3-7908-1513-6

Vol. 110. E. Fink 1. 112. Y. Jin
Advanced Fuzzy Systems Design and Applications, 2003
ISBN 3-7908-1523-3

Vol. 111. P.S. Szcepaniak, J. Segovia, J. Kacprzyk and L.A. Zadeh (Eds.)
Intelligent Exploration of the Web, 2003
ISBN 3-7908-1529-2

Vol. 112. Y. Jin
Advanced Fuzzy Systems Design and Applications, 2003
ISBN 3-7908-1537-3

Vol. 113. A. Abraham, L.C. Jain and J. Kacprzyk (Eds.)
Recent Advances in Intelligent Paradigms and Applications", 2003
ISBN 3-7908-1538-1

Vol. 114. M. Fitting and E. Orowska (Eds.)
Beyond Two: Theory and Applications of Multiple Valued Logic, 2003
ISBN 3-7908-1541-1

Vol. 115. J.J. Buckley
Fuzzy Probabilities, 2003
ISBN 3-7908-1542-X

Vol. 116. C. Zhou, D. Maravall and D. Ruan (Eds.)
Autonomous Robotic Systems, 2003
ISBN 3-7908-1546-2

Vol 117. O. Castillo, P. Melin
Soft Computing and Fractal Theory for Intelligent Manufacturing, 2003
ISBN 3-7908-1547-0

Vol. 118. M. Wygralak
Cardinalities of Fuzzy Sets, 2003
ISBN 3-540-00337-1

Vol. 119. Karmeshu (Ed.)
Entropy Measures, Maximum Entropy Principle and Emerging Applications, 2003
ISBN 3-540-00242-1

Vol. 120. H.M. Cartwright, L.M. Sztandera (Eds.)
Soft Computing Approaches in Chemistry, 2003
ISBN 3-540-00245-6

Vol. 121. J. Lee (Ed.)
Software Engineering with Computational Intelligence, 2003
ISBN 3-540-00472-6

Vol. 122. M. Nachtegael, D. Van der Weken, D. Van de Ville and E.E. Kerre (Eds.)
Fuzzy Filters for Image Processing, 2003
ISBN 3-540-00465-3

Vol. 123. V. Torra (Ed.)
Information Fusion in Data Mining, 2003
ISBN 3-540-00676-1

Vol. 124. X. Yu, J. Kacprzyk (Eds.)
Applied Decision Support with Soft Computing, 2003
ISBN 3-540-02491-3

Vol. 125. M. Inuiguchi, S. Hirano and S. Tsumoto (Eds.)
Rough Set Theory and Granular Computing, 2003
ISBN 3-540-00574-9

Vol. 126. J.-L. Verdegay (Ed.)
Fuzzy Sets Based Heuristics for Optimization, 2003
ISBN 3-540-00551-X

Vol 127. L. Reznik, V. Kreinovich (Eds.)
Soft Computing in Measurement and Information Acquisition, 2003
ISBN 3-540-00246-4

J. Casillas
O. Cordón
F. Herrera
L. Magdalena (Eds.)

Interpretability Issues in Fuzzy Modeling

Springer

Dr. Jorge Casillas
E-mail: casillas@decsai.ugr.es
Dr. Oscar Cordón
E-mail: ocordon@decsai.ugr.es
Dr. Francisco Herrera
E-mail: herrera@decsai.ugr.es
Dpto. Ciencias de la Computación
e Inteligencia Artificial
Escuela Técnica Superior
de Ingeniería Informática
Universidad de Granada
E - 18071 Granada
Spain

Dr. Luis Magdalena
E-mail: llayos@mat.upm.es
Dpto. Matemáticas Aplicadas
a las Tecnologías de la Información
Escuela Técnica Superior de Ingenieros
de Telecomunicación
Universidad Politécnica de Madrid
E - 28040 Madrid
Spain

DOI 10.1007/978-3-540-37057-4

Library of Congress Cataloging-in-Publication-Data applied for

A catalog record for this book is available from the Library of Congress.

Bibliographic information published by Die Deutsche Bibliothek
Die Deutsche Bibliothek lists this publication in the Deutsche Nationalbibliographie;
detailed bibliographic data is available in the internet at <http://dnb.ddb.de>.

http://www.springer.de

Originally published by Springer-Verlag Berlin Heidelberg New York in 2003
MyCopy version of the original edition 2003

Typesetting: camera-ready by editors
Cover design: E. Kirchner, Springer-Verlag, Heidelberg
Printed on acid free paper 62/3020/M - 5 4 3 2 1 0
www.springer.com/mycopy

Foreword

When I accepted the editors' invitation to write this foreword, I assumed that it would have been an easy task. At that time I did not realize the monumental effort that went into the organization and compilation of these chapters, the depth of each contribution, and the thoroughness with which the book's theme had been covered.

A foreword usually tries to impress upon the reader the importance of the book's main topic, placing the work within a comparative framework, and identifying the new trends or ideas that are pushing the state-of-the-art. While doing this, one also tries to relate the book's main theme to some personal experience that will help the reader understand the usefulness and applicability of the various contributions. I will do my best to achieve at least some of these lofty goals.

The need for trading off interpretability and accuracy is intrinsic to the use of fuzzy systems. Before the advent of soft computing, and in particular of fuzzy logic, accuracy was the main concern of model builders, since interpretability was practically a lost cause. In a recent article in which I reviewed hybrid Soft Computing (SC) systems and compared them with more traditional approaches [1], I remarked that the main reason for the popularity of soft computing was the synergy derived from its components. In fact, SC's main characteristic is its intrinsic capability to create hybrid systems that are based on the integration of constituent technologies. This integration provides complementary reasoning and searching methods that allow us to combine domain knowledge and empirical data to develop flexible computing tools and solve complex problems.

Soft Computing provides a different paradigm in terms of representation and methodologies, which facilitates these integration attempts. For instance, in classical control theory the problem of developing models is usually decomposed into system identification (or system structure) and parameter estimation. The former determines the order of the differential equations, while the latter determines its coefficients. In these traditional approaches, the main goal is the construction of accurate models, within the assumptions used for the model construction. However, the models' interpretability is very limited, given the rigidity of the underlying representation language.

The equation "**model = structure + parameters**"[1], followed by the traditional approaches to model building, does not change with the advent of soft computing. However, with soft computing we have a much richer repertoire to represent the structure, to tune the parameters, and to iterate this process. This repertoire enables us to choose among different trade-

[1] It is understood that the search method used to postulate the structures and find the parameter values is an important and implicit part of the above equation, and needs to be chosen carefully for efficient model construction.

offs between the model's interpretability and accuracy. For instance, one approach aimed at maintaining the model's transparency might start with knowledge-derived linguistic models, where the domain knowledge is translated into an initial structure and parameters. Then the model's accuracy could be improved by using global or local data-driven search methods to tune the structure and/or the parameters. An alternative approach aimed at building more accurate models might start with data-driven search methods. Then, we could embed domain knowledge into the search operators to control or limit the search space, or to maintain the model's interpretability. Post-processing approaches could also be used to extract more explicit structural information from the models.

This book provides a comprehensive yet detailed review of all these approaches. In the introduction the reader will find a general framework, within which these approaches can be compared, and a description of alternative methods for achieving different balances between models' interpretability and accuracy. The book is mainly focused on the achievement of the mentioned tradeoff by improving the interpretability in fuzzy modeling. It delves with the use of flexible rule structures to improve legibility, the issues of complexity reduction in linguistic or precise fuzzy models, the interpretability constraints in Takagi-Sugeno-Kang models, the use of measures to assess the interpretability loss, and the applicability of fuzzy rule-based models to interpret black-box models.

These topics are germane to many applications and resonate with recent issues that I have addressed. Therefore, I would like to illustrate the pervasiveness of this book's main theme by relating it to a personal experience. By virtue of working in an industrial research center, I am constantly faced with the constraints derived from real-world problems. There are situations in which the use of black-box models is not acceptable, due to legal or compliance reasons. On the other hand, the same situations require a degree of accuracy that is usually prohibitive for purely transparent models.

An example of such a situation is the automation of the insurance underwriting process, which consists in evaluating an applicant's medical and personal information to assess his/her potential risk and determine the appropriate rate class corresponding to such risk. To address this problem, we need to maintain full accountability of the model decisions, i.e. full transparency. This legal requirement, imposed by the states insurance commissioners, is necessary since the insurance companies need to notify their customers and explain to them the reasons for issuing policies that are not at the most competitive rates. Yet, the model must also be extremely accurate to avoid underestimating the applicants' risk, which would decrease the company's profitability, or overestimating it, which would reduce the company's competitive position in the market.

We solved this problem by creating several hybrid SC models, some of them transparent, for use in production, and some of them opaque, for use in quality assurance. The commonalities among these models are the tight integration of knowledge and data, leveraged in their construction, and the loose integration of their outputs, exploited in their off-line use. In different parts of this project we strived to achieve different balances between interpretability and accuracy. This project exemplifies the pervasiveness of the theme and highlights the timeliness of this book, which fills a void in the technical literature and describes a topic of extreme relevance and applicability.

Piero P. Bonissone
General Electric Global Research Center
Schenectady, New York, 12308, USA

[1] "Hybrid Soft Computing Systems: Industrial and Commercial Applications", P. P. Bonissone, Y-T Chen, K. Goebel and P. S. Khedkar, Proceedings of the IEEE, pp 1641-1667, vol. 87, no. 9, September 1999.

Preface

System modeling with fuzzy rule-based systems, i.e. fuzzy modeling, usually comes with two contradictory requirements in the obtained model: the *interpretability*, capability to express the behavior of the real system in an understandable way, and the *accuracy*, capability to faithfully represent the real system.

Obtaining high degrees of interpretability and accuracy is a contradictory purpose and, in practice, one of the two properties prevails over the other. While linguistic fuzzy modeling (mainly developed by linguistic fuzzy systems) is focused on the interpretability, precise fuzzy modeling (mainly developed by Takagi-Sugeno-Kang fuzzy systems) is focused on the accuracy.

The relatively easy design of fuzzy systems, their attractive advantages, and their emergent proliferation have made fuzzy modeling suffer a deviation from the seminal purpose directed towards exploiting the descriptive power of the concept of a linguistic variable. Instead, in the last few years, the prevailing research in fuzzy modeling has focused on increasing the accuracy as much as possible, paying little attention to the interpretability of the final model.

Nevertheless, a new tendency in the fuzzy modeling scientific community that looks for a good balance between interpretability and accuracy is increasing in importance. This searching of the desired trade-off is usually performed from two different perspectives, mainly using mechanisms to improve the interpretability of accurate fuzzy models, or to improve the accuracy of linguistic fuzzy models with a good interpretability. From both perspectives, a tendency emerges as one of the most important issues: the interpretability improvements. This book aims at present a state-of-the-art on the recent proposals that address it.

More specifically, the book presents the following structure. Section 1 introduces an overview of the different interpretability improvement mechanisms existing in the recent literature. Section 2 proposes interpretability improvement tools that consider alternative, more legible fuzzy rule structures. Sections 3 and 4, devoted to linguistic and precise fuzzy modeling respectively, are composed of a set of contributions showing how to improve the interpretability by, mainly, decreasing the complexity of the fuzzy models as a consequence of the reduction in the number of rules, variables, linguistic terms, etc. Section 5 shows several contributions that improve the interpretability of Takagi-Sugeno-Kang fuzzy systems by imposing constraints to their parameters. Section 6 collects a set of contributions that mainly propose new criteria to assess the interpretability loss. Finally, Section 7 presents two proposals that allow a translation from black-box models to fuzzy models, thus improving the interpretability of the former ones.

We believe that this volume presents an up-to-date state of the current research that will be useful for non expert readers, whatever their background, to easily get some knowledge about this area of research. Besides, it will also support those specialists who wish to discover the latest results as well as the latest trends in research work in fuzzy modeling.

Finally, we would like to express our most sincere gratitude to Springer-Verlag (Heidelberg, Germany) and in particular to Prof. J. Kacprzyk, for having given us the opportunity to prepare the text and for having supported and encouraged us throughout its preparation. We would also like to acknowledge our gratitude to all those who have contributed to the books by producing the papers that we consider to be of the highest quality. We also like to mention the somehow obscure and altruistic, though absolutely essential, task carried out by a group of referees (all the contributions have been reviewed by two of them), who, through their comments, suggestions, and criticisms, have contributed to raising the quality of this edited book.

Granada and Madrid (Spain)
January 2003

Jorge Casillas, Oscar Cordón,
Francisco Herrera, and Luis Magdalena

Table of Contents

6. ASSESSMENTS ON THE INTERPRETABILITY LOSS

7. INTERPRETATION OF BLACK-BOX MODELS AS FUZZY RULE-BASED MODELS

SECTION 1

OVERVIEW

Interpretability Improvements to Find the Balance Interpretability-Accuracy in Fuzzy Modeling: An Overview

Jorge Casillas[1], Oscar Cordón[1], Francisco Herrera[1], and Luis Magdalena[2]

[1] Department of Computer Science and Artificial Intelligence,
University of Granada, E-18071 Granada, Spain
e-mail: {casillas,ocordon,herrera}@decsai.ugr.es

[2] Department of Mathematics Applied to Information Technologies,
Technical University of Madrid, E-28040 Madrid, Spain
e-mail: llayos@mat.upm.es

Abstract. System modeling with fuzzy rule-based systems (FRBSs), i.e. fuzzy modeling (FM), usually comes with two contradictory requirements in the obtained model: the *interpretability*, capability to express the behavior of the real system in an understandable way, and the *accuracy*, capability to faithfully represent the real system. While linguistic FM (mainly developed by linguistic FRBSs) is focused on the interpretability, precise FM (mainly developed by Takagi-Sugeno-Kang FRBSs) is focused on the accuracy. Since both criteria are of vital importance in system modeling, the balance between them has started to pay attention in the fuzzy community in the last few years.

The chapter analyzes mechanisms to find this balance by improving the interpretability in linguistic FM: selecting input variables, reducing the fuzzy rule set, using more descriptive expressions, or performing linguistic approximation; and in precise FM: reducing the fuzzy rule set, reducing the number of fuzzy sets, or exploiting the local description of the rules.

1 Introduction

System modeling is the action and effect of approaching to a model, i.e., to a theoretical scheme that simplifies a real system or complex reality with the aim of easing its understanding. Thanks to these models, the real system can be explained, controlled, simulated, predicted, and even improved. The development of *reliable* and *comprehensible* models is the main objective in system modeling. If not so, the model loses its usefulness.

There are at least three different paradigms in system modeling. The most traditional approach is the *white box modeling*, which assumes that a thorough knowledge of the system's nature and a suitable mathematical scheme to represent it are available. As opposed to it, the *black box modeling* [60] is performed entirely from data using no additional a priori knowledge and considering a sufficiently general structure. Whereas the white box modeling has serious difficulties when complex and poorly understood systems are considered, the black box modeling deals with structures and associated parameters that usually do not have any physical significance [2]. Therefore, generally the former approach does not adequately obtain reliable models while the latter one does not adequately obtain comprehensible models.

A third, intermediate approach arises as a combination of the said paradigms, the *grey box modeling* [28], where certain known parts of the system are modeled considering the prior understood and the unknown or less certain parts are identified with black box procedures. With this approach, the mentioned disadvantages are palliated and a better balance between reliability and comprehensibility is attained.

Nowadays, one of the most successful tools to develop grey box models is *fuzzy modeling* (FM) [41], which is an approach used to model a system making use of a descriptive language based on fuzzy logic with fuzzy predicates [63]. FM usually considers model structures (fuzzy systems) in the form of fuzzy rule-based systems (FRBSs) and constructs them by means of different parametric system identification techniques. Fuzzy systems have demonstrated their ability for control [17], modeling [49], or classification [12] in a huge number of applications. The keys for their success and interest are the ability to incorporate human expert knowledge – which is the information mostly provided for many real-world systems and is described by vague and imprecise statements – and the facility to express the behavior of the system with a language easily interpretable by human beings. These interesting advantages allow them to be even used as mechanisms to interpret black box models such as neural networks [11].

As a system modeling discipline, FM is mainly characterized by two features that assess the quality of the obtained fuzzy models:

- *Interpretability* — It refers to the capability of the fuzzy model to express the behavior of the system in a understandable way. This is a subjective property that depends on several factors, mainly the model structure, the number of input variables, the number of fuzzy rules, the number of linguistic terms, and the shape of the fuzzy sets. With the term interpretability we englobe different criteria appeared in the literature such as *compactness*, *completeness*, *consistency*, or *transparency*.
- *Accuracy* — It refers to the capability of the fuzzy model to faithfully represent the modeled system. The closer the model to the system, the higher its accuracy. As closeness we understand the similarity between the responses of the real system and the fuzzy model. This is why the term approximation is also used to express the accuracy, being a fuzzy model a fuzzy function approximation model.

As Zadeh stated in its *Principle of Incompatibility* [75], "*as the complexity of a system increases, our ability to make precise and yet significant statements about its behavior diminishes until a threshold is reached beyond which precision and significance (or relevance) become almost mutually exclusive characteristics.*"

Therefore, to obtain high degrees of interpretability and accuracy is a contradictory purpose and, in practice, one of the two properties prevails over the other one. Depending on what requirement is mainly pursued, the FM field may be divided into two different areas:

- *Linguistic fuzzy modeling (LFM)* — The main objective is to obtain fuzzy models with a good interpretability.
- *Precise fuzzy modeling (PFM)* — The main objective is to obtain fuzzy models with a good accuracy.

The relatively easy design of fuzzy systems, their attractive advantages, and their emergent proliferation have made FM to suffer a deviation from the seminal purpose directed towards exploiting the descriptive power of the concept of a linguistic variable [75,76]. Instead, in the last few years, the prevailing research in FM has focused on increasing the accuracy as much as possible paying little attention to the interpretability of the final model.

Nevertheless, a new tendency in the FM scientific community that looks for a good balance between interpretability and accuracy is increasing in importance [3,9,54,65]. The aim of this chapter is to review some of the recent proposals that attempt to address this issue using mechanisms to improve the interpretability of fuzzy models.

The chapter is organized as follows. Section 2 analyzes the different existing lines of research related to the improvement of interpretability and accuracy to find a good balance in FM, Sect. 3 introduces the most important kinds of FRBSs used to improve their interpretability, Sect. 4 shows how to improve the interpretability of linguistic fuzzy models, Sect. 5 introduces tools to improve the interpretability of precise fuzzy models and, finally, Sect. 6 points out some conclusions.

2 Major Lines of Work

The two main objectives to be addressed in the FM field are *interpretability* and *accuracy*. Of course, the ideal thing would be to satisfy both criteria to a high degree but, since they are contradictory issues, it is generally not possible. In this case, more priority is given to one of them (defined by the problem nature), leaving the other one in the background. Hence, two FM approaches arise depending on the main objective to be considered: LFM (interpretability) and PFM (accuracy).

Regardless of the approach, a common scheme is found in the existing literature to perform the FM:

1. Firstly, the main objective (interpretability or accuracy) is tackled defining a specific model structure to be used, thus setting the FM approach.
2. Then, the modeling components (model structure and/or modeling process) are improved by means of different mechanisms to define the desired ratio interpretability-accuracy.

This procedure results in four different possibilities (see Fig. 1): LFM with improved interpretability, LFM with improved accuracy, PFM with improved interpretability, and PFM with improved accuracy.

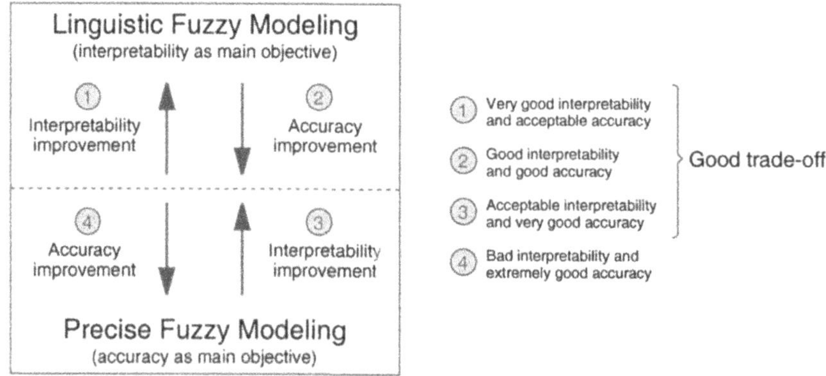

Fig. 1. Improvements of interpretability and accuracy in fuzzy modeling

Although historically more priority has been given to the accuracy, currently the search of a good balance between both criteria is increasing in importance. Indeed, a significative effort is being performed by several researchers proposing *improvement mechanisms* to compensate for the initial difference. Among the four said lines of work, clearly this philosophy is pursued by two of them: *LFM with improved accuracy* and *PFM with improved interpretability* (approaches 2 and 3 in Fig. 1, respectively).

Moreover, another interesting proposal is *LFM with improved interpretability* (approach 1 in Fig. 1). Although LFM uses a model structure with a high description power by itself, there are some problems (curse of dimensionality, excessive number of input variables or fuzzy rules, garbled fuzzy sets, etc.) that make it not to be as interpretable as desired and the need of interpretability improvements to restore the searched balance is justified.

Finally, the modus operandi of obtaining more accuracy in PFM (approach 4 in Fig. 1) does not pay attention to the comprehensibility of the model and acts close to black box techniques. This approach does not follow the original objective of FM and does not profit from the advantages that distinguish it from other modeling techniques. Although the approach is useful when only accuracy is required, it goes away from the aim of the present book.

This chapter is devoted to review different interpretability improvements that have been proposed to attain the desired balance. Thus, Sects. 4 and 5 show some mechanisms found in the recent literature to do so. In [27], an overview from a different point of view is explored by analyzing the interpretability of several proposals instead of considering how the balance interpretability-accuracy is achieved.

3 Types of Fuzzy Rule-Based Systems

Before presenting the search of a balance interpretability-accuracy in FM, it seems that there is need to introduce the different kinds of FRBSs usually employed. It is a significant aspect to consider since depending on the rule structure used, an FRBS has itself a specific capability of description and approximation. The section is only focused on the FRBS types usually considered to improve their interpretability for the sake of a good trade-off.

3.1 Linguistic Fuzzy Rule-Based System

Also known as Mamdani-type FRBS [44,45], the linguistic FRBS constitutes the main tool to develop LFM. A crucial reason why this approach is worth considering is that it may remain verbally interpretable, playing the concept of linguistic variable [76] a central role. Linguistic FRBSs are formed by linguistic rules with the following structure:

IF X_1 is A_1 and ... and X_n is A_n
THEN Y_1 is B_1 and ... and Y_m is B_m ,

with X_i and Y_j being input and output linguistic variables respectively, and with A_i and B_j being linguistic labels with fuzzy sets associated defining their meaning. These linguistic labels will be taken from a global *semantic* defining the set of possible fuzzy sets used for each variable (Fig. 2 shows an example with triangular membership functions). This structure provides a natural framework to include expert knowledge in the form of fuzzy rules.

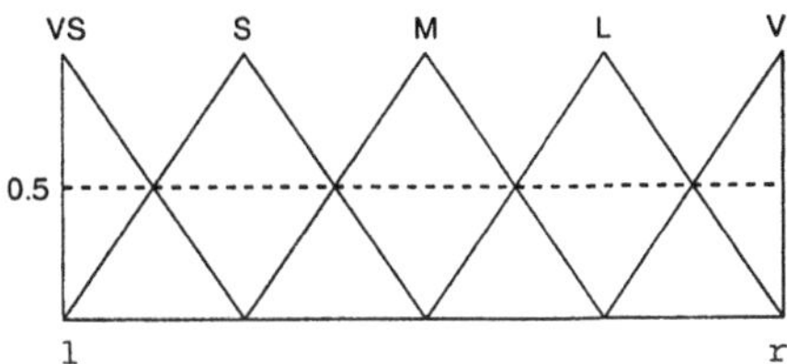

Fig. 2. Graphical representation of an example of the semantic considered for a variable, standing VS for *very small*, S for *small*, M for *medium*, L for *large*, and VL for *very large*, with $[l, r]$ being the corresponding variable domain

In these systems, the knowledge base (KB) – the component of the FRBS that stores the knowledge about the problem being solved – is composed of:

- the *rule base* (RB), constituted by the collection of linguistic rules themselves joined by means of the connective *also*, and
- the *data base* (DB), containing the term sets and the membership functions defining their semantics.

3.2 Takagi-Sugeno-Kang-type Fuzzy Rule-Based Systems

The system proposed by Takagi, Sugeno, and Kang [62,64] (shortly called TSK) differs from the linguistic one in the use of a different consequent structure. While linguistic rules consider a linguistic variable in the consequent, TSK-type fuzzy rules are based on representing the output variables as polynomial functions of the input variables, i.e.,

IF X_1 is A_1 and ... and X_n is A_n
THEN $Y_1 = p_1(X_1, \ldots, X_n)$ and ... and $Y_m = p_m(X_1, \ldots, X_n)$,

with $p_j(\cdot)$ being the polynomial function defined for the jth output variable. Using this fuzzy rule structure, the human interpretation on the action suggested by each rule is garbled but, on the contrary, the approximation capability is significantly increased. For this reason, TSK-type FRBSs are very useful in PFM.

3.3 Other Kinds of Fuzzy Rule-Based Systems

Other model structures may be considered apart from the two main types mentioned, some examples follow:

- The *singleton FRBS*, where the rule consequent takes a single real-valued number, may be considered as a particular case of the linguistic FRBS (the consequent is a fuzzy set where the membership function is one for a specific value and zero for the remaining ones) or of the TSK-type FRBS (the polynomial function of the consequent is a constant). Since the single consequent seems to be more easily interpretable than a polynomial function, the singleton FRBS may be used to develop LFM. Nevertheless, compared with the linguistic FRBS, the fact of having a different consequent value for each rule (no global semantic is used for the output variable) worsens the interpretability.
- The *fuzzy rule-based classification system*, which is an automatic classification system that uses fuzzy rules as knowledge representation tool. The classical fuzzy classification rule structure is the one that have a class label in the consequent part instead of the mentioned fuzzy set. Other alternative representations that consider a certainty degree for each rule or that include all the possible class labels with their corresponding certainty degrees in the consequent part are usually also considered.

- The *approximate FRBS*, which differs from the linguistic one in the direct use of fuzzy variable [1,7,15,36,61]. Each fuzzy rule thus presents its own semantic, i.e., the variables take different fuzzy sets as values and not linguistic terms from a global term set. The fuzzy rule structure is then as follows:

  IF X_1 is $\widehat{A}_1$ and ... and X_n is $\widehat{A}_n$
  THEN Y_1 is $\widehat{B}_1$ and ... and Y_m is $\widehat{B}_m$,

  with $\widehat{A}_i$ and $\widehat{B}_j$ being fuzzy sets. Since no global semantic is used in approximate FRBSs, these fuzzy sets can not be interpreted. This more flexible structure allows the model to be more accurate, being very appropriate to develop PFM.
  Other names have been proposed by different authors to designate approximate FRBSs. Among others, we may find FRBSs with *local fuzzy sets* [7], *rule-based* FRBSs [14], or *scatter-partitioning* FRBSs [22].

Moreover, different extensions to the FRBSs such as adding rule weights [20,74] or using disjunctive forms [13,42,24] have also been proposed.

4 Improving the Interpretability in Linguistic Fuzzy Modeling

A possibility to achieve a good trade-off between interpretability and accuracy is to perform an LFM process trying to obtain accurate initial models, and subsequently applying a process to improve the interpretability of the obtained model even at the expense of losing certain accuracy. To generate these accurate initial models in LFM, a large number of input variables, a great variety of linguistic terms, oversized RBs, or illegible fuzzy sets are usually considered. This section analyzes different mechanisms to improve the interpretability in these kinds of models. Moreover, there is always the chance of indirectly improving the accuracy in terms of generalization capability when removing the existing redundancies and inconsistencies.

4.1 Selecting Input Variables in the Model and/or in the Linguistic Rules

When managing high-dimensional problems with a large number of input variables, the RB suffers from an exponential growth in its size due to the homogeneous partitioning of the input and output spaces caused by the use of linguistic variables [4] and, therefore, a good interpretability is not guaranteed. Moreover, with an excessive number of input variables, every linguistic rule also loses part of its description ability since the understanding of the condition to activate the rule comes more difficult. A solution to these disadvantages is to make an input variable selection process that reduces the number of variables used by the model.

Basically, we may distinguish between two variable selection processes:

- *Selecting input variables in the model* — This simplification task involves selecting a subset of input variables to be used in the model or, similarly, removing those input variables that do not significantly contribute to the FRBS performance.
  Usually, this variable (or feature) selection has been applied to LFM in classification [8,30,35,40,58,59], where a large number of variables is frequently tackled. Nevertheless, in [34] an interesting contribution is proposed for linguistic FRBSs including the variable selection within a more complex deriving process (with rule generation, DB tuning, and rule selection).
- *Selecting input variables in the linguistic rules* — Other innovative approach involves the selection of a subset of input variables for each rule. In this case, the antecedent length of each rule is variable avoiding the need of using all the variables involved in the system. It does not mean that a specific input variable ignored in a rule could not be used in another one. To manage with this structure in the inference process, a special linguistic term with a membership function with a value one in all the domain may be assigned to the ignored variables.
  In [35], this input variable selection at RB level is considered together with a global selection of the variables and a merging of the rules. On the other hand, the methods proposed in [10,26,43,68] obtain the most significative input variable for each rule during the learning process, instead of making an a priori variable selection.

It is important to emphasize that the fact of removing some input variables may cause the existence of several rules with identical antecedents (mainly when the selection is performed a posteriori) that could imply an inconsistence. In this case, the most usual solution is to merge these rules. This process is explained in the following section.

4.2 Selecting/Merging Linguistic Rules

Sometimes, an RB with an excessive size must be used to reach an acceptable accuracy degree in linguistic FRBSs. However, this effect is often caused by a deficient RB learning process (sometimes advisedly) with tendency to generate too many rules. Thus, in an RB we may find *redundant rules*, which do not contain relevant information and whose actions are covered by other rules; *erroneous rules*, which are wrong defined and distort the FRBS performance; and *conflictive rules*, which perturb the FRBS performance when coexist with others. Besides worsening the accuracy, an excessive number of rules makes difficult to understand the model behavior.

To face this problem, an *RB reduction process* can be developed by merging rules and/or selecting a subset of rules from a given RB to achieve the goal of minimizing the number of rules used while maintaining (or even improving) the FRBS performance. Indeed, depending on the criteria considered to reduce the RB, this process can be considered as a mechanism to improve not only the interpretability but also the accuracy.

The RB reduction is generally applied as a postprocessing stage, once an initial RB has been derived. We may distinguish between two approaches to reduce the fuzzy rule set size in order to obtain a *compact* RB:

- *Selecting linguistic rules* — It involves obtaining an optimized subset of rules from a previous RB by selecting some of them. We may find several methods to do so with different search algorithms in the specialized literature [15,16,23,29,32,33,38].
  In [39], an interesting heuristic rule selection procedure is proposed where, by means of statistical measures, a relevance factor is computed for each fuzzy rule composing the linguistic FRBSs to subsequently select the most relevant ones. The philosophy of ordering the rules with respect to an importance criterion and selecting a subset of them seems similar to the orthogonal transformation-methods used for TSK-type FRBSs [72] (explained in Sect. 5.1). Another heuristic rule selection procedure is proposed in [67].
- *Merging linguistic rules* — It is an alternative approach that reduces the RB by merging the existing linguistic rules. In [35], the authors propose to merge neighboring rules, i.e., linguistic rules where the linguistic terms used by the same variable in each rule are adjacent. The merge is performed in three different ways: using a new fuzzy set that groups the adjacent linguistic terms, merging the adjacent fuzzy sets if they are very similar, or giving the set of rules in disjunctive normal form. Another proposal is presented in [31], where a special consideration to the merging order is made.
  From a different point of view, the RB may be reduced by using a disjunctive form for the fuzzy rules that groups several rules within a more general expression, thus easing the interpretability. This approach is explained in next section.

4.3 Alternative Linguistic Rule Expressions

Another possibility to improve the interpretability of linguistic FRBSs is to use an alternative model structure to give a higher descriptive power to each rule. With this extended description we may represent a linguistic fuzzy model in a more compact structure with minor accuracy loss. To do that, the linguistic fuzzy rule expression is extended to make it more flexible. Some examples are shown in the following:

- *Disjunctive normal form (DNF)* — The DNF-type fuzzy rule has the following form [24]:

  IF X_1 is $\widetilde{A_1}$ and ... and X_n is $\widetilde{A_n}$ **THEN** Y is B

  where each input variable X_i takes as a value a set of linguistic terms $\widetilde{A_i} = \{A_{i1} \; or \; \ldots \; or \; A_{il_i}\}$, whose members are joined by a disjunctive operator, whilst the output variable remains a usual linguistic variable with a single label associated.

  This structure uses a more compact description that improves the interpretability. Moreover, the structure is a natural support to allow the absence of some input variables in each rule (simply making $\widetilde{A_i}$ be the whole set of linguistic terms). Several learning methods have been proposed following this rule structure [10,24–26,35,42,43,68].

- *Exception rules* — Another interesting possibility to represent a more interpretable and compact description is the use of exceptions [37]. In [6], the authors make a fine-tuning of the meaning of each linguistic rule by excluding a local region of its firing region. This consideration is especially useful when structures with multiple fuzzy input subspaces (e.g., the DNF one) are considered. An example follows:

  IF X_1 is {Big or Small} and X_2 is {Medium or Small} **THEN** Y is Big
  except **IF** X_1 is Small and X_2 is Medium

- *Union-rule configuration* — In [13], an attempt to palliate the *curse of dimensionality* problem (exponential growth in the number of rules when a large number of input variables are considered) is proposed by converting a multiple-input-variable linguistic rule into single-input-variable linguistic rules connected by the disjunction operator.

4.4 Linguistic Approximation

The linguistic approximation [19] lies in finding a linguistic description that represents a given fuzzy set. Given a linguistic FRBS where the DB has been automatically obtained or optimized, this procedure may be used in LFM to find an interpretation of the involved fuzzy sets to improve the comprehensibility of the model. During this linguistic approximation, a certain accuracy loss is assumed.

The linguistic approximation is usually performed with linguistic terms and sometimes linguistic modifiers are used as well. Some examples of methods that improve the interpretability of the model with linguistic approximation are [18,46,63].

5 Improving the Interpretability in Precise Fuzzy Modeling

The birth of more flexible FRBSs such as TSK or approximate ones also entails the eruption of PFM since the new structures allow the FM to achieve more accurate fuzzy models. This fact causes a shift from the seminal intent of FM and the modeling tasks with these kinds of FRBSs increasingly become black box processes.

Fortunately, nowadays there is a sense shared by several researches in the way of rescuing the good interpretability advantages offered by fuzzy systems. This interpretability consideration is usually attained by reducing the complexity of the model. On the other hand, approaches that improve the local description of the TSK-type fuzzy rules are also proposed. In the following subsections, some specific proposals are reviewed.

5.1 Ordering/Selecting TSK-type Fuzzy Rules

As mentioned in Sect. 4, an efficient way to improve the interpretability in FM is to select a subset of significative fuzzy rules that represent in a more compactly way the system to be modeled. Moreover, this selection of important rules has the interesting advantage of reducing the possible redundancy existing in the RB, thus improving the generalization capability of the system, i.e., its accuracy.

Recently, one of the most successful approaches to make such rule selection in TSK-type FRBS has been proposed by obtaining a subset of important fuzzy rules considering orthogonal transformations [47,48,55,57,66,69–72]. This mechanism is used to give an importance degree to each fuzzy rule, thus obtaining an ordering of them. Once they are sorted, the selection is achieved using only the most promising ones.

Given a previously defined RB, let us assume we have a matrix that allocates the firing degree of each rule for each training example considered:

$$F = \begin{bmatrix} f_{11} & f_{21} & \cdots & f_{r1} \\ f_{12} & f_{22} & \cdots & f_{r2} \\ \vdots & \vdots & \vdots & \vdots \\ f_{1N} & f_{2N} & \cdots & f_{rN} \end{bmatrix}$$

with f_{ij} being the normalized firing degree of the i-th rule when the j-th example is used, r the number of rules, and N the data set size. The relevance of each rule (column) may be analyzed by means of orthogonal transformations of this matrix.

The two orthogonal transformations and the most usual extensions to select a subset of TSK-type fuzzy rules are:

- *Orthogonal least-squares methods* (OLS) — The OLS-based method (whose first application to fuzzy rule selection was proposed in [66]) transforms the columns of the firing matrix F into a set of orthogonal basis vectors. With them, the individual error reduction ratio of each rule may be easily computed, thus defining its importance in the whole set of possible rules. Its main interest in system modeling is that it considers the output contribution of the rules to sort them. However, since the OLS-based method is guided by the approximation capabilities of the rules (fitting error) without paying attention to the premise structures, it is possible to give a high importance to redundant fuzzy rules with high firing degrees thanks to their contributions to the output [72].
  In [55], this drawback is faced considering the dependency between the current rule to be selected and the set of rules previously selected. If the firing vector corresponding to the current fuzzy rule is (or nearly is) a linear combination of the firing vectors corresponding to the previously selected rules, a low importance degree is assigned.
  In [47], another improvement to the basic OLS approach is made to consider the RB redundancy. Each time a new rule is selected, its similarity to the previously rules is analyzed. If it significantly differs from the others, the rule is added. Otherwise, the previously selected rule being similar to the current one is properly updated considering the latter.
- *Singular value decomposition and QR with column pivoting methods* (SVD-QR) — The first application of the SVD-QR method to the selection of the most important fuzzy rules was proposed in [48]. Firstly, the SVD algorithm obtains a factorization of the firing matrix F into a product of three matrices. The obtained information will determine the rules to be considered to construct the reduced RB. Then, QR with column pivoting is applied to determine the most important fuzzy rules.
  In [70], this method is improved disregarding the QR process to order the selected rules and obtaining this information directly from the SVD result (from the singular values). The main advantage of this improvement is its simplicity in terms of implementation and computational time consumption.
  In [55], the authors support the opinion that methods based on SVD fails to produce an importance ordering and they propose an optional solution only considering the QR process.

Generally, the final objective of orthogonal transformation-based methods is to order the candidate fuzzy rules for subsequently selecting the most important ones. Usually, the number of selected rules is established as a rule of thumb. However, in [57,71], statistical information criteria are used to automatically decide the number of rules to reduce the human intervention and consider a proper trade-off between simplicity of the model (interpretability) and data approximation (accuracy). Moreover, the use of orthogonal transformation has been used to perform other kinds of FM tasks with TSK-type FRBSs, such as the estimation of the consequent model parameters [47,73].

On the other hand, an interesting approach different from the orthogonal transformation one is proposed in [20] to make the TSK-type fuzzy rule selection. In this case, the model structure is extended incorporating an intensity parameter to each rule that allows the method to give different importance degrees during the inference process. These parameters could be considered as rule weights in linguistic FRBSs. This improvement makes the model more flexible giving it a higher approximation capability. To select the rules, the method considers an initial oversized RB and an iterative algorithm progressively removes the most redundant fuzzy rules.

Finally, we should say that an RB reduction is also indirectly attained when the fuzzy sets are merged or removed. The following section focuses on this approach.

5.2 Merging/Removing Fuzzy Sets in Precise Fuzzy Rule-Based Systems

Other successful way to obtain precise FRBSs (basically TSK or approximate ones) with a better interpretability is to reduce the number of rules by merging the fuzzy sets involved in the system. In this section, two different approaches are introduced in the following. While the former one tries to simplify TSK-type FRBSs, the latter one starts from a linguistic FRBSs with an excessive number of rules and fuzzy sets and generates a compact approximate FRBSs that is mostly equivalent.

The interpretability of TSK-type FRBSs may be improved by removing those fuzzy sets that, after an automatic adaptation and/or acquisition, do not significatively contribute to the model behavior. There are two effects caused by the fuzzy sets composing an FRBS that make the model unnecessarily more complex [52]:

- *Redundancy* — It refers to the coexistence of similar fuzzy sets representing compatible concepts. With these kinds of fuzzy sets, the model becomes more complex and difficult to be understandable (the distinguishability property [65] is not met).
- *Irrelevancy* — It is given when fuzzy sets with a constant membership degree equal to one, or close to it, are used. These kinds of fuzzy sets do not furnish relevant information.

To automatically detect these undesired fuzzy sets, the use of similarity measures between fuzzy sets has been proposed [51–53,56]. To properly use these measures for a RB reduction process, several properties - such as a similarity value of zero for nonoverlapping fuzzy sets, a value greater than zero for overlapping fuzzy sets, a value equal to one for equal fuzzy sets, and a measure independent of the scaling domain - must be satisfied [52]. For example, the following similarity measure between the fuzzy sets A and B for a discrete domain is recommended to reduce the RB:

$$S(A,B) = \frac{\sum_{j=1}^{m} Min(\mu_A(x_j), \mu_B(x_j))}{\sum_{j=1}^{m} Max(\mu_A(x_j), \mu_B(x_j))}.$$

The process to simplify the model consists of two steps:

1. *Merging/removing fuzzy sets* — Those fuzzy sets with a high degree of similarity are merged to a unique fuzzy set that represent the collection of similar fuzzy sets. On the other hand, those irrelevant fuzzy sets – i.e., the ones with a high similarity degree to the universal set (a fuzzy set with a membership degree constantly one) – are removed. This step is called RB simplification [53].
2. *Merging fuzzy rules* — Moreover, it is interesting to mention that the fact of reducing the number of fuzzy sets in a variable fuzzy partition might result in rules with equal antecedents that can also be merged. This step is called RB reduction [53].

Hence, the precise fuzzy model go through an interpretability improvement (or complexity reduction) process that make it less complex (more compact) and more easily interpretable (more transparent).

Another interesting approach that also merges fuzzy sets is proposed in [61]. In this case, linguistic FRBSs with a bad interpretability are transformed into compact approximate FRBSs to develop PFM. Firstly, an iterative algorithm generates fuzzy partitions with a number of fuzzy sets large enough to achieve the desired accuracy degree, thus deliberately generating a linguistic RB with an excessive size. Subsequently, a merging process reduces it without losing accuracy by combining linguistic fuzzy rules that have adjacent fuzzy sets, thus obtaining a set of approximate fuzzy rules where each one has its own semantic.

5.3 Exploiting the Local Description of TSK-type Fuzzy Rules

In system modeling, a TSK-type FRBS is usually considered as the combination of simple models (the rules) that describe local behaviors of the system to be modeled. Hence, insofar as each TSK-type fuzzy rule is either forced to have a smoother consequent polynomial function or to develop an isolated action, the interpretability will be improved. Several contributions follow these approaches to make TSK models more comprehensible:

- *Smoothing the consequent polynomial function* — For example, in [21] the author proposes a method that imposes several constraints to the weights involved in the polynomial function of each rule consequent:

$$w_0 = 0, \quad \sum_{j=1}^{n} w_j = 1, \quad w_j \geq 0 \; (j = 1, \ldots, n)$$

with w_j being the weight of the rule consequent polynomial function corresponding to the j-th input variable, w_0 the independent term, and n the number of input variables. Thanks to this, a convex combination of the input variables is performed, thus contributing to a better understanding of the model. Like in the previously mentioned contribution [20], a parameter associated with each rule is used to modulate its action.

Another approach that softens the consequent polynomial functions is proposed in [73]. To do that, an objective function that properly combines two criteria during the regression algorithm is used. On the one hand, the classical mean square error is considered as global measure error to evaluate the quality of the rule set, thus favoring the cooperation. On the other hand, a local error measure is used to induce competition among the rules. While the former criterion increases the accuracy, the latter allows the rules to describe each region better, thus improving the interpretability. Figure 3 shows two models with different description capabilities depending on the used polynomial functions.

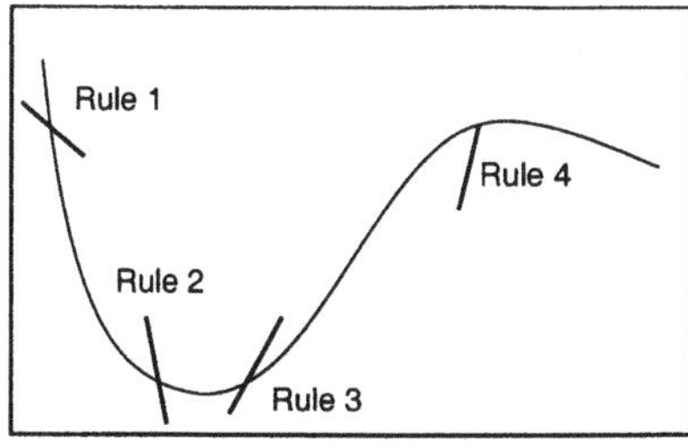

(a) Model with garbled rules

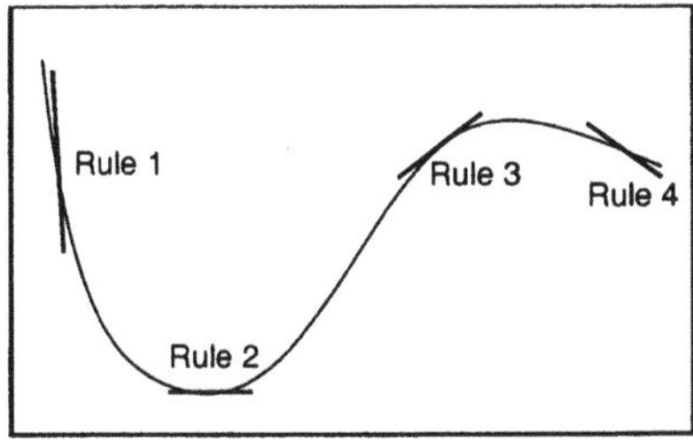

(b) Model with understandable rules

Fig. 3. The local interpretability of a TSK-type fuzzy model may be improved with smooth consequent polynomial functions

- *Isolating the fuzzy rule actions* — In [50], an study concludes that the description of each TSK-type fuzzy rule is improved when the overlapping between adjacent input fuzzy sets is reduced. This is because the performance region of a particular rule is more clearly defined by avoiding that other rules having a high firing degree in such an area. This approach is an alternative proposal to improve the local description of the TSK-type rules by designing the DB instead of the RB.

The proposal in [5] tries to englobe the two said philosophies for improving the local interpretability of TSK-type FRBSs. On the one hand, the use of a special type of membership function based on splines improves the local behavior of the fuzzy system by only firing the most immediate rules to the given input vector. On the other hand, the consequent polynomial structure is modified to interpret the coefficients as a Taylor series expansion around the center of the corresponding rule (i.e., the vector containing the vertex of each membership function considered for each input variable in the fuzzy rule).

6 Concluding Remarks

The FM research developed in the last two decades was mainly focused on exploiting the flexibility of FM to obtain the maximum accuracy. During this evolution, the derivation methods were improved, the components to be designed were extended, and new model structures were proposed. This search of the accuracy usually set aside the interpretability of the obtained models.

However, we should remember the initial philosophy of fuzzy set theory directed to serve the bridge between the human understanding and the machine processing. In this challenge, the faculty of fuzzy models to express the behavior of the real system in a comprehensible manner acquires a great importance. This is why the current tendency in FM tries to find a better balance between interpretability and accuracy.

This equilibrium is attained from different perspectives. One of the things that attracts the eye is the fact that it is frequently performed by means of previous existing extensions, but used in a more rational and moderate way. Other times, however, new approaches explicitly proposed are considered. This chapter was aimed to present an introduction to the different tends recently proposed in the specialized literature to improve the interpretability degree of the fuzzy models with the objective of finding the desired trade-off.

The remaining 26 chapters contained in this volume are excellent works of research in the FM approach studied in this chapter and they properly represent the existing state-of-the-art.

References

1. R. Alcalá, J. Casillas, O. Cordón, and F. Herrera. Building fuzzy graphs: features and taxonomy of learning for non-grid-oriented fuzzy rule-based systems. To appear in *Journal of Intelligent and Fuzzy Systems*. Draft version available at `http://decsai.ugr.es/~casillas/`.
2. R. Babuška. *Fuzzy modeling for control*. Kluwer Academic, Norwell, MA, USA, 1998.
3. R. Babuška, H. Bersini, D.A. Linkens, D. Nauck, G. Tselentis, and O. Wolkenhauer. Future prospects for fuzzy systems and technology. ERUDIT Newsletter Vol. 6, No. 1. Aachen, Germany, 2000. Available at `http://www.erudit.de/erudit/newsletters/news_61/page5.htm`.
4. A. Bastian. How to handle the flexibility of linguistic variables with applications. *International Journal of Uncertainty, Fuzziness and Knowledge-Based Systems*, 2(4):463–484, 1994.

5. M. Bikdash. A highly interpretable form of Sugeno inference systems. *IEEE Transactions on Fuzzy Systems*, 7(6):686–696, 1999.
6. P. Carmona, J.L. Castro, and J.M. Zurita. Learning maximal structure fuzzy rules with exceptions. In *Proceedings of the 2nd International Conference in Fuzzy Logic and Technology*, pages 113–117, Leicester, UK, 2001.
7. B. Carse, T.C. Fogarty, and A. Munro. Evolving fuzzy rule based controllers using genetic algorithms. *Fuzzy Sets and Systems*, 80:273–294, 1996.
8. J. Casillas, O. Cordón, M.J. del Jesus, and F. Herrera. Genetic feature selection in a fuzzy rule-based classification system learning process for high dimensional problems. *Information Sciences*, 136(1-4):169–191, 2001.
9. J. Casillas, O. Cordón, and F. Herrera. Can linguistic modeling be as accurate as fuzzy modeling without losing its description to a high degree? Technical Report #DECSAI-00-01-20, Department of Computer Science and Artificial Intelligence, University of Granada, Granada, Spain, 2000. Available at `http://decsai.ugr.es/~casillas/`.
10. J.L. Castro, J.J. Castro-Schez, and J.M. Zurita. Learning maximal structure rules in fuzzy logic for knowledge acquisition in expert systems. *Fuzzy Sets and Systems*, 101(3):331–342, 1999.
11. J.L. Castro, C.J. Mantas, and J.M. Benítez. Interpretation of artificial neural networks by means of fuzzy rules. *IEEE Transactions on Neural Networks*, 13(1):101–116, 2002.
12. Z. Chi, H. Yan, and T. Pham. *Fuzzy algorithms with application to image processing and pattern recognition.* World Scientific, Singapore, 1996.
13. W.E. Combs and J.E. Andrews. Combinatorial rule explosion eliminated by a fuzzy rule configuration. *IEEE Transactions on Fuzzy Systems*, 6(1):1–11, 1998.
14. M.G. Cooper and J.J. Vidal. Genetic design of fuzzy controllers: the cart and jointed pole problem. In *Proceedings of the 3rd IEEE International Conference on Fuzzy Systems*, pages 1332–1337, Piscataway, NJ, USA, 1994.
15. O. Cordón and F. Herrera. A three-stage evolutionary process for learning descriptive and approximate fuzzy logic controller knowledge bases from examples. *International Journal of Approximate Reasoning*, 17(4):369–407, 1997.
16. O. Cordón and F. Herrera. A proposal for improving the accuracy of linguistic modeling. *IEEE Transactions on Fuzzy Systems*, 8(3):335–344, 2000.
17. D. Driankov, H. Hellendoorn, and M. Reinfrank. *An introduction to fuzzy control.* Springer-Verlag, Heidelberg, Germany, 1993.
18. A. Dvořák. On linguistic approximation in the frame of fuzzy logic deduction. *Soft Computing*, 3(2):111–116, 1999.
19. F. Eshragh and E.H. Mamdani. A general approach to linguistic approximation. In E.H. Mamdani and B.R. Gaines, editors, *Fuzzy Reasoning and its Applications*, pages 169–187. Academic Press, London, UK, 1981.
20. A. Fiordaliso. Autostructuration of fuzzy systems by rules sensitivity analysis. *Fuzzy Sets and Systems*, 118(2):281–296, 2001.
21. A. Fiordaliso. A constrained Takagi-Sugeno fuzzy system that allows for better interpretation and analysis. *Fuzzy Sets and Systems*, 118(2):307–318, 2001.
22. B. Fritzke. Incremental neuro-fuzzy systems. In B. Bosacchi, J.C. Bezdek, and D.B. Fogel, editors, *Proceedings of the International Society for Optical Engineering: Applications of Soft Computing*, volume 3165, pages 86–97, 1997.
23. A.F. Gómez-Skarmeta and F. Jiménez. Fuzzy modeling with hybrid systems. *Fuzzy Sets and Systems*, 104(2):199–208, 1999.

24. A. González and R. Pérez. Completeness and consistency conditions for learning fuzzy rules. *Fuzzy Sets and Systems*, 96(1):37–51, 1998.
25. A. González and R. Pérez. SLAVE: a genetic learning system based on an iterative approach. *IEEE Transactions on Fuzzy Systems*, 7(2):176–191, 1999.
26. A. González and R. Pérez. Selection of relevant features in a fuzzy genetic learning algorithm. *IEEE Transactions on Systems, Man, and Cybernetics—Part B: Cybernetics*, 31(3):417–425, 2001.
27. S. Guillaume. Designing fuzzy inference systems from data: an interpretability-oriented review. *IEEE Transactions on Fuzzy Systems*, 9(3):426–443, 2001.
28. K.M. Hangos, editor. *Special issue on grey box modelling*, volume 9(6) of *International Journal of Adaptive Control and Signal Processing*. John Wiley & Sons, New York, NY, USA, 1995.
29. F. Herrera, M. Lozano, and J.L. Verdegay. A learning process for fuzzy control rules using genetic algorithms. *Fuzzy Sets and Systems*, 100:143–158, 1998.
30. T.-P. Hong and J.-B. Chen. Finding relevant attributes and membership functions. *Fuzzy Sets and Systems*, 103(3):389–404, 1999.
31. T.-P. Hong and C.-Y. Lee. Effect of merging order on performance of fuzzy induction. *Intelligent Data Analysis*, 3(2):139–151, 1999.
32. H. Ishibuchi, T. Murata, and I.B. Türkşen. Single-objective and two-objective genetic algorithms for selecting linguistic rules for pattern classification problems. *Fuzzy Sets and Systems*, 89(2):135–150, 1997.
33. H. Ishibuchi, K. Nozaki, N. Yamamoto, and H. Tanaka. Selecting fuzzy if-then rules for classification problems using genetic algorithms. *IEEE Transactions on Fuzzy Systems*, 3(3):260–270, 1995.
34. Y. Jin. Fuzzy modeling of high-dimensional systems: complexity reduction and interpretability improvement. *IEEE Transactions on Fuzzy Systems*, 8(2):212–221, 2000.
35. A. Klose, A. Nurnberger, and D. Nauck. Some approaches to improve the interpretability of neuro-fuzzy classifiers. In *Proceedings of the 6th European Congress on Intelligent Techniques and Soft Computing*, pages 629–633, Aachen, Germany, 1998.
36. L. Koczy. Fuzzy if ... then rule models and their transformation one another. *IEEE Transactions on Systems, Man, and Cybernetics*, 26(5):621–637, 1996.
37. A. Krone and H. Kiendl. Automatic generation of positive and negative rules for two-way fuzzy controllers. In *Proceedings of the Second European Congress on Intelligent Techniques and Soft Computing*, volume 1, pages 438–447, Aachen, Germany, 1994. Verlag Mainz.
38. A. Krone, P. Krause, and T. Slawinski. A new rule reduction method for finding interpretable and small rule bases in high dimensional search spaces. In *Proceedings of the 9th IEEE International Conference on Fuzzy Systems*, pages 693–699, San Antonio, TX, USA, 2000.
39. A. Krone and H. Taeger. Data-based fuzzy rule test for fuzzy modelling. *Fuzzy Sets and Systems*, 123(3):343–358, 2001.
40. H.-M. Lee, C.-M. Chen, J.-M. Chen, and Y.-L. Jou. An efficient fuzzy classifier with feature selection based on fuzzy entropy. *IEEE Transactions on Systems, Man, and Cybernetics—Part B: Cybernetics*, 31(3):426–432, 2001.
41. P. Lindskog. Fuzzy identification from a grey box modeling point of view. In H. Hellendoorn and D. Driankov, editors, *Fuzzy model identification*, pages 3–50. Springer-Verlag, Heidelberg, Germany, 1997.

42. L. Magdalena. Adapting the gain of an FLC with genetic algorithms. *International Journal of Approximate Reasoning*, 17(4):327–349, 1997.
43. L. Magdalena and F. Monasterio-Huelin. A fuzzy logic controller with learning through the evolution of its knowledge base. *International Journal of Approximate Reasoning*, 16(3):335–358, 1997.
44. E.H. Mamdani. Applications of fuzzy algorithms for control a simple dynamic plant. *Proceedings of the IEE 121*, 12:1585–1588, 1974.
45. E.H. Mamdani and S. Assilian. An experiment in linguistic systhesis with fuzzy logic controller. *International Journal of Man-Machine Studies*, 7:1–13, 1975.
46. J.G. Marín-Blázquez, Q. Shen, and A.F. Gómez-Skarmeta. From approximative to descriptive models. In *Proceedings of the 9th IEEE International Conference on Fuzzy Systems*, pages 829–834, San Antonio, TX, USA, 2000.
47. P.A. Mastorocostas, J.B. Theocharis, and V.S. Petridis. A constrained orthogonal least-squares method for generating TSK fuzzy models: application to short-term load forecasting. *Fuzzy Sets and Systems*, 118(2):215–233, 2001.
48. G.C. Mouzouris and J.M. Mendel. A singular-value-QR decomposition based method for training fuzzy logic systems in uncertain environments. *Journal of Intelligent and Fuzzy Systems*, 5:367–374, 1997.
49. W. Pedrycz, editor. *Fuzzy modelling: paradigms and practice.* Kluwer Academic, Norwell, MA, USA, 1996.
50. A. Riid and E. Rüstern. Interpretability versus adaptability in fuzzy systems. *Proceedings of the Estonian Academy of Sciences. Engineering*, 49(2):76–95, 2000.
51. H. Roubos and M. Setnes. Compact and transparent fuzzy models and classifiers through iterative complexity reduction. *IEEE Transactions on Fuzzy Systems*, 9(4):516–524, 2001.
52. M. Setnes, R. Babuška, U. Kaymak, and H.R. van Nauta Lemke. Similarity measures in fuzzy rule base simplification. *IEEE Transactions on Systems, Man, and Cybernetics—Part B: Cybernetics*, 28(3):376–386, 1998.
53. M. Setnes, R. Babuška, and H.B. Verbruggen. Complexity reduction in fuzzy modeling. *Mathematics and Computers in Simulation*, 46(5-6):509–518, 1998.
54. M. Setnes, R. Babuška, and H.B. Verbruggen. Rule-based modeling: precision and transparency. *IEEE Transactions on Systems, Man, and Cybernetics—Part C: Applications and Reviews*, 28(1):165–169, 1998.
55. M. Setnes and H. Hellendoorn. Orthogonal transforms for ordering and reduction of fuzzy rules. In *Proceedings of the 9th IEEE International Conference on Fuzzy Systems*, pages 700–705, San Antonio, TX, USA, 2000.
56. M. Setnes and H. Roubos. GA-fuzzy modeling and classification: complexity and performance. *IEEE Transactions on Fuzzy Systems*, 8(5):509–522, 2000.
57. L.I.U. Shi-Rong and Y.U. Jin-Shou. Model construction optimization for a class of fuzzy models. *Chinese Journal of Computers*, 24(2):164–172, 2001.
58. R. Silipo and M. Berthold. Discriminative power of input features in a fuzzy model. In D. Hand, J. Kok, and M. Berthold, editors, *Advances in Intelligent Data Analysis (IDA-99)*, volume LNCS 1642, pages 87–98. Springer-Verlag, Heidelberg, Germany, 1999.
59. R. Silipo and M. Berthold. Input features' impact on fuzzy decision processes. *IEEE Transactions on Systems, Man, and Cybernetics—Part B: Cybernetics*, 30(6):821–834, 2000.
60. T. Söderström and P. Stoica. *System identification.* Prentice Hall, Engleweed Cliffs, NJ, USA, 1989.

61. T. Sudkamp, J. Knapp, and A. Knapp. Refine and merge: generating small bases from training data. In *Proceedings of the 9th IFSA World Congress and the 20th NAFIPS International Conference*, pages 197–202, Vancouver, Canada, 2001.
62. M. Sugeno and G.T. Kang. Structure identification of fuzzy model. *Fuzzy Sets and Systems*, 28:15–33, 1988.
63. M. Sugeno and T. Yasukawa. A fuzzy-logic-based approach to qualitative modeling. *IEEE Transactions on Fuzzy Systems*, 1(1):7–31, 1993.
64. T. Takagi and M. Sugeno. Fuzzy identification of systems and its application to modeling and control. *IEEE Transactions on Systems, Man, and Cybernetics*, 15:116–132, 1985.
65. J. Valente de Oliveira. Semantic constraints for membership function optimization. *IEEE Transactions on Systems, Man, and Cybernetics—Part A: Systems and Humans*, 29(1):128–138, 1999.
66. L.-X. Wang and J.M. Mendel. Fuzzy basis functions, universal approximation, and orthogonal least squares learning. *IEEE Transactions on Neural Networks*, 3:807–814, 1992.
67. X. Wang and J. Hong. Learning optimization in simplifying fuzzy rules. *Fuzzy Sets and Systems*, 106(3):349–356, 1999.
68. N. Xiong and L. Litz. Fuzzy modeling based on premise optimization. In *Proceedings of the 9th IEEE International Conference on Fuzzy Systems*, pages 859–864, San Antonio, TX, USA, 2000.
69. Y. Yam, P. Baranyi, and C.-T. Yang. Reduction of fuzzy rule base via singular value decomposition. *IEEE Transactions on Fuzzy Systems*, 7(2):120–132, 1999.
70. J. Yen and L. Wang. An SVD-based fuzzy model reduction strategy. In *Proceedings of the 5th IEEE International Conference on Fuzzy Systems*, pages 835–841, New Orleans, LA, USA, 1996.
71. J. Yen and L. Wang. Application of statistical information criteria for optimal fuzzy model construction. *IEEE Transactions on Fuzzy Systems*, 6(3):362–372, 1998.
72. J. Yen and L. Wang. Simplifying fuzzy rule-based models using orthogonal transformation methods. *IEEE Transactions on Systems, Man, and Cybernetics—Part B: Cybernetics*, 29(1):13–24, 1999.
73. J. Yen, L. Wang, and C.W. Gillespie. Improving the interpretability of TSK fuzzy models by combining global learning and local learning. *IEEE Transactions on Fuzzy Systems*, 6(4):530–537, 1998.
74. W. Yu and Z. Bien. Design of fuzzy logic controller with inconsistent rule base. *Journal of Intelligent and Fuzzy Systems*, 2:147–159, 1994.
75. L.A. Zadeh. Outline of a new approach to the analysis of complex systems and desision processes. *IEEE Transactions on Systems, Man, and Cybernetics*, 3:28–44, 1973.
76. L.A. Zadeh. The concept of a linguistic variable and its application to approximate reasoning. Parts I, II and III. *Information Science*, 8, 8, 9:199–249, 301–357, 43–80, 1975.

SECTION 2

IMPROVING THE INTERPRETABILITY WITH FLEXIBLE RULE STRUCTURES

Regaining Comprehensibility of Approximative Fuzzy Models via the Use of Linguistic Hedges

Javier G. Marín-Blázquez and Qiang Shen

Centre for Intelligent Systems and their Applications
Division of Informatics, The University of Edinburgh
80 South Bridge, Edinburgh EH1 1HN, UK

Abstract. This chapter presents an effective and efficient approach for translating rules that use approximative sets to rules that use descriptive sets and linguistic hedges of predefined meaning. Following this approach, descriptive models can take advantage of any existing approach to approximative modelling which is generally efficient and accurate, whilst employing rules that are comprehensible to human users. This allows the comprehensibility of approximative models to be restored.

Although trapezoidal fuzzy sets, including triangular ones, are most commonly used in fuzzy modelling for computational simplicity, applications of conventional linguistic hedges over such sets typically fail to result in significant changes of the sets definition. In particular, the full membership part of a trapezoid membership function does not change at all. This does not help in many modelling tasks as intended. Therefore, this chapter also presents an improved version of more effective hedges specifically devised for trapezoidal fuzzy sets, including three which do not appear in the literature. Simulation results are provided to demonstrate the advantages of utilising the revised and newly introduced hedges for assisting fuzzy modelling, in comparison to the use of conventional ones.

1 Introduction

Computing with words is a fundamental contribution of fuzzy logic [32]. This is feasible via the utilisation of linguistic variables which are variables whose values can be words rather than numbers. These words can be interpreted as semantic labels to the fuzzy sets employed within the fuzzy models [31]. Thus, human comprehensible computer representation of the domain problems concerned can be created when desired.

A great majority of fuzzy rule bases and/or fuzzy sets used in fuzzy models are, created and tuned to best fit the data available. They are not encoded to keep the meaning of the semantic labels, following the so-called *approximative approach* (or precise fuzzy modelling). Note that the word *approximative* is herein used instead of approximate to mirror the word *descriptive* in descriptive modelling which is itself an approximate approach. The approximative approach is, under minor restrictions, equivalent to neural networks [8] and, like them, offers little explanatory power over the inferences drawn. In fact, important automatic model generation techniques such as clustering, variations of hillclimbers and genetic algorithms are all data-driven. When employed unrestrictedly for fuzzy modelling, they tend to create fuzzy sets that fit the data extremely well but that usually lack features which are considered important to make it easy for human users to interpret the resulting model and its reasoning [29].

Opposing this stands the *descriptive approach* (or linguistic fuzzy modelling). From this approach point of view, semantics are as important as accuracy. The definition of the fuzzy sets is human given. For pure descriptive modelling no changes to such definition are allowed. As humans can not process efficiently more than 7±2 different entities in the short term memory [16], a linguistic variable of practical use should not have more than such a number of possible different labels. However, this constraint leads to coarse partitions of the underlying value ranges of the linguistic variables and hence affects the accuracy of the model to be built. The descriptive sets used induce a fuzzy grid in the product space of the domain variables. As few or even no modifications are permitted, the grid, and the hyperboxes delimited by it, is almost fixed. The approximative approaches solve this problem by changing the definitions of the sets, and therefore the hyperboxes themselves. Yet, in so doing, they generally ruin the underlying semantics. The current literature either pays little attention to this semantic ruining effect from using an approximative model (some even regarding such models as descriptive), or maintains a descriptive model and accepts high modelling errors.

There also exist the so-called *pseudo-descriptive* approaches to improve model flexibility. These approaches include approximative methods where certain restrictions are imposed to allow for some descriptive features [17]. Other pseudo-descriptive approaches [20–22] try to regain transparency by aggregating approximative sets of a rule set until some descriptive features are obtained. The attachment of linguistic labels and hence interpretability is done *a posteriori*. The resulting models use computer-defined words instead of words given by humans. The fact that the sets are modified or generated by computer implies a lack of certainty in obtaining a real descriptive model. It is for this reason that have been referred to as pseudo-descriptive.

One of the important disadvantages of descriptive modelling is that the best rules are usually generated by an exhaustive search in the given data space. This can only be done for a small number of variables/labels due to the potential combinatorial explosion. Even with a manageable size of variables/labels, the automatic generation of descriptive rules is generally either very slow, or very inaccurate, whilst many existing approximative methods are able to find very accurate rules rapidly. Thus, the question becomes if there exists a way to generate descriptive rules using a variation of these approximative methods.

A translation procedure has been proposed to obtain descriptive explanations of approximative models [26]. It works by mapping each approximative rule into a close descriptive rule. However, this procedure adds on additional runtime cost and the explanation generated may not be sufficiently accurate due to the one-to-one approximate translation. This chapter proposes to perform a one-to-many rule translation using a heuristic method and to perform a fine tuning using a systematic search through evolutionary computing. The search will be guided by *functional equivalence* rather than by similarity between the antecedent approximative and the predefined descriptive fuzzy sets. To allow more flexible modelling results, such that predefined fuzzy sets may be modified in their effects within the resulting descriptive rules without changing ther definition, linguistic hedges are used [12,15].

For computational simplicity, and easiness of being encoded into chips, piecewise linear fuzzy sets have been widely used in the literature. This kind of sets, be they triangular, trapezoidal or shouldered, can be defined with parameters that are easy to understand, and hence they are also very user friendly and popular [23]. However, conventional definitions of hedges, such as those of applying powers to the membership values of the original set [5,9], do not result in significant changes on trapezoid fuzzy sets. In particular, the full membership part of a trapezoid membership function does not get changed at all. This causes a conflict with the intended use of the corresponding hedges in the present work, that is, to ensure highly accurate translations from approximative to descriptive models. This chapter, therefore, presents an improved version of more effective hedges specifically devised for trapezoidal fuzzy sets, including triangular and shouldered ones. These hedges are to be applied to dilate or concentrate a given descriptive set by shrinking or expanding its constituent parts. The chapter also introduces three new hedges, named UPPER, LOWER and MID, which do not appear in the literature. Simulation results on well known classification problems are provided to demonstrate the advantages of utilising the revised and newly introduced hedges for assisting fuzzy modelling. These simulations have been done in comparison to the use of conventional ones.

The rest of the chapter is organised as follows. Section 2 proposes a set of useful linguistic hedges, which differ from those conventionally employed in the literature. Section 3 presents the translation process, covering two methods, one based on the use of heuristics and the other on a genetic algorithm. Section 4 reports on typical experimental results for the translation procedure, demonstrating the potential of the present proposal. Section 5 describes further results, showing the superiority of the newly proposed hedges in use. The chapter is concluded in section 6, with further work pointed out.

2 Hedges

A linguistic hedge modifies the shape of a fuzzy set's definition, causing changes in the membership function [5]. That is, a hedge transforms one fuzzy set into another. The meaning of the transformed set can be extracted from that of the original set and that embedded in the hedge applied. The definition of hedges has more to do with common sense knowledge in a domain than with mathematical theories. The following work reflects this observation.

2.1 Revised Hedges

A trapezoidal membership function, characterised by 4 parameters (a, b, c, d), consists of three consecutive segments as illustrated in figure 1. The application of dilation/concentration hedges should increase/decrease the size of these segments and, therefore, be implemented with a modification kept in proportion to the center of the full membership segment (m). This means that the center of gravity of the original fuzzy set and that of the modified will be different. In so doing, one set is always included in the other due to the piecewise linearity of the set membership definition. For concentration the modified set is included in the original and for dilation the original set is included in the modified. This makes more sense than if the shrinking/expansion is made with regarding to the center of gravity (since it would otherwise produce shapes that may break the intuition of using the dilation/concentration hedges). For example, figure 1(i) shows the case where the hedge VERY is applied such that the center of gravity is preserved. This breaks the dilation principle over the grey region where $\mu_{VERY \cdot S}(x) > \mu_S(x)$. However, shrinking the original relatively to the center of the full membership segment (m), as shown in figure 1(ii), guarantees that $\mu_{VERY \cdot S}(x) \leq \mu_S(x)$. The following formalises these ideas and redefines the concentration and dilation hedges.

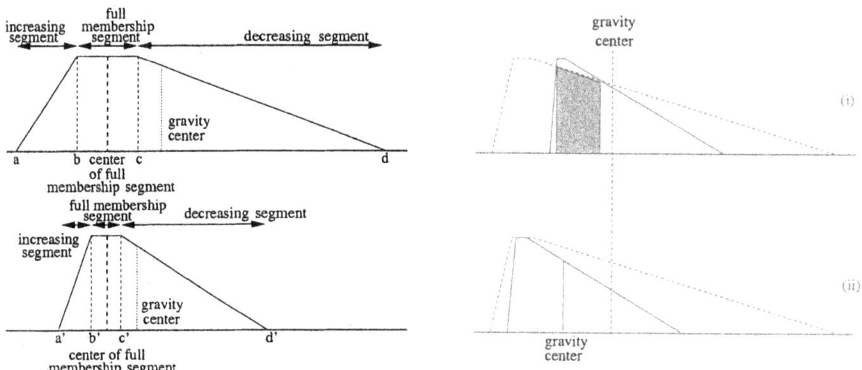

Fig. 1. Parts of a trapezoidal set and application of the hedge VERY (left), and how it affects (right): (i) center of gravity and (ii) center of full membership segment.

Concentration Concentration hedges reduce the size of segments or each part of the membership values of an original set. For a given trapezoidal fuzzy set S with a membership function $\mu_S(x)$, the set modified by a concentration hedge CON should comply with $\forall x \in X, \mu_{CON \cdot S}(x) \leq \mu(x)$. The parameters of the modified set are therefore defined to be calculated by:

$$\begin{array}{c} m = \frac{b+c}{2} \\ b' = m - ((m-b) * \beta) \; c' = m + ((c-m) * \beta) \\ a' = b' - ((b-a) * \beta) \;\; d' = c' + ((d-c) * \beta) \end{array} \tag{1}$$

where β controls the shrinking degree of a concentration hedge. In order to reduce the set effectively β must satisfy $0 < \beta < 1$. In particular, the commonly used hedge terms $MORE$, $VERY$ (see figure 1) and $EXTREMELY$ are defined as follows:

- $MORE$ reduces the segments to $\frac{2}{3}$ of the original size ($\beta = \frac{2}{3}$).
- $VERY$ reduces the segments by half ($\beta = \frac{1}{2}$).
- $EXTREMELY$ reduces the segments to $\frac{1}{8}$ of the original size ($\beta = \frac{1}{8}$).

Dilation Dilation hedges increase the size of segments or each part of the membership values of a fuzzy set. As opposite to a concentration hedge, a dilation hedge DIL should comply with the following intuition: $\forall x \in X, \mu_{DIL \cdot S}(x) \geq \mu_S(x)$. The parameters of the modified set are calculated as concentration hedges, but this time the factors will be greater than one:

- $GREATLY$ increases the segments by 2 times the original size ($\beta = 2$).
- $LESS$ increases the segments by $\frac{3}{2}$ of the original size ($\beta = \frac{3}{2}$).

It is easy to see that the pair $MORE - LESS$ and the pair $VERY - GREATLY$ are complementary; they cancel each other as they express exactly opposite concentration-dilation concepts. A basic optimisation is to remove these pairs if they appear in a model. No hedge was found in the literature that matches the opposite of $EXTREMELY$ (perhaps $REMOTELY$ could be a candidate for this), but including such a dilation hedge is as simple as making $\beta = 8$.

2.2 Newly Introduced Detailisation Hedges

Existing hedges modify a given fuzzy set but cannot create sets which have a focus on certain parts of the original without destroying the original set itself. In practical applications, there are cases where fuzzy models may well be better represented to suit the problem at hand if a loosely predefined fuzzy set could be described with different focuses, say on its lower, middle or upper part. Having recognised this, a new type of three hedges is proposed here, called detailisation ones. Jointly, they allow decomposing the original set into three, whilst keeping the order of these decomposed sets the same as the order of the elements belonging to the full membership segment of the original.

Given a trapezoidal set S the resulting three sets are denoted by $LOWER \cdot S$, $MID \cdot S$ and $UPPER \cdot S$, arranged in an increasing order, and they are defined as follows:

$$LOWER \cdot S \begin{cases} a' = a & b' = b \\ c' = b + \frac{c-b}{3} & d' = b + \frac{2*(c-b)}{3} \end{cases} \tag{2}$$

$$MID \cdot S \begin{cases} a' = b & b' = b + \frac{c-b}{3} \\ c' = b + \frac{2*(c-b)}{3} & d' = c \end{cases} \tag{3}$$

$$UPPER \cdot S \begin{cases} a' = b + \frac{c-b}{3} & b' = b + \frac{2*(c-b)}{3} \\ c' = c & d' = d \end{cases} \tag{4}$$

2.3 Traditional Hedges

To ease comparison with the use of conventional hedges, a brief summary of classical dilation/concentration hedges is given below, together with another type of hedge that may be applied to restrict extreme values of linguistic variables.

Dilation/Concentration As indicated previously, traditional hedges usually use powers of the normal membership functions for dilation and concentration. In particular the traditional form of such hedges is:

$$\mu_{INT \cdot S}(x) = \mu_S^e(x), \; e > 1 \tag{5}$$

$$\mu_{DIL \cdot S}(x) = \mu_S^e(x), \; 0 \leq e < 1 \tag{6}$$

The particular e may vary. In this chapter the values used for the e of the traditional hedges will be the reciprocal of the β factors of the newly proposed hedges. For example, for the hedge $EXTREMELY$ the value of β is $\frac{1}{8}$ so the exponent e for the same $EXTREMELY$ hedge uses in the traditional way is $e = 8$. Any traditional concentration or dilation hedge can be converted to the newly proposed ones by using β as the reciprocal of the exponent e.

Restriction Restriction hedges [5], $ABOVE$ and $BELOW$, are applied to variables where fuzzy values are ordered. The set modified by applying the $ABOVE$ hedge denotes the set which is "greater than" the original set and that modified by the $BELOW$ hedge represents the set which is "less than" the original. The resulting sets are therefore shouldered ones, the left shouldered for $BELOW$ and the right shouldered for $ABOVE$. Their membership functions are defined as follows:

$$\mu_{ABOVE \cdot S}(x) = \begin{cases} x < c & 0 \\ c \leq x < d & 1 - \mu_S(x) \\ x \geq d & 1 \end{cases} \tag{7}$$

$$\mu_{BELOW \cdot S}(x) = \begin{cases} x < a & 1 \\ a \leq x < b & 1 - \mu_S(x) \\ x \geq b & 0 \end{cases} \tag{8}$$

Note that the application of restriction hedges to some fuzzy sets may make no sense. This includes cases where the hedge $ABOVE$ is to be applied to a right shouldered set (or any set whose $\mu(x) = 1$ when $x \to \infty$) and similar ones where $BELOW$ is used to modify a left shouldered set. Such nonsense hedge-set combinations are disallowed in modelling.

2.4 The NOT operator

Although NOT is not a hedge but a logical operator, in terms of its application effects it may be viewed as a hedge. This is because the application of this operator to a fuzzy set also changes the shape of the membership function of that set (as $\mu_{NOT \cdot S} = 1 - \mu_S$). For this reason, it will be treated similarly as any other hedge hereafter.

In applying the above hedges, for any original shouldered membership function, it is assumed that the original fuzzy set is trapezoidal with its extreme parameters (a and b for the left shouldered and c and d for the right shouldered) being equal to a certain domain maximum/minimum value. These limit values can be determined using domain knowledge (e.g. by identifying the maximum and minimum values in the training data set with, perhaps, a safe offset of, say, $\pm 10\%$). If these maximum/minimum values get updated (for example, when new training data becomes available), the same hedge applied to a shouldered set before or after the update may yield different modified sets. In the event that the new maximum/minimum value exceeds the corresponding safe offset, modifications made to shouldered sets must then be recomputed using the new limit values.

To illustrate and compare the effects of applying the hedges conventionally used and presently proposed, figure 2 shows the results before and after a hedge is used to modify an irregular piecewise linear fuzzy set.

3 Mapping approximative into descriptive rules

The aim is to find an effective and efficient way to translate rules that use approximative sets to rules that use descriptive sets and hedges. The translated rules will be equivalent to the original or the closest equivalence possible with the advantage of having (or regaining) meaningful interpretation.

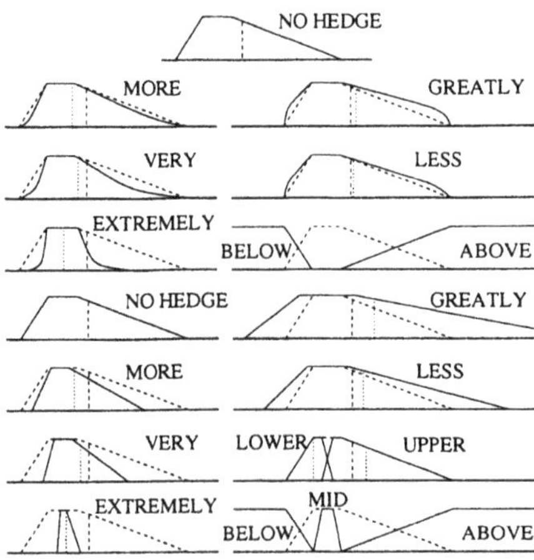

Fig. 2. Traditional hedges (left) and newly proposed ones (right) applied to an irregular trapezoid and how they change the center of gravity.

Many different methods exist for automatically generating approximative rules. Broadly speaking, they can be categorised into three groups: clustering methods [1,3], partition methods [4,10] and search methods [6,7,18,24]. For the the present work it does not matter which technique is used to generate the original approximative rules. What is required is a set of approximative rules, and the definition of the descriptive fuzzy sets and linguistic hedges.

To perform the translation, two different methods are proposed here. One is based on a heuristic search. The second method uses evolutionary computation techniques. As evolutionary search usually works better if a good start point is identified [27], the first method will be employed as the generator of the initial population for the genetic method to perform a finer-grain search.

3.1 Heuristic approach

This approach is based on hyperbox intersection. Descriptive rules, that is, hyperboxes defined on descriptive sets shown as white boxes in figure 3, are created if they intersect with an approximative rule (that is, the hyperbox defined by the antecedent approximative sets like the shadowed box in figure 3). This produces a crude translation and so it is in fact used as the generator of the initial population for the evolutionary method to be presented later.

Basic Method The idea is to build a layered graph to represent degrees of intersection, using a similarity measure. Each layer of the graph consists of a certain number of nodes, each of which represents the degree of intersection between one of the approximative sets of the antecedent and one of the descriptive sets of the same variable involved. The degree of intersection between two sets S_1 and S_2 is hereafter called the *Similarity Value* (I) of the two, which is defined by:

$$I_{S_1 S_2} = \frac{A(S_1 \cap S_2)}{\max(A(S_2), A(S_1))} \tag{9}$$

with $A(Set)$ denoting the area of the set Set. Clearly, this value may vary from zero for no intersection to one for equality. When intersection is null the corresponding node is removed from the graph. Although this particular definition is utilised in this work, empirical results have shown that other similarity metrics proposed in the literature [30,33,34] may be adopted to take its place without major disruption in the mapping results. However, this definition has proven to be computationally simple and performance-wise robust.

Having finished the building of the graph all possible paths are checked. A T-Norm is used to evaluate each path, yielding an overall coverage degree of the resulting descriptive rule over the original approximative one. The *Coverage Degree* (CD) of a rule R_1 over another R_2 is defined by:

$$CD_{R_1 R_2} = T\text{-}Norm(I_{S_{i1}, S_{i2}})$$

where $i = 1, \ldots, n$, with n being the number of antecedent variables. A standard T-norm like min can be used for the implementation of this evaluation.

All valid paths will then become rules and the set of rules obtained will jointly be a translation of the original approximative rule. To reduce the search an I-threshold may be used to prune the graph by removing nodes of a low I value. This will help to reduce the number of possible paths and hence speed up the entire graph creation and evaluation process.

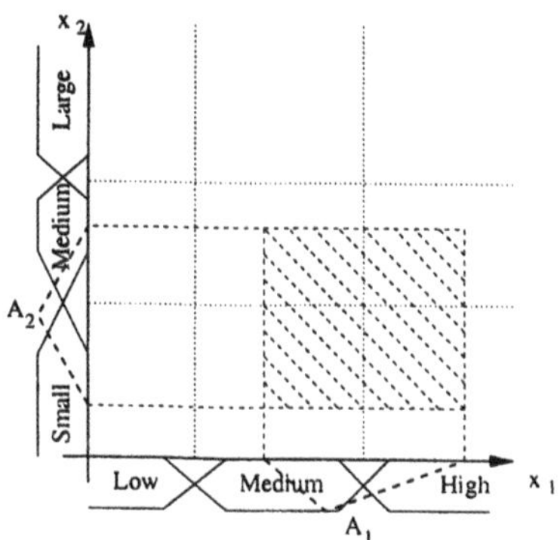

Fig. 3. Approximative rule and descriptive sets.

For example, the approximative rule given in figure 3 (IF x_1 IS S_1 AND x_2 is S_2 THEN OUTPUT A) leads to the generated graph as shown in figure

4 (without using any I-threshold) with $I_{High,A_1} = 0.43$, etc. Using the T-Norm min, for instance, the following four CDs are obtained with respect to the four resulting descriptive rules: $min(0.43, 0.5) = 0.43$ (High-Medium), $min(0.43, 0.69) = 0.43$ (High-Small), $min(0.62, 0.5) = 0.5$ (Medium-Medium) and $min(0.62, 0.69) = 0.62$ (Medium-Small).

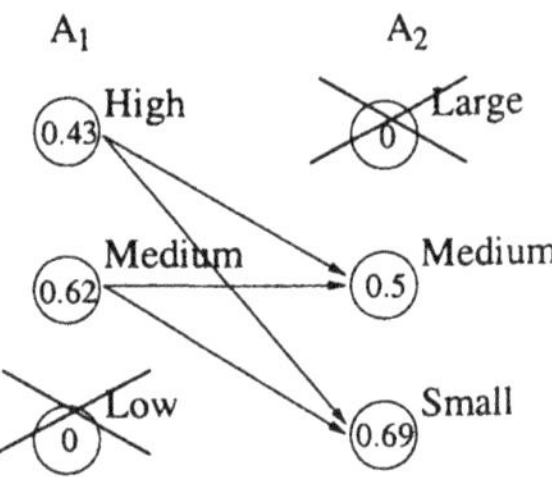

Fig. 4. Graph generation.

Thus, the resultant set of descriptive rules which collectively form the translation of the given approximative rule are:

R_1: IF x_1 IS High AND x_2 IS Medium THEN OUTPUT A, CD=0.43
R_2: IF x_1 IS High AND x_1 IS Small THEN OUTPUT A, CD=0.43
R_3: IF x_1 IS Medium AND x_1 IS Medium THEN OUTPUT A, CD=0.5
R_4: IF x_1 IS Medium AND x_1 IS Small THEN OUTPUT A, CD=0.62

After this, all rules with a CD lower than a prescribed CD-threshold may be discarded. (This offers another degree of freedom for reducing the complexity of the resulting model in addition to the use of I-threshold for node-pruning). Note that if the T-Norm used is min the same result can be obtained if, before creating the paths and hence the rules, all nodes with an $I_{L_{ij}}$ less than the CD-threshold are removed from the graph. This will reduce the number of possible paths and hence speed up the entire graph creating and evaluation process.

Extended Method A posible improvement of the method above is to include hedges. In so doing, the number of nodes will, however, increase dramatically. This is because the label of a node can now be any combination of a descriptive set and a hedge used to modify the set. Even if nodes with an I value below the I-threshold are eliminated this may still result in an important increase of the nodes in the graph. To reduce such increases, for each layer a similarity measure between any pair of nodes is calculated. A node that is very similar to some others can then be eliminated. For this elimination two different heuristic methods are proposed here.

The first ensures that all nodes will not be similar among themselves above a certain threshold (termed S-Threshold). It works via identifying the best I value node that has a similarity value above the S-Threshold to other relevant nodes. All these other nodes are removed as they have a smaller I value. After this, the process iterates with the node of the next highest I value which possesses a similarity value still above the S-Threshold to the remaining nodes.

The second method ensures a particular number of nodes per layer. It works by, for each layer, sorting the similarity values of every pair of nodes and removing, in order of high to low similarity, the node of a pair with a smaller I value, until reaching an allowed number of nodes remaining.

3.2 GA-based approach

This approach relies on the concept of *functional equivalence*. A method following this approach will search for a set of descriptive rules that collectively behave like the original approximative rule to be translated. That is, for the data that is covered by the approximative rule, the obtained descriptive rules will fire with at least the same firing strength. Meanwhile, for data that would not cause the original approximative rules to fire, the resultant descriptive rules will either not fire or fire if their consequents comply with the desired output. The search mechanism is herein implemented by a GA.

Training Sets and Objective Functions One strategy for translation is that each approximative rule is translated independently. This assumes that the result of rule firings always takes the consequent that is obtained by executing the rule which has the highest rule firing strength. Therefore, a training subset of data for each original rule needs to be generated from the global training set. Such rule training subsets are obtained via the following preprocessing process.

Given a set $\mathcal{X}$ of training examples, for each example $x_i \in \mathcal{X}$ the firing strength $AFS_R(x_i)$ of the approximative rule R is determined. If $AFS_R(x_i) > 0$ it would be desirable that, if the consequent of this rule is the same as the desired consequent, the resulting translated descriptive rules will fire for such an example with a strength $DFS_R(x_i)$ equal or greater than $AFS_R(x_i)$. This kind of example will be referred to as type one.

If $AFS_R(x_i) > 0$ but the consequent does not match the desired, then it would be desirable that the firing strengths of the resulting descriptive rules $DFS_R(x_i)$ will be less than, or at worst equal to $AFS_R(x_i)$. This kind of example will be referred to as type two.

Furthermore, if $AFS_R(x_i) = 0$ then, if the example x_i is of the same desired consequent as that of the original rule, it is not selected to form the training subset (as this example is expected to be covered by other rules). If, however, the consequent is different, the firing strength of the resulting descriptive rules should be zero. This last kind of example will be referred to as type three.

In summary, $\forall x_i \in \mathcal{X}$, the preprocessing aims to allow the GA to enforce:

$$\overbrace{\begin{array}{lll}\text{same cons.} & DFS_R(x_i) \geq AFS_R(x_i) & \text{Type 1} \\ \text{diff. cons.} & DFS_R(x_i) \leq AFS_R(x_i) & \text{Type 2}\end{array}}^{AFS_R(x_i) > 0}$$

$$\overbrace{\begin{array}{ll}\text{same cons.} & \text{ignore } x_i \\ \text{diff. cons.} & DFS_R(x_i) = 0 \ \ \text{Type 3}\end{array}}^{AFS_R(x_i) = 0}$$

In general, a given approximative rule may not be covered by just one descriptive rule. It may happen to obtain several descriptive rules which, collectively, form the functionally equivalent, or the closest possible, to the original. For this reason, a new objective is needed to minimise the number of descriptive rules used to represent the given approximative rule. Further, given the goal of genetic search being to obtain a model that behaves like the original approximative one, if the original model is inaccurate the translated one will generally also be inaccurate. To overcome in part this potential drawback, an extra objective of reducing the model error is also included. Thus, the GA will be guided by several objectives as listed below:

Objectives to minimise
$\sum_{x_i \in T_1} \max(0, AFS(x_i) - DFS(x_i))$
$\sum_{x_i \in T_2} \max(0, DFS(x_i) - AFS(x_i))$
$\sum_{x_i \in T_3} DFS(x_i)$
Number of Rules
Model error

There exist in the genetic algorithms literature several approaches to deal with multiple objectives. The one adopted in this work is called "Sum of Weighted Global Ratios" [2]. This method independently normalises each objective first with respect to the best and worst value ever found for it. Then, each objective is weighted and added together into a final fitness value.

Note that for specific application domains, there may be further more objectives requiring consideration (see section 4). Such domain-particular objectives can be readily incorporated with the above mentioned general ones to define the fitness function.

The individual translation strategy described above has the drawback that, when the individual translations are put together to form the final translated descriptive rule set, the independently translated rules may interfere with each other. Although a close fit of the descriptive rules to the approximative ones may help resolve the problem, this cannot be guaranteed. It is therefore interesting to consider possible alternative translation strategies.

Instead of translating individual approximative rules one by one, a possible alternative, valid for problems with a limited range of output values (such as classification problems), is to translate one group of all the approximative rules regarding a single output value at a time. In this so-called group strategy an AFS value is calculated as the firing strength of the entire subset of rules concerning the same output value, which is defined by the strongest firing strength of all the original approximative rules that characterize the same class. Thus, the translation can be done class by class, instead of rule by rule.

For completeness, another version of the GA search strategy is also included here, and termed the global strategy, where all rules for all classes are represented together in each chromosome. That is, a chromosome is itself a whole translation of the given approximative model.

Genetic Representation and Genetic Engine The genetic chromosome representation is based on the work in [11]. The specific codification and use of this DNA-like representation is explained in [14]. This representation resembles very closely the way in which real DNA chromosomes are represented. In particular, a chromosome is a sequence of single-valued cells, each of which is termed a locus and represents the basic information building block. Within a chromosome certain sub-sequences of loci represent genes, each of which denotes an instruction that is used, in conjunction with zero or more other genes, to build a fuzzy rule. Genes are separated by specific sequences of loci that act as the delimiting boundaries, or START and END marks. Any other loci between genes are considered junk and are therefore ignored. (Note that those junk sequences are nonetheless helpful since they lessen the disruptive effect of genetic operators.)

This representation allows a variable number of rules and a variable number of conditions in each rule. The same variable may even appear several times in the same antecedent. The length of the chromosomes is also variable. Overall, each chromosome represents an entire collection of instructions needed to construct a fuzzy rule set, following the general approach known as Pittsburgh style GAs [25]. Each population therefore consists of sets of emerging translated rulesets. For GAs implementing either individual or group strategy the chromosomes do not encode consequents (as all of them share the same consequent anyway). However, for use of the global strategy the consequents are also encoded.

The GA adopted here is a steady-state one. The selection of the two parents is done for one by linear ranking and for the other by random choice. Each child replaces a random member of the worst half of the population. The search stops when the best half of the population does not improve for a prescribed number of generations. Diversity is maintained thanks to the random replacement within the worst half of the population.

After a translation process terminates, those possible descriptive rules that did not fire with the training data available are eliminated. Note that there may exist cases where the eliminated rules cover certain training data, but such data is already covered by other rules with higher firing strength, so they did not ever fire. In experimental studies, to be reported next, not all available data is used for training in order to exploit part of the data to check for possible overfitting.

Three different mutation and four different crossover operators have been implemented in order to investigate what combinations may lead to a good translation. The use of a steady-state GA states that a crossover must always take place. To improve this without destroying the underlying GA operation mechanism a special no-crossover operator is included to allow mutations to be applied without a crossover. Each crossover has a fixed minimum chance of being applied (with a total of 50% shared among all possible crossovers). If applying a certain crossover has so far resulted in a successful new member then its chance to be applied in future generations is increased at the expense of the rest. Similarly, if it is unsuccessful its chance for future use will decrease. In general, during the execution of a GA there will be more unsuccessful tries than successful ones. Hence, the rate at which a particular crossover is applied should decrease slowlier than the speed to increase the rate of applying a successful crossover. Empirically, the inclusion of this dynamic schema helps improve significantly the performance of the GA employed.

To be self-contained, a brief description of the mutation and crossover operators implemented is given below:

- **Flipping loci** This mutation operator changes the contents of random loci in the chromosome, being a low level mutation that modifies the basic instructions given to build the whole descriptive ruleset.
- **Flipping hedges/sets** This mutation operator works by changing certain randomly identified sets and/or hedges in the antecedent of the emerging rules, thereby being a high level mutation.
- **Adding/Removing a rule** This mutation operator removes or adds a rule to the emerging fuzzy rule set by removing from or inserting into the gene a new instruction sequence.
- **Rulemixing crossover** This shuffles the rules of two rule sets by cutting the two parent chromosomes into chunks of individual rules (individual genes) and rearranging randomly these chunks into two new chromosomes.
- **One point crossover** This classical operator cuts the parent chromosomes into two pieces in random loci and then swaps the resulting halves.
- **Two point crossover** This is similar to the one point crossover in cutting the parent chromosomes but interchanges the sections of the chromosomes contained between two random cutting points.
- **Uniform crossover** This interchanges the loci of two chromosomes randomly with the number of loci to be interchanged being also random.

4 Results for the translation procedure

4.1 Set-up

To demonstrate the proposed approach at work, typical classification problems are used here. The benchmark classification problems used are selected from [28], including the Breast Cancer, Diabetes, New Thyroid, Wine, and Iris datasets. Table 1 summarises the set-ups of these datasets. Classification problems require the maximisation of the number of correctly classified samples whilst minimising the number of incorrectly or not covered cases. This introduces the following new objectives, in addition to those general ones given in section 3.2.

Additional classification objectives
Maximum number of correctly classified
Minimum number of incorrectly classified
Minimum number of not covered

Table 1. Classification problems

Name	No. of Inputs	No. of Output Classes	No. of Samples
Breast Cancer	9	2	683
Diabetes	8	2	768
Iris	4	3	150
New Thyroid	5	3	215
Wine	13	3	178

A neuro-fuzzy approximative rule induction algorithm ANFIS [9] using bell-shaped approximative sets, and which has been optimized for classificatin, were trained for both problems. The resulting approximative ruleset is employed as the original set of rules for translation.

The first step of the translation is performed using the extended heuristic method with the two methods outlined in 3.1. As indicated before, the output of the heuristic method is used to act as the generator of the initial population for the GA that performs finer search for the final descriptive rules. To ensure the readability and understandability of the resulting descriptive ruleset, the maximum number of hedges (including the *NOT* operator) allowed to be applied to a given set is limited to 2 (no such limit is needed in theory though).

For comparison, the pure descriptive induction algorithm as given in [13], which is a form of exhaustive search with different parameter settings, is also tested. In particular, this algorithm (referred to as Lozowski's algorithm hereafter) has a parameter that trades off between the model accuracy and the size of learned ruleset. It determines the minimum difference between the firing strengths of any given rules that have the same antecedent but different class values. In the present investigation, this parameter is set up with reference to the number of rules that the heuristic method has generated to ease comparison. Also, for comparison purposes, results obtained by running the standard C4.5 algorithm [19] are included.

Fuzzification was carried out proportionally with respect to the size of the universe of discourse of the individual variables for each problem considered. That is, for each variable, the distance between its maximum and minimum value within the data set is divided such that all of them approximately cover an equal range of the underlying real values (with soft boundaries of course). The fuzzy sets resulting from such a partition are regarded as the given descriptive sets. This is not necessary in practical applications when this can be done by users or experts. For simplicity, each variable is allowed to take 3 linguistic labels.

As the GA executions are computationally affordable for the present investigation, it is worthy to execute the genetic search for as many times as possible in order to obtain a number of different translations. The figures to be presented below will show the mean error of the translated rulesets depending on the number of runs allowed for the GA. Such error measures were obtained using a bootstrapping of 1000 samples over 100 real runs. Finally, to avoid possible overfitting each dataset has been separated into a training set containing 75% of all the given data and a test set comprising the remaining 25%.

4.2 Results

First of all, it is interesting to investigate the effects of using different translation strategies. Figure 5 collects the results of such experiments on the Diabetes problem, as an example (results on other problems are very similar). The group strategy gave better results than the global and individual ones during the training phase. In testing, all the three strategies performed very similarly overall but difference exists in terms of the number of rules needed to achieve the similar test results. Individual translation gave more (sometimes many more) rules. This is so because the GA tried to produce good translations locally. The pressure on the search exerted by the number of rules objective on individual strategy was not so high as the pressure by the same objective in carrying out group or global translations, where more data points were involved. It may be the case that, when using group or global strategies, several approximative rules can be covered by just one descriptive rule. This may produce potential savings in the number of rules generated.

As group and global strategies allow for more data to be covered there is more pressure to reduce the number of rules using either of these strategies than the individual strategy. At the same time the more general strategies have more evaluations to run so this difference in pressure can, to a certain extent, be compensated. The strong point of the individual strategy rests in the fact that it leads to very short and compact rules as can be seen in figure 6.

In order to reduce the size of the resultant ruleset the pressure on the number of rules objective can be used. This makes it easier to understand the general characteristics of the problem under consideration. However, in

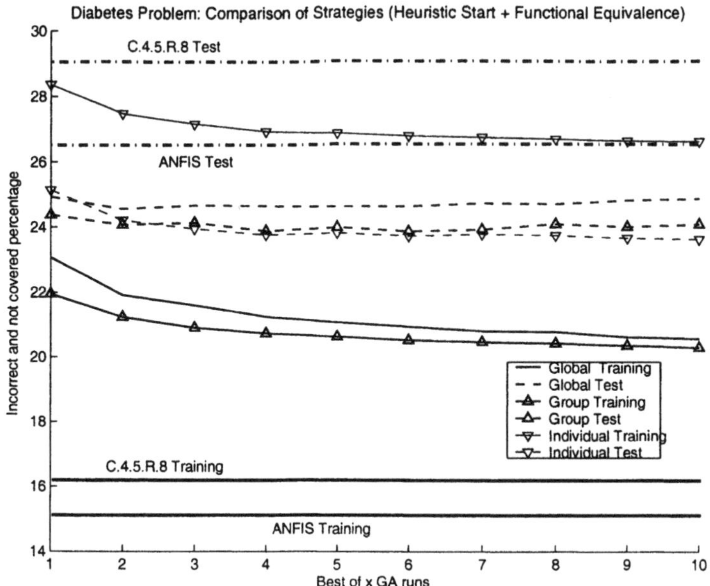

Fig. 5. Classification error vs translation strategy for the Diabetes problem.

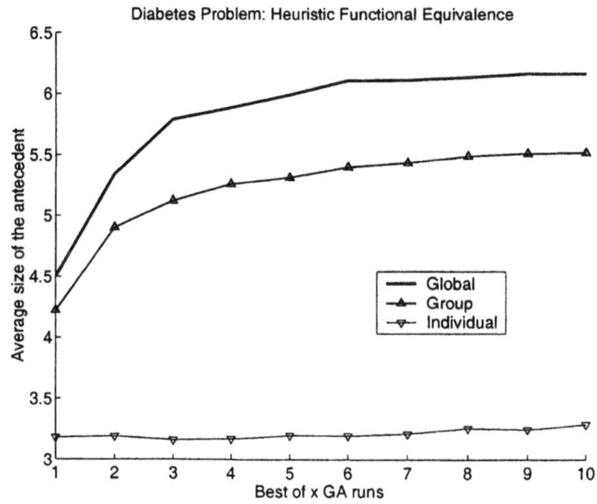

Fig. 6. Mean size of antecedents vs translation strategy.

general, a reduced number of rules is likely to result in a reduced classification rate (compare figures 7 and 8 or 5 and 9).

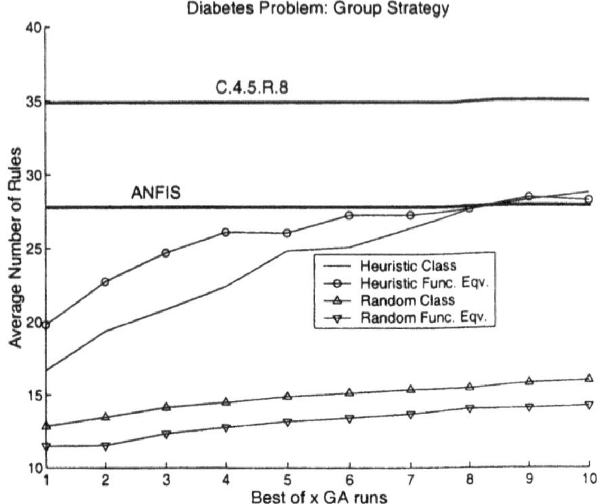

Fig. 7. Number of descriptive rules using the group strategy for the Diabetes problem.

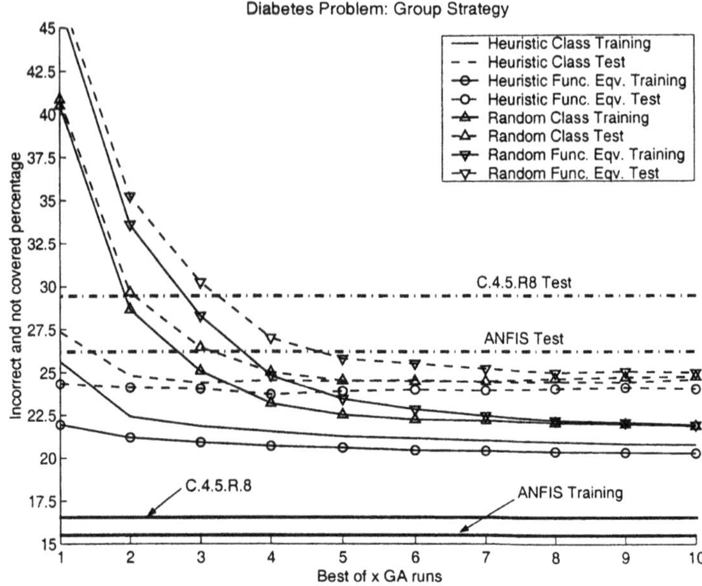

Fig. 8. Classification error for the Diabetes problem in using the group strategy.

As different translation strategies lead to very similar testing performances of the resulting descriptive rulesets, only those results obtained by

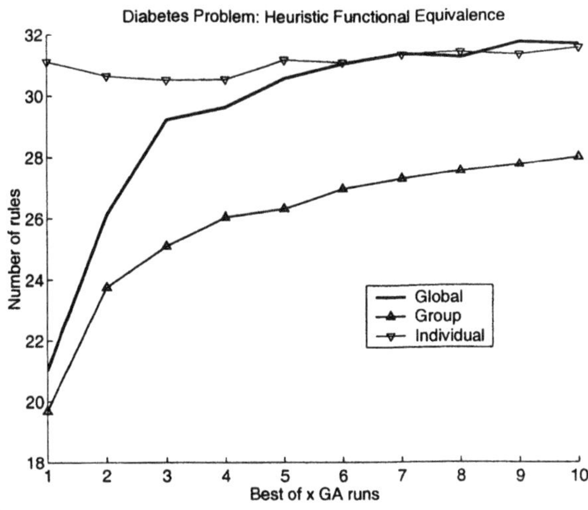

Fig. 9. Number of rules vs translation strategy for the Diabetes problem.

the use of group strategy are hereafter presented. Tables 2 and 3 give the results of runs with respect to the following (where Trn, Tst, and Rul respectively stand for the classification error percentage over training data, that over testing data, and the number of descriptive rules obtained):

a) Starting from a heuristic translation and guiding the GA by classification objectives only.
b) Starting from a heuristic translation and guiding the GA by *functional equivalence* and classification objectives.
c) Starting from randomly generated rules and guiding the GA by classification objectives only.
d) Starting from randomly generated rules and guiding the GA by *functional equivalence* and classification objectives.

The most interesting points to compare are those given in the columns concerning items b) and c) above. The former shows the result of what is suggested in this work, and the later shows that of pure data-driven search (no heuristic translation nor approximative model provided). In order to illustrate the impact of the *number of rules*objective upon the translation, the results on the Diabetes problem were collected with two different weights set for this objective.

To support the comparison, results of applying a descriptive ruleset which is obtained by the initial heuristic translation alone, and those of using C4.5, ANFIS and Lozowski's algorithm are also provided as given in table 4. The work presented in this chapter performs well and does so consistently in testing. The accuracy of translated descriptive models is close to that achievable by the optimized ANFIS (comparing table 3 and the middle column of table 4), and generally outperforms the other pure descriptive modelling techniques tested (comparing table 3 and columns 3 and 4 of table 4).

Figure 8 shows a graphical comparison, for the Diabetes problem, of the translation results with respect to the number of GA runs. It reveals the difference between different ways of guiding the GA, either starting from a set of descriptive rules obtained by the heuristic translation of the approximative rules, or from a set of randomly generated descriptive rules. Graphs plotting the classification errors for the different classification problems appear to be almost the same in their general tendency. They only differ in the actual values of the classification error and in how many GA runs are needed to reach a state where running more GAs will not improve the result. Such graphs are therefore omitted here.

Table 2. Mean results of translations for one GA run only

Problem	Heuristic Class			Heur Fun. Eqv.			Random Class			Rand Fun. Eqv.		
	Trn	Tst	Rul	Trn	Tst	Rul	Trn	Tst	Rul	Trn	Tst	Rul
Breast Cancer	2.9	7.3	9.5	1.1	5.6	8.5	14.9	18.3	9.8	10.0	14.0	9.1
Diabetes (Norm)	25.0	26.8	16.9	21.9	24.5	19.8	41.0	41.5	12.8	46.2	46.8	11.5
Diabetes (High)	28.7	29.7	8.3	25.3	27.2	9.1	48.5	48.6	9.4	46.4	46.4	3.4
Iris	1.3	4.6	5.4	1.1	4.4	5.3	5.2	9.9	5.2	3.7	6.2	5.1
New Thyroid	12.3	13.0	7.2	7.0	7.9	6.2	54.6	55.3	7.2	45.9	46.9	6.1
Wine	3.8	10.3	10.3	1.5	5.7	10.8	45.4	47.5	10.2	36.4	39.7	10.6

Table 3. Mean results of the best translation out of 5 GA runs

Problem	Heuristic Class			Heur Fun. Eqv.			Random Class			Rand Fun. Eqv.		
	Trn	Tst	Rul	Trn	Tst	Rul	Trn	Tst	Rul	Trn	Tst	Rul
Breast Cancer	0.8	5.7	9.3	0.7	5.6	8.3	0.8	5.7	9.1	0.7	5.7	9.0
Diabetes (Norm)	21.2	24.4	24.0	20.5	23.8	26.7	24.6	27.2	14.0	23.5	25.4	15.7
Diabetes (High)	22.7	26.0	9.8	21.4	24.2	10	25.2	26.6	8.2	25.1	27.8	9.1
Iris	0.49	3.98	5.4	0.43	3.43	4.7	0.9	5.0	5.4	0.8	4.6	5.5
New Thyroid	6.4	7.1	7.5	5.8	6.6	6.1	11.3	13.3	7.3	9.3	11.4	6.2
Wine	0.3	6.5	10.8	0.3	3.2	9.8	6.8	13.1	9.7	2.7	9.21	10.4

This general trend shows that a GA-based translation starting from random rules produces systematically worse results than that starting from the

Table 4. Results of initial translation and of other modelling methods

Problem	C4.5			ANFIS			Lozowski			Heuristic		
	Trn	Tst	Rul	Trn	Tst	Rul	Trn	Tst	Rul	Trn	Tst	Rul
Breast Cancer	1.2	7.6	29	0.4	4.1	18	10.3	15.2	74	25.7	23.9	94
Diabetes	16.1	29.2	35	15.7	26.6	28	38.2	39.6	65	33.5	32.8	43
Iris	1.8	2.6	5	0.8	2.6	4	4.5	10.5	11	8.9	11.6	9
New Thyroid	1.9	7.4	13	3.1	1.8	4	82.6	83.3	4	25.5	25.9	4
Wine	0.8	2.2	17	0	2.2	6	39.8	44.4	21	30.1	22.2	23

heuristic translation. In addition, the results achievable with a heuristic translation start are far more stable than those obtained with a random start. This supports the need for generating such heuristic crude translation first. For simpler problems (e.g., Breast Cancer) both starting points may eventually lead to similar classification accuracies, yet several runs are needed to ensure this similarity in results. Nevertheless, as shown in table 2, if only a single run is executed (for example, due to computational resource limit) the results obtained using a randomly generated initial population would be much worse than those obtained using the initial population produced by the heuristic method. It clearly pays off to generate the heuristic translation first and to use such a descriptive rule set to act as the generator for the initial population of the GA.

Guidance for the GA by classification error only is also, in general, rather unstable, when compared to the use of *functional equivalence* guidance. This comparison is shown in figure 10 and 11 , where results of 100 runs on classifying the Thyroid problem, with a heuristic start and using the group strategy, are depicted. Following the guidance by classification error alone produces rather poor runs with high error peaks (which actually happens in all tested problems). This instability of the results can be partially overcome with a higher number of GA runs, though those extra runs are not always affordable computationally. Even if extra GA runs were affordable it would still be better to use the criterion of *functional equivalence* over all the runs.

The above results of a) starting from a heuristic translation and b) making use of *functional equivalence* reveal an important point. That is, generating an approximative model first (to get required *functional equivalence* objectives) and then implementing a heuristic translation (to act as the initial population generator for the GA) significantly improves the performance of the final descriptive ruleset.

A positive side effect of the way that the rules are encoded in the GA is that the size of the antecedent, i.e. the number of conditions in the antecedent, is variable. This implies that there may be an implicit attribute reduction going along with the GA-based optimization as shown in figures 6 and 12. This reduction is more evident in high dimensional problems. If desired, the reduc-

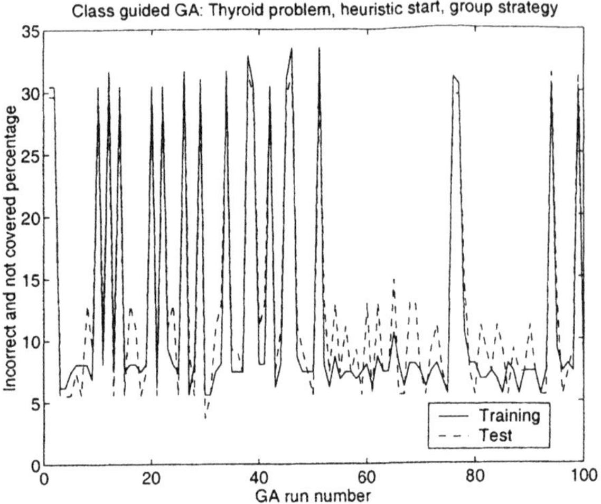

Fig. 10. 100 runs for a class guided GA (Thyroid group).

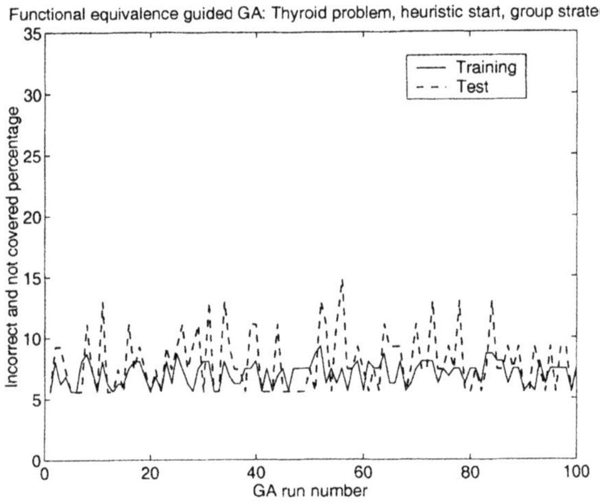

Fig. 11. 100 runs for a functional equivalence guided GA (Thyroid group).

tion of antecedent conditions may even be explicitly introduced as another optimization objective.

Finally, it is worth noting that the performance of Lozowski's algorithm, which is a pure descriptive modelling method based in exhaustive search, never gets close to that of ANFIS or C4.5. In all the problems tested this algorithm gives poorer results than the approximative modelling methods used and yet requires more rules even for reaching such less desirable results.

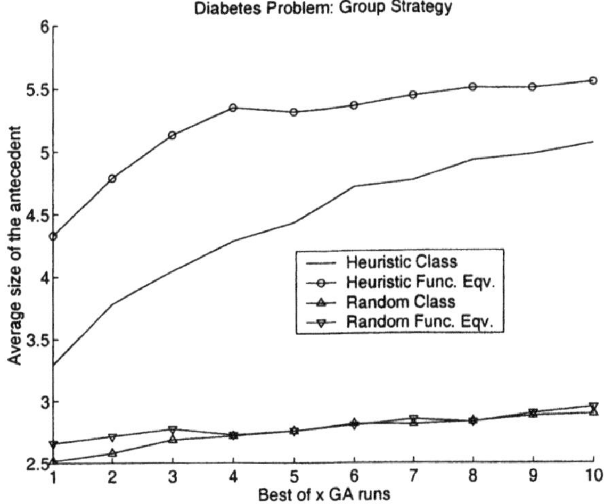

Fig. 12. Mean size of the antecedents for the Diabetes problem in using the group strategy.

Compared just to the heuristic translation Lozowski's algorithm only defeats it for the Breast Cancer and Iris problems, but its corresponding translated models contain a considerably higher number of rules. However, to be fair with this comparison it must be remembered that Lozowski's algorithm makes no use of hedges.

5 Experiments for redefined trapezoidal hedges

5.1 Set-up

To test the performance of the conventional and present implementations of hedges the work described in section 3 was used to carry out data-driven modelling (though any other descriptive modelling technique may be employed as an alternative for this purpose). Again, ANFIS was used to obtain the original approximative fuzzy model. Also, to ensure the readability of the resulting descriptive models, it is disallowed for any consecutive use of more than two linguistic hedges over a given fuzzy set within the experiments.

Although, in theory, the translation process works by having a heuristic translation to serve as the generator of the initial population for the GA, to avoid any possible advantage taking by this initial heuristic population (that has been found to produce better results when used as tables 2 and 3 have shown) random rules are used to initialise the GA. To minimise the possibility that a particular random ruleset may potentially benefit one or another group of hedges ten different random rulesets have been generated, and GA runs were started using one of these rulesets at random.

The version of the GA used has no special features designed for the proposed new hedges. Detailed parameters of the GA were set the same across all experiments and so no advantage may be taken by the present work. Again, the figures to be presented below will show the mean error of the translated rulesets depending on the number of runs allowed for the GA.

5.2 Results

Table 5 lists the results of using the original ANFIS models and those of applying the descriptive models obtained by employing conventional linguistic hedges. As compared to these results, table 6 lists the outcome of applying the hedges defined in this work. Note that, Trn, Tst and Rul in these tables respectively stand for error on training data, error on testing data and number of rules produced. To provide a fair comparison between the use of new hedges and that of the traditional ones (which lack the novel detailisation hedges and therefore may be argued of having less variety), experiments without involving the three new hedges $LOWER$, MID and $UPPER$ were also carried out and their results are also reported in table 6.

Table 5. Results of using ANFIS model and of employing traditional hedges

Problem	ANFIS			Traditional					
				1 run			5 runs		
	Trn	Tst	Rul	Trn	Tst	Rul	Trn	Tst	Rul
Breast Cancer	0.4	4.1	18	3.9	7.0	7	2.1	6	8
Diabetes	15.7	26.6	28	31.3	31.7	7.5	27.6	27.6	7.9
Iris	0.8	2.6	4	14.1	15.1	3.5	7.9	7.0	3.5
New Thyroid	3.1	1.8	4	17.8	17.9	4.1	9.5	10.7	4.1
Wine	0	2.2	6	15.2	20.6	7.7	10.2	17	8.4

Table 6. Results of using revised hedges (left) and of including detailisation ones (right)

Problem	Without detailisation hedges						With detailisation hedges					
	1 run			5 runs			1 run			5 runs		
	Trn	Tst	Rul	Trn	Tst	Rul	Trn	Tst	Rul	Trn	Tst	Rul
Breast Cancer	2.4	6.8	8.7	1.7	6.4	9	2.5	6.6	9	1.6	6.2	8.8
Diabetes	29	26.7	9.2	25.8	23.7	8.6	28.5	25.7	9.8	24.7	22.6	9.5
Iris	4.1	4.8	4.4	2.5	5.4	3.9	3.6	3.5	4.3	2.3	3.2	4.3
New Thyroid	7.5	12.9	5.9	5.1	7.7	5.9	6.2	9.3	6.3	3.7	6.2	6.2
Wine	7.8	10.7	7.8	4.2	8.6	8.0	6.6	9.1	7.3	3.4	6.4	7.2

When comparing the results of a descriptive model with those of the original approximative model obtained by ANFIS, the performance of any descriptive model does not appear to be impressive. However, this is not the point of this investigation; much better descriptive models can be acquired when an initial heuristic translation is employed. The intention of the present experiments is to show the differences between applications of different sets of hedges.

The results leave little doubt about the clear advantage of the use of the redefined hedges. Even better performance is reached when the newly introduced hedges are exploited. Either way, the use of hedges implemented in the present work significantly outperforms the use of traditional ones. This result is obtained consistently across all tested problems. The performance improvement is particularly dramatic for the last three problem cases (namely, Iris, New Thyroid and Wine). In addition to these results, figure 13 shows in greater detail the evolution of the error measured as the number of GA runs allowed for getting the best model is increased. This highlights further the success of the present research.

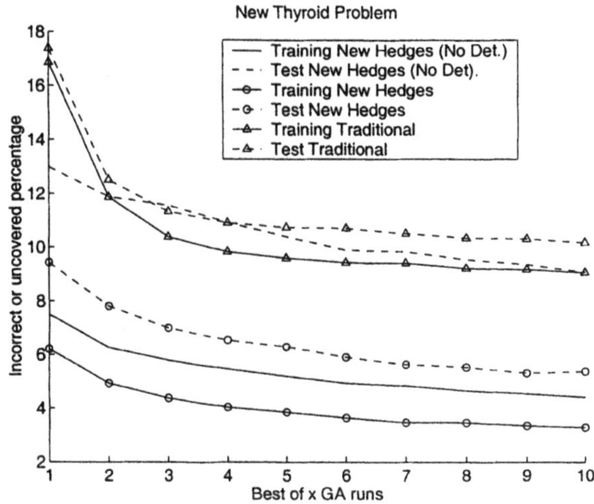

Fig. 13. Classification error for New Thyroid problem using the best of several GA runs.

6 Conclusions

A major disadvantage of several existing methods for building descriptive fuzzy systems is that the generation of fuzzy rules is usually made via an exhaustive search throughout the input product space. In addition, the rules produced by pure descriptive methods usually have a low accuracy. However, approximative rule generation methods can be very fast and accurate, though they tend to have difficulties in interpreting the underlying meaning of the data being modeled and in facilitating understanding of the inferences drawn from the resultant models.

In order to generate accurate rules that possess the desirable property of being readily comprehensible to human users and to create such rules in an efficient way, a translation technique from approximative to descriptive rules has been proposed here. The translation is enabled by the use of linguistic hedges to change implicitly the prescribed, meaningful descriptive fuzzy sets, such that the modified fuzzy sets will closely resemble the original approximative ones in function. The approximative rules may be themselves created by any standard approximative modelling technique. The modification process is implemented via a *functional equivalence* guided search. Novel hedges that are useful for supporting such translations are defined in this work, for trapezoidal fuzzy sets, including triangular and shouldered ones. A heuristic algorithm has been presented, which implements a preliminary translation that is employed to act as the generator for the initial population used by a GA to perform the fine grain modifications.

The results obtained so far have demonstrated that the proposed approach does not decrease significantly the accuracy attainable by the original optimized approximative models, and that the descriptive rules obtained are interpretable by humans. It outperforms exhaustive search methods for pure descriptive rule generation in terms of search efficiency, as searches for descriptive rulesets are herein heuristically guided. In terms of classification accuracy, it also outperforms pure descriptive methods which do not use hedges, whilst it is impractical to run exhaustive search with hedges, as that would complicate even more the usually already huge search space.

The work on the revised implementation of, and newly introduced, hedges has been compared to the traditional implementation of hedges, again using a series of well known classification problems as test cases. The results are overwhelmingly in favour of the new implementation. Although the experiments carried out were focused on developing descriptive fuzzy classifiers, there is no obvious theoretical limitation that may invalidate the achieved results for other application problems.

A number of possible improvements to the translation process implemented herein can be done, however. The GA was poorly optimised within the present work, as no important attempts to optimize it were made. Optimisations to the individual rules may be done periodically during the genetic search. Furthermore, the entire implementation could be integrated with an on-line descriptive modeller that could include micro-tuning of the original fuzzy descriptive partition. Here, by micro-tuning it means a tuning where the modified sets are constrained to keep a very high similarity with the original ones, such that the underlying meaning of the prescribed fuzzy sets is not disrupted. The modeller might also allow adaptation to new unforeseen data thorough vicinity detection for close but uncovered points, and through generation and translation of new approximative rules for far away points.

A further possible component of improvement is to build into the GA evaluation function another objective that would minimise the number of hedges present in the descriptive rules that are to be created. This is interesting as conditions with less or no hedges are easier to read. Work along these lines is currently being carried out at Edinburgh.

Another further issue to consider is how the present work on the hedges may be extended to modifying other types of fuzzy sets, in an effort to make it more generally applicable. Also, even just for trapezoidal membership functions, it would be interesting to investigate whether the current hedge definitions could be refined such that they capture better common sense understanding of the corresponding linguistic terms. Relevant viewpoints from cognitive studies may well help with this further development.

Acknowledgements

The first author is partly supported by the Fundación Marín-Blázquez, Spain. Whilst taking full responsibility for the views expressed here, both authors are very grateful to Professor A. F. Gómez Skarmeta of the University of Murcia, Spain, Professor P. Ross of Napier University, UK, and to Alexios Chouchoulas for helpful discussions and assistance in the research reported.

References

1. Shigeo Abe. Fuzzy function approximators with ellipsoidal regions. *IEEE Transactions on Systems, Man and Cybernetics-Part B: Cybernetics*, 29(4):654–661, August 1999.
2. P. J. Bentley. *Evolutionary Design by Computers*. Academic Press Ltd, London, 1999.
3. James C. Bezdek and Sankar K. Pal, editors. *Fuzzy Models for Pattern Recognition: Methods that Search for Structures in Data*. IEEE, Pascataway, 1992.
4. C. L. Chen, S. H. Hsu, C. T. Hsieh, and W. K. Lin. Generating crisp-type fuzzy models from operating data. In *Proceedings of the 7th IEEE International Conference on Fuzzy Systems*, pages 686–691, 1998.
5. Earl Cox. *The Fuzzy Systems Handbook*. AP Professional, 1994.
6. A. F. Gómez Skarmeta and F. Jimenez. Generating and tuning fuzzy rules using hybrid systems. In *Proceedings of the 6th IEEE International Conference on Fuzzy Systems*, volume 1, pages 247–252, 1997.
7. Isao Hayashi, Hiroyoshi Nomura, Hisayo Yamasaki, and Noboru Wakami. Construction of fuzzy inference rules by neural network driven fuzzy reasoning with learning functions. *International Journal of Approximate Reasoning*, 6:241–266, February 1992.

8. J.-S. R. Jang, Y. C. Lee, and C.-T. Sun. Functional equivalence between radial basis function networks and fuzzy inference systems. *IEEE Transactions on Neural Networks*, 4(1):156–159, January 1993.
9. J.-S. R. Jang, C.-T. Sun, and E. Mizutani. *Neuro-Fuzzy and Soft Computing.* Matlab Curriculum. Prentice Hall, 1997.
10. C. J. Kim. An algorithmic approach for fuzzy inference. *IEEE Transactions on Fuzzy Systems*, 5(4):585–598, November 1997.
11. D. D. Leich. *A New Genetic Algorithm for the Evolution of Fuzzy Systems.* PhD thesis, Dept. of Egineering Science, University of Oxford, 1995.
12. Bin-Da Liu, Chuen-Yau Chen, and Ju-Ying Tsao. Desing of adaptive fuzzy logic controller based on linguistic-hedge concepts and genetic algorithms. *Transactions on Systems, Man, and Cybernetics – Part B: Cybernetics*, 31(1):32–53, February 2001.
13. A. Lozowski, T. J. Cholewo, and J. M. Zurada. Crisp rule extraction from perceptron network classifiers. In *Proceedings of International Conference on Neural Networks*, volume Plenary, Panel and Special Sessions, pages 94–99, Washington, D.C., 1996.
14. J. Gómez Marín-Blázquez and Q. Shen. Towards the generation of descriptive fuzzy rules via approximative modelling. In *Proceedings of 6th UK Workshop on Fuzzy Systems*, pages 1–14, 1999.
15. J. Gómez Marín-Blázquez, Q. Shen, and A. F. Gómez Skármeta. From approximative to descriptive models. In *Proceedings of the 9th IEEE International Conference on Fuzzy Systems*, pages 829–834, May 2000.
16. G. A. Miller, E. Galanter, and K.H. Pribam. *Plans and Structure of Behaviour.* Holt, Rinehart and Winston, New York, 1960.
17. D. Nauck, F. Klawonn, and R. Kruse. *Foundations of Neuro-Fuzzy Systems.* John Wiley & sons, 1997.
18. H. Nomura, I. Hayashi, and W. Noboru. A Learning Method of Fuzzy Inference Rules by Descent Method. *Proceedings of the 1st IEEE International Conference on Fuzzy Systems*, pages 203–210, 1992.
19. J. R. Quinlan. *C4.5: Programs for machine learning.* Morgan Kaufmann, San Mateo, CA, 1993.
20. J.A. Roubos, M. Setnes, and J. Abonyi. Learning fuzzy classification rules from data. In *Recent Advances in Soft Computing (RASC2000)*, pages 6–12, Leicester, UK, 2000.
21. M. Setnes, R. Babuska, and H. B. Verbruggen. Rule-based modeling: Precision and transparency. *IEEE Transactions on Systems, Man and Cybernetics - Part c: Applications and Reviews*, 28(1):165–169, Feb. 1998.
22. M. Setnes, R. Babuska, and H. B. Verbruggen. Transparent fuzzy modelling. *International Journal of Human-Computer Studies*, 49(2):159–179, 1998.
23. Q. Shen and R. Leitch. Fuzzy qualitative simulation. *IEEE Transactions on Systems, Man and Cybernetics*, 23(4):1038–1061, 1993.
24. Patrick K. Simpson. Fuzzy min-max neural networks-part 1: Classification. *IEEE Transactions on Neural Networks*, 3(5):776–786, September 1992.
25. S. F. Smith. *A learning system based on genetic algorithms.* PhD thesis, University of Pittsburgh, 1980.
26. M. Sugeno and T. Yasukawa. A fuzzy logic based approach to qualitative modeling. *IEEE Transactions on Fuzzy Systems*, 1(1):7–31, August 1993.

27. P. D. Surry and N. J. Radcliffe. Inoculation to initialise evolutionary search. In T. C. Fogarty, editor, *Proceedings of the 3rd AISB workshop on Evolutionary Computation*. Springer-Verlag (LNCS), 1996.
28. Uci machine learning databases. Available on web: `http://ftp.ics.uci.edu/pub/machine-learning-databases/`.
29. J. Valente de Oliveira. Semantic constrains for membership function optimization. *IEEE Transactions on Systems, Man and Cybernetics - Part A: Systems and Humans*, 29(1):128–138, Jan. 1999.
30. D. S. Yeung and C. C. Tsang. A comparative study on similarity-based fuzzy reasoning methods. *IEEE Transactions on Systems, Man, and Cybernetics - Part B: Cybernetics*, 27(2):216–227, April 1997.
31. Lofti A. Zadeh. The concept of a linguistic variable and its application to approximate reasoning i. *Information Sciences*, 8:199–249, 1975.
32. Lofti A. Zadeh. Fuzzy logic = computing with words. *IEEE Transactions on Fuzzy Systems*, 4(2):103–111, May 1996.
33. Lotfi A. Zadeh. Similarity relations and fuzzy orderings. In R. R. Yager, S. Ovchinnikov, R. M. Tong, and H. T. Nguyen, editors, *Fuzzy Sets and Applications: Selected Papers by L.A. Zadeh*, pages 81–104, New York, 1987. John Wiley & Sons, Inc.
34. R. Zwick, E. Carlstein, and D. V. Budescu. Measures of similarity among fuzzy concepts: A comparative analysis. *International Journal of Approximate Reasoning*, 1:221–242, 1987.

Identifying Flexible Structured Premises for Mining Concise Fuzzy Knowledge

N. Xiong[1] and L. Litz

Institute of Process Automation, University of Kaiserslautern
D-67663 Kaiserslautern, Germany

Abstract. Data mining attains growing importance to ease the knowledge-acquisition bottleneck. This chapter discusses the issue of extracting, from accumulated data, a compact fuzzy rule base, to reach concise yet highly generalizing knowledge. For this purpose, we establish flexible structured premises of rules, allowing for not only canonical AND combinations of input fuzzy sets but also OR connectives of linguistic terms as well as incomplete compositions of input variables in the premise constitution. The later two forms of premises are beneficial for rule number reduction, as they achieve bigger coverage of the input space compared with the first premise form. A genetic-based search algorithm is utilized to explore optimal premise structure in combination with parameters of fuzzy set membership functions. Simulation results on several data sets are given to demonstrate the merits and characteristics of the presented method.

1 Introduction

Data analysis and knowledge discovery attains growing importance to establish intelligent knowledge-based systems, particularly when no explicit human expertise is available for the underlying problem domain. Extracting valuable knowledge from experiences/examples has many practical utilities, including behavioral cloning of human skill, gaining insights into complex processes that are hard to be described with mathematical models, identifying prospective customers in commerce, as well as obtaining user behavior on the internet to mention only a few cases. The term data mining reflects the general purpose of efficient discovery of useful knowledge and information which are hidden somewhere in the accumulated data sets.

Fuzzy logic [29] merits to be incorporated in data mining processes to achieve linguistically interpretable results. Rules based on fuzzy logic well represent vagueness and imprecision prevalent in human statement and opinion. Fuzzy rule based systems [30], working upon if-then expressions of linguistic terms, appear closer to the way of human reasoning and in many real applications are thus more

---

[1] N. Xiong is currently with the Center for Autonomous Systems, Royal Institute of Technology, SE-10044 Stockholm, Sweden. E-mail: n.xiong@ieee.org. The work was performed at the University of Kaiserslautern.

powerful and flexible for knowledge treatment than traditional crisp counterparts. This motivates our paper to adopt fuzzy linguistic rules as the form of knowledge to be mined.

Construction of fuzzy rules involves specification of antecedents and determination of corresponding conclusions. Rule conditions appear here a critical issue since they determine the structure of the knowledge base. Traditionally, a rule set can be built by utilizing all AND combinations of input fuzzy sets as rule premises and then determining their consequent counterparts, see ([3], [13], [17], [19], [22], [25]) as examples. However, some drawbacks may arise by doing so, as the number of rules will increase exponentially with the number of inputs. The computational efficiency associated with fuzzy logic is lost and the robustness decreases when using a large amount of rules [4]. Moreover, a rule set of huge size could be tough for human users to check and understand the content of it.

This chapter aims to achieve a compact rule base by means of premise learning. Flexible structured premises of rules are introduced, allowing for not only incomplete composition of input variables but also OR-connectives of linguistic terms. Compared with the commonly used premise structure in the form of canonical AND connections, flexible premises of rules enable bigger coverage of input domain and therefore make a reduced rule number possible. This contributes to mining concise (fuzzy) knowledge which is beneficial in various respects in that it saves memory size, enhances computational efficiency, as well as makes itself easier for human understanding.

2 Fuzzy Rules Based on Flexible Structured Premises

2.1 Flexible Structured Premises versus Elementary Premises

Suppose a fuzzy system with $X=(x_1, x_2, ..., x_n)$ as its inputs and y as its output. Each input x_i (i=1...n) has q[i] linguistic terms denoted as $A(i,1)$, $A(i,2)$, ..., $A(i, q[i])$. $p(\bullet)$ is an integer function mapping from $\{1,2, ...,s(s\leq n)\}$ to $\{1,2,....,n\}$ satisfying $\forall\ x\neq y,\ p(x)\neq p(y)$. A flexible structured rule premise is formulated as follows:

Definition 1: A flexible structured premise is a compound fuzzy proposition of the form as:

$$[x_{p(1)} = \bigcup_{j\in D(1)} A(p(1),j)] \ \ and \ \ [x_{p(2)} = \bigcup_{j\in D(2)} A(p(2),j)] \ \ and \cdots \\ and \ \ [x_{p(s)} = \bigcup_{j\in D(s)} A(p(s),j)] \tag{1}$$

where $D(k)\subset\{1, 2, \ldots, q[p(k)]\}, \quad k \in\{1,2,\cdots,s\}$

Clearly the premise formulated by (1) can be regarded as a conjunction of simple propositions, each of which corresponds to OR connectives of fuzzy sets defined for the input variable involved in it. If such a premise includes simple propositions for all inputs (e.g. $s = n$), we say that this condition has a complete composition, otherwise its composition is incomplete.

Definition 2: The premise expressed in (1) is an elementary one if the following conditions hold:

$$\text{(a) } s = n;$$
$$\text{(b) } \forall \quad k \in \{1,2,\cdots,n\}, \qquad \|D(k)\| = 1$$

Elementary rule premises defined above have complete compositions and contain one and only one linguistic term for every input variable. They can be regarded as a degraded case of the general form of flexible premises in (1). Using canonical AND connectives of linguistic terms, elementary premises are especially suitable to describe simple systems with low input dimensions.

For rule based modeling of complex problems with high input dimensions, flexible premises of the general form are preferable as they enable bigger coverage of the input domain compared with elementary ones. The main purpose of this chapter is to establish such premises optimally for building a compact and easily understandable rule base. A flexible antecedent with incomplete composition or containing OR connectives of linguistic terms can replace a set of related elementary premises, as illustrated in the following two examples.

Example 1: The antecedent '(x1=NZ or PZ) and (x2=NZ or PZ)' shown in Figure 1 covers four elementary premises listed below:

1) If (x1=NZ) and (x2=NZ) 2) If (x1=NZ) and (x2=PZ)
3) If (x1=PZ) and (x2=NZ) 4) If (x1=PZ) and (x2=PZ)

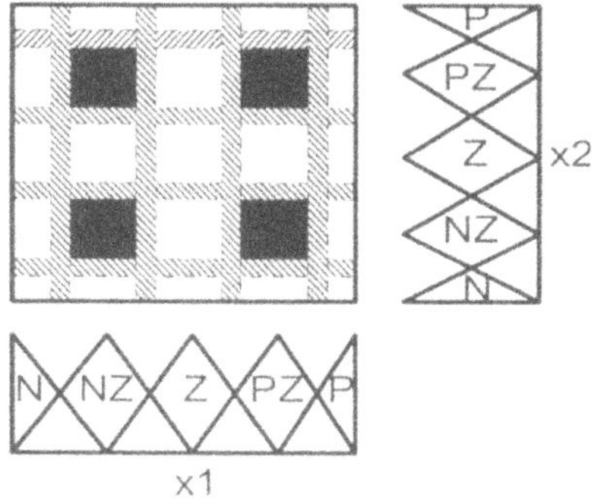

Fig. 1. Rule premise in Example 1

Example 2: The incompletely composed antecedent '(x1: don't care) and (x2=Z)' shown in Figure 2 covers the following group of elementary premises:

1) If (x1=N) and (x2=Z) 2) If (x1=NZ) and (x2=Z)
3) If (x1=Z) and (x2=Z) 4) If (x1=PZ) and (x2=Z)
5) If (x1=P) and (x2=Z)

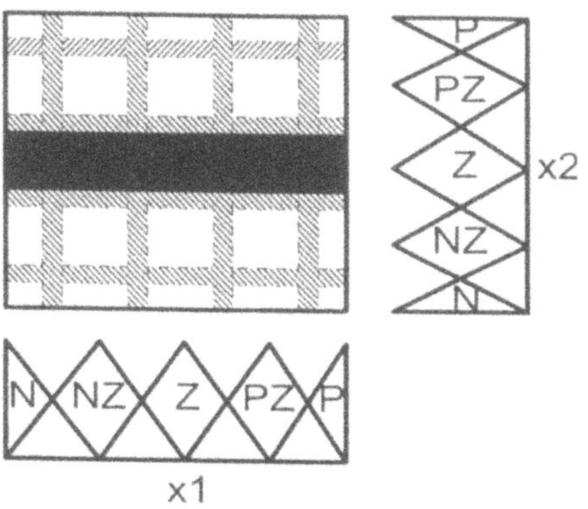

Fig. 2. Rule premise in Example 2

Rule interestingness is an important aspect of knowledge representation for neuro-fuzzy systems [24]. Informally speaking, the more general the antecedent description and the more specific the consequent part are, the more interesting and informative the rule is. The premises in the above two examples are more interesting than the elementary premises covered by them in the sense that the input situations covered by the elementary ones are relatively much narrower. In view of this, the introduction of flexible premise structure helps enhancing interestingness of the fuzzy knowledge to be mined.

2.2 Determining the Consequence under a Flexible Premise

By representing the flexible premise expressed in (1) with symbol $\tilde{A}$ with the definition as

$$\begin{aligned}\tilde{A} = & [x_{p(1)} \text{ is } \bigcup_{j \in D(1)} A(p(1), j)] \text{ and } [x_{p(2)} \text{ is } \bigcup_{j \in D(2)} A(p(2), j)] \\ & \text{and} \cdots \text{and } [x_{p(s)} \text{ is } \bigcup_{j \in D(s)} A(p(s), j)]\end{aligned} \tag{2}$$

a flexibly conditioned fuzzy rule can be abbreviated as "*If* $\tilde{A}$ *Then* $\tilde{B}$". The condition $\tilde{A}$ includes several simple fuzzy propositions each of which corresponds to OR connectives of fuzzy sets defined for the input variable involved in it. A simple proposition in $\tilde{A}$ determines the membership grade of a certain input variable with respect to the rule condition. If an input variable does not appear in $\tilde{A}$, it is considered to have a full membership grade with this rule condition. Supposing an object $e = (x_1, x_2, \ldots, x_n, y)$ with x_i $(i=1\ldots n)$ as input values and y as its output value, the membership grade of input x_i with respect to condition $\tilde{A}$ can be expressed as follows:

$$\mu_{\tilde{A}}(x_i) = \begin{cases} 1.0 & if \ \ p^{-1}(i) = \phi \\ \bigcup\limits_{j \in D(p^{-1}(i))} \mu_{A(i,j)}(x_i) & otherwise \end{cases} \tag{3}$$

Further $\tilde{A}$ is a compound fuzzy proposition consisting of a conjunction of simple propositions. So we define the degree of match $\mu_{\tilde{A}}(e)$ between $\tilde{A}$ and the object e as a T-norm as:

$$\mu_{\tilde{A}}(e) = t(\mu_{\tilde{A}}(x_1), \mu_{\tilde{A}}(x_2), \cdots, \mu_{\tilde{A}}(x_n)) \tag{4}$$

Likewise the degree of match between conclusion $\tilde{B}$ and object e is given by $\mu_{\tilde{B}}(e) = \mu_{\tilde{B}}(y)$.

On the other hand, the compound proposition $\tilde{A}$ can be viewed as a discrete fuzzy subset on the training set $U=\{e_1, e_2, \ldots, e_M\}$. Each object in the training set belongs to $\tilde{A}$ with a membership degree. The value of this membership grade is equal to the degree to which $\tilde{A}$ is satisfied by the inputs of the object. Thus we write:

$$\tilde{A} = \left\{ \frac{e_1}{\mu_{\tilde{A}}(e_1)}, \frac{e_2}{\mu_{\tilde{A}}(e_2)}, \cdots, \frac{e_M}{\mu_{\tilde{A}}(e_M)} \right\} \tag{5}$$

$$\mu_{\tilde{A}}(e_k) == t(\mu_{\tilde{A}}(x_{k1}), \mu_{\tilde{A}}(x_{k2}), \cdots, \mu_{\tilde{A}}(x_{kn})), \quad k = 1 \cdots M \tag{6}$$

Here $x_{k1}, x_{k2}, \ldots, x_{kn}$ are input values of object e_k in the training set. Similarly the conclusion $\tilde{B}$ is also treated as a discrete fuzzy subset on the training set. Each object in the training set belongs to $\tilde{B}$ with a membership grade, which equals the degree to which $\tilde{B}$ is satisfied by the output of the object. Therefore the fuzzy subset for rule conclusion can be expressed as

$$\tilde{B} = \left\{ \frac{e_1}{\mu_{\tilde{B}}(e_1)}, \frac{e_2}{\mu_{\tilde{B}}(e_2)}, \cdots, \frac{e_M}{\mu_{\tilde{B}}(e_M)} \right\} = \left\{ \frac{e_1}{\mu_{\tilde{B}}(y_1)}, \frac{e_2}{\mu_{\tilde{B}}(y_2)}, \cdots, \frac{e_M}{\mu_{\tilde{B}}(y_M)} \right\} \tag{7}$$

where y_k (k=1...M) denotes the output value of object e_k.

The rule "*If* $\tilde{A}$ *Then* $\tilde{B}$" corresponds to an implication of $\tilde{A} \Rightarrow \tilde{B}$, which is equivalent to the proposition that $\tilde{A}$ is $\tilde{B}$'s subset, i.e. $\tilde{A} \subseteq \tilde{B}$. Further, the relation $\tilde{A} \subseteq \tilde{B}$ holds if and only if $\mu_{\tilde{B}}(e) \geq \mu_{\tilde{A}}(e)$ for all $e \in U$. If there exists some $e \in U$ such that $\mu_{\tilde{A}}(e) > \mu_{\tilde{B}}(e)$, the rule will not be completely satisfied but still may be partially true. Rule confidence is thus reflected by the degree to which $\tilde{A}$ is a subset of $\tilde{B}$. With this view point, the measure of subsethood [12] of $\tilde{A}$ in $\tilde{B}$ is utilized as the confidence value of the rule. So we obtain

$$confidence(\tilde{A} \Rightarrow \tilde{B}) = \frac{\|\tilde{A} \cap \tilde{B}\|}{\|\tilde{A}\|} = \frac{\sum_{e \in U} (\mu_{\tilde{A}}(e) \wedge \mu_{\tilde{B}}(e))}{\sum_{e \in U} \mu_{\tilde{A}}(e)} \tag{8}$$

where $\|\tilde{A}\|$ and $\|\tilde{A} \cap \tilde{B}\|$ indicate the cardinality measures of the sets $\tilde{A}$ and $\tilde{A} \cap \tilde{B}$ respectively.

Given condition $\tilde{A}$, we choose a linguistic term $\tilde{B}$ from $\{B(1), B(2), \ldots, B(N_Y)\}$ such that the confidence value of the corresponding rule reaches its maximum. This leads to a two-step procedure for selecting proper consequence as follows.

Step 1: Calculate the voting strength (VS) for each linguistic output based on all members of the training set $e \in U$

$$VS[\tilde{B}] = \sum_{e \in U} \mu_{\tilde{A}}(e) \wedge \mu_{\tilde{B}}(e), \qquad \tilde{B} \in \{B(1), B(2), \cdots, B(N_Y)\} \tag{9}$$

Notice that the voting strength defined above can be considered as the accumulated evidence of implication $\tilde{A} \Rightarrow \tilde{B}$ found in the training data.

Step 2: Determine rule consequence as the linguistic output B^* that has the maximal voting strength, i.e.,

$$VS[B^*] = \max\{VS[B(1)], VS[B(2)], \cdots, VS[B(N_Y)]\} \tag{10}$$

Finally, it bears mentioning that the above procedure also applies to selecting consequent classes of fuzzy classification rules under minor revision. For doing this, we merely need to define the consequence B of a fuzzy classification rule "*If* $\tilde{A}$ *Then B*" as a crisp subset on the training set with its membership grade as

$$\mu_B(e_i) = \begin{cases} 1 & \text{if } class(e_i) = B \\ 0 & otherwise \end{cases} \qquad i = 1 \cdots M \tag{11}$$

such that the voting strength of a candidate class B under condition $\tilde{A}$ turns to

$$VS[B] = \sum_{class(e)=B} \mu_{\tilde{A}}(e) \tag{12}$$

2.3 Compatibility of a Rule Base Employing Flexible Premises

The introduction of the flexible structure of rule premises can lead to knowledge incompatibility. A simple example showing this is the following rule base:

If x_1 is A(1,2) then B_1
If x_2 is A(2,2) then B_2
If x_1 is A(1,1) and x_2 is [A(2,1) or A(2,2)] then B_3
If x_1 is [A(1,1) or A(1,3)] and x_2 is [A(2,1) or A(2,3)] then B_4

which leads to ambiguous conclusions as shown in Table 1. From this table, we can see that ambiguous conclusions exist for three input situations including

$A(1,1) \wedge A(2,1)$, $A(1,1) \wedge A(2,2)$, and $A(1,2) \wedge A(2,2)$. Human users will therefore be confused about outcomes in these circumstances.

Table 1. Example of incompabilities in a rule base

$x_1 \backslash x_2$	A(2,1)	A(2,2)	A(2,3)
A(1,1)	B_3, B_4	B_2, B_3	B_4
A(1,2)	B_1	B_1, B_2	B_1
A(1,3)	B_4	B_2	B_4

Generally speaking, linguistic overlapping between conditions of two rules leads to a conflict between them as long as the two rules suggest distinct output fuzzy sets in their conclusions. Since linguistic incompatibility between rules undermine knowledge interpretation and human understanding, we can not disregard this point in the procedure of premise learning. In order to give a numerical assessment of knowledge compatibility in a rule base, several definitions are set up below.

Definition 3: The input pattern (IP) of a flexible rule premise in (1) is the following set of linguistic term connections:

$$IP = \{A(1,m_1) \cap A(2,m_2) \cap \cdots \cap A(n,m_n) | m_1 \in F(1), m_2 \in F(2), \cdots m_n \in F(n)\} \tag{13}$$

where $F(i)$ is a set of integers defined as follows:

$$F(i) = \begin{cases} D(p^{-1}(i)) \ \ if \ \ p^{-1}(i) \neq \phi \\ \{1,2,\cdots,q[i]\} \ \ in \ \ other \ \ cases \end{cases} \quad (i=1,2,\ldots,n) \tag{14}$$

A conflict occurs in the rule base if there exist two rules which have overlapping input patterns in conditions but different linguistic outputs in conclusions. We divide all the rules in a rule base into several clusters according to rule conclusions such that rules in the same cluster suggest the same linguistic output. The number of clusters in a rule base is equal to the number of output fuzzy sets. The formal description of the input pattern of a rule cluster is given in Definition 4.

Definition 4: The input pattern of a rule cluster Clu is a union of input patterns for the premises of the rules that belong to it:

$$IP(Clu) = \bigcup_{r \in Clu} IP(r) \tag{15}$$

It is clear that there is no disagreement inside any cluster. However, between two different clusters, there may exist conflict if the intersection between their input patterns is not empty. The size of this intersection set reflects the conflicting scale between two clusters.

Definition 5: The conflicting scale (CS) between two clusters Clu_1 *and* Clu_2 is the cardinality measure of the intersection set between their input patterns:

$$CS(Clu_1, Clu_2) = \|IP(Clu_1) \cap IP(Clu_2)\| \quad (16)$$

Definition 6: The conflicting area (CA) in the whole knowledge base is the sum of conflicting scales between any two distinct clusters in it:

$$CA = \sum_{i=1}^{Ny-1} \sum_{j=i+1}^{Ny} CS(Clu_i, Clu_j) \quad (17)$$

where N_y represents the number of rule clusters.

Finally, we consider a numerical index to evaluate the compatibility of a rule base. This index is designed to be in inverse proportion to the conflicting area. In the case of no conflict, the compatibility index reaches its maximal value '1'. Otherwise it decreases linearly with the increment of conflicting area until its minimal value '0' is reached. The calculation of the compatibility index λ_c is given in the following formula:

$$\lambda_c = \begin{cases} 1 - \gamma \cdot CA & if \quad \gamma \cdot CA < 1; \\ 0 & if \quad \gamma \cdot CA \geq 1. \end{cases} \quad (18)$$

where $\gamma \in (0, 1]$ denotes the decreasing rate of λ_c. The value of γ should be determined in terms of particular application requirements.

3 Fuzzy Knowledge Mining by Means of Premise Learning

3.1 A General Perspective

We intend to develop a premise learning method based on genetic algorithms [8] to achieve concise fuzzy knowledge from numerical examples. The upper limit of the rule number is predetermined by user in advance. It can be considered as an estimation of the sufficient amount of rules to achieve a satisfactory accuracy. The actual rule number is adapted automatically within this specified limit during the running of the genetic algorithm (GA). The overall structure for fuzzy knowledge mining consists of two loops: the external loop and the internal loop, as illustrated in Fig. 3. In the external loop, the genetic algorithm is used to search in the combinatorial space for optimal structure of premises as well as to optimize parameters of fuzzy set membership functions simultaneously. The internal loop, serving the purpose of fitness evaluation for premise learning, is tasked to determine consequences of individual rules under given preconditions. This function is prerequisite to derive model errors and the rule base compatibility value required by the GA in the external loop.

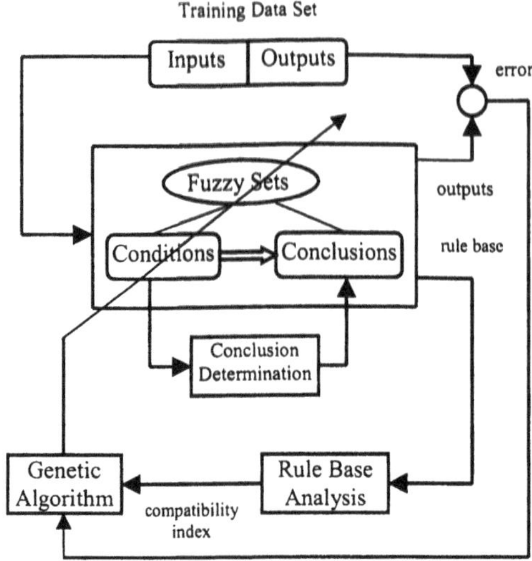

Fig. 3. The overall structure for fuzzy knowledge mining

The basic philosophy behind the proposed data mining structure is to obtain fuzzy knowledge exhibiting not only good predictive properties but also strong descriptive capabilities. In addition to minimization of errors between desired and predicted outputs, the GA makes efforts to optimize the compatibility index of the rule base. Compatibility in interpreting rules is enforced by incorporation of the compatibility index into the evaluation function of the GA. Consequently, an economical and compatible fuzzy knowledge base can be expected; it has high generalization ability and is also easy for human inspection and understanding.

In investigation about fuzzy knowledge mining, we have to distinguish two basic types of fuzzy rule based systems: Mamdani and Sugeno models. Owing to limited scope, this chapter only considers Mamdani fuzzy rules employing output fuzzy sets in conclusions and fuzzy classification rules taking object classes as consequences. Construction of Sugeno fuzzy systems based on premise learning has been tackled in [26-27].

3.2 Encoding of Flexible Premises

In order to apply GAs to a premise learning problem, we must have an suitable representation of the space of premise descriptions that are generally hybrid in nature (containing both continuous and discrete parameters), have semantic and interpretable constraints, and can be of varying length and complexity. The key point here lies in the mapping of flexible premises into linear strings. In the sequel we discuss a specific form of strings used in the chapter for such purpose.

3.2.1 Binary Coding of Premise Structure

From Definition 1, we can see that flexible premise of a rule is characterized by sets $D(k) \subset \{1, 2, ..., q[p(k)]\}$. This fact suggests that a binary string be a suitable scheme for representing the structure of a premise, as inclusion or exclusion of an integer in the sets $D(k)$ can be declared binary. For input variable x_i (i=1, 2, ..., n) with $q[i]$ linguistic values, a segment consisting of $q[i]$ binary bits is required to encode the conjunctive term for this variable. Every bit of the segment corresponds to a linguistic value with bit '1' for presence and bit '0' for absence of its fuzzy set in forming the conjunctive term. For instance, assume that x_i has three linguistic values {low, middle, high}, the segments '010' and '100' correspond to the conjunctive terms 'x_i=middle' and 'x_i=low' respectively. Similarly, the conjunctive term with internal disjunction 'x_i=middle or high' is represented by the segment '011'. Notice that a segment including all '1's matches any value of the corresponding input and is thus equivalent to "dropping" that conjunctive term (i.e. the input is irrelevant for that premise). By composing the binary segments for all individual inputs, the premise structure of a rule can be encoded into a binary group P_r=(seg(x_1), seg(x_2), ..., seg(x_n)), where seg(x_i) (i=1....n) denotes the segment of the conjunctive term for input variable x_i.

Example 3: Assume four input variables x_1, x_2, x_3 and x_4 with x_i={low, middle, high} (i=1....4). The premise structure: (x_1=low or high) and (x_3=middle) and (x_4=middle or high) corresponds to the following binary code:

seg(x_1)=101, seg(x_2)=111, seg(x_3)=010, seg(x_4)=011,

P_r=[seg(x_1), seg(x_2), seg(x_3), seg(x_4)]=[101 111 010 011]

Further, since a knowledge base consists of more than one rule, the coding of premise structure for the whole rule set is realized through merging binary groups of all individual rule premises in an head-to-tail manner. Let N_m be the maximal number of rules possible to appear in the rule base, the binary code (CB) for premise structure of all rules as a whole is written as:

$$CB = [P_r(1),\ P_r(2),\cdots,\ P_r(N_m)] \tag{19}$$

where $P_r(i)$ (i=1, 2, ..., N_m) indicates the binary group for the premise of the ith possible rule.

Invalid Premises Encoded. It is worthy noting that either of the following two cases in a binary group leads to an invalid rule premise encoded:

1) All bits in the group are equal to one, meaning that all input variables are irrelevant for the premise. A rule condition like this is absurd and doesn't have any real meaning at all.

2) The group contains an all-zero segment. The corresponding conjunctive term thus includes no linguistic value in it and matches nothing. This means that any rule premise containing such a pattern will not match any points in the input space.

Variable Sized Rule Base Implied. Rules with invalid premises are meaningless or play no role in the fuzzy reasoning. Therefore they should be removed from the knowledge base. If all the binary groups P_r(i) in (19) correspond to valid premises, the rule base has exactly N_m rules in it. Otherwise the size of the rule base can be reduced according to invalid premises detected in (19). It is clear that the exact size of the rule base is not determined by human in advance. Rather, the actual rule number is implicated in the binary code CB, which is to be learned by the GA. In this respect, we note that adjustment of the size of the rule set is possible within a constraint of prescribed maximal rule amount.

To illustrate the representation of premise structure more concretely, consider the following example of binary code with three binary groups:

x_1	x_2	x_1	x_2	x_1	x_2
101	1111	010	0101	000	0100

Since the third binary group contains an all-zero segment, an invalid rule premise is implied such that the corresponding rule base contains in fact only two rules. Suppose that the input x_1 has linguistic values {low, medium, high} and x_2 has linguistic values {zero, small, medium, large}, this rule set is equivalent to:

If (x_1 is low or high) then ......
If (x_1 is medium) and (x_2 is small or large) then ......

3.2.2 Integer-Coding of Fuzzy Set Membership Functions

Triangular membership functions are adopted to define k (k: a prespecified number) fuzzy sets of a variable x. For the sake of optimal interfaces [15], we give certain constraints to the membership functions $\mu_j(x)$ (j=1, 2, ..., k) defined on the whole universe of discourse $[X_{min}, X_{max}]$ such that 1) at each point $x \in [X_{min}, X_{max}]$ there exist only two membership functions which are non-null; 2) two neighboring membership functions overlap at a 0.5 degree of coverage; 3) the right boundary X_{max} has a full membership grade to the most right fuzzy set and the left boundary X_{min} has a full membership degree to the most left fuzzy set. Let C_j's be the centers of the triangular membership functions $\mu_j(x)$ (j=1, 2, ..., k) which are numbered in an increasing order from left to right, that is $X_{min} = C_1 < C_2 < \cdots < C_k = X_{max}$. Between any two consecutive centers C_j and C_{j+1} there are only two nonzero membership functions which are defined by the equations:

$$\mu_j(x) = -\frac{1}{C_{j+1} - C_j} \cdot x + \frac{C_{j+1}}{C_{j+1} - C_j} \tag{20}$$

$$\mu_{j+1}(x) = \frac{1}{C_{j+1} - C_j} \cdot x - \frac{C_j}{C_{j+1} - C_j} \tag{21}$$

Under the above restrictions on membership functions, the corresponding fuzzy sets form a fuzzy partition of the universe of discourse in the strict sense, i.e.

$$\forall_{x \in [X_{min}, X_{max}]} \mu_1(x) + \mu_2(x) + \cdots + \mu_k(x) = 1 \tag{22}$$

Actually C_2, C_3, ..., C_{k-1} correspond to the endpoints of membership functions as shown in Fig. 4; they are critical parameters determining shapes and locations of membership functions to partition the universe of discourse. Exact definition of these triangular membership functions is thus equivalent to the specification of positions of their endpoints. In this respect we have transformed the problem of coding membership functions to the one of representing some important parameters.

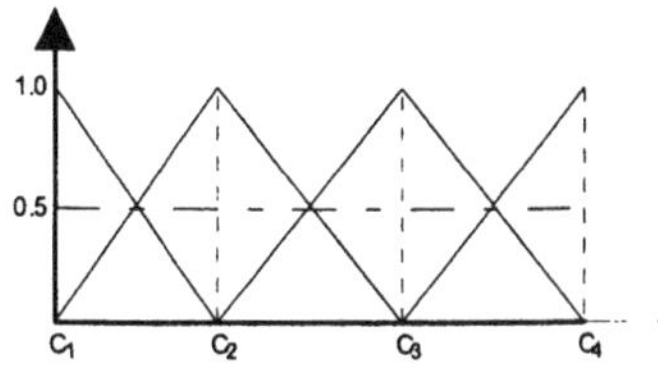

Fig. 4. Triangular membership functions adopted

Let a parameter C_j (j=2,...,k-1) be mapped by an integer N_j in the interval {0,1, ... N_{max}}, the relationship between the parameter value C_j and the integer N_j is below:

$$C_j = X_{min} + \frac{N_j}{N_{max}}(X_{max} - X_{min}) \tag{23}$$

We quantize every endpoint of the membership functions with an integer, and the composition of such integers results in an integer-chain: M_f=(N_2, N_3, ..., N_{k-1}), depicting all fuzzy subsets corresponding to the variable. For instance, the endpoints of the six membership functions in Fig. 5 are represented by integers and the six fuzzy sets are thus characterized by the integer-chain: M_f=(15, 25, 45, 70). The elements in the chain M_f are ordered from small to big with correspondence to endpoints from left to right.

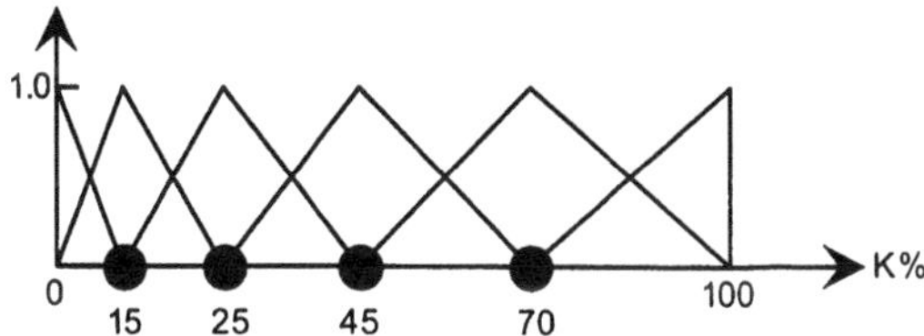

Fig. 5. Six fuzzy sets characterized by four integers

Further, the integer-chains for membership functions of all individual variables are merged together to produce an integer-code to represent information of fuzzy sets as a whole. Suppose there are n_v input/output variables in the fuzzy system and $M_f(i)$ (i=1,2, ...,n_v) is the integer-chain for membership functions of the ith variable, the integer-code (CI) for the fuzzy sets of the whole system is given by:

$$CI = [M_f(1),\ M_f(2),\ \cdots,\ M_f(n_v)] \tag{24}$$

3.2.3 Hybrid Coding of Conditions of Rules

Surface structure and fuzzy set membership functions are two essential aspects to describe rule conditions of a fuzzy knowledge base. Because of the inherent relationship between premise structure and membership functions, it is preferable that the both of them be optimized simultaneously. For this purpose, we devise a hybrid string (HS) which consists of the binary code in (19) and the integer-code in (24) as its two substrings. Figure 6 illustrates the constitution of such a hybrid string. The substring in left is the binary code for premise structure of the rule base. N_m denotes the upper limit of the total number of rules in the rule base and $P_r(j)$ (j=1....N_m) is the binary group corresponding to the premise structure of the jth possible rule. The substring in right is the integer-code representing membership functions of all input (and output) fuzzy sets, on which the surface structure of rule conditions depends. $M_f(1)$, $M_f(2)$, ..., $M_f(n_v)$ indicate the integer-chains for membership functions of the variables 1, 2, ..., n_v respectively. Such a hybrid string is treated as a chromosome in the GA population and it corresponds to a possible solution to rule premises of a fuzzy knowledge base. The initial population of hybrid strings can be randomly created. Through evolutionary process based on genetic operators, the quality of individuals in the population is expected to be gradually improved from generation to generation.

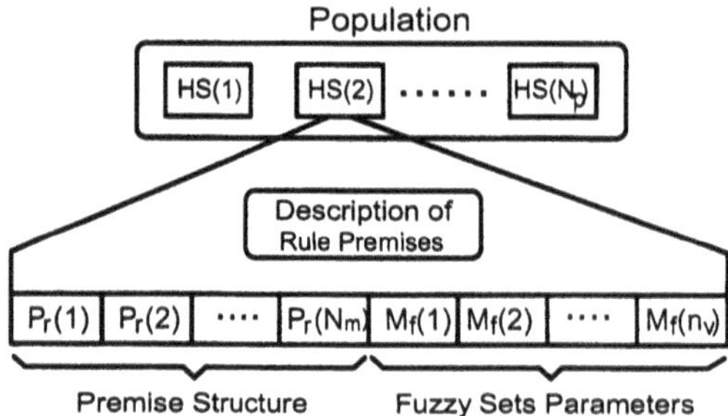

Fig. 6. A hybrid string in the population

3.3 Fitness Evaluation

Evaluation of a (hybrid) string is based upon the performance of the fuzzy knowledge base resulting from it. On one side, we require the model to be built predict system behavior as accurately as possible. On the other side, a compatible knowledge base free from internal contradictory is desirable and interesting. According to the above two criteria, the cost function for a hybrid string HS is constructed below:

$$Cost(HS) = err(HS) + \alpha_C[1 - \lambda_C(HS)] \tag{25}$$

Here $err(HS)$ and $\lambda_C(HS)$ indicate the error measure (on the training data) and the compatibility index of the knowledge base corresponding to the string HS. α_C is a penalty factor determined by users. As the compatibility measure of the rule base is now directly incorporated into the evaluation function, the GA searches for solutions with not only best accuracy but also minimal linguistic contradictory.

To acquire system outputs, consequences of fuzzy rules must first be selected according to their preconditions. This is the reason why the procedure of conclusion determination is nested in premise identification. The method of selecting the optimal conclusion under a fixed antecedent has been described in Section 2.2.

3.4 Genetic Operators

Genetic operators modify individuals within a population to create new individuals for testing and evaluation. Historically crossover, mutation and reproduction have been the most important and best understood genetic operators. One of the goals in the chapter is to establish a premise learning mechanism that could exploit these fundamental operators. We have achieved this goal with the hybrid string representation described previously, involving both premise structure and membership functions of fuzzy sets.

By the operation of crossover, parent strings (old premise descriptions) mix and exchange their attributes through a random process, so that offspring (new premise descriptions) with even higher fitness than current individuals can be generated. Owing to the distinct nature between the two substrings, it is preferable that the attributes in both substrings be mixed and exchanged separately. Therefore a special three-point crossover is devised here. One breakpoint for this operation is fixed to be the splitting point between both substrings, and the other two breakpoints are randomly selected within the two substrings respectively. At breakpoints the parent bits are alternatively passed on to the offspring. This means that offspring get bits from one of the parents until a breakpoint is encountered, at which they switch and take bits from the other parent.

Example 4: Consider two chromosomes in the following:

$$HS_1 = (b_1^1, b_1^2, b_1^3, b_1^4, b_1^5, b_1^6, b_1^7, b_1^8, b_1^9, b_1^{10}, b_1^{11}, b_1^{12} \mid c_1^1, c_1^2, c_1^3, c_1^4, c_1^5, c_1^6)$$

$$HS_2 = (b_2^1, b_2^2, b_2^3, b_2^4, b_2^5, b_2^6, b_2^7, b_2^8, b_2^9, b_2^{10}, b_2^{11}, b_2^{12} \mid c_2^1, c_2^2, c_2^3, c_2^4, c_2^5, c_2^6)$$

Both HS_1 and HS_2 consist of two substrings $(b_i^1, b_i^2, \cdots, b_i^{12})$, $(c_i^1, c_i^2, \cdots, c_i^6)$ (i=1,2) representing the premise structure and the parameters of fuzzy set membership functions respectively. The position between b_i^{12} and c_i^1 is the splitting point between two substrings. Selecting the other two breakpoints for the crossover operator as the position between b_i^5 and b_i^6 and the position between c_i^4 and c_i^5 respectively, we obtain the two child strings as follows:

$$HS_3 = (b_1^1, b_1^2, b_1^3, b_1^4, b_1^5, b_2^6, b_2^7, b_2^8, b_2^9, b_2^{10}, b_2^{11}, b_2^{12} \mid c_1^1, c_1^2, c_1^3, c_1^4, c_2^5, c_2^6)$$
$$HS_4 = (b_2^1, b_2^2, b_2^3, b_2^4, b_2^5, b_1^6, b_1^7, b_1^8, b_1^9, b_1^{10}, b_1^{11}, b_1^{12} \mid c_2^1, c_2^2, c_2^3, c_2^4, c_1^5, c_1^6)$$

Clearly this three-point crossover used here is equivalent to two one-point crossovers operating on both substrings separately.

Mutation is a local operator that transforms the bits of a GA construct, so as to increase the variability of population. Because of the distinct substrings used, different mutation schemes are needed to suit their purposes.

Membership Function Mutation. Since parameters of fuzzy set membership functions are essentially continuous, a small mutation with high probability is more meaningful. Therefore it is so designed that each element in the integer-substrings for membership functions undergo a disturbance. The magnitude of this disturbance is determined by a Gaussian random variable $N(0, \delta)$ with mean zero and variance δ.

Premise Structure Mutation. Bit mutation is applied on the binary substrings codifying surface structure of rule conditions. This is a random operation that occasionally occurs with a small probability (typically 0.01-0.05). Each bit of the binary code is flipped if a probability test is satisfied, i.e. a randomly generated real number is smaller than the prespecified probability.

Parent selection is a routine emulating the mechanism of survival of the better fitness in nature. It is expected that a better hybrid string will produce a higher number of offspring and thus has a higher chance of surviving in the subsequent generation. The selection rate of a hybrid string *HS*, *prob(HS)*, is determined by

$$prob(HS) = \frac{S - Cost(HS)}{(N_p - 1) \cdot S} \tag{26}$$

where N_P denotes the population size and S is the sum of cost values of all individuals in the population. Finally a new generation is generated according to selective breeding [14] as follows:

1) A set of offspring is created by selecting parents from current generation and applying genetic operators for recombination. Each integer-chain in the offspring then undergoes a reordering operation to maintain the proper order required for codifying membership functions of every input variable.
2) The individuals in the offspring set are evaluated in terms of accuracy and compatibility of the associated fuzzy system using the cost function in (25).
3) The best N_P individuals from the current population and the offspring set are chosen to form the next generation (N_P is the population size).

4 Tests and Results

Three examples are given in this section to demonstrate the capability of the proposed approach to induce compact fuzzy knowledge while maintaining excellent modeling accuracy. The first test was to set up a Mamdani fuzzy model for a system with continuous outputs. The latter two experiments were dedicated to establish fuzzy rule-based systems for pattern classification.

4.1 Modeling of Gas Furnace Data

The proposed approach was first applied to the well-known problem of modeling a gas furnace system introduced by Box and Jenkins [2]. This data set consists of 296 input-output measurements from a gas furnace system with a single input u being the gas flow rate and a single output y being the CO_2 concentration in outlet gas. The sampling interval is nine seconds. We chose $u(t\text{-}4)$ and $y(t\text{-}1)$ as inputs for the fuzzy model to predict the CO_2 concentration at time instant t. Such choice of inputs was motivated by the identified fact [20] that $u(t\text{-}4)$ and $y(t\text{-}1)$ are the most important factors affecting $y(t)$ in the underlying gas furnace system.

Before model building, we prescribed that both variables (gas flow rate and CO2 concentration) have five fuzzy sets. The linguistic terms like NB (negative big), NS (negative small), Z (zero), PS (positive small), and PB (positive big) were associated with the gas flow rate, while other five linguistic values including VL (very low), LL (little low), NM (normal), LH (little high), and VH (very high) were used to depict the CO_2 concentration. The corresponding membership functions for all fuzzy subsets are illustrated in Figs.7-8. It was requested that the sum of membership grades for every variable be always equal to one, so that only nine critical points ($a1$, $a2$, $a3$, $a4$ in Fig.7; $c1$, $c2$, $c3$, $c4$, $c5$ in Fig.8) were to be adapted.

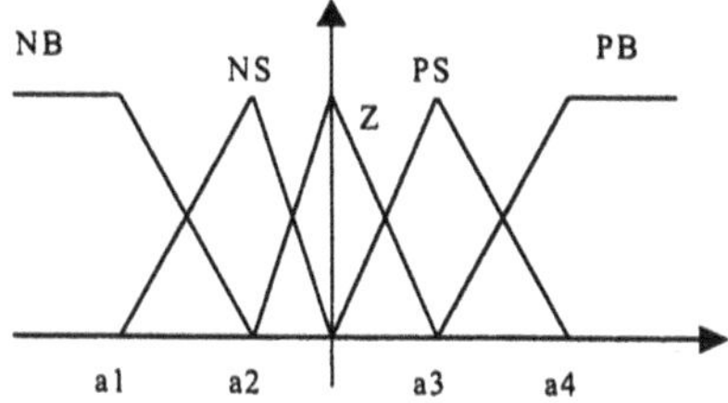

Fig. 7. Fuzzy sets for the gas flow rate

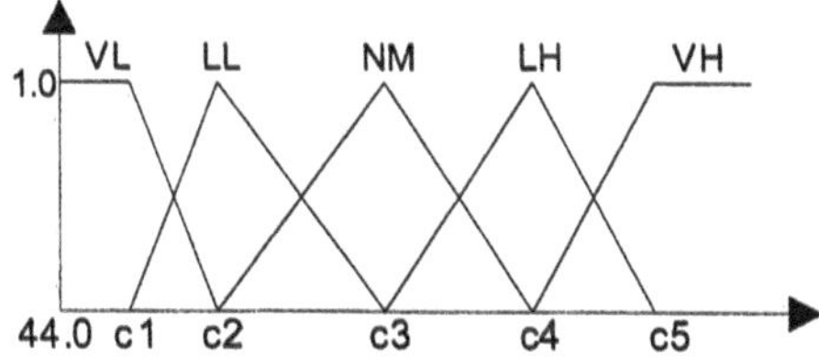

Fig. 8. Fuzzy sets for the CO_2 concentration

Assuming the upper limit of the number of required (Mamdani) rules to be 18, genetic algorithm was used to identify the structure of premises of these possible rules as well as to optimize the parameters ($a1$, $a2$, $a3$, $a4$, $c1$, $c2$, $c3$, $c4$, $c5$) of the fuzzy set membership functions concurrently. As a result of the genetic search, a compact fuzzy model was decoded from the best hybrid string found. Since five premises encoded in the binary sub-string are actually invalid, the obtained knowledge base contains only 13 rules in it. Further there is no linguistic incompatibility between the rules.

Selecting maximum operation as *s-Norm* and algebraic product as *t-Norm*, this fuzzy model has a mean-squared-error (MSE) of 0.338 on the training data. Fig. 9 shows the outputs from the fuzzy model in comparison with the actual outputs of the process.

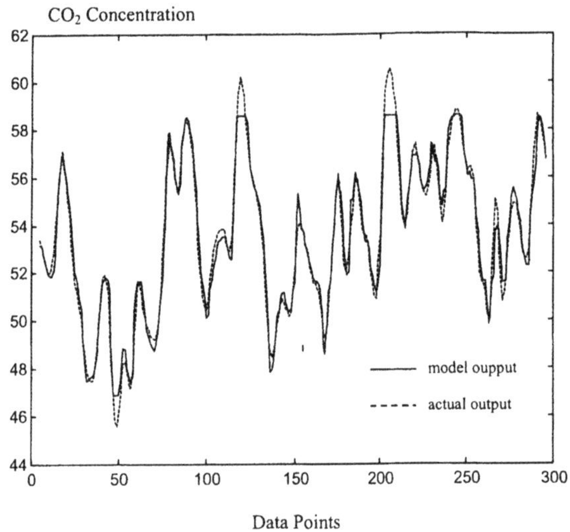

Fig. 9. Outputs from the Mamdani fuzzy model

Mamdani fuzzy models on the gas furnace data were studied as an interesting benchmark problem in the literature. A comparison between our result and those obtained by others is given in Table 2, in terms of model accuracy, the number of input fuzzy sets as well as the number of Mamdani rules required. The MSE here reflects the predicting ability of the models generated, while the other two indexes indicate model's complexity and description capability. Obviously our fuzzy model achieves much better accuracy than those reported in [21] and [16]. It also exhibits remarkable advantage over the model by [28] in requiring significantly fewer fuzzy rules but offering very similar numerical accuracy.

Table 2. Comparison with other Mamdani models

Model	Fuzzy Model Inputs	Number of Input Membership Functions	Number of Rules	MSE
Tong [21]	y(t-1), u(t-4)	13	19	0.469
Pedrycz [16]	y(t-1), u(t-4)	10	25	0.776
Xu and Lu [28]	y(t-1), u(t-4)	10	25	0.328
Our model	y(t-1), u(t-4)	10	13	0.338

4.2 Classification of Iris data

We also applied the proposed approach to mine classification rules from the Iris Data [7]. The task was to classify three species of iris (setosa, versicolor and virginica) by four-dimensional attribute vectors consisting of sepal length (x_1), sepal width(x_2), petal length (x_3) and petal width(x_4). There are 50 samples of each class in this data set. By randomly taking 10 patterns from each class as test data, the total data set was divided into training set (80%) and test set (20%). We built the fuzzy classification system based on the training data and then verified its performance according to the test dada that were not used for learning.

Each attribute of the classification system was assigned with three linguistic terms: *short, middle and long*. By normalizing attribute values of each attribute into the interval between zero and one, the membership functions of input fuzzy sets are depicted in Figure 10. The upper limit of the rule number in the rule base was set to *16*, meaning that *16* rules were supposed to be sufficient to achieve desirable classification accuracy. The GA was put into work to search for the premise structure of possible rules and to optimize the parameters (corresponding to the circle in Fig. 10) of the input fuzzy sets at the same time. As a result of the search process, 10 rule premises were identified as invalid so that the rule base in fact contains only six rules in it. Again, the size of the rule base in this example was not exactly prescribed by human. Instead it was automatically adjusted by the learning algorithm within an upper limit.

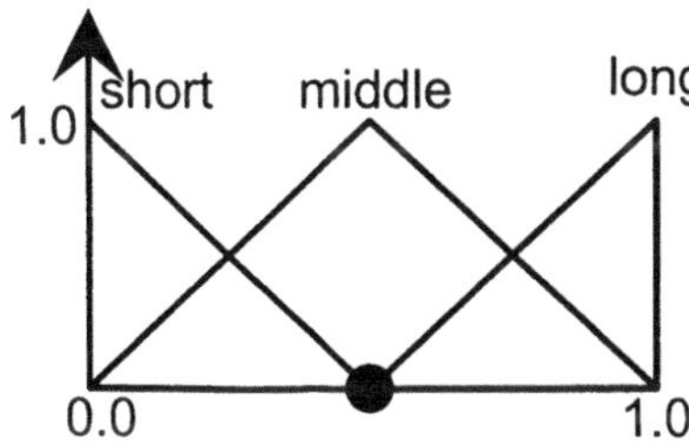

Fig. 10. The membership functions of the attributes for iris data

The fuzzy classification system learned is completely compatible and has very high accuracy despite its quite small number of rules. On the training set it correctly classifies *119* patterns (i.e. *99.2%* of the *120* patterns). On the test data *30* patterns (i.e. *100%* of the *30* patterns) are properly classified.

The results of some other machine learning algorithms on the Iris Data are given in Table 3 for comparison. It is striking that our model, despite its very small rule number, achieves the best accuracy among those yielded by other algorithms in the table.

Table 3. Accuracy of the other five algorithms on the Iris Data

Algorithms	Setosa	Viginica	Versicolor	Average
Hirsh [9]	100%	93.33%	94.00%	95.78%
Aha [1]	100%	93.50%	91.13%	94.87%
Dasarathy [6]	100%	98%	86%	94.67%
C4 [18]	100%	91.07%	90.61%	93.89%
Hong [10]	100%	94%	94%	96.00%

4.3 Classification of Cancer Data

In order to examine the capability of the proposed method for high dimensional data mining problems, we used the cancer data as the third example for case study. This data set consisting of 683 samples from two classes with 9 continuous attributes is available from the University of California, Irvine, database (available via anonymous ftp from *ftp.ics.uci.edu/pub/machine-learning-databases*). The task was to classify character of breast cancers (either benign or malignant) according to their attribute values. By randomly taking 100 patterns as test data, the total data set was divided into training set (85.4%) and test set (14.6%). We built the fuzzy classification system based on the training data and then verified its performance according to the test dada that were not used for learning.

Each attribute x_i (i=1, 2, ..., 9) of the classification system was assigned with three fuzzy subsets: $A(i,1)$, $A(i,2)$ and $A(i,3)$, whose membership functions are depicted in Fig. 11. At first the maximal number of rules was supposed to be to 15, implying that 15 rules were assumed to be adequate to achieve satisfying classification accuracy. The GA was used to search for the premise structure of possible rules and to optimize the nine parameters (corresponding to the circle in Fig.11) of the input fuzzy sets at the same time. As a result of the search process, 11 premises from the best string are invalid such that the rule base learned contains only four rules in it.

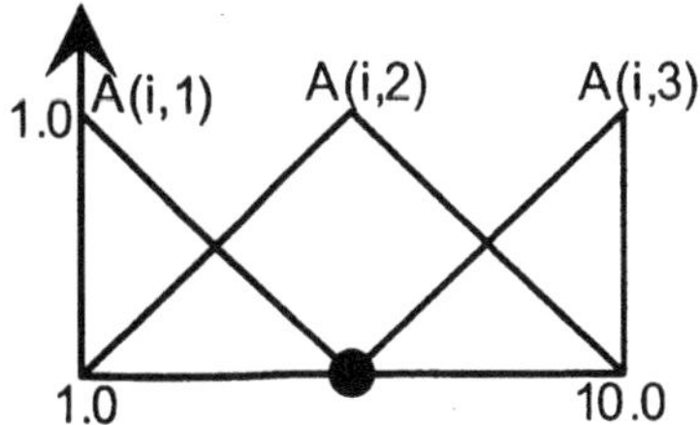

Fig. 11. The three membership functions of attribute x_i for cancer data

The fuzzy classification system learned from the GA has a very good accuracy in spite of the low number of rules. On the training set it correctly classifies 573 patterns (i.e. 98.2847% of the 583 patterns). On the test data 91 patterns (i.e. 91% of the 100 patterns) are properly classified. Moreover, there is no linguistic incompatibility between the four rules in the knowledge base.

Cancer data were used in [23] for evaluating various data mining methods. The following results were obtained from using random resampling procedure and reported in [23] as the performances of ten non-fuzzy classification methods.

Best classification rate on test data among ten methods: 77.1%;
Average classification rate on test data over ten methods: 70.9%;
Worst classification rate on test data among ten methods: 65.3%.

In [11] the cancer data was classified with fuzzy technique. They generated 36 (${}_9C_7$=36) fuzzy rule-based classification systems by partitioning seven of the nine axes of the pattern space into two triangular fuzzy sets. The remaining attributes were treated as irrelevant features in each fuzzy rule-based classification system. For instance, one of such fuzzy classification systems was derived based on the 2×2×2×2×2×2×2×1×1 fuzzy grid. On average, a classification rate of 74.7% was achieved on test data by applying random resampling procedures.

From the results reported in [23] and [11], we see that the 91% classification rate acquired by our approach is evidently superior to the results by all the nine non-fuzzy classification methods as well as the scheme based on fuzzy grid. Moreover the fuzzy grid method used in [11] led to the requirement of 128 rules, while the fuzzy classification system learned here contains only four rules but provides much better accuracy. This fact demonstrates the power and effectiveness of the proposed method for mining concise and accurate knowledge, in particular when the number of attributes for the system is high.

5 Conclusion and Discussion

A key issue in constructing fuzzy rule sets is to specify rule conditions, which determine the structure of a fuzzy knowledge base. Taking into account all canonical AND connectives of input fuzzy sets as rule antecedents can not scale well up to high-dimensional problems with multiple input variables, since the number of rules generated increases exponentially with the input dimension. It is argued in this chapter that flexible premises of rules can be identified automatically via machine learning technique. GA is used for this purpose to yield appropriate premise structure of various rules as well as to optimize membership functions of fuzzy sets at the same time. A rule number reduction is enabled by this genetic-based premise learning, as flexible premises are encoded in the GA population and they can achieve bigger coverage of the input domain compared with those based on grid-type fuzzy partitions.

The outcome of this work should be a concise fuzzy knowledge base that exhibits high generalizing capability and enhances human understanding. Further, there is no or little inconsistency in the rule base. Linguistic incompatibility between rules is strongly discouraged in the genetic search process by incorporating rule base's compatibility index into the evaluation function of the GA.

The proposed approach also enables a variable size of the rule base. Before running of the algorithm, human users are not required to specify the exact number of rules in the knowledge base. What is needed is only a 'guess' about how many rules appear sufficient for the problem at hand. The GA adjusts the actual rule number within an upper limit specified by human in advance.

In relevance to genetic fuzzy systems [5], our current work falls into the category of Pittsburgh-learning systems by treating the whole set of premises as an individual in the GA population. The other two fundamental modes for evolving fuzzy systems are Michigan-learning and Iterative-learning, as presented in [5]. However, there is still much to be investigated about the relative merits of the three genetic learning strategies to identify flexible structured premises in mining concise (fuzzy) knowledge. Exploring potentials of Michigan-learning and iterative-learning approaches for premise optimization would be intriguing for future research.

References

[1] D. W. Aha and D. Kibler. Noise-tolerant instance-based learning algorithms. In *Proc. 11th Internat. Joint Conf. on Artificial Intelligence*, pages 794-799, Detroit, MI, 1989.

[2] G. E. P. Box and G. M. Jenkins. Time Series Analysis. In *Forecasting and Control*, San Francisco, CA: Holden Day, 1970.

[3] Y. Cho et al. Autogeneration of fuzzy rules and membership functions for fuzzy modelling using rough set theory. *IEE Proc.-Control Theory and Applications*, 145: 437- 442, 1998.

[4] O. Cordon, F. Herrera and M. Lozano. On the bidirectional integration of fuzzy logic and genetic algorithms. In *Proc. 2nd Online Workshop Evolutionary Computation (WEC2)*, pages 13-17, Nagoya, Japan, 1996.

[5] O. Cordon, F. Herrera, F. Hoffman and L. Magdalena. *Genetic fuzzy systems: Evolutionary tuning and learning of fuzzy knowledge bases*. World Scientific, Singapore, 2001.

[6] B. V. Dasarathy. Noising around the neighbourhood: a new system structure and classification rule for recognition in partially exposed environments. *IEEE Trans. Pattern Analysis and Machine Intelligence*, 2: 67-71, 1980.

[7] R. A. Fisher. The use of multiple measurements in taxonomic problems. *Ann. Eugen*, 7: 179-188, 1936.

[8] D. E. Goldberg. *Genetic algorithms in search, optimization and machine learning*. Addison-Wesley, New York, 1989.

[9] H. Hirsh. Generalizing version spaces. *Machine Learning*, 17:5-46, 1994.

[10] T. P. Hong and S. S. Tseng. A generalized version space learning algorithm for noisy and uncertain data. *IEEE Trans. Knowledge Data Eng.*, 9: 336-340, 1997.

[11] H. Ishibuchi, et al. Voting in fuzzy rule-based systems for pattern classification problems. *Fuzzy Sets and systems*, 103: 223-238, 1999.

[12] B. Kosko. *Neural Networks and Fuzzy Systems*. Prentice-Hall, Englewood Cliffs, NJ., 1992.

[13] J. Liska and S. S. Melsheimer. Design of fuzzy logic systems for nonlinear process identification. In *Proceedings of the American Control Conference*, pages 976-980, Maryland, 1994.

[14] D. A. Linkens and H. O. Nyongesa. Genetic algorithms for fuzzy control. Part 1: Offline system development and application. *IEE Proc.-Control Theory and Applications*, 142(3): 161-176, 1995.

[15] J. V. Oliveira. Semantic constraints for membership function optimization. *IEEE Trans. Fuzzy Systems*, 29: 128-138, 1999.

[16] W. Pedrycz. An identification algorithm in fuzzy relational systems. *Fuzzy Sets and Systems*, 13: 153-167, 1984.

[17] H-P. Preuß and J. Ockel. Lernverfahren für Fuzzy-Regler. *Automatisierungstechnische Praxis*, **37**: 10-20, 1995.

[18] J. R. Quinlan. *C4.5: Programs for Machine Learning*. Morgan Kaufmann, 1993.

[19] A. Rahmoun and M. Benmohamed. Genetic algorithm based methodology to generate automatically optimal fuzzy systems. *IEE Proc.-Control Theory and Applications*, 145: 583-586, 1998.

[20] M. Sugeno and T. A. Yasukawa. A fuzzy-logic-based approach to qualitative modelling. *IEEE Trans. Fuzzy Systems*, **1**: 7-31, 1993.

[21] R. M. Tong. The evaluation of fuzzy models derived from experimental data. *Fuzzy Sets and Systems*, 4: 1-12, 1980.

[22] L.-X. Wang and J. M. Mendel. Generating fuzzy rules by learning from examples. *IEEE Trans. Systems, Man, and Cybernetics*, 22: 1414-1427, 1992.

[23] S. M. Weiss and C.A. Kulikowski. *Computer systems that learn*. Morgan Kaufmann, San Mateo, 1991.

[24] T. Wittmann, J. Ruhland and M. Eichholz. Enhancing rule interestingness for neuro-fuzzy systems. In *Principles of Data Mining and Knowledge Discovery*, pages 242-250. Spring Verlag, 1999.

[25] N. Xiong. Fuzzy tuning of PID controllers based on genetic algorithm. *Systems Science*, 25(3): 65-75, 1999.

[26] N. Xiong and L. Litz. Fuzzy modeling based on premise optimization.In *Proceedings of the 9th IEEE International Conference on Fuzzy Systems*, pages 859-864, San Antonio, 2000.
[27] N. Xiong. Evolutionary learning of rule premises for fuzzy modelling. *International Journal of Systems Science*, 32(9): 1109-1118, 2001.
[28] C. W. Xu, and Y. Z. Lu. Fuzzy model identification and self-learning for dynamic systems. *IEEE Trans. Syst., Man, Cybern.*, **17:** 683-689, 1987.
[29] L. A. Zadeh. Fuzzy sets. *Information Control* , 8: 338-353, 1965.
[30] L. A. Zadeh. Outline of a new approach to the analysis of complex systems and decision processes. *IEEE Transactions on Systems, Man, Cybernetics*, 3: 28-44, 1973.

SECTION 3

COMPLEXITY REDUCTION IN LINGUISTIC FUZZY MODELS

A Multiobjective Genetic Learning Process for joint Feature Selection and Granularity and Contexts Learning in Fuzzy Rule-Based Classification Systems

Oscar Cordón[1], María José Del Jesus[2], Francisco Herrera[1], Luis Magdalena[3], and Pedro Villar[4]

[1] Department of Computer Science and Artificial Intelligence,
University of Granada, 18071 Granada, Spain
e-mails: {ocordon,herrera}@decsai.ugr.es
[2] Department of Computer Science,
University of Jaén, 23071 Jaén, Spain
e-mail: mjjesus@ujaen.es
[3] Departmento de Matemática Aplicada,
Universidad Politécnica de Madrid, 28040 Madrid, Spain
e-mail: llayos@mat.upm.es
[4] Department of Computer Science,
University of Vigo, 32001 Ourense, Spain
e-mails: {amlopez,pvillar}@uvigo.es

Abstract. In this contribution, we propose a genetic process to select an appropiate set of features in a Fuzzy Rule-Based Classification System (FRBCS) and to automatically learn the whole Data Base definition using a non linear scaling function to adapt the fuzzy partition contexts and determining an appropiate granularity for each of them. An ad-hoc data covering learning method is considered to obtain the Rule Base. The method uses a multiobjective genetic algorithm in order to obtain a good trade-off between accuracy and interpretability.
Keywords: Fuzzy Rule-Based Classification Systems, Feature Selection, Granularity, Fuzzy Context Adaptation, Genetic Algorithms.

1 Introduction

An FRBCS presents two main components: the Fuzzy Reasoning Method and the Knowledge Base (KB). The KB is composed of the Rule Base (RB) constituted by the collection of fuzzy rules, and of the Data Base (DB), containing the membership functions of the fuzzy partitions associated to the linguistic variables. The composition of the KB of an FRBCS directly depends on the problem being solved. If there is no expert information about that problem, an automatic learning process must be used to derive the KB from examples.

[1] This research has been supported by CICYT under project PB98-1319.

Although, there is a large quantity of RB learning methods proposed in the specialised literature [6,9,20,31,38], among others, there is not much information about the way to derive the DB and most of these RB learning methods need of the existence of a previous definition for it. The usual way to proceed involves choosing a number of linguistic terms (granularity) for each linguistic variable, which is normally the same for all of them, and building the fuzzy partition by a uniform partitioning of the variable domain into this number of terms. This operation mode makes the granularity and fuzzy set definitions have a significant influence on the FRBCS performance. In fact, some studies in Fuzzy Rule-Based Systems have shown the importance of the choice of the semantics in the DB [14]. In this contribution we propose a process to learn the number of labels for each variable and the form of each fuzzy membership function in non-uniform fuzzy partitions, using a non-linear scaling function that defines different areas in the variable working range where the FRBCS has a higher or a lower relative sensibility, i.e., the fuzzy partition contexts.

Moreover, high dimensionality problems present a new difficulty to obtain FRBCSs with good behaviour: the large number of features, that can originate an RB with a high number of rules, thus presenting a low degree of interpretability and a possible over-fitting. This problem can be tackled from a double perspective:

- Via the compactness and reduction of the rule set, minimising the number of fuzzy rules included in it. Unnecessary rules can be eliminated with the aim of having a more cooperative rule set in order to obtain an FRBCS with better performance.
- Via a feature selection process that reduces the number of features used by the FRBCS.

Rule reduction methods have been formulated using Neural Networks, clustering techniques, orthogonal transformation methods and similarity measures [42], as well as using GA-based rule selection processes to get a cooperative set of rules from a candidate rule set [9,29].

Notice that, for high dimensional problems and problems where a high number of instances is available, it is difficult for the latter approaches to get small rule sets, and therefore the system comprehensibility may not be as good as desired. For high dimensionality classification problems, a feature selection method that determines the most relevant variables before or during the FRBCS inductive learning process, must be considered [3,37]. It increases the efficiency and accuracy of the learning and classification stages.

Our goal is to develop a genetic process for feature selection and whole DB learning (granularity and membership functions for each variable) to obtain FRBCSs composed of a compact set of comprehensible fuzzy rules with high classification ability. This method uses a multiobjective GA and considers a simple generation method to derive the RB.

To carry out this task, this paper is organised as follows. In Section 2, the preliminaries will be briefly introduced: FRBCS components, feature selection, DB learning, fuzzy context adaptation and multiobjective optimisation. In Section 3 we will expose the characteristics of our proposal for the FRBCS design. Some expermiental results will be shown in Section 4. In the last section, some conclusions will be pointed out.

2 Preliminaries

2.1 Fuzzy Rule-Based Classification Systems

An FRBCS is an automatic classification system that uses fuzzy rules as knowledge representation tool. Two different components are distinguished within it:

1. The **KB**, composed of:
   - **DB**, which contains the fuzzy set definitions related to the labels used in the fuzzy rules. So, the DB components for every variable are the number of linguistic terms (granularity) and the membership function shape of each term.
   - **RB**, comprised by a set of fuzzy rules that in this work are considered to have the following structure:

     $$R_k: \text{ If } X_1 \text{ is } A_1^k \text{ and } \ldots \text{ and } X_N \text{ is } A_N^k \\ \text{then } Y \text{ is } C_j \text{ with } r^k$$

     where $X_1, \ldots, X_N$ are the input variables considered in the problem (features) and $A_1^k, \ldots, A_N^k$ are linguistic labels employed to represent the values of the variables.
     This kind of fuzzy rules represents, in the antecedent part, a subspace of the complete search space defined by means of a linguistic label for each considered variable and, in the consequent part, a class label (C_j) and a certainty degree (r^k). This numerical value indicates the degree of certainty of the classification in that class for the examples belonging to the fuzzy subspace delimited by the antecedent part.
2. The **Fuzzy Reasoning Method (FRM)**, an inference procedure which, combining the information provided by the fuzzy rules related to the example to classify, determines the class to which it belongs to.

The majority of FRBCSs (see [6,20] among others) use the classical FRM that classifies a new example with the consequent of the fuzzy rule having the highest degree of association. Another family of FRMs that uses the information provided by all the rules compatible with the example (or a subset of them) have been developed [6,10,30]. In this work, we use two different FRMs, each one belonging to one of the said groups: maximum and normalised sum.

2.2 Feature Selection Process

The main objective of any feature selection process is to reduce the dimensionality of the problem for the supervised inductive learning process. This fact implies that the feature selection algorithm must determine the best set of features for its design.

There are two kinds of feature selection algorithms:

- *Filter feature selection algorithms* [33], which remove the irrelevant characteristics without using a learning algorithm (e.g. by means of class separability measures). They are efficient processes but, on the other hand, the feature subsets obtained by them may not be the best ones for a specific learning process because of the exclusion of the heuristic and the bias of the learning process in the selection procedure [32].
- *Wrapper feature selection algorithms* [32,33]. This kind of feature selection algorithms selects feature subsets by means of the evaluation of each candidate subset with the precision estimation obtained by the learning algorithm. In this form, they obtain feature subsets with the best behaviour in the classifier design. Their problem is their inefficiency since the classifier has to be built for the evaluation of each candidate feature subset.

In our proposal we will use a wrapper feature selection algorithm which utilises the precision estimation provided by an efficient fuzzy rule generation process (Wang and Mendel's fuzzy rule generation process) and considers a GA as search algorithm. Within the DB derivation, the granularity learning will provide us an additional way to select features when the number of linguistic labels assigned to a specific variable is only one (we will explain this in a further section).

2.3 DB Learning

As said, the derivation of the DB highly influences the FRBCS performance. Some approaches have been proposed to improve the FRBCS behaviour by means of a tuning process once the RB has been derived [1,9]. However, these tuning processes only adjust the shapes of the membership functions and not the number of linguistic terms in each fuzzy partition, which remains fixed from the begining of the design process.

The methods that try to learn appropiate DB components per variable usually work in collaboration with an RB derivation method. A DB generation process wraps an RB learning one working as follows: each time a DB has been obtained by the DB definition process, the RB generation method is used to derive the rules, and some type of error measure is used to validate the whole KB obtained. The approaches proposed in [13–15] use Simulated Annealing and GAs to learn an appropiate DB in a Fuzzy Rule-Based System. The method proposed in [28] considers a GA to design an FRBCS, working in the said way.

2.4 Context Adaptation in Fuzzy Processing

The use of contextual transformation functions to adjust membership functions in Fuzzy Systems can be a good way to find the fuzzy sets that best represent the linguistic terms they are associated with.

Considering concepts representable by fuzzy sets, the main effect of a context can be related to some sort of filtering. That is, the same base concept can be perceived in different situations, provided it is filtered to suit the context particulars [23].

The scaling functions map the input variables onto the universe of discourse of the fuzzy sets definitions. From a linguistic point of view, the scaling function can be interpreted as a sort of context information. While the membership functions describe the relative semantics (context-independent) of the linguistic variables contained in the rules, the union of the scaling functions and the membership functions generates the absolute semantics of the linguistic variables (context-dependent through the scaling functions).

Different results may be obtained when modifying the scaling function working on a certain variable, obtaining two kinds of contexts, linear and non-linear. In a non-linear context, the membership function shape can be adjusted. This is not possible when using linear context adaptation.

In [23], several details about fuzzy context adaptation can be found, such us a formalisation of the context adaptation procedure, the formal requirements for the scaling function and some context adaptation approaches.

The next proposals are examples of Fuzzy Rule-Based System learning methods based on context adaptation:

- The approach proposed in [39] adjusts the membership functions of the input variables and the gain for the output variable in the framework of Fuzzy Control.
- In [23], it is presented a genetic learning method for the parameters of a determinated non-linear scaling function, in order to find an adequate fuzzy partition per variable. A similar study based on Neural Networks is proposed in [40].
- The approach proposed in [34] uses a GA to adapt the whole KB definition of a Fuzzy Controller. For the learning of the DB, the GA evolves the working ranges of the linguistic variables and the membership functions shapes, by means a non-linear scaling function that produces high sensibility either for the medium values or for the extreme values of the linguistic variables. This approach is refined in [35] by adding a new parameter to the non-linear scaling function that may produce high sensibility in only one of the working range limits of the linguistic variables.

The main contribution of our method, apart from including the feature selection, is that it also evolves the number of labels per linguistic variable, an information that has not been considered to be relevant for the context adaptation methods mentioned before. The fuzzy partition granularity of a linguistic variable can also be viewed as a sort of context information. Considering a specific label set for a variable, the scaling functions can adapt the membership functions to a determinated real situation. However, in other contexts, some labels can result irrelevant, that is, they can contribute nothing and even can cause confusion. In other cases, it would be necessary to add new labels to appropiately differenciate the values of the variable.

2.5 Multiobjective optimisation

Many real-world problems involve simultaneous optimisation of multiple objectives. In principle, multiobjective optimisation is very different from single-objective optimisation. The latter attempts to obtain the best solution (the global minimum or the global maximum depending on the problem). However, in the case of multiple objectives, it may no exist one solution better than the rest with respect to all objectives.

In a typical multiobjective optimisation problem, there is a set of solutions that are superior to the remainder in the search space when all the objectives are considered (the *Pareto* set), but are inferior to other solutions in the search space of only several of them. These solutions are known as *non-dominated solutions* [5,24], while the remaining solutions are known as *dominated solutions.* Since none of the solutions in the non-dominated set is absolutely better than the other ones, all of them are equally acceptable as regards the satisfaction of all the objectives.

Matematically, the concept of *Pareto-optimality* or *non-dominance* is defined as follows. Let us consider, without loss of generality, a multiobjective minimisation problem with m parameters (decision variables) and n objectives:

$$Minimise f(x) = (f_1(x), f_2(x), \dots, f_n(x)), \quad being\ x = (x_1, x_2, \dots, x_m) \in X$$

A decision vector $a \in X$ dominates $b \in X$ if, and only if:

$$\forall i \in 1, 2, \dots, n \mid f_i(a) \leq f_i(b) \quad \wedge \quad \exists j \in 1, 2, , \dots, n \mid f_j(a) < f_j(b)$$

Any vector that is not dominated by any other is said to be *Pareto-optimal* or *non-dominated.*

A common difficulty with multiobjective optimisation is the appearance of an *objective conflict* [24], i.e., none of the feasible solutions allows simultaneous optimal values for all objectives. Mathematically, the best way of acting is keeping those solutions with the least objective conflict. These solutions can be viewed as points in the search space that are optimally placed from the individual optimum of each objective. However, such solutions may not satisfy a decision-maker because he or she may want a solution that satisfies some associated priorities of the objectives. To find such points, the classical multiobjective methods scalarise the objective vector reducing it to a scalar optimisation problem.

The three classical methods most commonly used to solve the multioptimality are the *objective weighting* method, the method of *distance functions* and the method based on the *min-max formulation* [7]. However, these classical techniques have serious drawbacks mainly due to objectives are combined to form an unique objective:

- An only Pareto-optimal solution is usually obtained. It would be desirable that the method offers different alternatives.
- If some objectives are noisy or have a discontinuous variable space, the methods may not work properly.
- If the objective functions are not deterministic, the setting of a weight vector or a demand level may become even more difficult.
- There is much sensivity and dependency towards weights.

In order to solve these problems, in the last few years some advanced multiobjective optimisation techniques have been proposed. Many of then are based on GAs [8,16]. The following subsection will be focused on multiobjective GAs.

Multiobjective Genetic Algorithms. Thanks to the use of a population of solutions, GAs can simultaneously search many Pareto-optimal solutions. For this reason, GAs have been recognised to be possibly well-suited to multiobjective optimisation. Furthermore, the ability to handle complex problems, involving features such as discontinuities, multimodality, disjoint feasible spaces, and noisy function evalutions, reinforces the potencial effectiveness of GAs in multiobjective search and optimisation.

Generally, the multiobjective approaches only differ from the rest of GAs in the fitness function and/or in the selection operator. The evolutionary approaches in multiobjective optimisation can been clasified in three groups: plain aggregating approaches, population-based non-Pareto approaches, and Pareto-based approaches [8,16].

The first group constitutes the extension of classical methods to GAs. The objectives are artificially combined, or aggregated, into a scalar function according to some understanding of the problem, and then the GA is applied. Practically, all the clasical aggregation approaches can be used with GAs. Optimising a combination of the objectives has the advantage of producing a single compromise solution. The problem is that, if the optimal solution can not be accepted, new runs of the GA may be required until a suitable solution is found.

The population-based non-Pareto approaches allow to exploit the special characteristics of GAs. A non-dominated individual set is obtained instead of obtaining only one of them. In order to do this, the selection operator is changed. Generally, the best individuals according to each of the objectives are selected, and then these partial results are combined to obtain the new population. An example of multiobjective GA of this group is Vector Evaluated Genetic Algorithm (VEGA) [41].

Finally, the Pareto-based approaches attempt to promote the generation of multiple non-dominated solutions, as the former group, but directly making use of the Pareto-optimality definition. In this approach, to calculate the probability of reproduction of each individual, the solutions are compared by means of the dominance relation. However, although the Pareto-based ranking correctly assigns all non-dominated individuals the same fitness, it does not guarantee that the Pareto set is uniformly sampled. When multiple equivalent optima exist, finite populations tend to converge to only one of them, due to stochastic errors in the selection process. This phenomenom is known as *genetic drift* [18]. Since preservation of diversity is crucial in the field of multiobjective optimisation, several multiobjective GAs have incorporated the *niche* and *species* concepts for the purpose of favouring such behaviour. Examples of Pareto-based multiobjective GAs are: Niched Pareto Genetic Algorithm (NPGA) [27], Non-dominated Sorting Genetic Algorithm (NSGA) [43] and Multiple Objective Genetic Algorithm (MOGA) [17]. In this contribution we will consider the latter.

3 Multiobjective Genetic Algorithm for Feature Selection and DB Learning

In this section, we propose a new learning approach to automatically generate the KB of an FRBCS composed of two methods with different goals:

- A genetic learning process for the DB that allows us to define:
  - The relevant variables for the classification process (feature selection).
  - The number of labels for each variable (granularity learning).
  - The form of each fuzzy membership function in non-uniform fuzzy partitions, using a non-linear scaling function that defines different areas in the variable working range where the FRBCS has a higher or a lower relative sensibility.
- A quick *ad hoc data-driven method* that derives the fuzzy classification rules considering the DB previously obtained. In this work we use the extension of Wang and Mendel's fuzzy rule generation method [45] for classification problems [6], but other efficient generation methods can be considered.

The main purpose of our KB design process is to obtain FRBCSs with good accuracy and high interpretability. Unfortunately, it is not easy to achieve these two objectives at the same time. Normally, FRBCSs with good performance have a high number of selected variables and also a high number of rules, thus presenting a low degree of readability. On the other hand, the KB design methods sometimes lead to a certain over-fitting to the training data set used for the learning process.

To avoid these problems, our genetic process uses a multiobjective GA with two goals:

- Design an accurate KB. This objective is performed by minimising the classification error percentage over the training data set.
- Design a compact and interpretable KB. This objective is performed by penalising FRBCSs with a large number of selected features and high granularity.

The next two subsections explain in detail the learning process proposed. The first one deals with the context adaptation proposal and the other one describes the main components of the GA.

3.1 Context adaptation proposal

All the components of the DB will be adapted throughout a genetic process. Since it is interesting to reduce the dimensionality of the search space for that process, the use of non-linear scaling functions is conditioned by the necessity of using parameterised functions with a reduced number of parameters. We consider the scaling funtion proposed in [13], that has a single sensibility parameter called a ($a \in \mathbb{R}$). The function used is:

$$f : [-1,1] \rightarrow [-1,1] \qquad f(x) = sign(x) \cdot |x|^a \,, \quad with \;\; a > 0$$

The final result is a value in $[-1,1]$ where the parameter a produces uniform sensibility ($a = 1$), higher sensibility for center values ($a > 1$), or higher sensibility for extreme values ($a < 1$). In this paper, triangular membership functions are considered due to their simplicity. So, the non-linear scaling function will only be applied on the three definition points of the membership function (which is equal to transform the scaling function in a continuous piece-wise linear function), in order to make easier the structure of the generated DB and to simplify the defuzzification process. Figure 1 shows a graphical representation of the three possibilities.

We should note that the previous scaling function is recommended to be used with symmetrical variables since it causes symetrical effects around the center point of the interval. For example, it can not produce higher sensibility in only one of the working range extents. In the method presented in this paper, we add a new parameter (called S) to the non-linear scaling function as described also in [13]. S is a parameter in $\{0,1\}$ to distinguish between non-linearities with symmetric shape (lower sensibility for middle or for extreme values, Figure 1) and asymmetric shape (lower sensibility for the lowest or for the highest values, Figure 2).

Therefore, the context adaptation process involves three or four steps to build the fuzzy partition associated to each variable depending on the value of the parameter S:

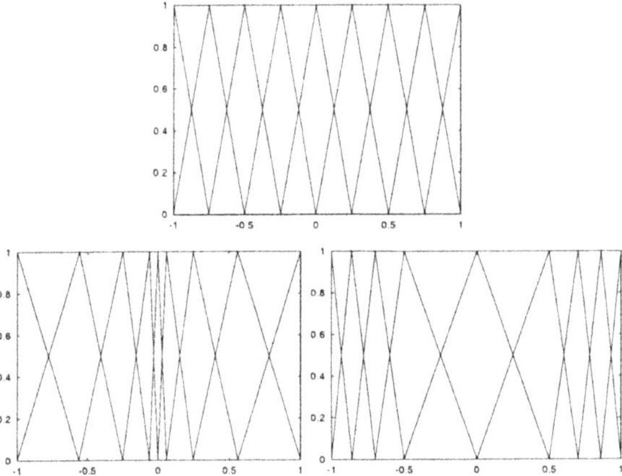

Fig. 1. Fuzzy partitions with $a = 1$ (top), $a > 1$ (down left), and $a < 1$ (down right)

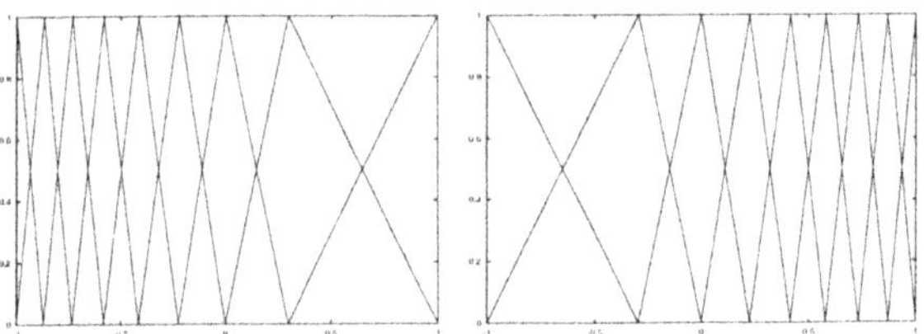

Fig. 2. Fuzzy partitions with $S = 1$ (left with $a > 1$ and right with $a < 1$)

- The first step builds a uniform fuzzy partition considering the number of labels of the variable and its working range.
- If $S = 0$ or $a = 1$:
  - The second step produces a linear mapping of the fuzzy partition from the domain interval to interval $[-1, 1]$.
  - The third step introduces the non-linearity through the non-linear scaling function ($f(x) = sign(x) \cdot |x|^a$), that maintains the extremes of the interval unchanged ($[-1, 1]$).
- If $S = 1$ and $a \neq 1$:
  - The second step produces a linear mapping of the fuzzy partition from the domain interval to interval $[0, 1]$.
  - The third step introduces the non-linearity through the non-linear scaling function. With the aim of permitting a symmetric effect in the two extremes, we adapt the non-linear function when $a > 1$. So, the scaling functions considered are:
    * $f' : [0, 1] \to [0, 1] \qquad f'(x) = |x|^a, \quad when\ a < 1$
    * $f'' : [0, 1] \to [0, 1] \qquad f''(x) = 1 - (1 - |x|)^{1/a}, \quad when\ a > 1$

- In the fourth step, a second linear mapping transforming the resulting interval $[0,1]$ into $[-1,1]$ is applied.

The overall result is a non-linear mapping from the variable domain to [-1,1].

3.2 Genetic Learning Process

GAs [19,36] are search and optimisation techniques based on a formalisation of natural genetics. The genetic process starts with a population of solutions called chromosomes, that constitutes the first generation ($G(0)$), and undergoes evolution over it. While a certain termination condition is not met, each chromosome is evaluated by means of an evaluation function (a fitness value is assigned to the chromosome) and a new population is created ($G(t+1)$), by applying a set of genetic operators to the individuals of generation $G(t)$. Different proposals that use GAs in order to design FRBSs are contained in [12,25].

The main questions when using GAs are: how to code each solution, how to evaluate these solutions and how to create new solutions from existing ones. Moreover, it is relatively important the choice of the initial population, because we can obtain good solutions more quickly if an appropiate initial gene pool is chosen.

In our process, each chromosome represents a complete DB definition by encoding the said parameters. To evaluate a cromosome, we use an ad hoc data-driven method to learn the RB considering the DB contained in it, generating a complete KB, and then the accuracy of the FRBS obtained on a training data set is measured.

The main purpose of our KB design process is to obtain FRBCSs with good accuracy and high interpretability. Unfortunately, it is not easy to achieve these two objectives at the same time. Normally, FRBCSs with good performance have a high number of selected variables and also a high number of rules, thus presenting a low degree of readability (for more information about interpretability and accuracy in fuzzy systems, refer to [4]). On the other hand, KB design methods sometimes lead to a certain overfitting of the training data set used for the learning process.

To avoid these problems, our genetic process jointly considers two objectives:

1. Minimise the classification error percentage over the training data set.
2. Design a compact and interpretable KB. This objective is performed by penalising FRBCSs with a large number of selected features and/or with high granularities.

The next sections describe the main components of the genetic learning process.

Encoding the DB. The two main DB components of the selected variables are the number of linguistic terms and the membership functions that define their semantics. Therefore, each chromosome will be composed of three parts:

- Relevant variables (C_1): For a classification problem with N variables, the selected features are stored into a binary coded array of length N. In this array, an 1 indicates that the corresponding variable is selected for the FRBCS.
- Number of labels (C_2): The number of labels per variable is stored into an integer array of length N. In this contribution, the possible values considered are taken from the set $\{1, \ldots, 5\}$. We should note that the granularity learning allows us another way of feature selection: if a variable is assigned only to one label, it has no influence in the RB, so it will not be considered as a relevant variable. A similar double-level feature selection process has been previously considered in genetic learning processes of FRBCSs such as SLAVE [21].
- Sensibility parameters (C_3): An array of lenght $N \times 2$, where the sensibility parameters (a,S) are stored for each variable. In our case, the range considered for the parameter a is the interval $(0, 8]$.

If v_i is the bit that represents whether the variable i is selected and l_i is the granularity of variable i, a graphical representation of the chromosome is shown next:

$$C_1 = (v_1, \ldots, v_N) \quad C_2 = (l_1, \ldots, l_N)$$

$$C_3 = (a_1, \ldots, a_N, S_1, \ldots, S_N)$$

$$C = C_1 C_2 C_3$$

Initial Gene Pool. The initial population is composed of four parts. The generation of the initial gene pool is described next:

- In the first group all the chromosomes have all the features selected, that is, $C_1 = (1, 1, 1, \ldots, 1)$, and each one of them has the same granularity in all its variables. This group is composed of $\#val$ chromosomes, with $\#val$ being the cardinality of the significant term set, in our case $\#val = 4$, corresponding to the four possibilities for the number of labels, $2 \ldots 5$, (a single label is not considered because the variable would not be selected). For each value of the number of labels, one individual is created.
- The second part is composed of $\#val \times 4$ chromosomes and each one of them has the same granularity in all its variables. For each possible number of labels, four individuals are created with a different percentage of randomly selected variables (75%, 50%, 25% and 10%).

- The third group has five subgroups, each of them with the different percentages for the selected variables considered in the previous groups (100%, 75%, 50%, 25% and 10%), and all of the chromosomes with a randomly selected granularity per variable. In the experiments, all of these five subgroups have 10 chromosomes.
- The fourth part is composed of the remaining chromosomes, whose components are randomly selected.

The aim of generating the initial population in these four different groups is to sample it to achieve an appropiate diversity. Although GAs have proven to be robust and get good solutions starting from randomly generated populations (group 4), a quicker convergence can be obtained using the knowledge available about the problem to sample the population in a biased way.

Evaluating the chromosome. There are three steps that must be done to evaluate each chromosome:

1. Generate the fuzzy partitions for all the input variables using the information contained in the chromosome as introduced in Section 3.1
2. Generate the RB by running a fuzzy rule learning method considering the DB obtained in the previous step.
3. Calculate the two values of the evaluation function, that correspond to the two objectives considered:
   - CPE: classification percentage error over the training set.
   - $S\_V \cdot A\_Gr$: with S_V being the number of selected variables and A_Gr being the averaged granularity of the selected variables.

Genetic operators. A set of genetic operators is applied to the genetic code of the DBs contained in $G(t)$, to obtain $G(t+1)$. Due to the special nature of the chromosomes involved in this DB definition process, the design of genetic operators able to deal with it becomes a main task. Since there is a strong relationship among the three chromosome parts, operators working cooperatively in C_1, C_2 and C_3 are required in order to make best use of the representation tackled.

Taking into account these aspects, the following operators are considered:

Selection. We have used the selection mechanism of MOGA [17], which is based on the definition of Pareto-optimality. Taking this idea as a base, MOGA assigns the same selection probability to all non-dominated solutions in the current population. The method involves dividing the population into several classes depending on the number of individuals dominating the members of each class.

Therefore, the selection scheme of our multiobjective GA involves the following five steps:

1. Each individual is assigned a rank equal to the number of individuals dominating it plus one (chromosomes encoding non-dominated solutions receive rank 1).
2. The population is increasingly sorted according to that rank.
3. Each individual is assigned a selection probability which depends on its ranking in the population, with lower ranking receiving lesser probabilities.
4. The selection probability of each equivalence class (group of chromosomes with the same rank, i.e., which are non-dominated among them) is averaged.
5. The new population is created by following the Baker's stochastic universal sampling [2].

Crossover. As regards the recombination process, two different crossover operators are considered depending on the two parents' scope:

- *Crossover when both parents have the same selected variables and equal granularity level per variable:* If the two parents have the same values in C_1 and C_2, the genetic search has located a promising space zone that has to be adequatelly exploitated. This task is developed by applying the max-min-arithmetical (MMA) crossover operator in the chromosome part based on real-coding scheme (parameters a_i) and obviously by maintaining the parent C_1 and C_2 values in the offspring. This crossover operator is proposed in [26] and works in the way shown below.
  If $C_v^t = (c_1, ..., c_k, ..., c_H)$ and $C_w^t = (c'_1, ..., c'_k, ..., c'_H)$ are to be crossed, the following offspring is generated (with $d \in [0, 1]$)
  $$C_1^{t+1} = dC_w^t + (1 - d)C_v^t$$
  $$C_2^{t+1} = dC_v^t + (1 - d)C_w^t$$
  $$C_3^{t+1} \text{ with } c_{3k}^{t+1} = \min\{c_k, c'_k\}$$
  $$C_4^{t+1} \text{ with } c_{4k}^{t+1} = \max\{c_k, c'_k\}$$
  In our proposal, eight chromosomes are generated, the previous four with $S = 0$ and the same four with $S = 1$. This operator uses a parameter d which is either a constant, or a variable whose value depends on the age of the population. The resulting descendants are the two best of the eight aforesaid individuals.
- *Crossover when the parents encode different selected variables or granularity levels:* This second case highly recommends the use of the information encoded by the parents to explore the search space in order to discover new promising zones. So, a standard crossover operator is applied over the three parts of the chromosome. This operator performs as follows: a crossover point p is randomly generated in C_1 and the two parents are crossed at the p-th variable in all the chromosome parts, thereby producing two meaningful descendants.

Let us look at an example in order to clarify the standard crossover application. Consider:

$$C_t = (v_1, \ldots, v_p, v_{p+1}, \ldots, v_N, l_1, \ldots, l_p, l_{p+1}, \ldots, l_N,$$
$$a_1, \ldots, a_p, a_{p+1}, \ldots, a_N, S_1, \ldots, S_p, S_{p+1}, \ldots, S_N)$$

$$C_t' = (v_1', \ldots, v_p', v_{p+1}', \ldots, v_N', l_1', \ldots, l_p', l_{p+1}', \ldots, l_N',$$
$$a_1', \ldots, a_p', a_{p+1}', \ldots, a_N', S_1', \ldots, S_p', S_{p+1}', \ldots, S_N')$$

as the individuals to be crossed at point p, the two resulting offsprings are:

$$C_t = (v_1, \ldots, v_p, v_{p+1}', \ldots, v_N', l_1, \ldots, l_p, l_{p+1}', \ldots, l_N',$$
$$a_1, \ldots, a_p, a_{p+1}', \ldots, a_N', S_1, \ldots, S_p, S_{p+1}', \ldots, S_N')$$

$$C_t' = (v_1', \ldots, v_p', v_{p+1}, \ldots, v_N, l_1', \ldots, l_p', l_{p+1}, \ldots, l_N,$$
$$a_1', \ldots, a_p', a_{p+1}, \ldots, a_N, S_1', \ldots, S_p', S_{p+1}, \ldots, S_N)$$

Hence, the complete recombination process will allow the GA to follow an adequate exploration-exploitation balance in the genetic search. The expected behavior consists of an initial phase where a high number of standard crossovers and a very small of MMA ones (equal to zero in the great majority of the cases) are developed. The genetic search will perform a wide exploration in this first stage, locating the promising zones and sampling the population individuals at them in several runs. At this moment, a new phase starts, characterised by the increase of the exploitation of these zones and the decrease of the space exploration. Therefore, the number of MMA crossovers rises a lot and the application of the standard crossover decreases. This way to perform an appropiate exploration-exploitation balance in the search was succesfully applied in [11].

Mutation. Three different operators are used, each one of them acting on different chromosome parts. A brief description of them is given below:

- *Mutation on C_1 and on the second part of C_3 (parameters S_i)*: As these parts of the chromosome are binary coded, a simple binary mutation is developed, flipping the value of the gene.
- *Mutation on C_2*: The mutation operator selected for the granularity levels is similar to the one proposed by Thrift in [44]. A local modification is developed by changing the number of labels of the variable to the immediately upper or lower value (the decision is made at random). When the value to be changed is the lowest (1) or the highest one, the only possible change is developed.

- *Mutation on first part of C_3 (parameters a_i)*: As this part is based on a real-coding scheme, Michalewicz's non-uniform mutation operator is employed [36]. This operator performs as follows:
  If $C_v^t = (c_1, ..., c_k, ..., c_H)$ is a chromosome and the gene c_k was selected for mutation (the domain of c_k is $[c_{kl}, c_{kr}]$), the result is a vector $C_v^{t+1} = (c_1, ..., c'_k, ..., c_H)$, with $k \in 1, ..., H$, and

$$c'_k = \begin{cases} c_k + \triangle(t, c_{kr} - c_k) \text{ if } e = 0, \\ c_k - \triangle(t, c_k - c_{kl}) \text{ if } e = 1 \end{cases}$$

  where e is a random number that may have a value of zero or one, and the function $\triangle(t, y)$ returns a value in the range $[0, y]$ such that the probability of $\triangle(t, y)$ being close to 0 increases as t increases:

$$\triangle(t, y) = y(1 - r^{(1-\frac{t}{T})^b})$$

  where r is a random number in the interval $[0, 1]$, T is the maximum number of generations and b is a parameter chosen by the user, which determines the degree of dependency with the number of iterations. This property causes this operator to make an uniform search in the initial space when t is small, and a very local one in later stages.

4 Experimentation

Three different benchmarks have been considered in the experiments:

- The *Sonar* data set [22], which has 208 instances of a sonar objective classification problem. Each one of these instances is described by 60 features to discriminate between a sonar output corresponding to a cylindrical metal or an approximately cylindrical rock. The training set contains 104 elements and the test set contains 104 elements, randomly selected from the whole data set.
- The *Wisconsin Breast Cancer* data set, donated by Olvi Magasarian, which currently contains 698 instances of a diagnostic problem, described by means of 9 features to distinguish between malignant and benign breast tumor. The training and data sets contain 349 examples ramdomly selected from the whole data set.
- The *glass identification* data set, from the USA Forensic Science Service, which has 214 instances of 6 types of glass, defined in terms of their oxide content (i.e. Na, Fe, K, etc.) by means of 9 features. The training set contains 107 elements and the test set contains 107 elements, randomly selected from the whole data set.

For every benchmark, the method were run four times for each one of the two FRMs considered with different initial seeds for the GA random number

Table 1. Parameter values

Parameter	Value
Granularity values	$\{1,\ldots,5\}$
Population size	200
Crossover probability	0.6
Mutation probability	0.2
Parameter b (non-uniform mutation)	5
Parameter d (MMA crossover)	0.35
Number of generations	$\{100, 500\}$

generator. Table 1 shows the parameter values considered for the experiments developed.

The experiments provided a wide range of FRBCSs with a significant reduction on the feature number and different results over the prediction ability and the interpretability. Some of these solutions have been rejected because its high CPE over the test data set since, of course, this value is not used during the learning process. Table 2 shows a brief set of good solutions for the two FRMs considered. The best results found with the Wang and Mendel's RB generation method considering all the features selected and the same number of labels for each one of them are also shown in the top line of each FRM. The table contains the following columns:

- **Method**: KB design method used, Wang and Mendel (**WM**) and the multiobjective genetic algorithm proposed (**MGA**).
- **S_V**: Number of selected variables.
- **A_Gr**: Average of the granularity considered for the selected variables.
- **N_R**: The number of rules of the FRBCS.
- **% Tra**: Classification percentage error obtained in the training data set.
- **% Tst**: Classification percentage error obtained in the test data set.

As it can be observed, the proposed method achieves a significant reduction in the number of variables selected (about the 90% of the original number of features, or even more in some cases) even with an important increase of the generalisation capability (classification rate over the test data set). Besides, many solutions also present a significant decrease in the number of rules, reducing the complexity of the KB. Therefore, our multiobjective GA provides a wide set of solutions that permit an adequate choice depending on the main goal required: good performance or high degree of interpretability.

5 Conclusions

This contribution has proposed a multiobjective genetic process for jointly performing feature selection and DB components learning, which is combined with an efficient fuzzy classification rule generation method to obtain

Benchmark	FRM = Maximum						FRM = Normalised sum					
	Method	S_V	A_Gr	N_R	% Tra	% Tst	Method	S_V	A_Gr	N_R	% Tra	% Tst
Sonar	WM	60	3	104	0.96	23.1	WM	60	3	104	2.88	25.9
	MGA	8	3.5	93	0.96	16.3	MGA	7	4.4	101	0.0	15.3
		6	4.1	90	1.92	17.3		6	4.1	93	5.77	15.3
		5	3.6	75	10.5	17.3		4	3.7	63	10.5	16.3
		4	3.0	44	17.3	20.2		3	3.6	34	20.2	21.1
		3	4.0	33	17.3	22.1		3	3.0	24	22.1	25.0
		2	4.0	14	24.1	24.1		2	3.0	9	26.9	26.9
Cancer	WM	9	3	160	2.29	6.88	WM	9	3	160	2.01	5.73
	MGA	4	4.0	85	1.43	4.58	MGA	4	4.2	88	1.71	5.73
		4	3.2	64	2.01	4.29		4	2.7	27	2.57	5.15
		3	4.3	57	2.01	4.58		3	2.6	18	3.15	4.01
		3	3.3	34	3.15	4.58		3	2.3	13	3.43	4.01
		3	3.0	22	3.15	6.01		2	3.5	11	4.01	4.01
		2	3.5	14	4.01	6.31		2	3.0	8	4.01	5.44
Glass	WM	9	5	72	17.8	39.2	WM	9	3	52	30.8	44.9
	MGA	4	4.5	50	15.8	42.1	MGA	4	4.2	30	21.5	33.6
		4	3.2	38	22.4	34.5		3	4.3	35	25.2	35.5
		3	4.0	31	25.2	35.5		3	3.3	21	27.1	33.6
		3	3.0	23	31.7	32.7		2	5.0	28	27.1	34.5
		2	2.5	15	36.4	34.5		2	3.0	14	34.5	41.1

Table 2. Best results obtained

the complete KB for a descriptive FRBCS. Our method achieves an important reduction of the relevant variables selected for the final system and also adapts the fuzzy partition of each variable to the problem being solved. So, we can conclude that the proposed method allows us to significantly enhance the interpretability and accuracy of the FRBCSs generated. We have used a simple RB generation algorithm but another more accurate one can be used, having in mind its run time. Our future work will focus on improving the performance of the multiobjective GA by using a niching technique or employing a co-evolutive GA and on comparing the results with other feature selection approaches.

References

1. S. Abe, M.-S. Lan, and R. Thawonmas. Tuning of a fuzzy classifier derived from data. *International Journal of Approximate Reasoning*, 12:1–24, 1996.
2. J. E. Baker. Reducing bias and inefficiency in the selection algorithms. In *Proc. of the Second International Conference on Genetic Algorithms (ICGA'87)*, pages 14–21, Hillsdale, 1987.
3. J. Casillas, O. Cordón, M. J. del Jesus, and F. Herrera. Genetic feature selection in a fuzzy rule-based classification system learning process for high dimensional problems. *Information Sciences*, 136(1-4):135–157, 2001.

4. J. Casillas, O. Cordón, F. Herrera, and L. Magdalena (Eds.). *Fuzzy Modeling and the Interpretability-Accuracy Trade-Off. Part I: Interpretability Issues. Part II: Accuracy Improvements Preserving Interpretability*. Physica-Verlag. in press.
5. V. Chankong and Y.Y. Haimes. *Multiobjective decision making theory and methodology*. North-Holland, 1983.
6. Z. Chi, H. Yan, and T. Pham. *Fuzzy algorithms with applications to image processing and pattern recognition*. World Scientific, 1996.
7. C.A. Coello. A comprehensive survey of evolutionary-based multiobjective optimization techniques. *Knowledge and Information Systems. An International Journal*, 1(3):269–308, 1999.
8. C.A. Coello, D.A. Van Veldhuizen, and G.B. Lamont. *Evolutionary algorithms for solving multi-objective problems*. Kluwer Academic Publishers, 2002.
9. O. Cordón, M. J. del Jesus, and F. Herrera. Genetic learning of fuzzy rule-based classification systems co-operating with fuzzy reasoning methods. *International Journal of Intelligent Systems*, 13(10/11):1025–1053, 1998.
10. O. Cordón, M. J. del Jesus, and F. Herrera. A proposal on reasoning methods in fuzzy rule-based classification systems. *International Journal of Approximate Reasoning*, 20(1):21–45, 1999.
11. O. Cordón and F. Herrera. Hybridizing genetic algorithms with sharing scheme and evolution strategies for designing approximate fuzzy rule-based systems. *Fuzzy Sets and Systems*, 118(2):235–255, 2000.
12. O. Cordón, F. Herrera, F. Hoffmann, and L. Magdalena. *Genetic fuzzy systems. Evolutionary Tuning and Learning of Fuzzy Knowledge Bases*. Kluwer Academic Publishers, 2001.
13. O. Cordón, F. Herrera, L. Magdalena, and P. Villar. A genetic learning process for the scaling factors, granularity and contexts of the fuzzy rule-based system data base. *Information Sciences*, 136(1-4):85–107, 2001.
14. O. Cordón, F. Herrera, and P. Villar. Analysis and guidelines to obtain a good uniform fuzzy partition granularity for fuzzy rule-based systems using simulated annealing. *International Journal of Approximate Reasoning*, 25(3):187–216, 2000.
15. O. Cordón, F. Herrera, and P. Villar. Generating the knowledge base of a fuzzy rule-based system by the genetic learning of the data base. *IEEE Transactions on Fuzzy Systems*, 9(4):667–675, 2001.
16. K. Deb. *Multi-objective optimization using evolutionary algorithms*. John Wiley & Sons, 2001.
17. C.M. Fonseca and P.J. Fleming. Genetic algorithms for multiobjective optimization: Formulation, discussion and generalization. In S. Forrest, editor, *Proc. of the Fifth International Conference on Genetic Algorithms (ICGA'93)*, pages 416–423. Morgan Kaufmann, 1993.
18. C.M. Fonseca and P.J. Fleming. An overview of evolutionary algorithms in multiobjective optimization. *Evolutionary Computation*, 3(1):1–16, 1995.
19. D.E. Goldberg. *Genetic Algorithms in Search, Optimizacion & Machine Learning*. Addison Wesley, 1989.
20. A. González and R. Pérez. SLAVE: A genetic learning system based on an iterative approach. *IEEE Transactions on Fuzzy Systems*, 7(2):176–191, 1999.
21. A. González and R. Pérez. Selection of relevant features in a fuzzy genetic learning algorithm. *IEEE Transactions on Systems, Man and Cybernetics - Part B: Cybernetics*, 31(3):417–425, 2001.

22. R. P. Gorman and T. J. Sejnowski. Analysis of hidden units in a layered network trained to classify sonar targets. *Neural Networks*, 1:75–89, 1988.
23. R. Gudwin, F. Gomide, and W. Pedrycz. Context adaptation in fuzzy processing and genetic algorithms. *International Journal of Intelligent Systems*, 13:929–948, 1998.
24. A.E. Hans. Multicriteria optimization for highly accurate systems. In W. Stadler, editor, *Multicriteria optimization in engineering and sciences*, pages 309–352. Plenum Press, 1988.
25. F. Herrera and J.L. Verdegay (eds.). *Genetic Algorithms and Soft Computing*. Physica-Verlag, 1996.
26. F. Herrera, M. Lozano, and J.L. Verdegay. Fuzzy connectives based crossover operators to model genetic algorihtms population diversity. *Fuzzy Sets and Systems*, 92(1):21–30, 1997.
27. J. Horn, N. Nafpliotis, and D.E. Goldberg. A niched pareto genetic algorithm for multiobjective optimization. In *Proc. First Conf. on Evolutionary Computation*, pages 82–87, Piscataway, 1994.
28. H. Ishibuchi and T. Murata. A genetic-algorithm-based fuzzy partition method for pattern classification problems. In F. Herrera and J.L. Verdegay, editors, *Genetic Algorithms and Soft Computing*, pages 555–578. Physica-Verlag, 1996.
29. H. Ishibuchi, T. Murata, and I. B Turksen. Single-objective and two-objective genetic algorithms for selecting linguistic rules for pattern classification problems. *Fuzzy Sets and Systems*, 89:135–150, 1997.
30. H. Ishibuchi, T. Nakashima, and T. Morisawa. Voting in fuzzy rule-based systems for pattern classification problems. *Fuzzy Sets and Systems*, 103:223–238, 1999.
31. H. Ishibuchi, K. Nozaki, and H. Tanaka. Construction of fuzzy classification systems with rectangular fuzzy rules using genetic algorithms. *Fuzzy Sets and Systems*, 65:237–253, 1994.
32. R. Kohavi and G.H. John. Wrappers for feature subset selection. *Artificial Intelligence*, 97:273–324, 1997.
33. Huan Liu and Hiroshi Motoda. *Feature selection for knowledge discovery and data mining*. Kluwer Academic Publishers, 1998.
34. L. Magdalena. Adapting the gain of an FLC with genetic algorithms. *International Journal of Approximate Reasoning*, 17(4):327–349, 1997.
35. L. Magdalena and J.R. Velasco. Evolutionary based learning of fuzzy controllers. In W. Pedrycz, editor, *Fuzzy Evolutionary Computation*, pages 249–268. Kluwer Academic, 1997.
36. Z. Michalewicz. *Genetic Algorithms + Data Structures = Evolution Program*. Springer-Verlag, 1996.
37. T. Nakashima, T. Morisawa, and H. Ishibuchi. Input selection in fuzzy rule-based classification systems. In *Proc. of the Sixth IEEE International Conference on Fuzzy Systems (FUZZ-IEEE'97)*, pages 1457–1462, 1997.
38. D. Nauck and R. Kruse. A neuro-fuzzy method to learn fuzzy classification rules from data. *Fuzzy Sets and Systems*, 89:277–288, 1997.
39. Y. Oh and D.J. Park. Self-tuning fuzzy controller with variable universe of discourse. In *Proc. IEEE International Conference on System, Man and Cybernetics*, pages 2628–2632, Vancouver, 1995.
40. W. Pedrycz, R. Gudwin, and F. Gomide. Nonlinear context adaptation in the calibration of fuzzy sets. *Fuzzy Sets and Systems*, 88:91–97, 1997.

41. J.D. Schaffer. Multiple objective optimization with vector evaluated genetic algorithms. In J.J Grefenstette, editor, *Genetic algorithms and their applications: Proc. of the 1st Int. Conf. on Genetic Algorithms*, pages 93–100. Lawrence Erlbaum, 1985.
42. M. Setnes, R. Babuska, U. Kaymak, and H.R. van Nauta-Lemke. Similarity measures in fuzzy rule base simplification. *IEEE Transctions on Systems, Man and Cybernetics - Part B: Cybernetics*, 28:376–386, 1998.
43. N. Srinivas and D. Kalyanmoy. Multiobjective optimization using nondominated sorting in genetic algorithms. *Evolutionary Computation*, 2(3):221–248, 1994.
44. P. Thrift. Fuzzy logic synthesis with genetic algorithms. In *Proc. Fourth International Conference on Genetic Algorithms (ICGA'91)*, pages 509–513, 1991.
45. L. X. Wang and J. M. Mendel. Generating fuzzy rules by learning from examples. *IEEE Transactions on Systems, Man, and Cybernetics*, 25(2):353–361, 1992.

Extracting Linguistic Fuzzy Models from Numerical Data-AFRELI Algorithm

Jairo Espinosa[1] and Joos Vandewalle[2]

[1] IPCOS N.V. Technologielaan 11,0101
B-3001 Leuven, Belgium
[2] Katholieke Universiteit Leuven, Kasteelpark 10,
B-3001 Leuven, Belgium

Abstract. This paper discusses the concepts of linguistic integrity and interpretability. The concepts are used as a framework to design an algorithm to construct linguistic fuzzy models from Numerical Data. The constructed model combines prior knowledge (if present) and numerical information. Two algorithms are presented in this chapter. The main algorithm is the Autonomous Fuzzy Rule Extractor with Linguistic Integrity (AFRELI). This algorithm is complemented with the use of the FuZion algorithm created to merge consecutive membership functions while guaranteeing the distinguishability between fuzzy sets. Examples of function approximations and modeling of industrial data are presented as application examples.

1 Introduction

The use of models is an essential element of human behavior. When a human being predicts the impact of his actions, it is using a model. Causality is a paramount assumption that makes models useful. Causality is reflected in language as IF-THEN rules (IF **cause-happens** THEN **a consequence is foreseen**). A set of IF-THEN rules is a linguistic representation of a mental model created inside the brain. New instrumentation and data acquisition system have expanded the capacity of human beings beyond the five senses. This expanded sensorial capacity has been accompanied with an increase in the storage capacity but the capacity of the human brain to interact with this information remains limited. This situation motivates the development of computer techniques that can extract the "knowledge" and represent it in a linguistic way using IF-THEN rules.
There is a trade-off between numerical accuracy and linguistic interpretability. This trade off is a consequence of a limitation of the human brain to represent a limited number of categories in a given domain. A consequence of this limitation is reflected in language. The number of linguistic labels that a human being can generate to represent categories in a given domain is limited to as much as nine and it will be typically seven.
On the other hand, the numerical accuracy is important in the implementation of policies and control actions based on the information given by a model. This issue of accuracy is especially relevant when the models are used

in a dynamic way where the predicted value is fed back to extend the prediction over a long term horizon.
Several methods for model construction have been proposed in the literature. Initially, the algorithms based on gradient descent techniques, most of the original neurofuzzy models [1] [13] make use of these techniques. The goal of the optimization has been to minimize the numerical error without paying attention to the interpretability. The use of this technique generates, in some cases, fuzzy sets with "too much" or "none" overlap making the interpretation of the model a difficult task.
To overcome the drawback of the initial selection of the fuzzy sets, several methods have been proposed, some of them based on local error approximation [2] [3] and others based on clustering techniques [4] [5] [6]. These methods generate multidimensional fuzzy sets and project them into the input spaces. The projections also exhibit unsatisfactory overlap making the interpretation and the labeling of the fuzzy sets a difficult task.
This chapter presents the AFRELI algorithm (**A**utonomous **F**uzzy **R**ule **E**xtractor with **L**inguistic **I**ntegrity); the algorithm is able to fit input-output data while maintaining the semantic integrity of the rule base. Jang et. al mentioned the issue of interpretability at the end of chapter 12 in [7]. There they only mention some basic ideas about how to constrain the optimization of the ANFIS scheme to preserve the interpretability. In [8] Valente presents a formulation of some constraints to guarantee semantic integrity. For his formulation, he assumes a number of sets on each input domain and then constrains the optimization of the membership functions to guarantee the integrity of the resulting fuzzy sets. The trade-off between precision and transparency is mentioned in [10] together with some examples where the consequences of the rules use two terms; one numeric and one linguistic. The AFRELI algorithm presented in this chapter uses clustering and projection techniques to find a good initial position for the fuzzy sets in the input domains. A FuZion algorithm is introduced to reduce the complexity of the projected fuzzy sets. A rule base is constructed using the reduced representation of the fuzzy sets and the consequences are initialized and calculated with a method that improves generalization and avoids the lack of excitation in some rules. Finally, the consequences of the rules are represented by two fuzzy sets with different strength. The number of terms in the consequences of the fuzzy rules is reduced again using the FuZion algorithm.
The next section presents a discussion on the selection of the structure of the fuzzy model, the next two sections present the AFRELI and the FuZion algorithms and the chapter is closed with some examples including an application to the chemical process industry.

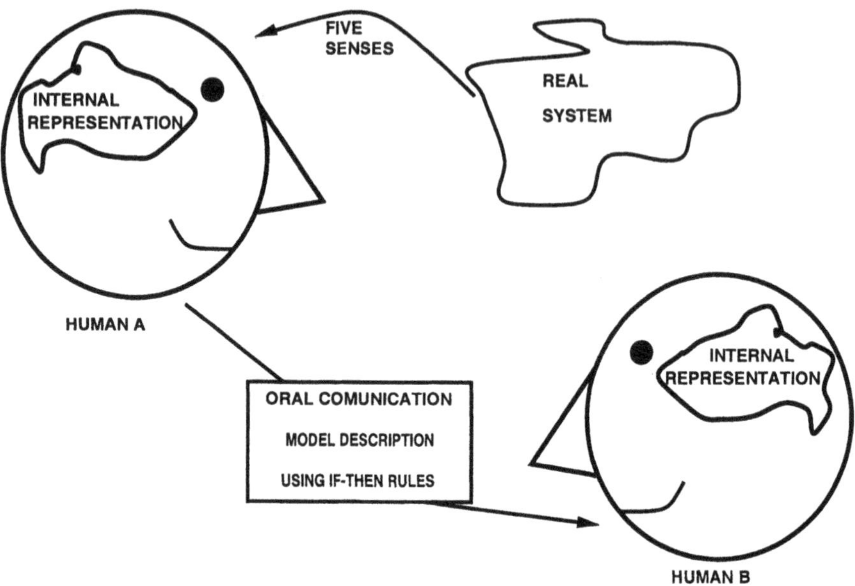

Fig. 1. *Traditional Knowledge Acquisition*

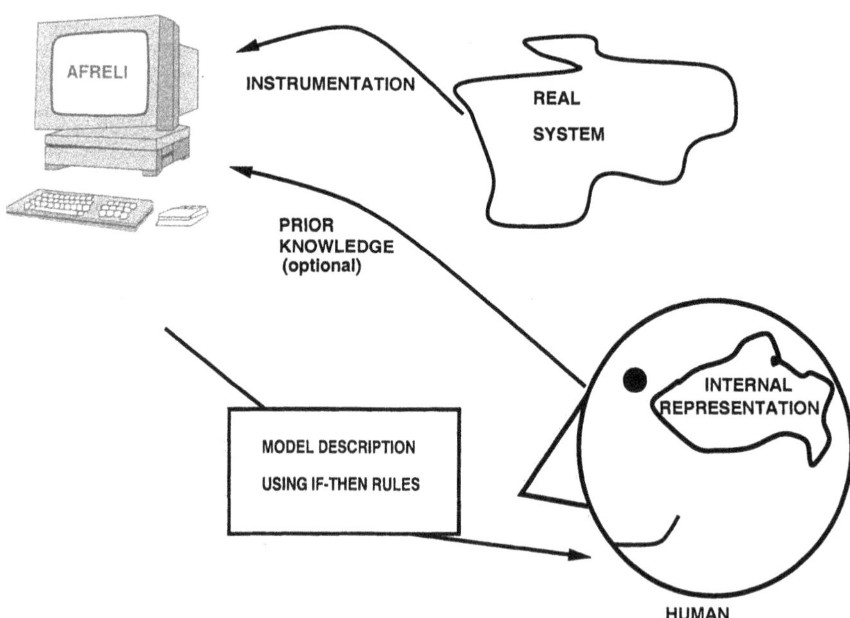

Fig. 2. *Knowledge Acquisition using AFRELI*

2 Structure of the fuzzy model

The high number of degrees of freedom in a fuzzy inference system (shape and number of membership functions, T-norms, aggregation methods, etc) gives high flexibility to the fuzzy system but also demands systematic criteria to select these parameters. Some parameters can be fixed by taking into account the following issues: optimal interface design [12] and semantic integrity [14].

- Optimal interface design
  - *Error-free Reconstruction*: In a fuzzy system a numerical value is converted into a linguistic value by means of fuzzification. A defuzzification method should guarantee that this linguistic value can be reconstructed in the same numerical value

$$\forall x \in [a, b]: \qquad \mathcal{L}^{-1}[\mathcal{L}(x)] = x \tag{1}$$

    where $[a, b]$ is the universe of discourse, $\mathcal{L}$ is the fuzzification process and $\mathcal{L}^{-1}$ is the defuzzification process. The use of triangular membership functions with overlap $\frac{1}{2}$ and centroid defuzzification will satisfy this requirement (see proof: [12]). Polynomial membership functions with overlap $\frac{1}{2}$ also are used but a new defuzzification process must be designed to guarantee an error-free reconstruction.
- Semantic integrity: This property guarantees that the fuzzy sets represent a linguistic concept. The conditions for semantic integrity are:
  - *Distinguishability:* Each linguistic label should have semantic meaning and the fuzzy set should clearly define a range in the universe of discourse. Therefore, the membership functions should be clearly different. The assumption of the overlap equal to $\frac{1}{2}$ makes sure that the support of each fuzzy set will be different. A minimum distance between the modal values of the membership functions makes sure that the membership functions can be distinguished. The modal value of a membership function is defined as the α-cut with $\alpha = 1$

$$m_i = \mu_{i(\alpha=1)}(x), \quad i = 1, \ldots, q \tag{2}$$

  - *Justifiable Number of Elements:* The number of sets on each domain should be compatible with the number of "quantifiers" that a human being can handle. This number should not exceed the limit of 7 ± 2 distinct terms [15]. The choice of the shape of the membership functions does not guarantee this property. To assure that this requirement is fulfilled the FuZion algorithm is presented further in this paper. This algorithm reduces the number of sets on each input or output domain.
  - *Coverage:* Any element from the universe of discourse should belong to at least one of the fuzzy sets. This concept is also mentioned in the literature as ϵ completeness [13].
  - *Normalization:* Due to the fact that each linguistic label has semantic meaning, at least one of the values in the universe of discourse should have a membership degree equal to one. In other words, all the fuzzy sets should be normal.

Based on these criteria the selected *membership functions* will be triangular and normal $(\mu_1(x), \mu_2(x), \ldots, \mu_q(x))$ with a specific overlap of $\frac{1}{2}$. It means that the height of the intersection of two successive fuzzy sets is

$$\text{hgt}(\mu_i \cap \mu_{i\pm1}) = \frac{1}{2}. \tag{3}$$

The choice of the AND and the OR operation will be conditioned by the need of constructing a continuous and differentiable nonlinear map. This property is important if optimization of the antecedent terms is needed in this case AND and OR operations using *product* and *probabilistic sum* will be preferred because their derivatives are continuous.

3 The AFRELI algorithm

The AFRELI (Automatic Fuzzy Rule Extractor with Linguistic Integrity) [16] [11] is an algorithm designed to obtain a good compromise between numerical approximation and linguistic meaning. This particular trade-off has been referenced for long time in science (for a compilation of remarks see: [9]). The main steps of this algorithm are:

- Clustering
- Projection
- Reduction of terms in the antecedents(FuZion see section 4)
- Consequence Calculation
- (optional step) Further antecedent optimization
- Reduction of terms in the consequences and rule modification (FuZion see section 4)

The AFRELI algorithm proceeds as follows:

1. Collect N points from the inputs $(X = \{x^1, \ldots, x^N\})$ and the output $(Y = \{y^1, \ldots, y^N\})$

$$x^k = \begin{bmatrix} x_1^k \\ \vdots \\ x_p^k \end{bmatrix} \tag{4}$$

where $x^k \in \Re^p$ and $y^k \in \Re$ represent the inputs and the output on instant k and construct the *feature vectors*

$$U^k = \begin{bmatrix} x_1^k \\ \vdots \\ x_p^k \\ y^k \end{bmatrix} \tag{5}$$

$u^k \in \Re^{p+1}$.

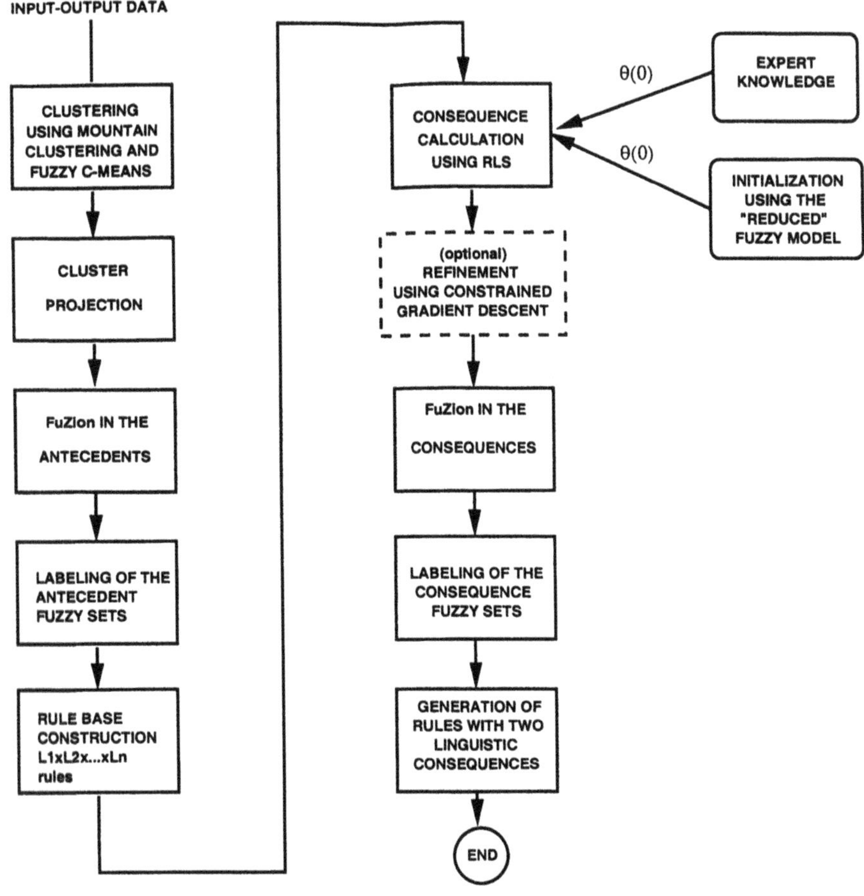

Fig. 3. *Flow Diagram of AFRELI algorithm*

2. Using the N feature vectors find C clusters. Apply the mountain clustering method [5] [6] to initialize the centers and to obtain the number of clusters (C). Refine the clusters using the fuzzy C-means algorithm [17].

$$Uc = \begin{pmatrix} \tilde{x}_1^1 & \tilde{x}_1^2 & \dots & \tilde{x}_1^C \\ \vdots & \vdots & \ddots & \vdots \\ \tilde{x}_p^1 & \tilde{x}_p^2 & \dots & \tilde{x}_p^C \\ \tilde{y}^1 & \tilde{y}^2 & \dots & \tilde{y}^C \end{pmatrix} \tag{6}$$

with $Uc \in \Re^{p+1 \times C}$ and $\tilde{x}_i^j$ represents the i-th coordinate of the j-th cluster.

It is very important to remark that the use of *mountain clustering* will be limited to low dimensional problems (up to four or five dimensions). Its inherent advantage is that it can guide good initial points and number of clusters. For high dimensional problems, the alternative is to use only fuzzy c-means and overestimate the number of clusters; the subsequent steps (FuZion) will reduce the number of terms.

3. Project the C prototypes of the clusters into the input spaces, by converting the projected value of each prototype into the modal value of a triangular membership function.

$$m_j^i = \tilde{x}_i^j \tag{7}$$

where $i = 1, \ldots, p$, $j = 1, \ldots, C$

4. Sort the modal values on each domain such that:

$$m_j^i \leq m_{j+1}^i \qquad \forall i = 1, \ldots, p \tag{8}$$

5. Add two more modal values to each input to guarantee full coverage of the input space.

$$m_0^i = \min_{k=1,\ldots,N} x_i^k \tag{9}$$

$$m_{C+1}^i = \max_{k=1,\ldots,N} x_i^k \tag{10}$$

6. Construct the triangular membership functions with overlap of $\frac{1}{2}$ as:

$$\mu_j^i(x_i^k) = \max\left[0, \min\left(\frac{x_i^k - m_{j-1}^i}{m_j^i - m_{j-1}^i}, \frac{x_i^k - m_{j+1}^i}{m_j^i - m_{j+1}^i}\right)\right] \tag{11}$$

where: $j = 1, \ldots, C$, and the trapezoidal membership functions at the extremes of each universe of discourse

$$\mu_0^i(x_i^k) = \max\left[0, \min\left(\frac{x_i^k - m_1^j}{m_0^j - m_1^j}, 1\right)\right] \tag{12}$$

$$\mu_{C+1}^j(x_i^k) = \max\left[0, \min\left(\frac{x_i^k - m_C^j}{m_{C+1}^j - m_C^j}, 1\right)\right] \tag{13}$$

7. Apply FuZion algorithm (see section 4) to reduce the number of membership functions on each input domain. This algorithm does a somewhat one-dimensional clustering among the modal values of the fuzzy sets.
8. Associate linguistic labels (i.e.BIG, MEDIUM, SMALL, etc.) to the resulting membership functions. This association will depend on the type of variable and the criteria of the designer. In fact the association of a fuzzy set with a label will be the result of the agreement between the fuzzy set proposed by the algorithm and the "sense" that this set creates in the mind of the user.

9. Construct the rule base with all the possible antecedents (all possible permutations) using rules of the form:

$$\cdots$$
$$\text{IF } x_1^k \text{ is } \mu_l^1 \text{ AND } x_2^k \text{ is } \mu_l^2 \text{ AND} \ldots \text{AND } x_p^k \text{ is } \mu_l^p \text{ THEN } \hat{y}_k = \bar{y}_l$$
$$\cdots$$

   Equivalently, the evaluation of the antecedents of each rule can be expressed in terms of the operators *min* and *product* . Using *min* operator:

$$\mu_l(x^k) = \min\{\mu_l^1(x_1^k), \mu_l^2(x_2^k), \ldots, \mu_l^p(x_p^k)\} \tag{14}$$

   Using *product* operator:

$$\mu_l(x^k) = \mu_l^1(x_1^k) \cdot \mu_l^2(x_2^k) \cdot \ldots \cdot \mu_l^p(x_p^k) \tag{15}$$

   Observe that if the number of fuzzy sets on the input i is L_i and there are p inputs, the number of rules will be:

$$L_1 \times L_2 \times \ldots \times L_p$$

   This structure guarantees a complete description of the system in the space interval U, because every possible condition will be represented in the rule base. Observe that the number of rules will grow very fast as the number of input increases. This fact is a limitation in the sense that the comprehension of a set of rules with a large number of antecedents is difficult. In addition, the storage problem generated by a large number of terms to be kept in the computer's memory. On the other hand, it does not represent a limitation in terms of execution time because the use of the described type of triangular membership functions will guarantee that at most 2^p rules will be evaluated during the inference process.
10. Propagate the N values of the inputs and calculate the consequences of the rules as singletons ($\bar{y}_l$). The calculation of these singletons is very critical to guarantee a good generalization in the model. The fact that the rules cover "all the possible cases" generates a poorly conditioned Least Squares problem. For this reason we propose to initialize the consequences of the rules to the value given by a linear model or a "rich excited" reduced fuzzy model (see [20] section 2.3) and continue the tuning of the singletons with Recursive Least Squares (RLS). The use of RLS will guarantee that only those rules that has been excited will be tuned, the other will remain around their initialization value. Because the "reduced" model is the best multi-linear model that can be built with the given data, there is a guarantee that the "full" fuzzy model will be at least as good as the best multi-linear model.

11. (**Optional Step**) If further refinement is required to improve the approximation, constrained gradient descent methods can be applied to improve

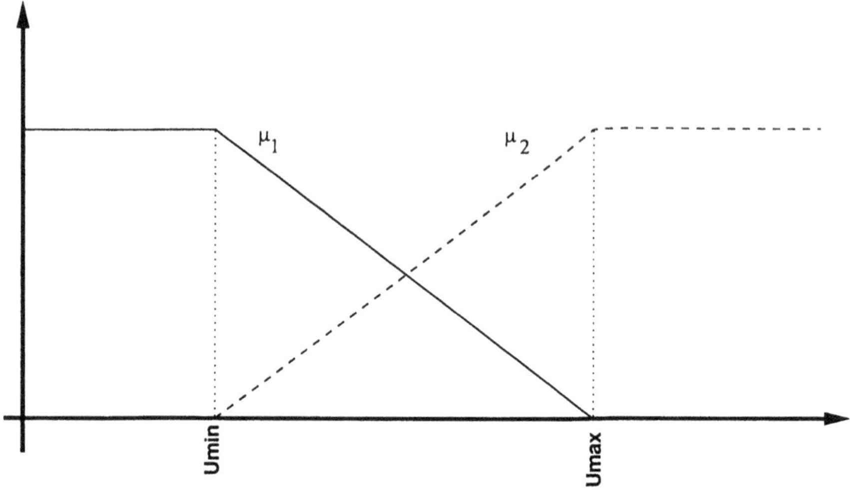

Fig. 4. *Input membership functions for the smallest fuzzy model*

the location of the modal values of the antecedent membership functions. The main constraint applied in the optimization phase is the "distinguishability" constraint that is represented as the minimum acceptable distance between two consecutive modal values. The use of gradient descent methods will move the system parameters towards a "local minimum" close to the initial values. Because the improvement obtained by this step will not be very significant this step is considered *optional* and only will be recommended when the numeric performance of the model does not satisfy the requirements of the user.

12. Because the singletons in the consequences are crisp sets, the linguistic meaning of the rules will be lost. The next step is to convert the singletons of the consequences to triangular membership functions with overlap $\frac{1}{2}$ and modal values equal to the position of the singleton $\bar{y}_l$. Consider the vector $\tilde{Y}$ whose entries are the L consequences of the rules but sorted in such a way that:

$$\tilde{y}_1 \leq \tilde{y}_2 \leq \ldots \leq \tilde{y}_L \tag{16}$$

The triangular membership function of the i-th consequence is:

$$\mu_i^{\tilde{y}}(y) = \max\left[0, \min\left(\frac{y - \tilde{y}_{i-1}}{\tilde{y}_i - \tilde{y}_{i-1}}, \frac{y - \tilde{y}_{i+1}}{\tilde{y}_i - \tilde{y}_{i+1}}\right)\right] \tag{17}$$

and the two membership functions of the extremes:

$$\mu_1^{\tilde{y}}(y) = \max\left[0, \min\left(\frac{y - 2\tilde{y}_1 + \tilde{y}_2}{-\tilde{y}_1 + \tilde{y}_2}, \frac{y - \tilde{y}_2}{\tilde{y}_1 - \tilde{y}_2}\right)\right] \tag{18}$$

$$\mu_L^{\tilde{y}}(y) = \max\left[0, \min\left(\frac{y - \tilde{y}_{L-1}}{\tilde{y}_L - \tilde{y}_{L-1}}, \frac{y - 2\tilde{y}_L + \tilde{y}_{L-1}}{-\tilde{y}_L + \tilde{y}_{L-1}}\right)\right] \tag{19}$$

This description of the outer membership functions guarantees that their centers of gravity will be exactly on their modal values. This guarantees that the condition of error-free reconstruction for optimal interface will be achieved.

13. Apply FuZion algorithm (see section 4) to reduce the number of membership functions in the output universe. The FuZion process reduces groups of neighboring singletons to triangular membership functions whose modal values are representative for a group of singletons. It is optimal in a sense that the modal value of the "FuZioned" membership function is placed at the mean value of the neighboring singletons.
14. Associate linguistic labels to the resulting membership functions.
15. With the partition of the output universe, fuzzify the values of the singletons. Observe that each singleton will have a membership degree in at least one set and in as much as two.
16. Relate the fuzzified values with the corresponding rule. It means that each rule will have one consequence or two weighted consequences where the weights are the non-zero membership values of the fuzzified singleton. This description of the consequences of the rules using two linguistic fuzzy sets and two strength values improves the interpretability of the consequences compared when only one singleton describes the consequence. The advantage of this description is that interpretability is gained without a cost in numerical precision. This strategy was independently proposed previously in [18] and [19].

4 The FuZion algorithm

The FuZion algorithm is a routine that merges consecutive triangular membership functions when their modal values are "too close" to each other. This merging process is needed to preserve the *distinguishability* and the *justifiable number of elements* on each domain guaranteeing the semantic integrity. A fundamental parameter of this algorithm is the minimum acceptable distance between modal values and it is given by M. The FuZion algorithm goes as follows:

1. Take the triangular membership functions $\mu_1(x), \mu_2(x), \ldots, \mu_Q(x)$ with $\frac{1}{2}$ overlap, and the modal values

$$m_i = \mu_{i(\alpha=1)}(x), \quad i = 1, \ldots, Q \tag{20}$$

with:

$$m_1 \leq m_2 \leq \ldots \leq m_Q \tag{21}$$

2. Define the minimum distance acceptable M between the modal values.
3. Calculate the difference between successive modal values as:

$$d_j = m_{j+1} - m_j, \quad j = 1, \ldots, Q-1 \tag{22}$$

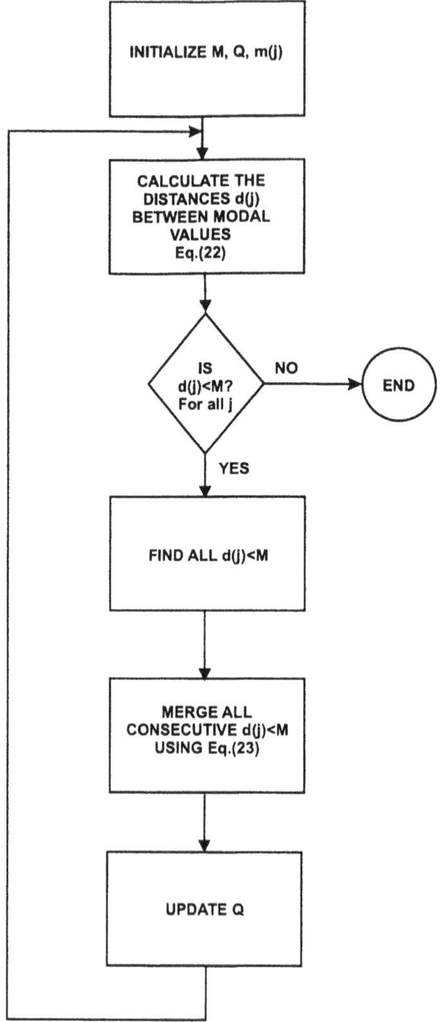

Fig. 5. *Flow Diagram of FuZion algorithm*

4. While $\exists d_j < M$ do
5. Find all the differences smaller than M.
6. Merge all the modal values corresponding to consecutive differences smaller than M using (23).

$$m_{new} = \frac{\sum_{i=a}^{b} m_i}{D} \tag{23}$$

$$D = b - a + 1 \tag{24}$$

where a and b are respectively the index of the first and the last modal value of the fusioned sequence and D is the number of merged membership functions.

7. Update Q
8. Calculate the difference between the new successive modal values as:

$$d_j = m_{j+1} - m_j, \quad j = 1, \ldots, Q-1 \tag{25}$$

9. end while
10. end

5 Examples

The present section shows two examples of applications of the AFRELI and FuZion algorithm, the first example is the approximation of a nonlinear static map and the second is the prediction of a physical property in a chemical process.

Example 1. Modeling a two input nonlinear function In this example, we consider the function:

$$f(x,y) = \sin\left(\frac{\pi x}{10}\right) \sin\left(\frac{\pi y}{10}\right) \tag{26}$$

The steps applied to the current example will be numbered using the same numbering as the one used in the FuZion algorithm in section 4.

- **Step 1** 441 points regularly distributed were selected from the interval $[-10, 10] \times [-10, 10]$. The graph of the function is shown in figure 7
- **Step 2** Using mountain clustering and fuzzy C-means algorithm 26 clusters were found and are shown in figure 8 represented with 'x'.
- **Step 3** After the clusters were found their center values were projected into the input domains as shown in figure 9.
- **Steps 4,5,6** The modal values were sorted and two more modal values were added to each input domain on -10 and 10. The triangular membership functions were constructed. Figure 10 shows the projected membership functions.
- **Steps 7,8** The FuZion algorithm was applied with M equal to the 10 % of the universe of discourse on each domain; observe that with this value of M five (5) membership functions were generated as shown in figure 11. In figure 12, it can be observed that the obtained membership functions (seven (7)) when the M parameter in the FuZion algorithm is chosen equal to 7 % of the universe of discourse. It is clear that the smaller the value of M the larger the number of membership functions. Linguistic labels were associated to the membership functions as shown in figure 11.

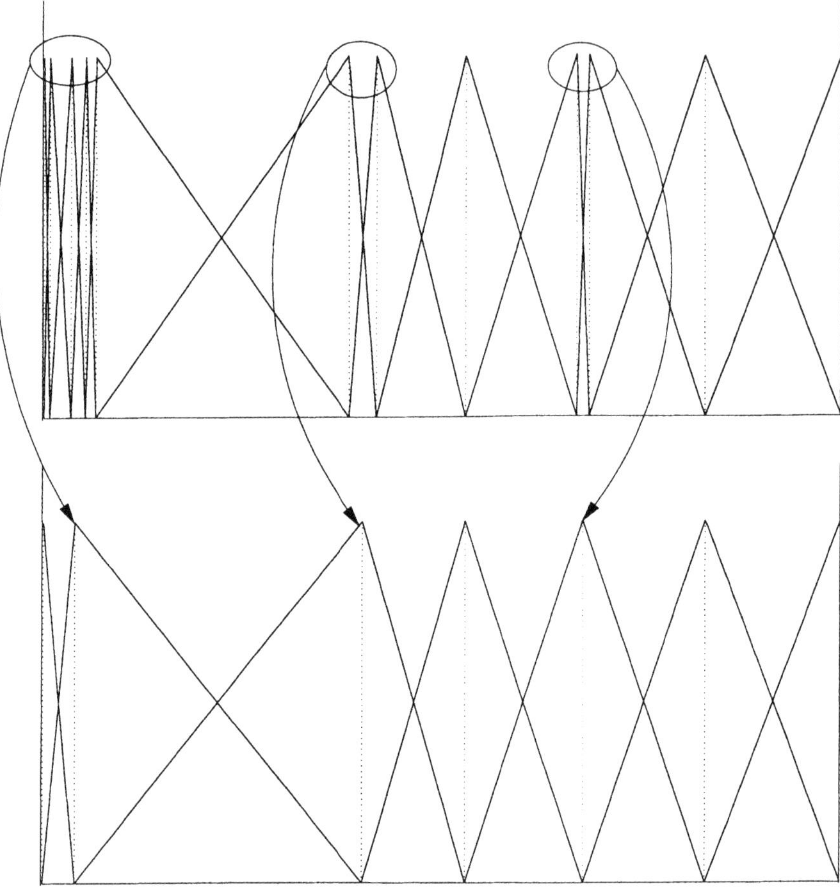

Fig. 6. *Effect of the FuZion algorithm*

- **Step 9** A rule base was generated by combining all the membership functions present on each domain ($5 \times 5 = 25$ rules).
- **Step 10** The 25 singletons of the consequences were calculated using RLS. The output membership functions are shown in figure 13(a).
- **Step 11** The optional step was not applied because the approximation was considered acceptable.
- **Step 12** The singletons were converted into 25 triangular membership functions.
- **Steps 13,14** The FuZion algorithm was applied to the 25 output triangular membership functions with M equal to the 10 % of the universe of

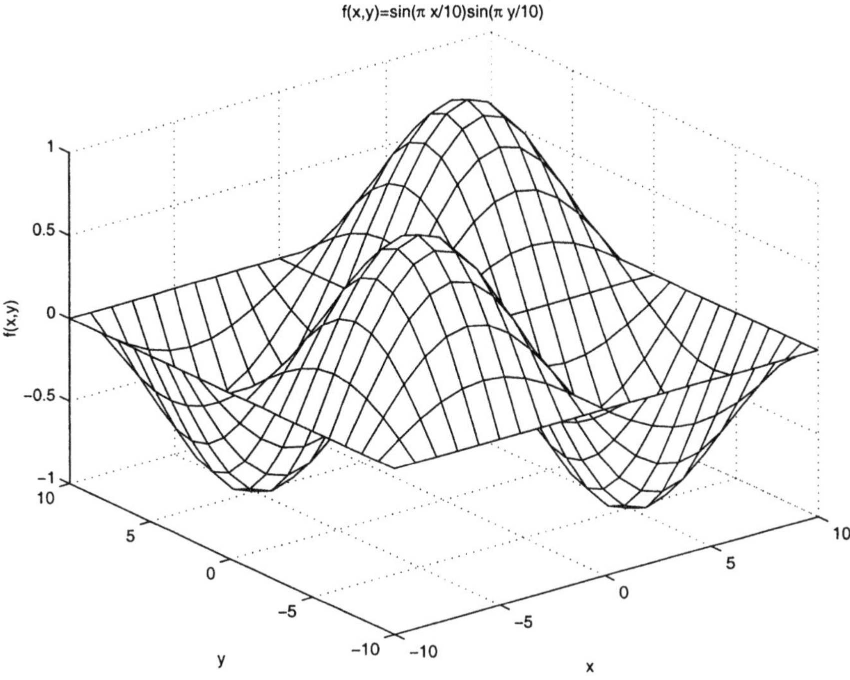

Fig. 7. *Example 1:Function* $f(x,y) = \sin\left(\frac{\pi x}{10}\right)\sin\left(\frac{\pi y}{10}\right)$

discourse. Three membership functions were obtained and they received their linguistic values as shown in figure 13(b).

- **Steps 15,16** The 25 singletons were fuzzified using the three membership functions obtained in the previous step. All the non-zero membership values were associated to the consequences of the rules as shown in the following list.

1. **IF** x is Negative Large AND y is Negative Large **THEN** z is Negative with strength 0.01 AND Zero with strength 0.99
2. **IF** x is Negative Medium AND y is Negative Large **THEN** z is Zero with strength 0.92 AND Positive with strength 0.08
3. **IF** x is Zero AND y is Negative Large **THEN** z is Negative with strength 0.01 AND Zero with strength 0.99
4. **IF** x is Positive Medium AND y is Negative Large **THEN** z is Negative with strength 0.1 AND Zero with strength 0.9
5. **IF** x is Positive Large AND y is Negative Large **THEN** z is Negative with strength 0.03 AND Zero with strength 0.97
6. **IF** x is Negative Large AND y is Negative Medium **THEN** z is Zero with strength 0.96 AND Positive with strength 0.04
7. **IF** x is Negative Medium AND y is Negative Medium **THEN** z is Zero with strength 0.01 AND Positive with strength 0.99

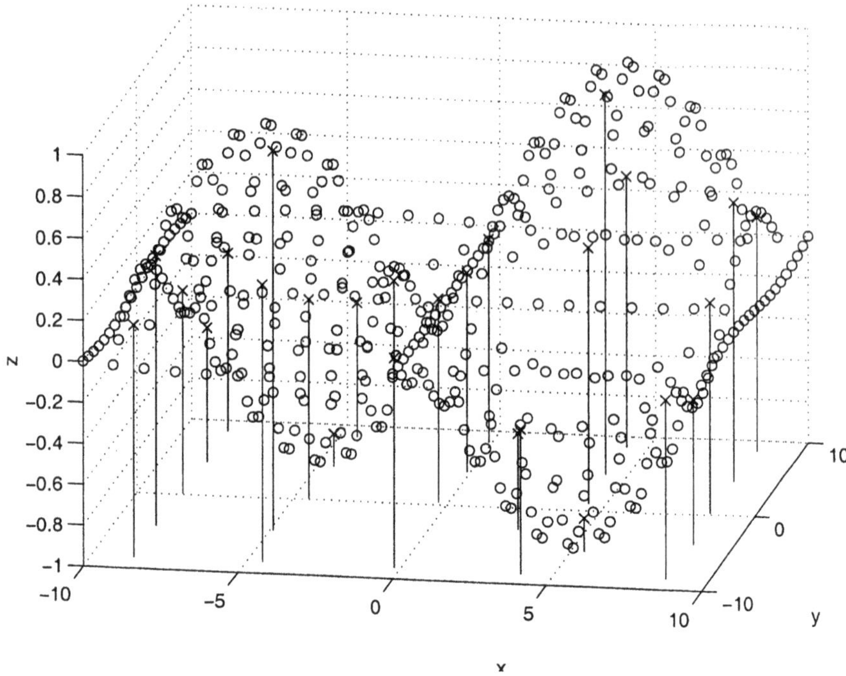

Fig. 8. *Example 1:Extracted clusters ('x') from the data (o)*

8. **IF** x is Zero AND y is Negative Medium **THEN** z is Zero with strength 0.92 AND Positive with strength 0.08
9. **IF** x is Positive Medium AND y is Negative Medium **THEN** z is Negative with strength 0.99 AND Zero with strength 0.01
10. **IF** x is Positive Large AND y is Negative Medium **THEN** z is Negative with strength 0.1 AND Zero with strength 0.90
11. **IF** x is Negative Large AND y is Zero **THEN** z is Negative with strength 0.02 AND Zero with strength 0.98
12. **IF** x is Negative Medium AND y is Zero **THEN** z is Negative with strength 0.11 AND Zero with strength 0.89
13. **IF** x is Zero AND y is Zero **THEN** z is Negative with strength 0.03 AND Zero with strength 0.97
14. **IF** x is Positive Medium AND y is Zero **THEN** z is Zero with strength 0.92 AND Positive with strength 0.08
15. **IF** x is Positive Large AND y is Zero **THEN** z is Negative with strength 0.01 AND Zero with strength 0.99
16. **IF** x is Negative Large AND y is Positive Medium **THEN** z is Negative with strength 0.07 AND Zero with strength 0.93
17. **IF** x is Negative Medium AND y is Positive Medium **THEN** z is Negative with strength 1
18. **IF** x is Zero AND y is Positive Medium **THEN** z is Negative with strength 0.11 AND Zero with strength 0.89

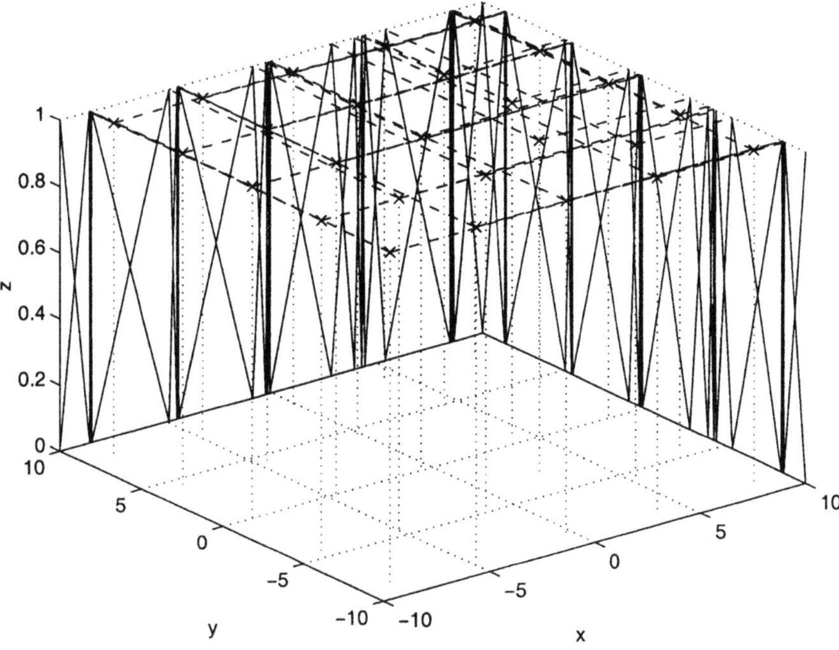

Fig. 9. *Example 1:Projection of the centers of the clusters*

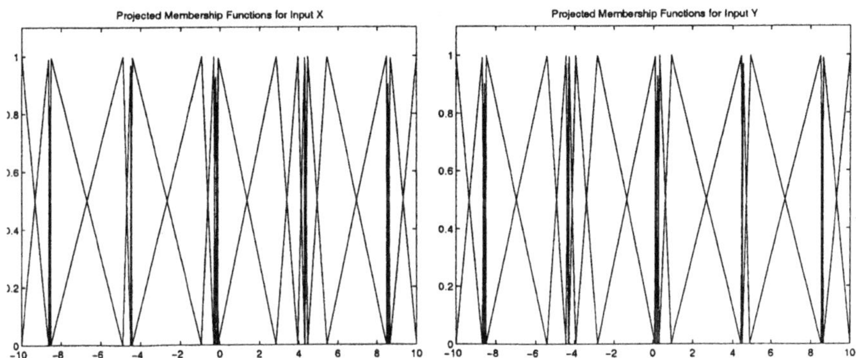

Fig. 10. *Example 1:Projected membership functions*

19. **IF** x is Positive Medium AND y is Positive Medium **THEN** z is Positive with strength 1
20. **IF** x is Positive Large AND y is Positive Medium **THEN** z is Zero with strength 0.93 AND Positive 0.07
21. **IF** x is Negative Large AND y is Positive Large **THEN** z is Negative with strength 0.02 AND Zero with strength 0.98

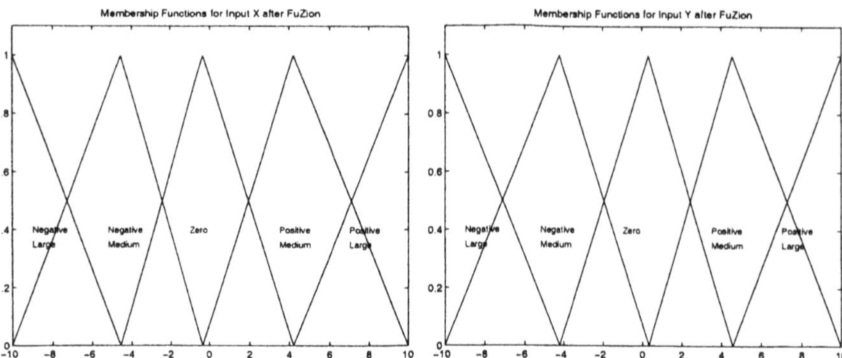

Fig. 11. *Example 1:Membership functions after FuZion with M =10%*

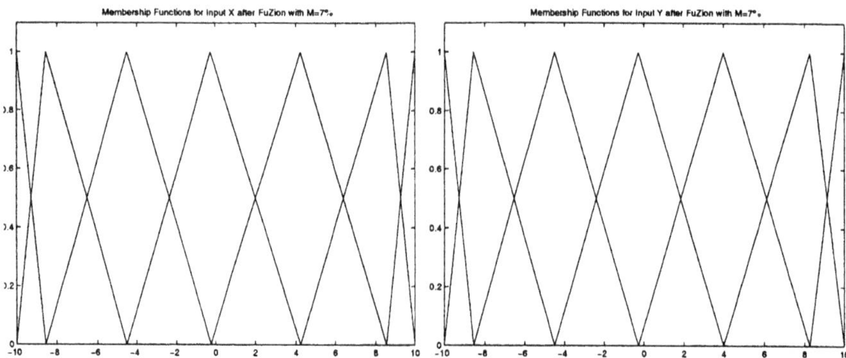

Fig. 12. *Example 1:Membership functions after FuZion with M =7%*

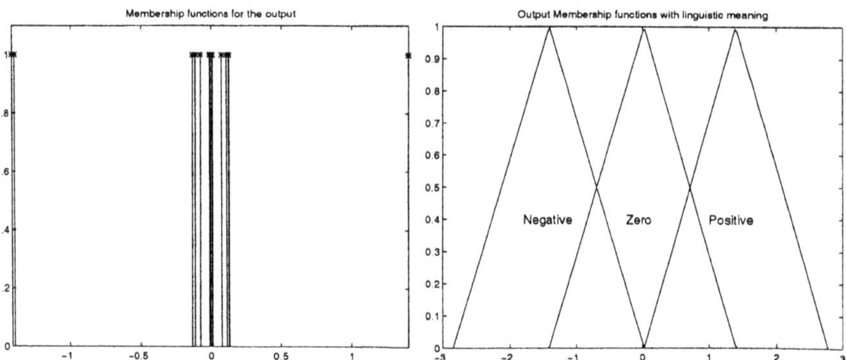

Fig. 13. *Example 1:(a) Singletons (b) Membership functions with linguistic meaning*

22. **IF** x is Negative Medium AND y is Positive Large **THEN** z is Negative with strength 0.07 AND Zero with strength 0.93

23. **IF** x is Zero AND y is Positive Large **THEN** z is Negative with strength 0.02 AND Zero with strength 0.98
24. **IF** x is Positive Medium AND y is Positive Large **THEN** z is Zero with strength 0.96 AND Positive with strength 0.04
25. **IF** x is Positive Large AND y is Positive Large **THEN** z is Negative with strength 0.01 AND Zero with strength 0.99

Observe that the obtained rules exhibit a clear dominance of one of the consequences. When this happens it will be possible to eliminate the consequence with the small strength without a major impact on the numerical approximation. However, this step is a decision that must be left to the designer because it is case dependent. Figure 14 shows the identified surface. Observe that the main features of the function were captured by the fuzzy system.

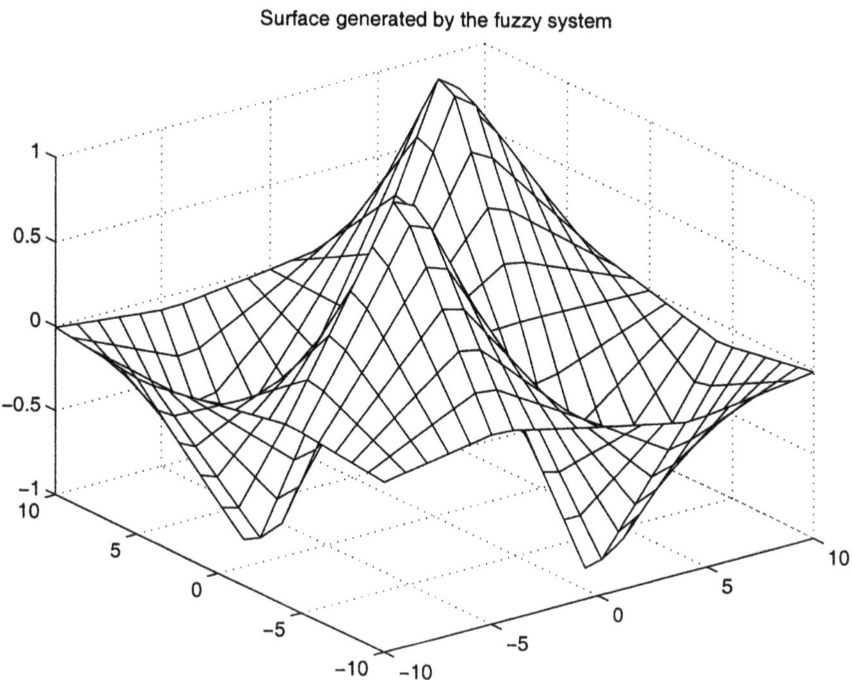

Fig. 14. *Example 1:Surface Generated by the Fuzzy System*

Example 2. Modeling of a High Density Polyethylene (HDPE) Reactor
In this example the purpose is to predict the density of the Polyethylene produced in a gas phase HDPE reactor. For this purpose three signals are collected and preprocessed to eliminate dynamic information. Finally 254

samples were selected and from this set two subsets were chosen, one for training (178 samples) and one for validation (76 samples). The input signals are C_4/C_2 ratio, H_2/C_2 ratio and Product Outflow (see figure 15), and the output signal is the Density of the PolyEthylene (see figure 16).

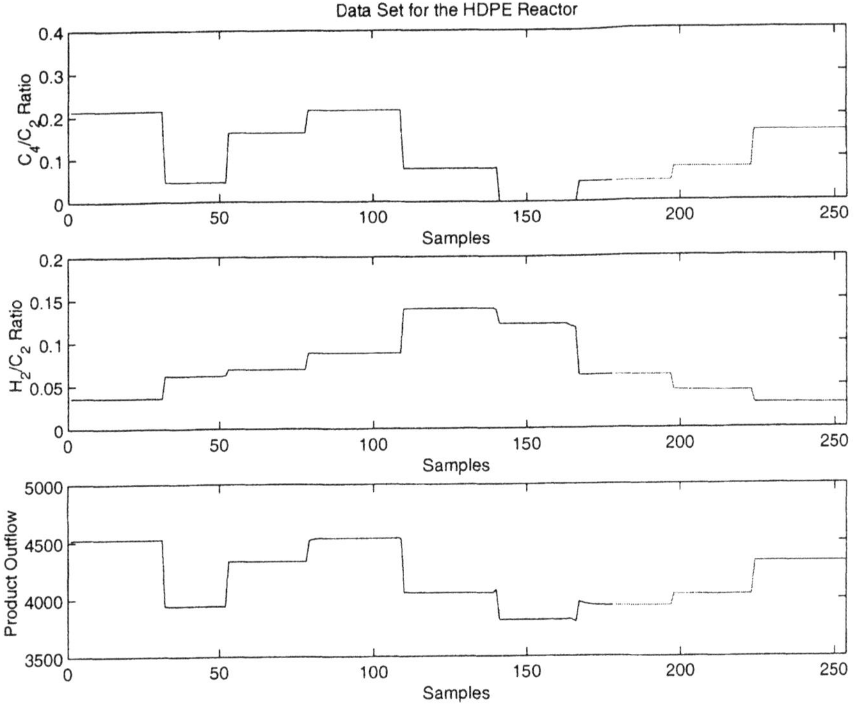

Fig. 15. *Example 2:Signals used to predict the density in the HDPE reactor*

The training set was clustered using mountain clustering ([6])with a grid of five(5) divisions per dimension. From this procedure, 6 clusters were selected as the most important candidates and they were refined using fuzzy C-means ([17]) (**step 2**). These clusters were projected into input and the membership functions were constructed (**steps 3,4,5,6**). Figure 17 shows the projected membership functions. The FuZion algorithm was applied to the input domains with M equal to 10 % of the universe of discourse (**steps 7,8**). Figure 18 shows the membership functions after the FuZion algorithm was applied. A rule base of 60 rules ($4 \times 5 \times 3$) was constructed and the consequences of the rules were calculated (**steps 9,10**). In figure 19(a) the singleton consequences are represented.
The singletons were converted into triangular membership functions and re-

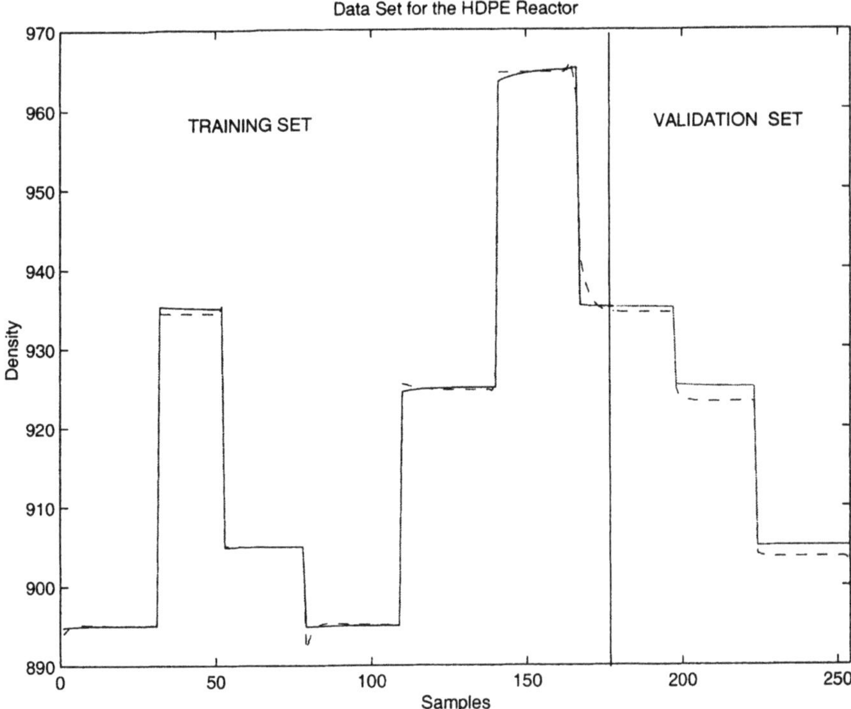

Fig. 16. *Example 2:Data for density in the HDPE reactor Original Data (-) Predicted data (- -)*

duced by means of the FuZion algorithm to only six membership functions (see figure 19(b)) (**steps 12,13,14**).
The new membership functions are associated with the rules (**steps 15,16**). Some of the obtained rules are:

- **IF** C4-C2 Ratio is Small AND H2-C2 Ratio is Very Small AND Prod.Outflow is Small **THEN** Density is Medium Low with strength 0.9 AND Low with strength 0.1
- **IF** C4-C2 Ratio is Small AND H2-C2 Ratio is Very Small AND Prod.Outflow is Large **THEN** Density is Very High with strength 0.99 AND High with strength 0.01
- **IF** C4-C2 Ratio is Small AND H2-C2 Ratio is Very Large AND Prod.Outflow is Small **THEN** Density is Medium Low with strength 0.89 AND Medium High with strength 0.11
- **IF** C4-C2 Ratio is Small AND H2-C2 Ratio is Very Large AND Prod.Outflow is Large **THEN** Density is Very High with strength 0.99 AND High with strength 0.01

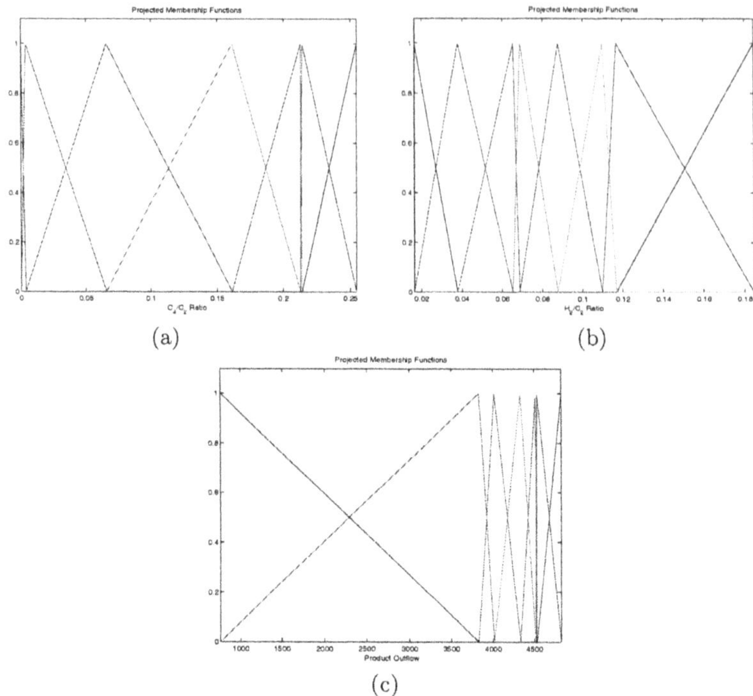

Fig. 17. *Example 2: Membership functions after cluster projection. (a) C_4/C_2 ratio, (b) H_2/C_2 ratio, (c) Product Outflow*

- **IF** C4-C2 Ratio is Very Large AND H2-C2 Ratio is Very Small AND Prod.Outflow is Small **THEN** Density is Very Low with strength 0.6 AND Low with strength 0.4
- **IF** C4-C2 Ratio is Very Large AND H2-C2 Ratio is Very Small AND Prod.Outflow is Large **THEN** Density is High with strength 0.95 AND Very High with strength 0.05
- **IF** C4-C2 Ratio is Very Large AND H2-C2 Ratio is Very Large AND Prod.Outflow is Small **THEN** Density is Very Low with strength 1
- **IF** C4-C2 Ratio is Very Large AND H2-C2 Ratio is Very Large AND Prod.Outflow is Large **THEN** Density is Medium High with strength 0.92 AND Medium Low with strength 0.08

Finally, figure 16 shows the prediction of the fuzzy model, observe in figure 15 that the conditions of the validation set are different to the ones presented in the training set, however the prediction is still very good for the validation set.

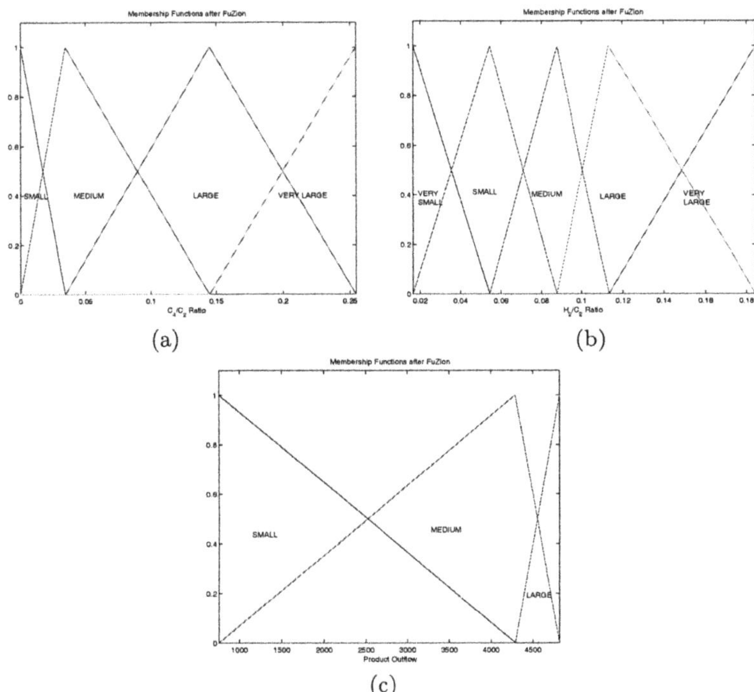

Fig. 18. *Example 2: Membership functions after FuZion. (a) C_4/C_2 ratio, (b) H_2/C_2 ratio, (c) Product Outflow*

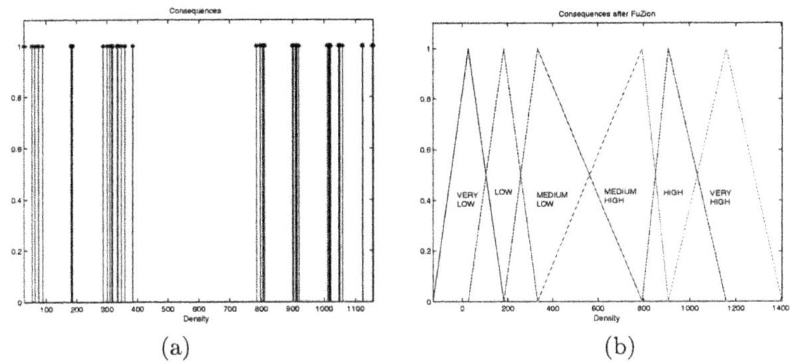

Fig. 19. *Example 2: Output Membership functions. (a) Singletons (b) After FuZion*

6 Closing Remarks

The AFRELI algorithm is an algorithm based on clustering methods to locate the rules. The clusters are projected into the input space and converted into fuzzy sets. The FuZion algorithm reduces the large number of projected fuzzy sets to preserve the linguistic integrity. With the reduced number of fuzzy sets a combinatorial set of IF-THEN rules is constructed using only AND operators among the antecedents. The rule base will cover every possible case in the compact set defined by the universes of discourse of the input domains. This is an important advantage because the system will be able to predict values of the approximated function, even if there were no similar values in the data set used to construct the model. The generalization of this method for the case of rules badly excited during training will be as good as the generalization given by a multilinear model of dimension p where p is the number of inputs.
However, such large number of rules present some problems: i) a large set of rules with a large set of antecedents is difficult to understand and analyze; ii) a large set of rules demands large memory storage; for instance, a rule base created for a system with five inputs and six membership functions on each input will demand approximately 30 Kbytes of memory and a similar system with 10 inputs will demand 230 Mbytes of memory. On the other hand the evaluation will not grow that fast, for the first system only $2^5 = 32$ rules need to be evaluated, for the second system $2^{10} = 1024$ rules.

7 Conclusions

The AFRELI algorithm in combination with the FuZion algorithm guarantees a good trade-off between numerical accuracy and interpretability. The method exploits some successful elements proposed in other methods to reduce the complexity of the model construction.
The algorithm automatically generates the fuzzy sets and the interactive labeling process (with intervention of the designer) guarantees an agreement between the fuzzy set and the assigned label.
The method generates a rule base covering all the possible cases, this guarantees the completeness of the rule base, but the associated drawback is the exponential growth of the rule base as the number of inputs increases. However, this is only a storage problem because the description of the fuzzy sets guarantees that only 2^p rules (p number of inputs) are activated on each inference. This makes the inference process fast because only a limited number of rules are evaluated.
The numerical accuracy of the algorithm is directly related with the choices of the M parameter governing the FuZion algorithm, and the choices in the clustering algorithm. When the number of inputs is large, the clustering will be limited to use Fuzzy C-means algorithm with an overestimated number

of clusters. The poor performance of the Mountain Clustering method with large dimensions motivated this modification.
Some improvements of the numerical performance of the model can be obtained by making a "fine" tuning of the parameters of the antecedents by means of constrained gradient descent techniques.

Acknowledgements

This research was supported by grants from several funding agencies and sources : Research Council KUL : Concerted Action GOA-Mefisto 666 (Mathematical Engineering), Flemish Government : Fund for Scientific Research Flanders (G.0240.99 (multilinear algebra), G.0407.02 (support vector machines), research communities ICCoS, ANMMM), IWT; Belgian Federal Government: DWTC (IUAP IV-02 (1996-2001), and IUAP V-10-29 (2002-2006) : Dynamical Systems and Control : Computation, Identification & Modelling).

References

1. Wang, L. X. (1994) Adaptive Fuzzy Systems and Control. 1st. Ed., Prentice Hall.
2. Jang, J. S. R. (1994) Structure Determination in Fuzzy Modeling: A Fuzzy CART Approach. Proceedings of IEEE international conference on fuzzy systems
3. Tan S., Vandewalle J. (1995) An On-line Structural and Parametric Scheme for Fuzzy Modelling. Proc. of the 6th International Fuzzy Systems Association World Congress IFSA-95. Sao Paulo, 189–192
4. Sugeno M., Yasukawa T. (1993) A Fuzzy Logic Based Approach to Qualitative Modeling. IEEE Trans. on Fuzzy Systems **1**, 7-31
5. Yager R., Filev D. (1994) Essentials of fuzzy modeling and control.1st. Ed. John Wiley & Sons. New York
6. Lori N., Costa Branco P. J. (1995) Autonomous Mountain-Clustering Method Applied to Fuzzy Systems Modeling. In: Dagli C. H., Akay M., Philip C. L., Chen, Fernández B., Ghosh, J. (Eds.): Intelligent Engineering Systems Through Artificial Neural Networks, Smart Engineering Systems: Fuzzy Logic and Evolutionary Programming. ASME Press. New York, 311-316
7. Jang J. S. R., Sun C. T., Mizutani E. (1997) Neuro Fuzzy and Soft Computing. Prentice Hall International. USA.
8. Valente de Oliveira J. (1999) Semantic Constraints for Membership Function Optimization. IEEE Trans. Systems Man and Cybernetics-Part A **1**
9. Jang, J. S. R (1998) Fuzzy Logic Toolbox, User's Guide, V-2.0. Mathworks. USA.
10. Setnes M., Babuska R., Verbruggen H. (1998) Rule-Based Modeling: Precision and Transparency. IEEE Trans. Systems Man and Cybernetics-Part C **1**
11. Espinosa J., Vandewalle J. (1998) Fuzzy Modeling and Identification, Using AFRELI and FuZion Algorithms. Proceedings of the 5th. International Conference on Soft Computing IIZUKA-98. Iizuka-Japan, 535–540

12. Pedrycz W. (1994) Why Triangular Membership Functions?. Fuzzy Sets and Systems. **64**, 21-30
13. Jang J. S. R. (1992) Neuro-Fuzzy Modeling: Architectures, Analyses and Applications. University of Berkeley-California
14. Pedrycz W., Valente de Oliveira J.(1996) Optimization of fuzzy models. IEEE Trans. Syst.,Man,Cybern. Part B. **26** 627-636
15. Broadbent D. (1975) The Magic Number Seven After Fifteen Years. In: Kennedy A., Wilkes A. Studies in Long Term Memory **3** Addison Wesley. New York
16. Espinosa J., Vandewalle J. (2000) Constructing Fuzzy Models with Linguistic Integrity from Numerical Data-AFRELI Algorithm. IEEE Trans. on Fuzzy Systems.**8** 591-600
17. Bezdek J. C. (1976) A Physical Interpretation of Fuzzy ISODATA. IEEE Trans. Syst.,Man,Cybern. 387-389
18. Nozaki K., Ishibuchi H., Tanaka H. (1997) A Simple but Powerful Heuristic Method for Generating Fuzzy Rules from Numerical Data. Fuzzy Sets and Systems.**86** 251-270
19. Espinosa J., Vandewalle J. (1998) Fuzzy Modeling with Linguistic Integrity. Proceedings of the International Workshop on Advanced Black Box Techniques for Nonlinear Modeling. Leuven - Belgium. 197–202
20. Espinosa J. (2001) Fuzzy Modeling and Control. Ph.D.Thesis. Katholieke Universiteit Leuven. Leuven Belgium

Constrained Optimization of Fuzzy Decision Trees

Pierre-Yves Glorennec
pierre-yves.glorennec@irisa.fr

INSA de RENNES/IRISA
F-35043 Rennes Cedex, France

Abstract. This paper proposes to build and optimize Takagi-Sugeno-like fuzzy regression trees with constraints aiming at preserving the interpretability of the rules. For this purpose, we state five requirements and deduce some conditions such as membership functions shared by the rules and strong fuzzy partitions on input variable domains. The membership functions are automatically placed thanks to an evolutionary strategy. We propose also a heuristics in order to find a sub-optimal structure of the tree.

1 Presentation

Different approaches can be used for modeling a non-linear input/output relationship: Neural Networks, Support Vector Machine, Fuzzy Inference Systems... These approaches generally lead to comparable results, with their own advantages and drawbacks. The main interest in using a Fuzzy Inference System (FIS) lies in its capability to represent an expert human knowledge, with interpretable rules written in natural language. This interpretability makes possible to incorporate the available *a priori* knowledge and to easily initialize parameters with realistic values. Moreover, if interpretability is guaranteed throughout the learning stage, it is possible to do an *a posteriori* examination and validation of the optimized rule base.

The first FIS were designed and tuned by experts, using some "rule of thumb" and trial and error. For this reason, the applications were restricted to simple systems (typically: two inputs, one output). A Self-Organizing Controller (SOC) was proposed by Procyk and Mamdani in 1979 [15], but the real development of optimization techniques started in the beginning of the 90's, see *e.g.* [9]. Now, it is possible to automatically extract knowledge from data [5], but the application are often restricted to low dimensional input spaces (less than four or five inputs). Two families of optimization techniques are usually used according to the available feedback:

- Supervised Learning, if there are input/output pairs recorded experimentally;
- Reinforcement Learning, when the only feedback from environment is a simple scalar signal understood as reward or punishment.

The sequel of this chapter is devoted to supervised learning only, with the following goals:

- increase the dimension of the input space, in order to deal with complex problems,
- increase the accuracy while keeping interpretability of the rule base.

In section 2, we show some consequences of unconstrained learning and deduce five requirements for interpretability and readability. We justify our choice of Fuzzy Decision Trees (FDT) in section 3, claiming that they are a natural and useful extension of both FIS and crisp Decision Trees. A method for building a FDT is shown in section 4 and its optimization thanks to an Evolutionary Strategy in section 5. The usefulness of the proposed method is illustrated with a known subjective evaluation problem in section 6.

2 Requirements for interpretability and large input spaces

We consider a Takagi-Sugeno model of order zero, [21], for two reasons:

- these FIS have the universal approximation property [2], and therefore they can be used for any non-linear input/output mapping;
- the input/output mapping is described by a very simple analytical equation, see Eq. 2, allowing easy tuning of parameters.

Takagi-Sugeno FIS can have as many real conclusions as the number of rules, thus reducing interpretability, but they have better accuracy than Mamdani FIS. Moreover, it is possible to drastically reduce the number of different values in the conclusion part of the rules with a limited loss of accuracy and simultaneously an increase in legibility, see [6]. In this model the rules have the form:

$$\textbf{if } x_1 \text{ is } A_1^i \ldots \textbf{ and } x_n \text{ is } A_n^i \textbf{ then } y = b^i, \text{ for } i = 1 \text{ to } N \tag{1}$$

where:

- $(A_j^i)_{j=1,n}$ are fuzzy sets characterized by linguistic labels (e.g. "small", "medium", "positive large",...) and by membership functions, quantifying the membership degree of x to A_j^i. We suppose that all the membership functions are entirely defined by some parameters to be tuned.
- b^i is a real value to be tuned.

Let $y : \mathbf{x} \rightarrow y(\mathbf{x}) = FIS(\mathbf{x})$ be the input/output mapping performed by the set of rules. For a given input vector $\mathbf{x}$, the output is obtained by:

$$y(\mathbf{x}) = \frac{\sum_{i=1}^{N} \alpha_i(\mathbf{x}) \times b^i}{\sum_{j=1}^{N} \alpha_j(\mathbf{x})} = \sum_{i=1}^{N} \frac{\alpha_i(\mathbf{x})}{\sum_{j=1}^{N} \alpha_j(\mathbf{x})} \times b^i \tag{2}$$

where $\alpha_i(\mathbf{x})$ is the firing strength of rule i for an input vector $\mathbf{x}$, given a conjunction operator *And*:

$$\alpha_i(\mathbf{x}) = And(\mu_{A_1^i}(x_1), \ldots, \mu_{A_n^i}(x_n)) \tag{3}$$

From equations (1) and (2), we see that there is an immediate passage from a linguistic to an analytic form of the rule base.

2.1 How to optimize ?

Takagi-Sugeno FIS are non-linear non-parametric analytical models, see [19]. Therefore, it is possible to use different regression algorithms in order to minimize some criterion error. Algorithms derived from gradient descent are very popular because they provide simple updating equations. The general form is the following. Let $\mathbf{x} \rightarrow S_\theta(\mathbf{x})$ be a FIS entirely described by the vector $\theta = (p_1, \ldots, p_K)$, where the p_k are the parameters defining the membership functions and conclusions. The criterion to minimize is, for example,

$$F_\theta = \frac{1}{2} \sum_x (S_\theta(\mathbf{x}) - d(\mathbf{x}))^2 \tag{4}$$

where $d(\mathbf{x})$ is the reference output for input $\mathbf{x}$. Then, if F_θ is differentiable *w.r.t.* p_k, the updating equation is:

$$\triangle p_k = -\epsilon \frac{\partial F_\theta}{\partial p_k} \tag{5}$$

Unfortunately, when using only equation 5, we have no guarantee that the parameters retain any physical meaning throughout the learning stage. This method is used, *e.g.* in [11], with real efficiency, but the optimized rules are not always interpretable, leading to a paradoxical passage from knowledge to ignorance or incoherence (from a linguistic point of view).

A (not so) caricatural example of unconstrained optimization is illustrated in Figure 1 where we can see:

- *empty overlap of membership functions*, Figure 1-a, which can occur when the learning points are not dense in the input space.
- *loss of semantic*, Figure 1-a, where the linguistic labels A_1 and A_2 are inverted: the "linguistic" ordering relation is no longer true.
- *"brouhaha"*, Figure 1-b. At the point x^*, all the rules are fired and, for example, the rule "if x is A_1" cannot be examined separately.

2.2 Five requirements

To avoid these problems and according to our goals, we can state five requirements.

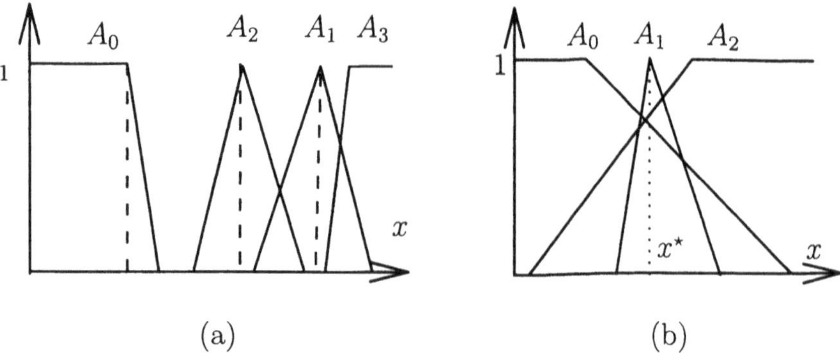

Fig. 1. Semantic problems with unconstrained optimization, from [4].

R_1- **Completeness**

Let $x \in U$ be a variable and (A_j) are fuzzy sets defined on the domain U of x. If $\delta > 0$ is a given completeness degree, the completeness condition is:

$$\forall x \in U, \exists j \text{ such that } \mu_{A_j}(x) \geq \delta > 0 \tag{6}$$

R_2- **Semantic ordering relation**

Preserving the semantic ordering relation[1] involves a condition on membership functions:

$$(\forall i < j)(\max_{a_i \in \mathcal{N}(A_i)} a_i < \min_{a_j \in \mathcal{N}(A_j)} a_j) \tag{7}$$

where $\mathcal{N}(A) = \{x/\mu_A(x) = 1\}$ is the kernel of fuzzy set A.

R_3- **Prototype**

For examination and validation, each rule needs at least a prototype defined by:

$$(\forall i, 1 < i < N)(\exists \mathbf{x}_i^\star)(\alpha_i(\mathbf{x}_i^\star) \gg \sum_{j \neq i} \alpha_j(\mathbf{x}_i^\star)) \tag{8}$$

For the input vector $\mathbf{x}_i^\star$, the contribution of all the other rules is negligible: the rule i can be examined separately. More precisely, the *prototypicality degree* of $\mathbf{x}$ for rule i is:

$$Prot(\mathbf{x}|i) = \alpha_i(\mathbf{x}) - \sum_{j \neq i} \alpha_j(\mathbf{x}) \leq 1 \tag{9}$$

R_4- **Coherence**

The predicate "x is A" must have the same meaning in all the rules where it appears, for all x and A.

[1] *e.g.* "low" is less then "medium" which is less than "large".

R_5- **Low complexity**
The number of fuzzy labels in the premise part of the rules must be as low as possible in case of large dimensional input spaces.

2.3 Choices compatible with the requirements

The previous requirements naturally lead to the following choices (although other choices are possible, see [7]).

C_1- **Strong Fuzzy Partitions**
A strong fuzzy partition on the universe U of a variable x is defined by:

$$(\forall x)(\sum_i \mu_{A_i}(x) = 1) \tag{10}$$

where A_i are fuzzy subsets on U. Such a constraint is easily achieved using trapezoid or triangular membership functions crossing at the grade 0.5, cf Figure 2.

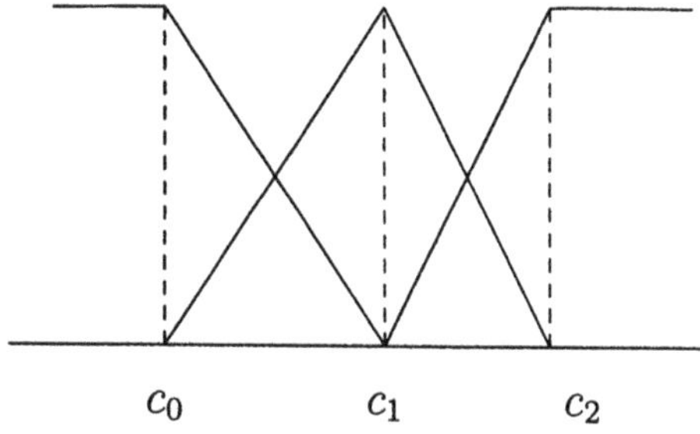

Fig. 2. Strong fuzzy partition using triangular membership functions

With the notations of Figure 2, the partition is entirely defined by the values of c_0, c_1 and c_2. Requirement R_1 is fulfilled with $\delta = 0.5$ and requirement R_2 implies:

$$c_0 < c_1 < c_2 \tag{11}$$

A modal value, a_j^i, for the fuzzy set A_j^i, is defined by $\mu_{A_j^i}(a_j^i) = 1$. $\mathbf{m}_i$ is a modal vector for rule i if its components are modal values for the fuzzy sets in the antecedent part of the rule. When using triangular strong fuzzy partitions, if $\mathbf{m}_i$ is a modal vector for rule i we have:

$$(\forall j \neq i)((\alpha_i(\mathbf{m}_i) = 1) \wedge (\alpha_j(\mathbf{m}_i) = 0)) \tag{12}$$

which is equivalent to:

$$Prot(\mathbf{x}|i) = 1 \tag{13}$$

and moreover we have:

$$y(\mathbf{m}_i) = b^i \tag{14}$$

Therefore requirement R_3 is fulfiled.

C_2- **Evolutionary methods**
Gradient descent-based methods are not compatible with requirements R_1 to R_3 nor with the structure of strong fuzzy partitions. On the contrary, these constraints can be easily incorporated into evolutionary methods: Genetic Algorithms or Evolutionary Strategies.

C_3- **Shared membership functions**
The domain of input j is partitioned into m_j fuzzy sets, $\mathcal{A}_j^1, \ldots, \mathcal{A}_j^{m_j}$, with their associated linguistic labels. With the notations of equation (1), we have necessarily:

$$(\forall i, 1 \leq i \leq N)(A_j^i \in \{\mathcal{A}_j^1, \ldots, \mathcal{A}_j^{m_j}\}) \quad (15)$$

By this way, requirement R_4 (coherence) is fulfiled, but this choice can lead to possible combinatorial explosion when taking all the combinations of antecedent labels. In this case, the total number of rules is $\prod_{j=1}^{n} m_j$. For practical applications, in case of large input space, it is necessary to choose an appropriate model avoiding combinatorial explosion by reduction of the number and the size of the rules (*i.e.* the number of fuzzy labels in the premise part). This appropriate model can be a FDT.

C_4- **Fuzzy Decision Trees**
Inducing a FDT leads naturally to a small set of incomplete rules:

- at each node, it is possible to evaluate quantitatively the information gain (see section 4) and to decide to stop or not the expansion of the tree before the complete examination of all the attributes;
- moreover, the pruning process eliminates subtrees and shorten the way from the root to a leaf.

Both the number of rules and the number of input variables taken into account are reduced, satisfying requirement R_5.

3 Fuzzy Decision Trees

3.1 Induction of crisp Decision Trees

Crisp Decision Trees (DT) are powerful tools for classification or regression problems, [8,12,17]. They proceed by successive partitions of a learning set into subsets with growing homogeneity. For a classification problem, in the ideal case, the training vectors associated with a terminal node (or leaf) belong to the same class. At each node we have to choose the "best" test for partitioning, according to a measure of effectiveness in classifying: the information gain.

With a set of training examples, $X = \{(\mathbf{x}, d(\mathbf{x}))\}$, where $\mathbf{x} = (x_1, \ldots, x_n)$ is an input vector[2] and $d(\mathbf{x})$ is the class number, it is possible to induce a DT. Let p_k be the probability of class k in X. The information content of X is measured by the entropy defined by[3]:

$$H(X) = -\sum_k p_k \log(p_k) \tag{16}$$

For two classes, with probabilities p and $1 - p$, this function has the form depicted in Figure 3. The minimal values, $H(X) = 0$, correspond to $p = 0$ or $p = 1$, when the elements of X belong to the same class. The maximal value occurs when $p = 0.5$. The goal is to divide X into subsets minimizing entropy.

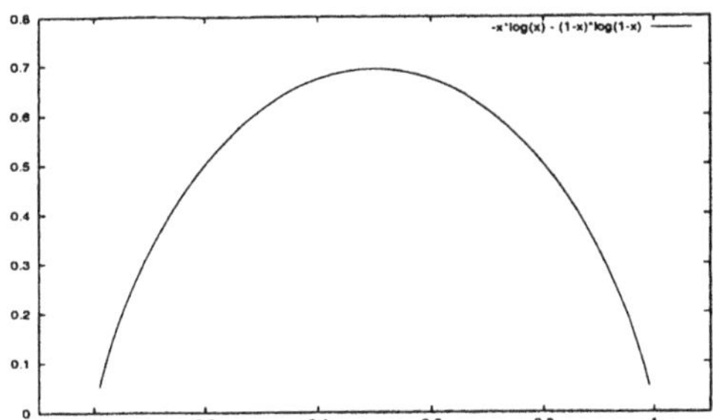

Fig. 3. The entropy function for two classes

The information gain from a variable at a node is the expected decrease in entropy by partitioning the examples (subset of X) according to the modalities of this variable. Therefore, for each node, we have to choose between the remaining variables (*i.e.* variables not chosen in a previous node) the one that has the maximum information gain.

3.2 FDT extend crisp Decision Trees

DT can be used with boolean, multi-valued or continuous data. The case of continuous data is tricky and is the origin of instability in classification. In the following example, illustrated in Figure 4 and Table 1, we can see that four quasi-identical situations for temperature (T) and wind velocity (V) give three different decisions. The reason is that a boolean test (*e.g.* T $<$ 20.6°C) introduce an arbitrary threshold in the variation domain of a continuous variable: such a threshold makes the test very sensitive to noisy data.

[2] The variables x_i are often called "attributes".

[3] This definition of entropy, in Eq. 16, was proposed by Shannon [18].

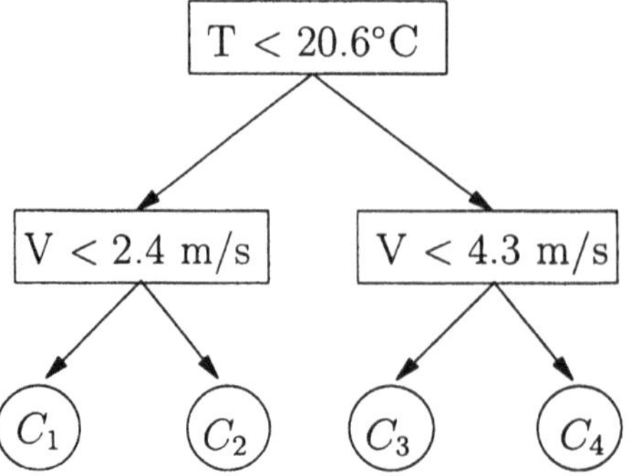

Fig. 4. Boolean tests with continuous values

Table 1. Four quasi-identical situations leading to three leaves

V \ T	20.5°C	20.7°C
2.2 m/s	C_1	C_3
2.4 m/s	C_2	C_3

Fuzzy representation allows to overcome this drawback, because a membership degree is associated to a test, instead of a boolean value. In Figure 5, the linguistic variable x has three fuzzy labels, therefore three nodes are created. For an actual value of x, x_0, the branches issued from the node will be valued by the membership degrees $\mu_{A_i}(x_0)$, $i = 1$ to 3.

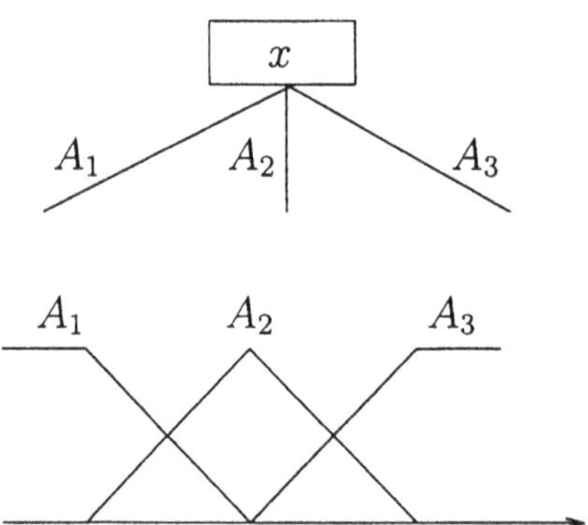

Fig. 5. Fuzzy Decision/Regression Tree

Two architectures are possible: binary FDT, where a node has at most two children, and more general FDT, where the number of children corresponds to the number of fuzzy label of the linguistic variable chosen at the node, as in Figure 5. We develop a methodology for general FDT.

3.3 FDT extend FIS

FDT are systems of fuzzy rules: each branch, from the root to a leaf, forms the premise part and the value associated to a leaf forms the conclusion of a fuzzy rule. Let c_L be this conclusion and $\alpha_L(\mathbf{x})$ the truth value of the rule.

- The truth value is computed with all the membership degrees along the branch for an input vector $\mathbf{x}$, see Equation 6.
- As in a FIS, see [5], c_L is calculated by:

$$c_L = \frac{\sum_{X_\epsilon} \alpha_L(\mathbf{x}) d(\mathbf{x})}{\sum_{X_\epsilon} \alpha_L(\mathbf{x})} \tag{17}$$

  where X_ϵ is a subset of the training set, X, for a given threshold ϵ:

$$\mathbf{x} \in X_\epsilon \Leftrightarrow \alpha_L(\mathbf{x}) > \epsilon \tag{18}$$

Therefore, by analogy with Equation 2, the inferred output of a Takagi-Sugeno-like FDT is:

$$FDT(\mathbf{x}) = \frac{\sum_L \alpha_L(\mathbf{x}) c_L}{\sum_L \alpha_L(\mathbf{x})} \tag{19}$$

The main differences between FIS and FDT are:

- in a FIS, the order of variables in the premise part has no importance; on the contrary, in a FDT, the variables are hierarchized and the way from the root to a leaf considers them from the most informative to the less informative;
- a FDT allows graphical representation which is both more readable and more informative when the number of input variables is large;
- in a FDT, the most important variables are automatically selected thanks to the information gain, with two consequences:
  - the dimension of the input space can be high, but the number of used inputs can be low,
  - the induction of a FDT naturally generates incomplete rules.

Despite from these differences, we want to emphasize that FIS and FDT correspond to the same reasoning scheme.

4 Induction of FDT

4.1 Notations

We consider the following general problem: let $f : (x_1, \ldots, x_n) \to y = f(x_1, \ldots, x_n)$ be an unknown input/output mapping, x_1 to x_n being potential explicative variables and y the output. Among the inputs, some are important, some are redundant, other are not very informative. Building a FDT consists in:

- hierarchizing the input variables according to their informative power,
- evaluating the utility to consider or not some non-informative variable (on-line or off-line by pruning).

We need a learning set $E = \{(\mathbf{x}^i, y^i); \mathbf{x}^i = (x_1^i, \ldots, x_n^i),\ y^i \in \mathcal{R}$, for $i = 1$ to $P\}$, where x_j is a linguistic variable, $j = 1$ to n, with m_j membership functions[4], $(A_{jk})_{j=1}^{m_k}$, defining a strong fuzzy partition of the domain of x_j. Without loss of generality, we choose the product as T-norm: $And(x, y) = x \times y$. With these choices, we have the property:

$$\sum_L \alpha_L(\mathbf{x}) \equiv 1 \tag{20}$$

The demonstration is straightforward. With this property, Equations 19 becomes:

$$FDT(\mathbf{x}) = \sum_L \alpha_L(\mathbf{x}) c_L \tag{21}$$

Let $(B_k)_{k=1}^{K}$ be a set of membership functions on the output domain. Without any loss of generality, we can suppose that we have triangular strong fuzzy partitions. To emphasize the analogy with crisp decision trees, B_k is regarded as a class, for $k = 1$ to K. Let:

- $\hat{\mu}_k(\mathbf{x}^i) \stackrel{\text{def}}{=} \mu_{B_k}(y^i)$: membership degree of $\mathbf{x}^i$ at class k.
- $\alpha_N(\mathbf{x}^i)$: membership degree of $\mathbf{x}^i$ at node N, calculated from the root to the node. If N is the root, $\alpha_N(\mathbf{x}^i) = 1$ for all i. If N is a leaf, then $\alpha_N(\mathbf{x}^i)$ is the product of membership degrees along the path from the root to the node:
$$\alpha_N(\mathbf{x}^i) = \begin{cases} 1 & \text{if } N \text{ is the root} \\ \alpha_M(\mathbf{x}^i) \times \mu_C(x_m^i) & \text{elsewhere} \end{cases} \tag{22}$$
with notations of Figure 6, where M is the parent of N, x_m the variable attached to M and C is a fuzzy set on domain of x_m.

[4] The same notation is used for both membership function and linguistic label.

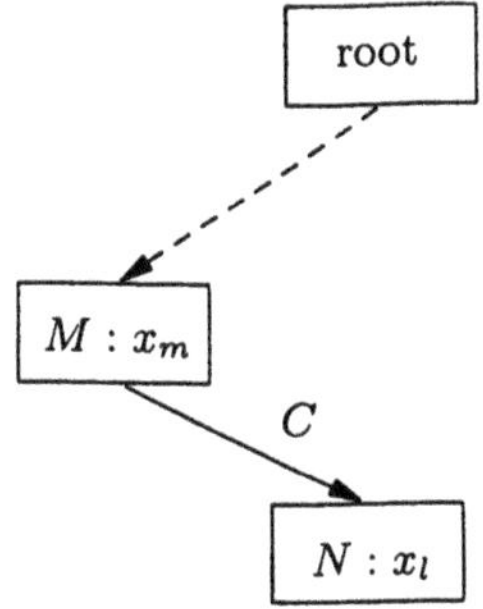

Fig. 6. Membership to node N

4.2 Information Gain

The *representation* of class k for the label A_{lj} of the variable x_l at node N, $r(k,l,j,N)$, plays a key role. It is defined by:

$$r(k,l,j,N) \stackrel{\text{def}}{=} \sum_{i=1}^{P} \hat{\mu}_k(\mathbf{x}^i) \times \mu_{A_{lj}}(x_l^i) \times \alpha_N(\mathbf{x}^i) \tag{23}$$

The information gain brought by variable x_l , with modalities $(A_{lj})_j$, at node N, is:

$$\begin{aligned} G(x_l,N) = &- \sum_k p_k \log(p_k) \\ &- \sum_j w_j \{ - \sum_k \frac{r(k,l,j,N)}{\sum_k r(k,l,j,N)} \log(\frac{r(k,l,j,N)}{\sum_k r(k,l,j,N)}) \} \end{aligned} \tag{24}$$

with:

$$p_k = \frac{\sum_j r(k,l,j,N)}{\sum_{i=1}^{P} \alpha_N(\mathbf{x}^i)} \tag{25}$$

$$w_j = \frac{\sum_k r(k,l,j,N)}{\sum_{i=1}^{P} \alpha_N(\mathbf{x}^i)} \tag{26}$$

4.3 Context of a variable

From Equations 22 and 23, we see that the information gain for variable x_l is strongly dependent of its position in the tree. The predicates previously used from the root to a node form the *context*: in different contexts, the same variable has different information gains.

In Figure 7, the context of x_l is not the same in the two paths. As a consequence, a variable can be very discriminant in a specific context only

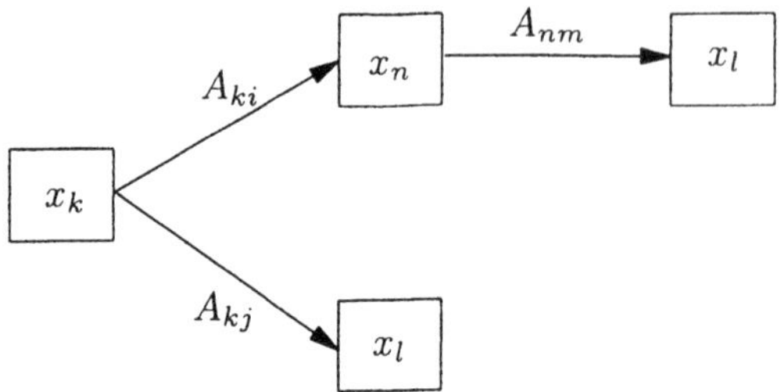

Fig. 7. Two different contexts for x_l.

(high information gain) and be insignificant elsewhere. This remark highlights the difficulty of input variable selection in regression problems. The relevance of a subset of input variables cannot be directly estimated at the root.

4.4 Entropy at the root

At the root we have $\alpha_N(\mathbf{x}^i) = 1$ for all i. Therefore, Equation 25 can be simplified and becomes:

$$\begin{aligned} p_k &= \frac{1}{P}\sum_j r(k,l,j,N) \\ &= \frac{1}{P}\sum_j\sum_{i=1}^{P}\hat{\mu}_k(\mathbf{x}^i) \times \mu_{A_{lj}}(x_l^i) \\ &= \frac{1}{P}\sum_i \hat{\mu}_k(\mathbf{x}^i)\sum_j \mu_{A_{lj}}(x_l^i) = \frac{1}{P}\sum_i \hat{\mu}_k(\mathbf{x}^i) \end{aligned} \tag{27}$$

thanks to the strong fuzzy partition ($\sum_j \mu_{A_{lj}}(x_l^i) = 1.$). The entropy is calculated with this expression of p_k, see Equation 16.

Suppose now that we have a crisp classification problem with:

$$\hat{\mu}_k(\mathbf{x}^i) = \begin{cases} 1 \text{ if } \mathbf{x}^i \text{ is in class } k \\ 0 \text{ elsewhere} \end{cases} \tag{28}$$

In this case, $p_k = \frac{n_k}{P}$, where n_k is the number of training examples belonging to class k. We have the same expression of entropy that in a crisp DT.

5 Constrained Optimization

The crux of FDT is that the user has to fix *a priori* the number of membership functions and place them by some "rule of thumb" on each input domain. These choices have direct consequences on the value of the different information gains, as already seen in section 4.3. Therefore, the user has to initialize by hand an important number of parameters, that is a difficult task.

We propose a "divide and conquer" strategy in order to reduce the complexity of the problem:

- for a given structure (number of membership functions), find the optimal placement using an evolutionary algorithm,
- increase the number of membership functions according to a heuristics and find the new optimal placement.

The proposed method has the following general form.

1. Begin with only two membership functions per input.
2. Optimize the placement.
3. Add one membership function for one input.
4. Optimize the placement for this input.
5. Accept or reject (*w.r.t.* some criterion).
6. Go to step 3 or quit.

5.1 Automatic placement of membership functions

In a FIS, the usual performance measure is the mean squared error (MSE) defined by MSE $= \frac{F_\theta}{P}$, see Equation 4, where P is the size of the training set. Here, θ can be restricted to input membership function parameters only, because the conclusions are obtained using Equation 17. But this method is computationally expensive because we have to re-build the tree after each modification of θ (the information gains are modified). An alternative consists in finding the placement of membership functions that maximizes the information gain. The major advantage of this choice is that we have to deal with low dimensional search spaces, variable per variable, in a top-down process:

- at the root: compute the maximum information gain for all variables; select the best variable and create a number of children corresponding to the number of fuzzy labels;
- at a node: compute the maximum information gain for only the non already used variables.

Because we have strong fuzzy partitions, we have chosen an evolutionary method capable to integrate such a constraint. The principle is depicted in Figure 8: a set of membership functions on an input variable domain corresponds to a string and *vice-versa*. Each new string generated by the evolutionary method is converted into a strong fuzzy partition and the information gain is computed (Equation 24). This information gain is the fitness value of the string.

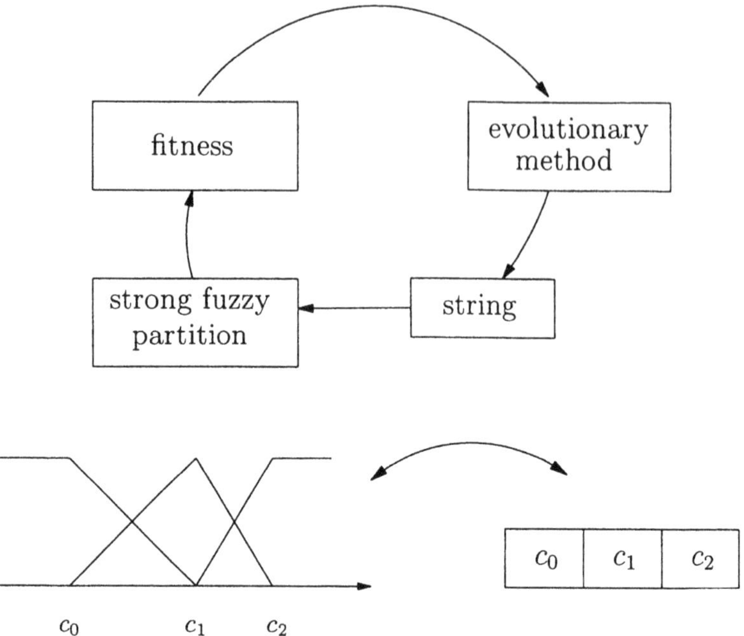

Fig. 8. Principle of the method.

An efficient algorithm, corresponding to a mono-agent evolutionary strategy, was proposed by Solis and Wetts [20]. It is an hill climbing method with memorization of the previous success thanks to a bias vector B. A mutation operator pays a key role in the stochastic search. This is equivalent to the moving of modal values in a fuzzy partition, (see Figure 9), abiding by the constraint of equation (11). A simplified version of Solis and Wetts's algorithm is shown in Table 2.

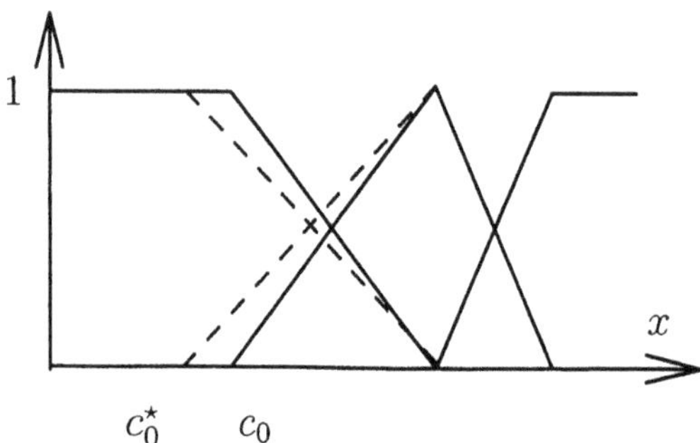

Fig. 9. Effects of a mutation: a biased Gaussian noise is added to c_0. The new point, $c_0^\star$, is used to rebuild the strong fuzzy partition.

Table 2. Solis and Wetts's algorithm

1. Choose $M^{(0)}$. Let $B^{(0)} = 0$ and $k = 0$.
2. Calculate $E(M^{(k)})$
3. Generate a Gaussian vector, $g^{(k)}$, with $g^{(k)} = B^{(k)} + \mathcal{N}(0, \sigma)$.
4. **if** $E(M^{(k)} + g^{(k)}) < E(M^{(k)})$
   then $M^{(k+1)} = M^{(k)} + g^{(k)}$
   $B^{(k+1)} = 0.4g^{(k)} + 0.2B^{(k)}$
   else if $E(M^{(k)} - g^{(k)}) < E(M^{(k)})$
   then $M^{(k+1)} = M^{(k)} - g^{(k)}$
   $B^{(k+1)} = B^{(k)} - 0.4g^{(k)}$
   else $M^{(k+1)} = M^{(k)}$
   $B^{(k+1)} = 0.5B^{(k)}$
5. **if** $k >$ MaxIter or error $<$ threshold
   then stop
   else $k = k + 1$; go to 3.

Implementation The user has to fix:

- the number, n, of membership functions,
- the size of the interval search (minimum and maximum values for the variable): c_0 = minimum, c_{n-1} = maximum.

The input domain is automatically divided into $(n - 1)$ equal parts and a triangular strong fuzzy partition is generated (see Figure 2 where $n = 3$). The values $c_0, \ldots, c_{n-1}$ form the initial string. Therefore, the search does not begin randomly but from a reasonable initial value, speeding up learning.

The other parameters are:

- the maximum number of iterations: generally between 20 and 40,
- a seed for the random generator,
- the value of the standard deviation, σ.

These parameters are not very sensitive and a sub-optimal solution is quickly found.

5.2 Heuristic for the number of membership functions per input

In order to build compact and readable FDT, it would be better to have as less membership functions as possible. But, in this case, the performance of the tree can be poor. To solve this dilemma, we propose a controlled growing, using two indicators:

- the MSE, computed with a training and a test set,

- a threshold for the gain in information, θ.

The principle consists in accepting to add one membership function for one input *only* when the corresponding improvement in information gain is greater than the threshold.

- A complete FDT is built, allowing to add a membership function only once for each variable. The conclusion part is computed with Equation 17.
- The MSE is calculated. If the performance is poor, the threshold is decreased and a new FDT is built. Conversely, if the number of generated rules is too large and the MSE is reasonably low, the threshold is increased.

The growing process stops when the MSE is low and/or the size of the tree reaches a maximum limit.

The heuristics is described in Table 3.

Table 3. Structural optimization algorithm: find the number of membership functions (MF) on each input domain

1. Choose a threshold, β.
2. Initialize with 2 MF per variable and optimize the placement.
3. For each variable, $Flag = 0$.
4. current node $\leftarrow$ root.
5. At current node:
   search for variable, x^*, with maximum gain; $IG = \text{Gain}(x^*)$
   if $Flag = 0$,
      add one MF on x^* domain
      optimize the placement, get IG_{new}
      if $IG_{new} - IG > \beta \times IG$ then accept the new MF
         $Flag = 1$
      else reject
   create children (nodes or leaves)
6. If not ended, change current node and go to 5.

It is always possible to increase the accuracy of the tree after its structural optimization, by a stochastic gradient descent method limited to the conclusion parts of the rules.

- Firstly, the conclusions are initialized[5] using Equation 17.

[5] Notice that in many cases, this initialization gives good results. Moreover, we have the guarantee that the conclusion belongs to the convex envelope of possible solutions.

- After that, they are incrementally updated by:

$$\triangle C_L = -\epsilon(FDT(\mathbf{x}) - d(\mathbf{x}))\alpha_L(\mathbf{x}) \tag{29}$$

where ϵ is a learning rate.

6 Example: Rice Taste Evaluation

Subjective evaluation is a difficult problem when the number of relevant variables is large. In most cases, a group of experts evaluates a product according to a set of characteristics. They give marks which can be:

- boolean values (*e.g.* bad, good),
- continuous values (*e.g.* between 0 (bad) to 1 (good)),
- linguistic values (*e.g.* very bad, bad, medium, good, very good).

Moreover, the tests suffer from:

- non-repeatability: the same expert can give different marks for the same product at different tests,
- large dispersion of marks in some cases, for some characteristics.

Rice taste evaluation is an example of such sensory tests. This problem was firstly proposed by [10] and recently revisited by [3,7]. In this problem, a group of 24 Japanese experts had to evaluate rice plates according to five characteristics: flavor, appearance, taste, stickiness and toughness.

The data set is a file with 105 lines, each line containing the five marks given by an expert and his/her overall evaluation. All the variables are normalized and lie in the interval [0,1]. The data set is randomly divided into two parts: 75 examples for learning, 30 for testing. For this problem, Cordon and Herrera [3] have the results in Table 4 for ten different data partitions. The first column, MF, gives the number of membership functions per input; the second, #R stands for the average number of rules, the other columns give the average of MSE for the training and the test sets.

Table 4. Results of Cordon and Herrera [3]: mean results for ten different data partitions.

MF	#R	MSE_{learn}	MSE_{test}
2	5	0.00341	0.00398
3	12.2	0.00185	0.00290

Our goals are:

- extract from data an accurate and interpretable model of the underlying relationship between the five chosen characteristics and the overall evaluation,
- provide some insight on the reasoning process performed by the experts,
- sort the characteristics from the most important to the less one.

6.1 Information gain at the root

A first induction of FDT is made with the minimum structure: only two membership functions per input. We take six classes on the output domain (see Section 4.1). Table 5 shows the effects of optimal placement of membership functions on the information gain. The table gives also the values of the modal points c_0 and c_1 after optimization (the initial values were $c_0 = 0$ and $c_1 = 1$).

Table 5. Automatic placement of membership functions. The information gain, IG, is increased.

input	IG before	IG after	c_0	c_1
flavor	0.05	0.10	0.07	0.80
appearance	0.06	0.09	0.08	0.83
taste	0.08	0.14	0.03	0.81
stickiness	0.12	0.19	0.12	0.88
toughness	0.03	0.07	0.15	0.86

The first criterion is stickiness[6] and the second is the taste. At the root, the toughness is the less informative variable, but in the context "*rice is not sticky* **and** *taste is bad* **and** *appearance is good*", this variable is more important than flavor, see Figure 10.

6.2 Induction of two FDT

We follow the method described in the previous sections. In the presented cases, the conclusion parts are calculated with Equation 17 and the gradient descent method (Equation 29) is not used.

The first induced FDT is the minimal tree (two labels/variable). Without optimal placement of membership functions on input domains, the performances of the inducted tree are MSE = 0.0037 on the training set and MSE = 0.0052 on the test set. Then, the optimization method is used and produce a tree with a comparable structure but better performances, see Table 6. It

[6] Imagine you are eating rice with chopsticks...

has 13 rules and is depicted in Figure 10. Its growing was stopped before the complete development because the information gain of the unused variables was low. All the variables appear but the premises have only three or four predicates. We have incomplete rules and not all the 32 possible combinations, therefore the rule base is easily understandable.

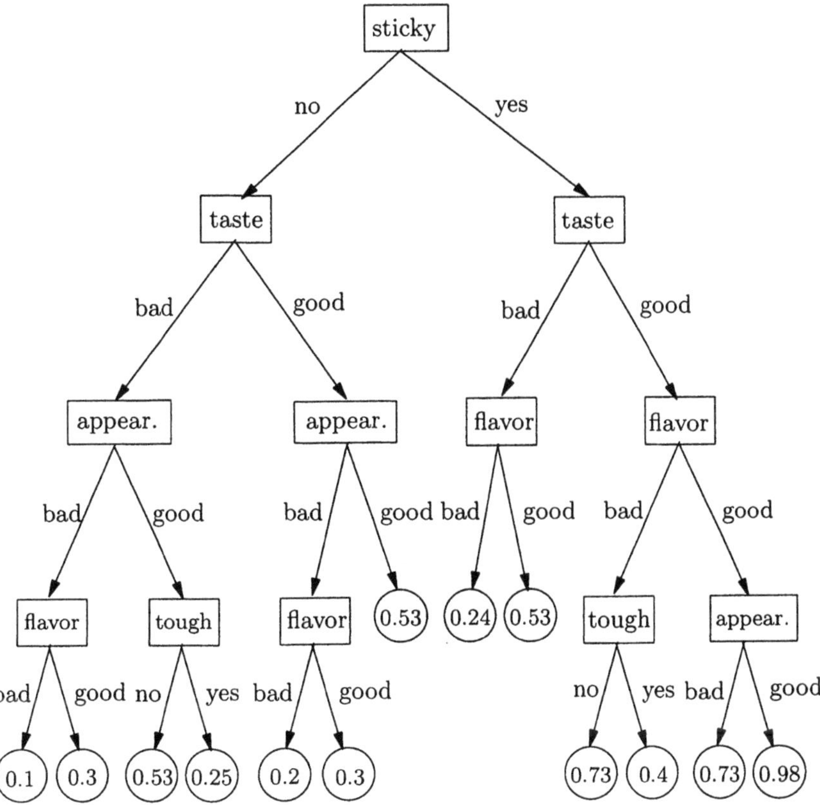

Fig. 10. FDT (13 rules) with 2 labels per variable. The marks are in the leaves.

One membership function is now added to the variable in the root (stickiness) with a significant increase in information gain: from IG = 0.19 with two labels to IG = 0.48 after optimization with 3 labels. The structural optimization starts but the other attempts to add a membership function to another input failed. Finally, the tree, depicted in Figure 11, has 10 rules and is simpler that the first one, because the rules have 2 or 3 predicates only. Continuing the development of the tree do not improve the performances.

Table 6 shows that the results are comparable.

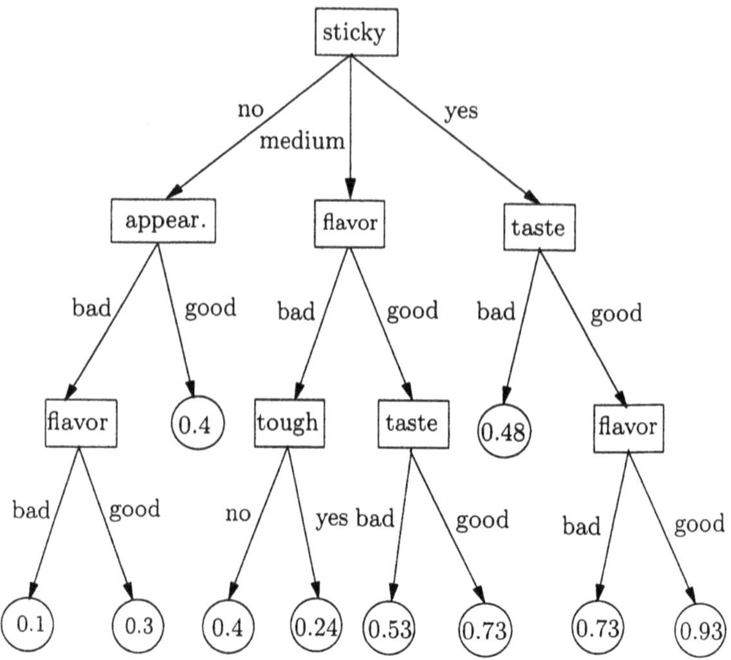

Fig. 11. FDT with 3 labels for stickiness (10 rules). The marks are in the leaves.

Table 6. Performances of the two FDT on the rice evaluation problem, mean of MSE using ten randomly selected data set for learning and testing. The 13-rules tree has two lines: the first, (a), without optimization of membership functions, the second, (b), with optimization.

FDT	MSE_{train}	MSE_{test}
13 rules (a)	0.0037	0.0052
13 rules (b)	0.0015	0.0018
10 rules	0.0017	0.0022

A *a posteriori* study of the two trees brings some useful indications.

- In both cases, the right subtrees are comparable, but the 10-rules tree narrows the concept "rice is sticky", associated with marks greater than 0.48, and give results with few tests. On the contrary, this concept is

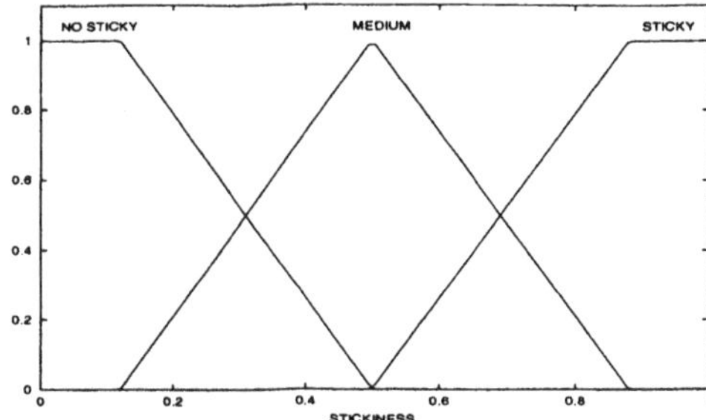

Fig. 12. Optimized membership functions for Stickiness, in the 10-rules tree.

wider in the 13-rules tree which therefore needs additional variables in order to discriminate between marks belonging to [0.24; 0.98].

- In the minimal structure, we have $2^5 = 32$ potential rules (although 13 are used only). Adding one membership function leads to $3 \times 2^4 = 48$ potential rules but increases significantly the information gain. Really, the complexity is reduced because the different subtrees have better characterizations and finally the number of rules is lower than expected.

7 Conclusion

The main drawbacks of Fuzzy inference systems are, firstly, a number of rules increasing exponentially when the number of inputs increases and, secondly, the dilemma between accuracy and interpretability.

- Tuning by hand a set of fuzzy rules is very hard when the number of inputs is larger than two or three. Automatic optimization methods are needed, but most of then (principally the gradient descent-based methods) can lead to a loss of semantic and can change initial interpretable rules into a sort of (often efficient) black box. Our claim is that introducing some natural constraints in the tuning process allows to improve the accuracy and to preserve the interpretability.
- The human brain can manage complex situations, involving a lot of simultaneous informations. Two mechanisms are used:
  - a coarse categorization and hierarchization of inputs, leading to fast recognition of contexts, where useless variables are neglected (a context can be thought of as a low dimensional space that the human brain can manage efficiency);
  - an interpolating scheme do deal with several contexts according to their activation level.

These mechanisms are reproduced by FDT which combine coarse categorization (thanks to the fuzzy partitions), hierarchization (thanks to the information gain) and interpolation (thanks to fuzzy inference). The proposed optimization method links these mechanisms and makes possible to use FDT for modeling complex problems.

References

1. Boyen X., Wehenkel L. (1996) Automatic Induction of Continuous Decision Trees. Proc. of the 6th Int. Conf. IPMU, pp. 419-424.
2. Castro J.L. (1995) Fuzzy logic controllers are universal approximators. IEEE Trans. on SMC 25-4, pp. 629-635.
3. Cordon O., Herrera F. (2000) A proposal for Improving the Accuracy of Linguistic Modeling. IEEE Trans. on Fuzzy Systems 8:3, pp. 335-344.
4. Glorennec PY. (1996) Constrained Optimization of Fuzzy Inference Systems using an Evolutionary Method. *In* Genetic Algorithms and Soft Computing, F. Herrera and J.L. Verdegay (Eds), Coll. Studies in Fuzziness, Physica-Verlag, pp. 349-368.
5. Glorennec P.Y. (1999) Algorithmes d'apprentissage pour systèmes d'inférence floue. Editions Hermès, ISBN 2-7462-0044-9.
6. Glorennec P.Y. (2002) Mamdani and Takagi-Sugeno Inference: an Unified View. Proc. of the Int. Conf. IPMU, july 2002.
7. Guillaume S. (2001) Induction de règles floues interprétables. Thèse à l'INSA de Toulouse.
8. Haskell R.E. (1998) Neuro-Fuzzy Classification and Regression Trees. Proc. of the 3th Int. Conf. on Applications of Fuzzy Systems and Soft Computing, Wiesbaden.
9. Horikawa S.I., Furuhashi T., Okuma S., Uchkawa Y. (1990) A fuzzy controller using a neural network. Proc. of Iizuka'90, Iizuka, Japan, pp. 103-106.
10. Ishibuchi H., Nozaki K., Tanaka H., Hosaka Y., Matsuda M. (1994) Empirical study on learning in fuzzy systems by rice test analysis. Fuzzy Sets and Systems 64, pp. 129-144.
11. Jang J.-S. R. (1993) Anfis: Adaptive-network-based fuzzy inference systems. IEEE Trans. on SMC 23 (03), pp. 665-685.
12. Janikow C. (1995) A genetic algorithm for optimizing fuzzy decision trees. Proc. of the 6th Int. Conf. on Genetic Algorithms, pp. 421-428.
13. Kajitani Y., Kuwata K., Katayama R., Nishida Y. (1991) An automatic fuzzy modeling with constraints of membership functions and a model determination for neuro and fuzzy model by plural performance indices. In Proc. IFES'91, pp. 586-597.
14. Marsala C. (1998) Apprentissage inductif en présence de données imprécises : construction et utilisation d'arbres de décision flous. Thèse à l'Université Paris 6.
15. Procyk T., Mamdani E. (1979) A linguistic self organizing process controller. Automatica 15-1, pp. 15-30.
16. Pedrycz W., Valente de Oliveira J. (1996) Optimization of fuzzy models. IEEE Transactions on Systems, Man, and Cybernetics 26:4, pp. 627-636.
17. Quinlan J.R. (1986) Induction of Decision Trees. Machine Learning 1(1), pp. 81-106.

18. Shannon C.E. (1948) The Mathematical Theory of Communication. University of Illinois Press, Shannon and Weaver Eds.
19. Sjöberg J., Zhang Q., Ljung L., Benveniste A., Delyon B., Glorennec P.Y., Hjalmarsonn H., Judiski A. (1995) Nonlinear Black-box Modeling in System Identification : a Unified Overview. Automatica 31-12, pp. 1691-1723.
20. Solis F., Wetts J. (1981) Minimization by random search techniques. Mathematics of Operation Research, Vol. 6, pp. 19-30.
21. Takagi T., Sugeno M. (1985) Fuzzy identification of systems and its applications to modeling and control. IEEE Trans. on SMC 15 (1), pp. 116-132.
22. Umano M., Okamoto H., Hatono I., Tamura H., Kawachi F., Umedzu S., Kinoshita J. (1994) Fuzzy decision trees by fuzzy ID3 algorithm and its application to diagnosis systems. Proc. of the 3th IEEE Conf. on Fuzzy Systems, Orlando, pp. 2113-2118.
23. Valente de Oliveira J. (1999) Semantic constraints for membership function optimization. IEEE Transactions on Systems, Man, and Cybernetics-Part A: Systems and Humans 29:1, pp. 128-138.
24. Wehenkel L. (1997) Discretization of continuous attributes for supervised learning. Variance evaluation and variance reductions. Proc. of the 7th IFSA Congress, Prague, pp. 381-388.
25. Yuan Y., Shaw M.J. (1995) Induction of Fuzzy Decision Trees. Fuzzy Sets and Systems 69, pp. 125-139.

A new method for inducing a set of interpretable fuzzy partitions and fuzzy inference systems from data

Serge Guillaume[1] and Brigitte Charnomordic[2]

[1] Cemagref, 361 rue Jean-François Breton, 34033 Montpellier, France
[2] INRA, 2 Place Viala, 34060 Montpellier, France

Abstract. To improve the interpretability of a fuzzy rule base generated from data, three conditions are necessary: semantic integrity must be respected, the number of rules should be small, and incomplete rules have to be handled. An incomplete rule is a rule defined only by a few variables. The presence of incomplete rules reflects the fact that all the variables do not have the same importance for all rules.

We propose a new method for learning a fuzzy rule base. A hierarchy of fuzzy partitions is generated for each input variable, based on a special metric suitable for fuzzy partitioning, before being used in building fuzzy inference systems of increasing complexity. Then the rule base is simplified. We introduce an intermediate selection level represented by a group of rules. The simplification tolerates some loss of accuracy while being guided by indices complementary to the usual numerical performance index.

The proposed approach is applied to a rice quality evaluation problem.

1 Introduction

This chapter introduces a new method for inducing a set of interpretable rules from data. Although being generic, it has been designed for dealing with complex systems, such as food processes, usually composed of a sequence of several unit operations. However these processes cannot be described by a simple decomposition, as their overall behavior shows new properties which did not appear in unit operations. For such systems, the influence of the input variables is not the same for all the regions of the input space, it depends on the context [9]. Food process control mostly relies on expert knowledge due to a lack of mathematical modeling. Knowledge induction from data will benefit to human-computer cooperation, for building decision support systems for example, under the restriction that induced rules are meaningful for the expert.

To improve the interpretability of a fuzzy rule base generated from data [8], three conditions have to be fulfilled. First semantic integrity should be respected within the partition. The partition size must be kept reasonably small, and each fuzzy set is to be assigned a readable linguistic label. Secondly the number of rules should be small. The third condition is specific to complex systems with a large number of input variables: incomplete rules have to be handled. An incomplete rule is a rule defined only by a few variables. The presence of incomplete rules reflects the fact that all the variables do not have the same importance for all rules. Incomplete rules naturally appear when formalizing expert knowledge and are easier to interpret than complete rules.

For semantic reasons we will prefer fuzzy rule conclusions (Mamdani type) over the linear combinations of inputs used in precise fuzzy modeling (TSK rule bases).

The method includes fuzzy partitioning and yields a small set of FRBSs. Each step pays attention to the interpretability conditions stated above.

The first step of the procedure generates a hierarchy of fuzzy partitions of different sizes for each input variable. We use an ascending method for which a merging criterion is needed. This criterion is based on the definition of a special metric distance suitable for fuzzy partitioning.

The second step generates several FRBSs of increasing complexity. The refinement procedure is based upon accuracy improvement. It uses the previously found hierarchies of fuzzy partitions. The trade-off between complexity and accuracy is user controlled.

The final step consists in rule base simplification. Between global variable removal and selection done at rule level, we introduce an intermediate selection level. It allows for evaluating the influence of a given variable within a context defined by the other inputs, and represented by a group of rules. The context implementation requires a rule distance function. The simplification stage tolerates some loss of accuracy while being guided by new indices complementary to the usual numerical performance index. According to the degree of simplification several FRBSs are designed.

This chapter is organized as follows: section 2 introduces the distance adapted to fuzzy partitioning, which is used in section 3 to build hierarchical fuzzy partitions. Section 4 describes the fuzzy rule base generation from the family of partitions. The procedure for simplifying a rule base in order to get incomplete rules is given in Section 5. The proposed approach is applied, in section 6, to a rice quality evaluation problem and the results are compared to those previously obtained by other teams. Finally section 7 contains some concluding remarks.

2 A distance metric suitable for fuzzy partitioning

For our needs, the hierarchical fuzzy partitioning that will be described in the next section, only a dissimilarity function is required. Nevertheless, we will show that the proposed function fulfills the triangle inequality, in order not to limit its use to this application.

Basic definitions A function d is a dissimilarity if

$$\forall\ q,t \qquad \begin{cases} d(q,t) \geq 0 \\ d(q,q) = 0 \\ d(q,t) = d(t,q) \end{cases} \tag{1}$$

A dissimilarity is proper if:

$$d(q,t) = 0 \quad \Rightarrow \quad q = t \tag{2}$$

A semi distance is a dissimilarity which verifies the triangle inequality:

$$\forall\ q,s,t \qquad d(q,t) \leq d(q,s) + d(s,t) \tag{3}$$

A proper semi distance is called a distance.

Fuzzy specificities The training dataset E is a collection of n multiple input-single output numerical data pairs $(x_k, y_k), k = 1, 2, \ldots, n$ where x_k is a p dimensional input vector $x_k^1, x_k^2, \ldots, x_k^p$ and y_k is a one-dimensional output vector. Consider two data points with respective x_q^j, x_t^j coordinates in the jth dimension. Due to the fuzzification procedure, they can belong to several fuzzy sets with a non zero degree.

To alleviate the notations we will denote $\mu_q^f = \mu_j^f(x_q^j)$, and $d(q,t) = d(x_q^j, x_t^j)$ the pairwise distance between these two data points. Let m be the number of fuzzy sets in the partition.

$d(q,t)$ is defined only if $max(\mu_q^f) > 0\ and\ max(\mu_t^f) > 0$ for $f = 1, \ldots, m$.

Two non exclusive cases are to be distinguished:

1. x_q^j and x_t^j partially belong to the same fuzzy set f, $\mu_q^f > 0$, $\mu_t^f > 0$
2. x_q^j and x_t^j partially belong to two fuzzy sets f and g, $f \neq g$, and $\mu_q^f > 0$, $\mu_t^g > 0$,

We introduce the terms of internal distance in the first case, external distance in the second one.

The pairwise distance $d(q,t) = d(x_q^j, x_t^j)$ will take into account q and t memberships to the various fuzzy sets by combining the respective parts of internal and external distances.

2.1 Internal Distance

The membership degree complement $(1 - \mu_q^f)$ can be interpreted as the distance of x_q^j to the fuzzy set f. It measures the similarity of x_q^j to the fuzzy set prototypes that delimit the kernel. Let us recall that a prototype x is such that $\mu(x) = 1$. Given two data points with (x_q^j, x_t^j) coordinates, their internal distance is equal to the difference of their respective prototype similarities, which comes to:

$$d_{int}(q,t) = |\mu_q^f - \mu_t^f|$$

The proposed internal distance function d_{int} is a semi distance: counterexamples for property 2 are easy to find. Many distinct data pairs have an identical membership degree, yielding a zero internal distance, as illustrated in figure 1 for a triangular membership function.

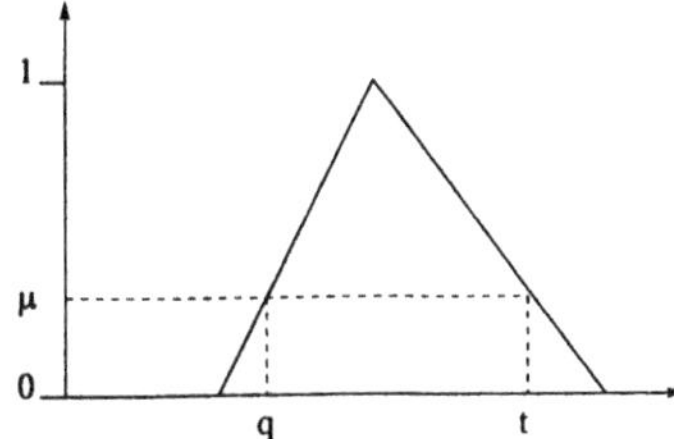

Fig. 1. A zero internal distance $d(q,t)$

2.2 External distance

The external distance is defined for two points q, t which belong to two fuzzy sets $f, g,\ f \neq g$. It must take account of each point location within its fuzzy set, and of the relative fuzzy set location within the fuzzy partition. A way of doing so is to define a prototype distance.

Distances are made independent of measurement units by scaling universes into the unit space.

We propose two different definitions of the distance $d_{prot}(f, g)$ between the prototypes of fuzzy sets f and g.

1. a numerical prototype distance $d_{prot}^{num}(f,g) = \sqrt{(c_f - c_g)^2}$, where c_f, c_g are the respective fuzzy set kernel locations. This definition corresponds to the kernel Euclidean distance. When dealing with fuzzy sets where the prototype is not unique, such as trapezoidal ones, several prototype distances are available. In this case, the shortest one is chosen.
2. a more symbolic prototype distance

$$d_{prot}^{sym}(f,g) = \frac{g - f}{m - 1} \tag{4}$$

   where m is the partition size, f and g are the indices of the fuzzy sets sorted in ascending order relatively to the kernel coordinates.

The external distance between two points which belong to f and g is defined as follows:

$$d_{ext}(q,t) = |\mu_q^f - \mu_t^g| + d_{prot}(f,g) + D_c \tag{5}$$

where $d_{prot}(f,g)$ is the distance between f and g fuzzy set prototypes, and D_c is a constant correction factor, which ensures the validity of the following semantic principle.

Semantic Principle Two points which mainly belong to the same fuzzy set will always be considered closer than others which mainly belong to distinct fuzzy sets.

The fuzzy set to which the point mainly belongs is the one for which the point membership degree is the maximum of all. In case of multiple maxima, the point is said to mainly belong to all of them.

The notion of main belonging could seem to be in contradiction with multiple membership. This is not the case. All membership degrees are considered for evaluating distances (see section 2.3), and main belonging is only used for setting the D_c constant. Its value depends on the practical implementation and it will be specified in section 3.1.

The semantic principle ensures that the distance will reflect the partition structure and preserve the fuzzy set label semantic.

Example of an external distance Within the partition illustrated on figure 2, the external distance is shown with single membership data points. The symbolic choice for the prototype distance makes the fuzzy set 3 at the same distance from 2 and 4, while the numerical choice puts it closer to 4. The symbolic distance is more faithful to the symbolic representation.

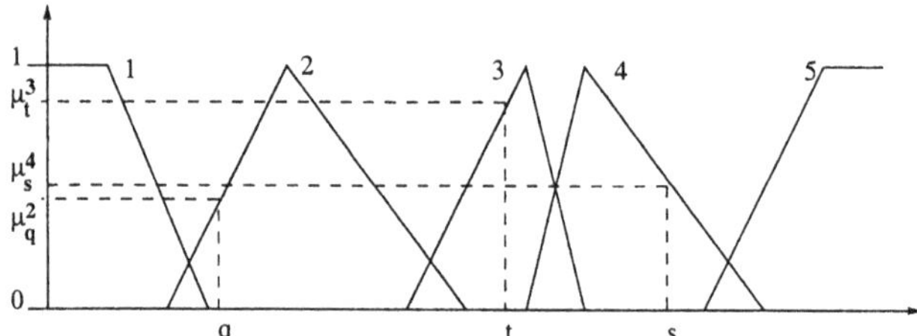

Fig. 2. An illustration of external distances

On figure 2, the external distances for the (q,t,s) triplet can be written as:

$$d_{ext}(q,t) = \mu_t^3 - \mu_q^2 + d_{prot}(2,3) + D_c$$
$$d_{ext}(t,s) = \mu_t^3 - \mu_s^4 + d_{prot}(3,4) + D_c$$
$$d_{ext}(q,s) = \mu_s^4 - \mu_q^2 + d_{prot}(2,4) + D_c$$

It can be shown that the external distance is a semi distance. Let us note that the external distance reduces to the prototype distance plus the correction factor, when points q,t have internal identical membership degrees.

2.3 Distance combination

To manage multiple membership, the pairwise distance $d(q,t)$ is taken as a combination of the internal and external distances. Let us denote $d_{f,g}(q,t)$ the partial (q,t) distance that represents respective memberships to f and g. It is an internal distance if $f = g$, an external distance otherwise. Let m be the partition size.

$d(q,t)$ results from the combination of at most m^2 distances:

$$d(q,t) = \frac{1}{\sum\limits_{f=1}^{m} \mu_q^f} \sum_{f=1}^{m} \left[\mu_q^f \frac{1}{\sum\limits_{g=1}^{m} \mu_t^g} \sum_{g=1}^{m} \left[\mu_t^g \; d_{f,g}(q,t) \right] \right] \tag{6}$$

Generally, the number of fuzzy sets for which μ_q and μ_t are different from zero is kept small.

The term $d(q,t)$, being a combination of semi distances, is a semi distance. Nevertheless to alleviate the notations we will refer to d as a distance.

3 Hierarchical fuzzy partitioning

The Hierarchical Fuzzy Partitioning method, which we will refer to as HFP, generates a collection of univariate fuzzy partitions from a multidimensional training dataset.

A univariate fuzzy partition is composed of m fuzzy sets, the fth fuzzy set for the jth input variable being defined by its membership function $\left(x, \mu_j^f(x)\right)$.

The procedure is carried independently over all dimensions. It has some similarities with Hierarchical Clustering which is widely used in Data Analysis [10]. Hierarchical Clustering makes clusters of multidimensional data pairs according to a given criterion. The starting point is a n-cluster partition, each cluster containing a single individual. The final partition obtained by recursive group aggregating is a one-cluster partition including all data pairs. At each stage the two "nearest" clusters are combined to form a bigger cluster. In our method, we wish to group fuzzy sets instead of data points. We start from an initial m-fuzzy set partition, and end with a single-fuzzy set partition. At each stage the procedure aggregates two fuzzy sets to form a new wider range one. Semantic integrity constraints are imposed at each stage.

This approach should not be confused with Fuzzy Clustering [4,1], which does not aggregate fuzzy sets. In Fuzzy Clustering, fuzzy rules are built from clusters of multidimensional data pairs. Each rule premisses has its own fuzzy sets, which are obtained by projecting the corresponding cluster onto each dimension. For a given dimension, the fuzzy partition results from the union of the fuzzy sets for all rules. The fuzzy set distinguishability, which is essential for semantic integrity, is not guaranteed and is even unlikely to be met.

3.1 Semantic considerations

The necessary conditions for interpretable fuzzy partitions have been studied by several authors [15,3,7,5]. Let us recall the main points :

- Distinguishability: Semantic integrity requires that the membership functions represent a linguistic concept.
- Justifiable number of fuzzy sets.
- Completeness: Each data point should belong at least to one fuzzy set.
- Normalization: All the fuzzy sets should be normal.
- Overlapping: All the fuzzy sets should significantly overlap.

We implement these constraints as follows:

$$\begin{cases} \forall x \sum_{f=1,2,\ldots,m} \mu_j^f(x) = 1 \\ \forall f \; \exists x \; \mu_j^f(x) = 1 \end{cases} \tag{7}$$

Equation 7 means that any point belongs at most to two fuzzy sets. We also have $\forall x \max_{f=1,2,\ldots,m} \mu_j^f(x) \geq \frac{1}{2}$.

Both for clarity in demonstrating the method and due to their specific properties [14] we choose all fuzzy sets of triangular shape, except at the domain edges, where they are semi trapezoidal.

A triangle fuzzy set f is defined by its breakpoints $left^f, c^f, right^f$. Conditions from equation 7 are implemented by choosing fuzzy set breakpoints as shown in figure 3. Three contiguous fuzzy sets f, g, h have such boundaries as:

$$\begin{cases} left^g = c^f \\ c^g = left^h = right^f \\ right^g = c^h \end{cases}$$

Fig. 3. A standardized fuzzy partition

D_c value One point q mainly belongs to a fuzzy set f if $\mu_q^f \geq 0.5$. Consequently the pairwise distance $d(q,t)$ will be mainly internal when both points q, t mainly belong to the same fuzzy set. These points must be closer than points whose distance is mainly external. The maximum value of a mainly internal distance is 0.5, from definition 7. We therefore set $D_c = 0.5$ in equation 5.

For a standardized fuzzy partition, as defined in equation 7, all denominators in equation 6 are equal to 1 and membership degrees can be non zero for at most two fuzzy sets for any given point.

3.2 Procedure

For a given dimension j, let us denote FP^m the HFP generated fuzzy partition of size m. We have:

$$FP^m = \{MF_j^{f/m}, f = 1, \ldots, m\} \tag{8}$$

where $MF_j^{f/m}$ defines the fth fuzzy set within the partition. $MF_j^{1/1}$ represents the universal set.

The HFP procedure can be summarized as a sequence of (m, $m-1, \ldots,$ 2) iterations in each dimension, as shown in algorithm 1. D_m is the criterion for merging two fuzzy sets. It will be given in section 3.4, equation 11.

Algorithm 1 Hierarchical fuzzy partitioning

1: base partition $= FP^m =$ initial partition of size m;
2: evaluate D_m; $s = 1$
3: **while** $m > 2$ **do**
4: **for** $1 \le s \le m-1$ **do**
5: merge fuzzy sets s and $s+1$
6: evaluate D_{m-1}^s
7: restore base partition
8: $s = s+1$
9: **end for**
10: select and store the partition FP^{m-1} for which:
11: $$D_{m-1} = argmax\left(\frac{D_{m-1}^s}{D_m}\right)$$
12: base partition $= FP^{m-1}$; $m = m-1$
13: **end while**

The following sections detail the steps of the procedure: initial fuzzy partition and fuzzy set merging.

3.3 Choice of the initial fuzzy partition

In all rigor the initial partition could include as many fuzzy sets as n, the number of pairs in the training dataset. The kth triangular membership function would then be centered on x_k^j. In practice, the initial partition size can be reduced. The goal is to accelerate the procedure without a loss of performance. We therefore form m clusters of so called unique c^l values. These unique values are determined by sorting the x_t^j, $t = 1, 2, \ldots, n$ values and setting an equality threshold tol. The Cl_f clusters are built through an iterative algorithm, where n_f is the number of data points in the cluster.

$$Cl_1 = \left\{x_t^j \quad | \quad x_t^j - x_1^j < tol\right\} \qquad \forall t = 1, \ldots, n$$
$$Cl_2 = \left\{x_t^j \quad | \quad x_t^j - x_{n_1+1}^j < tol\right\} \qquad \forall t = n_1 + 1, \ldots, n$$

The algorithm continues to form clusters until all data points are assigned to a cluster. The fth cluster center c_j^f is computed as:

$$c_j^f = \frac{\sum\limits_{i \in Cl_f} x_i^j}{n_f}$$

We therefore obtain a m fuzzy set partition, where each fuzzy set center is a cluster center.

3.4 Fuzzy set merging

The only part of the procedure which is depending upon membership function shape, is the fuzzy set merging. In this paper, it will be defined for triangular and trapezoidal shapes.

Triangular fuzzy sets: Each fuzzy set is assigned a weight equal to its cardinality $w^f = \sum\limits_{k=1}^{n} \mu_j^f(x_k^j)$.

Before the first merging, the weight of each fuzzy set in the initial partition is equal to its crisp cardinality, $w_f = n_f$, $\forall f = 1 \ldots m$. The merging of fuzzy sets labeled 2 and 3 is illustrated on figure 4. The resulting fuzzy set is labeled $2'$ and defined as follows:

$$\begin{cases} left^{2'} = c^1 \\ c^{2'} = \dfrac{w^2 c^2 + w^3 c^3}{w^2 + w^3} \\ right^{2'} = c^4 \end{cases} \tag{9}$$

The neighboring fuzzy sets 1 and 4 are turned into $1'$ and $3'$. Their left or right breakpoints are modified so that the fuzzy partition is kept standardized, as in equation 7

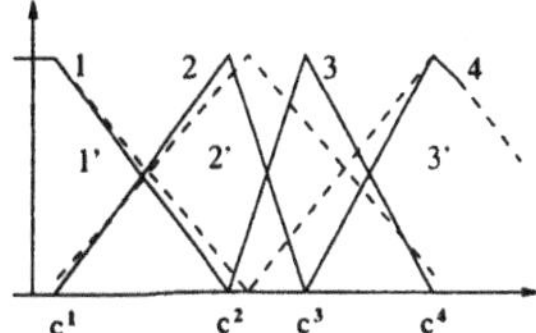

Fig. 4. Merging triangular fuzzy sets 2 and 3 results in $2^{'}$, $1 \Rightarrow 1^{'}$, $4 \Rightarrow 3^{'}$

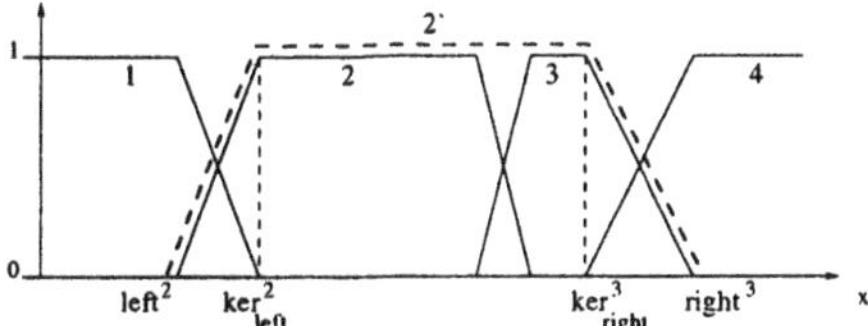

Fig. 5. Trapezoidal fuzzy set merging

Trapezoidal fuzzy sets: The merging is shown on figure 5.

The resulting fuzzy set is labeled $2^{'}$ and defined as follows:

$$\begin{cases} left^{2^{'}} = left^{2} \\ ker^{2^{'}}_{left} = ker^{2}_{left} \\ ker^{2^{'}}_{right} = ker^{3}_{right} \\ right^{2^{'}} = right^{3} \end{cases} \tag{10}$$

Merging criterion For a training data set, a given m fuzzy set partition can be characterized by the sum of pairwise distances over all the data points:

$$D_m = \frac{1}{n(n-1)} \sum_{q,t=1,2,\ldots,n,\ q \neq t} d(q,t) \tag{11}$$

where $d(q,t)$ is the pairwise distance defined in equation 6.

During the merging process, the number of fuzzy sets is reduced by one at each stage. Obviously, some external distances become internal distances, inducing a change on the D_m index.

The best merge at a given stage can be considered as the one that minimizes the variation of D_m. The underlying idea is to maintain as far as possible the homogeneity of the structure built at the previous stage.

Due to the fact that internal distances are smaller than external ones, the sum of distances decreases, except for some particular cases in the very first steps of the procedure.

Let us note that the merging algorithm has a reduced complexity. Assuming a m fuzzy set partition at a given step in a given dimension, the number of possible merges is equal to $m - 1$. The D_m index is computed on the prototypes resulting from the preliminary stage, whose number can be reasonably bounded according to the chosen tolerance tol.

Illustration We now apply this method to the well known Fisher iris data [6]. We will derive a family of partitions for the two petal characteristics, petal length and petal width. The iris are from three species Setosa, Virginica and Versicolor. The petal length and width histograms are plotted at the top of figure 6 with a different colormap for each species.

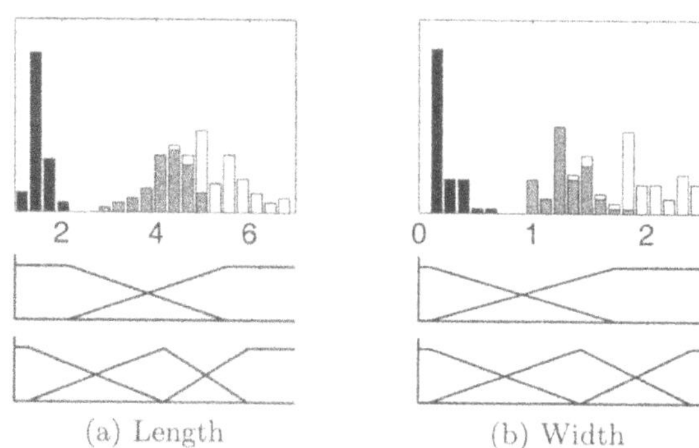

Fig. 6. Histograms and fuzzy partitions of *iris* petal features

The last steps of the procedure are reported in table 1, for triangular fuzzy sets and $tol = 0.01$, chosen equal to the numerical precision on the iris measurement data. Each row shows the fuzzy set centers for a given partition size. Each center corresponds to the mean of ten repeats, each training set being made up of 35 randomly chosen items of each species. The standard deviation is given into parentheses. A graphical display of the size 2 and 3 fuzzy partitions appears at the bottom of figure 6. There are three iris species, but the petal length histogram exhibits two modes. This distribution yields important variations on the fuzzy set centers of the smallest partitions.

The center coordinates can be compared with those obtained by a reference method. Table 2 shows the ones yielded by the $k - means$ algorithm [11], when the number of groups is set to 3 for petal width, and successively to 3 and 2 for petal length.

4 Fuzzy rule base generation

Our objective is to generate a few complete rule bases and to simplify them in order to improve their legibility. By a complete rule base we mean a set of rules where all rules include the same variables, and only differ by the fuzzy labels that appear in each rule. The whole procedure is summarized in figure 7. The present section deals with the generation part, and section 5 explains how to simplify a rule base.

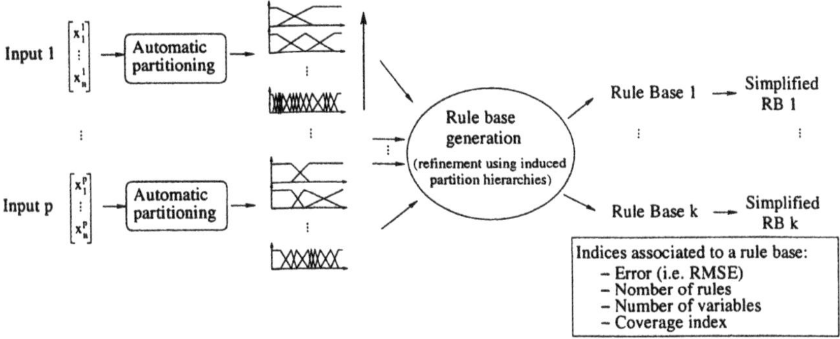

Fig. 7. Rule base generation and simplification

Partition size	Fuzzy set centers					
	Petal width					
2	0.10 (0.00)	1.73 (0.02)				
3	0.10 (0.00)	1.43 (0.03)	2.40 (0.10)			
4	0.10 (0.00)	0.97 (0.11)	1.78 (0.07)	2.40 (0.10)		
5	0.10 (0.00)	0.46 (0.16)	1.32 (0.08)	1.78 (0.07)	2.40 (0.10)	
6	0.10 (0.00)	0.37 (0.09)	1.21 (0.08)	1.49 (0.13)	1.90 (0.10)	2.49 (0.09)
	Petal length					
2	2.15 (0.76)	5.52 (0.45)				
3	1.31 (0.07)	4.19 (0.41)	5.99 (0.35)			
4	1.31 (0.07)	2.91 (0.95)	4.92 (0.23)	6.14 (0.39)		
5	1.29 (0.06)	1.76 (0.36)	4.30 (0.19)	5.03 (0.08)	6.14 (0.39)	
6	1.29 (0.06)	1.57 (0.04)	4.00 (0.20)	4.68 (0.15)	5.18 (0.16)	6.24 (0.44)

Table 1. Fuzzy set center evolution for the iris petal features

Feature	Groups	Centers
Petal width	3	0.25 1.32 2.06
Petal length	3	1.46 4.23 5.56
Petal length	2	1.49 4.93

Table 2. Group centers found by the k-means algorithm for the iris petal features

Generating complete rule bases is a refinement procedure using the fuzzy sets resulting from the HFP stage. We start by considering the simplest rule base, which has only one rule including a single fuzzy set $MF_j^{1/1}$ in each dimension $j = 1, \ldots, p$. The generation procedure builds new rule bases by refining the fuzzy partitions. The refinement algorithm is detailed in section 4.2. It calls a rule base generation algorithm described in section 4.3.

First we give the definition of some elements that will be used all along.

4.1 Definitions

Fuzzy partition notation In each dimension, the HFP procedure yields a family of partitions of decreasing size. For a given dimension j, let us denote $FP_j^{n_j}$ the HFP generated fuzzy partition of size n_j (defined in equation 8); n_j^{max} is the maximum size of the partition (see 3.3).

Performance index The numerical performance index is chosen as the root mean square error and calculated as:

$$Perf = RMSE = \frac{1}{n}\sqrt{\sum_{i=1}^{n} \|\widehat{y_i} - y_i\|^2} \tag{12}$$

where n is the sample size, y_i the observed output for the *ith* example, and $\widehat{y_i}$ the inferred output for the *ith* example.

Coverage index This index characterizes the representativity of the rule base relative to the training set, and as such, is complementary to the performance index. A rule potentially covers the subset of the multidimensional input space corresponding to the combination of the fuzzy sets composing its premisses. The *rth* rule will be activated by the *ith* example to a degree, called rule weight:

$$w^r(x_i) = \mu_{FS_1^a}(x_1) \wedge \mu_{FS_2^b}(x_2) \wedge \ldots \wedge \mu_{FS_p^q}(x_p) \tag{13}$$

where $\wedge$ is a T-norm operator for fuzzy set intersection.

The *ith* example will be considered as inactive or blank for a given rule r if $w^r(i) \leq \mu_{min}$, μ_{min} being a fixed threshold value.

We call E^r the subset of non blank examples for the *rth* rule. It is a subset of the learning sample E such as:

$$E^r = \{(x_i, y_i) \in E \quad | \quad w^r(x_i) > \mu_{min}\}. \tag{14}$$

The examples in E^r are sorted by descending order of w^r. They are said to fire the rule r.

Similarly, an example is said to be blank for the $(r_1, r_2, \dots, r_R)$ rule base if $\sum_{r=1}^{R} w^r(x_i) \leq \mu_{min}$. Due to the cumulated sum, some examples which are blank for all rules may not be blank for the rule base. This could be avoided by using a max operator.

The rule base coverage index CI is based on the ratio of blank examples to the sample size:

$$CI = 1 - \frac{\sum_{i=1}^{n} \delta(x_i)}{n} \qquad \delta(x_i) = \begin{cases} 1 \; if \; \sum_{r=1}^{R} w^r(x_i) \leq \mu_{min} \\ 0 \; otherwise. \end{cases} \tag{15}$$

4.2 Refinement procedure

The iterative algorithm is presented below.

Algorithm 2 Refinement procedure

1: iter = 1; $\forall j \; n_j = 1$
2: CALL *Rule Base Generation* (Algorithm 3)
3: **while** $iter \leq iter_{max}$ **do**
4: Store system as base system
5: **for** $1 \leq j \leq p$ **do**
6: **if** $n_j = n_j^{max}$ **then** next j (partition size limit reached for input j)
7: $n_j = n_j + 1$
8: CALL *Rule Base Generation* (Algorithm 3)
9: $Perf_j = Perf$
10: $n_j = n_j$ - 1
11: Restore base system
12: **end for**
13: s = argmin $\{Perf_j, \; j = 1, \dots, p\}$ (Select inputs to refine)
14: $S_i = s \cup \{j \mid (Perf_j/Perf_s) < \delta\}$
15: **for all** $j \in S_i$ **do**
16: **if** $n_j < n_j^{max}$ **then** $n_j = n_j + 1$ (Refine each selected input)
17: **end for**
18: CALL *Rule Base Generation* (Algorithm 3)
19: keep RB_{iter}
20: **if** $\forall j \; n_j = n_j^{max}$ **then** exit (no more inputs to refine)
21: iter = iter + 1
22: **end while**

The key idea is to introduce as many variables, described by a sufficient number of fuzzy sets, as necessary to get a good rule base. A good rule base represents a reasonable trade-off between complexity, in relationship with the number of rules, and accuracy, measured by the $RMSE$ performance index.

The refinement procedure is responsible for the selection of the variables or fuzzy sets to be introduced in the rule bases. The initial rule base is the simplest one possible (Algorithm 2, lines 1-2). The search loop (lines 5 to 12) builds up temporary rule bases. Each of them corresponds to adding to the initial rule base a fuzzy set in a given dimension. The selection of the dimensions to retain is done in lines 13-14, where we see that several input variables can be selected at one given step. Following this selection, a rule base to be kept is built up. It will serve as a base to reiterate the sequence (lines 3 to 22). Thus the result of the procedure is not a single rule base, but a series of rule bases $RB_1, RB_2, \ldots$. The choice of the best ones is left to the user, as several criteria are to be considered: performance, complexity, coverage index.

Remark: the iterative algorithm is not a greedy algorithm, contrary to other techniques. It does not implement all possible combinations of the fuzzy sets, but only a few chosen ones.

When necessary, the procedure calls a rule base generation algorithm, referred to as Algorithm 3, which is now detailed.

4.3 Generating a rule base

Characteristics of a rule base A rule base RB_k is stored as:

$$\{FP_j^{n_j}, j = 1, \ldots, p\,; R_r, r = 1, \ldots, r_k\,; Perf_k\,; CI_k\}$$

where $FP_j^{n_j}$ is the fuzzy partition for the jth dimension, R_r is one of the r_k rules, $Perf_k$ is the system performance defined in equation 12, CI_k is the coverage index defined in equation 15.

$FP_j^{n_j}$ is uniquely determined by its size n_j, the fuzzy set centers being the coordinates given by the HFP method.

The associated fuzzy inference system is completely defined by the rule base and the inference method.

Algorithm The rule generation is done by combining the fuzzy sets of the FP_j partitions for $j = 1, \ldots, p$, as described by Algorithm 3. The algorithm then removes the less influential rules and evaluates the rule conclusions.

We give detailed explanations about the way of evaluating the rule conclusions and inferring the system output.

Evaluating the rule conclusions We consider the case of a continuous output, and we wish each rule to yield a fuzzy conclusion, both for interpretability and robustness considerations. The use of a regular grid is the easiest way to obtain a fuzzy output partition. More sophisticated techniques could be used, as the HFP or the recourse to an expert choice.

Algorithm 3 Rule base generation

Require: $\{n_j \mid j = 1, ..., p\}$
1: get $FP_j^{n_j} \forall j = 1, ..., p$
2: Generate the $\prod_{j=1}^{p} n_j$ rules premisses
3: **for all** Rule $r \in RB$ **do**
4: $\quad CV_r = \sum_{k=1}^{n} w^r(x_k)$
5: $\quad$ **if** $CV_r < CV_t$ **then** remove rule r
6: **end for**
7: Evaluate the rule conclusions
8: Compute Perf and CI

To start with, a crisp conclusion is calculated for each rule:

$$C^r_{crisp} = \frac{\sum\limits_{i \in S^r} w^r(x_i) * y_i}{\sum\limits_{i \in S^r} w^r(x_i)} \tag{16}$$

$S^r \subset E^r$ is composed of the first n_s elements of E^r, the subset defined in equation 14. To ensure that the rule conclusion is representative of the rule best matching examples, n_s should be kept small ($n_s = 3$ for instance).

Then C^r_{crisp} is fuzzified, and the fuzzy conclusion C^r is chosen as the fuzzy set of maximal membership for C^r_{crisp}.

Calculating the fuzzy inference system output A simple defuzzification procedure is done, that aggregates through a weighted area technique the rule conclusions C^r. Finally, the inferred system output for the ith example is equal to:

$$\widehat{y}_i = \frac{\sum\limits_{r=1}^{R} \alpha^{C^r} \; area(C^r_\alpha)}{\sum\limits_{r=1}^{R} area(C^r_\alpha)}$$

where C^r_α is a trapezoidal fuzzy set whose support is the support of the fuzzy set C^r, and whose kernel is the α-cut of C^r, with $\alpha = w^r(x_i)$. The x-coordinate of the C^r_α centroid is denoted α^{C^r}.

The defuzzification method is illustrated on figure 8 when two rules are fired.

This value of $\widehat{y}_i$ will be used in the $RMSE$ calculation (equation 12).

We can now proceed with the essential work of simplifying further the rule bases.

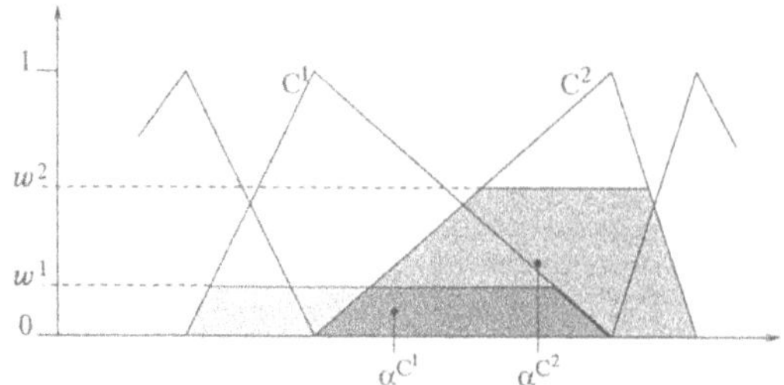

Fig. 8. Weighted area defuzzification

5 Simplifying a rule base

In the complete rule bases that we now have at our disposal, it is somewhat difficult to give a meaning to the rules. All variables equally appear in all rules. If we wish to interpret the rules as interaction rules, it is important to think of a means to highlight the strongest interactions and to erase the lightest ones. A good way to do so is to try out a simplification procedure leading to an incomplete rule base, where one or more variables appear in some rules, and not in all of them.

Basically we could then consider the variable absent from one incomplete rule and present in some others, as of no importance in the context of that incomplete rule. However, the semantic of incomplete rules can be viewed under two different angles. The first explanation would attribute the unimportantness of the absent variable to the fact that if could take any value in that context. The second one would, in a completely different way, state that the absent variable range, conditioned to the other variables in the premisses, could on the contrary be very specific. We will make more comments on this point when presenting the rice case in section 6.

The variable elimination in order to obtain incomplete rules could be undertaken at different levels. Many existing methods eliminate variables from the whole rule base, on the faith of overall indicators which could be misleading. Other techniques remove variables from one rule at a time, not considering the relationships that could exist between the rules. We will favor an intermediary selection level, which is an attempt to make up for these difficulties. This intermediary level is chosen as the level of a group of rules, and includes rules with a common context. Our main axis in the simplification procedure will be turned toward the merging of some rules into a more generic incomplete rule. We will now define the notions of rule distance and group of rules.

Rule distance The distance between two rules a and b depends on the number of variables that differ in the premisses, and on their labels.

$$d^r(a,b) = \sum_{j=1}^{p} d^j_{part}(a,b) \tag{17}$$

In the definition above, $d^j_{part}(a,b)$ is the partial distance of the rules for the jth variable. Let us call FS^a_j and FS^b_j the corresponding fuzzy set labels. The label ANY means that the corresponding variable is missing from the rule. We choose:

$$d^j_{part}(a,b) = 0 \quad if \quad \begin{cases} FS^a_j = FS^b_j \\ or\ FS^a_j = ANY \\ or\ FS^b_j = ANY \end{cases}$$

$$d^j_{part}(a,b) = 1 \quad otherwise$$

This definition could be thought of as over simple, as it does not take into account the meaning of the fuzzy set labels. Indeed for a set of ordered labels as *small, medium, large,* it would be fair to consider the distance between *small* and *large* to be twice as much as the distance between *small* and *medium.* Nevertheless this definition is well adapted to our simplification purposes. For a variable to be removed from a rule, the label should be of no importance for that rule. Therefore it should not interfere with the calculation of the rule distance.

Rule associated heterogeneity For a continuous numerical output, the standard deviation of the observed outputs σ^r in the E^r subset is calculated and compared to the standard deviation σ of the observed outputs for the whole sample E. The rule associated output heterogeneity is defined as:

$$H^r = \frac{\sigma^r}{\sigma} \tag{18}$$

Group of rules A group of rules is a set of rules whose premisses only differ by a single fuzzy set label, corresponding to the same v variable. With the definition of the rule distance, a group of rules becomes the set of all rules whose pairwise distance is smaller than or equal to one.

5.1 Merging a group of rules into a generic rule

The procedure consists of examining each group of rules to see if it can be replaced by a generic incomplete rule g. The generic incomplete rule is formed by assigning $v = ANY$ in each of the premisses of the rules which constitute the group. The new rules are all identical within the group, so they can be replaced by a single one.

The merging is guided by the performance index $RMSE$ defined in equation 12, but also by a careful examination of the output heterogeneity H^g in the examples firing g, i.e. those which pertain to E^g.

Why this preoccupation?

To gain in interpretability, we can tolerate a loss of performance, measured by an increase in $RMSE$. However we must be careful as to the consequences of the widening of the space potentially covered by the new rule. If the output heterogeneity associated to the new rule is too high, it is a sign of inconsistency, and the replacement should not be done. Rule inconsistency means that similar input values yield very different output values. It is incompatible with a good behavior of the rule base, and must be avoided. Nevertheless some heterogeneity is unavoidable, and even desirable. It keeps the rules from being too specific, and favors interpolation in the fuzzy rule conclusions.

The simplification procedure is an iterative one, described by algorithm 4. $Perf$ is the reference system performance index, $NewPerf$ is the current system performance index, and $PerfLoss = (NewPerf - Perf)/Perf$ is the relative loss of performance.

Group search privileges rules whose distance is equal to one. This is the case when the rule premisses only differ by one fuzzy set label, for one variable. If no group is found (lines 5 to 8) or no merging is valid (lines 14-15), then the group search is widened to include rules whose distance is zero. That can happen when a rule has a missing variable (label ANY).

Case of a rule belonging to more than one group If the rule belongs to at least one group which is to be replaced by a generic rule, the replacement will be done and the original rule will be removed from the base. Indeed the generic rule covers a wider multidimensional space than the original one.

5.2 Final steps in the simplification procedure

In the generated rule base, redundancy is likely to happen, as no redundancy control is made during the generation phase. To make up for this drawback, rules of the base issued from the simplification procedure are tentatively removed one after the other. The removal is done according to algorithm 5.

The last step consists of individually simplifying each of the remaining rules, by removing variables at the rule level, according to algorithm 6. This simplification level, which is dependent on the variable order, is actually of minor importance in our whole procedure. The main simplifications occur at the group merging level.

Algorithm 4 Merging a group of rules into a generic rule

1: Compute Perf; DistanceValue = 1
2: **loop**
3: Identify all the groups with a rule distance = DistanceValue
4: k = number of groups
5: **if** k = 0
6: **if** DistanceValue = 1 **then** set DistanceValue = 0
7: **else** exit
8: **end if**
9: **for** $1 \leq g \leq k$ **do**
10: Compute the group generic rule
11: **if** $H_g < H_{thresh}$ **then** merging is valid
12: Restore the rule base
13: **end for**
14: **if** DistanceValue = 1 and No valid merging **then** set DistanceValue = 0
15: **else** exit
16: Perform all valid mergings
17: Store system; Compute NewPerf and PerfLoss
18: **if** $PerfLoss < Loss_{thresh}$ **then** DistanceValue=1
19: **else** exit
20: **end loop**

Algorithm 5 Rule removal

1: Compute Perf
2: **for all** $r \in RB$ **do**
3: remove rule r
4: store the corresponding system
5: compute NewPerf and PerfLoss
6: **if** $PerfLoss > Loss_{thres}$ OR $CI < CI_{thres}$ **then** Restore system
7: **end for**

Algorithm 6 Variable removal

1: Compute Perf
2: **for all** $r \in RB$ **do**
3: **for all** $v \in r$ **do**
4: Set its label to ANY
5: Store system
6: Compute H^r and PerfLoss
7: **if** $H^r > H_{thresh}$ or $PerfLoss > Loss_{thresh}$ or $CI < CI_{thresh}$
8: **then** Restore system
9: **end for**
10: **end for**
11: **if** two rules are identical **then** remove one of them

6 A rice taste case study

In this section, our rule generation and simplification method is applied to real-world data. The data are sensory test data collected from a subjective evaluation of many different kinds of rice by plural panelists. We will show how we can build a fuzzy rule base which performs a non linear mapping of the input-output relation in the rice data, while keeping a reasonable number of interpretable rules. We will compare the results with previous work done by other researchers on the same data.

6.1 The rice data

The sample includes 105 items. The following five factors constitute the input data: flavor, appearance, taste, stickiness, toughness, and the output factor is the overall evaluation of quality.

Though sensory data, the rice data were transformed and coded into numerical continuous variables standardized into the unit space. Let us remark that a loss of information results from the coding, and that it would be interesting to try the fuzzy rule induction on the original qualitative data. Unfortunately, they are not available.

6.2 Previous work

These data have been first used for learning FRBSs in [12]. The paper only aimed to prove the ability of trainable fuzzy inference systems as approximators of non linear mappings on real world data, and was successful in doing so. All factors are split into regular grids, and no simplification is done on the rule base. For a good performance, the complete rule base corresponding to all combinations of the fuzzy sets in the premisses is generated, which yields a $2^5 = 32$ rule base with crisp conclusions for a two fuzzy set input partitioning.

In more recent work [13], the same authors added a linguistic representation of the rule consequences, so that some interpretation could be drawn from the rule base. Though this being an interesting attempt, to acquire linguistic knowledge from the 32 complete rule base is not obvious, even if each rule premisses and consequent part have linguistic labels.

Further work has been done on the same data in [2]. The authors use fuzzy rule consequences and a genetic algorithm selection. The six rules they obtain for a two fuzzy set input and output partition are shown in symbolic form in table 7. They are directly interpretable and will be discussed later in this section. The performance index defined in equation 19 is used in that work and will be mentioned in our comparison tables :

$$PI = \frac{1}{2n} \sum_{i=1}^{n} (y_i - \widehat{y_i})^2 \qquad (19)$$

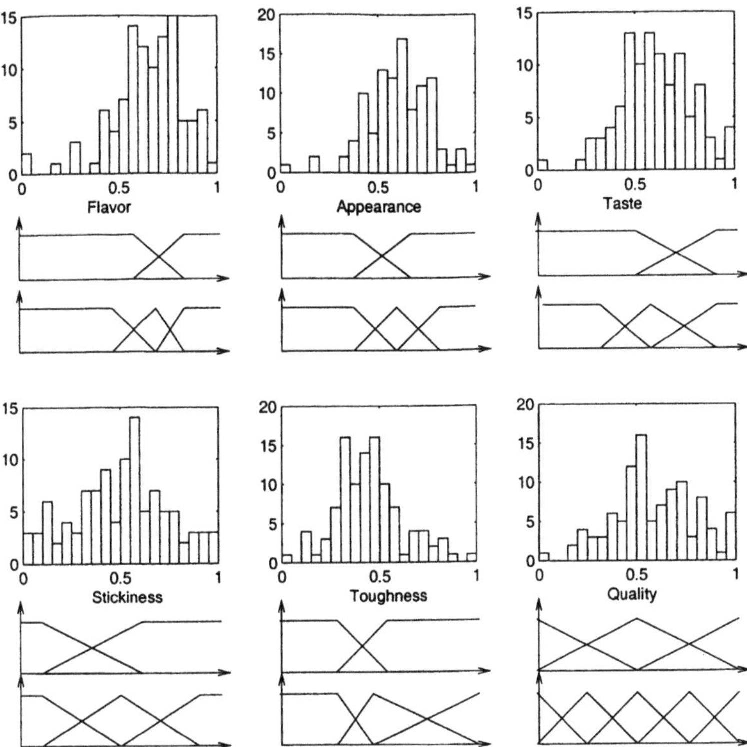

Fig. 9. Histograms and fuzzy partitions of the rice data variables

6.3 HFP, rule generation and simplification on the rice data

Our partitioning method was initialized with a unique value tolerance threshold $tol = 1\%$ as defined in section 3.3, and applied to the input variables using a numerical prototype distance.

The univariate histograms are shown on figure 9 for all variables. The data do not exhibit a well defined structure, and all variables basically have a Gaussian-like distribution. The same figure shows the size two and three HFP found fuzzy partitions below each input variable histogram. The output is partitioned using a three or five fuzzy set regular grid, as shown underneath the Quality histogram.

Table 3 shows the most interesting results of the rule generation and simplification procedure. It is split into two parts.

The left part gives the results of the rule generation method, which generated complete rule bases using the following parameters: $\mu_{min} = 0.3$ and $Card(E_r) = 1$. The column labels and meanings are as follows:

- $\#Out$: Number of fuzzy sets of the output partition,

#Out	Complete rule system FP	#MR	#R	RMSE	PI	MaxE	Simplified system #R	PI	MaxE	#D	#MD
3 FS	3 0 3 2 2	36	15	0.00776	0.00316	-0.330	8	0.00310	0.326	7	2
5_a	0 0 3 2 0	6	4	0.00655	0.00225	0.250					
5_b	2 0 3 3 0	18	10	0.00690	0.00250	0.255	8	0.00335	0.253	7	1
5_c	2 2 3 3 0	36	12	0.00766	0.00308	0.250	9	0.00220	0.250	11	2
5_d	2 2 4 4 0	64	19	0.00606	0.00193	0.250	10	0.00268	0.259	15	3

Table 3. Rice quality assessment: our results

- *FP*: Row vector giving the number of fuzzy sets retained for each variable,
- *#MR*: Maximum number of rules in the complete rule base,
- *#R*: Number of rules in the rule base,
- *RMSE*: Root mean square error over the training set according to equation 12,
- *PI*: Performance index over the training set according to equation 19, for comparison with results in [2],
- *MaxE*: Maximum error over the training set.

The first row corresponds to the best rule base obtained for a three fuzzy set output regular grid partition (denoted 3 FS), and the next ones, denoted $5_a, \ldots, 5_d$, to various steps of the rule generation for a five fuzzy set output regular grid partition.

The right part of the table shows the results for the corresponding simplified systems. The simplification procedure was applied using a heterogeneity threshold $H_{thresh} = 0.5$ (see definition 18). The second row is blank as there was no sense in simplifying in that case. The first three columns labeled *#R*, *#PI* and *MaxE* have the same meaning than in the left part of the table. The last two columns are:

- *#D*: Cumulated sum of the number of variables deleted by the simplification procedure over the rule base,
- *#MD*: Maximum number of variables deleted from one rule.

The coverage index is not mentioned, being equal to $CI = 1$ for all rule bases given in the table.

Examining the results of the generation procedure we see that they are on the whole better for a five fuzzy set output grid. The best numerically performing rule base is the one denoted 5_d. However it is a complex rule base including 19 rules and 4 variables, two of them being partitioned into 4 fuzzy sets. If we compare it with the rule base denoted 5_a, we notice a small decrease in performance, but a considerably reduced complexity. 5_a only has 4 rules altogether, with only two variables in the rule premisses, one partitioned into three fuzzy sets, the other one into two fuzzy sets. The maximum error is of the same order for 5_a and 5_d.

Rule bases 5_b and 5_c present an intermediary trade-off between complexity and accuracy.

Of course simplification is meaningless for 5_a. For the other systems, we note that simplification leads to a comparable number of rules, independently of the initial number of rules issued from the generation procedure. The $\#D$ and $\#MD$ values confirm this tendency, as their values increase with the initial rule base size.

In most cases, the numerical performance is a little lowered by simplification. It never drops abruptly, in some cases it is even improved. The maximum error does not significantly change. Figure 10 gives a graphical representation of the three FRBS accuracy. The fit can be considered as good, relatively to the nature of the rice data. It should be remembered that the *Quality* numerical value results from an aggregation of subjective imprecise assessments.

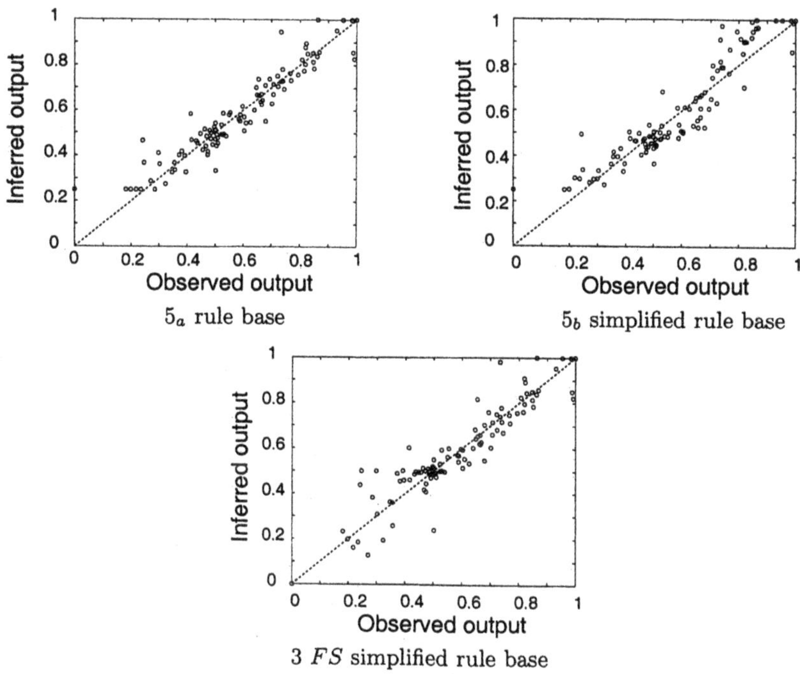

5_a rule base

5_b simplified rule base

3 *FS* simplified rule base

Fig. 10. Graphical representation of system prediction for rice assessment

6.4 Rule interpretation

The rules are expressed in linguistic terms in tables 4 and 5 for the rule bases denoted 5_a and 5_b. The linguistic terms are {*Low, High*}, {*Low, Average, High*} and {*Very low, Low, Average, High, Very High*} for a two, three and five fuzzy set partition respectively. An × in a case means that the corresponding variable is absent from the rule.

We first notice that out of the five labels used in the five fuzzy set output partition, only three appear in rule bases 5_a and 5_b.

Let us first comment the simplest rule base, shown in table 4.

Flavor	Appearance	Taste	Stickiness	Toughness	Quality
×	×	High	High	×	Very High
×	×	Low	Low	×	Low
×	×	Average	Low	×	Average
×	×	Average	High	×	Average

Table 4. 5_a rule base

Rule premisses include two variables only: *Taste* and *Stickiness.* It seems that experts mainly relied on these two factors to do their appraisal of the rice quality. The following interpretation can be given to the rules. If both *Taste* and *Stickiness* are good, then the overall quality is also good. Conversely, if they are both bad, then the overall quality is judged as bad. Finally, an average *Taste* seems to win the competition against *stickiness* in evaluating the quality.

Table 5 is more complex, as some of the rules include one more variable, *Flavor.* The first three rules show that, as soon as two of the three factors *Flavor, Taste or Stickiness* are satisfactory, the quality is judged as high. This takes us back to the semantic of incomplete rules mentioned at the beginning of this section. The absent factor *Flavor* from rule base 5_a means in this case that it takes *high* values, conditioned to *Taste* and *Stickiness* high values.

By comparing tables 4 and 5, we see that the first rule in table 4 is identical to the second rule in table 5. Rule 2 from table 4 is included into rule 8 from table 5. Rules 3 and 4 in table 4 can be related to rules 6 and 7 in table 5, if we remember that *Stickiness* is partitioned into two fuzzy sets for rule base 5_a and into three for rule base 5_b. This shows the consistency of the generation and simplification procedures.

Our last rule table, which corresponds to a three fuzzy set output regular grid, is table 6. It is more complex than the previous ones. Four factors appear in the rule base: *Flavor, Taste, Stickiness* and *Toughness.* This last factor is present in 5 rules out of 8, and its interpretation is delicate.

6 rules out of 8 have an *Average* fuzzy conclusion, thus probably making rule base 3 FS being less informative than rule bases 5_a or 5_d. It leads us to

Flavor	Appearance	Taste	Stickiness	Toughness	Quality
High	×	High	×	×	Very High
×	×	High	High	×	Very High
High	×	×	High	×	Very High
Low	×	×	Average	×	Average
Low	×	Average	×	×	Average
×	×	Average	Low	×	Low
×	×	Average	Average	×	Average
Low	×	Low	Low	×	Low

Table 5. 5_b simplified rule base

believe that a three fuzzy set regular grid partition is not well suited to the rice data.

Flavor	Appearance	Taste	Stickiness	Toughness	Quality
×	×	High	High	×	High
Average	×	Average	×	×	Average
Low	×	Average	×	High	Average
×	×	Average	Low	High	Average
High	×	Average	High	×	Average
Low	×	Low	Low	High	Low
Average	×	Low	Low	High	Average
Low	×	Low	Low	Low	Average

Table 6. $3FS$ simplified rule base

Comparison with other results on the rice data We can compare these rules with the fuzzy model in [2] for which the rule symbolic form, using the same linguistic labels as above, is given in table 7.

Flavor	Appearance	Taste	Stickiness	Toughness	Quality
High	High	High	High	High	High
High	High	High	High	Low	High
Low	High	High	Low	High	High
High	Low	Low	Low	High	Low
Low	Low	Low	Low	Low	Low
Low	High	Low	Low	High	Low

Table 7. Rules obtained by other researchers [2]

Let us first note these are complete rules. Nevertheless, the comparison with rule base 5_b is possible if we consider only the common factors, *Flavor*, *Taste* and *Stickiness*. The first two rules in 7 are identical and correspond to the first three rules in table 5. Similarly, the last two are identical and correspond to the last rule in table 5. The two middle rules state that *Taste* and *Stickiness* are more important than *Flavor*, which is also noticeable in table 4, for which the only two factors are *Taste* and *Stickiness*.

To complete the comparison, let us note that the numerical performance index obtained in [2] is $PI = 0.00285$, very close to the values reported in table 3.

To conclude on the rice data case study, we can say that the generation of incomplete rules facilitates the semantic interpretation of the rules, together with preserving a good accuracy. For sensory data, knowledge induction seems to be more relevant than a perfect fit: how many digits can be considered as significant in Quality evaluation?

7 Conclusion

In this chapter we presented a new method for generating interpretable fuzzy rules from data. There is no random effect or undeterministic choice all along the procedure, contrary to what happens in genetic algorithms for instance. At each stage, the final choice of the best partition or FRBS is left to the user, who can use his knowledge of the process or system to orient his decisions.

The approach can either be used in its whole: partitioning plus generation and simplification, or each component separately. The proposed partitioning method, called HFP, yields a hierarchy of univariate partitions which could advantageously replace regular grids for initializing fuzzy decision trees, or other algorithms requiring a predefined partition. Conversely, the FRBS generation method introduced in section 4 can rely upon any kind of fuzzy partition family: a HFP found partition, but also regular grids or partitions determined by other means, the only limitation being the respect of the semantic through some constraints.

The simplification step (section 5) can also be applied to various FRBS, not only to the ones resulting from our generation method. We explained the reasons for choosing an intermediate simplification level, half way between the global rule base level and the rule by rule one. The new rule heterogeneity and coverage indexes allow us to direct the simplification procedure and avoid inconsistency.

Results obtained for a real world case study on rice quality evaluation are promising and show the advantage of incomplete rules for interpretability.

References

1. J. C. Bezdek. *Pattern Recognition with Fuzzy Objective Functions Algorithms*. Plenum Press, New York, 1981.

2. Oscar Cordon and Francisco Herrera. A proposal for improving the accuracy of linguistic modeling. *IEEE Transactions on Fuzzy Systems*, 8 (3):335–344, June 2000.
3. J. Valente de Oliveira. Semantic constraints for membership functions optimization. *IEEE Transactions on Systems, Man and Cybernetics*, 29:128–138, 1999.
4. J. C. Dunn. A fuzzy relative of the isodata process and its use in detecting compact well-separated clusters. *J. Cybernetics*, 3 (3):32–57, 1973.
5. Jairo Espinosa and Joos Vandewalle. Constructing fuzzy models with linguistic integrity from numerical data-afreli algorithm. *IEEE Transactions on Fuzzy Systems*, 8 (5):591–600, October 2000.
6. Ronald A. Fisher. The use of multiple measurements in taxonomic problems. *Annals of Eugenics*, 7:179–188, 1936.
7. Pierre-Yves Glorennec. *Algorithmes d'apprentissage pour systèmes d'inférence floue*. Editions Hermès, Paris, 1999.
8. Serge Guillaume. Designing fuzzy inference systems from data: an interpretability-oriented review. *IEEE Transactions on Fuzzy Systems*, 9 (3):426–443, June 2001.
9. Serge Guillaume and Brigitte Charnomordic. Knowledge discovery for control purposes in food industry databases. *Fuzzy sets and Systems*, 122 (3):487–497, September 2001.
10. John A. Hartigan. *Clustering Algorithms*. Wiley, 1975.
11. John A. Hartigan and M.A. Wong. A k-means clustering algorithm. *Applied Statistics*, 28:100–108, 1979.
12. Hisao Ishibuchi, Ken Nozaki, Hideo Tanaka, Yukio Hosaka, and Masanori Matsuda. Empirical study on learning in fuzzy systems by rice test analysis. *Fuzzy Sets and Systems*, 64:129–144, 1994.
13. Ken Nozaki, Hisao Ishibuchi, and Hideo Tanaka. A simple but powerful heuristic method for generating fuzzy rules from numerical data. *Fuzzy Sets and Systems*, 86:251–270, 1997.
14. Witold Pedrycz. Why triangular membership functions? *Fuzzy sets and Systems*, 64 (1):21–30, 1994.
15. Enrique H. Ruspini. *Recent developments in fuzzy clustering.*, pages 133–147. Pergamon Press, New York, 1982.

A Feature Ranking Algorithm for Fuzzy Modelling Problems*

Domonkos Tikk[1,3], Tamás D. Gedeon[2], and Kok Wai Wong[2]

[1] Dept. of Telecommunications & Telematics,
Budapest University of Technology and Economics,
1117 Budapest, Magyar Tudósok Körútja 2., Hungary,
e-mail: tikk@ttt.bme.hu

[2] School of Information Technology, Murdoch University
South Street, Murdoch, 6150 W.A., Australia
e-mail: tgedeon@murdoch.edu.au

[3] Intelligent Integrated Systems Japanese–Hungarian Laboratory
1111 Budapest, Műegyetem rakpart 3., Hungary

Abstract. This paper presents a feature ranking method adapted to fuzzy modelling with output from a continuous range. Existing feature selection/ranking techniques are mostly suitable for classification problems, where the range of the output is discrete. These techniques result in a ranking of the input feature (variables). Our approach exploits an arbitrary fuzzy clustering of the model output data. Using these output clusters, similar feature ranking methods can be used as for classification, where the membership in a cluster (or class) will no longer be crisp, but a fuzzy value determined by the clustering. We propose the application of the Sequential Backward Selection (SBS) search method to determine the feature ranking by means of different criterion functions. We examined the proposed method and the criterion functions through a comparative analysis.

1 Introduction

It is well-known that traditional fuzzy rule-based systems (FRBSs) have exponential time and space complexity in terms of N, the number of variables [24]. The number of rules in a rule base increases exponentially with N. Thus the resulting model of the system is very large. In practice, if N exceeds the experimental limit of about 5–8 variables [24], the rule based system becomes intractable. Due to this fact, rule base reduction emerged as an important research field in the past decade including e.g. the topics of fuzzy rule interpolation methods (see e.g. [24,2,16,28]), hierarchical reasoning techniques [23,26], and other rule base reduction methods [3–5,7].

If the design of the modelled system is based on input-output data samples, a possible method of rule base reduction is the omission of those variables which have no relevant effect on the output. In pattern recognition and classification such methods are called feature selection [9]. Henceforth, when we

* This research was supported by the Australian Research Council, and by the Hungarian Scientific Research Fund (OTKA) Grants No. D034614, and T34212.

use the terms feature or variable in this paper, we refer to the same notion. In these contexts, having a finite number of classes (or clusters), the output of a sample data indicates which class the sample belongs to. Practically it means that outputs are selected from a finite set of labels or, equivalently, from a closed range of natural numbers.

Feature selection methods are of two main types: feature selection and ranking methods. The methods of the former type determine which input features are relevant in a given model, whilst the ones of the latter type result in a rank of importance. Feature ranking methods can be considered as preprocession of feature selection, because relevant features can be selected by taking the first k elements of the head of the feature ranking, and then, by optimizing the number of k, e.g. by a trial-and-error procedure. We aim at providing a reliable feature ranking method for fuzzy modelling problems.

On fuzzy modelling we mean the automatic design of a fuzzy system from a set of input-output data samples, where the sample data values (including the output) are real numbers. It means that opposing classification problems, the range of the output is continuous, so feature selection/ranking algorithms developed for pattern recognition or classification task can not be applied directly. Our goal is, therefore, to bridge the gap between existing feature selection/ranking methods and fuzzy modelling by the adaptation of the referred methods.

Classical feature selection/ranking methods need to be modified or the rule base has to be preprocessed before being applied to fuzzy modelling. A straightforward solution of this problem is the grouping of output data values. An obvious way of grouping is the clustering of output using some fuzzy clustering technique. A fuzzy clustering method divides the clustered space into various regions, called clusters, and determines a vector of membership degrees for each data, which indicates the grade to which the particular data belongs to the clusters. Because we cluster only the one dimensional output, the shape of the clusters (e.g. spherical or ellipsoid) is irrelevant, due to the fact that in our case clusters are interval. In our model we used fuzzy c-means (FCM) clustering ([6]; see Subsection 3.2), but other fuzzy clustering methods are also suitable for this purpose (e.g. subtractive clustering [21] or Gustafson–Kessel algorithm [17]).

Let us now briefly summarize our proposed method: first, we cluster the output data by FCM, then we apply an appropriately modified feature selection/ranking method. We adapt the interclass separability based feature selection/ranking method for this task. (The origin of this method is attributed to [12], who first applied the interclass distance concept to feature selection and extraction problems. Therefore this method is also known as Fischer's interclass separability method.)

This paper is organized as follows. In Section 2 we give a short overview of feature selection methods in fuzzy applications. In Section 3 we outline the interclass distance based feature selection problem in probabilistic case, and recall the FCM method. The main algorithm is described in section 4. In section 5 we analyze the proposed method on some sample data sets.

2 Feature selection methods in fuzzy application

Fuzzy modelling based on the clustering of output data was first proposed by Sugeno and Yasukawa [27]. For reducing the number of inputs they used the regularity criterion (RC) method [20]. RC creates a tree structure from the variables, where the nodes represent particular subsets of the entire variable set. The nodes are evaluated according to an objective function, and the evaluation process stops if a local minimum is found. We are also working on the automatic design of fuzzy modelling systems but we found the RC method unreliable: it is very sensitive to its parameters [29], therefore we decided to look for an alternative solution. Another deficiency of RC that it uses the fuzzy model itself to evaluate the nodes in the searching tree. Therefore, a preliminary fuzzy model should be built in advance.

Another solution was proposed to solve feature selection problem by Costa Branco *et al* [8]. They used the principal component analysis (PCA) method [22] for identifying the important variables in the fuzzy model of an electromechanical system. The PCA method transforms the input data matrix of (possibly highly) correlated variables to an orthogonal system, where the variables become uncorrelated. From the transformed system the variables having eigenvalues under a certain threshold can be omitted. However, by this transformation the meaning of the variables and hence the direct linguistic interpretability of the system is lost, which we consider one of the most important features of a fuzzy system.

Hong and Chen proposed a general learning algorithm which automatically derives fuzzy if–then rules and membership functions from a set a given training examples using a decision table [18]. They improved Hong's and Lee's algorithm [19] by first selecting relevant features and building an appropriate initial membership function in order to avoid very large rule bases. This method, although shows impressing performance benchmarks, is only suitable for finding relevant attributes for classification problems.

In the field of fuzzy classification another group of feature selection algorithms has been applied successfully [1,9,25] which are based on the interclass separability criterion. Let us briefly describe this method in the probabilistic case [10].

3 Preliminaries

3.1 The idea of interclass separability based classification

Let us take a data set consisting of N-dimensional vectors $\{\underline{x}_1, \ldots, \underline{x}_n\}$. Here N is the number of variables, features, or attributes. The vectors $\underline{x}_j$ $(j =$

$1,\ldots,n$) should be categorized into classes C_i ($i = 1,\ldots,C$) which possess *a priori* class probability P_i, and the cardinality of the classes is $|C_i| = n_i$.

Let us denote the features by f_k, $k = 1,\ldots,N$, and the original feature set of all the measured N features by $\mathcal{F}_N = \{f_k | k = 1,\ldots,N\}$. Analogously, denote by $\mathcal{F}_{N'}$, $N' < N$, the feature set where $N - N'$ features are removed from $\mathcal{F}_N$. Obviously, by deleting different features, we obtain different feature sets $\mathcal{F}_{N'}$.

Let us assume that the classes occupy different regions in the multi-dimensional feature space. Intuitively, more distant the classes are from each other, the better the chances of successful classification of input vectors. It is reasonable, therefore, to select as feature space that subspace of the original N-dimensional feature space in which classes are maximally separated in terms of certain distance function. Thus, we seek a feature set $\mathcal{F}$, which maximizes the average distances of C classes.

Let the matrix $\mathbf{X}_{\mathcal{F}_N}$ be formed by the vectors $\underline{x}_j$ ($j = 1,\ldots,n$) associated with the original feature set and vectors. Similarly, we denote by $\mathbf{X}_{\mathcal{F}_{N'}} = [\underline{x}'_1 \cdots \underline{x}'_n]$ the matrix associated with feature set $\mathcal{F}_{N'}$, and consisting of vectors obtained by deleting $N - N'$ features from vectors $\underline{x}_j$ ($j = 1,\ldots,n$). A general criterion function that maximizes the interclass distance, and thus can be exploited to rank features, is defined as [10]

$$J(\mathbf{X}_{\mathcal{F}_{N'}}) = \frac{1}{2}\sum_{i=1}^{C} P_i \sum_{j=1}^{C} P_j \frac{1}{n_i n_j} \sum_{k=1}^{n_i}\sum_{\ell=1}^{n_j} d(\underline{x}'_{ik}, \underline{x}'_{j\ell}) \tag{1}$$

which is the average distance between the elements of C classes. Here vectors $\underline{x}'_{ik}$, and $\underline{x}'_{j\ell}$ ($k = 1,\ldots,n_i$; $\ell = 1,\ldots,n_j$) are the elements of the ith and jth class, respectively, and $d(\underline{x}'_{ik}, \underline{x}'_{j\ell})$ denotes a distance metric, usually the square of the Euclidean norm, which is

$$d(\underline{x}'_k, \underline{x}'_\ell) = \sum_{j=1}^{N'} (x'_{kj} - x'_{\ell j})^2 = (\underline{x}'_k - \underline{x}'_\ell)^T (\underline{x}'_k - \underline{x}'_\ell), \tag{2}$$

where T denotes matrix transpose. The optimal feature set $\mathcal{F}$ satisfies

$$J(\mathbf{X}_{\mathcal{F}}) = \max_{\mathcal{F}_{N'}} J(\mathbf{X}_{N'}). \tag{3}$$

There are various choices for the criterion function. This issue is addressed in Subsection 4.3.

In order to rank the elements of the feature set, one may apply e.g. one of the following well-known search methods: Sequential Backward Selection (SBS), or Sequential Forward Selection (SFS) search method [10]. SBS is a simple top-down search procedure where one feature at a time is deleted from the current feature set. At each stage, the attribute to be removed from the feature set is selected from among the elements of the feature set so that the new shrunk set of features yields a minimum value of the criterion function used. SFS is the bottom-up counterpart of SBS search method, where the one feature at a time, having the largest effect on the criterion function, is added to the current feature set.

3.2 Fuzzy c-means (FCM) clustering algorithm

Let us now describe FCM algorithm proposed by Bezdek [6]. As we mentioned, we shall apply FCM method to group output values of a data set, where these values are from a continuous range.

Fuzzy clustering (in general) assigns a membership grade μ_{ij} to every vector $\underline{x}_j$ $(j = 1, \ldots, n)$ for every cluster i $(i = 1, \ldots, C)$, where C is the number of clusters. We require that

$$\sum_{i=1}^{C} \mu_{ij} = 1 \quad \forall j \in [1, n] \tag{4}$$

i.e. cluster membership degrees are normalized. We define a matrix $\mathbf{U}$ consisting of μ_{ij}. Our goal is to find an optimal C (according to an objective function) and to determine the matrix $\mathbf{U}$. Clusters are represented by their center, $\underline{v}_i$. In the presented algorithm $m > 1$ is an adjustable real valued parameter, the so-called fuzzifier, that may vary usually in the range of $[1.5, 3]$, and is set to 2 as default. The original algorithm is the following [6]:

1. Fix C, the number of clusters, set $\ell = 1$, and initialize $\mathbf{U}$ with $\mathbf{U}^{(1)}$.
2. Calculate the centers $\underline{v}_i$ of the fuzzy clusters as

$$\underline{v}_i = \sum_{j=1}^{n} (\mu_{ij})^m \underline{x}_j \Big/ \sum_{j=1}^{n} (\mu_{ij})^m \quad \forall i \in [1, C]. \tag{5}$$

   The distance of the jth vector from the ith cluster center is defined by

$$d_{ij} = \|\underline{x}_j - \underline{v}_i\|.$$

3. Calculate the new $\mathbf{U}^{(l)}$ for $\ell := \ell + 1$ as

$$\begin{aligned} I_j &:= \{i | 1 \leq i \leq C,\, d_{ij} = 0\} \\ \tilde{I}_j &:= \{1, \ldots, C\} - I_j \\ I_k &= \emptyset \Longrightarrow \mu_{ij} = \frac{1}{\sum_{k=1}^{C} (d_{ij}/d_{kj})^{2/(m-1)}} \\ I_k &\neq \emptyset \Longrightarrow \mu_{ij} = 0 \; \forall i \in \tilde{I}_j; \; \mu_{ij} = \frac{1}{|I_j|} \; \forall i \in I_j. \end{aligned}$$

4. If $\|\mathbf{U}^{(\ell-1)} - \mathbf{U}^{(\ell)}\| \leq \varepsilon$, where ε is a prescribed error, then stop; otherwise go to step 2.

The resulting FCM algorithm is able to recognize spherical clusters (clouds of points) of approximately the same size. FCM clustering reaches its limits for clusters of different shapes, sizes and densities. If these attributes of the clusters are very inhomogeneous, one can substitute FCM by an alternative clustering, e.g. Gustafson–Kessel method [17]. In our application we cluster one dimensional values, therefore the shape of clusters is homogeneous (interval).

There are many studies about FCM containing various objective functions to determine the optimal number of clusters [6,11,13]. We used the one proposed in [13], and also applied in [27], in which the goal is to minimize

$$S(C) = \sum_{j=1}^{n}\sum_{i=1}^{C}(\mu_{ij})^m \left(\|\underline{x}_j - \underline{v}_i\|^2 - \|\underline{v}_i - \underline{x}\|^2\right). \tag{6}$$

Here

$$\underline{x} = \frac{1}{n}\sum_{j=1}^{n}\underline{x}_j \tag{7}$$

denotes the averages of all input vectors.

4 The feature ranking on fuzzy clustered output (FRFCO) algorithm

4.1 Problem definition

Let us now turn to formalize our problem: feature ranking of fuzzy modelling systems. Now, the data set consists of pairs $(\underline{x}_j, y_j)$ $(j = 1, \ldots, n)$ determining the behavior of the modelled system. We intend to filter the features (variables), and keep only those ones, that have significant effect on the output. To achieve our purpose, we rank the features of $\mathcal{F}$ depending on their relevance to the determination of the output.

Let us assume that we already performed FCM clustering on the output values y_j $(j = 1, \ldots, n)$, and we obtained C as the optimal number of clusters by means of (6), and values μ_{ij} as membership grades in class C_i $(i = 1, \ldots, C)$, where condition (4) is satisfied. Thus, having classified the model output, instead of real values, we are now able to assign cluster membership grades to each input vector, where the membership grade μ_{ij} represents the distance (more precisely: closeness) of the original output value, y_j, from the representative cluster center v_i (see (5) with substitution $\underline{x}_j \to y_j$). Considering that y_j is the projection of the input vector $\underline{x}_j$ onto the output space, we can assign membership degree μ_{ij} to $\underline{x}_j$ as associated grade in the class C_i.

Now, on the analogy of the class probability in the probabilistic model, we can define the *fuzzy class frequency of occurrence* of a class C_i as the normalized sum of total membership grades assigned to the given class

$$P_i \sim F_i := \frac{1}{n}\sum_{j=1}^{n} \mu_{ij}$$

Observe that due to the condition (4), we have $\sum_{i=1}^{C} F_i = 1$. Similarly we can define *fuzzy class cardinality* on the analogy of class cardinality, n_i, by the following expression:

$$n_i \sim \sum_{j=1}^{n} \mu_{ij}.$$

We introduce matrices measuring between-class (interclass) and within-class (intraclass) averaged distances of input vectors $\underline{x}_j$, $(j = 1, \ldots, n)$:

$$\mathbf{Q}_b(\mathbf{X}_{\mathcal{F}}) = \sum_{i=1}^{C} F_i(\underline{v}_i - \underline{x})(\underline{v}_i - \underline{x})^T \tag{8}$$

$$\mathbf{Q}_i(\mathbf{X}_{\mathcal{F}}) = \frac{1}{\sum_{j=1}^{n} \mu_{ij}} \sum_{j=1}^{n} \mu_{ij}(\underline{x}_j - \underline{v}_i)(\underline{x}_j - \underline{v}_i)^T$$

$$\mathbf{Q}_w(\mathbf{X}_{\mathcal{F}}) = \sum_{i=1}^{C} \mathbf{Q}_i = \sum_{i=1}^{C} \frac{1}{\sum_{j=1}^{n} \mu_{ij}} \sum_{j=1}^{n} \mu_{ij}(\underline{x}_j - \underline{v}_i)(\underline{x}_j - \underline{v}_i)^T \tag{9}$$

$$\mathbf{Q}_t(\mathbf{X}_{\mathcal{F}}) = \mathbf{Q}_b + \mathbf{Q}_w = \sum_{i=1}^{C} \frac{1}{\sum_{j=1}^{n} \mu_{ij}} \sum_{j=1}^{n} \mu_{ij}(\underline{x}_j - \underline{x})(\underline{x}_j - \underline{x})^T \tag{10}$$

where

$$\underline{v}_i = \frac{1}{\sum_{j=1}^{n} \mu_{ij}} \sum_{j=1}^{n} \mu_{ij}\underline{x}_j \tag{11}$$

are the fuzzy centers of the ith cluster, and $\underline{x}$ is defined in (7).

Here (8) and (9) are called *fuzzy between-class (interclass)*, and *fuzzy within-class (intraclass) scatter matrices*, respectively, that sum up to the *total fuzzy scatter matrix* (10), fuzzy generalization of their crisp counterparts, see also [9]. (Scatter matrices are also known as covariance matrices.) The defined quantities are function of the matrix $\mathbf{X}_{\mathcal{F}}$ assigned to the feature set $\mathcal{F}$ in the sense that all vectors $\underline{x}_j$ contain exactly those features, that appear in the given feature set. When the context makes clear that we mean the quantities (8) and (9) associated to the feature set $\mathcal{F}$, we use simpler notation $\mathbf{Q}_b$ and $\mathbf{Q}_w$. In (9) and (10), membership grades μ_{ij} can be interpreted as a class weight of input vectors. We assume that $\mathbf{Q}_b$ and $\mathbf{Q}_w$ are nonsingular, positive semi-definite matrices.

The feature selection based on interclass separability concept is a trade-off between $\mathbf{Q}_b$ and $\mathbf{Q}_w$. Intuitively, the classes are well-separated, if the average interclass distance, $\mathbf{Q}_w$, is large, while the average intraclass distance, $\mathbf{Q}_b$, is low. Therefore, an appropriate solution to this problem is to find the feature set $\mathcal{F}$ which simultaneously maximizes (9) and minimizes (8).

Based on these quantities we can define appropriate criterion functions in Subsection 4.3. Needless to say, that the choice of the criterion function is problem dependent. Yet often, certain criterion functions possess better characteristic than others, therefore we will examine the effect of different criterion functions in Section 5, and introduce the FRFCO algorithm with a general criterion function.

4.2 The proposed algorithm

The proposed feature ranking algorithm is an instance of SBS search method. In our application, SBS method applies the interclass separability criterion function. In the ith step we delete temporarily a variable f_{temp} ($f_{\text{temp}} \in \mathcal{F}_{k-1}$), so we have feature set $\mathcal{F}_k = \{f | f \in \mathcal{F}_{k-1}, f \neq f_{\text{temp}}\}$ and input matrix $\mathbf{X}_{\mathcal{F}_k}$, where the starting feature set is $\mathcal{F}_0 = \mathcal{F}_N$, and we calculate the matrices $\mathbf{Q}_b(\mathbf{X}_{\mathcal{F}_k})$ and $\mathbf{Q}_w(\mathbf{X}_{\mathcal{F}_k})$ to be used in the criterion functions. This procedure is repeated for all the variables in $\mathcal{F}_{k-1}$. By means of an appropriate criterion function, the expression $\min_{f \in \mathcal{F}_{k-1}} J(\mathbf{X}_{\mathcal{F}_k})$ attains its minimum when the deviation between $\mathbf{Q}_b$ and $\mathbf{Q}_w$ is the least, i.e. when the most important variable is omitted. Then we remove the selected feature $f \in \mathcal{F}_{k-1}$ permanently, we can restart the algorithm with the updated feature set $\mathcal{F}_k$. The algorithm ends when the cardinality of the feature set is 1.

The FRFCO algorithm

1. Let $\mathcal{F}_0 := \{f_1, \ldots, f_N\}$, $k = 1$.
2. For all $f_{\text{temp}} \in \mathcal{F}_{k-1}$
   (a) Let $\mathcal{F}_k := \mathcal{F}_{k-1} - \{f_{\text{temp}}\}$, and update matrix $\mathbf{X}$ by deleting temporarily its f_{temp}th row, and vectors $\underline{v}_i$ (see (11)) and $\underline{x}$ (see (7)) by deleting temporarily their f_{temp}th element.
   (b) Calculate matrices $\mathbf{Q}_b(\mathbf{X}_{\mathcal{F}_k})$, $\mathbf{Q}_w(\mathbf{X}_{\mathcal{F}_k})$ and determine $J(\mathbf{X}_{\mathcal{F}_k})$.
3. Let $f_{\text{perm}} = \operatorname{argmin}_{f \in \mathcal{F}_{k-1}} J(\mathbf{X}_{\mathcal{F}_k})$, i.e. where J attains its minimal value. We obtain the final $\mathcal{F}_k$ by deleting permanently the variable f_{perm} from $\mathcal{F}_{k-1}$, and we update expressions $\mathbf{X}$, $\underline{v}_i$ and $\underline{x}$ appropriately.
4. If $k < N - 1$ then back to step 2, else stop.

The order of the deleted variables gives their rank of importance.

Remark 1. Note that f_{perm} can contain more than one variable. In such a case we delete all of them at a time.

Remark 2. Primarily, we apply the SBS search method, but we also run experiments with SFS search method for comparison, see Section 5.

4.3 Selection of criterion function

In [10] it is shown that (1) can be transformed with algebraic manipulation to

$$J_1(\mathbf{X}_{\mathcal{F}}) = \mathrm{tr}(\mathbf{Q}_w) + \mathrm{tr}(\mathbf{Q}_b) \tag{12}$$

where "tr" denotes the trace of a matrix, the sum of the diagonal elements. Recall that we intend to maximize $\mathbf{Q}_b$ and at the same time minimize $\mathbf{Q}_w$. It can be obtained by maximizing (12), however, in this case the effect of intraclass distance of input vectors is unchecked. Therefore, the magnitude of the criterion function (12) is not a good indicator of class separability.

A more realistic criterion function to maximize is

$$J_2(\mathbf{X}_{\mathcal{F}}) = \frac{\mathrm{tr}(\mathbf{Q}_b)}{\mathrm{tr}(\mathbf{Q}_w)} \tag{13}$$

which reflects more to the described intuitive notion. For more details see [9,10]. However, (13) is only suitable if the domain of variables are homogeneous or at least comparable, because if the dimensionality are different, e.g. one variable is in the linear domain and another is in the logarithmic, then the comparison is meaningless. Another drawback of J_2 was pointed out in [9], that is for a particular feature subset a class $\mathcal{C}_i$ is well scattered and a portion of $\mathcal{C}_i$ is overlapped with another class $\mathcal{C}_j$ but their centers are far away, then J_2 may be greater then for another feature subset, which separates the two classes in such a way that a single hyperplane may pass between them, but their centers are not so far apart.

Moreover, expression (13) ignores the effect on the separability of the correlated variables. This shortcoming can be overcome by preprocessing the scatter matrix $\mathbf{Q}_w$ by a suitable transformation $\mathbf{V}$ average covariance of the input vectors is the identity matrix: $\mathbf{V}^T\mathbf{Q}_w\mathbf{V} = \mathbf{I}$. It is easy to see that matrix $\mathbf{Q}_w^{1/2}$ is an appropriate choice for $\mathbf{V}$. In the transformed space criterion J_2 becomes

$$J_3(\mathbf{X}_{\mathcal{F}}) = \mathrm{tr}(\mathbf{Q}_w^{-1/2}\mathbf{Q}_b\mathbf{Q}_w^{1/2}) = \mathrm{tr}(\mathbf{Q}_w^{-1}\mathbf{Q}_b) = \sum_{k=1}^{N'} \lambda'_k \tag{14}$$

where λ'_k, $k = 1, \ldots, N'$ are the eigenvalues of matrix, $\mathbf{\Lambda}'$, i.e. the product matrix $\mathbf{Q}_w^{-1}\mathbf{Q}_b$, and N' is the size of the feature set $\mathcal{F}$. It can be shown [10], that J_3 corresponds to a separability measure which uses the quadratic metric with the average covariance of input vectors as scaling matrix. We can criticize J_3 from the point we mentioned for PCA, namely that it transforms the original features, hence the linguistic interpretability of the elements of a transformed feature set is questionable.

Another possible criterion function can be formulated by means of the determinants of scatter matrices. Determinant of scatter matrices has a geometrical interpretation [10], meaning the volume occupied by the elements of clusters in the multidimensional space. Intuitively, the greater the ratio of the determinants of between- and within-class matrices, or more conveniently, of the determinants of the total- and within-class matrices, the greater the spatial separation of classes. Hence, the function

$$J_4(\mathbf{X}_{\mathcal{F}}) = \det(\mathbf{Q}_t)/\det(\mathbf{Q}_w) \tag{15}$$

is a suitable indicator of class separability.

Let us now investigate J_4. Since matrices $\mathbf{Q}_t$ and $\mathbf{Q}_w$ are symmetric, there exists a matrix $\mathbf{V}$ that diagonalizes both of them:

$$\begin{aligned} \mathbf{V}^T\mathbf{Q}_t\mathbf{V} &= \mathbf{\Lambda}, \\ \mathbf{V}^T\mathbf{Q}_w\mathbf{V} &= \mathbf{I}. \end{aligned} \tag{16}$$

Because of (16), J_4 can be expressed in terms of diagonal elements of matrix $\mathbf{\Lambda}$ as:

$$J_4(\mathbf{X}_{\mathcal{F}}) = \frac{\det(\mathbf{V}^T\mathbf{Q}_t\mathbf{V})}{\det(\mathbf{V}^T\mathbf{Q}_w\mathbf{V})} = \sum_{k=1}^{N'} \lambda_k$$

Furthermore, from (16) follows

$$\mathbf{Q}_w^{-1}\mathbf{Q}_t\mathbf{V} = \mathbf{V}\mathbf{\Lambda} \tag{17}$$

so $\mathbf{\Lambda}$ is the matrix of eigenvalues of $\mathbf{Q}_w^{-1}\mathbf{Q}_t$. Due to (10), we have

$$\mathbf{Q}_w^{-1}\mathbf{Q}_t = \mathbf{I} + \mathbf{Q}_w^{-1}\mathbf{Q}_b$$

Now, from (14) and (17) we obtain

$$\mathbf{\Lambda}' = \mathbf{\Lambda} - \mathbf{I}$$

so $\lambda'_k = \lambda_k - 1$ and consequently J_4 can be expressed as

$$J_4(\mathbf{X}_{\mathcal{F}}) = \sum_{k=1}^{N'} (1 + \lambda'_k)$$

Comparing J_3 and J_4 we can state that the latter is likely to be more reliable in cases where the contributions of total interclass distance are distributed evenly over all the axes of the coordinate system. Thus, in contrast to J_3, it is unlikely, that J_4 would select a feature set which separates two classes well, but only at the expense of the ability of discriminate between all other classes. This situation can be characterized, in terms of eigenvalues λ_k, by the domination of a single eigenvalues in the criterion J_3.

We note that the value of J_4 can vary from very small to very large. This phenomenon can cause problems, as, on one side, a very small value can turn out to be zero if its order attains the accuracy of the underlying computer system, while on the other side, a very large value may result in loss of information, when rounding is involved. To overcome these possible drawbacks other criterion functions can be used.

Another problem can arise if any of the matrices appearing in (15) are singular, and hence the determinant is negative. This problem can be solved by taking the absolute value of the determinant.

The use of these functions does not require any modifications of the FR-FCO algorithm. We examine their effect on the ranking in Section 5.

We remark that the normalization of the input values can also solve the problem of computer system accuracy. The effect of this transformation on the ranking is out of the scope of this paper.

5 The comparative analysis of FRFCO

We applied the proposed method to several data sets. We compare our results to two other methods: the RC method [20] using a fixed setting and input contribution measure (ICM) analyzer (similar to sensibility analysis) [14]. RC divides the data into two groups. For this, we first ordered the data based on their output value, then put them in group A and alternately in group B according to this ordering. Here we remark again that the result obtained by RC is unstable as the RC is very sensitive to its parameters, i.e. how the two groups are selected and how the fuzzy submodels are built for the data [29]. The ICM makes use of a BPNN for training by using the given data set. It then varies the input variables one at a time to their minimum and maximum limit. The ICM of the input variable is then measured based on the effect of the input variable to the output.

For FCM we fixed the value of fuzzifier to the default ($m = 2$), and we use SBS search method unless stated otherwise.

5.1 Simple synthetic data set

First, we checked whether FRFCO gives consistent result on the synthetic and real data sets given in [27]. The synthetic data set contained 50 samples with 4 variables. The first two variables were obtained from the function

$$y = f(x_1, x_2) = (1 + x_1^{-2} + x_2^{-1.5}), \qquad 1 \leq x_1, x_2 \leq 5$$

and the last two were chosen randomly. The optimal number of clusters determined by (6) equals 6. In this case the determinants are all positive in each iteration. Table 1 shows the results obtained by FRFCO with various criterion functions with different search method and by other techiques.

On this synthetic sample our proposed method gives the correct ranking by using function J_3 or J_4 as the criterion function. Because all the determinants are positive we do not need to take their absolute value to solve singularity problem. The criterion function J_2 also finds the most important

Table 1. Ranking of the 4 input features of the synthetic data set from [27]

method	Ranking	Remarks
FRFCO with J_2	$\langle 2,3,1,4 \rangle$	
FRFCO with J_3	$\langle 2,1,3,4 \rangle$	
FRFCO with J_4	$\langle 2,1,3,4 \rangle$	
FRFCO with J_3	$\langle 2,1,4,3 \rangle$	SFS is used
FRFCO with J_4	$\langle 2,1,3,4 \rangle$	SFS is used
ICM with 4–8–1	$\langle 2,1,3,4 \rangle$	the contribution of each variable is:
architecture		$\langle 67.85, 29.57, 1.79, 0.79 \rangle$
RC	$\langle 1,2 \rangle$	Automatically pruned[a]

[a] The RC method automatically prunes the irrelevant variables.

variable, but fails to find the second one. When SFS search method is used the ranking is very similar, just the order of random variables is changed.

5.2 Real data set of a chemical plant

The second sample data set was the model of a chemical plant with 5 inputs [27]. The inputs were the following: x_1 – monomer concentration, x_2 – change of monomer concentration, x_3 – monomer flow rate, x_4 and x_5 – local temperature inside the plant. The output was the set point for monomer flow rate. 70 sample data were provided.

In [27] the first three variables were found important by means of the RC method. We have to admit that we could not generate this result with our RC implementation regardless of the applied parameter settings [29]. According to the ICM analyzer the third variable is the most important, then the first, while the remaining three were considered irrelevant. These results are compared with the FRFCO method in Table 2. The optimal number of clusters is again 6. All the determinants are positive during the elimination process.

The FRFCO gives the same result with all the criterion functions, and this also coincides with other techniques. Because the determinants are positive the rankings obtained with J_4 is correct, and absolute value function is need not to be applied to J_4. Criterion functions permute the order of the last three variables, but this is not very significant, because their contributions are similar and very low according to the ICM analyzer. Figure 1 shows the values of J_2 before the first deletion. Clearly, the difference between the first and the last two variables is insignificant and only due to rounding error of the computer system used. We remark that this value coincides with the one obtained by our RC implementation.

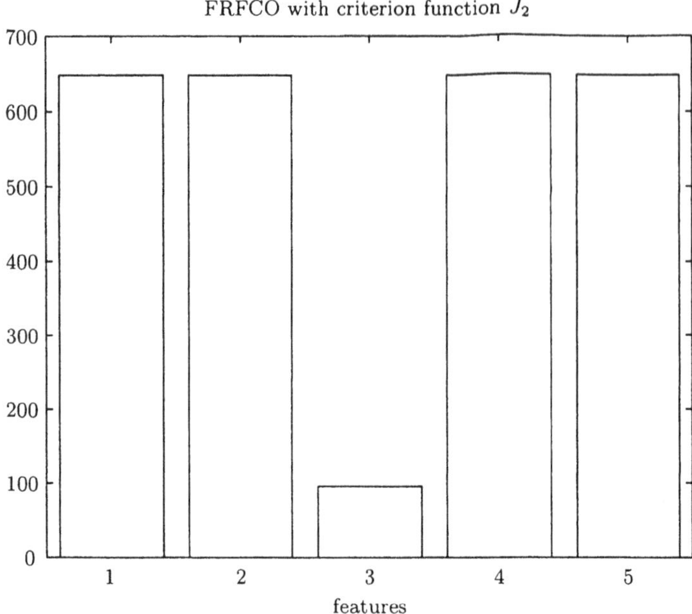

Fig. 1. The values of J_2 before the first elimination (on chemical plant data)

5.3 Eye-gaze data set

We also applied our technique to eye-gaze data set which was analyzed thoroughly by Gedeon in [15]. There, the author studied trained back propagation neural networks for ranking the 12 input features. The number of sample data

Table 2. Ranking of the 5 input features of the chemical plant data set from [27]

method	Ranking	Remarks
FRFCO with J_2	$\langle 3,1,5,2,4\rangle$	
FRFCO with J_3	$\langle 3,1,2,5,4\rangle$	
FRFCO with J_4	$\langle 3,1,4,5,2\rangle$	
ICM with 5–10–1	$\langle 3,1,2,5,4\rangle$	contribution of variables:
architecture		$\langle 75.94,22.17,1.10,0.78,0.01\rangle$
RC in [27]	$\langle 3,2,1\rangle$	Automatically pruned [a]
RC	$\langle 3\rangle$	Automatically pruned [a]

[a] The RC method automatically prunes the irrelevant variables.

was 909. The data set is described in details in [15]. We used the eye-gaze data set with its original output, but also with the network outputs of the best trained network. The optimal number of clusters was 2 in both cases. There are both negative and positive determinants during the iteration steps. The results (for both data sets) of the FRFCO with various criterion functions are compared with the result of the best trained network and the ICM analyzer and are depicted in Table 3. The first part of the table contains the results with the original data, while the second part with network output data.

Table 3. Ranking of the 12 input features of eye-gaze data set

method	Ranking	Remarks
best of [15]	$\langle 6,4,3,2,7,5,1,10,9,8,12,11\rangle$	
FRFCO with J_2	$\langle 7,1,8,12,3,2,11,10,4,5,9,6\rangle$	
FRFCO with J_4	$\langle 6,5,3,2,9,7,4,1,12,10,8,11\rangle$	
FRFCO with J_4'	$\langle 7,4,3,8,11,5,1,2,9,10,12,6\rangle$	
ICM with 12–24–1	$\langle 12,7,9,6,1,3,8,4,2,10,11,5\rangle$	contribution of variables:
architecture	$\langle 25.22,19.66,17.66,9.57,9.09,5.34,4.17,3.29,2.18,1.83,1.18,0.81\rangle$	
FRFCO with J_2	$\langle 7,12,1,3,2,8,10,11,9,5,6,4\rangle$	
FRFCO with J_4	$\langle 5,4,6,9,8,1,11,2,3,10,12,7\rangle$	
FRFCO with J_4'	$\langle 7,1,3,10,12,4,5,6,9,11,2,8\rangle$	
ICM with 12–24–1	$\langle 7,11,9,10,12,6,8,3,1,2,5,4\rangle$	contribution of variables:
architecture	$\langle 28.1,14.28,10.31,10.3,9.82,6.65,5.95,5.95,3.26,2.92,1.71,0.73\rangle$	

The results in Table 3 give many possibilities for comparison, from which we would like to emphasize the following. The results by J_2 are not in accordance with Gedeon's best considered one. The reason of this result partly is that the domains of variables vary. Due to the various signs of the determinants the use of the absolute value gives different results. Surprisingly, the result of J_4 without absolute value is more similar to the reference result of Gedeon than J_4 with absolute value.

The ICM analyzer also does not give correct results. It is in accordance with [15], where the similar sensitivity analysis was studied and it failed to find the correct or close-to-correct ranking. It may be argued that in this case we have only two clusters, and a small change in a variable can change the dominant cluster if the point is close to the boundary and the change applied ia towards the other cluster, but on the other hand, if a huge change applies in the opposite direction the dominant cluster remains the same. Note, that in this case the membership in the clusters is not fuzzy, but crisp!

Naturally, the results with the first data set is closer to Gedeon's results, and the evaluation with the network output the ranking gives a different result. Nevertheless, the six most significant variables coincide with Gedeon's result in 3 places, and the three most significant variables in 2 places.

5.4 Summary and hints for use

The best results were obtained in all the three examples with the use of the criterion function J_4. If the determinants are of both signs, then the use of absolute value function can improve its performance. When the domain of the features is homogeneous or at least comparable the J_2 criterion function also provides good results.

As the ranking does not specify how many features to use, the easiest way is to try it experimentally. For this we can start to build up fuzzy rule base models (e.g. with Sugeno and Yasukawa's method [27]) with a small number of top ranked features (one or two). Then they should be evaluated by a proper performance index function. If a local optimum is reached according to this index the fuzzy model structure is accepted. Finally, some parameter identification or tuning algorithms can be applied to improve the performance of the final model (see [25,27]).

6 Conclusion

We proposed in this paper a feature ranking algorithm adapted to fuzzy modelling with output from a continuous range. The main idea is to cluster the output data and to use the cluster-membership degrees as weights in the feature ranking method. Several criterion functions were proposed for determining the ranking. We applied our method to real world and synthetic data sets, and it was likely to find the proper or close-to-proper ranking. Finally, some hints for the use of the algorithm were presented.

Acknowledgement

The authors would like to thank the anonymous referees for their very valuable and constructive comments and help.

References

1. S. Abe, R. Thawonmas, and Y. Kobayashi. Feature selection by analyzing class regions approximated by ellipsoids. *IEEE Trans. on SMC, Part C*, 28(2), 1998. http://www.info.kochi-tech.ac.jp/ruck/paper.html.

2. P. Baranyi and L. T. Kóczy. A general and specialized solid cutting method for fuzzy rule interpolation. *BUSEFAL*, 67:13–22, 1996.
3. P. Baranyi, A. Martinovics, D. Tikk, L. T. Kóczy, and Y. Yam. A general extension of fuzzy SVD rule base reduction using arbitrary inference algorithm. In *Proc. of IEEE Int. Conf. on System Man and Cybernetics (IEEE-SMC'98)*, pages 2785–2790, San Diego, USA, 1998.
4. P. Baranyi and Y. Yam. Fuzzy rule base reduction. In D. Ruan and E. E. Kerre, editors, *Fuzzy IF-THEN Rules in Computational Intelligence: Theory and Applications*, pages 135–160. Kluwer, 2000.
5. P. Baranyi, Y. Yam, C. T. Yang, P. Várlaki, and P. Michelberger. Inference algorithm independent SVD fuzzy rule base complexity reduction. *International Journal of Advanced Computational Intelligence*, 5(1):22–30, 2001.
6. J. C. Bezdek. *Pattern Recognition with Fuzzy Objective Function Algorithms.* Plenum Press, New York, 1981.
7. J. Bruinzeel, V. Lacrose, A. Titli, and H. B. Verbruggen. Real time fuzzy control of complex systems using rule-base reduction methods. In *Proc. of the 2nd World Automation Congress (WAC'96)*, Montpellier, France, 1996.
8. P. J. Costa Branco, N. Lori, and J. A. Dente. New approaches on structure identification of fuzzy models: Case study in an electro-mechanical system. In T. Furuhashi and Y. Uchikawa, editors, *Fuzzy Logic, Neural Networks, and Evolutionary Computation*, pages 104–143. Springer-Verlag, Berlin, 1996.
9. R. K. De, N. R. Pal, and S. K. Pal. Feature analysis: Neural network and fuzzy set theoretic approaches. *Pattern Recognition*, 30(10):1579–1590, 1997.
10. P. A. Devijver and J. Kittler. *Pattern Recognition: A Statistical Approach.* Prentice Hall, London, 1982.
11. J. C. Dunn. Well-separated cluster and optimal fuzzy partition. *J. Cybern.*, 4:95–104, 1974.
12. R. A. Fischer. The use of multiple measurements in taxonomic problems. *Ann. Eugenics*, 7:179–188, 1936.
13. Y. Fukuyama and M. Sugeno. A new method of choosing the number of clusters for fuzzy c-means method. In *Proc. of the 5th Fuzzy System Symposium*, pages 247–250, 1989. (in Japanese).
14. C. C. Fung, K. W. Wong, and H. Crocker. Determining input contributions for a neural network based porosity prediction model. In *Proc. of the Eighth Australian Conference on Neural Network (ACNN97)*, pages 35–39, Melbourne, 1997.
15. T. D. Gedeon. Data mining of inputs: Analysing magnitude and functional measures. *International Journal of Neural Systems*, 8(2):209–218, 1997.
16. T. D. Gedeon and L. T. Kóczy. Conservation of fuzziness in rule interpolation. In *Proc. of the Symp. on New Trends in Control of Large Scale Systems*, volume 1, pages 13–19, Herľany, Slovakia, 1996.
17. D. E. Gustafson and W. C. Kessel. Fuzzy clustering with a covariance matrix. In *Proc. of the IEEE International Conf. of Fuzzy Systems*, pages 761–766, San Diego, 1979.
18. T.-P. Hong and J.-B. Chen. Finding relevant attributes and membership functions. *Fuzzy Sets and Systems*, 103(3):389–404, 1999.
19. T.-P. Hong and C.-Y. Lee. Induction of fuzzy rules and membership functions from training examples. *Fuzzy Sets and Systems*, 84:33–47, 1996.
20. J. Ihara. Group method of data handling towards a modeling of complex system IV. *Systems and Control*, 24:158–168, 1980. (In Japanese).

21. J-S. R. Jang, C.-T. Sun, and E. Mizutani. *Neuro-Fuzzy and Soft Computing: A Computational Approach to Learning and Machine Intelligence.* Prentice Hall, Upper Saddle River, NJ, 1997.
22. D. Kleinbaum, L. L. Kupper, and K. E. Muller. *Applied Regression Analysis and Other Multivariable Methods.* PWS-Kent, Boston, Mass., 2nd edition, 1988.
23. L. T. Kóczy and K. Hirota. Approximate inference in hierarchical structured rule bases. In *Proc. of 5th IFSA World Congress (IFSA'93)*, pages 1262–1265, Seoul, 1993.
24. L. T. Kóczy and K. Hirota. Size reduction by interpolation in fuzzy rule bases. *IEEE Trans. on SMC*, 27:14–25, 1997.
25. J.A. Roubos, M. Setnes, and J. Abonyi. Learning fuzzy classification rules from labeled data. *International Journal of Information Sciences*, submitted, July 2000. http://www.fmt.vein.hu/softcomp.
26. M. Sugeno, M. F. Griffin, and A. Bastian. Fuzzy hierarchical control of an unmanned helicopter. In *Proc. of the 5th IFSA World Congress (IFSA'93)*, pages 1262–1265, Seoul, 1993.
27. M. Sugeno and T. Yasukawa. A fuzzy logic based approach to qualitative modelling. *IEEE Trans. on Fuzzy Systems*, 1(1):7–31, 1993.
28. D. Tikk and P. Baranyi. Comprehensive analysis of a new fuzzy rule interpolation method. *IEEE Trans. on Fuzzy Systems*, 8(3):281–296, 2000.
29. D. Tikk, T. D. Gedeon, L. T. Kóczy, and Gy. Biró. Implementation details of problems in Sugeno and Yasukawa's qualitative modelling. Research Working Paper RWP-IT-02-2001, School of Information Technology, Murdoch University, Perth, W.A., 2001. p. 17.

Interpretability in Multidimensional Classification

Vincent Vanhoucke[1,2] and Rosaria Silipo[2]

[1] Stanford University, Stanford CA 94305, USA
[2] Nuance Communications, 1380 Willow Road, Menlo Park CA 94025, USA

Abstract. Generating rule-based models from data is an efficient way of inferring information from large datasets. In high-dimensional spaces, the complexity of the model itself can undermine the interpretability of this information. This chapter introduces metrics quantifying the information flow between inputs, feature dimensions and output classes. These metrics are used to estimate the contribution of individual input features to a fuzzy classification task without making explicit use of the data underlying the model. Application of these techniques to a speech classification problem shows that significant reduction in the model dimensionality can be achieved with minimal accuracy loss.

1 Introduction

1.1 Classification Algorithms and Interpretability

The increasing accessibility of data in many areas of technology has driven the development of many automatic classification and clustering algorithms. The use of these algorithms for classification and modeling purposes is in general well understood. However, the more complex these models are, the more they tend to obfuscate the relationships between the data and the classification output.

A more recent research trend focuses on investigating the data with the goal of getting some insights about the system that generated it. A classification algorithm is now required to provide a low error rate as well as an interpretable decision process [1, Chapter 1] [2].

The task of classification can be considered, in its canonical form, as the application of a one-to-one mapping from an input feature space into the space of output classes. In addition, it is of importance both when designing and using a classifier to be able to evaluate it on several grounds. Among these:

- the classifier's confidence in its decisions,
- the interpretability of the relationships between inputs and outputs,
- the sensitivity to the input features.

The earliest automatic classifiers that tried to answer the need for interpretability were based on fuzzy logic [3,4] [1, Chapter 8]. The representation of knowledge through rules and of the input space through linguistic values allows the user to easily apprehend how a given output class has been assigned to an input pattern.

Statistical decision trees, based on probabilistic observations, were also introduced to provide an interpretable statistical classification process [5,6]. A stepwise entropy maximization on a given subset of training data is used to determine the most informative split of the input space on one of the input dimensions. After recursively applying such split search, a tree can be built where the nodes correspond to the decisions and the leaves to the final classification results. A visual inspection of the tree can rapidly provide a summary about the whole classification process.

Statistical decision trees and fuzzy systems have also been combined to produce fuzzy decision trees (fuzzy ID3) [7]. On the basis of a fuzzy entropy definition, a fuzzy decision tree can be built in a similar way to statistical decision trees.

For input spaces with small dimensionality, fuzzy rules are easy to read. If the input space includes many features, or if the problem is complex and a very high number of fuzzy rules is generated, the decision process becomes much less transparent.

Whenever the classification task is complex and/or the input space has very high dimensionality, the visual inspection of the model can become intractable. In addition, there is a need for quantifiable measures to be associated with the interpretation process. A quantitative "interpreter" associated with a classification model can be much more powerful a tool for automated system analysis than a mechanism requiring visual inspection.

With the dramatic growth of the dimensionality of databases, even fuzzy models and decision trees can be impractical. It becomes necessary to make abstraction of the rule-based representation, and switch to a more compact description of the system based on information content. A more global measure of the system sensitivity to the input features can provide better insights into the classifier [8,9], using a much smaller set of descriptive variables.

1.2 Challenging Datasets

Large and high dimensional databases are increasingly becoming available to the community to challenge the way data analysis has been performed so far.

An example of a field that produces large size databases with high dimensionality is speech recognition. Research speech recognition systems compute up to 200 input features from a single frame of the speech signal. The OGI Corpus [10], for example, consists of one minute long speech segments spoken over commercial telephone lines of speakers in 12 different languages. The database contains a total of more than 1900 calls. Numerous corpora of similar or greater size are quite widespread and intensively exploited to train and analyze speech systems.

In the biomedical field, the recording of heterogeneous data is becoming more widespread. For example, for the electrocardiographic signal (ECG) the number of leads increased (up to 12) as well as the duration of each recording (up to 24 hours). Many physiological sources are now monitored simultaneously for longer and longer periods of time.

Many examples of this kind of biomedical data can be found on the PhysioBank web site [11]. PhysioBank is a large and growing archive of well-characterized digital recordings of physiologic signals and related data for use by the biomedical research community. PhysioBank currently includes databases of multi-parameter cardiopulmonary, neural, and other biomedical signals from healthy subjects and patients with a variety of conditions with major public health implications, including sudden cardiac death, congestive heart failure, epilepsy, gait disorders, sleep apnea, and aging. For example, the Apnea-ECG database consists of 70 ECG recordings with simultaneous respiration signals, each typically 8 hours long [12].

Similarly, bioinformatics work almost exclusively on extremely large databases. The National Center for Biotechnology Information (NCBI) [13] creates public databases on computational biology, genome data, and biomedical information, all of these aimed at the better understanding of molecular processes affecting human health and diseases [14].

In particular, GenBank [15] is the NIH genetic sequence database, an annotated collection of all publicly available DNA sequences. GenBank (at NCBI), together with the DNA DataBank of Japan (DDBJ) and the European Molecular Biology Laboratory (EMBL) comprise the International Nucleotide Sequence Database Collaboration. These three organizations exchange data on a daily basis. GenBank grows at an exponential rate, with the number of nucleotide bases doubling approximately every 14 months. Currently, GenBank contains more than 13 billion bases from over 100,000 species.

1.3 Information Measures

On large and/or high-dimensional databases, direct interpretation techniques, such as visual inspection of decision trees and representation of fuzzy rules on a two or three-dimensional space, are not informative. Alternative techniques have to be implemented to gain insights on the most significant aspects of the classification system.

One approach has been to view the system in a parallel coordinate environment. All input space coordinates are represented in different sections of a two-dimensional plane, allowing an unconventional but exhaustive view of the classification system [16–18]. Other techniques focus on the definition of a global measure characterizing one or more properties of the system.

An important part of a decision process consists of the measurement of how much each input feature contributes to the final outcome. Feature characterization can be used for several purposes:

- feature selection: removing superfluous inputs to simplify classification,
- robustness evaluation: analyzing the classification under noisy conditions,
- feature weighting: rebalancing of the feature impact on the decision.

Many information retrieval or data mining techniques, developed to allow some partial interpretation of the decision process, operate directly on the data [8], independently from the classification model. However, there are also reasons to treat classification and interpretation jointly.

First, techniques that operate directly on the data often actually create a model for it. A model trained for the purposes of classification will guarantee a better match to the problem at hand. Secondly, the classifier has in general orders of magnitude less degrees of freedom than the data, making the analysis simpler and more tractable computationally. As a consequence, techniques involving joint modeling and classification are getting a lot of attention [19,20].

Fuzzy systems, in particular, produce a computationally simple description of the input space and can be considered good candidates for joint classification and interpretation analysis. In Section 3, a joint classification and feature selection approach is proposed. Fuzzy rules are trained and subsequently analyzed in terms of the impact that input features have on the classification process.

To quantify the information content associated with the input features of a fuzzy model, a mutual information measure is applied (Section 2.5), being the relative difference between the intrinsic information available in the system before and after using a given input feature for the analysis [21] [1, chapter 8]. The measure of the information contained in a fuzzy model is derived solely on the basis of its fuzzy rules.

For the purposes of feature selection, the input features are ranked according to their impact on the fuzzy classification model and the least influencing features are discarded. The ranking provides useful insights on the relative importance of the input features for this fuzzy model. The feature selection also reduces the model size and possibly improves its performance, since noisy input features that do not contribute to or even confuse the classification process are identified and excluded from the analysis.

1.4 Feature Selection

Given a large dimensional input space and a classifier, the objective of feature selection [22] is to determine from the model which sets of features are meaningful to use and which can be discarded. There are several motivations for addressing this problem [23]:

- to reduce the number of rules to improve the interpretability of the model,
- to improve the accuracy of the model by removing potentially noisy superfluous features,

- to compress the model for complexity reduction and faster classification.

Features selection methods are usually referred to as either *filter* methods or *wrapper* methods. Filter methods quantify the importance of input features without taking into account the classification algorithm that will use the selected subset. Conversely, wrapper methods use the classification algorithm as the evaluation function to choose the relevant input features. The evaluation functions, for example, can be distance measures, information divergences, or the classifier's error rate [24].

Numerous criteria [25] have been proposed for selecting features, including correlation methods [26] and minimum description length [27]. A very extensive bibliography on feature selection can be found at [28].

In [21], for example, feature merit measures are defined on the basis of the entropy maximization theory of statistical decision trees. Another algorithm uses the ROC curves of the inductive algorithm to quantify the importance of the input features [29].

The main issue of feature selection is the intractability of an exhaustive search of the space of all possible subsets of features for high-dimensional problems. Various search strategies can be adopted [30,24], like sequential selection [31,32], or stochastic search strategies [33–35]. A general drawback of these techniques, however, is the amount of computation they require. These can involve a high number of training runs for different classifier configurations as well as decisions based on direct inspection of potentially large volumes of data.

1.5 Speech Classification

Automatic Speech Recognition (ASR) systems are classifiers that make use of various levels of linguistic information to decode spoken utterances into their textual transcription. Typical speech recognition systems consist of several data analysis blocks:

- Front-end: the speech signal is segmented into frames long of a few milliseconds, and passed trough an analysis transform from which features are extracted.
- Classifier (Acoustic Model): the features are submitted to a classifier that computes the probabilities of possible phonetic units.
- Decoder (Language Model): translates the probabilistic sequences of phonetic units into the most likely corresponding word.

ASR systems are usually very complex and involve a very high number of parameters derived from acoustic, phonetic and linguistic knowledge [36]. This structure usually does not provide much visibility into the workings of the system and the interpretability of ASR models is quite poor. As an example, any modification made to the front-end processing typically requires a retraining of the acoustic model, which can be an extremely time consuming process.

In commercial systems, typically between 20 and 50 spectral or time-domain features are extracted from the speech signal and used as input for the frame classification [36]. Due to the complexity of the system, it is rather difficult to estimate a priori the contribution of each input feature to the final result. For the same reason, any recursive search strategy aimed at reducing the dimensionality of the input space would involve a prohibitively large number of re-training and re-evaluation of the system performance.

The feature selection algorithm proposed in this chapter facilitates the task of investigating the input feature contributions. Because it uses fuzzy systems, and because it operates directly on the model while making abstraction of the data, it is computationally inexpensive and particularly suited for this class of problems.

The analysis in Section 4 focuses essentially on the influence of features on the acoustic model, with the assumption that a good frame classifier is the basis for an acceptable word recognition error rate. In particular, the phone classification task is decomposed into the recognition of simpler phonetic properties [37,38] from which phonetic units derive. This approach is more robust than direct classification of phones using a single all-purpose classifier. The phone labels can be recovered as a second step in the process by cross-analysis of all the phonetic properties of the speech frame [39].

2 Information Measures

2.1 Membership Degree and Information

A broad class of commonly used classifiers operate using mechanisms much more suitable to introspection than the simple specification of a functional mapping between inputs and outputs.

These classifiers operate by defining a set of membership functions that associate a given input pattern from domain D to an output class C in the set $\mathcal{C}$ of output classes by means of a membership degree $\mu_C : D \to [0,1]$. The membership degree summarizes by one scalar the information the model possesses about the class membership of a given input vector, with the assumption that the membership degree is proportional to the model's confidence in the association.

Classifying an input x amounts to finding the class that maximizes the membership degree:

$$C(x) = \arg\max_{C} \mu_C(x)$$

This simple representation of the information contained in the classifier adds a lot to its interpretability. This type of model contains information about all the possible class associations for any given input vector, as opposed to the output class only. A mathematically simple representation of the decision process is thus embedded into the model, and can be exploited for introspective analysis.

A consequence of this choice of a representation is that comparing relative membership degrees across regions of the feature and/or output space will provide measures of the amount of information contained in the model relative to any combination of inputs and outputs.

For example, the information contained in the model about any given input $\boldsymbol{x}$ is determined by the distribution of the membership degree of $\boldsymbol{x}$ across all possible classes. A very uneven distribution indicates that the model makes strong associations between this input and some of the classes, while a flat distribution indicates that the model has no information at all about the class membership of the input (Figure 1).

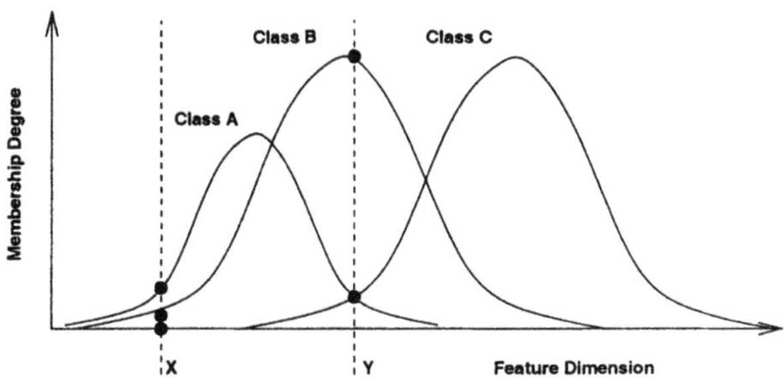

Fig. 1. Information as carried by membership functions: point X has very similar membership degrees across each class, implying that the model contains very little information about this input. The model is much more "informative" about the association of point Y with class B.

Measuring the degree of unevenness of a distribution $\Lambda = \{\lambda_k, k \in \mathcal{C}\}$, $(\sum \lambda_k = 1)$ has been tackled from different perspectives [21,40]. In particular, information theory introduced the concept of entropy [41], subsequently applied to rule-based systems [7]:

Entropy

$$I(\Lambda) = -\sum_k \lambda_k \log \lambda_k$$

The entropy is symmetric in its arguments, ensuring that no class is given more importance than any other. It is maximized when the distribution is even, and identically zero when any λ_k equals 1, effectively measuring the degree of randomness of the distribution.

By defining a normalized membership degree for class $C \in \mathcal{C}$ and a normalized information index as:

Normalized Membership Degree

$$\lambda_C(\boldsymbol{x}) = \frac{\mu_C(\boldsymbol{x})}{\sum_{k \in C} \mu_k(\boldsymbol{x})}$$

Normalized Information Index

$$\mathcal{I}(\boldsymbol{x}) = \frac{\log|\mathcal{C}| - I(\Lambda)}{\log|\mathcal{C}|}$$

$\mathcal{I}(\boldsymbol{x}) \in [0, 1]$ is a pointwise measure of the information the classifier provides about a given input $\boldsymbol{x}$. $\mathcal{I}$ in the vicinity of 1 implies that the classifier's confidence in the association between the input and the class is strong, while $\mathcal{I}$ in the vicinity of 0 indicates that the opposite.

2.2 Class-based Information Measure

Entropy can quantify the information a classifier provides about an input pattern. Similarly, information measures that relate to the output classes can be defined. A simple description of the impact of the classifier on a given class C over a domain D can be constructed using its average membership degree.

Average Membership Degree

$$V(C) = \frac{\int_D \mu_C(\boldsymbol{x})\, d\boldsymbol{x}}{\int_D d\boldsymbol{x}}$$

Note that the integration is carried over $d\boldsymbol{x}$, which is not necessarily uniform over D, if a prior on the feature distribution exists. Assuming normalized membership functions, a higher average membership degree to class C indicates a more uniformly distributed class over the input space. An output class represented by a membership function which takes value $+1$ everywhere on the input domain has average membership degree $+1$. A membership function with average value $V(C) = 0$ indicates an output class that is never related with any pattern of the input domain D.

2.3 Global Information Measure

In order to quantify the information contained in the whole classifier, all average membership degrees from the different membership functions should be considered together. Models with high informational content will have their average membership degrees spread across classes, while models with no information - i.e. all inputs map to a single class - will have their membership degrees unevenly distributed across classes. As previously, entropy metrics can be used to quantify this degree of unevenness. Define the relative average membership degree of class $C \in \mathcal{C}$ as follows:

Relative Average Membership Degree

$$v(C) = \frac{V(C)}{\sum_{k \in C} V(k)}$$

The information contained in the classifier over the domain D for the set of classes $\mathcal{C}$ can be computed as [1, chapter8] [9]:

Global Model Information

$$I(\mathcal{C}) = -\sum_{k \in \mathcal{C}} v(k) \log v(k)$$

2.4 Distance Between Models

In order to compare alternative linguistic descriptions of the same data, it is useful to define a metric on the model space. Consider a model m_1 described by the relative average membership degrees $v_1(k), k \in \mathcal{C}$ over the input domain D_1, and a second model m_2 described by $v_2(k), k \in \mathcal{C}$ over the input domain D_2. Define the divergence between models – or relative information – as:

Divergence between Models

$$\mathcal{D}(m_1 \| m_2) = \sum_{k \in \mathcal{C}} v_1(k) \log \frac{v_1(k)}{v_2(k)}$$

This divergence doesn't exactly define a metric on the model space, because it lacks the symmetry in its arguments:

$$\mathcal{D}(m_1 \| m_2) \neq \mathcal{D}(m_2 \| m_1)$$

However, it bears the other properties expected from a distance measure:

- Non-negativity: $D(m_1 \| m_2) \geq 0$
- Null kernel: $D(m_1 \| m_2) = 0 \iff \forall k \in \mathcal{C}, v_1(k) = v_2(k)$

The importance of this measure lies in the fact that there is no need for two models to share the same input space to be comparable. For example, define a hypothetical model U describing $\mathcal{C}$, which assigns equal relative average membership degrees $u(k) = 1/|\mathcal{C}|$ to each class. According to the definition of the global model information, this model has very high information, and indeed one can show that it is maximal over all models describing $\mathcal{C}$:

$$I_U(\mathcal{C}) = log|\mathcal{C}|$$

The global information of a given model m can now be reinterpreted in terms of its distance to this idealized model U:

$$I_m(\mathcal{C}) = I_U(\mathcal{C}) - \mathcal{D}(m\|U)$$

The topology induced by this divergence over the model space brings new dimensions to the analysis. Instead of simply evaluating models in terms of how they perform, they can now be compared in terms of the information they bring. Two otherwise good models can be redundant, and will not improve performance when combined. A model that is poor in isolation can be combined with models it is "distant" with, and improve significantly the performance.

2.5 Information Measure on Features

A measure of the information contained in input feature f can be derived from the global model information. For a given model m, define m_x as the model obtained by intersecting D with the hyperplane $\{f = x\}$. The conditional information contained in m at point x can be defined as:

$$I_m(\mathcal{C}|f = x) = I_{m_x}(\mathcal{C})$$

The information still available in the model after feature f has been exploited is thus:

Conditional Information

$$I(\mathcal{C}|f) = \int_x I(\mathcal{C}|f = x)\,dx$$

The difference between the information contained in the model before and after exploiting the feature f is defined as the mutual information:

Mutual Information

$$\mathcal{M}(\mathcal{C}, f) = I(\mathcal{C}) - I(\mathcal{C}|f)$$

The mutual information is an indicator of how much information was introduced in the model by using this feature. The less effective the input feature f is in the original model, the closer the remaining information $I(C|f)$ is to the original information $I(C)$, resulting in a lower mutual information. The input features producing the highest mutual informations are the most effective at reducing the total information contained in the model, and are deemed more informative for the analysis.

In order to compare distinct models, the relative mutual information - or information gain - is defined as:

Relative Mutual Information

$$g(\mathcal{C}, f) = \frac{\mathcal{M}(\mathcal{C}, f)}{I(\mathcal{C})} \in [0, 1]$$

The mutual information between a model m and a feature f can also be interpreted in terms of a divergence. Consider a model m_0 constructed by removing f from the feature set. The mutual information can be expressed as:

$$\mathcal{M}(\mathcal{C}, f) = \mathcal{D}(m \| m_0)$$

This means that evaluating the contribution of a feature to a model amounts to measuring the divergence between the model when exploiting and not exploiting the feature.

2.6 Linguistic Equivalence Classes

The difficult step in computing the mutual information between a feature and the model is the integration of the conditional information $\int_x I(\mathcal{C}|f = x)\, dx$. In some cases, the integration can be carried over explicitely by defining a collection of linguistic equivalence classes.

Linguistic equivalence classes are subsets of a feature space containing identical information. When such classes can be defined, conditioning the information on $f = x$ for any x in class L is equivalent to conditioning the information on the class itself:

$$I(\mathcal{C}|f = x) \equiv I(\mathcal{C}|f \in L)$$

When there is a collection $\mathcal{L}$ of linguistic equivalence classes for feature f, the computation of the conditional information becomes:

$$\begin{aligned} I(\mathcal{C}|f) &= \int_x I(\mathcal{C}|f = x)\, dx \\ &= \sum_{L \in \mathcal{L}} I(\mathcal{C}|f \in L) \int_{x \in L} dx \\ &= \sum_{L \in \mathcal{L}} I(\mathcal{C}|f \in L)\, \gamma_L \end{aligned}$$

The term $\gamma_L = \int_{x \in L} dx$ is a prior on the belonging of feature x to class L. Because outliers in the training data tend to induce their own classes, it is important to consider this weighting, even when there is no prior on the data in the probabilistic sense. For the purpose of downweighting outliers, the relative total membership in linguistic equivalence class L can be used as a heuristic:

$$\gamma_L = \frac{\int_{D \cap \{f \in L\}} \mu_C(\boldsymbol{x})\, d\boldsymbol{x}}{\int_D \mu_C(\boldsymbol{x})\, d\boldsymbol{x}}$$

3 Application to Fuzzy Rule Based Systems

3.1 Geometric Interpretation

Fuzzy rule-based systems represent linguistic associations using a superposition of unimodal membership functions that define a collection of fuzzy sets over the feature space (Figure 2). As a consequence, mathematical treatment of these rules translates into simple geometric manipulations.

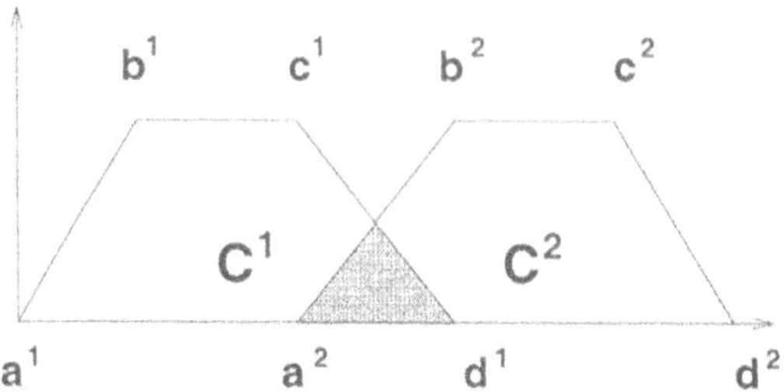

Fig. 2. Example of a class definition as a superposition of two fuzzy membership functions: the average membership degree of the class can be derived in a simple manner from their geometry.

The average membership degree to the union and the intersection of fuzzy sets derives from the min/max-definitions of intersection and union of fuzzy sets [3]:

$$V\left(\bigcup_{i=1}^{K} C^i\right) = \sum_{i=1}^{K}\left[V(C^i) - \sum_{j=i+1}^{K} V\left(C^i \cap C^j\right)\right] \tag{1}$$

If trapezoids are adopted as membership functions, the average membership degree of each fuzzy subset C^i can be computed from the trapezoid height h and the coordinate vectors of its vertices in the n-dimensional input space $< \boldsymbol{a}^i, \boldsymbol{b}^i, \boldsymbol{c}^i, \boldsymbol{d}^i >$:

$$V(C^i) = \frac{h\left[\prod_{j=1}^{n}\left(d_j^i - a_j^i\right) + \prod_{j=1}^{n}\left(c_j^i - b_j^i\right)\right]}{2\int_D dx} \tag{2}$$

Combining equations 1 and 2, all the class average membership degrees and information contents can be computed explicitely.

3.2 Linguistic Intervals

Linguistic equivalence classes, as defined in Section 2.6, can also be derived using the geometry of the membership functions. This leads to a computationally efficient algorithm to calculate the mutual information between any set of features and the fuzzy rule-based model.

Each intersection between trapezoids of different classes define a classification boundary. The boundary is located at:

- the intersection of their sides, if trapezoids overlap only on the sides,
- the middle point of their cores, if they overlap in their core areas,
- the middle point between the trapezoids, if they do not overlap anywhere.

Owing to the geometry of the trapezoids, each of these boundaries is a hyperplane orthogonal to one of the features in the feature set. The complete set of intersections in the model defines a partition of the input space into a finite collection of regions, each mapping a portion of the input space to a given class (Figure 3).

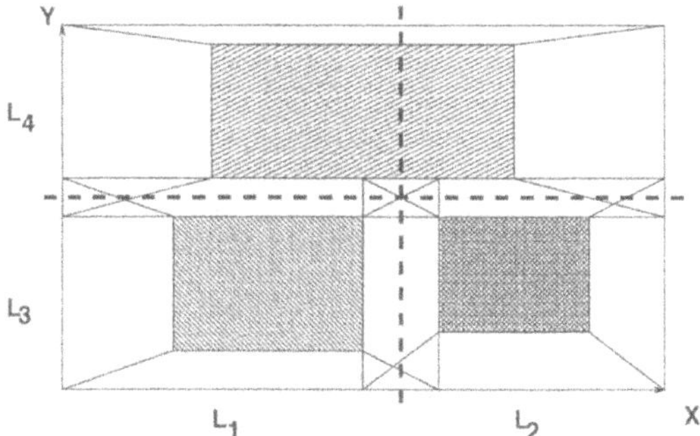

Fig. 3. Example of three classes defining a partition of the input space into regions separated by decision thresholds. Projected on the feature vectors, these decision thresholds define a set of linguistic intervals L_1 to L_4.

The projection of these decision boundaries along input feature f leads to the definition of a finite collection of thresholds that separate contiguous classes along this dimension. This set of thresholds summarizes all the information the feature is providing to the classifier: in between two of these thresholds, the membership association of the input is uniquely determined, regardless of the actual value of the input along f.

As a consequence, these intervals define linguistic equivalence classes over which the conditional information can be computed. The information content of each interval can be evaluated directly by intersecting the domain D with the slice $\{f \in L\}$ (Figure 4).

The complexity of the algorithm depends on the number of linguistic intervals, which is governed by the global geometry of the set of rules. The search for all the possible hyperplanes can be implemented efficiently using a partial ordering of the rules in the feature space, in order to discard quickly irrelevant rule combinations.

The complete algorithm [9,42] implementing the computation of the relative mutual information can be described as follows:

- Find all the k possible decision hyperplanes orthogonal to f by inspection of the possible rule combinations.
- Sort the hyperplanes by increasing order on f.
- For each of the $k+1$ linguistic intervals L_i
  - Build the class model by intersecting the rule set with $D \cap \{f \in L_i\}$.
  - Compute the linguistic interval information $I(C|f \in L_i)$.
- Average all the conditional informations into $I(C|f)$.
- Compute the relative mutual information $g(C, f)$.

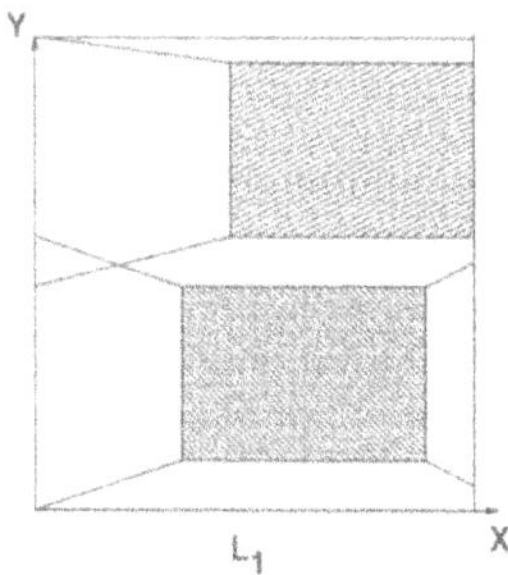

Fig. 4. Model conditioned on linguistic interval L_1: given the knowledge of the interval the input vector belongs to, feature X doesn't provide any additional decision information. The mutual information between feature X and the model is the difference between the information over the whole domain, and the sum of the informations left in intervals L_1 and L_2.

The constraint that the model only allows a finite set of decision boundaries orthogonal to a given feature is usually considered a limitation of such models, especially in high-dimensional spaces where natural clusters tend to be sphere-shaped. Here, this property is the key to make the conditional information computations tractable.

4 Application to Feature Selection: Classification of Phonetic Properties in Continuous Speech

4.1 Phonetic Properties

Phonetic properties have been used successfully in ASR [43,44], as well as in speech synthesis [45–47], as intermediate targets for general speech modeling and classification.

Each phone can be seen as the realization of a number of phonetic properties. A common example of such property is the distinction between *vowels*, *approximants* and *consonants*. Vowels in turn can be classified as *high*, *mid* or *low*, depending on the position of their first formant on the spectrum. On another dimension, they can be regarded as *front*, *central* or *back*, depending on the place where the vowel articulation occurs. As many as 28 different properties can be defined for American English phonemes.

Table 1. Example of characterization of phonetic units

	anterior	consonantal	liquid	sonorant	voiced	stop
/a/	n/a	-	-	+	+	n/a
/l/	+	+	+	+	+	-
/p/	+	+	-	-	-	+

In order to describe a complete spoken utterance in terms of phonetic properties, a *silence* class is also often introduced, broadening the definition of a phone. A complete characterization of phonetic units from an acoustic and articulatory perspective can be found in [48,49,37,38].

4.2 Speech Frame Classification

The goal of the experiments performed in this section is to classify speech frames as belonging or not to each phonetic property class.

The training data consists of 20000 American English utterances, collected over the telephone and sampled at 8 kHz. An additional 8000 utterances from distinct speakers in distinct environments are used for testing. All data is balanced by gender, microphone type and noise conditions. A speech recognizer is used to align the utterances with the corresponding phones using Viterbi alignments [50] on hand-labeled transcriptions (Figure 5). The phones are themselves labeled depending on whether they belong or not to each of the 29 phonetic classes that were defined[1].

The signal is segmented into 10 ms frames. Each phonetic class is assigned 1000 frames for training, and 1000 frames for testing. 12 Mel Filter Cepstral Coefficients (MFCC) [51] are extracted from the signal for each frame. A 39-dimensional input feature set - typical of ASR systems, is created from the MFCC (Table 2). A fuzzy classifier [52] is trained on this feature set for each of the binary classification tasks.

4.3 Feature Selection Based on Mutual Information

After the fuzzy models are trained on the available data and for each classification task, the resulting fuzzy models are analyzed in terms of relative mutual information, as described in Section 2.5 and Section 3.

[1] Class definition by Corey Miller, synthesizing data from [37], [45] and [48]

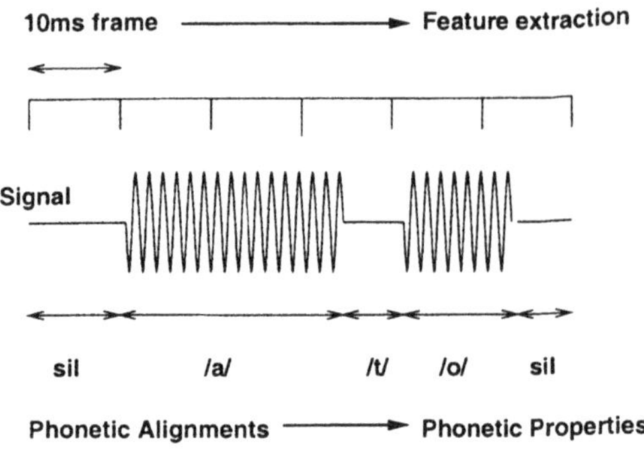

Fig. 5. Diagram of the signal processing involved: The signal is labeled phonetically in order to derive the phonetic properties, and feature extraction is performed on the signal frames to generate the inputs.

Table 2. MFCC feature vector structure

12	Cepstral Coefficients	c_1 to c_{12}
12	First Order Derivatives	∂c_1 to ∂c_{12}
12	Second Order Derivatives	$\partial^2 c_1$ to $\partial^2 c_{12}$
1	Frame Energy Coefficient	ϵ
1	First Order Energy Derivative	$\partial\epsilon$
1	Second Order Energy Derivative	$\partial^2\epsilon$

The 39 input features are ranked according to their relative mutual information. The input features with the lowest figure of merit are also the least relevant to the fuzzy model. Although the information contained in a feature is computed with respect to a given model, it is expected to be a stable measure of the feature discriminative quality for any model of the same class taking advantage of it.

The most natural way to take advantage of the mutual information for feature selection is to decide on a threshold below which the features are to be discarded. This selection can be performed in several ways.

The most obvious criterion is to choose an absolute threshold in $[0, 1]$. The value of the threshold is decided a priori. However, not all problems have the same distribution of information across the input features, and not all sets of fuzzy rules take advantage of the same input features. An absolute threshold might penalize those systems in which the information is distributed across many input features.

Another method involves defining the threshold as a proportion of the maximum normalized mutual information value. In this way, only the features with very low information with respect to the most informative ones will be discarded.

Finally, a last strategy consists of defining the threshold based on a percentile in the mutual information histogram. This method relies on the complete distribution of the mutual information across features to base its decision, and can be expected to be more robust across variations in the mutual information spread.

After removing the input features with normalized mutual information value below the threshold, the fuzzy classifier is retrained with the remaining input features to optimize its performance.

4.4 Speech vs. Silence

The first analyzed classification task consists of speech vs. silence detection.

Speech vs. silence discrimination is a much harder signal processing problem than one could suspect initially. The amount of variability characterizing silence is much greater than for any other acoustic class. At the same time oral stops such as /p/ and /b/ are mostly constituted of silence, and are characterized mostly through their effects on neighboring phonetic units. Fricatives such as /s/ and /ch/ have a broad spectrum very confusable with white noise, especially in low bandwidth signal.

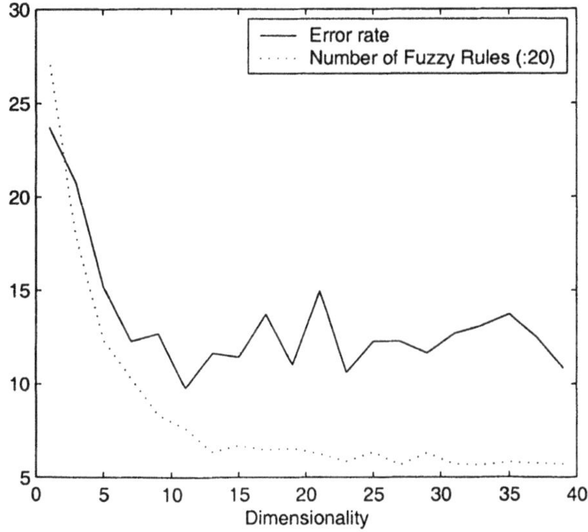

Fig. 6. Effects of input dimensionality reduction on number of rules and accuracy: speech vs. silence classification. The number of parameters (# rules × # dimensions) can be dramatically reduced while not degrading the classifier accuracy.

A great proportion of a speech signal is made of silence, which makes the correct recognition of silence very important for the good performance of an

ASR system. For this reason the classification of speech and silence is a very critical preliminary to any form of phonetic classification.

The classification error rate of the trained fuzzy model (Figure 6) is stable up until 10 input features are left. As expected, the energy ϵ is the predominant feature in this task ($g(\epsilon) = 0.63$). Because long silence segments are quite stationary, and because voiceless low-energy segments in the speech signal tend to be intertwined with louder segments, the first energy derivative $\partial\epsilon$ is also a very good indicator of the presence of speech ($g(\partial\epsilon) = 0.41$). Also in the highest scoring features are the lower cepstra c_1, c_2, c_3, c_5, the lower second derivatives $\partial^2 c_1$, $\partial^2 c_2$, $\partial^2 c_3$, and one first derivative ∂c_8. Conversely, the worst performing features are consistently related to the higher cepstra (c_6, c_8, c_12, ∂c_{11} and $\partial^2 c_{10}$).

4.5 Dental vs. Alveolar

Dental consonants are characterized by a constriction made against the upper teeth (examples: /dh/ as in "the", /f/), while for alveolars the constriction is made against the alveolar ridge (example: /d/). Distinguishing between the two classes can be quite difficult. Some strongly accented speakers sometimes pronounce some alveolars as dentals.

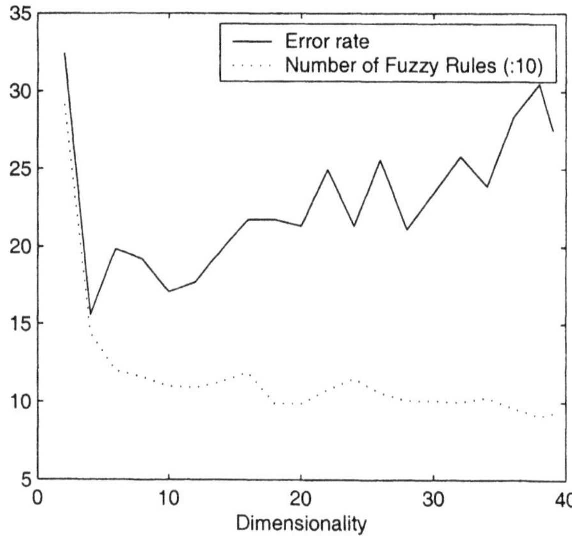

Fig. 7. Effects of input dimensionality reduction on number of rules and accuracy: dental vs. alveolar. The dimensionality reduction is associated with a dramatic improvement in accuracy

As shown in Figure 7, the reduction in dimensionality leads to a dramatic improvement in classification accuracy, due to the elimination of noisy and otherwise uninformative features. The features selected are high cepstra c_9, c_{11}, and the energy derivatives $\partial\epsilon$ and $\partial^2\epsilon$.

4.6 Oral Stops vs. Consonants

The classification of oral stops against all other consonants is a demonstration of the limitations of relying solely on the model to perform feature relevance analysis. In this situation, the amount of variability within each class (9 distinct phones are stops, while 14 are not) is too much for the classifier to handle given the amount of training data (Figure 8). The baseline error rate is about 40% using all 39 input features, showing that the system was unable to accurately learn the classification.

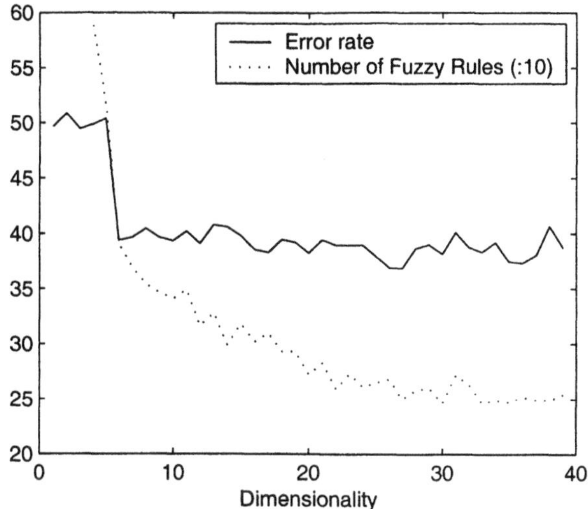

Fig. 8. Effects of dimensionality reduction on number of rules and accuracy: oral stops. The baseline model is extremely poor (about 40% error rate), and the analysis fails to recognize that one feature (the energy) is critical to the classification, and ranks it in 5th position.

The analysis ranks the following features in the top 6 in decreasing order of importance: $g(c_1) = 0.53$, $g(\partial c_3) = 0.51$, $g(\partial^2 c_9) = 0.50$, $g(\partial^2 c_4) = 0.44$, $g(\partial^2\epsilon) = 0.42$, $g(\epsilon) = 0.41$. As the error profile shows, this is inaccurate. With energy ϵ as a feature, the classifier performed at an error rate of about 40%. However, without ϵ, the error rate goes suddenly to 50%, which is as good as chance. This means that all the better scoring features were irrelevant in the absence of ϵ. Indeed, a model trained using the energy as sole feature still gives an error rate of about 42%.

This experiment stresses the fact that relying on the model implies that the model needs to be a good summary of the linguistic content in the data. If the model is not adequate in the first place, the analysis might not be able to distinguish limitations of the model from properties of the data.

Note that the total number of multidimensional rules in the model appears, in all the examples above, to be a reliable indicator of the performance to expect from a model. While the number of rules barely grows as non-meaningful dimensions are pruned out, it starts growing at a fast rate when meaningful features begin to be eliminated, as evidenced by the increase in error rate. This feature can be used as a trigger to determine when a pruned model's performance is to be questioned.

4.7 Threshold Selection

In feature selection, the choice of a threshold determines the tradeoff between input dimensionality reduction and error rate. The problem is to be able to determine the appropriate threshold directly from the mutual information of the different features, without having to retrain models.

Several selection strategies can be considered:

1. an *absolute threshold* on the mutual information,
2. a *relative threshold* with respect to the largest mutual information,
3. a *percentile threshold* based on the mutual information histogram.

When the error rates at all the possible thresholds are known, it is possible to pick the most adequate one by manual inspection of the tradeoff curves, and label the features as informative or not.

Figure 9 compares the three automated strategies in terms of precision and recall against selection of the threshold by visual inspection:

- *recall* measures the percentage of the informative features that were selected by a given strategy,
- *precision* measures what percentage of the selected features were actually informative, according to the manually selected threshold.

Were the selection to be optimal, both these statistics would be at 100%.

Strategy 3 outperforms strategies 1 and 2. Setting a hard threshold on the mutual information is brittle, since it doesn't prevent the system from having no feature selected at all. Simply normalizing the mutual information spread fixes only that issue, without improving the selectivity. On the other hand, taking advantage of the complete distribution of the mutual information across features is robust to any kind of mutual information spread and provides a practical metric for an automatic tuning of the threshold.

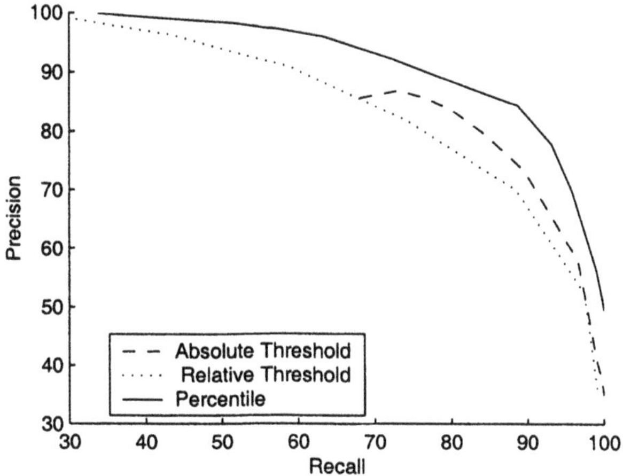

Fig. 9. Precision and recall of 3 automated threshold selection strategies, averaged across the 29 classification tasks. The optimal tradeoff lies at the upper right corner of the graph. The best method uses a feature selection threshold based on the histogram of the mutual information across all features.

4.8 Results

Figure 10 shows the average accuracy of the classifier when a global percentile threshold is used across all classification tasks.

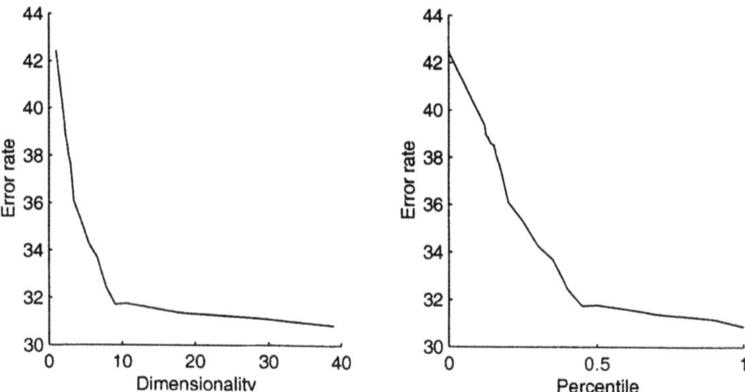

Fig. 10. Error rate vs. dimensionality across the 29 classification tasks (left). The dimensionality reduction corresponds to a threshold selection based on a percentile of the mutual information across all features (right).

The overall dimensionality can be divided by 4 on average across the 29 tasks with less than 1% loss in accuracy, by selecting the 50% percentile in the mutual information histogram. The average number of scalar parameters in the models goes from 3274 down to 1031, i.e. a 68% reduction in complexity.

Globally, no feature is consistently deemed irrelevant across all classification tasks. The average normalized mutual information for all features is 0.2 ± 0.05, which indicates that on average all features are relevant to the task. This conclusion is supported by the very widespread use made of these features across the ASR community. In noise-free environments for example, information contained in the first and second order derivatives is of great importance, because it compensates for a lack of temporal structure inherent to the typical Hidden Markov Models [53] that are generally used. On the other hand, derivatives, especially of the higher cepstra, are also the most subject to distortion by noise, downgrading their relative importance in very noisy environments.

5 Conclusion

Information theory provides an extensive set of tools aimed at the meta-analysis of models. These tools operate on the model parameters rather than on the data directly. This leads to efficient and robust algorithms describing the information flow from features to output classes. In many situations, such indirect methods are preferable to the direct analysis of the data, especially when it comes to evaluating the relevance of a large number of parameters governing a classification model.

In this chapter, a figure of merit for feature relevance, based on a fuzzy classification of the input space, was presented. Due to the low computational load of fuzzy systems, the information contained in a set of fuzzy rules can easily be quantified. This figure of merit is based on an estimation of the mutual information, measured as a difference between the information contained in the fuzzy model before and after a given input feature is used for classification.

A feature selection algorithm is implemented, which ranks input features according to their mutual information, and discards all features deemed irrelevant by a threshold criterion. Several strategies are investigated to define the optimal threshold for this feature selection process. The best criterion is based on discarding all features above a given percentile in the mutual information histogram across inputs.

The feature selection algorithm is shown to perform well on challenging real data in high dimensions. In particular, it has been used to evaluate the contribution of commonly used speech input features to the classification of speech segments into phonetic properties, achieving an average fourfold reduction in the dimensionality with minimal accuracy cost. Future work includes benchmarking its performance against alternative feature selection techniques on similar data.

6 Acknowledgments

The authors wish to thank Michael Berthold for giving us access to his fuzzy trainer, as well as Corey Miller, Chai-Shune Hsu and Francoise Beaufays for providing invaluable data and advice.

References

1. M. Berthold, D. Hand (Eds), *Intelligent Data Analysis, An Introduction*, Springer-Verlag, 1999
2. R. Silipo and G. Di Fatta, Learning to reason about data: the spring school on intelligent data analysis, *Intelligent Data Analysis journal*, Vol. 5, #5, to appear, 2001
3. L.A. Zadeh. Fuzzy logic and approximate reasoning. *Synthese*, Vol. 30, pp. 407-428, 1975
4. L.A. Zadeh, A fuzzy-algorithmic approach to the definition of complex or imprecise concepts, *International Journal on Man-Machine Studies*, Vol 8, pp. 249-291, 1976.
5. J.R. Quinlan, Induction of decision trees, in *Machine Learning*, pp. 81-106, 1986
6. J.R. Quinlan, *C4.5: Programs for Machine Learning*, Morgan Kaufmann Publishers, 1993
7. C.Z. Janikow, Fuzzy decision trees: issues and methods, *IEEE Transactions on Systems, Man, and Cybernetics*, part B, Vol 28, pp. 1-14, 1998
8. H. Liu and H. Motoda. *Feature Extraction, Construction and Selection: A Data Mining Perspective*, Kluwer Academic Publishers, 1998
9. R. Silipo and M. R. Berthold. Input features' impact on fuzzy decision processes. In *IEEE Transactions on Systems, Man, and Cybernetics*, part B, Vol. 30, #6, p. 821, December 2000
10. Center for Spoken Language Understanding, Dept. of Computer Science Engineering, Oregon Graduate Institute, Corvallis, "Stories Corpus", Release 1.0, 1995
11. PhysioBank: `http://www.physionet.org/resources.html`
12. M.R. Jarvis and P.P. Mitra. Sleep apnea classification based on frequency of heart-rate variability, In *Proceedings of Computers in Cardiology*, 2000
13. National Center for Biotechnology Information: `http://www.ncbi.nlm.nih.gov`
14. M. McClelland et al. Complete genome sequence of Salmonella enterica serovar Typhimurium LT2, Nature 413 (6858), pp. 852-856, 2001
15. GenBank: `http://www.ncbi.nlm.nih.gov/Database/`
16. A. Inselberg and B. Dimsdale. Multidimensional iines I: representation, *SIAM J. Applied Math*, Vol. 54 (2), pp. 559-577, 1994
17. A. Inselberg and B. Dimsdale. Multidimensional lines II: representation, *SIAM J. Applied Math*, Vol. 54 (2), pp. 578-596, 1994
18. M. R. Berthold and L. O. Hall. Visualizing fuzzy points in parallel coordinates, *Technical Report UCB/CSD-99-1082*, University of California at Berkeley, 1999

19. J. Li, R.M. Gray and R.A. Olshen. Joint image compression and classification with vector quantization and a two dimensional hidden Markov model. In *Proceedings of the 1999 IEEE Data Compression Conference (DCC)*, pp. 23-32, Snowbird, Utah, March 1999
20. N. Chaddha, K. Perlmutter and R.M. Gray. Joint image classification and compression using hierarchical table-lookup vector quantization. In *Proceedings of the 1996 IEEE Data Compression Conference (DCC)*, J.A. Storer and M. Cohn, editors, IEEE Computer Society Press, Snowbird, Utah, March 1996
21. C. Apte, S.J. Hong, J.R.M. Hosking, J. Lepre, E. Pednault, and B.K. Rosen. Decomposition of heterogeneous classification problems. *Intelligent Data Analysis Journal*, Vol 2, #2, 1998
22. A. Webb. *Statistical Pattern Recognition*, Chapter 8: Feature selection and extraction, Arnold, London, 1999
23. S. Salzberg. Improving classification methods via feature selection. *John Hopkins Technical Report*, 1992
24. M. Dash and H. Liu. Feature selection for classification. *Intelligent Data Analysis*, 1, (3), 1997
25. Y. Yang and J.O. Pedersen. A comparative study on feature selection in text categorization. In *Procccedings of the 14th International Conference on Machine Learning*, pp. 412-420, Morgan Kaufmann, 1997.
26. M.A. Hall. Correlation-based feature subset selection for machine learning. PhD dissertation, Department of Computer Science, University of Waikato, 1999
27. B. Pfahringer. Compression-based feature subset selection. In *Proceedings of the IJCAI-95 Workshop on Data Engineering for Inductive Learning*, pp. 109-119, Montreal, Canada, 1995
28. Bibliography compiled by P. Turney, maintained by O. Boz `http://home.ptd.net/ olcay/feature-selection.html`
29. K. Koumpis and S. Renals. The role of prosody in a voicemail summarization system. In *Proceedings of the ISCA Workshop on Prosody in Speech Recognition and Understanding*, pp. 87-92, 2001
30. G. John, R. Kohavi and K. Pfleger. Irrelevant features and the subset selection problem. In *Machine Learning: Proceedings of the Eleventh International Conference (ICML-94)*, pp. 121-129, New Brunswick, NJ, Morgan Kaufmann, 1994
31. D.W. Aha and R.L. Bankert. A comparative evaluation of sequential feature selection algorithms. In *Artificial Intelligence and Statistics V*, D. Fisher and J.-H. Lenz, editors, Springer-Verlag, New York, NY, 1996
32. D.W. Aha and R.L. Bankert. Feature selection for case-based classification of cloud types: An empirical comparison. In *Proceedings of the 1994 AAAI Workshop on Case-Based Reasoning*, pp. 106-112, AAAI Press, Seattle, WA, 1994
33. H. Almuallim and T.G. Dietterich. Learning with many irrelevant features. In *Proceedings of the Ninth National Conference on Artificial Intelligence*, pp. 547-552, Menlo Park, CA, AAAI Press, 1991
34. K. Kira and L.A. Rendell. A practical approach to feature selection. In *Proceedings of the Ninth International Workshop on Machine Learning*, pp. 249-256, Aberdeen, Scotland, Morgan Kaufmann, 1992

35. J. Casillas, O. Cordón, M. J. del Jesus, F. Herrera. Genetic Feature Selection in a Fuzzy Rule-Based Classification System Learning Process for High Dimensional Problems. *Information Science*, #136, pp. 169-191, 2001
36. F. Jelinek. *Statistical Methods for Speech Recognition*, MIT Press, Cambridge, 1998
37. P. Ladefoged. *A course in phonetics*, 1975
38. P. Ladefoged and I. Maddieson. *The sounds of the world's languages*, 1996
39. S. Chang, S. Greenberg and M. Wester. An elitist approach to articulatory-acoustic feature classification. In *Proceedings of Eurospeech 2001*, 2001
40. I. Kononenko. On biases in estimating multivalued attributes. In *Proceedings of the 14th International Joint Conference on Artificial Intelligence*, pp. 1034-1040, 1995
41. T.M. Cover and J.A. Thomas. *Elements of Information Theory*, John Wiley & Sons, New York, NY, USA, 1991
42. R. Silipo and M.R. Berthold. Discriminative power of input features in a fuzzy model. In *Proceedings of the Third International Symposium on Intelligent Data Analysis (IDA-99)*, D. Hand, J. Kok, M. Berthold, editors, *Advances in Intelligent Data Analysis" (IDA-99), Lecture Notes in Computer Science*, LNCS 1642, pp. 87-98, Springer-Verlag, 1999
43. S. King, T. Stephenson, S Isard, P. Taylor and A. Strachan. Speech recognition via phonetically featured syllables. In *Proceedings of the 5th International Conference on Spoken Language Processing (ICSLP'98)*, 1998
44. M. Richardson, J. Bilmes and C. Diorio. Hidden-articulator markov models for speech recognition In *Proceedings of ISCA ASR2000*, pp. 133-137, Paris, France, 2000
45. C.A. Miller, *Pronunciation Modeling in Speech Synthesis*, University of Pennsylvania, 1998
46. T. Sejnowski and C.R. Rosenberg. NETtalk: a parallel network that learns to read aloud. *Johns Hopkins University Technical Report*, JHU/EECS-86/01, 1986
47. T. Sejnowski and C.R. Rosenberg. Parallel networks that learn to pronounce English text. *Complex Systems*, #1, pp. 145-168, 1987
48. N. Chomsky and M. Halle. *The Sound Pattern of English*, MIT Press, 1968
49. L.M. Hyman. *Phonology: Theory and Analysis*, 1975
50. G. D. Forney. The Viterbi Algorithm. In *Proceedings of the IEEE*, pp. 268-278, 1973
51. S. B. Davis and P. Mermelstein. Comparison of parametric representations for monosyllabic word recognition in continuously spoken sentences. *IEEE Transactions on Acoustics, Speech, and Signal Processing*, vol. ASSP-28, #4, p. 357, 1980
52. K.P. Huber, M.R. Berthold. Building precise classifiers with automatic rule extraction. In *IEEE International Conference on Neural Networks*, Vol. 3, pp. 177-184, 1995
53. X. D. Huang, Y. Ariki and M. A. Jack. *Hidden Markov Models for Speech Recognition*, Edinburgh University Press, Edinburgh, 1990.

SECTION 4

COMPLEXITY REDUCTION IN PRECISE FUZZY MODELS

Interpretable Semi-Mechanistic Fuzzy Models by Clustering, OLS and FIS Model Reduction

Janos Abonyi[1], Hans Roubos[2], Robert Babuska[2], and Ferenc Szeifert[1]

[1] University of Veszprem, Department of Process Engineering, P.O. Box 158, H-8201, Hungary
[2] Delft University of Technology, Department of Information Technology and Systems, Systems and Control Engineering, P.O. Box 5031 2600 GA Delft, The Netherlands

Summary. A semi-mechanistic fuzzy modeling technique is proposed to obtain compact and transparent process models based on small data-sets. Semi-mechanistic models are hybrid models that consist of a white box structure based on mechanistic relationships and black-box substructures to model less defined parts. First, it is shown that certain type of white-box models can be efficiently incorporated into a Takagi-Sugeno fuzzy rule structure. Next, the proposed models are identified from learning data and special attention is paid to transparency and accuracy aspects. The approach is based on a combination of (i) prior knowledge-based model structures, (ii) fuzzy clustering, (iii) orthogonal least-squares, and (iv) the modified Fisher's interclass separability method. For the identification of the semi-mechanistic fuzzy model, a new fuzzy clustering method is proposed, i.e., clustering is achieved by the simultaneous identification of fuzzy sets defined on some of the scheduling variables and identification of the parameters of the local semi-mechanistic submodels. Subsequently, model reduction is applied to make the TS models as compact as possible, i.e., the most relevant consequent variables are selected by an orthogonal least squares method, and the modified Fisher's interclass separability criteria is used for selection of relevant antecedent (scheduling) variables. The overall procedure is demonstrated by the development of a semi-mechanistic model for a biochemical process. Although the results do not carry over directly to other engineering fields, the main ideas and conclusions, will certainly hold for other application areas as well.

1 Introduction

Fuzzy modeling and identification from process data proved to be effective for approximation of uncertain nonlinear processes [1]. The most frequently applied Takagi-Sugeno (TS) model decomposes the input space of the nonlinear model into fuzzy subspaces where the system is described by simple linear regression models [2]. Different approaches have been proposed to obtain such TS-fuzzy models from data. Most approaches, however, utilize only the function approximation capabilities of fuzzy systems, and little attention is paid to the qualitative aspects. This makes them less suited for applications in which emphasis is not only on accuracy, but also on interpretability, computational complexity and maintainability [3]. Furthermore, completely data-driven black-box identification techniques often yield unrealistic and non-interpretable models. This is typically due to an insufficient information content of the identification data and due to overparameterization of the models. Another disadvantage in process modeling is the non-scalability of black box models, i.e., one has to collect new training-data when the process is modified.

Due to the given drawbacks, combinations of *a priori* knowledge with black-box modeling techniques is gaining considerable interest. Two different approaches can be distinguished: grey-box modeling and semi-mechanistic modeling. In a grey-box model, *a priori* knowledge or information enters the black-box model as e.g. constraints on the model parameters or variables, the smoothness of the system behavior, or the open-loop stability [4–6]. For example, Lindskog and Ljung [5] applied combinations or (nonlinear) transformations of the input signals corresponding to physical variables and used the resulting signals in a black-box model. A major drawback of this approach is that it may suffer from the same drawbacks as the black-box model, i.e. no extrapolation is possible and time-varying processes remain a problem. On the other hand, transparency properties of fuzzy systems proved to be useful in the context of *grey-box* modeling, because it allows to effectively combine different types of information, namely linguistic knowledge, first-principle knowledge and information from data. For instance, if the model developer has *a priori* knowledge about steady-state or gain-independent dynamic behaviour of the process, a Hybrid Fuzzy Convolution Model [7] can be used as a combination of a fuzzy model and *a priori* knowledge based impulse response model of the system [8]. For the incorporation of prior knowledge into data-driven identification of dynamic fuzzy models of the Takagi-Sugeno type, a constrained identification algorithm has been developed [9]. This approach is based on a transformation of the *a priori* knowledge about stability, bounds on the stationary gains, and the settling time of the process into linear inequalities on the parameter set of the fuzzy model, similar to [4],[10]. This constrained identification is useful, because the TS fuzzy model is often overparameterized, hence explicit regularization, like penalties on non-smooth behaviour of the model and application of prior knowledge based parameter constraints can dramatically improve the robustness of the identification algorithm, eventually leading to more accurate parameter estimates [11].

One may also start by deriving a model based on first-principles and then include black-box elements as parts of the white-box model frame [12–16]. This modeling approach is usually denoted as *hybrid-modeling* or *semi-mechanistic modeling*. The latter term is used in the sequel because the first one is rather confusing with other methods. Johansen [16] describes various techniques to incorporate knowledge into black-box models. They use a local model structure and formulates the global identification as a nonlinear optimization problem. Thompson and Kramer [14] use the so-called paral-

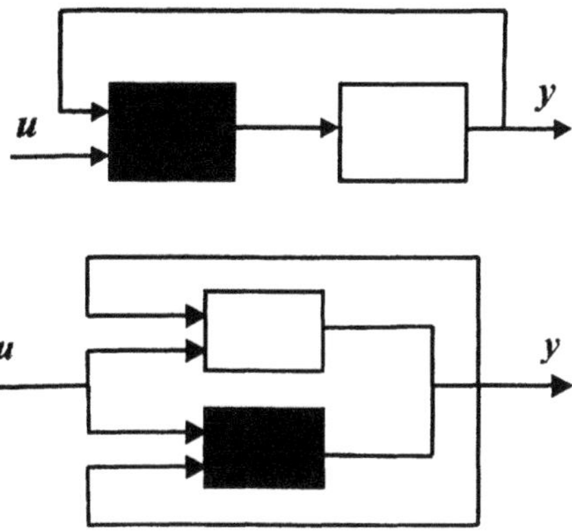

Fig. 1. Serial and parallel combinations of white and black box models.

lel approach of semi-mechanistic modeling (Fig. 1). Here, the error between the data and the white-box model is represented by a neural network. They also describe the serial approach where a neural network is used to model unknown parameters (Fig. 1). Several authors applied hybrid models for the modeling biotechnological systems [13,15,17,18]. Fuzzy models can also be incorporated into semi-physical models as linguistically interpretable black box elements [19]. A more complex semi-physical model was designed as a combination of a first-principles model, an artificial neural network and a fuzzy expert system for modeling a fed-batch cultivation [20]. The application of the fuzzy system was motivated by the fact that bioprocesses are often divided up into different phases (operating regimes) in which different mechanisms dominate. In many practical cases, the phase boundaries are not clear-cut boarders that can be described by crisp time instants. Hence, it is straightforward to use fuzzy techniques to determine in which phase the process is at a given time.

In this chapter, we also focus on semi-mechanistic models where a fuzzy model is used to represent, in a mechanistic sense, difficult-to-model parts of oure system. It will be show that fuzzy models can be efficiently incorporated into the semi-mechanistic modeling environment and we show also that this approach provides interpretable and accurate submodels. The proposed method is an extension to earlier work that was focused on TS-fuzzy model identification. In general, the bottleneck in data-driven identification of the fuzzy model is the identification of the model structure that requires nonlinear optimization. For this purpose often heuristic, data-driven approaches, like fuzzy clustering methods are applied, mainly for determining the rule antecedents of TS models [21]. In this chapter we propose a more advanced clustering method that pursues a further step in accomplishing the total parameter and structure identification of TS models. The clusters are represented by fuzzy sets and local semi-mechanistic models. The unknown parameters of the model are identified by expectation maximization (EM) [22] similarly to our new clustering algorithm: the modified Gath-Geva clustering [23].

Next, the obtained model is reduced by reducing the amount of antecedent variables and also the amount of consequent variables. Using too many antecedent variables may result in difficulties in the prediction and interpretability capabilities of the model due to redundancy, non-informative features and noise. Hence, selection of the scheduling variables is usually necessary. For this purpose, we modify our method that is based on Fischer interclass separability method and have been developed for feature selection of fuzzy classifiers [24]. Others have focused on reducing the antecedent by similarity analysis of the fuzzy sets [3,25], however this method is not very suitable for feature selection. Reduction of the consequent space is approached by an Orthogonal Least Squares (OLS) method. The application of orthogonal transforms for the reduction of the number of rules has received much attention in recent literature [26,27]. These methods evaluate the output contribution of the rules to obtain an importance ordering. In 1999 Yen and Wang investigated various techniques such as orthogonal least-squares, eigenvalue decomposition, SVD-QR with column pivoting method, total least square method and direct SVD method [26]. SVD based fuzzy approximation technique was initiated by Yam in 1997 [28], which directly finds the minimal number of rules from sampled values. Shortly after, this technique was introduced as SVD reduction of the rules and structure decomposition in [28]. For modeling purposes, the OLS is the most appropriate tool [26]. In this chapter, OLS is applied for a different purpose; the selection of the most relevant input and consequent variables based on the OLS analysis of the local models of the clusters.

The chapter is organized as follows. In Section 2, the semi-mechanistic model structure is presented. Section 3 describes the new clustering algorithm that allows for the direct identification of the models. Techniques for model reduction will be also given in this part. Section 4 presents an example, where the proposed approach is described by the development of a semi-mechanistic model for a biochemical process. Conclusions are given in Section 5.

2 Structure of the Semi-Mechanistic Fuzzy Model

2.1 Semi-Mechanistic modeling Approach

Generally, white box models of process systems are formulated by macroscopic balance equations, for instance, mass or energy balances. These balances are based on conservation principle that leads to differential equations written as

$$\begin{bmatrix}\text{accumulation}\\ \text{of } x_i\end{bmatrix} = \begin{bmatrix}\text{inflow}\\ \text{of } x_i\end{bmatrix} - \begin{bmatrix}\text{outflow}\\ \text{of } x_i\end{bmatrix} + \begin{bmatrix}\text{amount of}\\ x_i\text{generated}\end{bmatrix} + \begin{bmatrix}\text{amount of}\\ x_i\text{consumed}\end{bmatrix} \quad (1)$$

where x_i is a certain quantity, for example mass or energy.

Equation (1) can be simply formulated as commonly used state-space model of process systems given by

$$\begin{aligned} \dot{\boldsymbol{x}} &= f(\boldsymbol{x}, \boldsymbol{u}) \\ \boldsymbol{y} &= g(\boldsymbol{x}) \end{aligned} \quad (2)$$

where $\boldsymbol{x} = [x_1, \ldots, x_n]^T$ represents the state, and $\boldsymbol{y} = [y_1, \ldots, y_m]^T$ represents the output of the system, and f and g are nonlinear functions defined by the balance equations.

Besides the prediction properties of white box models, they also have the capability to explain the underlying mechanistic relationships of the process. These models are in general to a certain extent applicable independently of the process scale. Furthermore, these models are easy to modify or extend by changing parts of the model. White-box modeling is generally supported by experiments to get some of the parameters. In that case, even after scaling, new experiments are necessary because some of the parameters are difficult to express in a simple way as functions of process characteristics. In a practical situation, the construction of white-box models can be difficult because the underlaying mechanisms may be not completely clear, experimental results obtained in the laboratory do not carry over to practice, or parts of the white-box models are in fact not known.

In general, not all of the terms in (1) are exactly or even partially known. Hence, when we do not have detailed first principles knowledge about the process, the control law should be determined using approximations of functions of f and g. In semi-mechanistic modeling black-box models, like neural networks are used to represent the otherwise difficult-to-obtain parts of the model. The semi-mechanistic modeling strategy is be combined quite naturally with the general structure of white-box models in process systems, since this structure is usually based on macroscopic balances (e.g. mass, energy or momentum). These balances specify the dynamics of the relevant state variables and contain different rate terms. Some of these terms are directly associated with manipulated or measured variables (e.g. in- and out-going flows) and do not have to be modeled any further. In contrast, some rate terms (e.g. reaction rate) have a static mathematical relation with one or more state variables which should be modeled in order to obtain a fully specified model. These terms are then considered as inaccurately known terms. They can be modeled in a white-box way if a static mathematical relation for the rate terms can be based on easy obtainable first principles. If this is not possible, they can be modeled in a black-box way with a nonlinear modeling technique. In the latter case, one obtains the semi-mechanistic model configuration. One of the advantages of the semi-mechanistic modeling strategy is that it seems to be more promising with respect to extrapolation properties [18]. The resulting semi-mechanistic, also called hybrid model can be formulated similarly to (2)

$$\begin{aligned} \dot{\boldsymbol{x}} &= f_{FP}(\boldsymbol{x}, \boldsymbol{u}, \boldsymbol{\theta}, f_{BB}(\boldsymbol{x}, \boldsymbol{u}, \boldsymbol{\theta})) \\ \boldsymbol{y} &= g(\boldsymbol{x}) \end{aligned} \tag{3}$$

where f_{FP} represents the first-principle part of the model and f_{BB} the black-box part parameterized by the θ parameter vector.

Usually, the reaction rates (the amount of generated and consumed materials in chemical reactions) and the thermal effects (e.g. reaction heat) are especially difficult to model, but the first two transport terms (inlet and outlet flows) in (1) can be obtained easily and accurately. Hence, in the modeling phase it turns out which parts of the first principles model are easier and which are more laborious to obtain and often we can get the following hybrid model structure

$$\dot{\boldsymbol{x}} = f_{FP1}(\boldsymbol{x}, \boldsymbol{u}) + f_{FP2}(\boldsymbol{x}) \odot f_{BB}(\boldsymbol{x}, \boldsymbol{\theta}) \tag{4}$$

where $f_{FP2}(\boldsymbol{x}) \odot f_{BB}(\boldsymbol{x}, \boldsymbol{\theta})$ represents the reaction and thermal effects of the system that can be modelled as a function of the measured state variables, where $\odot$ denotes element-wise product.

Example 1: In this chapter, we use an example process from the biochemical industry. Here, most processes are carried out in a batch or fed-batch reactor, e.g., production of beer, penicillin and bakers yeast. These types of processes are characterized by their time-varying nature and their recipe-based processing. Furthermore, the measurement of all relevant state-variables is either impossible or very expensive. The on-line control of these processes is therefore usually limited to the setpoint-based control of several process-parameters like temperature, pH, and gasflows. The global process behavior is controlled by process operators that apply standard recipes, stating what to do at a certain time instant. These recipes are usually based on the available process experience and in some cases these are fine-tuned by by neural-networks. This method highly depends on the gained expertise and large amounts of data. On the other hand, when accurate process models can be developed based on small amounts of data, one may use these to calculate model-based recipes or control strategies. Besides, one can apply these models to predict process behavior at different scales or optimize the process lay-out, e.g., bioreactor geometry, batch scheduling, etc.

Modeling of batch and fed-batch processes is difficult due to the in general nonlinear characteristics and time-varying dynamics. Moreover, often reaction mechanisms and orders are unknown and have to be postulated and experimentally verified. Fortunately, however, these processes can be partly described by mass-balances that are easily obtained from general process knowledge. These mass balances describe the mass flows and dilution effects. The unknown parts, are the reaction kinetics, which are the conversion rates in the process. Based on this structure, we develop a hybrid model structure for a simulated fed-batch bioprocess. We use a simple model for biomass growth on a single substrate:

$$\begin{aligned} \frac{dx_1}{dt} &= \frac{u}{x_3}\left(x_{1,in} - x_1\right) - \sigma(\boldsymbol{x})x_2 \\ \frac{dx_2}{dt} &= -\frac{u}{x_3}\left(x_2\right) + \mu(\boldsymbol{x})x_2 \\ \frac{dx_3}{dt} &= u \end{aligned} \tag{5}$$

where x_1 is the concentration of substrate and x_2 is the biomass concentration of substrate, x_3 the volume of the liquid phase, $x_{1,in}$ is the substrate concentration the volumetric feed rate u, $\sigma(\boldsymbol{x})$ the specific substrate consumption rate and $\mu(\boldsymbol{x})$ the overall specific growth rate. Generally, the kinetic submodels $\sigma(\boldsymbol{x})$ and $\mu(\boldsymbol{x})$, are unknown, and therefore the following hybrid model of the process can be obtained:

$$\dot{\boldsymbol{x}} = \underbrace{\begin{bmatrix} \frac{u}{x_3}\left(x_{1,in} - x_1\right) \\ -\frac{u}{x_3}\left(x_2\right) \\ u \end{bmatrix}}_{f_{FP1}(\boldsymbol{x},\boldsymbol{u})} + \underbrace{\begin{bmatrix} -x_2 \\ x_2 \\ 0 \end{bmatrix}}_{f_{FP2}(\boldsymbol{x})} \odot \underbrace{\begin{bmatrix} \sigma(\boldsymbol{x}) \\ \mu(\boldsymbol{x}) \\ 0 \end{bmatrix}}_{f_{BB}(\boldsymbol{x})} \tag{6}$$

2.2 Operating Regime Based Semi-Mechanistic Model

In the given example, difficulties in modeling $\sigma(\boldsymbol{x})$ and $\mu(\boldsymbol{x})$ are due to lack of knowledge and understanding of relations between kinetic behavior and process states. To handle such uncertainties one can develop nonlinear data-based models. One specific class of nonlinear models with several special characteristics will be applied here, namely the *operating regime based model* [29]. Here, the global model is decomposed in several simple local models that are valid for predefined operating regions. For this, one should should examine what are the important states that can be applied for specification of operating points of the submodels and what are the most relevant effectors. This type of local understanding, in fact, is a key to identification of reliable local models based on a limited amount of data.

In the operating regime based modeling approach, $f_{BB}(\boldsymbol{x})$ is defined by local linear models:

$$\boldsymbol{A}_i\boldsymbol{x} + \boldsymbol{b}_i, \tag{7}$$

where the $\boldsymbol{A}_i$-matrix and $\boldsymbol{b}_i$-vector define the parameters of the i-th local linear model; $i = 1, \ldots, c$. The global black-box model f_{BB} is then given by:

$$f_{BB}(\boldsymbol{x}) = \sum_{i=1}^{c} \beta_i(\boldsymbol{z})\left(\boldsymbol{A}_i\boldsymbol{x} + \boldsymbol{b}_i\right), \tag{8}$$

where the $\beta_i(\boldsymbol{z})$-function describes the i-th operating region defined by the "scheduling" vector $\boldsymbol{z} = [z_1, \ldots, z_{nz}] \in \boldsymbol{x}$, that is usually a subset of the state variables; an example is given in Fig. 2 where overlapping operating regions of four local models are defined by two state variables. Integration of the

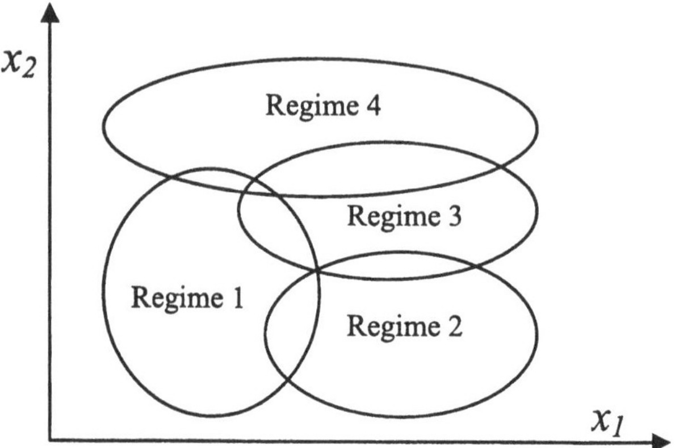

Fig. 2. Example of an operating regime based model decomposed into four regimes.

operating regime based model into hybrid model (4) results in the following formulation:

$$\dot{\boldsymbol{x}} = \sum_{i=1}^{c} \beta_i(\boldsymbol{z}) \left(f_{FP1}(\boldsymbol{x}, \boldsymbol{u}) + (\boldsymbol{A}_i \boldsymbol{x} + \boldsymbol{b}_i) \odot f_{FP2}(\boldsymbol{x}) \right) \tag{9}$$

Example I, Cont'd: We consider the process introduced in the previous section. The growth rate $\mu(\boldsymbol{x})$ and substrate consumption rate $\sigma(\boldsymbol{x})$ are probably defined by the biomass and substrate concentrations, however the relationship is unknown.Therefore, we can choose the scheduling vector as $\boldsymbol{z} = [x_1, x_2]^T$. This results in the semi-mechanistic:

$$\dot{\boldsymbol{x}} = \sum_{i=1}^{c} \beta_i(\boldsymbol{z}) \left(\begin{bmatrix} \frac{u}{x_3}(x_{1,in} - x_1) \\ -\frac{u}{x_3}(x_2) \\ u \end{bmatrix} + \right.$$

$$\left. \left(\underbrace{\begin{bmatrix} a_i^{1,1} & a_i^{1,2} & 0 \\ a_i^{2,1} & a_i^{2,2} & 0 \\ 0 & 0 & 0 \end{bmatrix}}_{\boldsymbol{A}_i} \begin{bmatrix} x_1 \\ x_2 \\ x_3 \end{bmatrix} + \underbrace{\begin{bmatrix} b_i^1 \\ b_i^2 \\ 1 \end{bmatrix}}_{\boldsymbol{b}_i} \right) \odot \begin{bmatrix} -x_2 \\ x_2 \\ 0 \end{bmatrix} \right) \tag{10}$$

2.3 Semi-Mechanistic Model Representation by Fuzzy Rules

The operating regime based modeling framework provides a method to accurately model nonlinear systems by an efficient interpolation of local linear models defined by operating regimes. In addition, it provides a transparent structure that can be represented by fuzzy rules, because the operating regimes can be given by fuzzy sets [30].

Such a fuzzy representation is appealing, since many systems show smooth changes in their dynamics as a function of the operating point, and the soft transition between the regimes introduced by the fuzzy set representation captures this feature in an elegant fashion. Hence, the entire global model can be conveniently represented by Takagi-Sugeno fuzzy rules [2]. Our previous hybrid model (9) can thus be given by multi-input multi-output (MIMO) Takagi-Sugeno fuzzy model rules:

$$\begin{aligned} R_i : \textbf{If } & z_1 \text{ is } A_{i,1} \textbf{ and } \ldots \textbf{ and } z_{nz} \text{ is } A_{i,nz} \textbf{ then} \\ & \dot{\boldsymbol{x}}_i = f_{FP1}(\boldsymbol{x}, \boldsymbol{u}) + (\boldsymbol{A}_i \boldsymbol{x} + \boldsymbol{b}_i) \odot f_{FP2}(\boldsymbol{x}), \;\; [w_i], \end{aligned} \tag{11}$$

where $A_{i,j}(z_j)$ is the i-th antecedent fuzzy set for the j-th input and $w_i = [0, 1]$ is the rule-weight that represents the desired impact of the i-th rule. The value of w_i is often chosen by the designer of the fuzzy system based on his or her belief in the goodness and accuracy of the i-th rule. When such knowledge is not available w_i is set as $w_i = 1, \forall i$.

Gaussian membership function are used in this chapter to represent the $A_{i,j}(z_j)$ fuzzy sets:

$$A_{i,j}(z_j) = \exp\left(-\frac{1}{2} \frac{(z_j - v_{i,j})^2}{\sigma_{i,j}^2} \right), \tag{12}$$

where $v_{i,j}$ represents the center and $\sigma_{i,j}^2$ the variance of the Gaussian function. The overall output of the MIMO TS-fuzzy model is inferred by computing a weighted average of the outputs of the consequent multivariable models:

$$\dot{\boldsymbol{x}} = \sum_{i=1}^{c} \beta_i(\boldsymbol{z}) \dot{\boldsymbol{x}}_i, \tag{13}$$

where c is the number of rules, and β_i the degree of fullfilment of the i-th rule, calculated by:

$$\beta_i(\boldsymbol{z}) = \frac{w_i \prod_{j=1}^{nz} A_{j,i}(z_j)}{\sum_i^c w_i \prod_{j=1}^{nz} A_{j,i}(z_j)}. \tag{14}$$

The fuzzy model can be formulated in the compact regression form:

$$\dot{\boldsymbol{x}}(k)^T = f_{FP1}(\boldsymbol{x}(k), \boldsymbol{u}(k))^T + \sum_{i=1}^{c} \beta_i(\boldsymbol{z}(k)) \boldsymbol{\phi}(k) \boldsymbol{\theta}_i^T, \tag{15}$$

where k denotes the index of the k-th data-pair, $\phi(k)$ the regressor matrix:

$$\phi(k) = [\boldsymbol{x}(k) \odot f_{FP2}(\boldsymbol{x}(k)) \;\; f_{FP2}(\boldsymbol{x}(k))],$$

and $\boldsymbol{\theta}_i$ the parameter matrix of the i-th local model:

$$\boldsymbol{\theta}_i = [\boldsymbol{A}_i, \boldsymbol{b}_i].$$

From (15) it is clear that the developed semi-mechanistic fuzzy model is linear in the consequent parameters of the fuzzy model. This allows for an effective identification of these parameters. It is interesting to see that the resulting model is a TS-fuzzy model with *semi-physical regressors*.

Next, we rewrite the MIMO fuzzy model as a set of MISO fuzzy models with same antecedents. This representation helps in the model analysis and interpretation because the input-output channels can be separately analyzed.

$$\begin{aligned} R_{i,j} &: \textbf{If } z_1 \text{ is } A_{i,1} \textbf{ and } \ldots \textbf{ and } z_{nz} \text{ is } A_{i,nz} \textbf{ then} \\ &\quad \dot{x}_{i,j}(k) = f_{Hyb,j}(\boldsymbol{x}(k), \boldsymbol{u}(k)), [w_i], \end{aligned} \tag{16}$$

where $j = 1, \ldots, n$ denotes the index of the state variable whose predicted value is calculated and $f_{Hyb,j}$ represents the corresponding rule consequent function of the proposed hybrid fuzzy model.

Note that this hybrid fuzzy model is much more parsimonious than the ordinary Takagi-Sugeno model of the process that is formulated as:

$$\begin{aligned} R_i &: \textbf{If } x_1 \text{ is } A_{i,1} \textbf{ and } \ldots \textbf{ and } x_n \text{ is } A_{i,n} \textbf{ then} \\ &\quad \dot{\boldsymbol{x}}_i = \widehat{\boldsymbol{A}}_i \boldsymbol{x} + \widehat{\boldsymbol{B}}_i \boldsymbol{u} + \widehat{\boldsymbol{c}}_i, \; [w_i]. \end{aligned} \tag{17}$$

The latter model has much more antacedent and consequent parameters because the sizes of $\widehat{\boldsymbol{A}}_i$, $\widehat{\boldsymbol{B}}_i$, and $\widehat{\boldsymbol{c}}_i$ are bigger than those of $\boldsymbol{A}_i$ and $\boldsymbol{b}_i$. Furthermore, the fuzzy model can be independently interpreted and analyzed from the first-principle part of the model since *the output of the fuzzy model has physical meaning!*

Example I, Cont'd: The operating regime based model of our bioreactor (10) can now be formulated by fuzzy rules of the form:

$$R_i : \textbf{If } x_1 \text{ is } A_{i,1} \textbf{ and } x_2 \text{ is } A_{i,2} \textbf{ then } \dot{\boldsymbol{x}} = \begin{bmatrix} \frac{u}{x_3}(x_{1,in} - x_1) \\ -\frac{u}{x_3}(x_2) \\ u \end{bmatrix} +$$

$$\left(\underbrace{\begin{bmatrix} a_i^{1,1} & a_i^{1,2} & 0 \\ a_i^{2,1} & a_i^{2,2} & 0 \\ 0 & 0 & 0 \end{bmatrix}}_{\boldsymbol{A}_i} \begin{bmatrix} x_1 \\ x_2 \\ x_3 \end{bmatrix} + \underbrace{\begin{bmatrix} b_i^1 \\ b_i^2 \\ 1 \end{bmatrix}}_{\boldsymbol{b}_i} \right) \odot \begin{bmatrix} -x_2 \\ x_2 \\ 0 \end{bmatrix}, [w_i]. \tag{18}$$

Now, the input-output channels can be separately analyzed, e.g., the fuzzy rule that describes the change of the first state of the system (substrate concentration), can be represented by a fuzzy model with rules:

$$R_i : \textbf{If } x_1 \text{ is } A_{i,1} \textbf{ and } x_2 \text{ is } A_{i,2} \textbf{ then } \dot{x}_{i,1}(k) = f_{Hyb,j}(\boldsymbol{x}(k), u(k)) = \frac{u(k)}{x_3(k)}(x_{1,in} - x_1(k)) + \left(a_i^{1,1}x_1(k) + a_i^{2,1}x_2(k) + b_i^1\right)(-x_2(k)) \quad (19)$$

The fuzzy models can now be independently interpreted and analyzed from the first-principle part of the model. For instance, the first output channel of the fuzzy bioprocess model can be written as:

$$R_i : \textbf{If } x_1 \text{ is } A_{i,1} \textbf{ and } x_2 \text{ is } A_{i,2} \textbf{ then } \sigma(\boldsymbol{x})_i = a_i^{1,1}x_1 + a_i^{1,2}x_2 + b_i^1, \ [w_i]. \quad (20)$$

3 Clustering-based Identification of the Semi-Mechanistic Fuzzy Model

The remaining part of this chapter deals with the identification of the presented hybrid structures. For this purpose we developed a clustering based identification method, with subsequent reduction of antecendent and consequent variables.

3.1 Problem Formulation

In general, measured input-output data are used in data-driven model identification. In many industrial cases it may be difficult to obtain large datasets. Moreover, the measurements may contain a high noise level and is often be asynchronous, which is often the case for bioprocesses. In this case, $\dot{\boldsymbol{x}}(k)$ cannot be be calculated by using the simple forward Euler discretization:

$$\dot{\boldsymbol{x}}(k) = \frac{(\boldsymbol{x}(k) - \boldsymbol{x}(k-1))}{\Delta T}. \quad (21)$$

Generation of synchronous data by an interpolation is often a good solution to deal with this problem. We applied Hermite interpolation to interpolate between the measured data and estimate the corresponding derivatives. This method is described in Appendix A.1.

After application of this spline-smoothing, and with the use of the first principle model parts, N semi-physical identification data pairs are obtained that can be arranged into the following matrices based on (15):

$$\boldsymbol{\phi} = [\phi(1)|\phi(2)|\cdots|\phi(N)]^T, \quad (22)$$

$$\boldsymbol{f_{FP1}} = [f_{FP1}(1)|f_{FP1}(2)|\cdots|f_{FP1}(N)]^T, \quad (23)$$

$$\dot{\boldsymbol{X}} = [\dot{\boldsymbol{x}}(1)|\dot{\boldsymbol{x}}(2)|\cdots|\dot{\boldsymbol{x}}(N)]^T. \quad (24)$$

3.2 Identification of the Consequent Parameters

The output of the proposed semi-mechanistic model is linear in the elements of the $\boldsymbol{A}_i$-consequent matrices and the $\boldsymbol{b}_i$-offset vector. Therefore, if the membership functions are given, these parameters can be estimated from input-output process data by a linear least-squares techniques based on the calculated degree of fulfillments:

$$\mathcal{B}_i = \begin{bmatrix} \beta_i(1) & 0 & \cdots & 0 \\ 0 & \beta_i(2) & \cdots & 0 \\ \vdots & \vdots & \ddots & \vdots \\ 0 & 0 & \cdots & \beta_i(N) \end{bmatrix}. \tag{25}$$

The local (weighted least squares for each rule) identification of the rule consequent parameters forces the local linear models to fit the systems separately and locally, i.e., resulting rule consequences are local linearizations of the nonlinear system [21]. Hence, in this chapter the parameters of the rule consequences are estimated separately by dividing the identification task into c weighted least-squares problems. By using this notation, the weighted least squares solution of $\boldsymbol{\theta}_i$ is:

$$\boldsymbol{\theta}_i = \left[\boldsymbol{\phi}^T \mathcal{B}_i \boldsymbol{\phi}\right]^{-1} \boldsymbol{\phi}^T \mathcal{B}_i \left(\dot{\boldsymbol{X}} - \boldsymbol{f_{FP1}}\right). \tag{26}$$

Using too many consequent variables results in difficulties in the prediction and interpretability capabilities of the fuzzy model due to redundancy, non-informative features and noise. To avoid these problems in the following the application of orthogonal least squares based model reduction is proposed.

3.3 Reduction of the Number of Consequent Parameters

The least squares identification of $\boldsymbol{\theta}_i$, (26) can be also formulated as:

$$\theta_i = \boldsymbol{B}^+ \left(\dot{\boldsymbol{X}} - \boldsymbol{f_{FP1}}\right) \sqrt{\mathcal{B}_i} \tag{27}$$

where $\boldsymbol{B}^+$ denotes the Moore-Penrose pseudo inverse of $\boldsymbol{\phi}^T \sqrt{\mathcal{B}_i}$.

The OLS method transforms the columns of $\boldsymbol{B}$ into a set of orthogonal basis vectors in order to inspect the individual contribution of each variable. To do this Gram-Schmidt orthogonalization of $\boldsymbol{B} = \boldsymbol{W}\boldsymbol{A}$ is used, where $\boldsymbol{W}$ is an orhogonal matrix $\boldsymbol{W}^T\boldsymbol{W} = \boldsymbol{I}$ and $\boldsymbol{A}$ is an upper triangular matrix with unity diagonal elements. If $\boldsymbol{w}_i$ denotes the i-th column of $\boldsymbol{W}$ and g_i is the corresponding element of the OLS solution vector $\boldsymbol{g} = \boldsymbol{A}\theta_i$, the output variance $(\left(\dot{\boldsymbol{X}} - \boldsymbol{f_{FP1}}\right)\sqrt{\mathcal{B}_i})^T(\left(\dot{\boldsymbol{X}} - \boldsymbol{f_{FP1}}\right)\sqrt{\mathcal{B}_i})/N$ can be explained by the regressors $\sum_{i=1}^{n_r} g_i \boldsymbol{w}_i^T \boldsymbol{w}_i / N$. Thus, the error reduction ratio, ϱ, due to an individual rule i can be expressed as

$$\varrho^i = \frac{g_i^2 \boldsymbol{w}_i^T \boldsymbol{w}_i}{\left(\dot{\boldsymbol{X}} - \boldsymbol{f_{FP1}}\right)\sqrt{\mathcal{B}_i})^T(\left(\dot{\boldsymbol{X}} - \boldsymbol{f_{FP1}}\right)\sqrt{\mathcal{B}_i})} \tag{28}$$

This ratio offers a simple mean for ordering the consequent variables, and can be easily used to select a subset of the inputs in a forward-regression manner.

3.4 Semi-Mechanistic Model-based Clustering

In the previous section, we showed that the unknown consequent parameters of the semi-mechanistic fuzzy model are linear in the output. Therefore, a (weighed) least squares method can be applied for the identification of these parameters when the antecedent membership functions (rule-weights) are given. Determination of the antecedent part of the fuzzy model, however, is more difficult because it requires nonlinear optimization. Hence, often heuristic approaches, like fuzzy clustering are applied for this purpose.

In fuzzy clustering, the objective is to partition the identification data into c clusters. Here, the fuzzy partition is represented by the $\boldsymbol{U} = [\mu_{i,k}]_{c \times N}$-matrix, where the $\mu_{i,k}$-element represents the degree of membership of each kth datapoint to the i-th cluster, and the clusters are parameterized by the η_i-set of parameters.

Different cluster shapes are obtained by clustering algorithm based on different prototypes, e.g., based on point or linear varieties (FCV) or with different distance measures. Often, the Gustafson-Kessel clustering algorithm, with hyper-elipsoid prototypes, is applied for fuzzy model identification [31]. However, a drawback of this algorithm is that mainly clusters with equal volumes are found. We apply the method introduced by Gath and Geva [32] who interpret the data as normally distributed random variables and assume a normal (Gaussian) distribution with excepted value $\boldsymbol{v}_i$ and covariance matrix $\boldsymbol{F}_i$ to be chosen for generating the datum with *a priori* probability $p(\eta_i)$. Hence, the Gath-Geva clustering algorithm is identical to the Expectation Maximization (EM) identification of mixture of Gaussians model.

A main disadvantage of clustering tools for the identification of interpretable Takagi-Sugeno fuzzy models is that the fuzzy covariance matrix is defined among all of the model inputs and the correlation among all the input variables. This causes difficulties for the extraction of membership functions from the clusters. This can be done by membership projection methods [33] or a transformed input approach [34]. Generally, however, this decreases the interpretability and accuracy of the initial fuzzy model based on the multivariable cluster representation. Furthermore, the decomposed univariate membership functions are defined on the domains of all input variables of the model $\boldsymbol{x}$ instead of only the domain of the scheduling variables, $\boldsymbol{z}$. This makes the model unnecessarily complex due to a large number of fuzzy sets which decreases the models transparency.

To avoid these problems, we propose a new clustering algorithm that generates hybrid fuzzy models in a direct approach. This algorithm obtains a set of clusters, that all contain an input distribution (the distribution of the scheduling variables, $p(\boldsymbol{z}(k)|\eta_i)$, a local hybrid model $\dot{\boldsymbol{x}}(k) =$

$f_{Hyb,i}(\boldsymbol{x}(k),\boldsymbol{u}(k))$, and an output distribution, $p(\dot{\boldsymbol{x}}(k)|\boldsymbol{x}(k),\boldsymbol{u}(k),\eta_i)$, where

$$\begin{aligned} p(\boldsymbol{x}(k),\boldsymbol{u}(k),\dot{\boldsymbol{x}}(k)) &= \sum_{i=1}^{c} p(\boldsymbol{x}(k),\boldsymbol{u}(k),\dot{\boldsymbol{x}}(k),\eta_i), \\ &= \sum_{i=1}^{c} p(\boldsymbol{x}(k),\boldsymbol{u}(k),\dot{\boldsymbol{x}}(k)|\eta_i)p(\eta_i), \\ &= \sum_{i=1}^{c} p(\dot{\boldsymbol{x}}(k)|\boldsymbol{x}(k),\boldsymbol{u}(k),\eta_i)p(\boldsymbol{z}(k)|\eta_i)p(\eta_i). \end{aligned} \tag{29}$$

The input distribution is parameterized as an unconditioned Gaussian [35] and defines the domain of influence of a cluster:

$$p(\boldsymbol{z}(k)|\eta_i) = \frac{\exp\left(-\frac{1}{2}(\boldsymbol{z}(k)-\boldsymbol{v}_i)^T(\boldsymbol{F}_i^{zz})^{-1}(\boldsymbol{z}(k)-\boldsymbol{v}_i)\right)}{(2\pi)^{\frac{n_z}{2}}\sqrt{|\boldsymbol{F}_i^{zz}|}}, \tag{30}$$

while the output distribution is taken to be:

$$p(\dot{\boldsymbol{x}}(k)|\boldsymbol{x}(k),\boldsymbol{u}(k),\eta_i) = \tag{31}$$

$$\frac{\exp\left(-(\dot{\boldsymbol{x}}(k)-f_{Hyb,i}(\boldsymbol{x}(k),\boldsymbol{u}(k)))^T(\boldsymbol{F}_i^{xx})^{-1}(\dot{\boldsymbol{x}}(k)-f_{Hyb,i}(\boldsymbol{x}(k),\boldsymbol{u}(k)))\right)}{(2\pi)^{\frac{n}{2}}\sqrt{|\boldsymbol{F}_i^{xx}|}}, \tag{32}$$

where $\boldsymbol{F}^{zz}$ and $\boldsymbol{F}^{xx}$ are covariance matrices calculated as:

$$\boldsymbol{F}_i^{zz} = \frac{\sum\limits_{k=1}^{N}(\boldsymbol{z}(k)-\boldsymbol{v}_i)(\boldsymbol{z}(k)-\boldsymbol{v}_i)^T p(\eta_i|\boldsymbol{x}(k),\boldsymbol{u}(k),\dot{\boldsymbol{x}}(k))}{\sum\limits_{k=1}^{N} p(\eta_i|\boldsymbol{x}(k),\boldsymbol{u}(k),\dot{\boldsymbol{x}}(k))}, \tag{33}$$

$$\boldsymbol{F}_i^{xx} = \frac{\sum\limits_{k=1}^{N}(\dot{\boldsymbol{x}}(k)-f_{Hyb,i}(\boldsymbol{x}(k),\boldsymbol{u}(k)))}{\sum\limits_{k=1}^{N} p(\eta_i|\boldsymbol{x}(k),\boldsymbol{u}(k),\dot{\boldsymbol{x}}(k))} \cdot \frac{(\dot{\boldsymbol{x}}(k)-f_{Hyb,i}(\boldsymbol{x}(k),\boldsymbol{u}(k)))^T p(\eta_i|\boldsymbol{x}(k),\boldsymbol{u}(k),\dot{\boldsymbol{x}}(k))}{\sum\limits_{k=1}^{N} p(\eta_i|\boldsymbol{x}(k),\boldsymbol{u}(k),\dot{\boldsymbol{x}}(k))}, \tag{34}$$

in which $p(\eta_i|\boldsymbol{x}(k),\boldsymbol{u}(k),\dot{\boldsymbol{x}}(k))$ are posterior probabilities that can be interpreted as the probability that a particular piece of data has been generated by a particular cluster. By using Bayes Theorem,

$$\begin{aligned} p(\eta_i|\boldsymbol{x}(k),\boldsymbol{u}(k),\dot{\boldsymbol{x}}(k)) &= \frac{p(\boldsymbol{x}(k),\boldsymbol{u}(k),\dot{\boldsymbol{x}}(k)|\eta_i)p(\eta_i)}{p(\boldsymbol{x}(k),\boldsymbol{u}(k),\dot{\boldsymbol{x}}(k))} \\ &= \frac{p(\boldsymbol{x}(k),\boldsymbol{u}(k),\dot{\boldsymbol{x}}(k)|\eta_i)p(\eta_i)}{\sum_{i=1}^{c} p(\dot{\boldsymbol{x}}(k)|\boldsymbol{x}(k),\boldsymbol{u}(k),\eta_i)p(\boldsymbol{z}(k)|\eta_i)p(\eta_i)}. \end{aligned} \tag{35}$$

When simplicity and interpretability of the model are important, then the F^{xx}-cluster weighted covariance matrix is reduced to its diagonal elements, resulting in the following density function:

$$p(z(k)|\eta_i) = \prod_{j=1}^{n} \frac{1}{\sqrt{2\pi\sigma_{i,j}^2}} \exp\left(-\frac{1}{2}\frac{\left(z_j(k) - v_{i,j}\right)^2}{\sigma_{i,j}^2}\right). \tag{36}$$

Identification of the model thus means determination of the $\eta_i = \{p(\eta_i), v_i, F_i^{zz}, \theta_i, F_i^{xx}\}$-parameters of the clusters, where θ_i represents the unknown parameters of the hybrid local models. In the appendix the Expectation Maximization (EM) identification of the model is that was reformulated in the form of fuzzy clustering, is presented.

Note that the application of (36) results in the direct determination of the parameters of the membership functions, as (36)is identical to (12), because the weight of the rule are determined as

$$w_i = \frac{p(\eta_i)}{(2\pi)^{n/2}\sqrt{|F_i^{xx}|}} \tag{37}$$

or equivalently,

$$w_i = \frac{p(\eta_i)}{\sqrt{2\pi\sigma_{i,j}^2}} \tag{38}$$

3.5 Reduction of the Number of Fuzzy Sets

In general, application of too many features and fuzzy sets decreases the models interpretability. Hence, feature selection is usually necessary. For this purpose, we modified Fischer's interclass separability method, that was applied for feature selection of labeled data in [24]. The method makes a importance ranking based on the interclass separability criterion which is obtained from statistical properties of the data. This criterion is based on the F_B between-class and the F_W within-class covariance matrices that sum up to the total covariance of the training data F_T, with:

$$\begin{aligned} F_W &= \sum_{i=1}^{c} p(\eta_i) F_i, \\ F_B &= \sum_{i=1}^{c} p(\eta_i)\left(v_i - v_0\right)^T\left(v_i - v_0\right), \\ v_0 &= \sum_{i=1}^{c} p(\eta_i) v_i. \end{aligned} \tag{39}$$

The feature interclass seperatibility selection criterion is then given as a trade-off between F_W and F_B:

$$J = \frac{\det F_B}{\det F_W}. \tag{40}$$

The importance of each feature is measured by leaving the feature out and calculating J for the reduced covariance matrices. Subsequent feature selection is done in an iterative way by leaving out the least needed feature in each iteration. Note that this method, in case of axis-parallel clusters (the fuzzy sets are directly obtained from such clusters), shows some similarities to similarity-based model reduction techniques [36].

4 Application Study

4.1 Process Description

We will demonstrate the proposed approach for identification of the previously presented bio-reactor example. Here, we assume that the *true* nonlinear kinetics are given by a previously published nonlinear process model. In this model, the two specific rates $\sigma(\boldsymbol{x})$ and $\mu(\boldsymbol{x})$ are inter-related by the following law:

$$\sigma(\boldsymbol{x}) = \frac{1}{Y_{x_2/x_1}}\mu(\boldsymbol{x}) + m, \tag{41}$$

where Y_{x_2/x_1} the biomass on substrate yield coefficient and m the specific maintenance demand, and the $\mu(\boldsymbol{x})$ specific growth rate was described by non-monotonic Haldane kinetics:

$$\mu(\boldsymbol{x}) = \mu_m \frac{x_1}{K_p + x_1 + \frac{x_1^2}{K_i}}, \tag{42}$$

where the K_p parameter relates the substrate level to the specific growth rate μ, together with the K_i inhibition parameter, which adds a negative substrate level effect. Nominal values of the model parameters are given in Table 1.

Table 1. Kinetic and operating parameter values

μ_m	2.1	h^-1
K_p	10	g/l
K_i	0.1	g/l
Y_{x_2/x_1}	0.47	$g\ DW/g$
m	0.29	$g\ DW/g\ h$
$x_{1,in}$	500	g/l
$x_3(0)$	7	l

The identification data has been gained from normal operation of the process and five batches have been simulated to obtain process data. Note that these type of laboratory experiments are in reality expensive and time-consuming, which makes such a low number of experiments very realistic.

Others, however, have used simulation studies with more than hundred simulated batches [37]. Each batch experiment was run for max. 16 *h.*, and all states were measured each 30 *min*. For every batch, a simple constant feeding strategy has been applied. In the five experiments we varied initial conditions $(x_1(0), x_2(0))$ and the feeding rate u. To make the simulation experiments more realistic, gaussian white noise on the measurements of the substrate concentration x_1 and on the biomass concentration x_2 was considered; variances of the measurement errors are given by $\sigma^2_{x_1} = 10^-2g^2/l^2$ and $\sigma^2_{x_2} = 6.25^-4g^2/l^2$, respectively.

In the next step, our spline-smoothing algorithm (see Appendix A-2) has been used to generate smoothed training data. Four knots were used to generate the splines; one at time zero and the others placed at 60%, 80% and 100% of the time till substrate concentration became zero, $x_1 = 0$. These knots were determined such that they provide a good fit at small substrate concentrations realized at the end of the batches where the effect of the process nonlinearity is the biggest. From every batch, 100 data pairs were sampled from the spline-interpolation and the change of the state variables was estimated from the analytic derivatives of the splines. This procedure resulted in 500 data points from the five batches.

4.2 Semi-Mechanistic Model of the Process

Based on the collected identification data, we first identified a standard MIMO TS model with eight rules. The previously presented fuzzy clustering algorithm has been applied without prior knowledge. To illustrate the complexity of the obtained fuzzy model, we show the first rule:

R_1 : **If** x_1 is $A_{1,1}$ **and** x_2 is $A_{1,2}$ **then**

$$\dot{x}_1 = \underbrace{\begin{bmatrix} -0.0875 & -0.4927 & 0.0569 \\ 0.0381 & 0.0937 & -0.0013 \\ 0.0000 & 0.0000 & 0.0000 \end{bmatrix}}_{\widehat{A}_1} \begin{bmatrix} x_1 \\ x_2 \\ x_3 \end{bmatrix} + \underbrace{\begin{bmatrix} 76.9633 \\ 0.9222 \\ 1.0000 \end{bmatrix}}_{\widehat{b}_1} u + \underbrace{\begin{bmatrix} -0.3093 \\ -0.0419 \\ 0 \end{bmatrix}}_{\widehat{c}_1}, \quad (43)$$

The fuzzy sets $A_{1,i}$ (Fig. 3) were obtained from the clusters, i.e, the clusters could be analytically projected and decomposed to membership functions defined on the individual features, because the proposed clustering method obtains axis-parallel clusters. It is clear that the fuzzy sets are more separated on the input x_1 than on x_2. Application of Fisher Inteclass separability therefore also showed that application of only the first input on antecedent part of the model was enough. This reflects our prior knowledge, since we know that the main source of the nonlinearity is given by the nonlinear relation between growth rate and substrate concentration x_1.

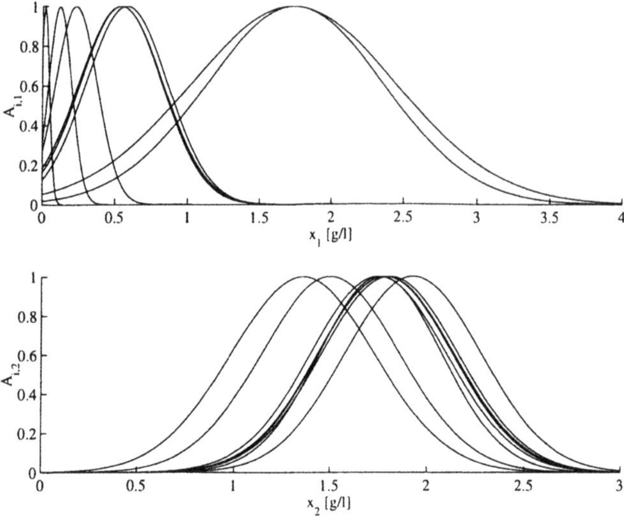

Fig. 3. Membership functions obtained by standard TS fuzzy modeling approach.

Second, we identified a eight rule semi-mechanistic model based on the presented theory. The obtained rules are similar to (18) and the membership functions are given by Fig. 4. The main difference between the semi-mechanistic and the standard TS models is that the semi-mechanistic model is better interpretable, i.e., its outputs are not the derivatives of the state variables but the *unknown* growth rates. For example, the specific growth rate has physical meaning that can be interpreted by the bio-engineers and is given by Fig. 5. Moreover, the application of the proposed clustering and model reduction tools resulted in a model where the growth rates are only the function of the substrate concentration, x_1 even in case that additional *prior* knowledge was considered to be unknown:

$$R_i \;:\; \textbf{If } x_1 \text{ is } A_{i,1} \textbf{ then}$$

$$\dot{\boldsymbol{x}}_i = \begin{bmatrix} \frac{u}{x_3}\left(x_{1,in} - x_1\right) \\ -\frac{u}{x_3}\left(x_2\right) \\ u \end{bmatrix} + \left(\underbrace{\begin{bmatrix} a_i^{1,1} & 0 & 0 \\ a_i^{2,1} & 0 & 0 \\ 0 & 0 & 0 \end{bmatrix}}_{\boldsymbol{A}_i} \begin{bmatrix} x_1 \\ x_2 \\ x_3 \end{bmatrix} + \underbrace{\begin{bmatrix} b_i^1 \\ b_i^2 \\ 1 \end{bmatrix}}_{\boldsymbol{b}_i} \right) \odot \begin{bmatrix} -x_2 \\ x_2 \\ 0 \end{bmatrix} \tag{44}$$

The fuzzy part of the obtained semi-mechanistic fuzzy model can be inspected separately, e.g.,:

$$R_i : \textbf{If } x_1 \text{ is } A_{i,1} \textbf{ then } \sigma(\boldsymbol{x})_i = a_i^{1,1} x_1 + b_i^1 . \tag{45}$$

The predicted growth rates obtained by this simple model are shown in Fig. 6.

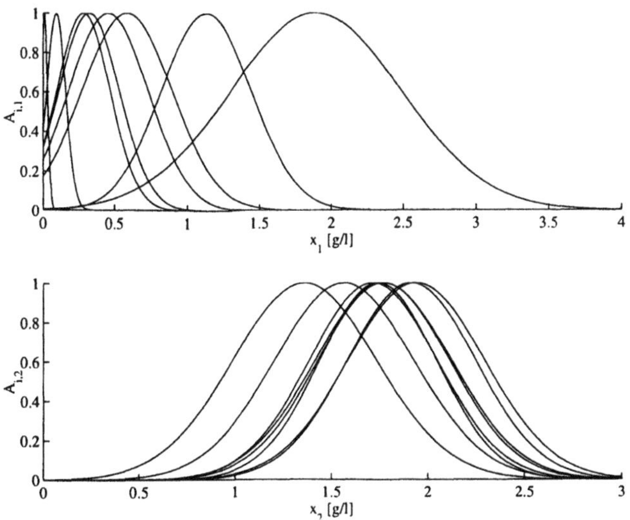

Fig. 4. Membership functions obtained by semi-mechanistic fuzzy modeling approach.

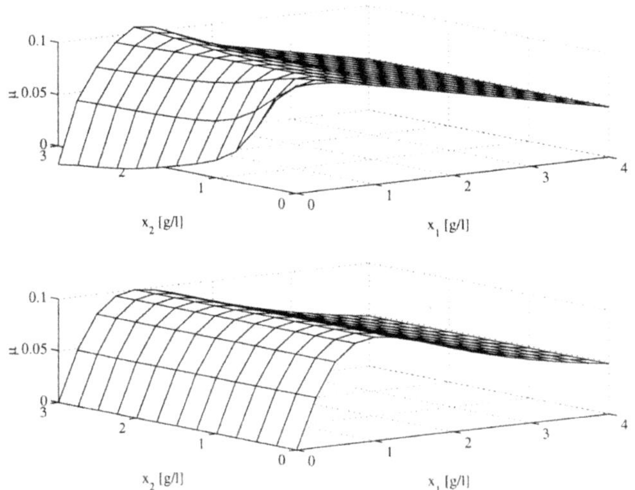

Fig. 5. Estimated (top) and "real" (bottom) growth rate $\mu(\boldsymbol{x})$.

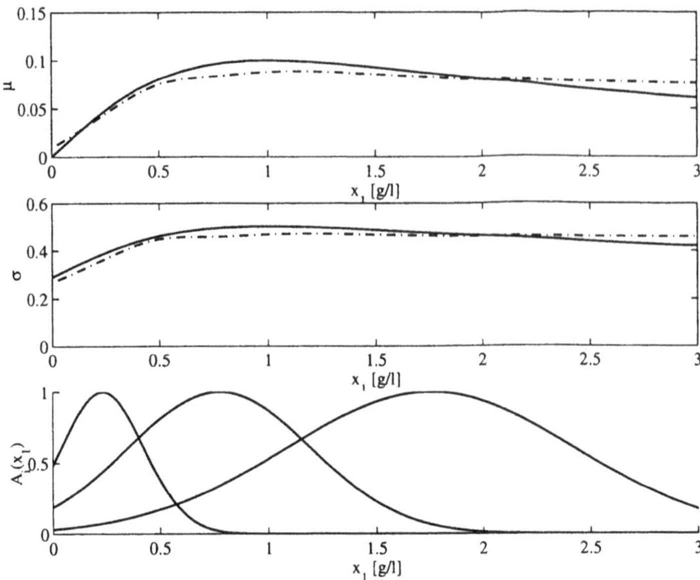

Fig. 6. True and estimated kinetic model and the operating regions of the local models.

Finally, the obtained models have been validated on a number of additional batches that were not used for identification. A typical ballistic prediction is shown in Fig. 7, and indicates that the prediction accuracy of the proposed model is excellent. It is observed that the performance of the proposed semi-mechanistic model is better than that of the complex TS fuzzy model which gives a worse prediction at low substrate concentrations where the effect of the nonlinearity is the biggest.

Concluding this example, we have shown that interpretable TS-fuzzy and semi-mechanistic fuzzy models can be generated by using fuzzy clustering and rule-base reduction methods. Moreover, in case of structured semi-mechanistic fuzzy models, one can extract simple and interpretable fuzzy submodels that describe the unknown physical effects of the modeled system.

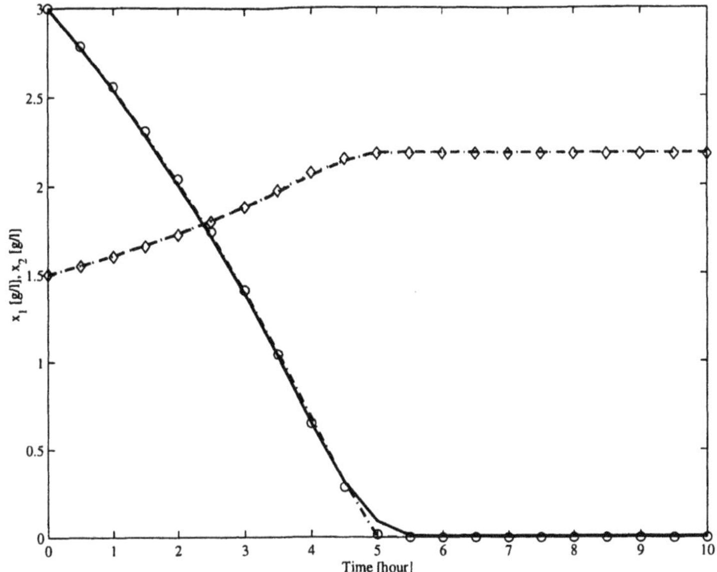

Fig. 7. Simulation (ballistic prediction) from correct initial states for a typical batch and the "true" measured variables, marked with circles *o*.

5 Conclusions

We have proposed a semi-mechanistic modeling strategy to obtain compact and transparent models of process systems based on small datasets. The semi-mechanistic fuzzy identification method is based on white-box modeling combined with fuzzy model parts. This method is suitable to model systems where a priori knowledge is available and experimental data is insufficient to make a conventional fuzzy model. First, for identification of the semi-mechanistic fuzzy model a new fuzzy clustering method is proposed. Here, clustering is achieved by the simultaneous identification of fuzzy sets defined on some of the scheduling variables and identification of the parameters of the local semi-mechanistic submodels. Second, application of orthogonal least squares and Fisher's interclass separability are proposed to keep the models as compact as possible. Next, the overall modeling procedure is described by the development of a model for a biochemical process and we showed that the proposed modeling approach was able to effectively identify compact and transparent models. The obtained semi-mechanistic fuzzy models are compared with classic TS fuzzy models and appeared to be more accurate and better interpretable.

6 Acknowledgements

The financial support of the Hungarian Ministry of Culture and Education (FKFP-0073/2001) and the Janos Bolyai Research Fellowship of the Hungarian Academy of Science is greatly appreciated.

References

1. H. Hellendoorn, D. Driankov (Eds.), Fuzzy Model Identification: Selected Approaches, Springer, Berlin, Germany, 1997.
2. T. Takagi, M. Sugeno, Fuzzy identification of systems and its application to modeling and control, IEEE Transactions on Systems, Man and Cybernetics 15 (1) (1985) 116–132.
3. M. Setnes, R. Babuška, H. Verbruggen, Rule-based modeling: Precision and transparency, IEEE Trans. SMC-C 28 (1998) 165–169.
4. H. Tulleken, Gray–box modelling and identification using physical knowledge and Bayesian techniques, Automatica 29 (1993) 285–308.
5. P. Lindskog, L. Ljung, Tools for semi-physical modeling, in: Proceedings IFAC SYSID, Vol. 3, Kopenhagen, Danmark, 1994, pp. 237–242.
6. T. Bohlin, A case study for grey box identification, Automatica 30(2) (1994) 307–318.
7. J. Abonyi, L. Nagy, F. Szeifert, Hybrid fuzzy convolution modelling and identification of chemical process systems, International Journal of Systems Science 31 (2000) 457–466.
8. J. Abonyi, A. Bodizs, L. Nagy, F. Szeifert, Hybrid fuzzy convolution model and its application in predictive control, Chemical Engineering Research and Design 78 (2000) 597–604.
9. J. Abonyi, R. Babuska, H. Verbruggen, F. Szeifert, Using a priori knowledge in fuzzy model identification, International Journal of Systems Science 31 (2000) 657–667.
10. W. Timmons, H. Chizeck, P. Katona, Parameter–constrained adaptive control, Ind. Eng. Chem. Res. 36 (1997) 4894–4905.
11. T. Johansen, Identification of non–linear systems using empirical data and a priori knowledge – an optimisation approach, Automatica 32 (1996) 337–356.
12. H. te Braake, Neural control of biotechnological processes, Ph.D. thesis, Delft University of Technology, Department of Electrical Engineering, Delft, The Netherlands (1997).
13. R. Simutis, A. Lübbert, Exploratory analysis of bioprocesses using artificial neural network-based methods, Biotechnology Progress 13 (1997) 479–487.
14. M. Thompson, M. Kramer, Modeling chemical processes using prior knowledge and neural networks, AIChE Journal 40 (1994) 1328–1340.
15. D. Psichogios, L. Ungar, A hybrid neural network – first principles approach to process modeling, AIChE J. 38 (1992) 1499–1511.
16. T. Johansen, Operating regime based process modeling and identification, Ph.D. thesis, Department of Engineering Cybernetics, Norwegian Institute of Technolgy, University of Trondheim, Norway (1994).
17. A. Tholudur, W. Ramirez, Optimization of fed-batch bioreactors using neural network parameter function models, Biotechnology Progress 12 (1996) 302–309.

18. H. van Can, H. te Braake, C. Hellinga, K. Luyben, J. Heijnen, Strategy for dynamic process modeling based on neural networks and macroscopic balances., AIChE Journal 42 (1996) 3403–3418.
19. R. Babuška, H. Verbruggen, H. van Can, Fuzzy modeling of enzymatic penicillin–G conversion, Engineering Applications of Artificial Intelligence 12 (1) (1999) 79–92.
20. J. Schubert, R. Simutis, M. Dors, I. H. ab A. Lubbert, Bioprocess optimization and control: Application of hybrid modelling, Journal of Biotechnology 35 (1994) 51–68.
21. R. Babuška, Fuzzy Modeling for Control, Kluwer Academic Publishers, Boston, 1998.
22. C. Bishop, Neural Networks for Pattern Recognition, Oxford University Press, 1995.
23. J. Abonyi, J. Roubos, M. Oosterom, F. Szeifert, Compact ts-fuzzy models through clustering and ols plus fis model reduction, in: Proc. of IEEE international conference on fuzzy systems, Sydney, Australia, 2001, p. to be published.
24. J. Roubos, M. Setnes, J. Abonyi, Learning fuzzy classification rules from data, in: R. John, R. Birkenhead (Eds.), Developments in Soft Computing, Springer, Physica Verlag, 2001.
25. J. Roubos, M. Setnes, Compact fuzzy models through complexity reduction and evolutionary optimization, in: Proc. of IEEE international conference on fuzzy systems, San Antonio, USA, 2000, pp. 762–767.
26. J. Yen, L. Wang., Simplifying fuzzy rule–based models using orthogonal transformation methods, IEEE Transaction on Systems, Man, and Cybernetics: Part B 29 (1999) 13–24.
27. M. Setnes, R. Babuška, U. Kaymak, H. R. van Nauta Lemke, Similarity measures in fuzzy rule base simplification, IEEE Transactions on Systems, Man and Cybernetics - Part B: Cybernetics 28 (3) (1998) 376–386.
28. Y. Yam, Fuzzy approximation via grid point sampling and singular value decomposition, IEEE Trans. on Systems, Man, and Cybertenics 27 (1997) 933–951.
29. R. Murray-Smith, T. Johansen (Eds.), Multiple Model Approaches to Nonlinear Modeling and Control, Taylor & Francis, London, UK, 1997.
30. R. Babuška, H. Verbruggen, Fuzzy set methods for local modeling and identification, in: R. Murray-Smith, T. A. Johansen (Eds.), Multiple Model Approaches to Nonlinear Modeling and Control, Taylor & Francis, London, UK, 1997, pp. 75–100.
31. D. Gustafson, W. Kessel, Fuzzy clustering with fuzzy covariance matrix, in: Proceedings of the IEEE CDC, San Diego, 1979, pp. 761–766.
32. I. Gath, A. Geva, Unsupervised optimal fuzzy clustering, IEEE Transactions on Pattern Analysis and Machine Intelligence 7 (1989) 773–781.
33. R. Babuška, H. Verbruggen, Constructing fuzzy models by product space clustering, in: H. Hellendoorn, D. Driankov (Eds.), Fuzzy Model Identification: Selected Approaches, Springer, Berlin, Germany, 1997, pp. 53–90.
34. E. Kim, M. Park, S. Kim, M. Park, A transformed input–domain approach to fuzzy modeling, IEEE Transactions on Fuzzy Systems 6 (1998) 596–604.
35. N. Gershenfeld, B. Schoner, E. Metois, Cluster-weighted modelling for time-series analysis, Nature 397 (1999) 329–332.

36. M. Setnes, R. Babuška, H. B. Verbruggen, Rule-based modeling: Precision and transparency, To appear in IEEE Transactions on Systems, Man and Cybernetics, Part A: Systems and Humans .
37. B. Foss, T. Johansen, A. Sorensen, Nonlinear predictive control using local models - applied to a batch process, Control Engineering Practice 3 (1995) 389–396.
38. J.-I. Horiuchi, M. Kamasawa, H. Miyakawa, M. Kishimoto, Phase control of fed-batch culture for α-amylase production based on culture phase identification using fuzzy inference, J. of Fermentation and Bioengineering 76 (1993) 207–212.

A Appendix

A.1 Hermite spline interpolation

Lets define a cubic spline for a knot sequence $\Xi, a = \xi_0 < \xi_1 < \ldots < \xi_N = b$. The cubic spline is then based on cubic polynomes of the form $S(x)$, given for each subinterval $[\xi_0, \xi_1]$, $[\xi_1, \xi_2], \ldots, [\xi_{N-1}]$. These polynomes are connected at the interior knots Ξ in such a way that $S(x)$ has *two* continuous derivatives on $[a, b]$. Every cubic splines can be written in the form:

$$S(x) = \alpha x^3 + \beta x^2 + \gamma x + \delta + \sum_{i=1}^{B-1} a_i |x - \xi_i|^3 . \tag{46}$$

Thus, $S(x)$ is simply a linear combination of the functions

$$x^3, x^2, x, 1, |x - \xi_1|^3, \ldots, |x - \xi_{N-1}|^3 . \tag{47}$$

An efficient notation for identification is given by Horiuchi [38], who defined the piecewise polynoms by a combination of the function-value and its first derivative at the knots:

$$\begin{aligned} S_i(x) = & \frac{dS(\xi_i)}{dx} \cdot \frac{(\xi_{i+1} - x)^2 (x - \xi_i)}{h_i^2} - \\ & \frac{dS(\xi_{i+1})}{dx} \cdot \frac{(x - \xi_i)^2 (\xi_{i+1} - x)}{h_i^2} + \\ & S(\xi_i) \cdot \frac{(\xi_{i+1} - x)^2 (2(x - \xi_i) + h_i)}{h_i^3} + \\ & S(\xi_{i+1}) \cdot \frac{(x - \xi_i)^2 (2(\xi_{i+1} - x) + h_i)}{h_i^3} , \end{aligned} \tag{48}$$

where $h_i = (\xi_{i+1} - \xi_i)$.
Denote:

$$a_i(x) = \frac{(\xi_{i+1} - x)^2 (x - \xi_i)}{h_i^2}, \tag{49}$$

$$b_i(x) = -\frac{(x - \xi_i)^2 (\xi_{i+1} - x)}{h_i^2}, \tag{50}$$

$$c_i(x) = \frac{(\xi_{i+1} - x)^2 (2(x - \xi_i) + h_i)}{h_i^3}, \tag{51}$$

$$d_i(x) = \frac{(x - \xi_i)^2 (2(\xi_{i+1} - x) + h_i)}{h_i^3}, \tag{52}$$

and

$$y_i = S(\xi_i), \tag{53}$$

$$y_i' = \frac{dS(\xi_i)}{dx}, \tag{54}$$

then (48) can be given as:

$$S_i(x) = y_i a_i(x) + y_{i+1} b_i(x) + y_i' c_i(x) + y_{i+1}' d_i(x). \tag{55}$$

Now, the unknown variables y_i and y_i' can be determined by a least square method. Define a quadratic objective function Q by:

$$Q(y_1, y_1', \dots, y_N, y_{N+1}') = \sum_{k=1}^{N} \left(S(x_k) - y_k\right)^2 . \tag{56}$$

The function Q can be minimized by solving the normal equations derived by partial differentiation of Eq. 56 with $\theta = [y_1, y_1', \dots, y_N, y_N']$:

Define for $\nu \in \{a, b, c, d\}$:

$$\nu_i = \sum_{k=p_i}^{q_i} \nu_1(x_k). \tag{57}$$

$$\frac{dQ}{d\theta} = 2 \left(\begin{bmatrix} \cdot & \cdot & \cdot & \cdot & \cdot & \cdot & \cdot & \cdot & \cdot \\ \cdot & \cdot & \cdot & \cdot & \cdot & \cdot & \cdot & \cdot & \cdot \\ c_1 d_1 & a_1 d_1 & d_1^2 & b_1 d_1 & \cdot & \cdot & \cdot & \cdot & -d_1 y_k \\ c_1 b_1 & a_1 b_1 & b_1 d_1^2 & b_1^2 & \cdot & \cdot & \cdot & \cdot & -b_1 y_k \\ \cdot & \cdot & \cdot & \cdot & \cdot & \cdot & \cdot & \cdot & \cdot \\ \cdot & \cdot & \cdot & \cdot & c_N d_N & a_N d_N & d_N^2 & b_N d_N & -d_N y_k \\ \cdot & \cdot & \cdot & \cdot & c_N b_N & a_N b_N & b_N d_N^2 & b_N^2 & -b_N y_k \end{bmatrix} + \begin{bmatrix} c_1^2 & a_1 c_1 & c_1 d_1 & b_1 c_1 & \cdot & \cdot & \cdot\cdot & -c_1 y_k \\ a_1 c_1 & a_1^2 & a_1 d_1 & a_1 b_1 & \cdot & \cdot & \cdot\cdot & -a_1 y_k \\ \cdot & \cdot & c_2^2 & a_2 c_2 & c_2 d_2 & b_2 c_2 & \cdot\cdot & -c_2 y_k \\ \cdot & \cdot & a_2 c_2 & a_2^2 & a_2 d_2 & a_2 b_2 & \cdot\cdot & -a_2 y_k \\ \cdot & \cdot & \cdot & \cdot & \cdot & \cdot & \cdot\cdot & \cdot \\ \cdot & \cdot & \cdot & \cdot & \cdot & \cdot & \cdot\cdot & \cdot \\ \cdot & \cdot & \cdot & \cdot & \cdot & \cdot & \cdot\cdot & \cdot \end{bmatrix} \right) \begin{bmatrix} \theta^T \\ 1 \end{bmatrix} \tag{58}$$

A.2 Proposed Clustering Algorithm

Clustering is based on minimization of the sum of weighted squared distances between the data pairs, $\boldsymbol{x}(k), \boldsymbol{u}(k), \dot{\boldsymbol{x}}(k)$ and the cluster prototypes, η_i. Here, the square of the D_{ik}^2 distances are weighted with the membership values μ_{ik}. Thus the objective function to be minimized by the clustering algorithm is formulated as:

$$J(\boldsymbol{Z}, \boldsymbol{U}, \eta) = \sum_{i=1}^{c} \sum_{j=1}^{N} (\mu_{i,k}) \, D_{i,k}^2, \tag{59}$$

where $\boldsymbol{Z}$ denotes the available training data, $\boldsymbol{U}$ the matrix of membership values, and η the set of cluster parameters to be identified. In order to obtain a fuzzy partitioning space, the membership values have to satisfy the following conditions:

$$U \in \mathbb{R}^{c \times N} | \mu_{i,k} \in [0,1], \; \forall i, k, \tag{60}$$

$$\sum_{i=1}^{c} \mu_{i,k} = 1, \; \forall k, \tag{61}$$

$$0 < \sum_{k=1}^{N} \mu_{i,k} < N, \; \forall i. \tag{62}$$

Minimization of (59) represents a non-linear optimization problem that is subject to constrains defined by (62) and can be solved by using a variety of methods. The most popular method, is the alternating optimization (AO), which is equivalent to the EM algorithm, and formulated as follows:

Initialization Given a set of data $\boldsymbol{Z}$ specify c, choose a weighting exponent $m = 2$ and a termination tolerance $\epsilon > 0$. Initialize the partition matrix, $\boldsymbol{U} = [\mu_{i,k}]_{c \times N}$, where $\mu_{i,k} = p(\eta_i | \boldsymbol{x}(k), \boldsymbol{u}(k), \dot{\boldsymbol{x}}(k))$ denotes the membership (posterior probability) that the $\boldsymbol{x}(k), \boldsymbol{u}(k), \dot{\boldsymbol{x}}(k)$ input-output data is generated by the i-th cluster.

Repeat for $l = 1, 2, \ldots$

Step 1 Calculate cluster parameters:

- Calculate centers of membership functions:

$$\boldsymbol{v}_i^{(l)} = \frac{\sum_{k=1}^{N} \mu_{i,k}^{(l-1)} \boldsymbol{z}(k)}{\sum_{k=1}^{N} \mu_{i,k}^{(l-1)}}, \; 1 \le i \le c. \tag{63}$$

- Calculate standard deviation of the Gaussian membership functions:

$$\sigma_{i,j}^{2\,(l)} = \frac{\sum_{k=1}^{N} \mu_{i,k}^{(l-1)} (z_j(k) - v_{j,k})^2}{\sum_{k=1}^{N} \mu_{i,k}^{(l-1)}}, \; 1 \le i \le c. \tag{64}$$

- Calculate the unknown parameters of the local models by a weighted least squares algorithm, where the weights of the data pairs are: $\mu_{i,k}^{(l-1)}$.
- Calculate covariance of the modeling errors of the local models (34).
- Calculate *a priori* probability of the cluster:

$$p(\eta_i) = \frac{1}{N} \sum_{k=1}^{N} \mu_{i,k}. \tag{65}$$

- Calculate rule weights:

$$w_i = p(\eta_i) \prod_{j=1}^{n} \frac{1}{\sqrt{2\pi\sigma_{i,j}^2}} \tag{66}$$

Step 2 Compute the distance measure $D_{i,k}^2(\boldsymbol{x}(k), \boldsymbol{u}(k), \dot{\boldsymbol{x}}(k), \eta_i)$. The distance measure consists of two terms. The first term is based on the geometrical distance between the cluster centers and the scheduling variable, $\boldsymbol{z}$, while the second is based on the performance of the local hybrid models:

$$\frac{1}{D_{i,k}^2(\boldsymbol{x}(k), \boldsymbol{u}(k), \dot{\boldsymbol{x}}(k), \eta_i)} = \tag{67}$$

$$w_i \prod_{j=1}^{n} z \exp\left(-\frac{1}{2}\frac{(z_{j,k}-v_{i,j})^2}{\sigma_{i,j}^2}\right) \cdot$$

$$\frac{\exp\left(-(\dot{\boldsymbol{x}}(k)-f_{Hyb,i}(\boldsymbol{x}(k),\boldsymbol{u}(k)))^T (\boldsymbol{F}_i^{xx})^{-1} (\dot{\boldsymbol{x}}(k)-f_{Hyb,i}(\boldsymbol{x}(k),\boldsymbol{u}(k)))\right)}{(2\pi)^{\frac{n}{2}} \sqrt{|\boldsymbol{F}_i^{xx}|}}$$

Step 3 Update the partition matrix:

$$\mu_{i,k}^{(l)} = \frac{1}{\sum_{j=1}^{c} \left(D_{i,k}(\boldsymbol{x}(k), \boldsymbol{u}(k), \dot{\boldsymbol{x}}(k), \eta_i) / D_{j,k}(\boldsymbol{x}(k), \boldsymbol{u}(k), \dot{\boldsymbol{x}}(k), \eta_j)\right)^{2/(m-1)}},$$

$$1 \leq i \leq c,\ 1 \leq k \leq N. \tag{68}$$

until $||\boldsymbol{U}^{(l)} - \boldsymbol{U}^{(l-1)}|| < \epsilon$.

Trade-off between approximation accuracy and complexity: TS controller design via HOSVD based complexity minimization*

Péter Baranyi[1,2], Yeung Yam[3], Domonkos Tikk[1,2], and Ron J. Patton[4]

[1] Dept. of Telecommunications & Telematics,
Budapest University of Technology and Economics,
1117 Budapest, Magyar Tudósok Körútja 2., Hungary,
e-mails: baranyi@ttt-202.ttt.bme.hu, tikk@ttt.bme.hu

[2] Intelligent Integrated Systems Japanese–Hungarian Laboratory
1111 Budapest, Műegyetem rakpart 3., Hungary

[3] Dept. Automation and Computer Aided Engineering
The Chinese University of Hong Kong, Shatin, N.T., Hong Kong
e-mail: yyam@acae.cuhk.edu.hk

[4] Control and Intelligent Systems Research Group, University of Hull
Cottingham Road, Hull, HU6 7RX, UK
e-mail: r.j.patton@eng.hull.ac.uk

Abstract. Higher order singular value decomposition (HOSVD) based complexity reduction method is proposed in this paper to tensor product based model approximation methods, especially, to *Takagi–Sugeno* (TS) fuzzy or polytopic models. The main motivation is that the TS model has exponentially growing computational complexity with the improvement of its approximation property through, as usually practiced, increasing the density of fuzzy partitions. The reduction technique proposed here is capable of pinpointing the contribution of each local linear model consequents of the fuzzy rules, which serves to remove the weakly contributing ones according to a given threshold. Reducing the number of local models leads directly to complexity reduction. The explicit form in this paper for the proposed reduction can also be applied on polytopic model approximation methods. A detailed illustrative example of a non-linear dynamic model is also given.

Keywords. TS fuzzy model, complexity reduction, singular value decomposition (SVD - HOSVD).

1 Introduction

As a result of the dramatic and continuing growth in computer power, and the advent of very powerful algorithms (and associated theory) for *convex optimization*, we can now solve very rapidly many convex optimization problems

* This research was supported by the Hungarian Scientific Research Fund (OTKA) Grants No. D034614, F30056, and T34212 and by the Hungarian Ministry of Education Grant No. FKFP 0180/2001. Péter Baranyi was supported by the Zoltán Magyary Scholarship.

for which no traditional "analytic" or "closed form" solution are known (or likely to exist). Indeed, the solution to many convex optimization problems can now be computed in a time frame comparable to that required to evaluate a "closed-form" solution for a similar problem. This fact has far-reaching implications for engineers: it changes our fundamental notion of what's being a solution to a problem. In the past, a "solution to a problem" generally meant a "closed-form" or "analytic" solution. There are "analytical solutions" to a few very special cases in the wide variety of problems in systems and control theory, but in general non-linear problems cannot be solved. Control engineering problem that reduces to solving two algebraic *Riccati* equations is now generally regarded as "solved". A control engineering problem that reduces to solving even a large number of convex algebraic Riccati inequalities (a problem which has no analytic solution) should also be regarded as "solved", even though there is no analytic solution. A number of problems that arise in the field of control, such as optimal matrix scaling, digital filter realization, interpolation problems for system identification, robustness analysis and state feedback synthesis via Lyapunov functions, can be reduced to a handful of standard *convex* and *quasiconvex problems* that involve *matrix inequalities*. Extremely efficient interior point algorithms have recently been developed for and tested on these standard problems; further development of algorithms for these standard problems is in an area of active research.

The notion of convex combination of a finite set of points gets considerable relevance in the context of dynamical systems if "points" become systems. Various model approximation techniques are based on the convex combination of local models. The combination is usually defined by the tensor product of basis functions, which express the local dominance of the local models, and by the array of local models. One of these approximation method is called *Takagi–Sugeno* (TS) fuzzy model approximation [?] (to be specific, we restrict our discussion to the most widely applied product inference based TS models), and is utilized to describe linear parametrically varying, or linear time variant systems where the parameters vary in time. In this regards, Polytopic Model Approximation (PMA) was actually introduced in the same explicit form as the TS models. The use of the PMA and TS fuzzy model techniques is motivated by the fact that they can easily be cast or recast as *convex optimization* problems that involve *Linear Matrix Inequalities* (LMIs). The controller design and stability analysis can, hence, be done in this framework [24,25]. Therefore, once the model approximation is given in the form of PMA or TS fuzzy model, the controller design and *Lyapunov* stability analysis of the non-linear system reduces to solving the LMI problem as outlined above.

Despite this advantage discussed above, the use of TS fuzzy models is practically limited. The reason of this is that though the TS fuzzy model possesses the universal approximation property in theory, it suffers from the problem of exponential complexity with regard to approximation accuracy. Here, we give a brief digression to illustrate the choice between approximation accuracy and complexity. In 1900, *D. Hilbert* listed 23 conjectures, hypotheses concerning unsolved problems he deemed as the most important to solve by mathematicians of the 20th century. The 13th conjecture states the existence of a continuous multivariable function, which cannot be decomposed as finite superposition of continuous functions of less variables. The hypothesis was disproved by *Arnold* in 1957 [1]. Moreover, in the same year, *Kolmogorov* [13] introduced a general representation theorem with a constructive proof, where the functions in the decomposition were one dimensional, but very complicated ones, therefore for practical purposes they were inappropriate to calculate with. Over the years, Kolmogorov's representation theorem was further improved by several authors (*Sprecher* [20] and *Lorentz* [18]). In 1980, *De Figueiredo* showed [10] that Kolmogorov's theorem could be generalized for multilayer feedforward neural networks, and these could, hence, be considered as universal approximators. From the late '80s several authors proved that different types of neural network possessed the universal approximation property (see, e.g. [6,9,11,15]). Similar results have been established from the early '90s in fuzzy theory. These results (see, e.g. [7,14,28]) claim that various fuzzy reasoning methods are capable of approximating arbitrary continuous function on a compact domain with any specified accuracy. As a result, soft-computing techniques were considered as universal approximators in general. Some papers focus on the proof of the universal approximation property of product inference operator based TS models [32–34]. Regarding the explicit form, the TS fuzzy model and PMA techniques share the same advantage.

In spite of these remarkable advantages, the neural network model as well as fuzzy approximation have exponential complexity in terms of the number of variables (see, e.g., *Kóczy* and *Hirota* [12]). It means that in the neural network context, the number of hidden neurons, or in the context of TS fuzzy model the number of local linear models (we say building units) grows exponentially as the approximation error tends to zero. This exponential increase cannot be eliminated, so the universal approximation property of these uncertainty based approaches cannot be exploited straightforwardly for practical purposes. Figure 1 shows the relation between the number of building units and approximation accuracy.

Moreover, for some fuzzy theory based approximation techniques (including TS product inference based approximation) it is shown, that if the number of the building units is bounded, the resulting set of functions is nowhere dense in the space of approximated functions, i.e. here in the space of continuous functions (*Moser* [19] and *Tikk* [26], 1999). According to the opinion of some researchers [26], analogous results should hold for most fuzzy and neural systems. The mutually contradicting results naturally pose the question as to what extent the model approximation is deemed accurate enough for the available computational cost. From the practical point of view, though, it is enough to achieve an "acceptably" good approximation, where the given problem determines the factor of acceptability in terms of the accuracy (ε).

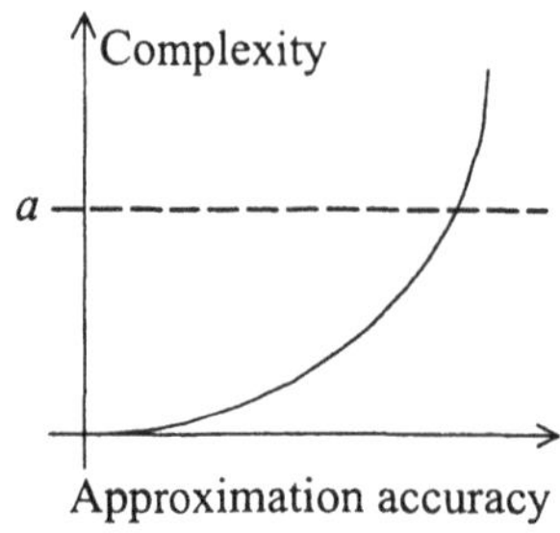

Fig. 1. Trade-off between complexity and approximation accuracy. "a" indicates the available computation capacity in real application.

The task is, hence, to find a possible trade-off between the specified accuracy and the number of building units.

Recently, several approaches applied orthogonal transformation methods to find the minimal number of building units in a given approximation. For instance, in 1999 *Yen* and *Wang* [31] investigated various techniques such as orthogonal least-squares, eigenvalue decomposition, SVD–QR with column pivoting method, total least square method, and direct SVD method. SVD based fuzzy approximation technique was initiated by *Yam* in 1997 [29], which directly finds the minimal number of building units from sampled values. Shortly after, this technique was introduced as SVD reduction of the building units and structure decomposition in [2–5,30]. An extension of *Yen* and *Wang's* work [31] to multi-dimensional cases may also be conducted in a similar fashion as the higher order SVD reduction technique proposed in the papers [2–5,29,30]. SVD is not merely used as a way of reduction of fuzzy rule bases. A brief enumeration of the potentials offered by SVD, of which some works were started by *Beltarmi* about 200 years ago, can be found in *Stewart* (1993) [21]. SVD is one of the most fruitful tools in linear algebra, belting its promising role in complexity reduction in general. The key idea of using SVD in complexity reduction is that the singular values can be applied to decompose a given system and indicate the degree of the significance of the decomposed parts. Reduction is conceptually obtained by the truncation of those parts which have weak or no contribution at all to the output according to the assigned singular values. This advantageous feature of SVD is used in this paper to extract a given model approximation, and discard those local linear models, namely, building units which have no significant role in the overall system according to a given approximation accuracy. However, reducing the number of building units does not imply the computational cost reduction in all cases since the computation also depends on the number of overlapping antecedent membership functions, see later. Therefore, as a subsequent aim, a detailed investigation is given in the aspect of the computational time reduction in this paper.

The concept of this paper is based on the above outlined ideas [2–5,29,30]. Presumably, the SVD technique in this paper as well as in the papers just referred can be replaced by other orthogonal techniques investigated in [31]. The present work constitutes a detailed investigation of the preliminary approaches outlined in the work [2], and gives a possible solution to the complexity problem described above. The algorithms proposed here are mostly developed in the papers [29,30], but are restructured in terms of tensor description in order to facilitate further developments. Concepts of HOSVD are investigated in tensor forms in the works of *Lathauwer et al.* (2000 and 2001) [16,17], *Comon* (1994) [8], and *Swami* and *Giannakis* (1996) [22].

2 Basic definitions

This section is devoted to introduce some elementary definitions and concepts utilized later in this paper. Before starting with the definitions, some comments are enumerated on the notation to be utilized. To facilitate the distinction between the types of given quantities, we follow the next rules for notation: scalar values are denoted by lower-case letters $\{a, b, \ldots\}$; column vectors and matrices are given by bold-face letters as $\{\mathbf{a}, \mathbf{b}, \ldots\}$ and $\{\mathbf{A}, \mathbf{B}, \ldots\}$ respectively. Tensors correspond to capital letters as $\{\mathcal{A}, \mathcal{B}, \ldots\}$, tensor $\mathcal{1}$ contains values 1 only. The transpose of matrix $\mathbf{A}$ is denoted as $\mathbf{A}^{\mathrm{T}}$. Subscript is consistently used for a lower order of a given structure. For example, an element of matrix $\mathbf{A}$ at row-column number i, j is symbolized as $(\mathbf{A})_{i,j} = a_{i,j}$. Systematically, the ith column vector of $\mathbf{A}$ is denoted as $\mathbf{a}_i$, i.e. $\mathbf{A} = \begin{bmatrix}\mathbf{a}_1 & \mathbf{a}_2 & \cdots\end{bmatrix}$. To enhance the overall readability characters $i, j, \ldots$ are in the meaning of indices (counters), $I, J, \ldots$ are reserved to denote the upper bounds for indices $i, j, \ldots$, unless stated otherwise. $\mathbf{R}^{I_1 \times I_2 \times \cdots \times I_N}$ is the vector space of real valued $(I_1 \times I_2 \times \cdots \times I_N)$-tensors. Letter N serves to denote the number of variables of the space where the coefficient matrices of the model are approximated. n denotes the index of a particular variable, $n = 1, \ldots, N$, while k denotes the set of indices of all other variables, $k = 1, \ldots, N$, $k \neq n$, unless stated otherwise. Finally, $\mathbf{I}$ denotes the identity matrix, and matrix $\mathbf{0}$ consists of zero values.

Definition 1. (n-mode matrix of tensor $\mathcal{A}$) *Assume an Nth order tensor $\mathcal{A} \in \mathbf{R}^{I_1 \times I_2 \times \ldots \times I_N}$. The n-mode matrix $\mathbf{A}_{(n)} \in \mathbf{R}^{I_n \times J}$, $J = \prod_k I_k$ contains all the vectors in the nth dimension of tensor $\mathcal{A}$. The ordering of the vectors is arbitrary in $\mathbf{A}_{(n)}$. This ordering shall, however, be consistently used later on. $(\mathbf{A}_{(n)})_j$ is called a jth n-mode vector.*

Note that any matrix of which the columns are given by n-mode vectors $(\mathbf{A}_{(n)})_j$ can readily be restored to become tensor $\mathcal{A}$. The restoration can be executed even in case when some rows of $\mathbf{A}_{(n)}$ are discarded since the value of I_n has no role in the ordering of $(\mathbf{A}_{(n)})_j$ [16,17].

Definition 2. (n-mode subtensor of tensor A) *Assume an Nth order tensor* $A \in \mathbf{R}^{I_1 \times I_2 \times \cdots \times I_N}$. *The n-mode subtensor* $A_{i_n=\alpha}$ *contains elements* $a_{i_1,i_2,\ldots,i_{n-1},\alpha,i_{n+1},\ldots,i_N}$.

Definition 3. (n-mode tensor partition) *Assume an Nth order tensor* $A \in \mathbf{R}^{I_1 \times I_2 \times \cdots \times I_N}$. *The n-mode partitions of tensor A are* B_l *(*$l = 1, \ldots, L$, L *is the number of partitions)* $B_l \in \mathbf{R}^{I_1 \times I_2 \times \cdots \times I_{n-1} \times J_l \times I_{n+1} \times \cdots \times I_N}$, *and the partition is denoted as* $A = \left[B_1 \; B_2 \; \cdots \; B_L\right]_n$, *where* $I_n = \sum_{l=1}^{L} J_l$.

Definition 4. (Scalar product) *The scalar product* $\langle A, B \rangle$ *of two tensors* $A, B \in \mathbf{R}^{I_1 \times I_2 \times \cdots \times I_N}$ *is defined as* $\langle A, B \rangle \overset{\text{def}}{=} \sum_{i_1} \sum_{i_2} \cdots \sum_{i_N} b_{i_1 i_2 \ldots i_N} a_{i_1 i_2 \ldots i_N}$.

Definition 5. (Orthogonality) *Tensors of which the scalar product equals* 0 *are mutually orthogonal.*

Definition 6. (Frobenius norm) *The Frobenius norm of a tensor A is given by* $\|A\| \overset{\text{def}}{=} \sqrt{\langle A, A \rangle}$.

Definition 7. (n-mode matrix-tensor product) *The n-mode product of tensor* $A \in \mathbf{R}^{I_1 \times I_2 \times \cdots \times I_N}$ *and a matrix* $\mathbf{U} \in \mathbf{R}^{J \times I_n}$, *as denoted by* $A \times_n \mathbf{U}$, *is an* $(I_1 \times I_2 \times \cdots \times I_{n-1} \times J \times I_{n+1} \times \cdots \times I_N)$*-tensor of which the entries are given by* $A \times_n \mathbf{U} = B$, *where* $B_{(n)} = \mathbf{U} \cdot \mathbf{A}_{(n)}$. *Let* $A \times_1 \mathbf{U}_1 \times_2 \mathbf{U}_2 \times \cdots \times_N \mathbf{U}_N$ *be denoted as* $A \bigotimes_{n=1}^{N} \mathbf{U}_n$ *for brevity.*

Theorem 1. (Matrix singular value decomposition (SVD)) *Every real valued* $(I_1 \times I_2)$*-matrix* $\mathbf{F}$ *can be written as the product of* $\mathbf{F} = \mathbf{U} \cdot \mathbf{S} \cdot \mathbf{V}^{\mathrm{T}} = \mathbf{S} \times_1 \mathbf{U} \times_2 \mathbf{V}$, *in which*

1. $\mathbf{U} = \left[\mathbf{u}_1 \; \mathbf{u}_2 \; \cdots \; \mathbf{u}_{I_1}\right]$ *is a unitary* $(I_1 \times I_1)$*-matrix,*
2. $\mathbf{V} = \left[\mathbf{v}_1 \; \mathbf{v}_2 \; \cdots \; \mathbf{v}_{I_2}\right]$ *is a unitary* $(I_2 \times I_2)$*-matrix,*
3. $\mathbf{S}$ *is an* $(I_1 \times I_2)$*-matrix with the properties of*

(i) *pseudodiagonality:* $\mathbf{S} = \mathrm{diag}(\sigma_1, \sigma_2, \ldots, \sigma_{\min(I_1,I_2)})$

(ii) *ordering:* $\sigma_1 \geq \sigma_2 \geq \cdots \geq \sigma_{\min(I_1,I_2)} \geq 0$. *The* σ_i *are the singular values of* $\mathbf{F}$ *and the vectors* $\mathbf{U}_i$ *and* $\mathbf{V}_i$ *are, respectively, an ith left and an ith right singular vector.*

There are major differences between matrices and higher-order tensors when rank properties are concerned. These differences can directly affect the way an SVD generalization look like. As a matter of fact, there is no unique way to generalize the rank concept. In this paper we restrict the description to n-mode rank only.

Definition 8. (n-mode rank of tensor) *The n-mode rank of A, denoted by* $R_n = \underset{n}{\mathrm{rank}}(A)$, *is the dimension of the vector space spanned by the n-mode vectors as* $\underset{n}{\mathrm{rank}}(A) = \mathrm{rank}(A_{(n)})$.

Theorem 2. (Higher Order SVD (HOSVD)) *Every tensor* $A \in \mathbf{R}^{I_1 \times I_2 \times \cdots \times I_N}$ *can be written as the product*

$$A = S \bigotimes_{n=1}^{N} \mathbf{U}_n \tag{1}$$

in which

1. $\mathbf{U}_n = \begin{bmatrix} \mathbf{u}_{1,n} & \mathbf{u}_{2,n} & \dots & \mathbf{u}_{I_N,n} \end{bmatrix}$ *is a unitary* $(I_N \times I_N)$*-matrix called n-mode singular matrix.*

2. *tensor* $S \in \mathbf{R}^{I_1 \times I_2 \times \ldots \times I_N}$ *of which the subtensors* $S_{i_n=\alpha}$ *have the properties of*

(i) *all-orthogonality: two subtensors* $S_{i_n=\alpha}$ *and* $S_{i_n=\beta}$ *are orthogonal for all possible values of* n, α *and* β : $\langle S_{i_n=\alpha}, S_{i_n=\beta} \rangle = 0$ *when* $\alpha \neq \beta$,

(ii) *ordering:* $\|S_{i_n=1}\| \geq \|S_{i_n=2}\| \geq \cdots \geq \|S_{i_n=I_n}\| \geq 0$ *for all possible values of* n.

The Frobenius-norm $\|S_{i_n=i}\|$, *symbolized by* $\sigma_i^{(n)}$, *are n-mode singular values of* A *and the vector* $\mathbf{u}_{i,n}$ *is an ith singular vector.* S *is called as core tensor.*

More detailed discussion of matrix SVD and HOSVD is given in [16,17]. A detailed algorithm of HOSVD is given in papers [2–5,29,30].

3 Takagi–Sugeno fuzzy model approximation

This section is intended to discuss the fundamental tensor product form of TS fuzzy model. Consider a parametrically varying dynamical system

$$\begin{aligned} \frac{d\mathbf{x}}{dt}(t) &= \mathbf{A}(p)\mathbf{x}(t) + \mathbf{B}(p)\mathbf{u}(t) \\ \mathbf{y}(t) &= \mathbf{C}(p)\mathbf{x}(t) + \mathbf{D}(p)\mathbf{u}(t) \end{aligned} \tag{2}$$

with input $\mathbf{u}(t)$, output $\mathbf{y}(t)$ and state $\mathbf{x}(t)$. Suppose that its system matrix

$$\mathbf{S}(p) = \begin{pmatrix} \mathbf{A}(p) & \mathbf{B}(p) \\ \mathbf{C}(p) & \mathbf{D}(p) \end{pmatrix} \tag{3}$$

is a parametrically varying object which for any parameter p can be written as a *convex combination* of the V system matrices $\mathbf{S}_1, \ldots, \mathbf{S}_V$ (as a matter of fact, p may depend on the state vector, as well, $\mathbf{S}(p)$ can, hence, be viewed as a time varying object if p is considered as $p(t)$). This means that there exists function $\alpha_v : \mathbf{R} \to [0,1]$ such that for any p we have that

$$\mathbf{S}(p) = \sum_{v=1}^{V} \alpha_v(p) \mathbf{S}_v$$

where $\sum_{v}^{V} \alpha_v(p) = 1$ and

$$\mathbf{S}_v = \begin{pmatrix} \mathbf{A}_v & \mathbf{B}_v \\ \mathbf{C}_v & \mathbf{D}_v \end{pmatrix}$$

are constant system matrices. In particular, this implies that the system matrices $\mathbf{S}(p)$ belong to convex hull of $\mathbf{S}_1, \ldots, \mathbf{S}_V$, i.e. $\mathbf{S}(p) \in \mathrm{co}(\mathbf{S}_1, \ldots, \mathbf{S}_V)$. Such models are also called polytopic linear differential inclusions and arise in the wide variety of modelling problems. Consequently, the system is approximated by a model, which consists of a number of local linear models assigned to regions defined by the membership function of antecedent fuzzy sets $\alpha_v(p)$. In multi-variable case, when the system is varying in a multi-dimensional vector space P, higher dimensional ordering is utilized to define the local linear models and the corresponding antecedent functions. The system matrix $\mathbf{S}(\mathbf{p})$, where $\mathbf{p} \in \mathbf{R}^N$, is approximated as:

$$\hat{\mathbf{S}}(\mathbf{p}) = \sum_{v_1=1}^{V_1} \sum_{v_2=2}^{V_2} \cdots \sum_{v_N=1}^{V_N} \prod_{n=1}^{N} \alpha_{n,v_n}(p_n) \mathbf{S}_{v_1,v_2,\ldots,v_N}, \tag{4}$$

where p_n are the elements of vector $\mathbf{p}$. The assignment between the local linear models and the antecedent sets in (4) is defined by the linguistic rules:

IF antecedent set α_{1,v_1} **AND** antecedent set α_{2,v_2} **AND** $\cdots$ **AND** antecedent set α_{N,v_N} **THEN** $\mathbf{S}_{v_1,v_2,\ldots,v_N}$

Along in the same line as above

$$\forall n, v : \alpha_{n,v}(p_n) \in [0,1] \tag{5}$$

and *Ruspini* partition $\forall n : \sum_{v=1}^{V_n} \alpha_{n,v}(p_n) = 1$ which implies that

$$1 = \sum_{v_1=1}^{V_1} \sum_{v_2=2}^{V_2} \cdots \sum_{v_N=1}^{V_N} \prod_{n=1}^{N} \alpha_{n,v_n}(p_n). \tag{6}$$

In order to have a general technique to mixed problems with various performance specification, consider a multi-channel system description as given by:

$$\begin{pmatrix} \dot{\mathbf{x}} \\ \mathbf{v}_1 \\ \vdots \\ \mathbf{v}_q \\ \mathbf{y} \end{pmatrix} = \begin{pmatrix} \mathbf{A}(\mathbf{p}) & \mathbf{B}_1(\mathbf{p}) & \cdots & \mathbf{B}_2(\mathbf{p}) & \mathbf{B}(\mathbf{p}) \\ \mathbf{C}_1(\mathbf{p}) & \mathbf{D}_1(\mathbf{p}) & \cdots & \mathbf{D}_{1q}(\mathbf{p}) & \mathbf{E}_1(\mathbf{p}) \\ \vdots & \vdots & \ddots & \vdots & \vdots \\ \mathbf{C}_q(\mathbf{p}) & \mathbf{D}_{1q}(\mathbf{p}) & \cdots & \mathbf{D}_{qq}(\mathbf{p}) & \mathbf{E}_q(\mathbf{p}) \\ \mathbf{C}(\mathbf{p}) & \mathbf{F}_1(\mathbf{p}) & \cdots & \mathbf{F}_q(\mathbf{p}) & \mathbf{D}(\mathbf{p}) \end{pmatrix} \begin{pmatrix} \mathbf{x} \\ \mathbf{w}_1 \\ \vdots \\ \mathbf{w}_q \\ \mathbf{u} \end{pmatrix} \tag{7}$$

where $\mathbf{w}_j \to \mathbf{v}_j$ are the channels on which we want to impose certain robustness and/or performance objectives. To facilitate further development, let the notation of (7) be simplified in a systematic form as:

$$\begin{pmatrix} \mathbf{z}_1 \\ \mathbf{z}_2 \\ \vdots \\ \mathbf{z}_K \end{pmatrix} = \begin{pmatrix} \mathbf{B}_{1,1}(\mathbf{p}) & \mathbf{B}_{1,2}(\mathbf{p}) & \cdots & \mathbf{B}_{1,L}(\mathbf{p}) \\ \mathbf{B}_{2,1}(\mathbf{p}) & \mathbf{B}_{2,2}(\mathbf{p}) & & \mathbf{B}_{2,L}(\mathbf{p}) \\ \vdots & & \ddots & \vdots \\ \mathbf{B}_{K,1}(\mathbf{p}) & \mathbf{B}_{K,2}(\mathbf{p}) & \cdots & \mathbf{B}_{K,L}(\mathbf{p}) \end{pmatrix} \begin{pmatrix} \mathbf{x}_1 \\ \mathbf{x}_2 \\ \vdots \\ \mathbf{x}_L \end{pmatrix} = \mathbf{S}(\mathbf{p}) \begin{pmatrix} \mathbf{x}_1 \\ \mathbf{x}_2 \\ \vdots \\ \mathbf{x}_L \end{pmatrix} \tag{8}$$

where $k = 1, \ldots, K$ and $l = 1, \ldots, L$, K being the number of rows in the model (8) (i.e. the number of equations describing the model), and L being the number of terms in the rows of the equations. For instance, $K = 2$ and $L = 2$ in (3). Vector $\mathbf{x}_l \in \mathbf{R}^{I_l}$ consists of the model input and state vectors, where I_l denotes the number of "input" elements in $\mathbf{x}_l$. Vector $\mathbf{z}_k \in \mathbf{R}^{O_k}$ contains the output values of the kth row in (8), where O_k denotes the number of "output" values in $\mathbf{z}_k$. This implies that the size of $\mathbf{B}_{k,l}(\mathbf{p})$ is $O_k \times I_l$. For example, describing (3) by (8) results in: $\mathbf{x}_1(t) = \mathbf{x}(t)$, $\mathbf{x}_2(t) = \mathbf{u}(t)$ and the outputs of the model are $\mathbf{z}_1(t) = \dot{\mathbf{x}}(t)$ and $\mathbf{z}_2(t) = \mathbf{y}(t)$. Coefficient matrices become: $\mathbf{B}_{1,1}(\mathbf{p}) = \mathbf{A}(\mathbf{p})$, $\mathbf{B}_{1,2}(\mathbf{p}) = \mathbf{B}(\mathbf{p})$, $\mathbf{B}_{2,1}(\mathbf{p}) = \mathbf{C}(\mathbf{p})$ and $\mathbf{B}_{2,2}(\mathbf{p}) = \mathbf{D}(\mathbf{p})$. Substitute (8) into (4) yields:

$$\hat{\mathbf{B}}_{k,l}(\mathbf{p}) = \sum_{v_1=1}^{V_1} \sum_{v_2=2}^{V_2} \cdots \sum_{v_N=1}^{V_N} \prod_{n=1}^{N} \alpha_{n,v_n}(p_n) \mathbf{B}_{v_1,v_2,\ldots,v_N,k,l}$$

which can be reformulated in terms of tensor product as:

$$\hat{\mathbf{B}}_{k,l}(\mathbf{p}) = \left(B_{k,l} \bigotimes_n \mathbf{m}_n(p_n) \right)_{(N+1)} \quad \text{or} \quad \mathbf{S}(\mathbf{p}) = \left(S \bigotimes_n \mathbf{m}_n(p_n) \right)_{(N+1)}$$

Here, the row vector $\mathbf{m}_n(p_n) \in \mathbf{R}^{V_n}$ contains the antecedent membership functions $\alpha_{n,v_n}(p_n)$, the $N + 2$ dimensional coefficient tensor $B_{k,l} \in \mathbf{R}^{V_1 \times V_2 \times \cdots \times V_N \times O_k \times I_l}$ is constructed from the matrices $\mathbf{B}_{v_1,v_2,\ldots,v_N,k,l} \in \mathbf{R}^{O_k \times I_l}$ and the tensor $S \in \mathbf{R}^{V_1 \times V_2 \times \cdots \times V_N \times \sum_k O_k \times \sum_l I_l}$ is constructed from system matrices $\mathbf{S}_{v_1,v_2,\ldots,v_N}$. The first N dimensions of $B_{k,l}$ are assigned to the dimensions of the parameter space P. The next two are assigned to the output and input vectors, respectively.

4 Complexity investigation

This section investigates the computation complexity of TS fuzzy model. The output values are calculated by (8) as:

$$\mathbf{z}_k = \left(\sum_l \left[B_{k,l} \bigotimes_n \mathbf{m}_n(p_n) \right] \times_{N+2} \mathbf{x}_l^{\mathrm{T}} \right)_{(N+1)} \tag{9}$$

Lemma 1. (Complexity explosion) *The computational complexity of TS fuzzy model techniques grows exponentially with the number of antecedent sets, the dimension of the parameter space and the size of the model coefficients. Specifically, in terms of the number of multiplications the computational requirement is characterized by:*

$$P = \prod_n V_n \left(\sum_k \sum_l O_k I_l + \sum_k O_k \right) + C_p \sum_n V_n, \tag{10}$$

where C_p indicates the number of multiplications within the calculation of a membership function.

To arrive at (10), one notes that calculating the output of one linear local model to a given input needs $\sum_k \sum_l O_k I_l$ multiplications. The number of the local linear models is $\prod_n V_n$. The outputs of the $\prod_n V_n$ local linear models are weighted by the antecedent membership functions, which implies $\prod_n V_n \cdot \sum_k O_k$ further multiplications. Finally, $C_p \sum_n V_n$ indicates the calculation of the membership functions, where C_p represents the number of multiplications in the calculation of one membership value. Consequently, (10) shows that increasing the density of the antecedent functions in pursue of good approximation leads to the explosion in the number of local linear models (building units) fully according to the paper by *Kóczy* and *Hirota* (1997) [12].

5 Key concept of HOSVD based reduction

This section briefly discusses the fundamentals of HOSVD in the sense of complexity reduction. Many reduction properties of the HOSVD of higher-order tensors have been investigated in related literatures. Here, let us shortly summarize those that have prominent roles in this paper. First, in multi-linear algebra as well as in matrix algebra, the Frobenius norm is unitary invariant. As a consequent, the squared Frobenius norm of a matrix can be generalized as equal to the sum of its squared singular values.

Property 1. (Approximation) *Let the HOSVD of A be given as in Theorem 2 and let the n-mode rank of A be equal to R_n. Define a tensor $\hat{A}$ by discarding singular values $\sigma^{(n)}_{I'_n+1}, \sigma^{(n)}_{I'_n+2}, \cdots, \sigma^{(n)}_{R_n}$ for given values of I'_n, i.e. set the corresponding parts of S equal to zero. Then we have:*

$$\left\| A - \hat{A} \right\|^2 \leq \sum_{n=1}^{N} \left(\sum_{i_n = I'_n+1}^{R_n} (\sigma^{(n)}_{i_n})^2 \right). \tag{11}$$

This property is the higher-order equivalent of the link between the SVD of a matrix and its best approximation in a least-squares sense, by a matrix of lower rank. The situation is, however, quite different for tensors. By discarding the smallest n-mode singular values, one obtains a tensor $\hat{A}$ with n-mode rank of I'_n, but this tensor is, in general, not the best possible approximation under the given n-mode rank constraints [16]. Nevertheless, the ordering implies that the main components of A are mainly concentrated in the part corresponding to low values of the indices. Consequently, if $\sigma^{(n)}_{I'_n} \gg \sigma^{(n)}_{I'_{n+1}}$, where I'_n actually corresponds to the numerical ranking of A, then the smaller n-mode singular values are not significant, justifying that they be discarded. In this sense, the $\hat{A}$ as obtained can still be considered as a good approximation of A. According to the present context, the following terms are therefore defined [29,30]:

Definition 9. (Exact/non-exact reduction) *Assume an Nth order tensor $A \in \mathbf{R}^{I_1 \times I_2 \times \cdots \times I_N}$.* ***Exact*** *reduced form $A = A^r \bigotimes_n \mathbf{U}_n$, where "r" denotes "reduced", is defined by the tensor $A^r \in \mathbf{R}^{I_1^r \times I_2^r \times \ldots \times I_N^r}$ and the n-mode singular matrices $\mathbf{U}_n \in \mathbf{R}^{I_n \times I_n^r}$, $\forall n : I_n^r \leq I_n$, as according to the results of Theorem 2 and discarding only the zero singular values and corresponding singular vectors.* ***Non-exact*** *reduced form $\hat{A} = A^r \otimes_n \mathbf{U}_n$ is obtained similarly but with some non-zero singular values and corresponding singular vectors discarded.*

6 SVD based complexity reduction of TS models

The main objective of the complexity reduction proposed in this section is twofold, as introduced via the following two methods. Method 1 is aimed to minimize values V_n, namely, to decrease the size of tensor $B_{k,l}$ in the first N dimension. This leads to the minimal number of local linear models. The reduction conducts HOSVD on tensor $B_{k,l}$ to root out linear dependencies by truncating zero or small singular values. In this discussion, exact reduction is first presented. Then, non-exact reduction is given to increase the effectiveness of the reduction by discarding some non-zero singular values in HOSVD. A comment on the reduction error for the process will be given in Remark 2 at the end of the section. Subsequently, the aim of the reduction in Method 2 is to decrease values O_k and I_l also appearing in the dominant term of (10). As the number of input and output values are defined by the problem at hand, O_k and I_l cannot be directly reduced. Similarly to [2], the key idea of reducing these values can be viewed as transforming the whole approximation to a smaller computational space. The input values are also projected in each state step of the modelled system, and the output values are calculated in the reduced computational space. Finally, the output values are transformed back to the original space. The reduction is based on executing SVD reduction to the coefficient matrices. As a matter of fact, exact reduction cannot be obtained in this step if the coefficient matrices are full in rank, which is usually guaranteed by modelling processes. Non-exact reduction is, however, still possible at the price of reduction error. Note that in this case the coefficient matrices computed by the reduced approximation will not be in full rank, which is not acceptable in various theorems of control design. In order to have a complete view, both reduction possibilities are discussed in the next part. First, let us characterize the concept and the goal of the reduction by the following theorem.

Theorem 3. (Complexity reduction) *Equation* (9) *can always be transformed into the following form:*

$$\mathbf{z}_k = \left(\sum_l \left[B^r_{k,l} \bigotimes_n \mathbf{m}^r_n(p_n) \right] \times_{N+1} \mathbf{A}_k \times_{N+2} \mathbf{x}_l^T \mathbf{C}_l \right)_{(N+1)}$$

which is equivalent to

$$\mathbf{z}_k = \mathbf{A}_k \left(\sum_l \left[B^r_{k,l} \bigotimes_n \mathbf{m}^r_n(p_n) \right] \times_{N+2} \mathbf{x}_l^T \mathbf{C}_l \right)_{(N+1)}, \tag{12}$$

and the size of $B^r_{k,l} \in \mathbf{R}^{V^r_1 \times V^r_2 \times \cdots \times V^r_N \times O^r_k \times I^r_l}$ *may be reduced as* $\forall n : V^r_n \leq V_n$, $O^r_k \leq O_k$ *and* $I^r_l \leq I_l$.

The quantity $\mathbf{m}^r_n(p_n) \in \mathbf{R}^{V^r_n}$ here constitutes the new antecedent functions. The number of the antecedent functions in the nth input dimension is V^r_n. $\mathbf{A}_k \in \mathbf{R}^{O_k \times O^r_k}$ and $\mathbf{C}_l \in \mathbf{R}^{I_l \times I^r_l}$ are applied to transform the inputs and the outputs between the reduced and the original computational space, see later at Method 2.

The proof of the Theorem 3 can readily be derived from the following Methods 1 and 2. Before going into details, let us have a brief digression, and represent the calculation of values $\mathbf{z}_k$ in respect to $\mathbf{x}_l$ in two different ways, similarly to [2]. Let tensor $G_k \in \mathbf{R}^{V_1 \times V_2 \times \cdots \times V_N \times O_k \times \left(\sum_l I_l\right)}$ be given by the form of $G_k = \left[B_{k,1}\ B_{k,2} \cdots B_{k,L}\right]_{N+2}$. The output value $\mathbf{z}_k$ of the approximation in respect to $\mathbf{x}_k$ is:

$$\mathbf{z}_k = \left(\left[G_k \bigotimes_n \mathbf{m}_n(p_n) \right] \times_{N+2} \left[\mathbf{x}_1^T\ \mathbf{x}_2^T \cdots \mathbf{x}_L^T\right] \right)_{(N+1)}.$$

The second way utilizes matrix $H_l \in \mathbf{R}^{V_1 \times V_2 \times \cdots \times V_N \times \left(\sum_k O_k\right) \times I_l}$ constructed by $H_l = \left[B_{1,l}\ B_{2,l} \cdots B_{K,l}\right]_{N+1}$. The output of the TS fuzzy model is:

$$\begin{bmatrix} \mathbf{z}_1 \\ \mathbf{z}_2 \\ \vdots \\ \mathbf{z}_K \end{bmatrix} = \left(\left[\left[H_1\ H_2 \cdots H_L\right]_{N+2} \bigotimes_n \mathbf{m}_n(p_n) \right] \times_{N+2} \left[\mathbf{x}_1^T\ \mathbf{x}_2^T \cdots \mathbf{x}_L^T\right] \right)_{(N+1)}. \tag{13}$$

The first method shows how to find the minimal number of local linear models.

Method 1 (Determination of the minimal values of $V_1, V_2, \ldots, V_N$)

Applying HOSVD to the $(N+2)$-dimensional system tensor

$$S = \left[G_1 \; G_2 \cdots G_K\right]_{N+1}$$

in such a way that the SVD is executed only for the dimensions $1 \ldots N$. This yields:

$$S = S^r \underset{n}{\otimes} \mathbf{T}_n \tag{14}$$

where "r" denotes "reduced". Tensors $B^r_{k,l} \in \mathbf{R}^{V_1^r \times V_2^r \times \ldots \times V_N^r \times O_l \times I_l}$ are found as the partitions of S^r

$$S^r = \left[G_1^r \; G_2^r \cdots G_K^r\right]_{N+1} \quad \text{and} \quad G_k^r = \left[B^r_{k,1} \; B^r_{k,2} \cdots B^r_{k,L}\right]_{N+2}.$$

If singular values are discarded, then the size of $B^r_{k,l} \in \mathbf{R}^{V_1^r \times V_2^r \times \ldots \times V_N^r \times O_l \times I_l}$ is less than the size of $B_{k,l} \in \mathbf{R}^{V_1 \times V_2 \times \ldots \times V_N \times O_l \times I_l}$, so, $\forall n : V_n^r \leq V_n$, which is the key point of the reduction. Thus, for (14) we obtain

$$B_{k,l} = B^r_{k,l} \underset{n}{\otimes} \mathbf{T}_n. \tag{15}$$

The new antecedent functions of the rules are constructed as

$$\mathbf{m}_n^r(p_n) = \mathbf{m}_n(p_n)\mathbf{T}_n. \tag{16}$$

Consequently, (9) can be written in the reduced form by substituting (15) and (16) into (9) which yields:

$$\mathbf{z}_k = \left(\sum_l \left[B^r_{k,l} \underset{n}{\otimes} \mathbf{m}_n^r(p_n)\right] \times_{N+2} \mathbf{x}_l^{\mathrm{T}}\right)_{(N+1)}$$

in full accordance with Theorem 3 of complexity reduction. On the other hand, Method 2 is aimed at decreasing O_k and I_l.

Method 2 (Determination of the minimal computational space)

1) Determination of matrices $\mathbf{A}_k$, namely, the reduction of O_k.

Let $\mathbf{R}_k = (G_k)_{(1)}$. Applying SVD to $\mathbf{R}_k$ and discarding the zero singular values yields:

$$\mathbf{R}_k = \mathbf{A}_k \cdot \mathbf{D}_k \cdot \mathbf{V}_k = \mathbf{A}_k \cdot \mathbf{R}'_k,$$

Matrix $\mathbf{R}'_k \in \mathbf{R}^{O_l^r \times \prod_n V_n \cdot \sum_l I_l}$ can be restored to tensor $G'_k \in \mathbf{R}^{V_1 \times V_2 \times \cdots \times V_N \times O_k^r \times \left(\sum_l I_l\right)}$.

2) Determination of matrices $\mathbf{C}_l$, namely, the reduction of I_l

Let tensor $H'_k \in \mathbf{R}^{V_1 \times V_2 \times \cdots \times V_N \times \sum_k O_k \times I_l}$ be constructed by tensors $B'_{k,l}$ are defined accordingly to the result G'_k of step 1), where

$$H'_l = \left[B'_{1,l}\ B'_{2,l}\ \cdots\ B'_{K,l}\right]_{N+1}, \quad \text{and} \quad G'_k = \left[B'_{k,1}\ B'_{k,2}\ \cdots\ B'_{k,L}\right]_{N+2}$$

Then let $\mathbf{M}_l = (H'_l)_{(N+2)}$ where upon executing SVD yields (zero singular values are discarded):

$$\mathbf{M}_l = \mathbf{C}_l \cdot \mathbf{D}'_l \cdot \mathbf{V}'_l = \mathbf{C}_l \cdot \mathbf{M}'_l.$$

Matrix $\mathbf{M}'_l$ defines tensors $B^r_{k,l} \in \mathbf{R}^{V_1 \times V_2 \times \cdots \times V_N \times O^r_k \times I^r_l}$ accordingly to $\mathbf{M}'_l = (H''_l)_{(N+2)}$ and $H''_l = \left[B^r_{1,l}\ B^r_{2,l}\ \cdots\ B^r_{K,l}\right]_{N+1}$.

The outcomes of Method 2 are $\mathbf{A}_k$ and $\mathbf{C}_l$. The quantity $\mathbf{C}_l$ is applied to transform the input values $\mathbf{x}_l$ to a reduced space as: $\mathbf{x}^r_l = \mathbf{C}^T_l \cdot \mathbf{x}_l$. The output is calculated in the reduced computational space as:

$$\mathbf{z}^r_k = \left(\sum_l \left[B^r_{k,l} \underset{n}{\otimes} \mathbf{m}_n(p_n)\right] \times_{N+2} (\mathbf{x}^r_l)^{\mathrm{T}}\right)_{(N+1)}.$$

The output $\mathbf{z}^r_k$ is projected to the original space by $\mathbf{z}_k = \mathbf{A}_k \cdot \mathbf{z}^r_k$, in full accordance with the Theorem 3 of complexity reduction.

The ordering of executing Method 1 and 2 is arbitrary. In the followings some important issues and interpretability problem of the results are discussed.

Remark 1. The antecedent functions in (16) obtained by Method 1 may not be interpretable as membership functions satisfying property (5), as the transformation using matrix $\mathbf{T}_n$ may result in functions with negative values. Another crucial point is that the resulted antecedent functions do not guarantee the Ruspini partition, which means that (6) may not be equal to 1. This fact would destroy the whole reduction concept since calculating (6) with the new antecedents may get far from 1. However, if only the saving of computational cost of final implementation is in purpose and the conditions (5) and (6) of the membership functions do not have to be accommodated then (12) is directly applicable. If the reduced form is for further studies in fuzzy theory and/or Lyapunov stability analysis, then the reduced antecedent functions should accommodate additional characterization pertaining to specific operations. This may require further transformations. To obtain matrices $\mathbf{T}_n$ in such a way that the reduced antecedent functions are bounded by $[0, 1]$ and hold (6), *Non-Negativeness* (NN) and *Sum Normalization* (SN) transformation techniques are developed by *Yam* in [29,30]. If SVD is accompanied by these transformations, then the resulted functions fulfil (5) and (6), and hence remain interpretable as antecedent fuzzy sets. This leads to the theoretically correct use of (9) and (8).

Remark 2. An advantage of the proposed algorithm is that it has error controllable property, i.e. if the HOSVD is executed in non-exact mode then the original and the reduced approximation differ, and the difference can be estimated during executing the reduction technique. In Section 5 it is shown that discarding non-zero singular values results in reduction error bounded by (11). Works [2–5,29,30] bound the maximum reduction error by the sum of the discarded singular values. As a matter of fact, the reduction errors of the proposed methods also depend on the type of the membership functions applied. In this regard various cases of antecedent functions are discussed in [5]. Generally speaking, it can be said that if the original antecedent sets hold (6) then the maximum final model approximation error is the sum of the discarded singular values, which can be controlled during executing Methods 1 and 2. For more details about the error bound of SVD reduction see works [2–5,29,30].

Remark 3. Method 1 may result in antecedent functions which cannot be analytically simplified, and hence their shapes are rather complicated and their computational loads may be greater than that of the original ones. However, it can be observed that C_p is not a dominant part in (10), which implies that this computational increase of the new antecedent is dispensable comparing to the exponential feature of the dominant term. In the worst case, the original functions can be calculated first, after which the values of the reduced membership functions are simply determined by (16) in each step of the system. Therefore the worst case is bounded by

$$P = \prod_n V_n^r \left(\sum_k \sum_l O_k^r I_k^r + \sum_k O_k^r \right) + C_p \sum_n V_n + \sum_n V_n V_n^r + \\ + \sum_k O_k O_k^r + \sum_l I_l I_l^r, \tag{17}$$

where the extra term $V_n V_n^r$ indicates the extra computational load of calculating the values of the antecedent functions in the nth input dimension. The terms $\sum_k O_k O_k^r$ and $\sum_l I_l I_l^r$ are the computation requirement of the transformation between the original and the reduced computational spaces. Consequently, the effectiveness of the reduction is

$$\eta = \frac{\prod_n V_n^r \left(\sum_k \sum_l O_k^r I_k^r + \sum_k O_k^r \right) + C_p \sum_n V_n + \sum_n V_n V_n^r + \sum_k O_k O_k^r + \sum_l I_l I_l^r}{\prod_n V_n \left(\sum_k \sum_l O_k I_k + \sum_k O_k \right) + C_p \sum_n V_n}.$$

In the case of a dense grid or higher dimensional parameter space the effectiveness of the reduction in the sense of computational complexity can be largely expressed as:

$$\eta \approx \frac{\prod_n V_n^r \left(\sum_k \sum_l O_k^r I_k^r + \sum_k O_k^r \right)}{\prod_n V_n \left(\sum_k \sum_l O_k I_k + \sum_k O_k \right)}.$$

Remark 4. Method 1 can be modified in such a way that reduction can be carried out for each row or column of (2) like in [2]. Furthermore, one antecedent system could be resulted for each coefficient tensor $B_{k,l}$. The advantage of the separately executed reduction of each $B_{k,l}$ in Method 1 is that the size of some $B^r_{k,l}$ may become less, while computing (14) the sizes of all partitions $B_{k,l}$ of S^r are the same in the first N dimension. This is due to the fact that the n-mode rank of tensor $B_{k,l}$ is less or equal to the n-mode rank of tensor S in (14). Consequently, replacing S in (14) with $B_{k,l}$ yields:

$$B_{k,l} = B^r_{k,l} \bigotimes_n \mathbf{T}_{n,k,l}.$$

According to (16), the new antecedent systems are then:

$$\mathbf{m}^r_{n,k,l}(p_n) = \mathbf{m}_n(p_n)\mathbf{T}_{n,k,l},$$

where the sets defined by $\mathbf{m}^r_{n,k,l}(p_n)$ are assigned to the approximation of $B_{k,l}$. Again, the benefit is that the size of each $B^r_{k,l}$ in the present modified case is less or equal to the common size of $B^r_{k,l}$ obtained in Method 1. As a matter of fact, the calculation cost of the antecedents may increase since different antecedent system should be calculated for each $B_{k,l}$. This extra calculation has not been included in the exponentially dominant part of (10) and (17). Such fine tuning of the reduction actually depends on the fact, and it is to be checked, whether performing the reduction for each coefficient tensor separately would yield a better computational reduction or not.

7 Example

This example is a controller design to a simple mass-spring-damper mechanical system. The controller is designed with the help of the proposed HOSVD based technique. In order to see the effectiveness of the proposed approach, the controller design is shown in an analytical way as well. Here, we chose a detailed demonstrative example from the works of *Tanaka et al.* [23], where a *Parallel Distributed Compensation* (PDC) controller design is performed via analytical derivation: Consider the non-linear mass-spring-damper mechanical system as depicted on Figure 2 and its dynamical equation,

$$m \cdot \ddot{x} + g(x, \dot{x}) + k(x) = \phi(\dot{x}) \cdot u \tag{18}$$

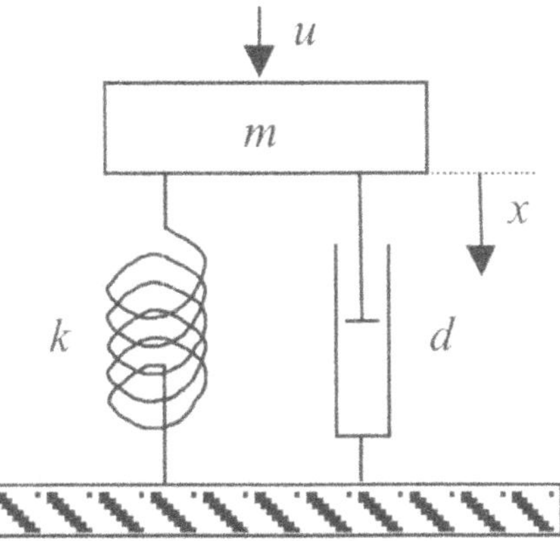

Fig. 2. Mass-spring-damper system

where m is the mass and u stands for the force. The function $k(x)$ is the non-linear or uncertain stiffness coefficient of the spring, $g(x, \dot{x})$ is the non-linear or uncertain term damping coefficient of the damper, and $\phi(\dot{x})$ is the non-linear input term. Assume that $g(x, \dot{x}) = d(c_1 x + c_2 \dot{x}^3)$, $k(x) = c_3 x + c_4 x^3$, and $\phi(\dot{x}) = 1 + c_5 \dot{x}^3$. Furthermore, assume that $x \in [-a, a]$, $\dot{x} \in [-b, b]$ and $a, b > 0$. The above parameters are set as follows [27]: $m = 1$, $d = 1$, $c_1 = 0.01$, $c_2 = 0.1$, $c_3 = 0.01$, $c_4 = 0.67$, $c_5 = 0$, $a = 1.5$, and $b = 1.5$. Equation (18) then becomes:

$$\ddot{x} = -0.1\dot{x}^3 - 0.02x - 0.67x^3 + u. \tag{19}$$

The non-linear terms are $-0.1\dot{x}^3$ and $-0.67x^3$.

7.1 Analytical design

Let us follow the detailed derivations of [23], and give the TS model form of (19). x and $\dot{x}$ have the following conditions:

$$\begin{cases} -1.5075x \leq -0.67x^3 \leq 0 \cdot x, & \text{if } x \geq 0 \\ 0 \cdot x \leq -0.67x^3 \leq -1.5075x, & \text{if } x < 0 \end{cases}$$

$$\begin{cases} -0.225\dot{x} \leq -0.1\dot{x}^3 \leq \dot{x} \cdot 0, & \text{if } x \geq 0 \\ 0 \cdot \dot{x} \leq -0.1\dot{x}^3 \leq -0.225\dot{x}, & \text{if } x < 0 \end{cases}.$$

This fact means that the non-linear terms can be represented by the upper and the lower bound.

$$\begin{aligned} -0.67x^3 &= f_{1,1}(x)x \cdot 0 - (1 - f_{1,1}(x)) \cdot 1.5075x \\ -0.1\dot{x}^3 &= f_{2,1}(\dot{x})\dot{x} \cdot 0 - (1 - f_{2,1}(\dot{x})) \cdot 0.225\dot{x}, \end{aligned}$$

where $f_{n,v_n}(\dot{x}) \in [0,1]$, $V_n = 2$. This leads to antecedent functions $f^a_{1,1}(x) = 1 - \frac{x^2}{2.25}$, (superscript "a" means that the function is obtained analytically), $f^a_{1,2}(x) = \frac{x^2}{2.25}$; $f^a_{2,1}(\dot{x}) = 1 - \frac{\dot{x}^2}{2.25}$; $f^a_{2,2}(\dot{x}) = \frac{\dot{x}^2}{2.25}$. The membership functions are depicted on Figure 4. Consequently, the following equality is obtained:

$$\ddot{x} = -0.1\dot{x}^3 - 0.02x - 0.67x^3 + u = \sum_{i=1}^{2}\sum_{j=1}^{2} f^a_{1,i}(x) f^a_{2,j}(\dot{x})\phi_{i,j}$$

where $\phi_{1,1} = -0.02x + u$, $\phi_{1,2} = -0.225\dot{x} - 0.02x + u$, $\phi_{2,1} = -1.5275x + u$ and $\phi_{2,2} = -0.225\dot{x} - 1.5275x + u$. This approximation in matrix representation takes the form:

$$\mathbf{A}(x,\dot{x}) = \sum_{i=1}^{2}\sum_{j=1}^{2} f^a_{1,i}(x) f^a_{2,j}(\dot{x})\mathbf{A}^a_{i,j}; \quad \mathbf{B}(x,\dot{x}) = \sum_{i=1}^{2}\sum_{j=1}^{2} f^a_{1,i}(x) f^a_{2,j}(\dot{x})\mathbf{B}^a_{i,j} \tag{20}$$

where

$$\mathbf{A}^a_{1,1} = \begin{bmatrix} 0 & -0.02 \\ 1 & 0 \end{bmatrix}, \quad \mathbf{B}^a_{1,1} = \begin{bmatrix} 1 \\ 0 \end{bmatrix}, \quad \mathbf{A}^a_{1,2} = \begin{bmatrix} -0.225 & -0.02 \\ 1 & 0 \end{bmatrix}, \quad \mathbf{B}^a_{1,2} = \begin{bmatrix} 1 \\ 0 \end{bmatrix},$$

$$\mathbf{A}^a_{2,1} = \begin{bmatrix} 0 & -1.5275 \\ 1 & 0 \end{bmatrix}, \quad \mathbf{B}^a_{2,1} = \begin{bmatrix} 1 \\ 0 \end{bmatrix}, \quad \mathbf{A}^a_{2,2} = \begin{bmatrix} -0.225 & -1.5275 \\ 1 & 0 \end{bmatrix}, \quad \mathbf{B}^a_{2,2} = \begin{bmatrix} 1 \\ 0 \end{bmatrix}.$$

Consequently, the fuzzy rules are constructed as:

$$\textbf{IF } f^a_{1,i}(x) \textbf{ AND } f^a_{2,j}(\dot{x}) \textbf{ THEN } \dot{\mathbf{x}}(t) = \mathbf{A}^a_{i,j}\mathbf{x}(t) + \mathbf{B}^a_{i,j}\mathbf{u}(t)$$

This implies that $\dot{\mathbf{x}}(t) = \sum_{i=1}^{2}\sum_{j=1}^{2} f^a_{1,i}(x) f^a_{2,j}(\dot{x})(\mathbf{A}^a_{i,j}\mathbf{x}(t) + \mathbf{B}^a_{i,j}\mathbf{u}(t))$. The resultant TS model with four linear local models exactly represents the nonlinear system given in (18) and (19). Note that the fuzzy system has the common $\mathbf{B}$. A fuzzy controller design via PDC determines four feedback gains as consequent parts. If we select the closed-loop eigenvalues to be $[-2\ -2]$, the following feedback gains are obtained to each local linear model:

$$\mathbf{K}^a_{1,1} = [4\ 3.98], \qquad \mathbf{K}^a_{1,2} = [3.775\ 3.98],$$
$$\mathbf{K}^a_{2,1} = [4\ 2.4725], \qquad \mathbf{K}^a_{2,2} = [3.775\ 2.4725].$$

Then the fuzzy controller takes the form:

$$\textbf{IF } f^a_{1,i}(x) \textbf{ AND } f^a_{2,j}(\dot{x}) \textbf{ THEN } \mathbf{u}(t) = -\mathbf{K}^a_{i,j}\mathbf{x}(t).$$

Thus,

$$\mathbf{u}(t) = -\sum_{i=1}^{2}\sum_{j=1}^{2} f^a_{1,i}(x) f^a_{2,j}(\dot{x})\mathbf{K}^a_{i,j}$$

Finally let the stability of the nonlinear system be analyzed. We apply the stability criteria C14, introduced in [23]. The criteria C14 is slightly specialized in Remark 4.2 of [23] according to the assumption that $\forall i,j : \mathbf{B}^a_{i,j} = \mathbf{B}^a$. Let us recall this simplified criteria:

Theorem 4. (Quadratical stability) *Equation*

$$\dot{\mathbf{x}}(t) = \sum_{i=1}^{I}\sum_{j=1}^{J} f_{1,i}(x) f_{2,j}(\dot{x}) \left(\mathbf{A}_{i,j}\mathbf{x}(t) + \mathbf{B}\mathbf{u}(t)\right)$$

where $\mathbf{u}(t) = -\sum_{i=1}^{2}\sum_{j=1}^{2} f_{1,i}(x) f_{2,j}(\dot{x})\mathbf{K}_{i,j}$ *is quadratically stable if and only if the following condition holds:*

$$\operatorname{Re}\lambda_i(\mathbf{H}) \neq 0 \quad where \quad \mathbf{H} = \begin{bmatrix} \mathbf{G} & \mathbf{0}_{n\times n} \\ \frac{r(r+1)}{2}\mathbf{I}_{n\times n} & -\mathbf{G}^{\mathrm{T}} \end{bmatrix}, \quad for \quad i = 1, 2, \ldots, 2\times n,$$

where $\mathbf{G} = \frac{1}{IJ}\sum_{i=1}^{I}\sum_{j=1}^{J}\left(\mathbf{A}_{i,j} - \mathbf{B}\mathbf{K}_{i,j}\right)$

In the present example $\mathbf{G} = \begin{bmatrix} -4 & -4 \\ 1 & 0 \end{bmatrix}$, which implies $\mathbf{H} = \begin{bmatrix} -4 & -4 & 0 & 0 \\ 1 & 0 & 0 & 0 \\ 10 & 0 & 4 & -1 \\ 0 & 10 & 4 & 0 \end{bmatrix}$.

The eigenvalues of $\mathbf{H}$ are -2.0, -2.0, 2.0 and 2.0. Therefore the designed fuzzy controller quadratically stabilizes the nonlinear system. We should remark here that the feedback system is linearized by the designed controller, because the fuzzy model has the common matrix $\mathbf{B}^a$ and $\mathbf{G} = \mathbf{A}^a_{i,j} - \mathbf{B}^a\mathbf{K}^a_{i,j}$.

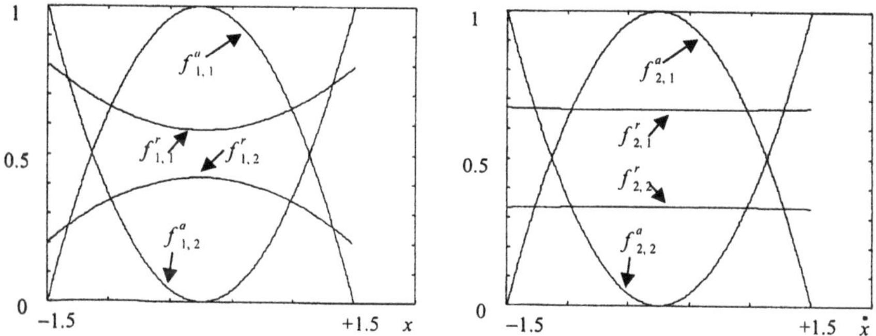

Fig. 3. Antecedent functions of the original approximation via analytical derivation and the antecedents extracted from training data by HOSVD reduction

7.2 Design via TS approximation and HOSVD reduction

In this part we assume that the analytical derivation from the model (18) to a TS fuzzy model (20) is unknown like in usual practical cases. Therefore, we utilize TS fuzzy approximation of the model over a dense approximation grid, namely, we simply sample the differential equations over the grid points. Note that this approximation can also be extracted by various fuzzy learning techniques from training data measured on the real system if the differential equations are unknown. The more dense approximation grid we use in the hope of resulting acceptable approximation accuracy, the more (exponentially growing) complex model we obtain. Therefore, in the next step we go about generating the minimal sized TS model, namely, finding the minimal number of linear local models, based on the proposed HOSVD technique. This may help us with reducing the complexity of the controller design and the resulted controller itself, which is the main purpose of this example. Then, we utilize PDC based controller design on the minimized model, which, hence, results in a controller with minimal complexity, and finally we check the stability of the designed system based on Theorem 4.

I) Approximation over a dense approximation grid

Let intervals $\dot{x}, x \in [-1.5, 1.5]$ be divided by 400 triangular shaped antecedent functions (or, in other words, by first order B-spline basis), see Figure 3.

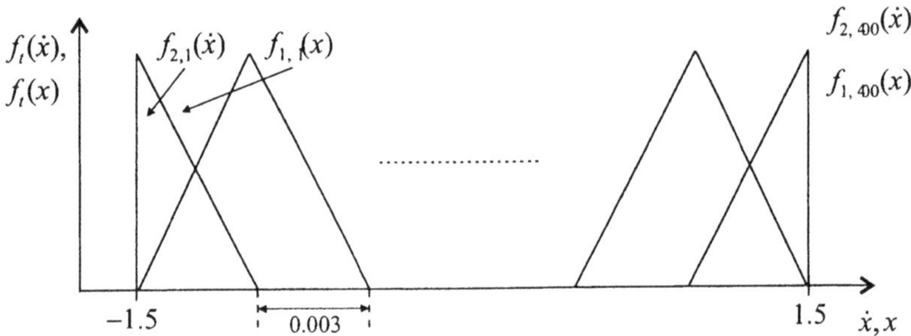

Fig. 4. Densely located antecedents to achieve a good approximation

Therefore the approximation is determined in the form as:

$$\hat{\mathbf{A}}(x, \dot{x}) = \sum_{i=1}^{400} \sum_{j=1}^{400} f_{1,i}(x) f_{2,j}(\dot{x}) \mathbf{A}_{i,j}; \quad \hat{\mathbf{B}}(x, \dot{x}) = \sum_{i=1}^{400} \sum_{j=1}^{400} f_{1,i}(x) f_{2,j}(\dot{x}) \mathbf{B}_{i,j} \tag{21}$$

We sample the differential equations over the approximation grid points defined by $x_i = -1.5 + (i-1)3/400$ and $\dot{x}_j = -1.5 + (j-1)3/400$. Thus the dense approximation becomes

$$\ddot{x} = \sum_{i=1}^{400} \sum_{j=1}^{400} f_{1,i}(x) f_{2,j}(\dot{x})(a_{i,j}\dot{x} + b_{i,j}x + c_{i,j}u), \tag{22}$$

where $a_{i,j} = -0.1(-1.5 + (j-1)3/400)^2$, $b_{i,j} = -0.02 - 0.67(-1.5 + (i-1) \cdot 3/400)^2$, and $c_{i,j} = 1$. The matrix form can easily be generated from (22), which results in

$$\hat{\mathbf{A}}(x, \dot{x}) = \sum_{i=1}^{400} \sum_{j=1}^{400} f_{1,i}(x) f_{2,j}(\dot{x}) \mathbf{A}_{i,j}; \quad \hat{\mathbf{B}}(x, \dot{x}) = \sum_{i=1}^{400} \sum_{j=1}^{400} f_{1,i}(x) f_{2,j}(\dot{x}) \mathbf{B}_{i,j}$$

This approximation employs $400 \times 400 = 160000$ linear local models.

II) Complexity reduction by Method 2

Executing Method 2 in exact mode on matrices $\mathbf{A}_{i,j}$, namely, on tensor $A \in \mathbf{R}^{400 \times 400 \times 2 \times 2}$ (note that matrices $\mathbf{B}_{i,j}$ are constant) results in two non-zero singular values to the first dimension such as $461.6404\ldots$ and $156.5663\ldots$, and after performing SN and NN transformation (see Remark 1) two non-zero singular values are obtained to the second dimension, such as $100.8708\ldots$ and $1.8970\ldots$. The resulted coefficient matrices are:

$$\begin{aligned}
\mathbf{A}^r_{1,1} &= \begin{bmatrix} -169.5952205449\ldots & -2.871864441639\ldots \\ 1 & 0 \end{bmatrix}, \\
\mathbf{A}^r_{1,2} &= \begin{bmatrix} 338.965358635779\ldots & -2.871864441665\ldots \\ 1 & 0 \end{bmatrix}, \\
\mathbf{A}^r_{2,1} &= \begin{bmatrix} -169.595220544832\ldots & 3.895957360524\ldots \\ 1 & 0 \end{bmatrix}, \\
\mathbf{A}^r_{2,2} &= \begin{bmatrix} 338.965358635644\ldots & 3.895957360558\ldots \\ 1 & 0 \end{bmatrix}.
\end{aligned} \tag{23}$$

This means that two antecedent sets on each dimension are sufficient for the same approximation, which is in full accordance with the analytical TS fuzzy model design. Furthermore, the resultant antecedent sets maintains the properties (5) and (6). The main conclusion is that the PDC design (or any further LMI analysis) can be restricted to the resulted four linear local models. Let us proceed further and determine the antecedent sets. The new membership functions inherit the piecewise linear property of the original triangular shaped membership functions. We approximate the break points of the pieces, (which are actually the elements in the columns of $\mathbf{T}_n$ [5]) by a polynomial fitting, which results in:

$$\begin{aligned}
f^r_{1,1}(x) &= \alpha_1 + \beta_1 x^2, & f^r_{1,2}(x) &= 1 - f^r_{1,1}(x), \\
f^r_{2,1}(\dot{x}) &= \alpha_2 + \beta_2 \dot{x}^2, & f^r_{2,2}(\dot{x}) &= 1 - f^r_{2,1}(\dot{x})
\end{aligned} \tag{24}$$

where $\alpha_1 = 0.5786141354\ldots$, $\beta_1 = 0.09899787843\ldots$, $\alpha_2 = 0.6665191376\ldots$ and $\beta_2 = 1.966334082796 \cdot 10^{-4}$. The membership functions are depicted on Figure 4. Let us take a brief digression here and show via linear transformations that the model obtained in (24) is a variant form of (20). The analytically derived antecedents can be transformed to the reduced sets as (in the following steps the equivalency is understood in numerical sense):

$$\mathbf{m}_1^r = \mathbf{m}_1^a \mathbf{T}_1 \quad \text{and} \quad \mathbf{m}_2^r = \mathbf{m}_2^a \mathbf{T}_2$$

where

$$\mathbf{m}_1^r(x) = \left[f_{1,1}^r(x)\ f_{1,2}^r(x)\right], \qquad \mathbf{m}_2^r(\dot{x}) = \left[f_{2,1}^r(\dot{x})\ f_{2,2}^r(\dot{x})\right],$$
$$\mathbf{m}_1^a(x) = \left[f_{1,1}^a(x)\ f_{1,2}^a(x)\right], \qquad \mathbf{m}_2^a(\dot{x}) = \left[f_{2,1}^a(\dot{x})\ f_{2,2}^a(\dot{x})\right].$$

The transformation matrices are:

$$\mathbf{T}_1 = \begin{bmatrix} 0.57861413538877\ldots & 0.42138586461123\ldots \\ 0.80135936185347\ldots & 0.19864063814653\ldots \end{bmatrix},$$
$$\mathbf{T}_2 = \begin{bmatrix} 0.66651913756641\ldots & 0.33348086243359\ldots \\ 0.66696156273504\ldots & 0.33303843726497\ldots \end{bmatrix}.$$

In the same way $\mathcal{A}^r = \mathcal{A}^a \times_1 \mathbf{T}_1^{-1} \times_2 \mathbf{T}_2^{-1}$, where coefficient tensors $\mathcal{A}^r \in \mathbf{R}^{2\times2\times2\times2}$ and $\mathcal{A}^a \in \mathbf{R}^{2\times2\times2\times2}$ are respectively constructed from matrices $\mathbf{A}_{i,j}^r$ and $\mathbf{A}_{i,j}^a$. Consequently,

$$\mathcal{A}^r \times_1 \mathbf{f}_1^r(x) \times_2 \mathbf{f}_2^r(x) = \mathcal{A}^a \times_1 \mathbf{T}_1^{-1} \times_2 \mathbf{T}_2^{-1} \times_1 (\mathbf{f}_1^a(x)\mathbf{T}_1) \times_2 (\mathbf{f}_2^a(x)\mathbf{T}_2) =$$
$$= \mathcal{A}^a \times_1 (\mathbf{f}_1^a(x)\mathbf{T}_1\mathbf{T}_1^{-1}) \times_2 (\mathbf{f}_2^a(x)\mathbf{T}_2\mathbf{T}_2^{-1}) = \mathcal{A}^a \times_1 \mathbf{f}_1^a(x) \times_2 \mathbf{f}_2^a(x).$$

We can conclude that the two models are equivalent with the model given by differential equations. Equivalency of the models is understood here in numerical sense, e.g. the difference between the outputs of the models to the same inputs is under $\varepsilon < 10^{-12}$. Figure 5 shows the response of the analytically derived and the reduced rule base in the case of step change. We can observe that the output signals are equivalent.

III) Controller design by PDC applying the reduced TS model

Along in the same line as in the above analytic design, let the closed-loop eigenvalues be $\left[-2\ -2\right]$. Thus, the feedback gains are:

$$\mathbf{K}_{1,1}^r = \left[-165.5952205449\ldots\ 1.128135558369\ldots\right],$$
$$\mathbf{K}_{1,2}^r = \left[342,965358635779\ldots\ 1.128135558319\ldots\right],$$
$$\mathbf{K}_{2,1}^r = \left[-165.595220544832\ldots\ 7.895957360534\ldots\right],$$
$$\mathbf{K}_{2,2}^r = \left[342.965358635644\ldots\ 7.89595736054\ldots\right]$$

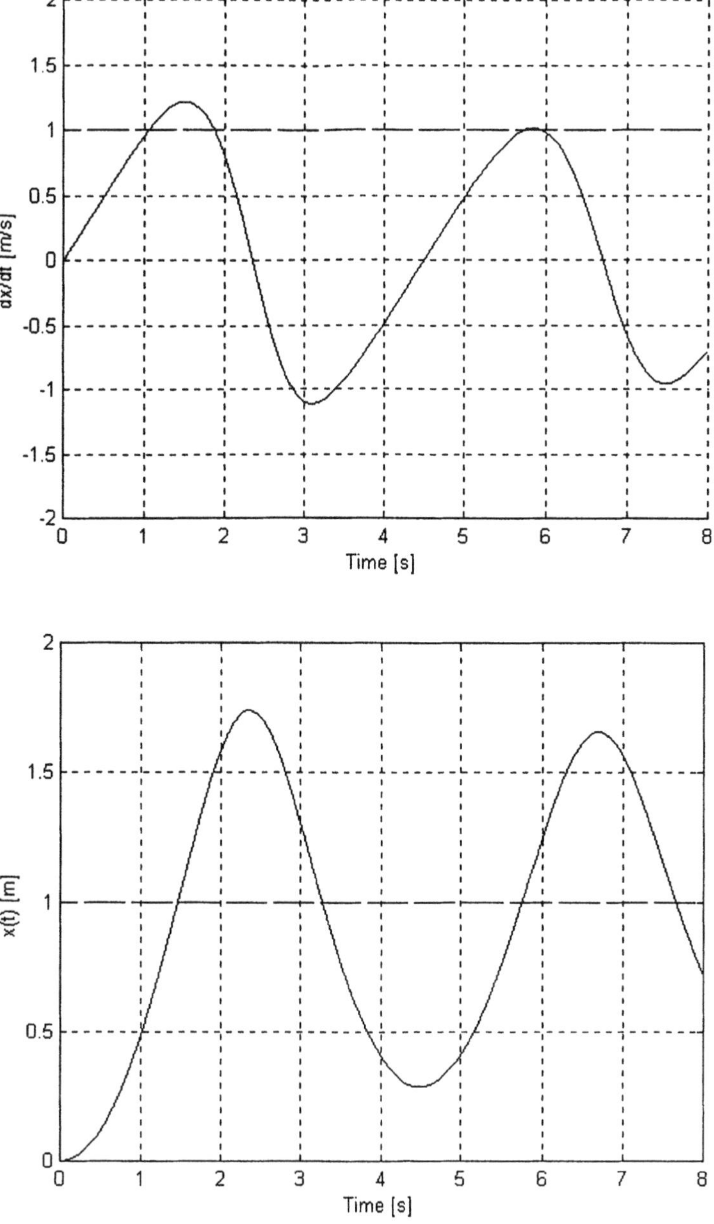

Fig. 5. System responses of the reduced (solid) and analytically derived (dash dotted) models to step input (dash line): a) shows the value $\dot{x}$ and b) shows the value x. The curves precisely cover each other on both figures.

Then computing $\mathbf{G} = \frac{1}{IJ} \sum_{i=1}^{I} \sum_{j=1}^{J} (\mathbf{A}_{i,j}^{r} - \mathbf{B}\mathbf{K}_{i,j}^{r})$ and $\mathbf{H}$, we obtain

$$\mathbf{G} = \begin{bmatrix} -4 & -4 \\ 1 & 0 \end{bmatrix} \quad \text{and} \quad \mathbf{H} = \begin{bmatrix} -4 & -4 & 0 & 0 \\ 1 & 0 & 0 & 0 \\ 10 & 0 & 4 & -1 \\ 0 & 10 & 4 & 0 \end{bmatrix}$$

that is the same result as in the case of the analytic design. We can, hence, say that the reduced controller with four rules quadratically stabilize the dynamical system.

7.3 Control results

In this part the control results of the complexity reduced controller is analyzed. Figure 6 shows the effect of the controller, where the stability point is achieved in about 8s. Figure 7 shows the output of the dynamical system without control in order to see the effectiveness of the controller. Figure 8 shows a case where white noise is added to signals u, $\dot{x}$ and x. The white noise has nonzero mean value, which is 10% of the original signal, and its maximum amplitude is 20% of the original signal.

7.4 Summary of the example

In point 7.1 the differential equations are analytically derived to a TS fuzzy model, in order to execute PDC controller design. In point 7.2 it is assumed the analytical derivation between the TS model and the differential equations is unknown and hard to solve. Therefore, the differential equations are sampled and approximated by TS model. In order to achieve numerically zero approximation error the approximation grid is chosen to be dense that leads to the high complexity of the TS model, the controller design (especially in the case of utilizing LMI based approaches) and the controller. Thus the reduction of the controller or the model is highly desired. The example shows that the proposed HOSVD based reduction method finds a minimal form of the TS model, which exists as proven by the analytic design. In the present example we show that the reduced model in 7.2 is capable of recovering the same structure embedded in the analytical model of 7.1. Finally the controller is determined from the reduced model. Without reduction the controller consists of 160000 local linear controllers (this large number is resulted by the fact that the most simple TS model construction and antecedents are utilized, using other technique this number would be much less, but the proposed HOSVD method finds the minimum number of rules). With the reduction its complexity is compressed to four local linear controllers. This implies that the computation effort of the controller is much reduced. As a matter of fact, in a general case we cannot always reduce the controller to

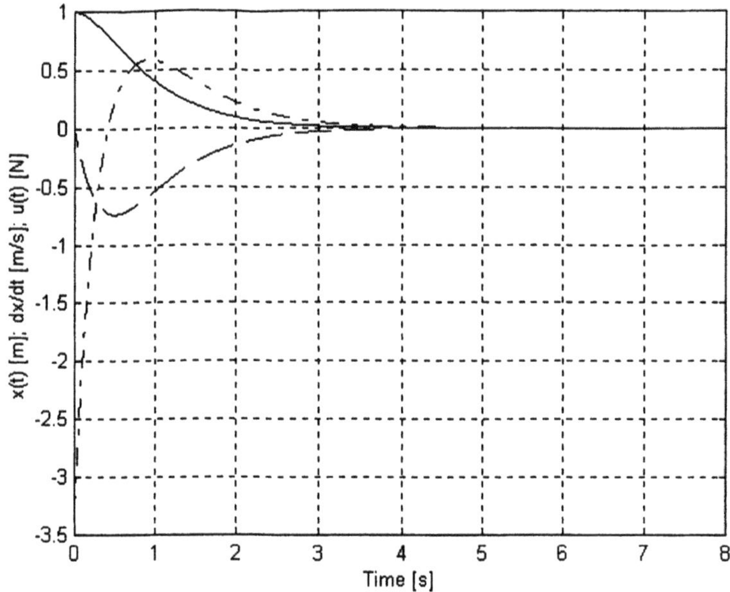

Fig. 6. Control result. $\dot{x}$ and x are depicted by dash and solid line respectively. The control signal is depicted by dash dot line.

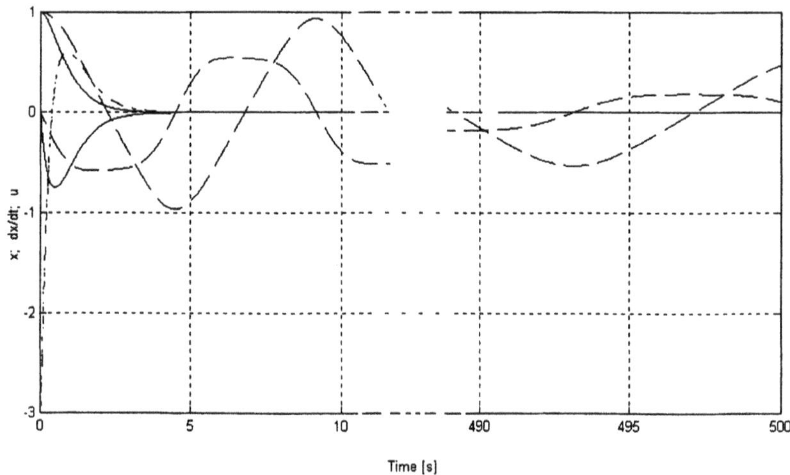

Fig. 7. Values $\dot{x}$ and x depicted by dash lines are uncontrolled. Values $\dot{x}$ (solid line) and x (solid line) are controlled by u (dash dot line).

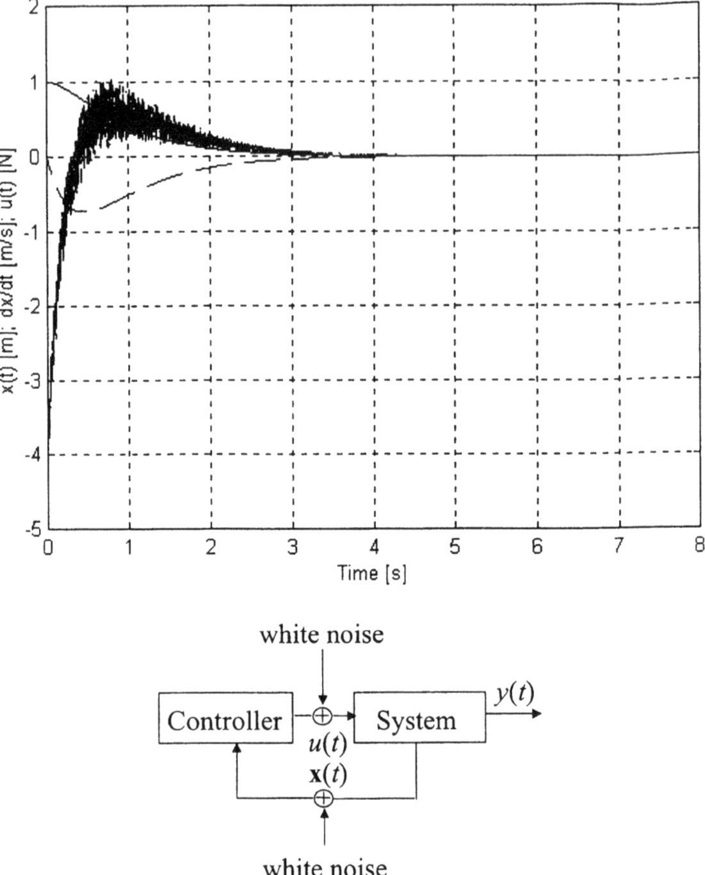

Fig. 8. Control result. $\dot{x}$ and x are depicted by dash and solid line respectively. The control signal is depicted by dash dot line.

four rules, but we can reduce the model as much as possible depending on the problem at hand. The same conclusion can be drawn in the case when even the differential equations are unknown, and the TS model is built up based on measurement data during training form training data. Since there is no mathematical framework to predefine the number of required liner local models employed in the learning process, there is no guarantee that the resulted TS model does not have redundancy. Furthermore, to be on the safe side, the number of the liner local models is usually overestimated. The proposed method in this case serves as a tool to root out linear dependent models and reduce the complexity of the controller.

8 Conclusions

In this paper we have argued that the identification of TS fuzzy models and controllers from training data needs to consider an important feature between data fitness and complexity. We emphasize the importance of these features by pointing out that TS fuzzy models and controllers with large number of local models may encounter the risk of having an approximation capable of fitting training data well, but incapable of running on satisfactory low computational cost. In order to help the developments of TS models and controllers to strive for balance between the two conflicting design objectives, we introduce a HOSVD based TS fuzzy model reduction technique. An other slightly different aspect of the proposed method is to find the minimal TS model representation (according to a given approximation error) of a given differential equations of a dynamic system. Using the proposed method, we have demonstrated a controller design to a dynamic system via HOSVD complexity minimization. This approach is expounded from single variable SVD based reduction technique of [2] to multi-variable cases.

References

1. V. I. Arnold. On functions of three variables. *Doklady Akademii Nauk USSR*, 114:679–681, 1957. (In Russian).
2. P. Baranyi, A. R. Várkonyi-Kóczy, Y. Yam, R. J. Patton, P. Michelberger, and M. Sugiyama. SVD-based reduction of TS model approximation. To appear in *IEEE Trans. Electronics.*
3. P. Baranyi and Y. Yam. Singular value-based approximation with non-singleton fuzzy rule base. In *Proc. of the 7th Int. Fuzzy Systems Association World Congress (IFSA'97)*, volume II, pages 127–132, Prague, Czech Republic, 1997.
4. P. Baranyi and Y. Yam. Singular value-based approximation with Takagi–Sugeno type fuzzy rule base. In *Proc. of the 6th IEEE Int. Conf. on Fuzzy Systems (FUZZ-IEEE'97)*, volume I, pages 265–270, Barcelona, Spain, 1997.
5. P. Baranyi and Y. Yam. Complexity reduction of a generalised rational form. In D. Ruan and E.E. Kerre, editors, *Fuzzy If-Then Rules in Computational Intelligence: Theory and Applications*, number SECS 553 in The Kluwer international series in engineering and computer science, pages 135–160. Kluwer Academic Publishers, Dordrecht, The Netherlands, 2000.
6. E. K. Blum and L. K. Li. Approximation theory and feedforward networks. *Neural Networks*, 4(4):511–515, 1991.
7. J. L. Castro. Fuzzy logic controllers are universal approximators. *IEEE Trans. on SMC*, 25:629–635, 1995.
8. P. Comon. Independent component analysis, a new concept? *Signal Processing, Special Issue on Higher Order Statistics*, 36:287–314, 1994.
9. G. Cybenko. Approximation by superposition of sigmoidal functions. *Mathematics of Control, Signals and Systems*, 2:303–314, 1989.
10. R. J. P. De Figueiredo. Implications and applications of Kolmogorov's superposition theorem. *IEEE Tr. Autom. Control*, pages 1227–1230, 1980.

11. K. Hornik, M. Stinchcombe, and H. White. Multilayer feedforward networks are universal approximators. *Neural Networks*, 2:359–366, 1989.
12. L. T. Kóczy and K. Hirota. Size reduction by interpolation in fuzzy rule bases. *IEEE Trans. on SMC*, 27:14–25, 1997.
13. A. N. Kolmogorov. On the representation of continuous functions of many variables by superpositions of continuous functions of one variable and addition. *Doklady Akademii Nauk USSR*, 114:953–956, 1957. (In Russian).
14. B. Kosko. Fuzzy systems as universal approximators. In *Proc. of the IEEE Int. Conf. on Fuzzy Systems*, pages 1153–1162, San Diego, 1992.
15. V. Kůrková. Kolmogorov's theorem and multilayer neural networks. *Neural Networks*, pages 501–506, 1992.
16. L. D. Lathauwer, B. D. Moor, and J. Vandewalle. A multi linear singular value decomposition. *SIAM Journal on Matrix Analysis and Applications*, 21(4):1253–1278, 2000.
17. L. D. Lathauwer, B. D. Moor, and J. Vandewalle. An introduction to independent component analysis. *Jour. Chemometrics*, 14:123–149, 2001.
18. G. G. Lorentz. *Approximation of functions.* Holt, Reinhard and Winston, New York, 1966.
19. B. Moser. Sugeno controllers with a bounded number of rules are nowhere dense. *Fuzzy Sets and Systems*, 104(2):269–277, 1999.
20. D. A. Sprecher. On the structure of continuous functions of several variables. *Trans. Amer. Math. Soc.*, 115:340–355, 1965.
21. G. W. Stewart. On the early history of the singular value decomposition. *SIAM Rev.*, 35(4):551–566, 1993.
22. A. Swami and G. Giannakis. Editorial. Higher-order statistics. *Signal Processing, Special Issue Higher Order Statistics*, 53(2–3):89–91, 1996.
23. K. Tanaka, T. Ikeda, and H. O. Wang. Robust stabilization of a class of uncertain nonlinear systems via fuzzy control: Quadratic stabilizability, H_∞ control theory, and Linear Matrix Inequalities. *IEEE Trans. Fuzzy Systems*, 4(1), 1996.
24. K. Tanaka, T. Ikeda, and H. O. Wang. Fuzzy regulators and fuzzy observers: relaxed stability conditions and lmi-based design. *IEEE Trans. Fuzzy Systems*, 31(2):250–265, 1998.
25. K. Tanaka and M. Sugeno. Stability analysis and design of fuzzy control systems. *Fuzzy Sets and Systems*, 45(2):135–156, 1992.
26. D. Tikk. On nowhere denseness of certain fuzzy controllers containing prerestricted number of rules. *Tatra Mountains Math. Publ.*, 16:369–377, 1999.
27. H. O. Wang, K. Tanaka, and M. F. P. Griffin. An approach to fuzzy control of non-linear systems: Stability and design issues. *IEEE Trans. Fuzzy Systems*, 4(1):14–23, 1996.
28. L. X. Wang. Fuzzy systems are universal approximators. In *Proc. of the IEEE Int. Conf. on Fuzzy Systems*, pages 1163–1169, San Diego, 1992.
29. Y. Yam. Fuzzy approximation via grid point sampling and singular value decomposition. *IEEE Trans. SMC*, 27:933–951, 1997.
30. Y. Yam, P. Baranyi, and C. T. Yang. Reduction of fuzzy rule base via singular value decomposition. *IEEE Trans. Fuzzy Systems*, 7(2):120–132, 1999.
31. J. Yen and L. Wang. Simplifying fuzzy rule-based models using orthogonal transformation methods. *IEEE Trans. SMC, Part B.*, 29(1):13–24, 1999.
32. H. Ying. General SISO Takagi–Sugeno fuzzy systems with linear rule consequents are universal approximators. *IEEE Trans. on FS*, 6(4):582–587, 1998.

33. H. Ying. Sufficient conditions on uniform approximation of multivariate functions by general Takagi–Sugeno fuzzy systems with linear rule consequents. *IEEE Trans. on SMC, Part A*, 28(4):515–520, 1998.
34. K. Zeng, N.-Y. Zhang, and W.-L. Xu. A comparative study on sufficient conditions for Takagi–Sugeno fuzzy systems as universal approximators. *IEEE Trans. on FS*, 8(6):773–780, 2000.

Simplification and reduction of fuzzy rules

Magne Setnes

Research & Development, Heineken Technical Services, PO Box 510,
2380 BB Zoeterwoude, The Netherlands. E-mail: magne.setnes@heineken.nl

Abstract. This chapter addresses rule base complexity in fuzzy models obtained from data. Data-driven fuzzy modeling is introduced, and two main approaches to complexity reduction in fuzzy rule-based models are presented: Similarity-driven rule base simplification, and rule reduction with orthogonal transforms.

1 Introduction

Most data-driven fuzzy modeling methods utilize only the function approximation capabilities of fuzzy systems, paying little attention to the qualitative aspects [1,2]. This makes them less suited for applications in which emphasis is not only on numerical properties, but also on interpretability, computational complexity and maintainability.

As reviewed in Chapter 1, recently methods have been proposed to improve the qualitative aspects of fuzzy models. This chapter describes two such approaches: Similarity-driven simplification and orthogonal transforms. The objective of similarity-driven simplification is to reduce redundant information present in the form of similar fuzzy *sets*. An application to modeling ecological data illustrates the approach. Orthogonal transforms can be used to reduce the number of *rules*. Some known methods are reviewed and improved, and a benchmark modeling problem is used for comparison.

2 Data-driven fuzzy modeling

Given observation data from an unknown system $y = f(\boldsymbol{x})$, data-driven modeling aim to construct a deterministic function $y = F(\boldsymbol{x})$ that can serve as an approximation of $f(\boldsymbol{x})$. In fuzzy modeling, F is represented by a collection of if–then rules. One particular rule-based model suitable for the approximation of a broad class of functions is the Takagi-Sugeno (TS) fuzzy model [3] which consists of a set of rules with the following structure :

$$R_i : \textbf{If } x_1 \text{ is } A_{i1} \textbf{ and } \ldots x_n \text{ is } A_{in} \textbf{ then } g_i = \boldsymbol{a}_i \boldsymbol{x} + b_i,\ i = 1, \ldots, M\,. \quad (1)$$

Here, R_i is the ith rule in the rule base, $\boldsymbol{x} = [x_1, \ldots, x_n]^T$ is the input (antecedent) variable and $A_{i1}, \ldots, A_{in}$ are fuzzy sets defined for the respective antecedent variable. The rule consequent g_i is an affine combination of the inputs with parameters $\boldsymbol{a}_i, b_i$, and each rule defines a hyperplane which locally approximates the real system's hypersurface. The output y of the model is a weighted sum of rule contributions:

$$y = \frac{\sum_{i=1}^{M} \beta_i g_i}{\sum_{i=1}^{M} \beta_i}, \tag{2}$$

where β_i is the degree of activation of the ith rule:

$$\beta_i = \Pi_{j=1}^{n} A_{ij}(x_j), \quad i = 1, 2, \ldots, M. \tag{3}$$

$A_{ij}(x_j)$ is the membership of x_j in the fuzzy set A_{ij}, i.e., the degree of match between the given fact and the proposition A_{ij} in the antecedent of rule i.

2.1 Parameter estimation

The TS model is usually identified in two steps. First, the fuzzy sets A_{ij} in the rule antecedents are determined, producing a partitioning of the input space. Then, when the rule antecedents are fixed, least-square (LS) estimation from observation data can be used to determine the consequent parameters, $\boldsymbol{a}_i$ and b_i, in two different ways. One solves M independent, or local, weighted LS problems, one for each rule. The other solves a global LS problem. Local LS gives more reliable local models, while global LS gives a minimal prediction error estimate (see [4] for a comparison).

Local learning The consequent parameters for each individual rule are obtained as a weighted least-square estimate. Let $\boldsymbol{\theta}_i^T = \left[\boldsymbol{a}_i^T, b_i\right]$, let X_e denote the matrix $[\mathrm{X}, \mathbf{1}]$ with rows $[\boldsymbol{x}_k, 1]$, and let W_i denote a diagonal matrix in $\mathbb{R}^{N \times N}$ having the degree of activation $\beta_i(\boldsymbol{x}_k)$ as its kth diagonal element. If the columns of X_e are linearly independent and $\beta_i(\boldsymbol{x}_k) > 0$ for $1 \le k \le N$, then the weighted LS solution of $\boldsymbol{y} = \mathrm{X}_e \boldsymbol{\theta}_i + \epsilon$ becomes

$$\boldsymbol{\theta}_i = \left[\mathrm{X}_e^T \mathrm{W}_i \mathrm{X}_e\right]^{-1} \mathrm{X}_e^T \mathrm{W}_i \boldsymbol{y}\,. \tag{4}$$

Global learning The contribution of all rules are considered simultaneously. Let X_e denote the matrix $[\mathrm{X}, \mathbf{1}]$. The activation of each rule is gathered in P_i which is a diagonal matrix in $\mathbb{R}^{N \times N}$ having the *normalized degree of activation* $p_i(\boldsymbol{x}_k)$ as its kth diagonal element

$$p_i(\boldsymbol{x}) = \frac{\beta_i}{\sum_{i'=1}^{M} \beta_{i'}}, \; i = 1, 2, \ldots, M. \tag{5}$$

Further, denote X' the matrix in $\mathbb{R}^{N \times MN}$ composed of matrices P_i and X_e

$$\mathrm{X}' = [\mathrm{P}_1 \mathrm{X}_e, \mathrm{P}_2 \mathrm{X}_e, \ldots, \mathrm{P}_M \mathrm{X}_e]\,. \tag{6}$$

Denote $\boldsymbol{\theta}'$ the vector in $\mathbb{R}^{M(n+1)}$ given by

$$\boldsymbol{\theta}' = \left[\boldsymbol{\theta}_1'^T, \boldsymbol{\theta}_2'^T, \ldots, \boldsymbol{\theta}_M'^T\right]^T \quad (7)$$

where $\boldsymbol{\theta}_i'^T = [\boldsymbol{a}_i^T, b_i]$ for $1 \leq i \leq M$. The resulting least-squares problem $\boldsymbol{y} = \mathrm{X}'\boldsymbol{\theta}' + \epsilon$, where ϵ is the approximation error, has the solution

$$\boldsymbol{\theta}' = \left[(\mathrm{X}')^T\mathrm{X}'\right]^{-1}(\mathrm{X}')^T\boldsymbol{y}. \quad (8)$$

From (7), the consequent parameters are given by $\boldsymbol{a}_i = [\theta'_{q+1}, \theta'_{q+2}, \ldots, \theta'_{q+n}]^T$, and $b_i = [\theta_{q+n+1}]$, where $q = (i-1)(n+1)$.

Example 1 We apply both local and global LS to estimate the consequents parameters in a TS model of a univariate function:

$$y(x) = 3e^{-x^2}\sin(\pi x) + \eta, \quad (9)$$

where η is Gaussian noise. 300 samples of $y(x)$ were obtained from (9) by random inputs $x \in [-3, 3]$. The input x is partitioned by seven equally distributed triangular fuzzy sets, giving a rule base of seven rules. The rule consequents are estimated both by local and by global LS. The results are shown in Fig. 1. Local learning gives consequents that better describe the behaviour of the modeled system in the regions where the rules are active. However, the overall approximation is better with global learning. □

2.2 Redundancy in data-driven modeling

The premise space partition of rule-based models obtained from data can contain overlapping and non-interpretable fuzzy sets. When the partition is obtained by clustering, this is caused by rules defined in the multidimensional premise that are overlapping in one or more dimensions. As a result, more membership functions will describe approximately the same concept in the final rule base. Another approach is to start with fuzzy sets defining a uniform grid in the premise space. The sets are then adapted using techniques like, e.g., steepest decent. During adaptation, fuzzy sets can move closer to each other and may become overlapping. Some sets may widen to cover the whole space, while others diminish to singletons. As such, an initially transparent model may become unreadable after adaptation.

Similar and redundant fuzzy sets and rules make a model unnecessarily complex and computationally demanding, and linguistic interpretation becomes difficult. In the following, two main methods for complexity reduction in fuzzy rule-based models are described.

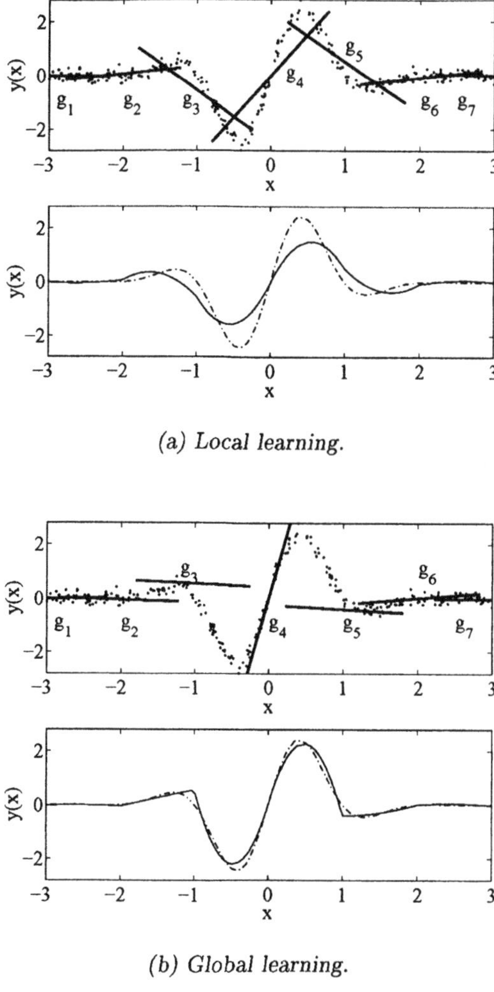

(a) Local learning.

(b) Global learning.

Fig. 1. Estimation of rule consequents using local and global learning.

3 Similarity-driven rule base simplification

Rule base simplification reduces redundant information in the form of *similar* fuzzy sets. Similar fuzzy sets are overlapping and describe almost the same region in the domain of some model variable. A similarity measure is used to identify such sets, which can be replaced by a common fuzzy set created by merging the similar ones. By specifying an acceptable level of similarity, the user can generate different models from the initial rule base. A complex but accurate model (high level of similarity accepted) can be useful for off-line simulation purposes. To validate and understand the basic concepts of the system, a transparent model is needed (little similarity accepted). A model with few fuzzy sets and rules also has a lower computational demand, making it more suitable for memory constrained and real-time applications.

3.1 Fuzzy set similarity

A similarity measure should generalize the classical crisp equality and capture a gradual transition between equality and non-equality [5]:

$$s = S(A, B) = \text{degree}(A = B), \quad s \in [0, 1], \tag{10}$$

where S is a similarity measure assigning a value s to the pair of fuzzy sets (A, B) that indicates the degree to which A and B are equal.

The fuzzy Jaccard index, Fig. 2, produces a smooth transition from equal ($S(A, B) = 1$) to completely non-equal fuzzy sets ($S(A, B) = 0$). It is based on the set-theoretic operations of intersection and union. Given fuzzy sets A_{lj} and A_{mj} defined on a discrete domain X_j by their membership functions $\mu_{A_{lj}}(x_j)$ and $\mu_{A_{mj}}(x_j)$, respectively, the Jaccard index can be written as

$$S(A_{lj}, A_{mj}) = \frac{|\min(\mu_{A_{lj}}(x_j), \mu_{A_{mj}}(x_j))|}{|\max(\mu_{A_{lj}}(x_j), \mu_{A_{mj}}(x_j))|}. \tag{11}$$

Here the cardinality of a set is given by $|\mu_A(x_j)| = \sum_{x_j \in X_j} \mu_A(x_j)$.

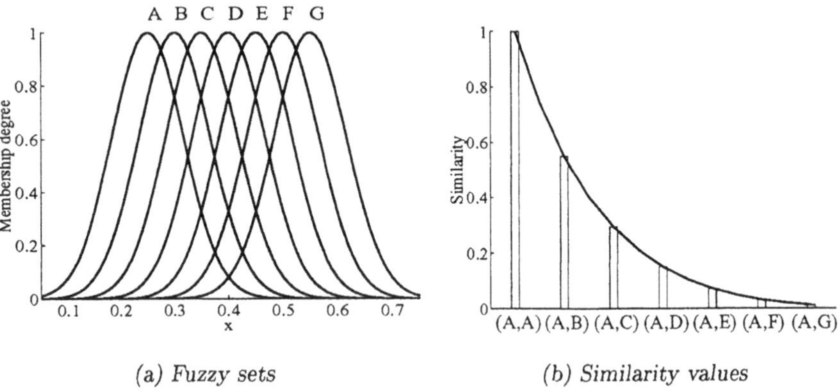

(a) *Fuzzy sets* (b) *Similarity values*

Fig. 2. Similarity computed for pairs of fuzzy sets $S(A, A)$, $S(A, B)$, ..., $S(A, G)$.

3.2 Simplification, rule- and feature reduction

By merging similar fuzzy sets, the number of fuzzy sets in the model decreases. In this way ***rule base simplification*** is achieved. The merging may result in equal rules. In the rule base, only one of the equal rules is needed, and ***rule base reduction*** is achieved. Hence, there is a difference between rule base simplification and rule base reduction. The former is the primary objective, and the latter may follow as a result. Fig. 3 illustrates the idea.

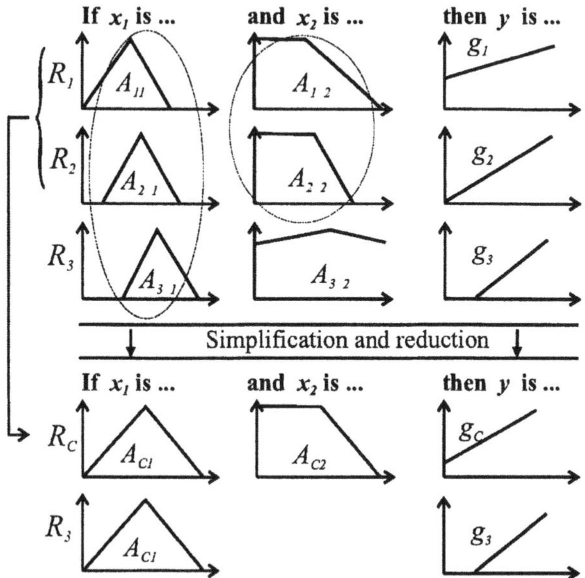

Fig. 3. Similar fuzzy sets, (A_{11}, A_{21}, A_{31}) and (A_{12}, A_{22}), are merged to produce the common sets A_{C1} and A_{C2}, respectively. Fuzzy set A_{31} is removed. The rules R_1 and R_2 get equal antecedents, and are replaced by the common rule R_C.

In the modeling of dynamic systems, when delayed samples of a variable are used as input, e.g., $x(k)$ and $x(k-1)$, there can be similarity in the partition of these inputs. The degree of firing of the rules can then be determined by one such input only. In this case, *dimensionality reduction* can be obtained for the rule base premise.

Merging fuzzy sets In [6] it was proposed to merge two fuzzy sets A and B by taking the support of $A \cup B$ as the support of the new fuzzy set C. This guarantees preservation of the coverage of the whole premise space when C replaces A and B in the premise of the rule base. Define a fuzzy set by a *parametric* membership function $\mu_A(x; a_1, a_2, a_3, a_4)$, $a_1 \leq a_2 \leq a_3 \leq a_4$:

$$\mu_A(x; a_1, a_2, a_3, a_4) = \begin{cases} 0, & x \leq a_1, \text{ or } x \geq a_4 \\ 1, & a_2 \leq x \leq a_3 \\ f(x), & f(x) \in (0,1), \text{otherwise} \end{cases} \quad (12)$$

Merging A and B gives a fuzzy set C where:

$$c_1 = \min(a_1, b_1), \quad (13)$$

$$c_2 = \lambda_2 a_2 + (1 - \lambda_2) b_2, \quad (14)$$

$$c_3 = \lambda_3 a_3 + (1 - \lambda_3) b_3, \quad (15)$$

$$c_4 = \max(a_4, b_4). \quad (16)$$

The parameters $\lambda_2, \lambda_3 \in [0,1]$ determines the influence of A and B, respectively, on the kernel of C. Averaging by $\lambda_2 = \lambda_3 = 0.5$ gives a tradeoff between contributions of the rules in which the fuzzy sets occur.

Removing fuzzy sets If a fuzzy set in the premise of a rule has a membership function $\mu(x) \approx 1, \forall x \in X$, it is similar to the universal set U and can be removed. As illustrated in Fig. 3, A_{32} can be removed and only A_{31} is necessary in the premise of rule R_3 to distinguish the associated region in the premise space.

A rule whose premise consists only of fuzzy sets similar to the universal set can be removed from the rule base. The activation of such rules is constant, and the contribution to the output can be accounted for by re-estimating the consequents of the remaining rules. The opposite may also occur. Singleton-like fuzzy sets have no similarity to the universal set (i.e. $S(A, U) \approx 0$). Rules with such fuzzy sets in their premise are also candidates to be removed from the rule base as they will never fire.

Removing inputs The similarity $\boldsymbol{S}_{lm}$ between the partitions of a pair of inputs (x_l, x_m) can be assessed by measuring the similarity between all corresponding pairs of fuzzy sets, $S(A_{il}, A_{im}), i = 1, 2, \ldots, M$, and take the minimum occurring similarity for each pair of inputs as the partition similarity:

$$\boldsymbol{S}_{lm} = \min_i S(A_{il}, A_{im}) \ , \ i = 1, 2, \ldots, M \, . \quad (17)$$

If the partition similarity $\boldsymbol{S}_{lm}$ is above an acceptable threshold given by the user, one of the two inputs, x_l or x_m, can be removed from the model's premise part. Note that it might still be necessary to keep all variables in the consequent part of the rule base.

Removing rules When the *premise* of $k \geq 2$ rules become equal, only one rule needs to remain in the rule base. The consequent parameters of the remaining rule must be re-estimated to account for the total contribution of the all the k rules it represents. This can be done by weighting the rule with k and let its consequent be an average of the consequents of all the k rules with equal premise parts [6,5].

If training data is available, the consequent parameters in the reduced rule base can be re-estimated as described in Section 2.1. This enables the consequent parameters of all rules to adapt to the new rule base and gives a more accurate result.

3.3 Rule base simplification algorithm

The algorithm is illustrated in Fig. 4 and consist of three parts: *Simplification*, *Dimensionality reduction*, and *Rule reduction*. The Jaccard index (11) determines the similarity between fuzzy sets in the rule base, and three thresholds are required: i) λ for merging similar fuzzy sets, ii) γ for removing fuzzy sets similar to the universal set, and iii) η for removing redundant input partitions. All thresholds take values in $[0, 1]$. The values of γ and η should be high, and typically $\gamma = 0.8$ and $\eta = 0.8$ have given good results in applications. The choice of a suitable threshold λ depends on the application. The threshold λ represents the degree to which overlap of fuzzy sets is tolerated. The lower the value of λ, the more fuzzy sets are combined, decreasing the term set of the model. In general, the numerical accuracy of the model decreases as λ decreases. However, if the model is overdetermined, the accuracy may improve. Since the algorithm does not require additional data acquisition or computationally expensive optimization, the user can easily investigate the effect of different thresholds, and the result best suiting the application can be used. For instance, for explaining the working of a system (operator training, expert validation), a comprehensible linguistic description is important. In such cases, it is reasonable to trade some accuracy for extra transparency and readability. This implies the use of a lower threshold λ than when aiming at applications like off-line simulation. To obtain rules sufficiently distinguishable to qualitatively describe the system, values of λ around 2/3 have proven to give good results in experiments.

3.4 Application example

We consider the problem of obtaining a model of algae growth from observation data. Algae contains Chlorophyll-a which makes photosynthesis possible. Algae growth models are used to predict, for instance, eutrophication, which can cause water quality problems in marine, estuarine and freshwater systems. Phosphate and nitrate accelerate algae growth in coastal and inland water, and high algae biomass cause visual pollution, bad odor and fish kills due to oxygen depletion. Toxic algae can threaten the ecological system and economic activities.

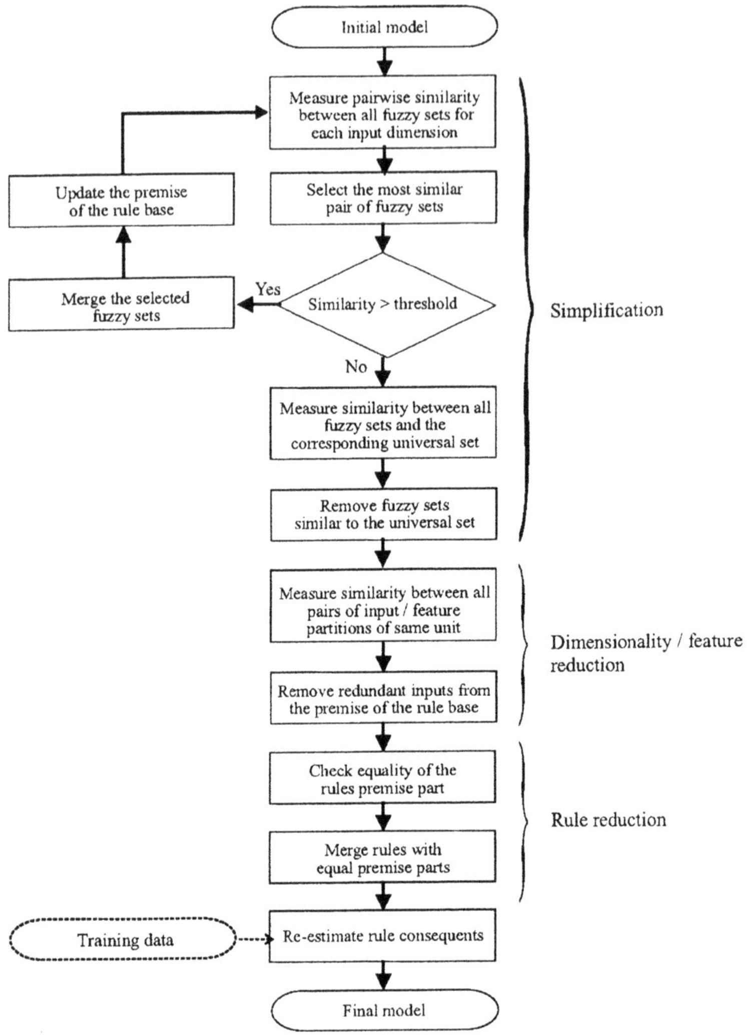

Fig. 4. The simplification algorithm.

Algae growth modeling We consider a data set of observations from nine different shallow lakes in the Netherlands, all with moderate to high levels of eutrophication. It contains measurements of the chlorophyll concentrations (output) and the states of biotic and abiotic parameters (inputs). The chlorophyll (Ch) is a measure of the algae biomass present in the ecosystem at a given moment. Because the ecosystem population is highly dynamic, the measurements are weekly or biweekly averages. In total 998 observations are available, obtained during different periods over several years.

The model's inputs are the energy input to the ecosystem (light) and the nutrient content of the lake. The temperature of the lake (T) acts as a catalyst for the algae growth. The light is represented by the equivalent light intensity (Ih), which results from the interaction of the sunlight that reaches the lake surface and the lake's depth. The duration of sunlight in a given day is accounted for by the day length (D). The three most important nutrients are: Nitrogen (N), Phosphorus (P) and Silicon (Si).

Table 1. Variables in the fuzzy model.

Variable	Parameter	Dimension
Input1, T	Temperature	[°C]
Input2, N	Nitrogen	[mg/l]
Input3, P	Phosphorus	[mg/l]
Input4, Si	Silicon	[mg/l]
Input5, D	Day length	[hour]
Input6, Ih	Light intensity	[watt/m^2]
Output, Ch	Chlorophyll-a	[μg/l]

The available observations are split into a training data set (688 samples) and an evaluation data set (310 samples). An initial TS fuzzy model with seven rules is identified from the observations in the training data set using the Gustafson–Kessels clustering algorithm [7]. The rules in the rule base have the form:

$$R_i : \textbf{If } \mathrm{T} \text{ is } A_{i\mathrm{T}} \textbf{ and } \mathrm{N} \text{ is } A_{i\mathrm{N}} \textbf{ and } \mathrm{P} \text{ is } A_{i\mathrm{P}} \textbf{ and } \mathrm{Si} \text{ is } A_{i\mathrm{Si}} \textbf{ and } \mathrm{D} \text{ is } A_{i\mathrm{D}} \textbf{ and } \mathrm{Ih} \text{ is } A_{i\mathrm{Ih}} \textbf{ then } \mathrm{Ch} = a_{i1}\mathrm{T} + a_{i2}\mathrm{N} + a_{i3}\mathrm{P} + a_{i4}\mathrm{Si} + a_{i5}\mathrm{D} + a_{i6}\mathrm{Ih} + b_i\,.$$

The fuzzy sets in the initial rule base are shown in Fig. 5. Many sets represent compatible concepts, making it difficult to assign meaningful qualitative labels to the different sets. This hampers the transparency of the model. To improve this, we apply the rule base simplification algorithm using the thresholds $\gamma = \lambda = \frac{2}{3}$. (The threshold η was 0.8, however, partition similarity is not an issue in this model.) The result is a simplified rule base consisting of five fuzzy rules that give a more qualitative description using only 10 fuzzy concepts. The simplified rule base is given in Table 2 and the fuzzy sets are shown in Fig. 6 with their linguistic labels.

Model inspection The reduced rule base consists of five rules and the rules do not all use the same antecedent variables. The two distinct regions for temperature, *cold* and *warm*, indicate strongly nonlinear influences of this variable. This agrees with the expert knowledge and experience. Furthermore, the variable "silicon" is removed from the premise, suggesting a more or less linear influence in the considered domain. Also, there is only one fuzzy set for the Nitrogen concentration. Nitrogen is a nutrient, and the obtained fuzzy

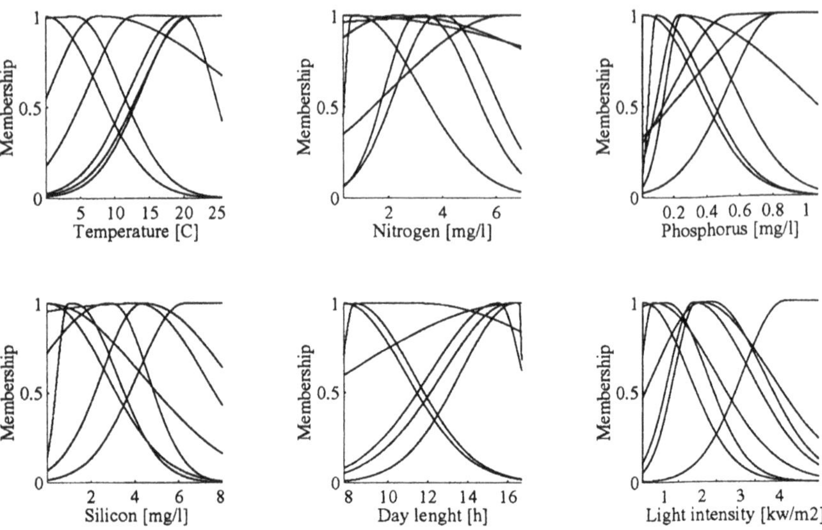

Fig. 5. Membership functions in the initial rule base.

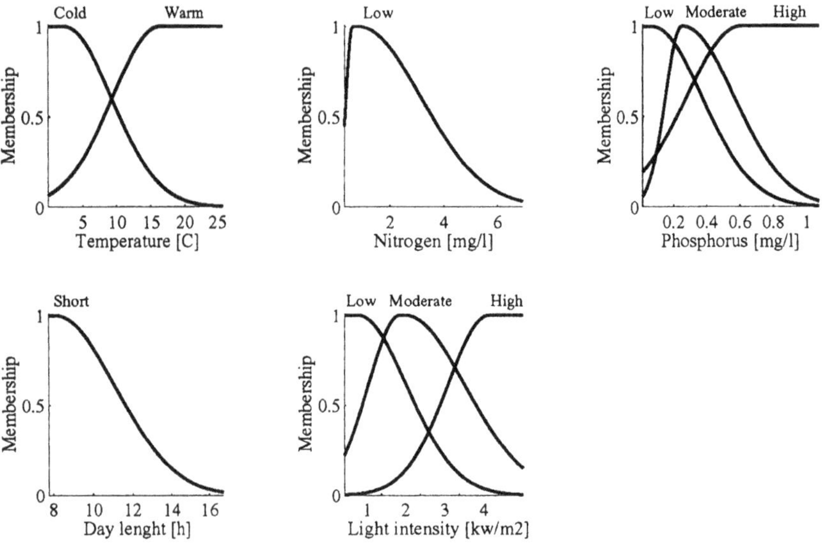

Fig. 6. Membership functions in the simplified rule base.

Table 2. The simplified model.

RULE	ANTECEDENT					CONSEQUENT
	Temp.	Nitrogen	Phosphorus	Day length	Light	Chlorophyll
R_1	Warm	–	High	–	Mod.	g_1
R_2	Cold	–	Low	Short	Low	g_2
R_3	Warm	Low	Low	–	High	g_3
R_4	Warm	–	High	–	Low	g_4
R_5	Cold	–	Mod.	Short	Low	g_5

set *low* is used in a rule to describe a situation in which there are favorable temperature and light conditions (rule R_3), but not enough nutrients for the algae.

For the variable "day length", only the fuzzy set *short* remains. It is used in the rules R_2 and R_5 which seem to cover the winter. The water is *cold* and the equivalent light intensity is generally *low* due to the small inclination angle of the sun light. Further, there seems to be a strong correlation between the three variables "temperature", "day length" and "light intensity". Knowledge about the climate in the Netherlands also supports this observation (e.g., high water temperature occurs in the summer, when the days are long and the light inclination angle is high).

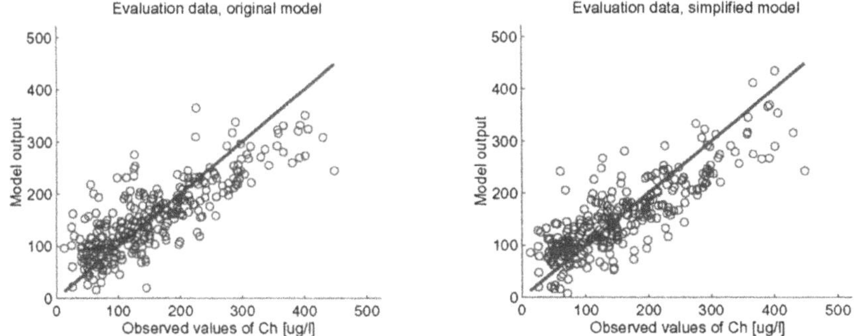

Fig. 7. Prediction performance of do not deteriorate as a result of the simplification.

The accuracy of the model has not changed much by the simplification. Figure 7 shows the evaluation of both the original and simplified model. The consequents of the simplified model have been re-estimated by means of global LS using the same training data as for the original model. It should be noted that the consequent structure of the remaining rules has not changed. All rule consequents are linear combinations of the six input variables in Table 1.

4 Orthogonal transforms for rule reduction

Orthogonal transforms methods for rule reduction [8–12] can be divided into two groups: the rank-revealing ones, and those that evaluate the individual contributions of the rules. Rank-revealing methods seek to determine the effective rank of the rule firing matrix from its singular values. In practice, this can be difficult and the ordering of the rules depends crucially on this difficult estimate. Methods that evaluate the output contribution of the rules to obtain an ordering, like the orthogonal least-squares approach (OLS) [13,8], are more attractive for systems modeling. However, OLS methods may assign high importance to rules that are correlated in the premise [11]. This can result in poor generalization capabilities of the model.

We first discuss rank-revealing reduction methods and advocate the use of the computationally simple pivoted QR decomposition which can produce a rule ordering without an estimate of the effective rank. Then, we extend the OLS method by introducing a simple detection for redundant rules. A benchmark example shows that the pivoted QR decomposition and the extended OLS algorithm give better results than other orthogonal methods.

4.1 Fuzzy modeling as a regression problem

Consider a TS model (1) with singleton consequents $g_i = b_i$. Using (5), the output (2) can be written as:

$$\hat{y} = \sum_{i=1}^{M} p_i(\boldsymbol{x}) b_i \,, \tag{18}$$

where $p_i(\boldsymbol{x})$ is the normalized degree of activation of the ith rule.

Given $N \gg M$ input-output data pairs $\{\boldsymbol{x}_k, y_k\}$, the model in (18) can be written as a linear regression problem

$$\boldsymbol{y} = \mathrm{P}\boldsymbol{\theta} + \boldsymbol{e} \tag{19}$$

where $\boldsymbol{y} = [y_1, y_2, \ldots, y_N]^T$ are the measured outputs, $\boldsymbol{\theta} = [b_1, b_2, \ldots, b_M]^T$ are the consequents of the M rules, and $\boldsymbol{e} = [e_1, e_2, \ldots, e_N]^T$ is a vector of approximation errors. The matrix $\mathrm{P} = [\boldsymbol{p}_1, \boldsymbol{p}_2, \ldots, \boldsymbol{p}_M] \in \mathbb{R}^{N \times M}$, where $\boldsymbol{p}_i = [p_{1i}, p_{2i}, \ldots, p_{Ni}]^T$, contains the firing strength of all the M rules for the N inputs.

4.2 Rule selection with rank-revealing methods

The least-squares solution to the overdetermined parameter estimation problem in (19) is given by

$$\mathrm{P}^T \mathrm{P} \boldsymbol{\theta}_{LS} = \mathrm{P}^T \boldsymbol{y} \,. \tag{20}$$

The solution $\boldsymbol{\theta}_{LS}$ is the one that minimizes the error $||\boldsymbol{y} - \mathrm{P}\boldsymbol{\theta}||$. Its determination requires the cross product matrix $\mathrm{P}^T\mathrm{P}$ to be invertible; P must have rank M. When P is *rank-deficient*, i.e., $r = \mathrm{rank}(\mathrm{P}) < M$, there will be no unique solution to the parameter estimation problem. A solution can be found via the *pseudo inverse* P^+ of P:

$$\boldsymbol{\theta}' = \mathrm{P}^+\boldsymbol{y}\,. \tag{21}$$

The pseudo inverse of P is obtained by singular value decomposition (SVD)

$$\mathrm{P} = \mathrm{U}\Sigma\mathrm{V}^T \tag{22}$$

where $\mathrm{U} \in \mathbb{R}^{N\times N}$ and $\mathrm{V} \in \mathbb{R}^{M\times M}$ are orthogonal matrices, and $\Sigma \in \mathbb{R}^{N\times M}$ is a diagonal matrix with the singular values $\sigma_1 \geq \sigma_2 \geq \ldots \geq \sigma_M \geq 0$ in decreasing order as diagonal elements. Given the SVD, the pseudo inverse is calculated as

$$\mathrm{P}^+ = \mathrm{V}\Sigma^+\mathrm{U}^T \tag{23}$$

where $\Sigma^+ \in \mathbb{R}^{M\times N}$ is a diagonal matrix with the reciprocals $1/\sigma_1, \ldots, 1/\sigma_r$ of the r *non zero* singular values as diagonal elements. The number of nonzero singular values σ in the SVD of P reveals the rank of P. Usually r is an estimate of the "effective rank" of P, identified by a (relative) gap $\sigma_r \gg \sigma_{r+1}$ in the singular values. The gap represents a natural point to reduce the dimensionality of the problem by setting the singular values below the gap to zero. Given an estimate $r \leq \mathrm{rank}(\mathrm{P})$, a solution is obtained that minimizes the error $||\boldsymbol{y} - \mathrm{P}'\boldsymbol{\theta}'||$ where P' is the closest matrix to P that has rank r:

$$\mathrm{P}' = \mathrm{U}\Sigma'\mathrm{V}^T\,. \tag{24}$$

Here, $\Sigma' \in \mathbb{R}^{N\times M}$ is a diagonal matrix extracted from Σ in (22) by taking the first r singular values $\sigma_1, \ldots, \sigma_r$ as diagonal elements. Indeed, if $r = M = \mathrm{rank}(\mathrm{P})$, then $\boldsymbol{\theta}' = \boldsymbol{\theta}_{LS}$ and $\mathrm{P}' = \mathrm{P}$. Replacing P with P' reduces redundancy since the redundant rules are associated with near zero singular values [10].

Rule selection with SVD Picking the most influential columns of P is known as *subset selection* [14]. An overview of subset selection methods for fuzzy models was given in [11]. The methods seek to replace $\mathrm{P}\boldsymbol{\theta}$ in (19) with $\mathrm{P}_r\boldsymbol{\theta}_r$ where $\mathrm{P}_r \in \mathbb{R}^{N\times r}$ consist of r columns picked from P. The natural way to determine r is to locate a gap in the singular values of the firing matrix P. One such approach is the *SVD-QR with pivoting* [15,10]. The algorithm works as follows:

1. Calculate the SVD of P as in (22) and estimate its effective rank $r \leq \mathrm{rank}(\mathrm{P})$ from Σ.
2. Calculate a permutation matrix Π such that the columns of the matrix $\mathrm{P}_r \in \mathbb{R}^{N\times r}$ in

$$\mathrm{P}\Pi = [\mathrm{P}_r, \mathrm{P}_{M-r}] \tag{25}$$

are independent ($\mathrm{rank}(\mathrm{P}_r) = r$).

3. Approximate $\boldsymbol{y}$ with $\mathrm{P}\boldsymbol{\theta}$ where

$$\boldsymbol{\theta} = \Pi \begin{bmatrix} \boldsymbol{\theta}_r \\ 0 \end{bmatrix}$$

and $\boldsymbol{\theta}_r \in \mathbb{R}^r$ minimizes $||\mathrm{P}_r\boldsymbol{\theta}_r - \boldsymbol{y}||$.

The actual rule selection is the calculation of the permutation matrix Π that extracts an independent subset of columns P_r from P, assumed to correspond to the most important rules. A heuristic solution to obtaining Π was given in [14] as computing the pivoted QR decomposition of the submatrix $[\mathrm{V}_{11}^T \mathrm{V}_{21}^T]$ of singular vectors extracted from V in (22) as

$$\mathrm{V} = \begin{bmatrix} \mathrm{V}_{11} & \mathrm{V}_{12} \\ \mathrm{V}_{21} & \mathrm{V}_{22} \end{bmatrix} \tag{26}$$

where $\mathrm{V}_{11} \in \mathbb{R}^{r\times r}$ and $\mathrm{V}_{21} \in \mathbb{R}^{(M-r)\times r}$. The pivoted QR decomposition determines a permutation matrix such that

$$\mathrm{Q}^T[\mathrm{V}_{11}^T \mathrm{V}_{21}^T]\Pi = [\mathrm{R}_{11}\mathrm{R}_{12}]\,.$$

$\mathrm{Q} \in \mathbb{R}^{r\times r}$ is orthogonal and R_{11} and R_{12} form an upper triangular matrix.

If the number of non-small singular values $r \leq \mathrm{rank(P)}$ can be properly determined, i.e., there is a well defined gap $\sigma_{r+1}(\mathrm{P}) \ll \sigma_r(\mathrm{P})$, then the subset selection performed by $\mathrm{P}\Pi = [\mathrm{P}_r\mathrm{P}_{M-r}]$ will produce a subset P_r containing the most important columns (rules) of P. However, often the singular values decrease smoothly without any clear gap. In such cases, r is determined by counting the number of (close to) zero singular values in the SVD of P, resulting in a rule reduction method for (almost) equal rules and rules that do not fire. In [16,4] it was proposed to use the size of the singular values as an ordering and to construct a reduced fuzzy model of any size $M_s \leq M$ by picking the M_s most important rules according to Π. This is different from subset selection, where a subset of r rules is extracted, and no further importance is associated with the order in which they are picked by the permutation matrix Π. It can be disputed whether Π produces an importance ordering. Moreover, the permutation is strongly dependent on the estimate of the effective rank r. This is illustrated in Example 2.

Rule selection with pivoted QR The pivoted QR (P-QR) decomposition can be applied directly to P to obtain a permutation matrix [14]. It skips the calculation of the SVD and does not need an estimate of r to obtain the permutation. For most practical cases, it will also reveal the rank of P

The QR decomposition of P is given by $\mathrm{P}\Pi = \mathrm{QR}$, where $\Pi \in \mathbb{R}^{M\times M}$ is a permutation matrix, $\mathrm{Q} \in \mathbb{R}^{N\times M}$ has orthonormal columns and $\mathrm{R} \in \mathbb{R}^{M\times M}$ is upper triangular. If P has full rank, then R is nonsingular (invertible). When P is (near) rank deficient, it is desirable to select the permutation matrix Π

such that the rank-deficiency is exhibited in R, having a small lower right block $R_{kk} \in \mathbb{R}^{k\times k}$ [17]:

$$R = \begin{bmatrix} R_{11} & R_{12} \\ 0 & R_{kk} \end{bmatrix}$$

It can be shown that for the $(M-k+1)$'th singular value of P, we have $\sigma_{M-k+1}(P) \leq \|R_{kk}\|_2$. Therefore, if $\|R_{kk}\|_2$ is small, then P has k small singular values.

The QR decomposition is uniquely determined by the permutation matrix Π, and many techniques have been proposed to compute it. The most well known is the column pivoting strategy [15] which in practice is very efficient in computing a triangular factor R with a small $\|R_{kk}\|$. The pivoting works such that the norm of the first column of R_{kk} dominates the norm of the other columns. The norm of this column is $|R(kk)|$, and the pivoting strategy can be regarded as a greedy algorithm to make the leading principal submatrices of R as well conditioned as possible by maximizing their trailing diagonals. Thus, the values $|R(kk)|$ on the diagonal of R, called the R values, are decreasing, and they tend to track the singular values $\sigma_k(P)$ well enough to expose gaps.

The pivoting algorithm favors columns of R with a high norm, related (through orthogonalization) to the norm of the columns of P. In regular regression problems, this is not necessarily an interesting observation. For a fuzzy rule base, however, the norms of the columns of P correspond to the firing strength and/or firing frequency of the rules. Thus, P-QR uses this to produce a rule ordering that picks first the most active and least redundant of the remaining rules.

Example 2 We illustrate the effect of the estimate of r on the "rule ordering" by the SVD-QR algorithm. We also show the ordering by P-QR.

A one-dimensional fuzzy partition of three trapezoid fuzzy sets A_1, A_2 and A_3 is considered. To this partition we add the three more or less redundant sets A_4, A_5 and A_6 as shown in Fig. 8a. An input data vector $x = [x_1, x_2, \ldots, x_{200}]^T$ of 200 observations evenly spaced in $[-3, 3]$ is constructed, and the 200×6 firing matrix P is calculated. Both the singular values and the R values of P are plotted in Fig. 8b, and it is seen that the R values track the singular values well. Further, we notice that P has two small (close to zero) singular values, but also a distinct gap after three values.

We now apply the SVD-QR algorithm several times with varying estimates of r. The results are reported in Table 3 together with those of the P-QR. The entries i in the table show the order in which the rules R_i are picked, with importance decreasing from top to bottom.

According to the distribution of the singular values, two good choices are $r = 4$ and $r = 3$. However, the resulting "importance ordering" of the SVD-QR differ completely for these two values. For $r = 4$, the algorithm assigns the highest importance to the two rules defined by the fuzzy sets A_2 and A_5 of which one is certainly redundant. In fact, the only reasonable ordering

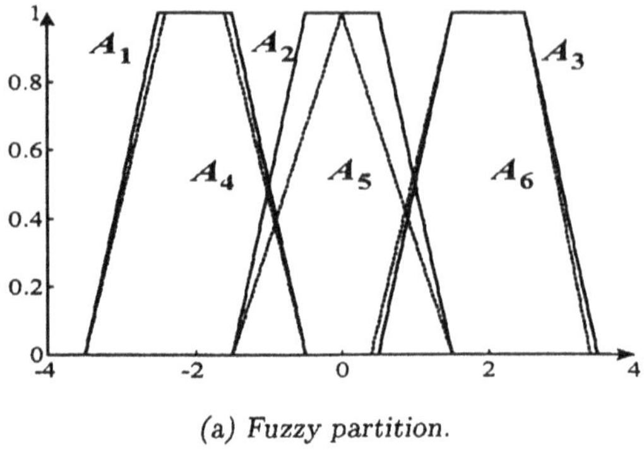

(a) Fuzzy partition.

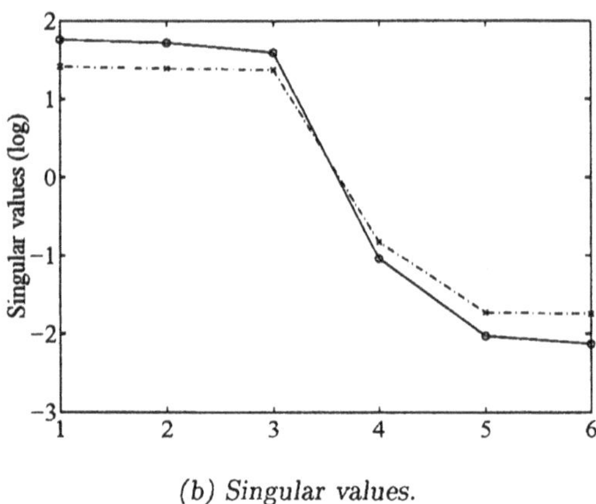

(b) Singular values.

Fig. 8. A redundant fuzzy partition of six fuzzy sets (a) and the singular values (o) and R values (x) of the 200×6 firing matrix P (b).

obtained with SVD-QR is the one obtained for $r = 3$. The P-QR does not need any estimate of r, and it produces a reasonable order. As for the singular values, the R values of the P-QR decomposition, see Fig. 8b, can help to determine the number of rules to pick. Due to the robust ordering, this estimate is not really necessary, and one can use other methods like information criteria [16] to pick the first M_s rules according to the permutation order produced by the pivoted QR decomposition. □

4.3 Orthogonal Least-Squares-based rule reduction

The orthogonal least-squares (OLS) method was first applied in [8] to select the most important fuzzy basis functions needed to approximate a data set.

Table 3. Rule ordering by P-QR and SVD-QR (most important at the top).

P-QR	SVD-QR					
	$r=6$	$r=5$	$r=4$	$r=3$	$r=2$	$r=1$
2	2	2	2	2	1	6
1	3	5	5	1	6	2
3	5	1	1	3	3	3
5	4	4	3	4	4	4
6	1	3	4	5	5	5
4	6	6	6	6	2	1

OLS transforms the columns of the firing matrix P into a set of orthogonal basis vectors in order to inspect the individual contribution of each rule. The Gram-Schmidt orthogonalization is used to perform the orthogonal decomposition $P = WA$, where W is an orthogonal matrix such that $W^T W = I$, and A is an upper-triangular matrix with unity diagonal elements. Substituting $P = WA$ into (19), we have $\boldsymbol{y} = WA\boldsymbol{\theta} + \boldsymbol{e} = W\boldsymbol{g} + \boldsymbol{e}$, where $\boldsymbol{g} = A\boldsymbol{\theta}$. Since the columns $\boldsymbol{w}_i$ of W are orthogonal, the sum of squares of $y(k)$ can be written as

$$\boldsymbol{y}^T\boldsymbol{y} = \sum_{i=1}^{M} g_i^2 \boldsymbol{w}_i^T \boldsymbol{w}_i + \boldsymbol{e}^T\boldsymbol{e}\,. \tag{27}$$

The part of the output variance $\boldsymbol{y}^T\boldsymbol{y}/N$ explained by the ith regressor is $\sum \boldsymbol{g}_i^2 \boldsymbol{w}_i^T \boldsymbol{w}_i/N$. An error reduction ratio due to an individual rule i can be defined as [13]:

$$[err]^i = \frac{g_i^2 \boldsymbol{w}_i^T \boldsymbol{w}_i}{\boldsymbol{y}^T\boldsymbol{y}}\,,\ 1 \le i \le M\,. \tag{28}$$

This ratio offers a simple means of ordering the rules, and was used in [8] to select a subset of important rules in a forward-regression manner.

The OLS method may sometimes produce an inappropriate subset of fuzzy rules [11]. The used error reduction ratio (28) minimizes the fitting error without paying attention to the premise partitioning. Thus, a rule that is rather redundant in the premise space can still be assigned a high importance because of its estimated contribution to the output. In [18] this problem was solved by introducing a check for $\boldsymbol{w}_k^T \boldsymbol{w}_k \le \epsilon$ when selecting the kth most important rule. Here $\epsilon > 0$ is some numerical approximation of zero. The relation $\boldsymbol{w}_i^T \boldsymbol{w}_i \le \epsilon$ implies that the corresponding column vector $\boldsymbol{p}_k$ is a linear combination of the column vectors corresponding to the previously selected $k-1$ rules [13]. An extended OLS algorithm is given below. It creates a vector $\boldsymbol{O} = [o_1, o_2, \ldots, o_M]^T$ of rule indices where the M rules are ordered in decreasing importance.

Algorithm 1 Extended OLS algorithm

Given the $N \times M$ firing matrix P *and a $N \times 1$ vector $\boldsymbol{y}$ of system outputs to be approximated.*

1. *Select the first vector* w_1 *of the orthogonal basis:*
   For $1 \leq i \leq M$,
   set $w_1^{(i)} = p_i$ *and calculate the corresponding element of the OLS solution vector*

$$g_1^{(i)} = \frac{\left(w_1^{(i)}\right)^T y}{\left(w_1^{(i)}\right)^T w_1^{(i)}}$$

   and the error-reduction ratio

$$[err]_1^{(i)} = \frac{\left(g_1^{(i)}\right)^2 \left(w_1^{(i)}\right)^T w_1^{(i)}}{y^T y}$$

   where $p_i = [p_{ki}, \ldots, p_{Ni}]^T$ *is given by the firing strength matrix* P.
   Find the rule with the largest error reduction ratio

$$o_1 = \arg\max_{1 \leq i \leq M} \left([err]_1^{(i)}\right)$$

   and select the first basis vector w_1 *and the first element* g_1 *of the OLS solution vector*

$$w_1 = w_1^{(o_1)} = p_{o_1},$$
$$g_1 = g_1^{(o_1)}.$$

2. *Select the next basis vectors* w_k*:*
   Repeat for $2 \leq k \leq M$*:*
   For $1 \leq i \leq M$, $i \neq o_1, \cdots, i \neq o_{k-1}$, *calculate*

$$\alpha_{jk}^{(i)} = \frac{w_j^T p_i}{w_j^T w_j}, 1 \leq j < k$$

$$w_k^{(i)} = p_i - \sum_{j=1}^{k-1} \alpha_{jk}^{(i)} w_j,$$

$$g_k^{(i)} = \frac{\left(w_k^{(i)}\right)^T y}{\left(w_k^{(i)}\right)^T w_k^{(i)}},$$

$$[err]_k^{(i)} = \frac{\left(g_k^{(i)}\right)^2 \left(w_k^{(i)}\right)^T w_k^{(i)}}{y^T y}.$$

   Find the remaining non-redundant rule with largest error reduction ratio

$$o_k = \begin{cases} \underset{1 \leq i \leq M,\, i \neq o_1, \cdots, i \neq o_{k-1}}{\arg\max} \left\{[err]_1^{(i)} \mid \left(w_k^{(i)}\right)^T w_k^{(i)} > \epsilon\right\}, & \text{if } \left(w_k^{(i)}\right)^T w_k^{(i)} > \epsilon \\ \underset{1 \leq i \leq M,\, i \neq o_1, \cdots, i \neq o_{k-1}}{\arg\max} \left([err]_1^{(i)}\right), & \text{otherwise} \end{cases}$$

and select the kth basis vector w_k and the kth element g_k of the OLS solution vector

$$w_k = w_k^{(o_k)},$$
$$g_k = g_k^{(o_k)}.$$

For rule selection, if a predetermined number of rules $M_S < M$ is to be selected, step two of the algorithm is ended at $k = M_S$. It is also possible to determine a stopping criterion based on, e.g., the approximation accuracy [9] or the relative contribution of the selected rules. The latter is proposed in [19,5], where a the OLS algorithm is generalized to models with functional consequents.

4.4 Benchmark example

We consider an example from [11] to illustrate the power of the pivoted QR decomposition and the extended OLS algorithm. The system under study is a second-order nonlinear plant

$$y(k) = f(y(k-1), y(k-2)) + u(k), \tag{29}$$

where

$$f(y(k-1), y(k-2)) = \frac{y(k-1)y(k-2)[y(k-1) - 0.5]}{1 + y^2(k-1) + y^2(k-2)}. \tag{30}$$

We want to approximate the nonlinear component f of the plant (the unforced system) with a fuzzy TS model. 1000 samples of identification data were obtained with a random input signal $u(k)$ uniformly distributed in $[-1.5, 1.5]$, followed by 200 samples of evaluation data obtained using a sinusoid input signal $u(k) = \sin(2\pi k/25), k = 1001, \ldots, 1200$. The input to the fuzzy model is $x_k = [y(k-1), y(k-2)]$, and 25 two-dimensional Gaussian membership functions [11] with the parameters shown in Table 4 are used to partition the input space as illustrated in Fig. 4.4

$$A_{ij}(x_j(k)) = \exp\left(-\frac{(x_j(k) - c_{ij})^2}{2\sigma_{ij}^2}\right), j = 1, 2, i = 1, \ldots, 25. \tag{31}$$

Each row in Table 4 is associated with one of the fuzzy rules in the rule base. The first two rows have equivalent membership function parameters. Thus, the first two rules in the rule base will always have nearly the same firing strengths. This implies that there is redundancy between the two rules and removing one of them will not affect the performance of the model. The same holds for rules 5 and 6 as well as rules 15 and 16. The Gaussian membership functions associated with the rules 10 and 20 are very narrow.

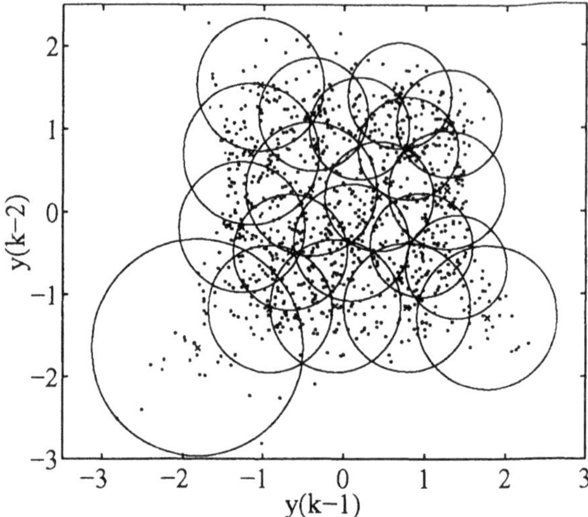

Fig. 9. The basis of the gaussian membership functions along with the input data. (The narrow gaussian membership functions cannot be recognized in this plot.)

This implies that the rules will virtually never fire, and they can thus be removed from the rule base.

From the first 1000 input data points $\boldsymbol{x}_k = [y(k-1), y(k-2)], k = 1, \ldots, 1000$, the 1000×25 normalized firing strength matrix P is calculated using (5). The singular values of the matrix are shown in Fig. 10 together with the R values of the pivoted QR decomposition. It is seen that the R values track the singular values with considerable fidelity.

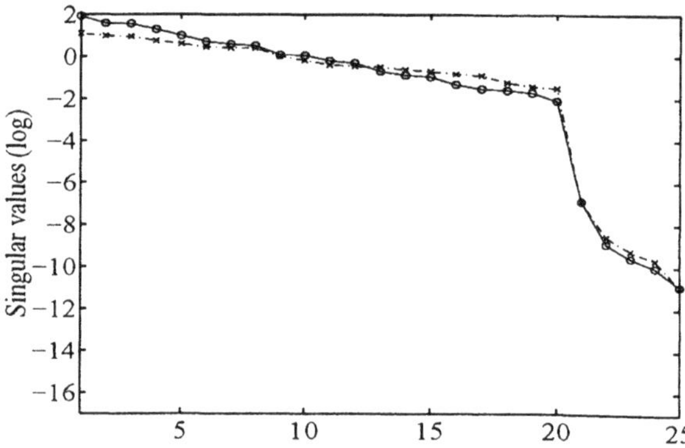

Fig. 10. Singular values (o) and R values (x) of the 1000×25 firing matrix P.

Table 4. Parameters of the Gaussian membership functions.

i	center c_{i1}	center c_{i2}	width σ_{i1}	width σ_{i2}
1	**0.0930**	**-0.3630**	**0.7095**	**0.7095**
2	**0.0933**	**-0.3632**	**0.7095**	**0.7095**
3	1.3828	-0.6617	0.6271	0.6271
4	-1.0414	1.5397	0.7969	0.7969
5	**-1.8130**	**-1.6470**	**1.3205**	**1.3205**
6	**-1.8125**	**-1.6469**	**1.3205**	**1.3205**
7	0.7776	-1.1555	0.7800	0.7800
8	0.1898	1.0142	0.6141	0.6141
9	-0.4052	0.2798	0.8099	0.8099
10	**-0.6613**	**-0.4846**	**0.0100**	**0.0100**
11	-0.6613	-0.4846	0.7051	0.7051
12	0.9529	-0.3965	0.6313	0.6313
13	0.7860	0.7723	0.6177	0.6177
14	0.4329	0.1910	0.6652	0.6652
15	**1.2940**	**1.0740**	**0.6474**	**0.6474**
16	**1.2942**	**1.0738**	**0.6474**	**0.6474**
17	0.6801	1.4083	0.6370	0.6370
18	1.2656	0.2698	0.7156	0.7156
19	-0.3846	1.1827	0.6772	0.6772
20	**-1.2642**	**-0.1808**	**0.0100**	**0.0100**
21	-1.2642	-0.1808	0.7907	0.7907
22	-0.9099	-1.1750	0.7728	0.7728
23	-0.1008	-1.1384	0.8046	0.8046
24	-1.1533	0.7037	0.8517	0.8517
25	1.7691	-1.2798	0.8746	0.8746

Rule subset selection The gap in the singular values in Fig. 10 indicates the presence of 5 near zero singular values. We now apply four different orthogonal transform-based methods to the problem. Table 5 shows the order in which the rules are picked from the rule base. The SVD-QR and the OLS method are the same as those studied for this problem in [11].

Table 5. Order in which the rules are picked (most important to the left).

P-QR	24	25	6	15	23	8	4	12	11	14	18	19	22	17	7	21	3	13	9	2	20	5	16	1	10
SVD-QR	25	4	19	7	3	24	8	13	23	14	21	17	22	18	12	9	11	2	5	16	15	10	20	6	1
E-OLS	5	24	25	16	8	21	23	11	3	22	7	9	19	4	14	18	1	17	13	12	2	6	10	15	20
OLS	5	14	15	16	8	21	23	11	3	22	6	7	15	19	4	9	17	13	18	12	1	2	14	10	20

The results show that both rank-revealing methods, the P-QR and the SVD-QR, pick as the least important rules three redundant ones and the two non-firing ones. In this case, the SVD-QR algorithm was executed with $r = 20$. We knew from Fig. 10 that there are $25 - 20 = 5$ non influential rules.

The P-QR method does not need this information, and still produces the correct subset of least important rules for this problem. Further, as concluded in [11], the OLS method correctly sorts out the non-firing rules (numbers 10 and 20), but fails to assign a low importance to one rule from each of the three pairs of redundant rules. This is not the case with the E-OLS method which successfully detects both the redundant as well as the non-firing rules.

Rule ordering According to each of the three methods, P-QR, SVD-QR and E-OLS, we make 20 models of increasing complexity with the rules picked in the order reported in Table 5. Since the ordering by the E-OLS method is similar to that of the OLS, only the models obtained with E-OLS are considered. The performance is evaluated on both the training (1000 samples) and evaluation (200 samples) data. For instance, according to the P-QR method, we first make a one-rule model consisting only of rule R_{24}, then a two-rule model with the rules R_{24} and R_{25}, etc., until we have a model of 20 rules. When $M_S \in [1, 20]$ rules are picked, the corresponding $1000 \times M_S$ firing matrix P_{M_S} is formed using the training data, and the rule consequents $\boldsymbol{\theta}$ are determined by solving the resulting global LS problem. To verify the methods, for each model complexity step of $M_S = 1, 2, \ldots, 20$ rules, 100 models are made with M_S different rules drawn at random from the total set of 25 rules. The average performance of the random models are recorded. The results are presented in Fig. 11.

In this experiment, the P-QR picks the rules such that they have good generalizing capabilities. It obtains an error on the evaluation data that is below that of the training data for a low number of rules ($r \geq 5$).

As expected, the E-OLS method fit the training data well with a low number of rules. Unlike the other methods, it uses information about the systems output and seek a best fit for the training data. From 7 rules on, the models show good performance also on the evaluation data.

The worst performing method is the SVD-QR. As expected, the order in which the rules are picked bears no proof of representing any importance order; neither with respect to fitting the training data, nor with respect to generalization capabilities. Its performance is qualitatively close to the random approach in which the redundant and non-firing rules were picked with the same probability as all the other rules.

5 Conclusions

Two main approaches to complexity reduction in fuzzy rule-based models are similarity-driven simplification and rule reduction by orthogonal transforms. The methods differs in their main objectives. Similarity-driven simplification seeks to reduce redundancy present in similar fuzzy sets, hereby reducing the term set of the model. Rule reduction may follow as a result. This serves two purposes: increase the transparency of the fuzzy model, and decrease

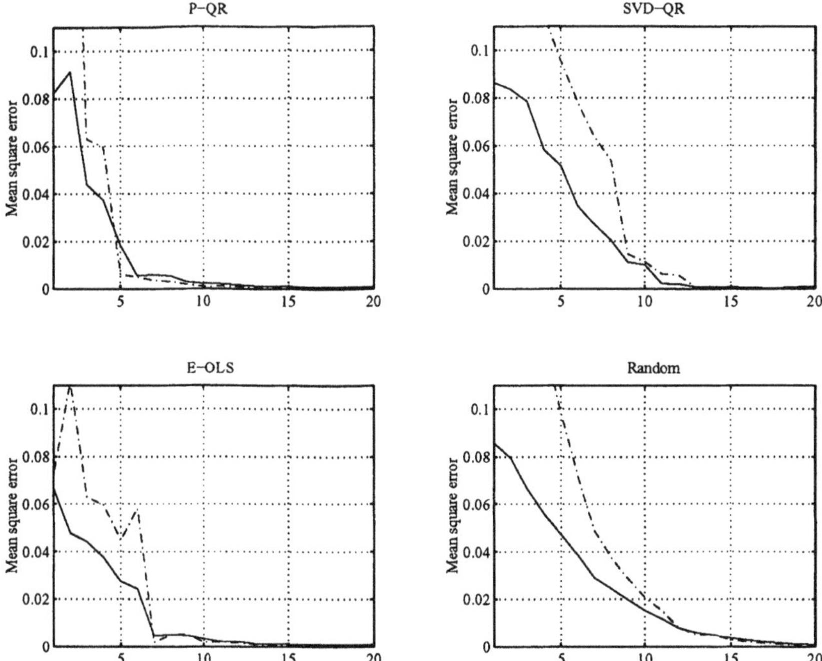

Fig. 11. Performance with increasing number of rules picked according to different methods. Solid line: error on training data. Dash-dot line: error on evaluation data.

the computational cost. Orthogonal transforms aim at rule-reduction only, by removing redundant an non-important rules.

Both approaches can be used together with other modeling tools to form a transparent fuzzy modeling scheme for data-driven modeling. This makes it possible to obtain inspectable and compact fuzzy rule-based models with qualitative rules directly from data [5,20,19,21,22]. Such models are often more suited than other function approximators for further analysis and for use in applications such as knowledge-based systems, controller design, process optimization, etc.

References

1. M. Setnes, R. Babuška, and H. B. Verbruggen. Rule-based modeling: Precision and transparency. *IEEE Trans. Systems, Man and Cybernetics – Part C: Applications and Reviews*, 28(1):165–169, 1998.
2. J. Valente de Oliveira. Semantic constraints for membership function optimization. *IEEE Trans. Systems, Man and Cybernetics – Part A: Systems and Humans*, 29(1):128–138, 1999.

3. T. Takagi and M. Sugeno. Fuzzy identification of systems and its applications to modelling and control. *IEEE Trans. Systems, Man, and Cybernetics*, 15:116–132, 1985.
4. J. Yen, L. Wang, and C.W. Gillespie. Improving the interpretability of TSK fuzzy models by combining global learning and local learning. *IEEE Trans. Fuzzy Systems*, 6(4):530–537, 1998.
5. M. Setnes. *Complexity reduction in fuzzy systems.* Ph.D. Thesis, Delft University of Technology, Dep. of El. Eng., Control Laboratory, Delft, the Netherlands, February 2001.
6. M. Setnes, R. Babuška, U. Kaymak, and H. R. van Nauta Lemke. Similarity measures in fuzzy rule base simplification. *IEEE Trans. Systems, Man and Cybernetics - Part B: Cybernetics*, 28(3):376–386, 1998.
7. R. Babuška. *Fuzzy Modeling for Control.* Kluwer Academic Pub., Boston, 1998.
8. L. X. Wang and J. M. Mendel. Fuzzy basis functions, universal approximation, and orthogonal least-squares learning. *IEEE Trans. Neural Networks*, 3(5):807–813, 1992.
9. J. Hohensohn and J. M. Mendel. Two-pass orthogonal least-squares algorithm to train and reduce fuzzy logic systems. In *Proceedings of FUZZ-IEEE'94*, pp 696–700, Orlando, USA, 1994.
10. G.C. Mouzouris and J.M. Mendel. Designing fuzzy logic systems for uncertain environments using a singular-value-qr decomposition method. In *Proceedings of FUZZ-IEEE'96*, pp 295–301, New Orleans, USA, 1996.
11. J. Yen and L. Wang. Simplifying fuzzy rule-based models using orthogonal transformation methods. *IEEE Trans. Systems, Man and Cybernetics - Part B: Cybernetics*, 29(1):13–24, 1999.
12. Y. Yam, P. Baranyi, and C-T. Yang. Reduction of fuzzy rule base via singular value decomposition. *IEEE Trans. Fuzzy Systems*, 7(2):120–132, 1999.
13. S. Chen, C.F.N. Cowan, and P.M. Grant. Orthogonal least squares learning algorithm for radial basis function networks. *IEEE Trans. Neural Networks*, 2(2):302–309, 1991.
14. G.H. Golub and C.F. van Loan. *Matrix Computations.* The Johns Hopkins University Press, London, 2 edition, 1989.
15. G.H. Golub. Numerical methods for solving least squares problems. *Numerical Methematics*, (7):206–216, 1965.
16. J. Yen and L. Wang. Application of statistical information criteria for optimal fuzzy model construction. *IEEE Trans. Fuzzy Systems*, 6(3):362–372, 1998.
17. G.W. Stewart. Rank degeneracy. *SIAM J. Sci. and Stat. Comput.*, 5(2):403–413, 1984.
18. M. Setnes and R. Babuška. Rule base reduction: some comments on the use of orthogonal transforms. *IEEE Trans. Systems, Man and Cybernetics - Part C: Applications and Reviews*, 31(2):199–206, 2001.
19. Magne Setnes. Supervised fuzzy clustering for rule extraction. *IEEE Trans. Fuzzy Systems*, 8(4):416–424, 2000.
20. M. Setnes, R. Babuška, and H. B. Verbruggen. Transparent fuzzy modeling. *International Journal of Human-Computer Studies*, 49(2):159–179, 1998.
21. M. Setnes and J.A. Roubos. GA-fuzzy modeling and classification: complexity and performance. *IEEE Trans. Fuzzy Systems*, 8(5):509–522, 2000.
22. M. Setnes and U. Kaymak. Fuzzy modeling of client preference from large data sets: an application to target selection in direct marketing. *IEEE Trans. Fuzzy Systems*, 9(1):153–163, 2001.

Effect of Rule Representation in Rule Base Reduction

Thomas Sudkamp, Aaron Knapp, and Jon Knapp

Wright State University, Department of Computer Science, Dayton OH, USA

Abstract. An objective of merging rules in rule bases designed for system modeling and function approximation is to increase the scope of the rules and enhance their interpretability. The effectiveness of rule merging depends upon the underlying system, the learning algorithm, and the type of rule. In this paper we examine the ability to merge rules using variations of Mamdani and Takagi-Sugeno-Kang style rules. The generation of the rule base is a two part process; initially a uniform partition of the input domain is used to construct a rule base that satisfies a prescribed precision bound on the training data. A greedy algorithm is then employed to merge adjacent regions while preserving the precision bound. The objective of the algorithm is to produce fuzzy models of acceptable precision with a small number of rules. A set of experiments has been performed to compare the effect of the rule representation on the ability to reduce the number of rules and on the precision of the resulting models.

1 Introduction

There are two primary strategies for reducing the number of rules in a fuzzy rule base: dimension reduction and rule merging. In dimension reduction, the determination of functional relationships between variables or the identification of variables that have minimal impact on the result are used to decrease the dimension of the input space [1,2]. Dimension reduction is frequently employed in classification problems in which the objects are defined by a large number of attributes. The objective of reducing the number of input variables is to facilitate the identification of the relationships between the variables and the classes of objects and to reduce the computational resources required. Rule merging is applied to models with a small number of inputs, primarily in system modeling and function approximation applications, to decrease the granularity of the rules and enhance the interpretability of the model [3]. Rule merging is accomplished by combining the regions of applicability of adjacent rules and constructing an appropriate consequent for the resulting rule. Rule merging algorithms have been developed for both Mamdani [4] and Takagi-Sugeno-Kang (TSK) style rules [5]. The objective of this study is to examine the effect of the form of the rule on the ability to merge regions and decrease the granularity of the rules.

In this work, rule merging is the second step in the generation of rule bases from training information. The learning procedure constructs an initial rule base that satisfies a prescribed bound on the error between the model and the training data. This rule base is generated using a uniform decomposable partition of the input space and a proximity based learning strategy. *Proximity based learning* describes a class of rule learning algorithms in which the domain decompositions are predetermined and the consequent of a rule is obtained from an analysis of the training data that occur in the support of the antecedent of the rule. Examples of proximity learning techniques can be found in [6–10].

In the generation of the initial rule base, regions in which the system has high variation may require a fine partition of the input space to satisfy the precision bound. Consequently, a uniform partition may produce many rules in regions of little variation where a small number of rules would suffice. After the generation of the initial rule base, a greedy strategy is employed to merge adjacent rules while maintaining the satisfaction of the precision bound. The technique for extending the scope of adjacent rules outlined in [11] provides the basis for the reduction. Enlarging the region of applicability of a rule may have two beneficial results. First, increasing the size of the support produces a rule with a greater degree of generalization, which guards against the problem of overfitting the data. The second benefit is that rules with larger granularity are more likely to admit a linguistic interpretation.

The objective of a learn-and-merge strategy is not to optimize the precision of the model on the training data, but rather to produce a model with a reasonably small set of rules that satisfies predetermined precision bound on the data. Ideally, the goal would be to identify the smallest set of rules satisfying the precision bound. However, the greedy merging strategy does not ensure the generation of a minimal size rule base.

In this paper we examine the effectiveness of rule merging using variations of Mamdani and TSK style rules. A set of experiments has been performed to compare the effect of the rule representation on the ability to reduce the number of rules and on the precision of the resulting models. Throughout this paper we will consider learning rules to model a two-input system with normalized input domains $U = [-1, 1]$ and $V = [-1, 1]$. The output will take values from the domain $W = [-1, 1]$.

2 Rule Representation

Algorithms for learning fuzzy rules from training data employ a local analysis of the training information to produce rules. The local regions are obtained by partitioning the input space into a set of fuzzy sets that define the antecedents of the rules. A fuzzy partition [12] of a two-dimensional input space $U \times V$ consists of a set of fuzzy sets $D_1, \ldots, D_t$ over $U \times V$ whose supports cover the space. A rule base consists of a rule whose antecedent has the form

$$\text{`if } X \times Y \text{ is } D_k \text{ then } \ldots\text{'},$$

for each fuzzy set D_k in the partition of $U \times V$.

A partition of a two-dimensional domain $U \times V$ is *decomposable* if there are fuzzy partitions $\{A_1, \ldots, A_n\}$ of U and $\{B_1, \ldots, B_m\}$ of V whose Cartesian product forms the partition of $U \times V$. That is, every fuzzy set D_k in the partition of $U \times V$ has the form $A_i \times B_j$ for some i and j. For linguistic interpretability and computational simplicity, fuzzy rule bases with multiple inputs frequently employ decomposable partitions. In this case, a rule with antecedent $D_k = A_i \times B_j$ is written 'if X is A_i and Y is B_j'.

Because of the simplicity and ease of computation, many partitions of one-dimensional spaces consist of triangular fuzzy sets [13]. In a triangular partition of a domain $U = [-1, 1]$, the fuzzy sets $A_1, \ldots, A_n$ are completely determined by the selection of a sequence of points $a_1 = -1, a_2, \ldots, a_n = 1$, where the point a_i is the center point of the triangular fuzzy set A_i. The support of A_i is the interval (a_{i-1}, a_{i+1}) and the membership function is

$$\mu_{A_i}(x) = \begin{cases} (x - a_{i-1})/(a_i - a_{i-1}) & \text{if } a_{i-1} \leq x \leq a_i \\ (-x + a_{i+1})/(a_{i+1} - a_i) & \text{if } a_i < x \leq a_{i+1} \\ 0 & \text{otherwise.} \end{cases}$$

Since a triangular partition of a one-dimensional space is completely determined by the center points of the membership functions, a decomposable partition of a two dimensional domain $U \times V$ is generated by the selection of membership function center points $\{a_1, \ldots, a_n\}$ and $\{b_1, \ldots, b_m\}$ of U and V respectively. The Cartesian product $\{a_1, \ldots, a_n\} \times \{b_1, \ldots, b_m\}$ defines a grid on $U \times V$. Figure 2 shows the set of grid points generated by decompositions consisting of $n = m = 12$ fuzzy sets.

The grid points divide the input domain into $(n-1)(m-1)$ rectangular regions, which will be used to produce the core and the support of the fuzzy sets in the antecedents of the rules. The precise manner in which the partitions of $U \times V$ are obtained from the grid points depends upon the form of the rule and the rule merging strategy.

The two major types of fuzzy rules differ in the form of the consequent. A two-dimensional Mamdani style rule [14,15] for a decomposable partition of the input space has the form

'if X is A_i and Y is B_j then Z is $C_{i,j}$',

where A_i and B_j are fuzzy sets from the partitions of the input domains U and V and $C_{i,j}$ is a fuzzy set over the output domain W. A TSK [16] rule has the form

'if X is A_i and Y is B_j then $z = g_{i,j}(x, y)$',

where $g_{i,j}$ is a function of the input (x, y). In a TSK rule base, all the consequent functions $g_{i,j}$ generally have a predefined parametric form.

A rule base consists of a rule for every fuzzy set in the partition of the input space. For a decomposable partition, this requires a rule for each pair of

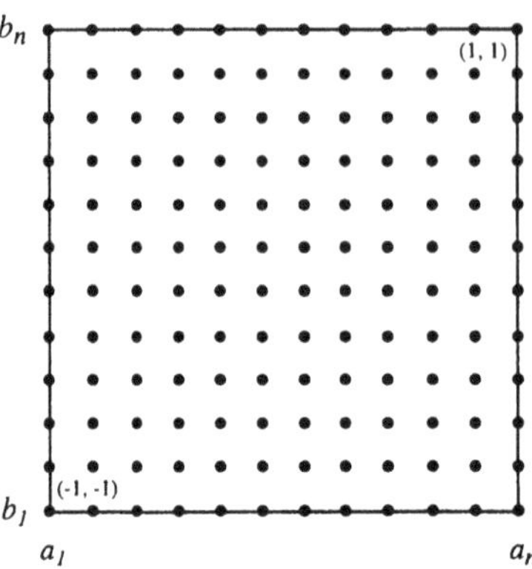

Fig. 1. Domain decomposition and grid points

fuzzy sets A_i and B_j. The rules and the rule aggregation technique combine to produce a model $\hat{f}$, a function $\hat{f} : U \times V \to W$. Mamdani and TSK rule bases produce the same models when the decomposition of the input space is generated from triangular partitions of the input domains, weighted averaging is used for rule aggregation, and the rules have form

'if X is A_i and Y is B_j then Z is $C_{i,j}$',
'if X is A_i and Y is B_j then $g_{i,j}(x,y) = c_{i,j}$',

where the constant $c_{i,j}$ in the consequent of the TSK rule is the center point of the fuzzy set $C_{i,j}$ in the consequent of the corresponding Mamdani rule.

When a rule base is defined in the preceding manner, the value of the model for input (x, y) is determined by at most four rules. For input $x \in [a_i, a_{i+1}]$ and $y \in [b_j, b_{j+1}]$, the rules

'if X is A_i and Y is B_j then Z is $C_{i,j}$',
'if X is A_i and Y is B_{j+1} then Z is $C_{i,j+1}$',
'if X is A_{i+1} and Y is B_j then Z is $C_{i+1,j}$',
'if X is A_{i+1} and Y is B_{j+1} then Z is $C_{i+1,j+1}$'

and weighted averaging combine to produce a function $\hat{f}_{i,j}$ over $[a_i, a_{i+1}] \times [b_j, b_{j+1}]$,

$$\begin{aligned}\hat{f}_{i,j}(x,y) &= \frac{\sum_{s=i}^{i+1}\sum_{t=j}^{j+1}\mu_{A_s}(x)\mu_{B_t}(y)c_{s,t}}{\sum_{s=i}^{i+1}\sum_{t=j}^{j+1}\mu_{A_s}(x)\mu_{B_t}(y)} \\ &= \frac{(a_{i+1}-x)(b_{j+1}-y)c_{i,j}}{(a_{i+1}-a_i)(b_{j+1}-b_j)} + \frac{(x-a_i)(b_{j+1}-y)c_{i+1,j}}{(a_{i+1}-a_i)(b_{j+1}-b_j)} \\ &\quad + \frac{(a_{i+1}-x)(y-b_{j+1})c_{i,i+1}}{(a_{i+1}-a_i)(b_{j+1}-b_j)} + \frac{(x-a_i)(y-b_j)c_{i+1,j+1}}{(a_{i+1}-a_i)(b_{j+1}-b_j)}. \end{aligned} \tag{1}$$

Equation 1 shows that the *local approximating function* $\hat{f}_{i,j}$ is completely determined by the grid points a_i, a_{i+1}, b_j, b_{j+1} that define the region and the values $c_{i,j}$, $c_{i+1,j}$, $c_{i,j+1}$, $c_{i+1,j+1}$.

The model $\hat{f}$ is constructed from the $(n-1)(m-1)$ local approximating functions $\hat{f}_{i,j}$, $1 \leq i \leq n-1$ and $1 \leq j \leq m-1$, where the domain of $\hat{f}_{i,j}$ is the rectangle $[a_i, a_{i+1}] \times [b_j, b_{j+1}]$. The fuzzy partition of the input domains produces a continuous transition between adjacent regions ensuring that the function $\hat{f}$ obtained from the local functions is continuous over $U \times V$.

Defining rule bases directly from Cartesian products of independent partitions of the input domains limits the ability of the rules to express the relationships between the variables [17]. This deficiency has provided the impetus for the use of clustering [18,19] and neural-fuzzy systems [20,21] to partition the input domains based on the distribution and values of the training data. To produce a dynamically generated partition within the framework of proximity learning and rule merging, it is necessary to have a rule representation that permits the antecedent to be determined by the training data rather than by an a priori partitioning of the input space.

The experimental results given in Section 4 will compare the ability to produce models using variations of the Mamdani and TSK style of rules presented above. These variations provide flexibility in the generation of local approximating functions to facilitate the reduction of the number of rules needed to define the model while maintaining the precision bound. The next two subsections describe the form of the rules using the grid points obtained from the domain decompositions. Rule bases consisting of Mamdani rules or TSK rules with constant consequents built from decomposable triangular partitions, as described above, will provide the baseline by which the variations will be judged.

2.1 Extended TSK Rules

The standard form of TSK rules with single point consequents is

$$\text{`if } X \text{ is } A_i \text{ and } Y \text{ is } B_j \text{ then } g_{i,j}(x,y) = c_{i,j}\text{'},$$

which we will refer to as type-1 rules. For a type-1 rule base with input $x \in [a_i, a_{i+1}]$ and $y \in [b_j, b_{j+1}]$, the local approximating function $\hat{f}_{i,j}(x,y)$ is determined by values of the function at the four corners of the rectangle defined by the point $(a_i, b_j, c_{i,j})$, $(a_i, b_{j+1}, c_{i,j+1})$, $(a_{i+1}, b_j, c_{i+1,j})$, and $(a_{i+1}, b_{j+1}, c_{i+1,j+1})$ as shown in Equation 1.

We will now define TSK rules that permit a single rule to produce a surface over a general rectangle formed by grid points in the input space. A rectangle in $U \times V$ is denoted by a pair of grid points $[(a_i, b_j), (a_r, b_s)]$ with $i < r$ and $j < s$. The points designate the corners of the rectangle; (a_i, b_j) is the lower left-hand corner of the rectangle and (a_r, b_s) is the upper right-hand corner.

A type-2 rule has the form

'if $X \times Y$ is $D = [(a_i, b_j), (a_r, b_s)]$ then $g_{i,j}(x, y) = [c_{i,j}, c_{i,s}, c_{r,j}, c_{r,s}]$',

where the subscript on the consequent function $g_{i,j}$ indicates the lower left-hand corner of the rectangle. The consequent of the rule specifies the values at the corners of the rectangular region; $(a_i, b_j, c_{i,j})$, $(a_i, b_s, c_{i,s})$, $(a_r, b_j, c_{r,j})$, and $(a_r, b_s, c_{r,s})$. Using the computation given in Equation 1 substituting a_r for a_{i+1}, b_s for b_{j+1}, $c_{i,s}$ for $c_{i,j+1}$, $c_{r,j}$ for $c_{i+1,j}$, and $c_{r,s}$ for $c_{i+1,j+1}$ produces the function $g_{i,j}(x, y)$.

A type-2 rule represents four standard TSK rules with constant consequents:

'if X is A_i and Y is B_j then $g_{i,j}(x, y) = c_{i,j}$',
'if X is A_i and Y is B_s then $g_{i,s}(x, y) = c_{i,s}$',
'if X is A_r and Y is B_j then $g_{r,j}(x, y) = c_{r,j}$',
'if X is A_r and Y is B_s then $g_{r,s}(x, y) = c_{r,s}$'.

The advantage of the type-2 rule is that the antecedent explicitly indicates the region of applicability of the rule and the rectangular regions of a type-2 rule base need not form a regular grid over the input space. Figure 2 shows the rectangular domain decomposition that resulted from merging regions formed by 6×6 decomposition. For this decomposition, a type-1 rule base consists of 36 rules, one for each original grid point. The type-2 rule base contains 15 rules representing 26 type-1 rules, one at each of the grid points occurring at intersections of rectangles in Figure 2.

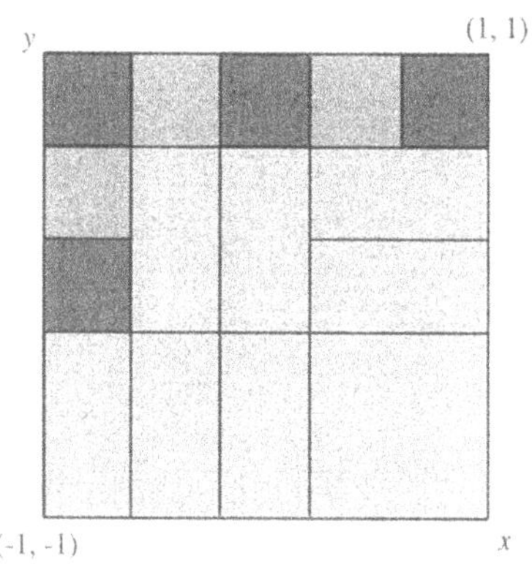

Fig. 2. Input domain after TSK rule merger

A type-2 rules have considerably more flexibility in constructing a model than type-1 rules. In a type-1 rule base, the value of an input $(x, y) \in [a_i, a_{i+1}] \times [b_j, b_{j+1}]$ is determined by the rules associated with the rectangle obtained from the grid points immediately surrounding it. A type-2 rule representing this relationship has the form

$$\text{`if } X \times Y \text{ is } D = [(a_i, b_j), (a_{i+1}, b_{j+1})] \text{ then}$$
$$g_{i,j}(x, y) = [c_{i,j}, c_{i,j+1}, c_{i+1,j}, c_{i+1,j+1}]\text{'}.$$

However, the rectangular region designated by a type-2 rule is not restricted to adjacent grid points. One artifact of the extended region is that the value for input (x, y) is obtained from the values of the corners of the rectangle in the antecedent, but these may not be the type-1 rules nearest (x, y). Consider the input (x, y) in Figure 3 that lies in the region indicated by a rule with antecedent 'if $X \times Y$ is $D = [(a_i, b_j), (a_r, b_s)]$'. The value $\hat{f}(x, y)$ is obtained from the points $(a_i, b_j, c_{i,j})$, $(a_i, b_s, c_{i,s})$, $(a_r, b_j, c_{r,j})$, and $(a_r, b_s, c_{r,s})$ although the grid point (a_k, b_s) is nearer to (x, y) than any of the corners of the rectangle D.

For the model defined by a type-2 rule base to be continuous, it is necessary that the values at the boundaries of adjacent rules are identical. In Figure 3, continuity requires that $\hat{f}_{i,j}(x, y) = \hat{f}_{k,s}(x, y)$ for all (x, b_k) with $x \in [a_k, a_s]$. The merging algorithm will ensure that the type-2 rule bases define continuous models.

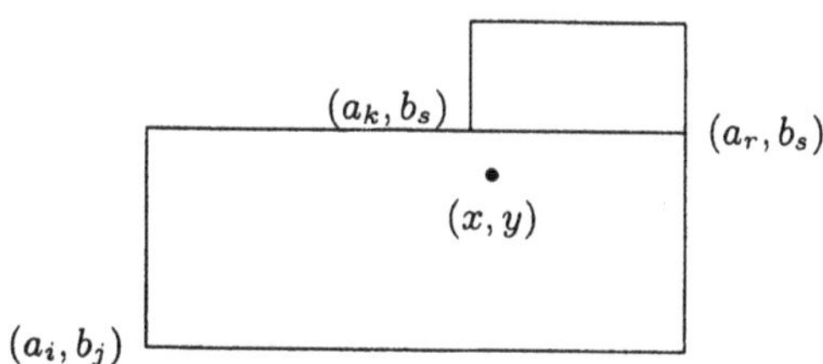

Fig. 3. Adjacent type-2 rules

2.2 Rectangular Core Rules

When the domain decomposition is defined by the Cartesian product of triangular fuzzy sets, the core of each rule is limited to a single point. Allowing the core to be a region of the input space permits the rule to define a surface that has a greater ability to match the training data. Expanding the core is accomplished by permitting the antecedent of a rule to be a rectangular fuzzy set obtained from the underlying set of grid points.

A rectangular fuzzy set is denoted by its core, which is a rectangle in $U \times V$ defined by a pair of grid points $[(a_i, b_j), (a_r, b_s)]$. The support of the fuzzy set is the rectangle specified by the grid points $[(a_{i-1}, b_{j-1}), (a_{r+1}, b_{s+1})]$. The outer rectangle in Figure 4 is the support of the rectangular fuzzy set $[(a_i, b_j), (a_r, b_s)]$ and the shaded region is the core. If the core of a fuzzy set abuts the boundary of the input domain, the support does not extend beyond the core in that direction. The region of the support not in the core provides the smooth transition between adjacent fuzzy sets in the decomposition of the input domain. For simplicity, we will refer to this region as the *buffer* of the fuzzy set.

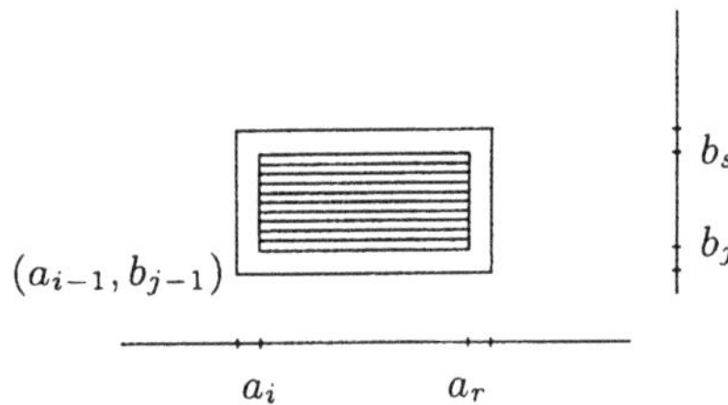

Fig. 4. Support of rectangular fuzzy set

Any point in the core of the fuzzy set $D = [(a_i, b_j), (a_r, b_s)]$ has membership value 1 in D. For a point (x, y) in the buffer, the membership value is determined by linear interpolation. Let (x_1, y_1) be the point on the boundary of the core that is nearest to (x, y). The line from (x_1, y_1) to (x, y) in $U \times V$ is extended until it intersects the boundary of the support at a point (x_2, y_2). The membership value $\mu_A(x, y)$ is the z value of the point (x, y, z) that lines on the line from $(x_1, y_1, 1)$ to $(x_2, y_2, 0)$. Figure 5 illustrates the generation of the membership values for a rectangular fuzzy set.

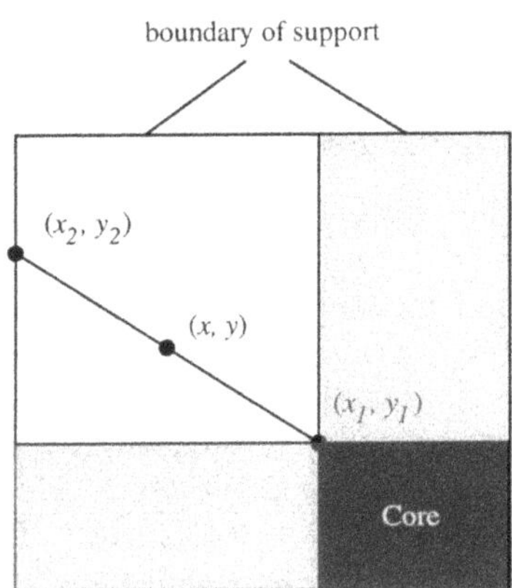

Fig. 5. Membership values of rectangular fuzzy sets

The general form of a rule with a rectangular fuzzy set antecedent, which we will refer to as a type-4 rule, is

$$\text{'if } X \times Y \text{ is } D = [(a_i, b_j), (a_r, b_s)] \text{ then } g_{i,j}(x, y) = [c_{i,j}, c_{i,s}, c_{r,j}, c_{r,s}]\text{'}. \quad (2)$$

This is a TSK style rule where the consequent defines a function over the support of D. The rule consequent specifies the values at the corners of the core; $(a_i, b_j, c_{i,j})$, $(a_i, b_s, c_{i,s})$, $(a_r, b_j, c_{r,j})$, and $(a_r, b_s, c_{r,s})$. Multi-linear interpolation from these corner points defines the function $g_{i,j}$ that provides the values for points that lie within in core of the antecedent. The 3-dimensional surface over the core obtained from the corner points is illustrated in Figure 6.

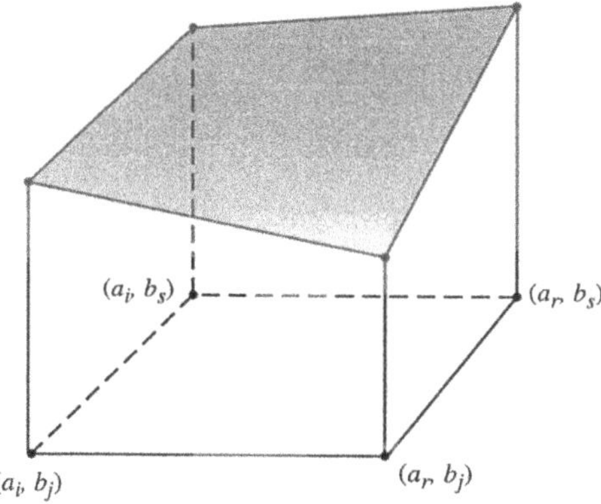

Fig. 6. Surface over core $[(a_i, b_j), (a_r, b_s)]$

The value $g_{i,j}(x, y)$ for a point (x, y) in the buffer of the antecedent of a rectangular core rule is obtained from the point (x_1, y_1) on the boundary of the core nearest (x, y): $g_{i,j}(x, y) = \mu_A(x, y) \cdot g_{i,j}(x_1, y_1)$. Multiplying by the membership function produces a transition from the surface over the core to the value 0 as shown in Figure 7.

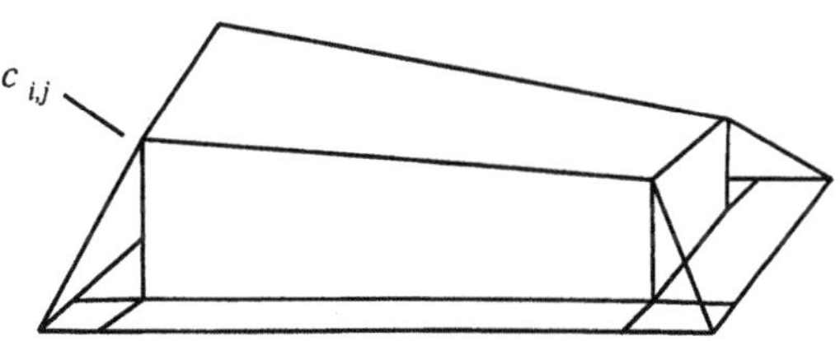

Fig. 7. Surface from type-3 rule

Figure 8 shows an example of rectangular regions partitioning the input space upon the completion of the rule merging algorithm described in Section 3.2. The shaded regions are the cores of the fuzzy sets in the antecedents of the rules and the dark regions are the buffers. The rule base was approximating the function $f(x,y) = x^2 - y^2$ and the initial domain decomposition produced a 50×50 grid. The merging clearly identified the relatively stable area of this surface near $(0,0)$, creating a large rule that that may be interpreted as "if the input is near $(0,0)$, then the output is near 0." In the partition of the domain by rectangular core rules, a point (x,y) in $U \times V$ is either in the core of a single rule or in a region buffer between adjacent rules.

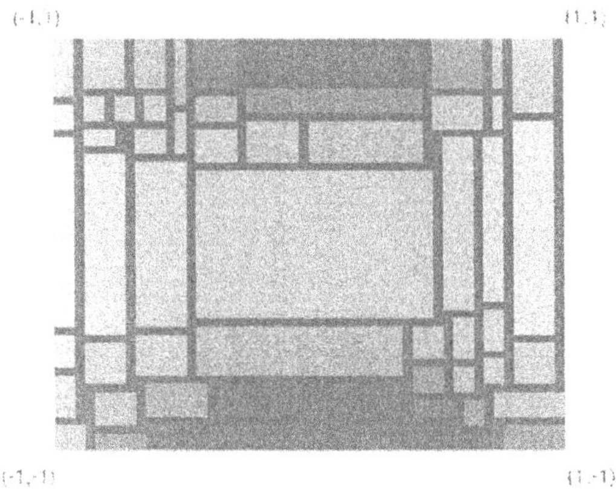

Fig. 8. Type-4 input domain partition by merging

The output value produced by a type-4 rule base with input (x,y) is the weighted average

$$\hat{f}(x,y) = \frac{\sum \mu_{D_t}(x,y) \cdot g_{i,j}(x,y)}{\sum \mu_{D_t}(x,y)}, \tag{3}$$

where the sum is over all rules in which (x,y) has nonzero support in the antecedent and the weight is the membership value. For an input in the core of a rule with antecedent 'if $X \times Y$ is $D = [(a_i, b_j), (a_r, b_s)]$', $\hat{f}(x,y)$ is simply $g_{i,j}(x,y)$. For a point (x,y) in a buffer area, the output is the weighted average of all fuzzy rules whose cores are adjacent to (x,y).

We will consider rule bases comprised of two types of rectangular core rules. Type-4 rules are general rectangular core rules. Rules of type-3 allow rectangular antecedents but require a constant value in the consequent:

'if $X \times Y$ is $D = [(a_i, b_j), (a_r, b_r)]$ then $g_{i,j}(x,y) = [c, c, c, c]$'.

The functions associated with a type-3 rule have a planar surface parallel to the U-V plane over the core of the rule. Rules of this type are more amenable to a linguistic interpretation than the more general type-4 rules since the consequent can be identified with the fuzzy set in a partition of the output domain in which c has maximal membership, the technique used by Wang and Mendel [22,6] to translate rules learned from training data to Mamdani style rules.

Type-1 rules may be considered a degenerate case of rectangular core rules. A type-1 rule may be written as a rectangular core rule by

$$\text{'if } X \times Y \text{ is } D = [(a_i, b_j), (a_i, b_j)] \text{ then } g_{i,j}(x, y) = [c, c, c, c]\text{'}.$$

That is, the rectangle in the antecedent reduces to the single point (a_i, b_j) and the consequent is a constant value c. A rule base with rules of this form is equivalent to a baseline Mamdani or TSK rule base described in the previous section; the function $\hat{f}$ defined by Equation 1 is identical to the weighted average in Equation 3.

3 Rule Base Generation

The construction of a rule base is a two step process that begins with the generation of an initial rule base that satisfies a specified precision bound followed by a refinement of the rule base by merging adjacent rules. Both of these phases depend upon the selection of the constant values in the consequent of a rule. The process begins by selecting points $\{a_1, \ldots, a_n\} \in U$ and $\{b_1, \ldots, b_m\} \in V$ that define the partitions of the input domains. The ordered pairs (a_i, b_j) form a grid over the space $U \times V$ as illustrated in Figure 2. We will first examine methods for generating rules from training data, followed by the presentation of the merging algorithms. Throughout this paper the initial partitions of the input domain consist of uniformly spaced grid points.

3.1 Consequent Selection

For a rule with antecedent 'if $X \times Y$ is $D = [(a_i, b_j), (a_r, b_s)]$', the learning process consists of selecting the constants in the consequent of the rule. A set $T = \{(x_s, y_s, z_s) \mid s = 1, \ldots, K\}$ of training examples, where $x_i \in U$ and $y_i \in V$ are input values and $z_i \in W$ is the associated response, provides the information needed for the generation of the rules.

For rules of types 1, 2, and 4, the value $c_{i,j}$ associated with grid point (a_i, b_j) is determined by the weighted average of training points whose projection onto the input space lies within a distance q of (a_i, b_j). The initial distance q is obtained from the spacing of the grid points: $q = .75 \min\{a_{i+1} - a_i, b_{i+1} - b_i\}$. Let (x_k, y_k, z_k) be a training point, $d_{i,j,k}$ be the distance from (x_k, y_k) to grid point (a_i, b_j), and $T_{i,j,q}$ be the set of training points with the distance $d_{i,j,k}$ less than or equal to q. The value associated with grid point (a_i, b_j) obtained from $T_{i,j,q}$ using weighted averaging is

$$c_{i,j} = \frac{\sum(1 - (d_{i,j,k}/q))z_k}{\sum(1 - (d_{i,j,k}/q))},$$

where the summation is taken over all training points in $T_{i,j,q}$.

Since training data may be sparsely distributed throughout the input space, it is possible that no training point lies within distance q from a grid point (a_i, b_j). When this occurs, the radius q defining the set $T_{i,j,q}$ is expanded incrementally until training information is found. The expansion of the radius is accomplished by adding $\min\{a_{i+1} - a_i, b_{i+1} - b_i\}/4$ to q after each unsuccessful search for training data.

In a type-3 rule, the single constant value c in the consequent is associated with an entire rectangular region and not a grid point. Consequently, the method of selection of the constant should not focus locally on the grid points at the corners of the core but rather should be concerned with matching the training data throughout the entire region. The value c is chosen to minimize the maximum error between the training data in the core and the surface $g(x,y) = c$ over the core region indicated by the rule antecedent. Thus, $c = (z_{\max} + z_{min})/2$ where $z_{\max}$ and $z_{\min}$ are the largest and smallest z values of training data that occur in the rectangle $D = [(a_i, b_j), (a_r, b_s)]$.

Several techniques have been examined for determining a value for an initial rectangle $D = [(a_i, b_j), (a_{i+1}, b_{j+1})]$ that contains no training data. These include a form of rule base completion [8,23] and selection of a value from an approximating surface generated by surrounding training data. The latter has proven more effective since the value is determined directly from training data rather than indirectly from derived rules. The process is illustrated in Figure 9. The nearest training point to the initial region in each of the four quadrants is located. A surface s is constructed from those points following the approach in Equation 1. The value assigned to the region is $s((a_i + a_{i+1})/2, (b_j + b_{j+1})/2))$; the value of s at the center of the rectangle D. The advantage of obtaining values using the approximation is that it provides a smooth set of auxiliary values when there are multiple adjacent initial regions with no training data.

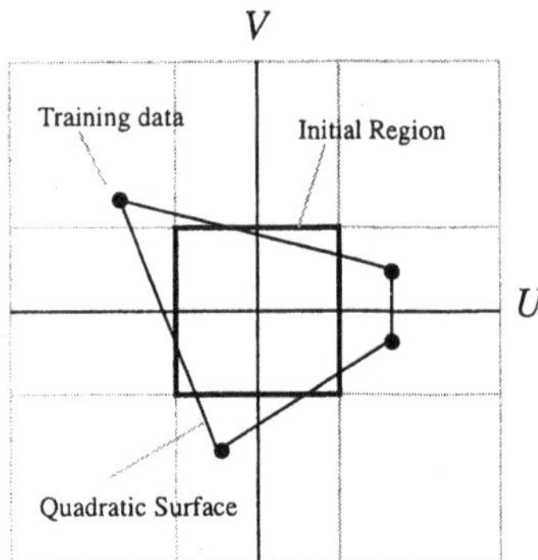

Fig. 9. Input domain partition by merging

3.2 Learning TSK Rule Bases

In this section we present the algorithm for generating type-1 and type-2 rule bases. The process begins by partitioning the input domains and constructing a type-1 rule base. If the resulting rule base satisfies a prescribed precision bound, the rule base serves as both the baseline by which the merging techniques will be judged and as the starting point for the generation of a type-2 rule base. Otherwise a new partition is selected and the generate-and-test procedure is repeated.

Throughout this presentation we will assume that partitions of input spaces have the same number of uniformly spaced fuzzy sets. Thus the grid formed by the partitions will be an $n \times n$ grid as shown in Figure 2. The refinement process initially sets n to 2 with grid points $a_1 = -1$, $a_2 = 1$ and $b_1 = -1$, $b_2 = 1$. The constant values $c_{i,j}$ for the type-1 rules are obtained as described in the previous section producing a model $\hat{f}$. The training data are then used to determine the precision of the model. The error at training point (x_t, y_t, z_t) is $|\hat{f}(x_t, y_t) - z_t|$. The type-1 rule base generation is completed if the maximum error over the training set lies within the specified precision bound. If not, the algorithm expands the number of regions in the grid by decomposing the input domains with $n+1$ points as shown in Figure 10. This generate-and-test procedure continues increasing n until a grid of size $n \times n$ is found that satisfies the precision bound.

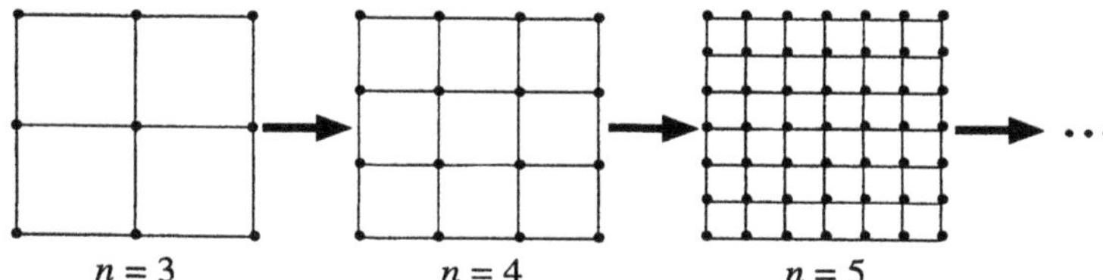

Fig. 10. Grid point refinement

The procedure for generating the type-1 rule base can be summarized as follows:

1. Input: training set T and error bound β.
2. Set $n = 2$.
3. Generate the type-1 rule base with n^2 rules defined by the $n \times n$ grid.
4. Calculate maximum error er over the set of training points.
5. If $er < \beta$ accept the rule base, else set $n = n + 1$ and go to 2.

As previously noted, this type-1 rule base defines a type-2 rule base with $(n-1)^2$ type-2 rules. The rectangular regions in the type-2 rule antecedents correspond to the regions defined by the domain decompositions.

Rules from the rule base with uniform partitions are combined to produce rules with rectangular antecedents that preserve the precision bound β. Merging begins with a region $[(a_i, b_j), (a_{i+1}, b_{j+1})]$ and attempts to construct a rule with a larger region of applicability as shown in Figure 11. Starting at the lower left corner of the region, regions are combined by expanding diagonally to produce rules with support $[(a_i, b_j), (a_{i+2}, b_{j+2})]$, $[(a_i, b_j), (a_{i+3}, b_{j+3})], \ldots$.

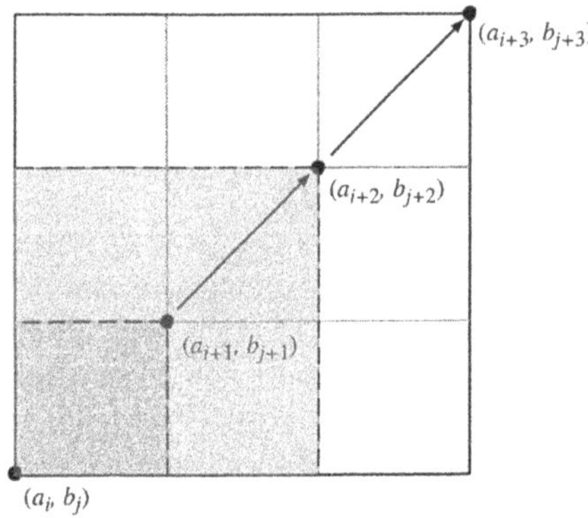

Fig. 11. Merging regions as a square

For type-2 rule bases, once a rule with enlarged support is proposed, the surfaces associated with adjacent rules may no longer be continuous. Figure 12 shows the relationship in the input domain between an expanded rule and the adjacent rules. Before expansion, the surface is continuous along the line segment $\overline{GH}$ since the values at points G and H are the same for the adjacent rules. When **S** is expanded, the values specified by the expanded rule at points G and H are no longer the values determined from the local training data but rather are obtained from the values at the corner points (a_i, b_{j+k}) and (a_{i+k}, b_{j+k}). To maintain continuity, the grid point values for the region **T** are modified to match the values generated by the rule over **S**. Since this alters the function associated with the region **T**, the modified rule over **T** must also be checked for compliance with the precision bound. If these conditions are met, the enlarged rule is accepted.

After the rule has been expanded as far as possible diagonally, expansions are attempted along the U and V axes. The region $[(a_i, b_j), (a_{i+k}, b_{j+k})]$ resulting from the diagonal expansion is then expanded vertically until halted by the precision bound, producing a rectangle $[(a_i, b_j), (a_{i+k}, b_{j+k+s})]$ as shown in Figure 13. In a like manner, $[(a_i, b_i), (a_{i+k}, b_{j+k})]$ is expanded horizontally producing $[(a_i, b_i), (a_{i+k+t}, b_{j+k})]$. The larger rectangle that preserves the precision bound is selected as the antecedent of the new rule.

The merging process is greedy; it begins with rectangle $[(a_0, b_0), (a_1, b_1)]$ and merges regions until the precision bound stops the process. Once an

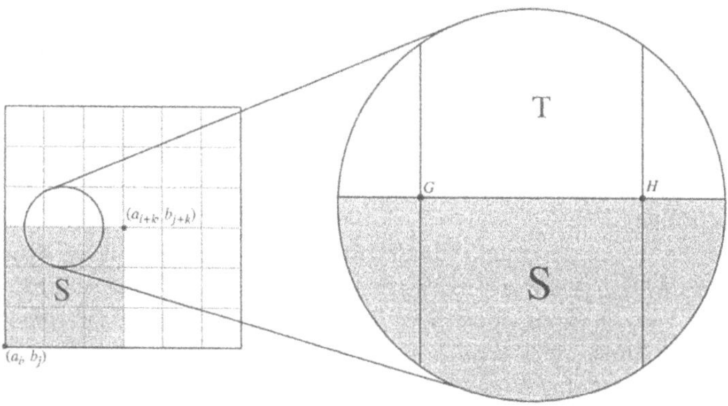

Fig. 12. Continuity considerations

expanded rule is accepted, the values at the grid points on the boundary of the rule are fixed and cannot by changed by subsequent expansions. The merging process continues left-to-right, bottom-to-top until the entire input domain has been considered.

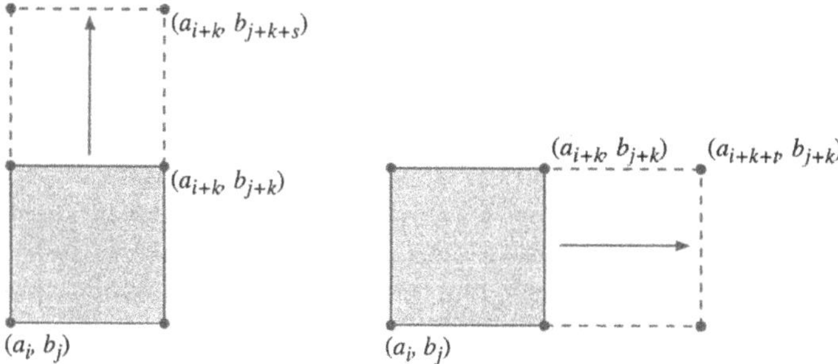

Fig. 13. Merging vertically and horizontally

3.3 Learning Rectangular Core Rule Bases

The learning process for type-3 and type-4 rule bases is similar to that of the type-2 rule bases. The first step is to find an input domain decomposition and construct a rule base that satisfies the precision bound. As before, the initial partitions are uniformly spaced $n \times n$ grids on the input domain.

The merging strategy is the same as for type-2 rule bases except for the incorporation of the buffer in the rule antecedent. The merging begins with rectangle $[(a_i, b_j), (a_{i+1}, b_{j+1})]$ and the region is expanded diagonally while preserving the precision bound as in Figure 11. Then horizontal and vertical extensions are considered. The largest expansion $[(a_i, b_j), (a_r, b_s)]$ is the core of the antecedent of the new rule and the rectangle $[(a_{i-1}, b_{j-1}), (a_{r+1}, b_{s+1})]$ is the support. The merging process is reinitiated with the first rectangle following the buffer. The process continues left-to-right, bottom-to-top until the entire domain is in the core of a single rule or in a buffer region between adjacent rules.

For rules bases of type-3 and type-3 rules, a merger is accepted if the training data within the core of the expanded rule satisfy the precision bound. Unlike the type-2 expansion, there is no need to check for the continuity of the model defined by a type-3 or type-4 rule base. The buffers define the transition between fuzzy sets in the domain decomposition which, in turn, ensures the continuity of the model.

3.4 Computational Issues

The algorithms and experimental results presented in this paper have considered rule generation for a system with two input dimensions. The techniques may be extended to higher dimensional systems but their applicability is limited by the associated growth in the number of rules required.

The generation of a rule base using the preceding algorithms is a two part process: the construction of the original set of rules by refinement and the rule reduction by merging. Refinement produces a grid with k^n grid points, where k is the number of input domains and $n = 2, 3, \ldots$ is the number of fuzzy sets in the partitions of the input domains. The proximity based algorithm used to learn to the rule consequents is straightforward, but is required for each of the k^n grid points. Thus the applicability of the refinement procedure is limited to problems with a small number of input variables, as is the case in many control, modeling, and function approximation applications which frequently have five or fewer inputs.

The greedy rule merging algorithm ensures that each hyper-rectangular region is considered at most once as the beginning of merging sequence. However, the number of adjacent regions that are examined for potential merging is dependent upon the dimension size. With k input dimensions, there are k adjacent regions that have not been previously allocated to a rule by the greedy strategy and are candidates for merging. However, the merging strategy is not dependent upon the generation of the original rule by the refinement method. It applicable to any rule base whose antecedents define hyper-rectangles in the input space.

4 Experimental Results

A set of experiments have been conducted to compare the effectiveness of the merging strategy on different rule types. An experiment begins with selection of a *target function* f, which represents the underlying system being modeled. A training set $T = \{(x_t, y_t, f(x_t, y_t)) \mid t = 1, \ldots, m\}$ is generated by randomly selecting m elements from the input space. After the selection of the target and the generation of the training set, the generate-and-test domain decomposition refinement procedure is used to find a type-1 rule base that satisfies the precision bound on the training set. The resulting rule base represents the minimal size type-1 rule base that satisfies the precision bound for a uniformly spaced $n \times n$ decomposition of the input domain. Rule bases of type-2, type-3, and type-4 are then constructed using the merging algorithm and compared with the baseline type-1 rule base.

Upon completion of the rule base generation, the function $\hat{f}(x, y)$ defined by the rule base is compared with the target function to determine the precision of the resulting model. The test set comprises 784 points uniformly distributed over the input domain $U \times V$. The experimental results consist of the average number of rules, the average of the average error, and the average maximum error over 20 iterations of the algorithm for each set of parameters and rule type. The algorithm was tested with a number of target functions, training set sizes, and precision bounds to evaluate the robustness of the merging strategy.

Before comparing the performance of the different rule types, we note several general characteristics exhibited by all of the learning strategies. As would be expected, reducing the precision bound increases the number of rules required satisfy the bound. With a fixed precision bound, the number of rules initially increases with the acquisition of training data. When sufficient training data are available, the number of rules declines until it stabilizes. Initially the acquisition of training data enhances the ability to find errors in more regions of the input domain, thereby requiring additional rules. When sufficient training data are available, there are few or no regions without the information needed to approximate each of the rules to a high degree of precision. The accurate generation of rules reduces the number needed. The point at which training data saturation occurs is dependent upon the precision bound and the target function.

We begin with a comparison of type-1 and type-2 rules. The sole difference between these rule bases is the merging, since the generation of the type-2 rule base begins with the completed type-1 rule base. Table 1 shows the effect of the training set size and the precision bound on the resulting rule base for the target function $f(x, y) = x^2 - y^2$. The column on the right gives the reduction of the number of rules due to rule merging.

The general observations about the relationships between the precision bound and the training set size described above are clearly evident in Table 1. For each precision bound, the number of rules initially increases with the acquisition of training data as the data provide information in previously uncovered areas of the input domain. For the type-1 rule bases, the maximum number of rules occurred with either 500 or 1000 training points. A training

		Type-1			Type-2			
Prec Bound	Training Examples	Rules	Ave Error	Max Error	Rules	Ave Error	Max Error	Percent Reduction
0.2	50	35.5	.100	.577	14.9	.111	.485	58%
	100	38.2	.076	.440	14.8	.084	.385	61%
	500	23.9	.039	.198	11.7	.058	.217	51%
	1000	13.2	.039	.171	12.4	.046	.188	6%
	10000	9.0	.042	.167	9.0	.043	.171	0%
0.1	50	153.9	.110	.607	43.7	.093	.563	71%
	100	163.5	.069	.520	53.7	.064	.395	67%
	500	193.5	.027	.249	42.0	.034	.147	78%
	1000	140.4	.020	.147	33.2	.032	.122	76%
	10000	25.0	.018	.080	25.0	.018	.081	0%
0.05	50	289.0	.109	.721	89.0	.088	.515	69%
	100	409.4	.073	.516	132.8	.062	.391	68%
	500	784.8	.029	.254	199.3	.021	.229	73%
	1000	873.9	.019	.183	175.5	.027	.149	80%
	10000	495.0	.006	.052	65.6	.015	.053	87%

Table 1. Type-1 vs type-2: target $f(x, y) = x^2 - y^2$

set with 10000 points saturates the input domain producing a rule base with significantly fewer rules.

The effectiveness of the merging for the rule base size is indicated by the rightmost column of the table. Merging is most effective with the strict precision bound. It is interesting to note the effect of the rule reduction on average and maximum error. For the .05 precision bound, three of the five reduced rule bases had better average error and four of the five better maximal error than their unreduced counterparts. This observation provides empirical support to the proposition that merging improves the generalization from the training data.

Table 2 shows the reduction in the size of the rule base over a number of target functions which vary in complexity. The results, given for a training set of 1000 points, illustrate the dependence of the reduction on the underlying surface. The planar surface is easily approximated by either type of rule. The 'wave-like' function $f(x, y) = .5(\sin(2\pi x))$ is well suited for approximation by type-2 rules but requires a large type-1 rule base. Neglecting the simple planar surface, a considerable reduction is obtained for each of the target functions with precision bounds .1 and .05 and the reduction has little or no impact on the error.

A comparison of type-3 and type-4 rule bases with the same target functions and 1000 training points is given in Table 3. As would be expected, the type-4 rules generally require fewer rules to satisfy the precision bound. An exception to this is the target function $f(x, y) = .5(\sin(2\pi x))$. The surface defined by this function is the cylindrical extension of the one-variable

		Type-1			Type-2			
Target $f(x,y) =$	Prec Bound	Rules	Ave Error	Max Error	Rules	Ave Error	Max Error	Percent Reduction
$(x+y/2)/2$	0.2	1.0	.023	.069	1.0	.022	.062	0%
	0.1	1.0	.023	.065	1.0	.022	.065	0%
	0.05	4.5	.013	.036	1.0	.013	.036	77%
$x^2 - y^2$	0.2	13.2	.039	.171	12.4	.046	.188	6%
	0.1	140.4	.020	.147	33.2	.032	.122	76%
	0.05	873.3	.019	.183	175.5	.019	.130	89%
$.5(\sin(2\pi x))$	0.2	68.3	.069	.225	7.5	.072	.196	89%
	0.1	306.4	.032	.233	52.1	.036	.147	83%
	0.05	1491.8	.025	.217	339.2	.019	.205	77%
$(\sin(\pi x) + \sin(\pi y))/2$	0.2	26.6	.054	.178	24.0	.058	.198	10%
	0.1	98.8	.027	.150	47.6	.037	.153	52%
	0.05	703.7	.018	.188	182.4	.021	.163	74%

Table 2. Type-1 vs type-2: 1000 training points

function $.5(\sin(2\pi x))$ to the input domain $U \times V$. The expansion of type-3 rule antecedents readily produces rectangular regions in the direction of the V axis. In this case, the flexibility of the type-4 rules and greedy nature of the algorithm combine to limit the ability to merge the same regions. The nonlinear surface generated by a type-4 rule allows the diagonal merging to extend beyond that of the type-3 rules. The smaller rectangles can be extended further along the V axis than larger rectangles, producing type-3 rule bases with fewer rules. This example demonstrates the effect of underlying system and the limitations imposed by the greedy strategy on the potential to merge rules.

The effect of additional training data on type-4 rules can be seen by comparing the final three columns in Tables 3 and 4. The rule bases in the former table were produced with 1000 training points and the latter with 10000. For target functions, $f(x,y) = x^2 - y^2$, $f(x,y) = .5(\sin(2\pi x))$, and $f(x,y) = (\sin(\pi x) + \sin(\pi y))/2$ the additional training data produced reductions of the rule base size by 78%, 86%, and 50%, respectively.

The two most general rule forms, type-2 and type-4, are compared in Table 4. The type-4 rules consistently produce smaller rule bases. With a large training set, the precision of the type-4 rule bases is also better. This indicates that the transitions between rules provided by the buffers, the gradual transitions between adjacent fuzzy sets in the input domain decomposition, provides a better approximation than that obtained using extended domain TSK rules.

No information is reported for type-2 rules and precision bound .05 for the target function $f(x,y) = .5(\sin(2\pi x))$. The type-2 rule base is obtained by merging regions in a type-1 rule base. The refinement process used to

Target	Prec Bound	Type-3 Rules	Type-3 Ave Error	Type-3 Max Error	Type-4 Rules	Type-4 Ave Error	Type-4 Max Error
$f(x,y) = (x + y/2)/2$	0.2	38.4	.093	.206	1.0	.011	.029
	0.1	93.7	.046	.151	1.0	.012	.039
	0.05	191.8	.022	.123	1.8	.012	.038
$f(x,y) = x^2 - y^2$	0.2	83.9	.080	.325	11.5	.066	.227
	0.1	177.4	.047	.335	33.2	.040	.174
	0.05	303.3	.030	.307	175.5	.027	.149
$f(x,y) = .5(\sin(2\pi x))$	0.2	17.3	.079	.358	12.3	.079	.242
	0.1	35.2	.046	.353	59.6	.043	.214
	0.05	82.5	.037	.374	131.9	.031	.234
$f(x,y) = (\sin(\pi x) + \sin(\pi y))/2$	0.2	82.7	.082	.320	16.4	.085	.228
	0.1	182.5	.046	.373	36.2	.045	.191
	0.05	314.8	.032	.424	105.5	.028	.182

Table 3. Type-3 vs type-4: 1000 training points

produce the initial type-1 rule base examined decompositions up to 150 by 150 grid points (22500 rules) without satisfying the bound, reflecting the high degree of change in a region of the surface.

Target	Prec Bound	Type-2 Rules	Type-2 Ave Error	Type-2 Max Error	Type-4 Rules	Type-4 Ave Error	Type-4 Max Error
$f(x,y) = (x + y/2)/2$	0.2	1.0	.022	.059	1.0	.004	.010
	0.1	1.0	.025	.063	1.0	.004	.010
	0.05	1.0	.011	.031	1.0	.005	.013
$f(x,y) = x^2 - y^2$	0.2	9.6	.043	.171	9.5	.017	.205
	0.1	25.0	.018	.081	15.0	.037	.106
	0.05	65.6	.015	.053	37.4	.020	.063
$f(x,y) = .5(\sin(2\pi x))$	0.2	6.0	.070	.144	5.1	.078	.180
	0.1	10.3	.033	.083	10.0	.042	.090
	0.05	-	-	-	20.5	.020	.070
$f(x,y) = (\sin(\pi x) + \sin(\pi y))/2$	0.2	25.0	.053	.182	17.1	.087	.201
	0.1	45.0	.033	.099	29.4	.044	.105
	0.05	100.0	.018	.053	52.3	.021	.059

Table 4. Type-2 vs type-4: 10000 training points

5 Conclusions

A suite of experiments has been performed to demonstrate the ability of a learn-and-merge strategy to generate fuzzy rules bases from training data. The incorporation of merging into proximity based rule generation decreases the granularity of the rule base while maintaining the ease of rule construction and the run time efficiency of rule bases produced using these techniques. Rule merging permits the partition of the input domain to conform to the training data and the type of the rule rather than be restricted to an a priori domain decomposition as in standard proximity based learning algorithms. We have shown that generalizing the antecedent of multi-dimensional rules significantly enhances the ability to reduce the number of fuzzy rules while maintaining equivalent precision in a model.

References

1. J. Yen and L. Wang, "An SVD-based fuzzy model reduction strategy," in *Proceedings of the Fifth IEEE International Conference on Fuzzy Systems*, (New Orleans), pp. 835–841, September 1996.
2. Y. Yam, P. Baranyi, and C.-T. Yang, "Reduction of fuzzy rule base via singular value decomposition," *IEEE Transactions on Fuzzy Systems*, vol. 7, no. 2, pp. 120–132, 1999.
3. M. Setnes, R. Babuska, U. Kaymak, and H. R. van Nauta Lemke, "Similarity measures in fuzzy rule base simplification," *IEEE Transactions on Systems, Man, and Cybernetics:B*, vol. 28, no. 3, pp. 376–386, 1998.
4. T. A. Sudkamp, A. Knapp, and J. Knapp, "A greedy approach to rule reduction in fuzzy models," in *Proceedings of the 2000 IEEE Conference on Systems, Man, and Cybernetics*, (Nashville), pp. 3716–3622, October 2000.
5. T. A. Sudkamp, J. Knapp, and A. Knapp, "Refine and merge: generating small rule bases from training data," in *Proceedings of the Joint Ninth IFSA World Congress and 20th NAFIPS International Conference*, (Vancouver), pp. 197–201, July 2001.
6. L. X. Wang and J. M. Mendel, "Generating fuzzy rules by learning from examples," *IEEE Transactions on Systems, Man, and Cybernetics*, vol. 22, pp. 1414–1427, 1992.
7. B. Kosko, *Neural Networks and Fuzzy Systems: A dynamical systems approach to machine intelligence*. Englewood Cliffs, NJ: Prentice Hall, 1992.
8. T. Sudkamp and R. J. Hammell II, "Interpolation, completion, and learning fuzzy rules," *IEEE Transactions on Systems, Man, and Cybernetics*, vol. 24, no. 2, pp. 332–342, 1994.
9. J. A. Dickerson and M. S. Lan, "Fuzzy rule extraction from numerical data for function approximation," *IEEE Transactions on Systems, Man, and Cybernetics*, vol. 26, pp. 119–129, 1995.
10. T. Thawonmas and S. Abe, "Function approximation based on fuzzy rules extracted from partitioned numerical data," *IEEE Transactions on Systems, Man, and Cybernetics:B*, vol. 29, no. 4, pp. 525–534, 1999.

11. T. A. Sudkamp and R. J. Hammell II, "Granularity and specificity in fuzzy function approximation," in *Proceedings of the NAFIPS-98*, pp. 105–109, 1998.
12. C. C. Lee, "Fuzzy logic in control systems: Part I," *IEEE Transactions on Systems, Man, and Cybernetics*, vol. 20, no. 2, pp. 404–418, 1990.
13. W. Pedrycz, "Why triangular membership functions?," *Fuzzy Sets and Systems*, vol. 64, pp. 21–30, 1994.
14. E. H. Mamdani and S. Assilian, "An experiment in linguistic synthesis with a fuzzy logic controller," *International Journal of Man-Machine Studies*, vol. 7, pp. 1–13, 1975.
15. E. H. Mamdani, "Advances in the linguistic synthesis of fuzzy controllers," *International Journal of Man-Machine Studies*, vol. 8, pp. 669–678, 1976.
16. T. Takagi and M. Sugeno, "Fuzzy identification of systems and its applications to modeling and control," *IEEE Transactions on Systems, Man, and Cybernetics*, vol. 15, pp. 329–346, 1985.
17. H. Takagi and I. Hayashi, "NN-driven fuzzy reasoning," *International Journal of Approximate Reasoning*, vol. 5, no. 3, pp. 191–212, 1991.
18. M. Delgado, A. F. Gomez-Skarmeta, and F. Martin, "Using fuzzy clusters to model fuzzy systems in a descriptive approach," in *Proceedings Information Processing Management Uncertainty Knowledge-Based Systems*, (Granada, Spain), pp. 564–568, July 1996.
19. S. Medasani, J. Kim, and R. Krishnapuram, "An overview of membership function generation for pattern recognition," *International Journal of Approximate Reasoning*, vol. 19, pp. 391–417, 1998.
20. J. R. Jang, "ANFIS: Adaptive-network based fuzzy inference system," *IEEE Transactions on Systems, Man, and Cybernetics*, vol. 23, pp. 665–684, 1993.
21. H. Ishibuchi, "Development of fuzzy neural networks," in *Fuzzy Modeling: Paradigms and Practices*, pp. 185–202, Norwell, MA: Kluwer Academic Publishers, 1996.
22. L. Wang and J. M. Mendel, "Generating fuzzy rules from numerical data, with applications," Tech. Rep. USC-SIPI-169, Signal and Image Processing Institute, University of Southern California, Los Angeles, CA 90089, 1991.
23. T. Sudkamp and R. J. Hammell II, "Rule base completion in fuzzy models," in *Fuzzy Modeling: Paradigms and Practice* (W. Pedrycz, ed.), pp. 313–330, Kluwer Academic Publishers, 1996.

Singular Value-Based Fuzzy Reduction With Relaxed Normality Condition

Yeung Yam[1], Chi Tin Yang[1], and Péter Baranyi[2]

[1] The Chinese University of Hong Kong, Shatin, N.T., Hong Kong
[2] Budapest University of Technology and Economics, Budapest, Hungary

Abstract. This work extends the results of a recent reduction method for fuzzy rule bases. The original approach conducts singular value decomposition (SVD) on the rule consequents and eliminates the weak and redundant components according to the magnitudes of the resulting singular values. The number of reduced rules as resulted depends on the number of singular values retained in the process. Conditions of sum normalization (SN), non-negativeness (NN) and Normality (NO) are imposed to ensure properly interpretable membership functions for the reduced rules. In this work, a new concept of relaxed Normality (RNO) condition is presented to enhance the interpretability of membership functions in situations where the NO condition cannot be strictly satisfied. The price to pay is an increase in the number of reduced rules and errors.

1 Introduction

The lack of well accepted theories in addressing the issues of design, optimality, reducibility, and partitioning of fuzzy rule base has been an on-going problem in the field of fuzzy system for some time. Without anything better, analysts often resort to over-parameterization as a mean to secure performance in the design of fuzzy rule bases. In addition, there may also be redundant, weakly-contributing, or even outright inconsistent components in any given set of fuzzy rules, whether it's generated from expert operators or by some learning or identification schemes. As a result, valuable computational time and storage may be lost in the actual implementation of the rule set as given, unnecessarily and sometimes with detrimental effects. A formal approach capable of capturing the essential elements in a data set or fuzzy rule base is hence highly desirable. Towards this end, a singular value-based approach to generate fuzzy approximator of a given function has been proposed [1]. The approach differs from previous works, e.g., [2]-[4], in that it pre-supposes no specific shapes for the membership functions. Rather, membership functions are characterized by the conditions of sum normalization (SN), non-negativeness (NN), and normality (NO). The approach calls for conducting singular value decomposition of a sample matrix over a rectangular grid, and then generates the membership functions and rule consequents by tailoring the orthogonal and singular value matrices to comply with the SN, NN, and NO conditions.

Numerous extensions of the approach in [1] have also been conducted. These include fuzzy rule reduction [5] [6], fuzzy identification with randomly scattered samples [7], and fuzzy interpolation of sparse rules [8] [9]. In all applications, it has been found that the NO condition is quite difficult to fully incorporate. To resolve this problem, a subdomain normalization technique to impose the NO condition at the cost of increasing the number of antecedent membership functions has also been proposed [10].

The present work constitutes another attempt to address this problem of NO incorporation. Here, we will first review the various concepts and procedures of the singular value-based approach. Generalization to high dimensional cases will also be discussed. Application of the approach to fuzzy reduction, including singleton and non-singleton cases, will be presented. Then, in situation where the NO condition cannot be fully satisfied, a new concept termed relaxed Normality (RNO) will be proposed to enhance the interpretability of membership functions. The RNO will result in an increase in the number of fuzzy rules. To consolidate the rules obtained as such, procedures to measure and merge similar membership functions, at the cost of increased errors, will be presented. The work also includes numerical examples to illustrate the various features of the procedures.

2 Basic Concepts

This section gives the basic concepts employed in the singular value-based approach.

Sum Normalization (SN): A matrix F is SN if the sums of its rows are all equal to 1, i.e.,

$$sum(F) = \begin{bmatrix} 1 \\ 1 \\ \vdots \\ 1 \end{bmatrix} \tag{1}$$

where $sum(F)$ denotes the column vector obtained by summing over the rows of matrix F.

Non-negativeness (NN): A matrix F is NN if each and every of its elements is non-negative, i.e., $F_{i,j} \geq 0$.

Normality (NO): A matrix F is NO if it is SN and NN and that each column contains the value 1 as its elements.

Note that a matrix F being SN and NN implies $0 \leq F_{i,j} \leq 1$ for its element $F_{i,j}$. The concepts of SN, NN, and NO as introduced are consistent with the usual definition of membership functions. The condition SN reflects the fact that the sum of membership degrees of a fuzzy variables is equal to 1, and NN reflects the fact that membership degree cannot be negative. The NO condition reflects the fact that membership functions should be localized, i.e., they take turns dominating the domain of the fuzzy variable, making it possible to have a meaningful assignment of linguistic labelings.

3 Singular Value-Based Procedures

This section outlines the singular value-based procedures of [1] to decompose a matrix into products involving SN and NN, and possibly NO, submatrices. The procedures is presented here with a 2-D matrix, and then extended to higher dimension in the next section.

Consider a 2-D matrix $\mathcal{F}$ and perform singular value decomposition,

$$\mathcal{F} = U\,\Sigma\,V^T, \tag{2}$$

where $\mathcal{F}$ is n_a by n_b, and U and V are n_a by n_a and n_b by n_b, respectively. Matrices U and V are orthogonal, i.e., $UU^T = I_{n_a \times n_a}$ and $VV^T = I_{n_b \times n_b}$, with $I_{s\times s}$ denoting the s by s identity matrix. For notation, we also let $O_{s\times t}$ denote the s by t matrix of zeros, and $\mathbf{1}_{s\times t}$ the s by t matrix of 1's. The n_a by n_b matrix Σ contains the singular values of $\mathcal{F}$ in decreasing magnitude. The maximum number of nonzero singular values is $n_{svd} = min(n_a, n_b)$. The singular values indicate the importance of the corresponding columns of U and V in the formation of $\mathcal{F}$. A close approximation to $\mathcal{F}$ can be obtained by keeping those components with large singular values.

Let $n^{(r)}$ be the number of singular values to keep. The corresponding approximation of $\mathcal{F}$ is therefore

$$\mathcal{F} \approx U^{(r)}\Sigma^{(r)}{V^{(r)}}^T, \tag{3}$$

where $U^{(r)}$ and $V^{(r)}$ contain the $n^{(r)}$ columns of U and V corresponding to the retained singular values in $\Sigma^{(r)}$. Approximation is exact if $n^{(r)} = n_{svd}$. Matrices $U^{(r)}$ and $V^{(r)}$ are in general neither SN nor NN. They can be converted into SN and NN matrices by the procedures described below.

3.1 Incorporating The SN Condition

We use the matrix U to state the following theorem.
Theorem 1: Let U be a n_a by n_a orthogonal matrix partitioned as $U = [\ U^{(r)} \mid U^{(d)}\ ]$, where $U^{(r)}$ and $U^{(d)}$ have, respectively, $n^{(r)}$ and $n^{(d)} = (n_a - n^{(r)})$ columns. Let Φ_a be any $n^{(r)}$ by $n^{(r)}$ matrix satisfying the constraint $\boldsymbol{sum}(\Phi_a) = \boldsymbol{sum}((U^{(r)})^T)$. Then,

if $\boldsymbol{sum}((U^{(d)})^T) = O_{n^{(d)}\times 1}$, the n_a by $n^{(r)}$ matrix $\mathcal{S}_a^{[1]}$ satisfies SN, with

$$\mathcal{S}_a^{[1]} = U^{(r)}\,\Phi_a, \tag{4}$$

and if $\boldsymbol{sum}((U^{(d)})^T) \neq O_{n^{(d)}\times 1}$, the n_a by $(n^{(r)}+1)$ matrix $\mathcal{S}_a^{[2]}$ satisfies SN, with

$$\mathcal{S}_a^{[2]} = [\ U^{(r)} \mid U^{(d)}\,\boldsymbol{sum}((U^{(d)})^T)\ ] \begin{bmatrix} \Phi_a & O_{n^{(r)}\times 1} \\ O_{1\times n^{(r)}} & 1 \end{bmatrix}. \tag{5}$$

Theorem 1 characterizes the extra column needed, if at all, to supplement the matrix $U^{(r)}$ in order to satisfy the SN condition. The proof can be found in [1] and is omitted here. Note that the theorem does not require Φ_a to be invertible. However, since our goal here is to have $U^{(r)}$ and $V^{(r)}$ for a close approximation of $\mathcal{F}$ in (3), an invertible Φ_a will be more efficient. Here, we adopt the following algorithm to construct Φ_a: if $\boldsymbol{sum}((U^{(r)})^T)$ does not contain zero elements, form matrix Φ_a as

$$\Phi_a = diag[\boldsymbol{sum}((U^{(r)})^T)]$$

and if $\boldsymbol{sum}((U^{(r)})^T)$ contains zero element(s), form Φ_a as

$$\Phi_a = I_{n^{(r)}\times n^{(r)}} + \left[O_{n^{(r)}\times(\widetilde{n}-1)}\middle|\boldsymbol{sum}((U^{(r)})^T) - \mathbf{1}_{n^{(r)}\times 1}\middle|O_{n^{(r)}\times(n^{(r)}-\widetilde{n})}\right]$$

and we need to ensure that the $\widetilde{n}^{th}$ entry of $\boldsymbol{sum}((U^{(r)})^T)$ is nonzero. After Φ_a is obtained, one then have

$$\begin{cases} For\ \boldsymbol{sum}((U^{(d)})^T) = O_{(n_a-n^{(r)})\times 1}, \\ \quad \mathcal{S}_a^{[1]} = U^{(r)}\Phi_a \\ \\ For\ \boldsymbol{sum}((U^{(d)})^T) \neq O_{(n_a-n^{(r)})\times 1}, \\ \quad \mathcal{S}_a^{[2]} = \left[U^{(r)}\middle|U^{(d)}\ \boldsymbol{sum}((U^{(d)})^T)\right]\begin{bmatrix} \Phi_a & O_{n^{(r)}\times 1} \\ O_{1\times n^{(r)}} & 1 \end{bmatrix} \end{cases} \tag{6}$$

With similar expressions for Φ_b, $\mathcal{S}_b^{[1]}$, and $\mathcal{S}_b^{[2]}$ when applying Theorem 1 to matrix V, (3) can be rewritten as

$$\mathcal{F} \approx U^{(r)}\Sigma^{(r)}V^{(r)^T} = \mathcal{S}_a^{[1]}(\Phi_a)^{-1}\Sigma^{(r)}(\Phi_b)^{-T}(\mathcal{S}_b^{[1]})^T \tag{7}$$

for the case $\boldsymbol{sum}((U^{(d)})^T) = O_{(n_a-n^{(r)})\times 1}$ and $\boldsymbol{sum}((V^{(d)})^T) = O_{(n_b-n^{(r)})\times 1}$. Expressions for cases of $\boldsymbol{sum}((U^{(d)})^T) \neq O_{(n_a-n^{(r)})\times 1}$ and/or $\boldsymbol{sum}((V^{(d)})^T) \neq O_{(n_b-n^{(r)})\times 1}$ follow in a straightforward manner. The matrices $\mathcal{S}_a^{[s]}$ and $\mathcal{S}_b^{[s]}$, $s = 1, 2$, are SN. They have either $n^{(r)}$ or $(n^{(r)} + 1)$ columns, depending on the case.

3.2 Incorporating The NN condition

Matrices $\mathcal{S}_a^{[s]}$ and $\mathcal{S}_b^{[s]}$ may contain negative elements and hence do not satisfy the NN condition. The following gives a set of procedures to generate from a matrix satisfying the SN condition another matrix with the same dimension satisfying both SN and NN conditions. Let $\mathcal{S}$ be a n_{row} by n_{col} matrix satisfying the SN condition. Then,

1. Look for the minimum element, $\min_{s,t} \mathcal{S}_{s,t}$, of $\mathcal{S}$. Set parameter

$$\zeta_{min} = \begin{cases} 1 & if\ \min_{s,t} \mathcal{S}_{s,t} \geq -1 \\ \frac{1}{|\min_{s,t} \mathcal{S}_{s,t}|} & otherwise \end{cases} \tag{8}$$

2. Form a n_{col} by n_{col} matrix $\mathcal{N}_\mathcal{S}$,

$$\mathcal{N}_\mathcal{S} = \frac{1}{(n_{col} + \zeta_{min})} \begin{bmatrix} (1+\zeta_{min}) & 1 & \cdots & 1 \\ 1 & (1+\zeta_{min}) & \cdots & 1 \\ \vdots & \vdots & \ddots & \vdots \\ 1 & 1 & \cdots & (1+\zeta_{min}) \end{bmatrix} \tag{9}$$

3. The matrix product $\mathcal{S}\mathcal{N}_\mathcal{S}$ satisfies the SN and NN conditions.

The subscript $\mathcal{S}$ in $\mathcal{N}_\mathcal{S}$ is adopted to stress the dependence of $\mathcal{N}_\mathcal{S}$ on $\mathcal{S}$. To see that $\mathcal{S}\mathcal{N}_\mathcal{S}$ is SN, one notes that $\boldsymbol{sum}(\mathcal{N}_\mathcal{S}) = \mathbf{1}_{n_{col}\times 1}$ and hence $\boldsymbol{sum}(\mathcal{S}\mathcal{N}_\mathcal{S}) = \mathcal{S}\mathbf{1}_{n_{col}\times 1} = \boldsymbol{sum}(\mathcal{S}) = \mathbf{1}_{n_{col}\times 1}$ as $\mathcal{S}$ is SN to begin with. To see that $\mathcal{S}\mathcal{N}_\mathcal{S}$ is also NN, one notes that the tth column of $\mathcal{S}\mathcal{N}_\mathcal{S}$ is given by the scalar $\frac{1}{(n_{col}+\zeta_{min})}$ multiplying a column vector which is the sum of $\mathbf{1}_{n_{row}\times 1}$ and ζ_{min} times the tth column of $\mathcal{S}$, and the parameter ζ_{min} ensures proper scaling to result in positive values for all entries.

Applying these steps to (7), one has

$$\mathcal{F} \approx \mathcal{S}_a^{[1]}(\Phi_a)^{-1}\Sigma^{(r)}(\Phi_b)^{-T}(\mathcal{S}_b^{[1]})^T = \widetilde{U}\widetilde{\mathcal{R}}\widetilde{V}^T, \tag{10}$$

with

$$\widetilde{U} = \mathcal{S}_a^{[1]}\mathcal{N}_{\mathcal{S}_a} \tag{11}$$

$$\widetilde{\mathcal{R}} = (\mathcal{N}_{\mathcal{S}_a})^{-1}(\Phi_a)^{-1}\Sigma^{(r)}(\Phi_b)^{-T}(\mathcal{N}_{\mathcal{S}_b})^{-T} \tag{12}$$

$$\widetilde{V} = \mathcal{S}_b^{[1]}\mathcal{N}_{\mathcal{S}_b} \tag{13}$$

The matrices $\widetilde{U}$ and $\widetilde{V}$ are n_a by $n^{(r)}$ and n_b by $n^{(r)}$, respectively. They satisfy the SN and NN conditions. The same procedures can be applied to other cases of $\boldsymbol{sum}((U^{(d)})^T) \neq O_{(n_a-n^{(r)})\times 1}$ and/or $\boldsymbol{sum}((V^{(d)})^T) \neq O_{(n_b-n^{(r)})\times 1}$. Again, the resulting SN and NN matrices $\widetilde{U}$ and $\widetilde{V}$ have $n^{(r)}$ or $(n^{(r)}+1)$ columns.

3.3 Incorporating the NO condition

Similar to what we have achieved for SN and NN, we here desire invertible matrices $\mathcal{Q}_a$ and $\mathcal{Q}_b$ of appropriate dimensions such that

$$\mathcal{F} \approx \widetilde{U}\widetilde{\mathcal{R}}\widetilde{V}^T = \widetilde{U}\mathcal{Q}_a(\mathcal{Q}_a)^{-1}\widetilde{\mathcal{R}}(\mathcal{Q}_b)^{-T}\mathcal{Q}_b^T\widetilde{V}^T \tag{14}$$

with matrix products $\widetilde{U}\mathcal{Q}_a$ and $\widetilde{V}\mathcal{Q}_b$ being SN, NN, and NO. However, while it is always possible to tailor the matrices to satisfy SN and NN, the same is not true for NO. Successful incorporation of the NO condition depends on the specific matrix at hand. Nonetheless, the following gives a set of tight bounding procedures which yields a NO matrix if possible, and a close-to-NO matrix otherwise. Take $\widetilde{U}$ as example and let its column dimension be n_{col}. Noting that each of the n_a rows of $\widetilde{U}$ corresponds to a point lying on a

n_{col}-dimensional space and since $\widetilde{U}$ is SN, these n_a points actually lie on a hyper-plane of $(n_{col} - 1)$-dimension. We present the following steps for tight bounding:

1. Project the n_a points in the n_{col}-dimensional space onto the $(n_{col} - 1)$-dimensional hyper-plane satisfying the SN condition. An efficient way to conduct the projection is to multiply $\widetilde{U}$ on the right by the n_{col} by n_{col} matrix

$$\begin{bmatrix} 1 & 0 & \cdots & 0 & 0 \\ 1 & 1 & \cdots & 0 & 0 \\ \vdots & \vdots & \ddots & \vdots & \vdots \\ 1 & 1 & \cdots & 1 & 0 \\ 1 & 1 & \cdots & 1 & 1 \end{bmatrix} \tag{15}$$

   The first column of the product will all be 1's as $\widetilde{U}$ is SN. The remaining $(n_{col} - 1)$ columns can be viewed as projected coordinates of the n_a points onto a $(n_{col} - 1)$-dimensional plane.
2. Obtain the convex hull of the n_a points on the $(n_{col} - 1)$-dimensional hyper-plane. Algorithms to treat convex hull problem in a general dimensional space are discussed in, e.g., [11] and [12].
3. Check the convex hull. If the convex hull has exactly n_{col} vertices, successful incorporation of the NO condition is possible. In this case the matrix $\mathcal{Q}_a$ can be obtained as inverse of the matrix containing the n_{col} rows of $\widetilde{U}$ associated with the convex hull. This is the NO case. If the convex hull has more than n_{col} vertices, however, determination of $\mathcal{Q}_a$ to strictly satisfying the NO condition is not possible. In this case, we have to search for a relaxed bounding with n_{col} vertices not all of which came from the n_a points of $\widetilde{U}$. The corresponding $\mathcal{Q}_a$ is then determined according to these n_{col} vertices. This is the close-to-NO case.

Carrying out the above procedures for $\widetilde{U}$ and $\widetilde{V}$, (14) now becomes,

$$\mathcal{F} \approx \widetilde{U}\tilde{\mathcal{R}}\widetilde{V}^{\,T} = \overline{U}\bar{\mathcal{R}}\overline{V}^{\,T} \tag{16}$$

where

$$\begin{aligned} \overline{U} &= \widetilde{U}\mathcal{Q}_a \\ \bar{\mathcal{R}} &= (\mathcal{Q}_a)^{-1}\tilde{\mathcal{R}}(\mathcal{Q}_b)^{-T} \\ \overline{V} &= \widetilde{V}\mathcal{Q}_b \end{aligned} \tag{17}$$

and $\overline{U}$ and $\overline{V}$ are SN, NN, and NO, or close-to-NO. Notice here that $\overline{U}$, $\bar{\mathcal{R}}$, and $\overline{V}$ in (17) are not unique. In fact, given that $\overline{U}$ and $\overline{V}$ are SN, NN, and NO, the choice of

$$\begin{aligned} \overline{U}' &= \overline{U}\,X_a \\ \bar{\mathcal{R}}' &= (X_a)^{-1}\,\bar{\mathcal{R}}\,(X_b)^{-T} \\ \overline{V}' &= \overline{V}\,X_b \end{aligned} \tag{18}$$

is another valid representation if the invertible matrices X_a and X_b are of appropriate dimensions and are SN, NN, and NO. Equation (18) actually defines a class of similarity transformed fuzzy rule bases.

3.4 Approximation Error

Noting that $U^{(r)}\Sigma^{(r)}V^{(r)^T} = \tilde{U}\tilde{\mathcal{R}}\tilde{V}^T = \overline{U}\bar{\mathcal{R}}\overline{V}^T$, the approximation error $(\mathcal{F} - \overline{U}\bar{\mathcal{R}}\overline{V}^T)$ due to the singular value-based approximation and SN, NN, and NO processing can thus be expressed as

$$|\mathcal{F} - \overline{U}\bar{\mathcal{R}}\overline{V}^T| = |U^{(d)}|\ \Sigma^{(d)}\ |V^{(d)}| \tag{19}$$

Since the columns of $U^{(d)}$ and $V^{(d)}$ all have Euclidean norm of unity, the absolute values of their elements must be bounded by 1. Thus,

$$|\mathcal{F} - \overline{U}\bar{\mathcal{R}}\overline{V}^T| \leq \mathbf{1}_{n_a \times n^{(d)}} \Sigma^{(d)} \mathbf{1}_{n_b \times n^{(d)}}^T = \left(\sum_{i=n^{(r)}+1}^{n_{svd}} \sigma_i \right) \mathbf{1}_{n_a \times n_b} \tag{20}$$

where σ_i denotes the ith singular value in Σ. Hence, the approximation error is bounded by the sum of singular values discarded in the process.

4 Generalization to Higher Dimensions

The previous section presents the singular value-based procedures to approximate $\mathcal{F}$ in term of SN, NN, and NO, or close-to-NO submatrices when $\mathcal{F}$ is 2-D. Expressing the result in terms of individual elements,

$$\mathcal{F}_{i,j} \approx \bar{\mathcal{F}}_{i,j} = \sum_{\bar{i}=1}^{\bar{n}_a} \sum_{\bar{j}=1}^{\bar{n}_b} \overline{U}_{i,\bar{i}} \overline{V}_{j,\bar{j}} \bar{\mathcal{R}}_{\bar{i},\bar{j}} \tag{21}$$

where $\bar{n}_a$ and $\bar{n}_b$ are, respectively, the numbers of columns of matrices $\overline{U}$ and $\overline{V}$. They have values of $n^{(r)}$ or $(n^{(r)} + 1)$, depending on the specific case at hand.

The procedures can be readily extended to matrix of higher dimension. Figure 1 depicts such procedures for the 3-D case. First, the n_a by n_b by n_c matrix $\mathcal{F}$ (step (i)) is spread in the j-index direction to form a 2-D n_a by $n_b n_c$ matrix $\mathcal{T}_a$ (Step (ii)). The procedures of Section 3 are then applied to $\mathcal{T}_a$ (step (iii)). This yield

$$\mathcal{T}_a \approx \overline{U}\mathcal{F}^{[a]}, \tag{22}$$

where $\overline{U}$ is SN, NN, and NO, or close-to-NO. The dimension of $\overline{U}$ is n_a by $\bar{n}_a$ where $\bar{n}_a = n_a^{(r)}$ or $\bar{n}_a = n_a^{(r)} + 1$, as the case may be. The $\bar{n}_a$ by $n_b\ n_c$ matrix $\mathcal{F}^{[a]}$ denotes the resultant matrix to the right of $\overline{U}$. One can now re-stack $\mathcal{F}^{[a]}$ to become a 3-D $\bar{n}_a$ by n_b by n_c matrix (step(iv)), which is then spread,

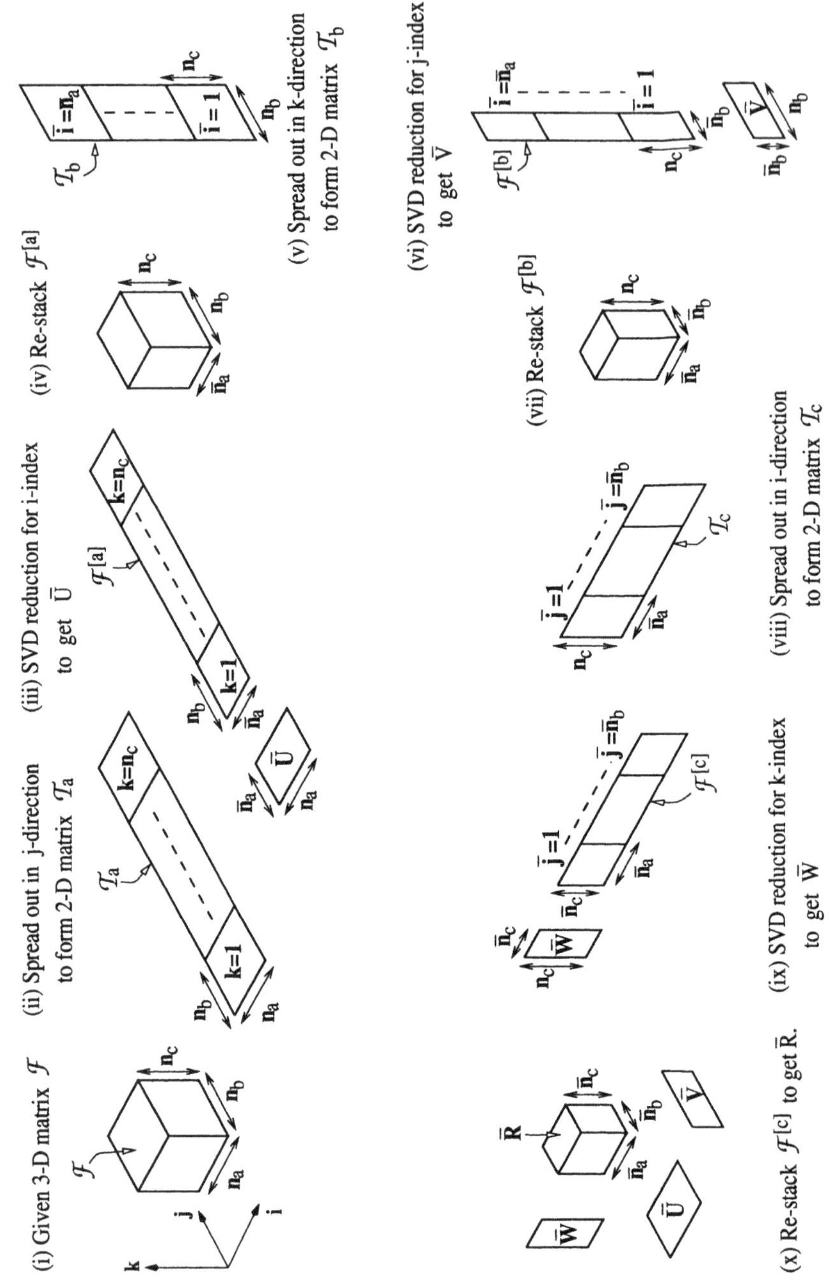

Fig. 1. Singular value-based procedures for 3-D matrix

this time in the k-index direction, to form a 2-D n_b by $n_c\bar{n}_a$ matrix $\mathcal{T}_b$ (step (v)). Again, applying the singular value-based procedures to $\mathcal{T}_b$ (step (vi)), one obtains

$$\mathcal{T}_b \approx \overline{V}\mathcal{F}^{[b]}, \tag{23}$$

where $\overline{V}$ is SN, NN, and NO, or close-to-NO. There are $\bar{n}_b = n_b^{(r)}$ or $\bar{n}_b = n_b^{(r)} + 1$ columns in $\overline{V}$. The matrix $\mathcal{F}^{[b]}$ is $\bar{n}_b$ by n_c $\bar{n}_a$. Re-stacking $\mathcal{F}^{[b]}$ into a $\bar{n}_b$ by $\bar{n}_b$ by n_c matrix (step(vii)) and spreading it in the i-index direction to form $\mathcal{T}_c$ (step(viii)), one has, after applying the singular value-based procedures to $\mathcal{T}_c$ (step(ix)),

$$\mathcal{T}_c \approx \overline{W}\mathcal{F}^{[c]}. \tag{24}$$

The matrix $\overline{W}$ is SN, NN, and NO, or close-to-NO, and there may be $\bar{n}_c = n_c^{(r)}$ or $\bar{n}_c = n_c^{(r)} + 1$ columns in $\overline{W}$. The matrix $\mathcal{F}^{[c]}$ is $\bar{n}_c$ by $\bar{n}_a$ $\bar{n}_b$. It can be re-stacked into a 3-D $\bar{n}_a$ by $\bar{n}_b$ by $\bar{n}_c$ matrix $\bar{\mathcal{R}}$ (step(x)). This completes the process of generating all necessary quantities. The resulting approximation to the elements of $\mathcal{F}$, $\mathcal{F}_{i,j,k}$, can be expressed as

$$\mathcal{F}_{i,j,k} \approx \bar{\mathcal{F}}_{i,j,k} = \sum_{\bar{i}=1}^{\bar{n}_a}\sum_{\bar{j}=1}^{\bar{n}_b}\sum_{\bar{k}=1}^{\bar{n}_c}\overline{U}_{i,\bar{i}}\overline{V}_{j,\bar{j}}\overline{W}_{k,\bar{k}}\bar{\mathcal{R}}_{\bar{i},\bar{j},\bar{k}} \tag{25}$$

Equation (25) is a 3-D generalization of (21). One can also obtain an approximation error bound for $|\mathcal{F}_{i,j,k} - \bar{\mathcal{F}}_{i,j,k}|$,

$$|\mathcal{F}_{i,j,k} - \bar{\mathcal{F}}_{i,j,k}| \leq \mathcal{E}_a + \mathcal{E}_b + \mathcal{E}_c. \tag{26}$$

where $\mathcal{E}_a$, $\mathcal{E}_b$, and $\mathcal{E}_c$ are sum of the discarded singular values of matrices $\mathcal{T}_a$, $\mathcal{T}_b$, and $\mathcal{T}_c$. They represent the errors induced in the process of generating matrices $\overline{U}$, $\overline{V}$, and $\overline{W}$. Interested readers are referred to [1] for the mathematical details of these results.

The above procedures structures a 3-D matrix so that the 2-D singular value-based procedures can be applied in stages. At each stage, approximation and conditioning of the matrix is conducted for one of the dimensions, and a certain singular value error is generated. With additional stages and proper indexing, the procedures can be extended to matrix of a general number of dimension.

5 Applications to Fuzzy Reduction

The singular value-based approach above has been applied to generate a set of fuzzy rules to closely approximate the outputs of a given function [1]. In this case, matrix $\mathcal{F}$ contains the sampled outputs of the given function at a set of input grid points. The approximation error then corresponds to the difference induced between outputs of the original function and the fuzzy approximator at the sampled points. Other applications have also been explored. These include fuzzy reduction [5] [6], fuzzy identification [7], and fuzzy interpolation of sparse rules [8] [9]. We will focus on fuzzy reduction in this section, and the application of the RNO concept and procedures to fuzzy reduction in the next.

We start with the singleton support reduction case. Consider a fuzzy rule base with 2 inputs, a and b, and a single output u,

$$\text{If } A_i(a) \text{ and } B_j(b) \Rightarrow \; u = r_{i,j}, \tag{27}$$

where $A_i(a)$ and $B_j(b)$, $i = 1, \ldots, n_a$, $j = 1, \ldots, n_b$, are the membership functions of variable a and b, respectively, and $r_{i,j}$ is the rule consequent of the (i,j)th fuzzy rule. To perform rule base reduction, one forms a n_a by n_b matrix $\mathcal{F}$ with $r_{i,j}$ as the (i,j) elements,

$$\mathcal{F} = \begin{bmatrix} r_{1,1} & r_{1,2} & \cdots & r_{1,n_b} \\ r_{2,1} & r_{2,2} & \cdots & r_{2,n_b} \\ \vdots & \vdots & \ddots & \vdots \\ r_{n_a,1} & r_{n_a,2} & \cdots & r_{n_a,n_b} \end{bmatrix}, \tag{28}$$

and then applies the singular value-based procedures to obtain

$$\mathcal{F} \approx \bar{\mathcal{F}} = \overline{U}\bar{\mathcal{R}}\overline{V}^T \tag{29}$$

where $\overline{U}$, $\bar{\mathcal{R}}$, and $\overline{V}$ are, respectively, n_a by $\bar{n}_a$, $\bar{n}_a$ by $\bar{n}_b$, and n_b by $\bar{n}_b$ matrices. A reduced rule base can then be obtained as:

$$\text{If } \bar{A}_{\bar{i}}(a) \text{ and } \bar{B}_{\bar{j}}(b) \Rightarrow \; u = \bar{r}_{\bar{i},\bar{j}},$$

with $\bar{i} = 1, \ldots, \bar{n}_a$ and $\bar{j} = 1, \ldots, \bar{n}_b$. Here, the new membership functions $\bar{A}_{\bar{i}}(a)$, $\bar{B}_{\bar{j}}(b)$ are given by

$$\bar{A}_{\bar{i}}(a) = \sum_{i=1}^{n_a} A_i(a)\overline{U}_{i,\bar{i}}, \tag{30}$$

$$\bar{B}_{\bar{j}}(b) = \sum_{j=1}^{n_b} B_j(b)\overline{V}_{j,\bar{j}}, \tag{31}$$

and the new rule consequent $\bar{r}_{\bar{i},\bar{j}}$ is given by the $(\bar{i},\bar{j})$ element of $\bar{\mathcal{R}}$,

$$\bar{r}_{\bar{i},\bar{j}} = \bar{\mathcal{R}}_{\bar{i},\bar{j}} \tag{32}$$

The reduced rule base has $\bar{n}_a\bar{n}_b$ fuzzy rules as compared to $n_a n_b$ of the original one. Since $\overline{U}$ is definitely SN and NN, it can be easily deduced (see [6], Theorem 1) that the membership functions $\{\bar{A}_{\bar{i}}(a), \bar{i} = 1, \ldots, \bar{n}_a\}$ are SN and NN if the original ones $\{A_i(a), i = 1, \dots, n_a\}$ are SN and NN, and if $\overline{U}$ is NO as well, the new membership functions are also NO if the original ones are NO. If $\overline{U}$ is only close-to-NO, however, $\{\bar{A}_{\bar{i}}(a), \bar{i} = 1, \dots, \bar{n}_a\}$ will only be SN and NN but not NO. Similar results apply to membership function $\{\bar{B}_{\bar{i}}(b), \bar{j} = 1, \dots, \bar{n}_b\}$.

The above scheme can be applied to any fuzzy rule base regardless of the adopted inference paradigm. Moreover, a compact expression for the output error bound can be obtained if PSG inference is adopted. Given that $a = a_o$, $b = b_o$, output error between the original and reduced rule base is given by

$$\mathcal{E}_{red} = \sum_{i=1}^{n_a} \sum_{j=1}^{n_b} A_i(a_o) B_j(b_o) r_{i,j} - \sum_{\bar{i}=1}^{\bar{n}_a} \sum_{\bar{j}=1}^{\bar{n}_b} \bar{A}_{\bar{i}}(a_o) \bar{B}_{\bar{j}}(b_o) \bar{r}_{\bar{i},\bar{j}}, \tag{33}$$

where the fact that membership functions $A_i(a)$, $B_j(b)$, $\bar{A}_{\bar{i}}(a)$, $\bar{B}_{\bar{j}}(b)$ satisfying the SN condition has been used. By (20) and (30)-(32),

$$|\mathcal{E}_{red}| \leq \sum_{i=n^{(r)}+1}^{n_{svd}} \sigma_i. \tag{34}$$

The output error incurred in the reduction process is hence bounded by the sum of the discarded singular values. If all nonzero singular values are included, the reduced rule base would yield the same output as the original one.

For the non-singleton support case, each fuzzy rules in (27) comes with an additional support factor $s_{i,j}$, which can be interpreted as a firing strength or a reliability coefficient. Given that $a = a_o$ and $b = b_o$, and assuming PSG inference, the inferred output now becomes

$$u = \frac{\sum_{i=1}^{n_a} \sum_{j=1}^{n_b} A_i(a_o) B_j(b_o) r_{i,j} s_{i,j}}{\sum_{i=1}^{n_a} \sum_{j=1}^{n_b} A_i(a_o) B_j(b_o) s_{i,j}}. \tag{35}$$

Equation (35) has a numerator part and a denominator part, as compared to just the numerator part in the singleton support case. To conduct reduction in this case, one forms a 3-D n_a by n_b by 2 matrix $\mathcal{F}$ with elements

$$\mathcal{F}_{i,j,1} = r_{i,j} s_{i,j} \ ; \quad \mathcal{F}_{i,j,2} = s_{i,j} \tag{36}$$

and then applies the procedures of the previous section as follows. With referring to Fig. 1, one first carries out steps (i) to (iii) to obtain a n_a by $\bar{n}_a$ matrix $\overline{U}$ for variable a, and then steps (iv) to (vi) to obtain a n_b by $\bar{n}_b$ matrix $\overline{V}$ for variable b. In the remaining steps (vii) to (ix), however, one keeps both singular values of $\mathcal{T}_c$ without any reduction, i.e., $\overline{W} = I_{2\times 2}$. As such, (25) becomes in this case

$$\mathcal{F}_{i,j,k} \approx \bar{\mathcal{F}}_{i,j,k} = \sum_{\bar{i}=1}^{\bar{n}_a} \sum_{\bar{j}=1}^{\bar{n}_b} \overline{U}_{i,\bar{i}} \overline{V}_{j,\bar{j}} \mathcal{R}_{\bar{i},\bar{j},k} \tag{37}$$

for $k = 1, 2$, where matrices $\overline{U}$ and $\overline{V}$ are SN, NN, and NO, or close-to-NO, and the matrix $\bar{\mathcal{R}}$ is $\bar{n}_a$ by $\bar{n}_b$ by 2. One can now write a reduced fuzzy rule base as

$$\text{If } \bar{A}_{\bar{i}}(a) \text{ and } \bar{B}_{\bar{j}}(b) \Rightarrow \ u = \bar{r}_{\bar{i},\bar{j}},$$

where $\bar{A}_{\bar{i}}(a) = \sum_{i=1}^{\bar{n}_a} A_i(a)\overline{U}_{i,\bar{i}}$ and $\bar{B}_{\bar{j}}(b) = \sum_{j=1}^{\bar{n}_b} B_j(b)\overline{V}_{j,\bar{j}}$ are the new membership functions. The new rule consequent $\bar{r}_{\bar{i},\bar{j}}$ and support factor $\bar{s}_{\bar{i},\bar{j}}$ are computed as

$$\bar{r}_{\bar{i},\bar{j}} = \bar{\mathcal{R}}_{\bar{i},\bar{j},1}/\bar{\mathcal{R}}_{\bar{i},\bar{j},2} \ ; \quad \bar{s}_{\bar{i},\bar{j}} = \bar{\mathcal{R}}_{\bar{i},\bar{j},2}$$

The reduced set has $\bar{n}_a\bar{n}_b$ fuzzy rules as compared to the original of $n_a n_b$ rules. Similar procedures can be applied to reducing Takagi-Sugeno-Kang (TSK) model [13] as well.

Table 1. Fuzzy rule consequent $r_{i,j}$ of the original rule base

	$j=1$	$j=2$	$j=3$	$j=4$	$j=5$	$j=6$	$j=7$	$j=8$	$j=9$
$i=1$	-5.00	-5.00	-5.00	-5.00	-5.00	-3.75	-2.5	-1.25	0
$i=2$	-5.00	-5.00	-5.00	-5.00	-3.75	-2.5	-1.25	0	1.25
$i=3$	-5.00	-5.00	-5.00	-3.75	-2.5	-1.25	0	1.25	2.5
$i=4$	-5.00	-5.00	-3.75	-2.5	-1.25	0	1.25	2.5	3.75
$i=5$	-5.00	-3.75	-2.5	-1.25	0	1.25	2.5	3.75	5.00
$i=6$	-3.75	-2.5	-1.25	0	1.25	2.5	3.75	5.00	5.00
$i=7$	-2.5	-1.25	0	1.25	2.5	3.75	5.00	5.00	5.00
$i=8$	-1.25	0	1.25	2.5	3.75	5.00	5.00	5.00	5.00
$i=9$	0	1.25	2.5	3.75	5.00	5.00	5.00	5.00	5.00

An numerical case study is now given to illustrate the reduction process.

Example 1: Consider the following example from [14] with a fuzzy rule base of input variables a and b. The membership functions of a are overlapping isosceles triangles of basewidth 0.08 centering at $a_1 = -0.16$ to $a_9 = +0.16$ at separations of 0.04, and those of b are also overlapping isosceles triangles but with basewidth of 0.6 centering at $b_1 = -1.2$ to $b_9 = +1.2$ at separations of 0.3. The original example in [14] utilizes Min-Max inference paradigm and assigns triangular membership functions to the fuzzy outputs. Here, the example is modified to using PSG inference and singleton rule consequents to coincide with our formulation. The fuzzy rule is then expressed as:

$$If\ A_i(a)\ and\ B_j(b) \Rightarrow \ u = r_{i,j},$$

where the values $r_{i,j}$s now correspond to the center points of the output membership functions in the original example and are as tabulated in Table 1. They form the matrix $\mathcal{F}$ on which reduction is to be conducted.

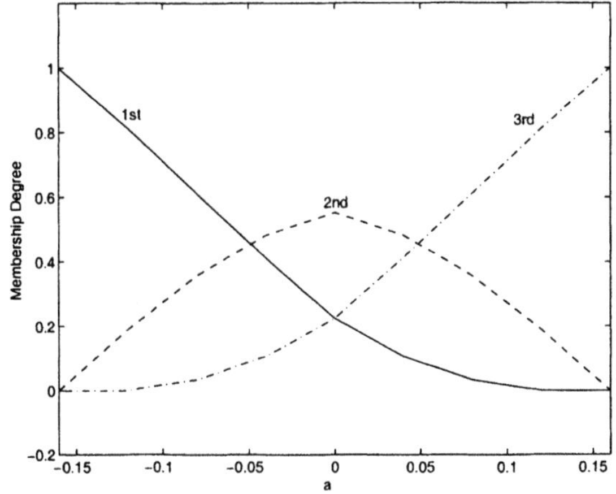

Fig. 2. Membership functions of a of the reduced rule base

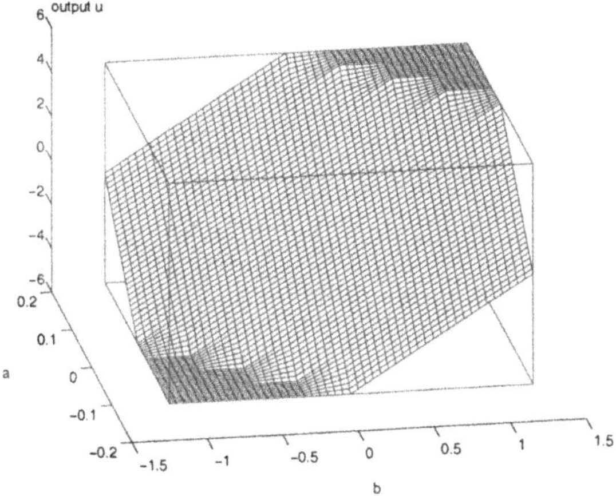

Fig. 3. PSG-inferred output of the original fuzzy rule base

Table 2. Fuzzy rule consequent $\bar{r}_{i,j}$ of the reduced rule base

	$\bar{j}=1$	$\bar{j}=2$	$\bar{j}=3$
$\bar{i}=1$	-4.7320	-6.6490	0
$\bar{i}=2$	-6.6490	0	6.6490
$\bar{i}=3$	0	6.6490	4.7320

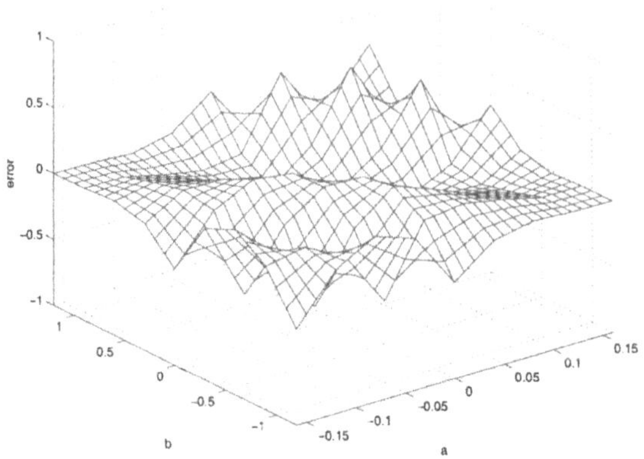

Fig. 4. Output error between the original and the reduced rule base

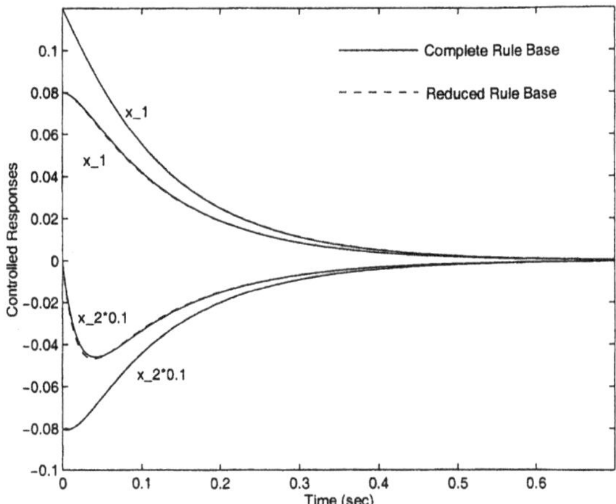

Fig. 5. Closed loop responses due to the original and the reduced rule base

The singular values of $\mathcal{F}$ are: {22.6767, 22.6767, 0.9835, 0.9835, 0.5523, 0.5523, 0.3433, 0.3433, 0}. Keeping the two largest singular values, the procedures yield a reduced fuzzy rule base with three membership functions for a and b. Figure 2 shows the membership function $\bar{A}_{\bar{i}}(a)$ for $\bar{i} = 1, 2, 3$. They are SN and NN but only close-to-NO. The membership functions of b are similar except for the domain of interest. The rule consequents $\bar{r}_{i,j}$ are tabulated in Table 2. The reduced rule base contains 9 rules compared to the original set of 81. To check how close the reduced rule set duplicates the output of the original set, the rule bases are applied to controlling an inverted pendulum as in [14]. The dynamics equation of the pendulum is

$$\begin{bmatrix} \dot{x}_1 \\ \dot{x}_2 \end{bmatrix} = \begin{bmatrix} 0 & 1 \\ 32.827 & 0 \end{bmatrix} \begin{bmatrix} x_1 \\ x_2 \end{bmatrix} + \begin{bmatrix} 0 \\ -14.954 \end{bmatrix} u, \tag{38}$$

where x_1 is the angular position, x_2 is the angular velocity, and u is the control input. Here, $a = x_1$ and $b = x_2$ are the fuzzy variables to infer u. Figures 3 and 4 show, respectively, the PSG-inferred output of the original rule base, and the output errors between the original and reduced rule base. It can be observed that the errors are reasonably small compared to the original outputs. Figure 5 shows the closed loop responses using the original and reduced rule bases for the initial conditions of $(0.08, 0)$ and $(0.12, -0.8)$. The response x_2 is scaled down by a factor of 0.1 in the figure. The closed loop responses are quite comparable to each other.

Of the three conditions on the membership functions, the NO condition is the most difficult to fully incorporate. As an alternative, we may have to accept a close-to-NO situation, as is the case of Fig. 2, where some but not all membership functions will be NO. In [10], a procedure to impose the NO condition over subdomains of the full range is introduced. The procedure enables possible incorporation of the NO condition to all membership functions. The price to pay is an increased in the number of membership functions for the fuzzy system. In the following section, we will introduce another approach to handle the situation where the NO condition cannot be fully satisfied. The approach makes use of a new concept called relaxed Normality (RNO) condition for membership functions.

6 The Relaxed Normality (RNO) Condition

As mentioned before, the NO condition is difficult to fully incorporate, especially when the number of membership functions is large. In this section, we introduce a milder version of the NO condition which can be satisfied by all membership functions, not just a few as in the close-to-NO case. The price to pay is, again, an increase in the number of rules.

Relaxed Normality (RNO): A matrix F is RNO if it is SN, NN, and there exists a constant c, $0 \leq c \leq 1$, such that for every column, the maximum value of its elements is equal to c, i.e.,

$$\max_i F_{i,j} = c \qquad \forall\, j \tag{39}$$

The NO condition before is just the RNO condition with the constant $c = 1$.

6.1 Incorporating the RNO condition

Consider the matrices $\widetilde{U}$ and $\widetilde{V}$ after the SN and NN conditions, we desire to have matrices $\mathcal{P}_a$ and $\mathcal{P}_b$ of appropriate dimensions such that

$$\widetilde{U} = \widehat{U}\mathcal{P}_a \tag{40}$$

$$\widetilde{V} = \widehat{V}\mathcal{P}_b \tag{41}$$

with matrices $\widehat{U}$ and $\widehat{V}$ being RNO, and that

$$\mathcal{F} \approx \widetilde{U}\widetilde{\mathcal{R}}\widetilde{V}^T = \widehat{U}\widehat{\mathcal{R}}\widehat{V}^T, \tag{42}$$

where

$$\widehat{\mathcal{R}} = \mathcal{P}_a \widetilde{\mathcal{R}} \mathcal{P}_b^T. \tag{43}$$

Notice that the formulation here is different from (14) for the incorporation of the NO condition. Here, $\mathcal{P}_a$ and $\mathcal{P}_b$ are actually not square matrices and hence their inverses do not exist. There is no unique way to find matrices $\mathcal{P}_a$ and $\mathcal{P}_b$ to have $\widehat{U}$ and $\widehat{V}$ RNO. One way to determine is the following. Take $\widetilde{U}$ as example and denote its n_{col} columns as

$$\widetilde{U} = [\widetilde{U}_1 \mid \widetilde{U}_2 \mid \ldots \mid \widetilde{U}_{n_{col}}]. \tag{44}$$

Then, for each $\widetilde{U}_i$, we add a complementary column and a transformation matrix T_i to form a two-column component $\widehat{U}_i$ that satisfies the SN, NN, and NO conditions:

$$\widehat{U}_i = \left[\widetilde{U}_i \;\; (\mathbf{1}_{n_a \times 1} - \widetilde{U}_i)\right] T_i. \tag{45}$$

The transformation matrix T_i is given by

$$T_i = \frac{1}{(\overline{f}_i - \underline{f}_i)} \begin{bmatrix} (1 - \underline{f}_i) & (\overline{f}_i - 1) \\ -\underline{f}_i & \overline{f}_i \end{bmatrix}, \tag{46}$$

where $\underline{f}_i$ and $\overline{f}_i$ are the minimum and maximum values, respectively, of the elements of column $\widetilde{U}_i$. As a result, we have

$$\widetilde{U} = \widehat{U}\mathcal{P}_a, \tag{47}$$

where

$$\widehat{U} = \frac{1}{n_{col}} \left[\widehat{U}_1 | \widehat{U}_2 | \cdots | \widehat{U}_{n_{col}}\right], \tag{48}$$

$$\mathcal{P}_a = n_{col} \begin{bmatrix} T_1^{-1} & O_{2\times2} & \cdots & O_{2\times2} \\ O_{2\times2} & T_2^{-1} & \cdots & O_{2\times2} \\ \vdots & \vdots & \ddots & \vdots \\ O_{2\times2} & O_{2\times2} & \cdots & T_{n_{col}}^{-1} \end{bmatrix} \left[\begin{array}{c|c|c|c|c} 1 & 0 & \cdots & \cdots & 0 \\ 0 & 0 & \cdots & \cdots & 0 \\ \hline 0 & 1 & \cdots & \cdots & 0 \\ 0 & 0 & \cdots & \cdots & 0 \\ \hline \vdots & \vdots & \ddots & & \vdots \\ \vdots & \vdots & & \ddots & \vdots \\ \hline 0 & 0 & \cdots & \cdots & 1 \\ 0 & 0 & \cdots & \cdots & 0 \end{array}\right]. \tag{49}$$

Notice that each $\widehat{U}_i$ is SN, NN, and NO for $i = 1, 2, ..., n_{col}$, and the overall matrix $\widehat{U}$ is SN, NN, and RNO with a constant $c = \frac{1}{n_{col}}$. As constructed, $\mathcal{P}_a$

is a $2n_{col}$ by n_{col} matrix, and $\widehat{U}$ is a n_a by $2n_{col}$ matrix. Hence, effectively, we have increased the number of membership functions by a factor of 2 for variable a, and similarly, for variable b. The total number of rules is hence increased 4 times. Notice also that up to now we still have $\widetilde{U}\widetilde{\mathcal{R}}\widetilde{V}^T = \widehat{U}\widehat{\mathcal{R}}\widehat{V}^T$. There is no additional error being introduced going from $\{\widetilde{U}, \widetilde{\mathcal{R}}, \widetilde{V}\}$ representation to $\{\widehat{U}, \widehat{\mathcal{R}}, \widehat{V}\}$ representation. The approximation error $|\mathcal{F} - \widehat{U}\widehat{\mathcal{R}}\widehat{V}^T|$ is still bounded by the sum of discarded singular values as in (20).

As a final point, here we initiate the RNO procedures with $\widetilde{U}$, $\widetilde{\mathcal{R}}$, and $\widetilde{V}$ after the SN and NN incorporation. If we so desire, the procedures can also start with $\overline{U}$, $\bar{\mathcal{R}}$, and $\overline{V}$ after the SN, NN and close-to-NO incorporation.

6.2 Similarity Measure

After the incorporation of the RNO condition, the number of membership functions is increased. Some of the resulting membership functions may be quite similar, however, and we may simplify the rule base by merging them into one membership function. The main objective of rule base simplification is to reduce the number of unequal membership functions used in the model. This will reduce the memory consumption in computer implementation of the model, and more directly, it will improve the semantic interpretation as the number of different labels or words needed to qualitatively describe the rule base at hand is reduced. To start, we first need a measure of similarity between fuzzy sets.

Many methods for measuring similarity have been reported in the literature [15], and some of them is based upon a distance measure. Similarity is seen as a sort of inverse of the distance. If the distance between two fuzzy sets is small, then their similarity is high. In this section, we consider a similarity measure between two fuzzy sets represented by, say, columns g_i and g_j of matrix $\widehat{U}$:

$$\mathcal{S}(g_i, g_j) = \frac{1}{1 + \mathcal{D}(g_i, g_j)} \tag{50}$$

where $\mathcal{D}(g_i, g_j)$ is a distance measure between g_i and g_j such that $\mathcal{D} \in [0, \infty)$, and therefore, $\mathcal{S} \in (0, 1]$. For example, one of the choices for the distance measure will be the 2-norm,

$$\mathcal{D}_2(g_i, g_j) = (\sum_{k=1}^{n_a} |g_{k,i} - g_{k,j}|^2)^{\frac{1}{2}} \tag{51}$$

and another will be the ∞-norm,

$$\mathcal{D}_\infty(g_i, g_j) = \max_{k=1,2,\ldots,n_a} |g_{k,i} - g_{k,j}| \tag{52}$$

where $g_{k,i}$ is the $(k, i)$the elements of $\widehat{U}$. Once we have a measure of the distance, we can get a similarity measure between two fuzzy sets. If the similarity is higher than a certain threshold, say, then we can merge the two columns into one for a more compact representation. In this work, we use the average of the two columns as the merged fuzzy set. At the same time, corresponding rows of $\widehat{R}$ should be merged by summing themselves together (or the columns of $\widehat{R}$ should be summed if it's the columns of $\widehat{V}$ being merged). Notice that merging, if it occurs, will come in even number. Referring to (45) and (48), if the columns of $\widehat{U}$ originated from $\widetilde{U}_i$ and $(1-\widetilde{U}_j)$ are considered very similar to each other, then the columns of $\widehat{U}$ originated from $(1-\widetilde{U}_i)$ and $\widetilde{U}_j$ should also be very similar to each other. Both pairs should be merged. As a result, $\widehat{U}$, which is originally RNO with constant $c = \frac{1}{n_{col}}$ will become RNO with constant $c = \frac{1}{(n_{col}-1)}$ after merging. The merging performed here will result in an error on top of that induced in the singular value approximation step. However, if the similarity is high, the error will be small.

The resulting $\widehat{U}$, $\widehat{R}$, and $\widehat{V}$ from the above manipulations yields the following rule base:

$$\text{If } \widehat{A}_{\hat{i}}(a) \text{ and } \widehat{B}_{\hat{j}}(b) \Rightarrow u = \widehat{r}_{\hat{i},\hat{j}}. \tag{53}$$

with indices $\hat{i}$ and $\hat{j}$ taking on appropriate integer values. Similar to (30)-(32), the new membership functions $\widehat{A}_{\hat{i}}(a)$ and $\widehat{B}_{\hat{j}}(b)$ are formed by linear combining the original membership functions $A_i(a)$ and $B_j(b)$ according to the columns of $\widehat{U}$ and $\widehat{V}$, and the rule consequents $\widehat{r}_{\hat{i},\hat{j}}$ are the $(\hat{i},\hat{j})$ element of $\widehat{R}$.

For the non-singleton support case, we note that (37) for $k = 1, 2$, can be rewritten as:

$$\mathcal{F}_1 \approx \overline{U}\bar{\mathcal{R}}_1\overline{V}^T; \quad \mathcal{F}_2 \approx \overline{U}\bar{\mathcal{R}}_2\overline{V}^T, \tag{54}$$

where $\mathcal{F}_1, \bar{\mathcal{R}}_1, \mathcal{F}_2$, and $\bar{\mathcal{R}}_2$ contain $\mathcal{F}_{i,j,1}, \bar{\mathcal{R}}_{\bar{i},\bar{j},1}, \mathcal{F}_{i,j,2}$, and $\bar{\mathcal{R}}_{\bar{i},\bar{j},2}$, respectively, as elements. The matrices $\overline{U}$ and $\overline{V}$ are SN, NN, and close-to-NO. Applying the RNO condition and similarity merging procedures to $\overline{U}$, $\overline{V}$, $\bar{\mathcal{R}}_1$, and $\bar{\mathcal{R}}_2$, we obtain matrices $\mathcal{P}_a$ and $\mathcal{P}_b$ to yield

$$\mathcal{F}_1 \approx \overline{U}\bar{\mathcal{R}}_1\overline{V}^T = \widehat{U}\widehat{\mathcal{R}}_1\widehat{V}^T; \quad \mathcal{F}_2 \approx \overline{U}\bar{\mathcal{R}}_2\overline{V}^T = \widehat{U}\widehat{\mathcal{R}}_2\widehat{V}^T, \tag{55}$$

with $\widehat{\mathcal{R}}_1 = \mathcal{P}_a\bar{\mathcal{R}}_1\mathcal{P}_b^T$, and $\widehat{\mathcal{R}}_2 = \mathcal{P}_a\bar{\mathcal{R}}_2\mathcal{P}_b^T$, and in this case $\overline{U} = \widehat{U}\mathcal{P}_a$ and $\overline{V} = \widehat{V}\mathcal{P}_b$ are RNO. Then, performing similarity merging on $\widehat{U}$ and $\widehat{V}$, and, correspondingly, on $\widehat{\mathcal{R}}_1$ and $\widehat{\mathcal{R}}_2$, we obtain a fuzzy rule base in the same form as (53), with membership functions $\widehat{A}_{\hat{i}}(a)$ and $\widehat{B}_{\hat{j}}(b)$ similarly obtained as linear combinations of $A_i(a)$ and $B_j(b)$ using $\widehat{U}$ and $\widehat{V}$. In this case, the rule consequent $\widehat{r}_{\hat{i},\hat{j}}$ and support factor $\widehat{s}_{\hat{i},\hat{j}}$ are given by

$$\widehat{r}_{\hat{i},\hat{j}} = \widehat{\mathcal{R}}_{\hat{i},\hat{j},1}/\widehat{\mathcal{R}}_{\hat{i},\hat{j},2}\ ; \quad \widehat{s}_{\hat{i},\hat{j}} = \widehat{\mathcal{R}}_{\hat{i},\hat{j},2}$$

where $\widehat{\mathcal{R}}_{\hat{i},\hat{j},1}$ and $\widehat{\mathcal{R}}_{\hat{i},\hat{j},2}$ are the $(\hat{i},\hat{j})$ elements of $\widehat{\mathcal{R}}_1$ and $\widehat{\mathcal{R}}_2$, respectively.

Two numerical examples are now given to illustrate the RNO procedures and performance.

Example 2: This example, originally from [16], was treated in [1] as a fuzzy approximation case study. Here, it is included to illustrate the induced errors in similarity merging. Consider the following function of two variables a and b:

$$f(a,b) = (1 + a^{-2} + b^{-1.5})^2, \qquad 1 \leq a, b \leq 5 \tag{56}$$

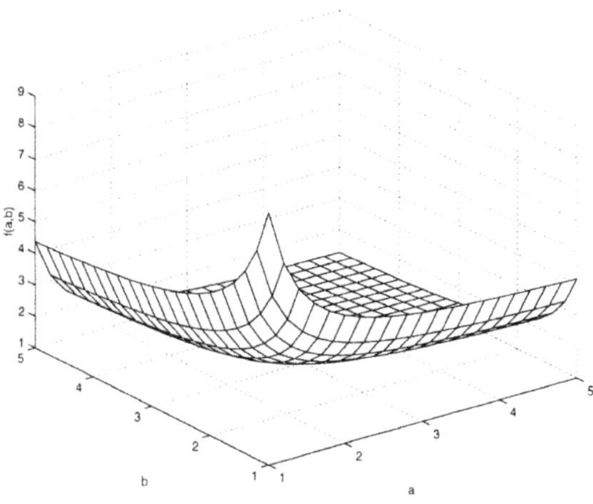

Fig. 6. Sampled outputs at grid points

Sampling the function at the rectangular grid of $\mathbf{A} \times \mathbf{B}$, with $\mathbf{A} = \mathbf{B} = [1, 1.2, 1.4, ..., 4.8, 5]$, we obtain the matrix $\mathcal{F}$ as shown in Fig. 6. Then, applying singular value decomposition to $\mathcal{F}$, we get 3 nonzero singular values: $\{55.7821,\ 3.2698,\ 0.0250\}$. Keeping all three singular values, and conducting the SN and NN procedures, we obtain the membership functions of a (given by the columns of $\widetilde{U}$) as shown in Fig. 7. The membership functions for b are similarly shaped. The resulting rule consequents are given by:

$$\widetilde{\mathcal{R}} = \begin{bmatrix} 9.0000 & 4.3658 & 5.9885 \\ 4.1616 & 1.2756 & 2.1538 \\ 5.7540 & 2.1579 & 3.3385 \end{bmatrix}. \tag{57}$$

As all nonzero singular values are being retained, the resulting SN and NN rule base in this case yields no error at the sampling grid points. There will be errors, however, in-between grid points due to fuzzy inferencing.

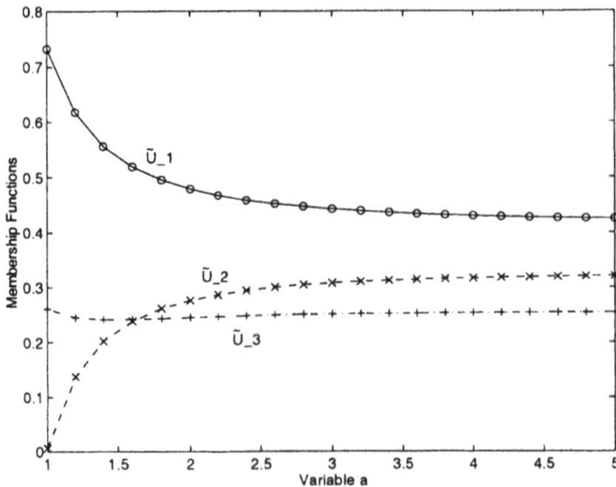

Fig. 7. Membership functions of a satisfying SN and NN conditions

Realizing that the membership functions cannot be made NO, we will try to incorporate the RNO condition. According to (44)-(49),

$$\widetilde{U} = \left[\widetilde{U}_1\ \widetilde{U}_2\ \widetilde{U}_3\right] = \widehat{U} \cdot 3 \cdot \begin{bmatrix} T_1^{-1} & O_{2\times 2} & O_{2\times 2} \\ O_{2\times 2} & T_2^{-1} & O_{2\times 2} \\ O_{2\times 2} & O_{2\times 2} & T_3^{-1} \end{bmatrix} \begin{bmatrix} 1 & 0 & 0 \\ 0 & 0 & 0 \\ \hline 0 & 1 & 0 \\ 0 & 0 & 0 \\ \hline 0 & 0 & 1 \\ 0 & 0 & 0 \end{bmatrix}$$

$$\widehat{U} = \frac{1}{3}\left[\widetilde{U}_1\ 1-\widetilde{U}_1\ \widetilde{U}_2\ 1-\widetilde{U}_2\ \widetilde{U}_3\ 1-\widetilde{U}_3\right] \begin{bmatrix} T_1 & O_{2\times 2} & O_{2\times 2} \\ O_{2\times 2} & T_2 & O_{2\times 2} \\ O_{2\times 2} & O_{2\times 2} & T_3 \end{bmatrix}$$
$$= [g_1\ g_2\ g_3\ g_4\ g_5\ g_6]$$

Figure 8 shows the membership functions of a after the RNO condition, in this case with $c = \frac{1}{3}$. Again, the RNO membership functions for b are roughly the same. The resulting rule consequents are

$$\widehat{\mathcal{R}} = 3^2 \times \begin{bmatrix} 14.7544 & 8.7969 & -0.0630 & -0.0063 & -4.0894 & -3.8200 \\ 8.5542 & 5.1002 & -0.0365 & -0.0037 & -2.3709 & -2.2147 \\ -0.0601 & -0.0359 & -0.8380 & -0.0845 & -0.2633 & -0.2459 \\ -0.0012 & -0.0007 & -0.0164 & -0.0017 & -0.0052 & -0.0048 \\ -3.9496 & -2.3548 & -0.2662 & -0.0268 & 2.3257 & 2.1725 \\ -3.6618 & -2.1833 & -0.2468 & -0.0249 & 2.1563 & 2.0142 \end{bmatrix}$$

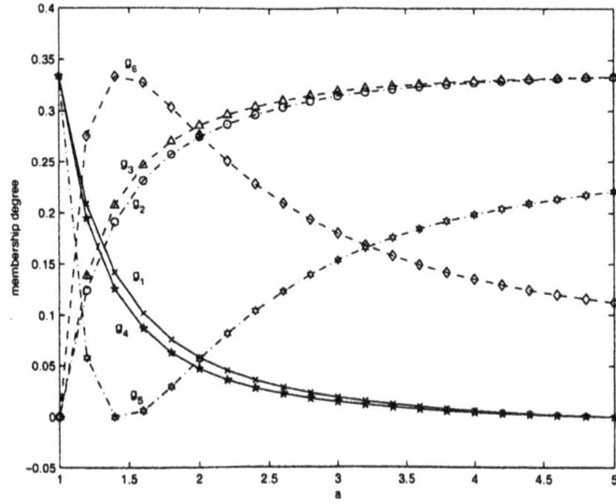

Fig. 8. Membership functions of a satisfying RNO with $c = \frac{1}{3}$

We now have six membership functions for each variable and a total of 36 rules. Some membership functions are very similar, however. The pairs of g_1 and g_4, and g_2 and g_3 have a 2-norm similarity measure of $\mathcal{S}_2(g_1, g_4) = \mathcal{S}_2(g_2, g_3) = 0.9038$. We can hence merge the pairs together and reduce the system to four membership functions for each variable. Merging is done by taking the average of the similar membership functions. Figure 9 shows the membership functions as resulted for variable a. They are now RNO with constant $c = \frac{1}{2}$. The corresponding rule consequent matrix becomes

$$\widehat{\mathcal{R}} = 2^2 \times \begin{bmatrix} 14.7452 & 8.7168 & -4.0945 & -3.8248 \\ 8.4059 & 4.1898 & -2.6342 & -2.4607 \\ -3.9764 & -2.6210 & 2.3257 & 2.1725 \\ -3.6867 & -2.4300 & 2.1563 & 2.0142 \end{bmatrix}$$

The errors due to the merging of membership functions are shown in Fig. 10. It can be observed that, comparing to Fig. 6, the errors are relatively small. This example is originally treated in [16] using qualitative modeling for the rule base identification. The identified system there has six fuzzy rules, each of the input variables and the output, however, has six different membership functions. The identified rules are hence sparse in nature. In our case, we have four membership functions for a, four for b, and a total number of sixteen dense rules. Linguistic labels can be readily assigned as:

g_1 is Small (S),
g_2 is more than Medium (m_M),
g_5 is Small or Medium Large (S_or_ML),
g_6 is more or less Medium Small (m_or_l_MS),

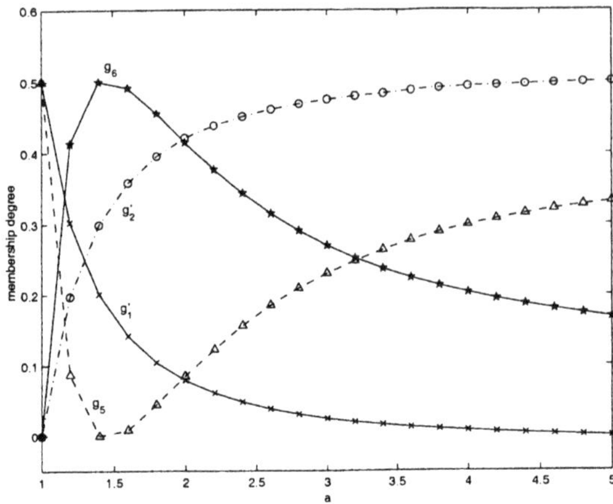

Fig. 9. Membership functions of a after similarity merging

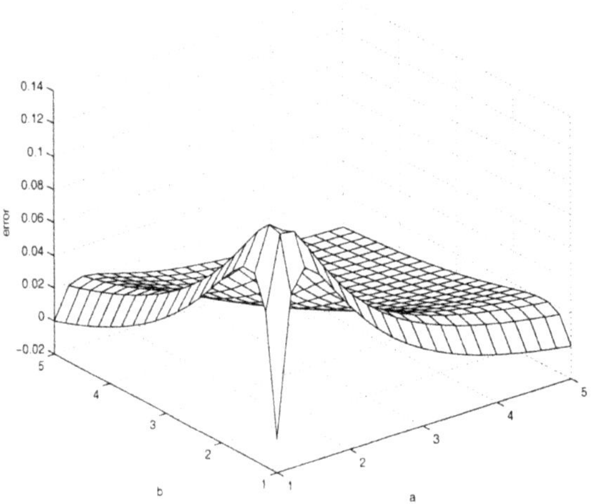

Fig. 10. Output errors due to similarity merging of membership functions

and similarly for the membership functions of b. Table 3 summaries the RNO fuzzy rule base as obtained.

Example 3: As a second example involving the RNO condition, we use the same inverted pendulum example from [14] as treated in Section 5 to illustrate the control performance of the resulting rule base. Recall that previously, the three membership functions of a after rule base reduction are shown in Fig. 2. We will initiate the RNO condition using these close-to-NO membership functions.

Table 3. Fuzzy rule base with RNO and similarity merging in Example 2

RC (MF a \ MF b)	S	m_M	S_or_ML	m_or_l_MS
S	58.9808	34.8672	−16.3780	−15.2992
m_M	33.6236	16.7592	−10.5368	−9.8428
S_or_ML	−15.9056	−10.4840	9.3028	8.6900
m_or_l_MS	−14.7468	−9.7200	8.6252	8.0568

Figure 11 shows the resulting membership functions of a after the RNO incorporation, in this case with $c = \frac{1}{3}$. Again, membership functions for b are similar except for the different domain of interest. The resulting rule consequents are

$$\widehat{\mathcal{R}} = 3^2 \times \begin{bmatrix} -4.7320 & 0 & -3.6731 & 0 & 0.0000 & 0 \\ 0 & 0 & 0 & 0 & 0 & 0 \\ -3.6731 & 0 & 0 & 0 & 3.6731 & 0 \\ 0 & 0 & 0 & 0 & 0 & 0 \\ 0 & 0 & 3.6731 & 0 & 4.7320 & 0 \\ 0 & 0 & 0 & 0 & -0.0000 & 0 \end{bmatrix},$$

which clearly shows that some rules are being added to satisfy the RNO condition.

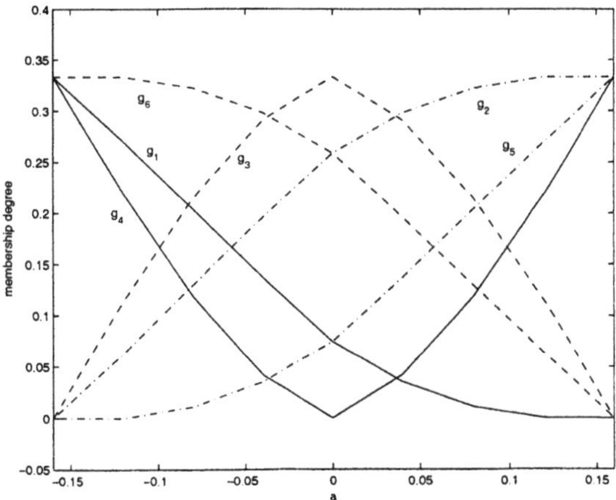

Fig. 11. Membership functions of a satisfying RNO with $c = \frac{1}{3}$

Now we have six membership functions for each variable and a total of 36 rules. In this case, we note that $\mathcal{S}_\infty(g_1, g_6) = \mathcal{S}_\infty(g_2, g_5) = 0.6442$. Although the similarity is not very high, we will merge the membership functions together and see how it will affect the control performance. Replacing the two membership functions by their average, $g'_1 = \frac{1}{2}(g_1 + g_6)$ and $g'_2 = \frac{1}{2}(g_2 + g_5)$, we obtain the membership functions as shown in Fig. 12 for variable a. Membership functions for b can be similarly processed. The rule consequents matrix is then

$$\hat{\hat{\mathcal{R}}} = 2^2 \times \begin{bmatrix} -4.7320 & -3.6731 & 0 & 0.0000 \\ -3.6731 & 0 & 0 & 3.6731 \\ 0 & 0 & 0 & 0 \\ 0 & 3.6731 & 0 & 4.7320 \end{bmatrix}.$$

Figure 13 shows the output errors of the RNO rule base with similarity merging as compared to the original rule base. Comparing to Fig. 4, it is noted that that additional errors induced in the similarity merging procedures do not actually increase significantly over those already after the singular value reduction, even though in this case we are merging membership functions with only moderate similarity.

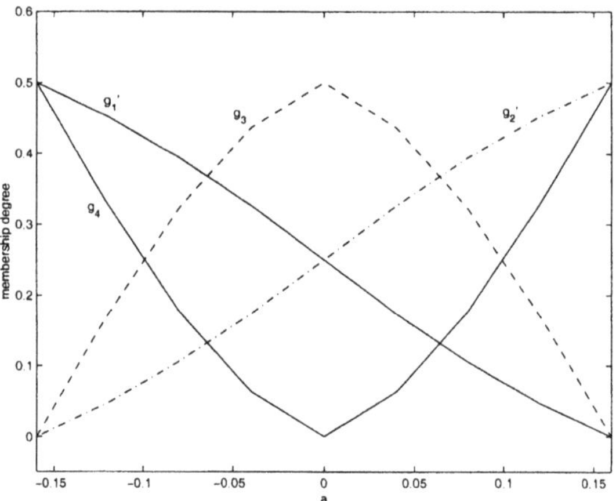

Fig. 12. Membership functions of a after RNO and similarity merging

Figure 14 compares the closed loop responses due to the original rule base and the reduced rule base after RNO and similarity merging. The initial conditions are same as before: $(0.08, 0)$ and $(0.12, -0.8)$. The responses are still quite reasonably close to each other.

In this example, we have four membership functions for a, four for b, and a total number of sixteen dense rules for the RNO and similarity merged rule

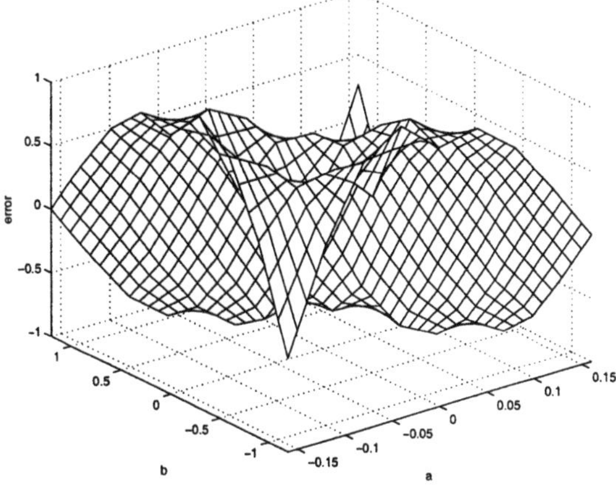

Fig. 13. Output error between the original rule base and the reduced rule base after RNO/similarity merging

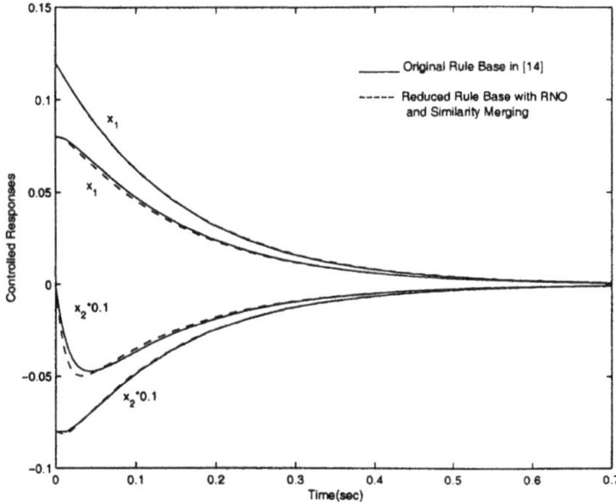

Fig. 14. Closed loop Responses due to the original rule base and the reduced rule base after RNO/similarity merging

base. Note that the rule consequents corresponding to the third membership functions of either variable a or b are all zeros. We can further eliminate these membership functions and get a total of nine fuzzy rules. However, then we would not have the SN condition. The advantage of the RNO condition is that all membership functions are now put on equal footing, and hence assignment of linguistic labelings can be applied uniformly on all of them. Here, we can assign linguistic labelings to the membership functions as

g_1' is Negative (N),
g_2' is About Zero (AZ),
g_3 is Not About Zero (NAZ),
g_4 is Positive (P),

and similarly to the membership functions of b. Notice that the labelings are obtained by mixing existing distinct labels. In reality, there are membership functions that required this kind of mixing labelings for linguistic approximation, see [16]. The resulting rule base is shown in Table 4.

Table 4. Fuzzy rule base after RNO and similarity merging in Example 3

RC / MF a \ MF b	N	AZ	NAZ	P
N	−18.9282	−14.6923	0	0
AZ	−14.6923	0	0	14.6923
NAZ	0	0	0	0
P	0	14.6923	0	18.9282

7 Conclusions

This chapter reviews the basic concepts and procedures of a recent singular value-based approach which can be applied for fuzzy approximation, identification, reduction, and interpolation. The approach calls for conducting singular value decomposition on a matrix and then tailors the relevant matrices to satisfy the SN, NN, and, possibly, NO conditions. For fuzzy approximation, the approach is applied to a sample matrix over a rectangular set of grid points. The tailored orthogonal matrix then yields the membership functions, with the accompanying singular values becoming the rule consequents. The PSG paradigm emerges naturally from the decomposition equation. Output error at the grid points is bounded by the sum of singular values discarded during the process. For fuzzy reduction, the approach is applied to the matrix containing the rule consequents. The tailored orthogonal matrices then become linear combinations to form the membership functions of the reduced rule set. This chapter presents the case of fuzzy reduction in details as a specific illustration of the approach. Generalization of the method to high dimensionality is also demonstrated.

Membership functions with the present approach can generally be made to satisfy the SN and NN conditions. Incorporation of the NO condition, however, is difficult, especially for cases where there is a large number of membership functions. A subdomain technique to impose the NO condition at subdivided ranges of the input domain has been proposed previously. Here, a new approach is introduced. The approach makes use of a milder version of the NO condition, termed the relaxed Normality (RNO) condition. All membership functions, not just a few as in the close-to-NO case, can be made to satisfy RNO. This puts all membership functions on equal footing, and assignment of linguistic labelings can be uniformly applied to all of them. The catch is that, under RNO, two or more membership functions may reach maximum membership degree of c at the same location in the input domain of the fuzzy variables, i.e., membership functions can be co-localized. The RNO will result in an increase in the number of fuzzy rules. To yield a more compact representation, a set of procedures to conduct similarity measure and merging of membership functions is also introduced, at the cost of increased output errors in the resulting rule base. The present work serves to establish a formulation to tradeoff the aspects of interpretability, number of rules, and induced errors in a rule base. Decision as to the best tradeoff point is left to individual analyst in consideration of his/her particular design and application. The present work also includes numerical examples to illustrate the fuzzy reduction approach and the subsequent RNO procedures.

References

1. Yam, Y. (1997) Fuzzy Approximation via Grid Point Sampling and Singular Value Decomposition. IEEE Transactions on Systems, Man, and Cybernetics. **Vol. 27, No. 6**, 933-951
2. Wang, L.X., Mendel, J.M. (1992) Fuzzy Basis Functions, Universal Approximation, and Orthogonal Least Square Learning. IEEE Transactions on Neural Networks. **Vol. 3, No. 5**, 807-814
3. Zeng, X.J., and Singh, M.G. (1994) Approximating Theory of Fuzzy System – SISO case. IEEE Transactions on Fuzzy Systems. **Vol. 2, No. 2**, 162-176
4. Zeng, X.J., and Singh, M.G. (1995) Approximating Theory of Fuzzy System – MIMO case. IEEE Transactions on Fuzzy Systems. **Vol. 3, No. 2**, 219-235
5. Baranyi, P., A. Martinovics, A., Kovacs, S., Tikk, D., Yam, Y., (1998) A General Extension of Fuzzy SVD Rule Base Reduction Using Arbitrary Inference Algorithm. Proceedings of the 1998 IEEE Conference on Systems, Man, and Cybernetics (IEEE SMC'98). 2785-2790
6. Yam, Y., Baranyi, P., Yang, C.T. (1999) Reduction of Fuzzy Rule Base via Singular Value Decomposition. IEEE Transactions on Fuzzy Systems. **Vol. 7, No. 2**, 120-132
7. Yang, C.T., Yam, Y. (1999) Fuzzy Identification using Scattered Samples and Singular Value Decomposition. Theory and Practice of Control and Systems, Eds. Tornambé, A., Conte, G., A.M. Perdon, A.M. World Scientific, 635-639
8. Baranyi, P., Yam, Y., Kóczy, L.T. (1997) Singular Value-Based Fuzzy Rule Interpolation. Proceedings of the IEEE International Conference on Intelligent Engineering Systems (IEEE INES97). Budapest, Hungary, 51-56

9. Baranyi, P., Kybic, J., Yam, Y., Kóczy, L.T. (1997) Extension of Singular Value-Based Rule Base Reduction to the General Fuzzy Rule Interpolation. Proceedings of the TEMPUS Symposium on Qualitative System Modelling, Qualitative Fault Diagnosis, and Fuzzy Logic and Control. Budapest and Miskolc, Hungary
10. Yam, Y. (1999) Fuzzy Identification with SVD and Subdomain Normality. Theory and Practice of Control and Systems. Eds. Tornambé, A., Conte, G., A.M. Perdon, A.M. World Scientific, 623-628
11. O'Rourke, J. (1994) Computational Geometry in C. Cambridge University Press
12. Edelsbrunner, H. (1987) Algorithms in Combinatorial Geometry. Springer-Verlag, Berlin, Germany
13. Takagi T., Sugeno, M. (1985) Fuzzy Identification of Systems and Its Applications to Modeling and Control. IEEE Transactions on Systems, Man, and Cybernetics. **Vol. 15**, **No. 1**, 116-132
14. Bouslama, F., Ichikawa, A. (1992) Application of Limit Fuzzy Controllers to Stability Analysis. Fuzzy Sets and Systems. **49**, 103-120
15. Setnes, M., Babuska, R., Verbruggen, H.B. (1998) Complexity Reduction in Fuzzy Modeling. Mathematics and Computers in Simulation. **46**, 507-516
16. Sugeno M., Yasukawa, T. (1993) A Fuzzy-Logic-Based Approach to Qualitative Modeling. IEEE Transactions on Fuzzy Systems. **Vol. 1**, **No. 1**, 7-31

SECTION 5

INTERPRETABILITY CONSTRAINTS IN TSK FUZZY RULE-BASED SYSTEMS

Interpretability, Complexity, and Modular Structure of Fuzzy Systems

Marwan Bikdash

Department of Electrical Engineering,
North Carolina A&T State University
Greensboro, NC 27411, USA
e-mail: bikdash@ncat.edu

Abstract. Zadeh's original motivation for fuzzy logic and the Fuzzy Rule-Based System (FRBS) was linguistic and hence possessed highly interpretable components. But as the complexity of a typical FRBS increases, it often becomes more like an uninterpretable neural network, and the Principle of Incompatibility predicts a degradation in interpretability for the same accuracy. This is particularly true of the so-called Takagi-Sugeno-Kang (TSK), or simply the Sugeno, approximator. We argue that imposing additional structure on the TSK system can significantly improve the tradeoff inherent in the Principle of Incompatibility. A promising structure was proposed recently in which the membership functions are local and sufficiently differentiable, and the consequent polynomials are rule-centered. This structure leads to the general interpretation that the consequent polynomials are Taylor series expansions. On this interpretation, a foundation for an algebra and a calculus of FRBSs can be built. We will illustrate these aspects of the proposed structure and discuss issues of modularity, functionality, and scalability of FRBSs.

1 Accuracy, Interpretability, and Complexity

There are two major traditions within the computational intelligence community regarding the nature of a Fuzzy Rule-Based System (FRBS). The first tradition is linguistic and hence inherently interpretable. The second is computational and hence inherently accurate.

1.1 Linguistic Tradition

Here the inference system is a codification of expert or intuitive but linguistically expressible knowledge. This is closer to Zadeh's original intent of introducing fuzzy sets [19,20]. Here the attributes or linguistic values of the system input variables have "meaning" and the Membership Functions (MFs) are a mathematical codification of that meaning. A classical example of this approach is the fuzzy-logic controller of [10] where the IF-THEN rule statements captured expert knowledge of what a good feedback controller of a certain chemical plant must be, and was then interfaced with the physical plant through meaningful fuzzification and defuzzification.

In this tradition, a typical rule is written as

$$R^k : \text{IF } x_1 \text{ is } A_1^k \text{ AND } x_2 \text{ is } A_2^k \text{ AND } \cdots \text{ AND } x_n \text{ is } A_n^k, \text{ THEN } y \text{ is } B^k. \tag{1}$$

In the antecedent, x_i is interpreted not as a crisp number but as a linguistic variable possessing many fuzzy attributes or linguistic values or terms. The collection of all such attributes or terms is called the term set. Each term is a fuzzy set which has a "meaning" codified through a MF, and usually labeled by a descriptive name. Examples of terms or attribute names include: "large", "small", "very reliable," and "about 15". In Eq. (1), A_i^k specifies which linguistic value or term of the linguistic variable x_i is being tested by rule k and B^k specifies which attribute of the output y is implied.

The exact meaning of every attribute is captured in the shape of its MF. The triangular function [4]

$$\mu_{triang}(x;\ L, g, U) = \begin{cases} 0, & x \leq L \\ \frac{x-L}{g-L}, & L \leq x \leq g \\ \frac{U-x}{U-g}, & g \leq x \leq U \\ 0, & U \leq x \end{cases} \tag{2}$$

is a well known example of a local MF. It is centered at $x = g$ and is nonzero over the support $[L, U]$. Examples of triangular MFs are shown in Figure 1. The MF labeled "3" can be interpreted as meaning "about 3" and is represented by the triangular MF $\mu_{triang}(x; 2, 3, 5)$. The three MFs denoted "3" through "5" are easy to interpret because they are local and well-ordered, have very similar shapes, and do not overlap excessively. In contrast, the dotted MF overlaps excessively with the other three and is generally harder to interpret for the same application. In general, local, similarly shaped MFs that do not overlap excessively usually have high interpretability, as illustrated in Figure 1b.

Every component of the rule is hence interpretable, and if the defuzzification process combining the outputs of the different rules is reasonably interpretable, then the whole fuzzy inference system is interpretable.

1.2 Computational Tradition

The other tradition emphasizes the computational aspects of fuzzy inference systems, often referred to as inference "engines", and is deeply connected with approximation theory. The Takagi-Sugeno-Kang (TSK) system, here referred to simply as the Sugeno Approximator (SA), is the prime example of this tradition. The universal approximation properties of these engines is often debated and emphasized [1, 5, 8, 2, 15, 17].

In its basic form, a linear Sugeno approximator is an FRBS mapping a crisp vector $x = [x_1, x_2, \cdots, x_n]^T \in \Re^n$ into $y \in \Re$. The output $Y_k(x)$ assigned to y by the kth rule is an affine function of the input vector:

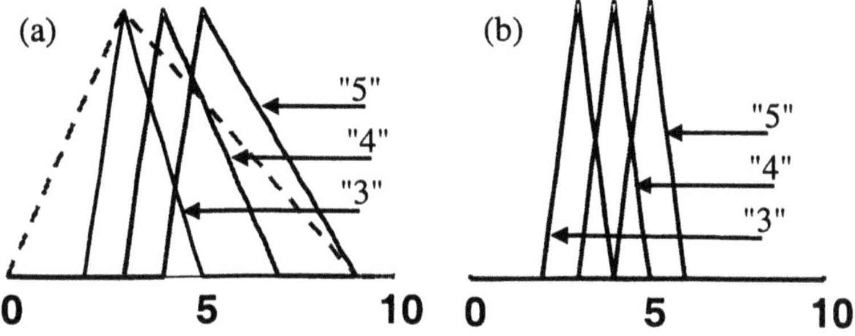

Fig. 1. (a) Three easily interpretable membership functions denoted "3", "4", and "5". Here "3" means "about 3". The dotted membership function is harder to interpret. (b) By reducing the overlap, the interpretability is improved.

$$\begin{aligned} R^k : \text{IF} \quad & x_1 \text{ is } A_1^k \text{ AND } x_2 \text{ is } A_2^k \text{ AND } \cdots \text{ AND } x_n \text{ is } A_n^k \\ \text{THEN } & y = y^k = Y_k(x) = c_0^k + c_1^k x_1 + \cdots + c_n^k x_n \end{aligned} \tag{3}$$

where $c_0^k, c_1^k, \cdots, c_n^k$ are the consequent coefficients of the kth fuzzy rule, but the antecedent has the same form as in Eq. (1) of the linguistic tradition. Hence

$$\begin{aligned} R^5 : \text{IF} \quad & x_1 \text{ is "3" AND } x_2 \text{ is "2" AND } \cdots \text{ AND } x_n \text{ is "4"} \\ \text{THEN } & y = Y^5(x) = 2 - x_1 + \cdots + 7x_n \end{aligned} \tag{4}$$

is read as: The fifth rule states that if x_1 is its third attribute, x_2 is its second attribute, ... and x_n is its fourth attribute, then $y = 2 - x_1 + \cdots + 7x_n$.

The consequent polynomial of the kth rule can be generalized to include higher-order terms [1]:

$$Y_k(x) = c_0^k + c^k x + \frac{1}{2} x^T W^k x + \cdots \tag{5}$$

where c^k is an n-dimensional row and W^k is an $n \times n$ matrix. This is the so-called "generalized Sugeno approximator."

Fuzzy systems arising in the computational tradition are often larger, supposedly representing more complicated systems, and the parameters defining the MFs or the consequent polynomials are often determined using numerically intensive procedures for learning from examples [16], supervised training [14] or some other parameter identification mechanism [13, 17]. Even the shapes of the MFs can become the subject of design, as in the ANFIS system, introduced by [4]. The resulting shapes can be rather wild-looking, and hence hard to interpret and label.

But even when the MFs are kept simple, and the antecedent of the rule remains the same as in the linguistic Zadeh-Mamdani tradition, the Sugeno approximator suffers from a lack of interpretability related to *the coefficients in the consequent polynomials.* What is the meaning or interpretation of the coefficients 2, −1, ⋯ , 7 in R^5 in Eq. (4)? Our main purpose in this chapter is to give an interpretation to these coefficients and to consider the implications of this interpretation.

1.3 A Qualified Principle of Incompatibility

Fuzzy engines often end up looking like neural networks or some other expression of the connectionist paradigm. Indeed, in the ANFIS systems, much of the notation, assumptions, and algorithms of neural networks is adopted [5]. For these reasons, Jang et al. [5, Chapter 13] discuss what they call the *Neuro-Fuzzy Interpretability Spectrum* which is a tradeoff between input-output mapping precision and the interpretability of the MFs. They seem to indicate that the higher the order of the system, the closer one must get to an uninterpretable neural network. Zadeh has stated it best [20] in a principle that relates three concepts: Complexity, interpretability, and accuracy:

Proposition 1. *(Zadeh's Principle of Incompatibility) As the complexity of a system increases, our ability to make precise and yet significant statements about its behavior diminishes until a threshold is reached beyond which precision and significance (or relevance) become almost mutually exclusive characteristics".*

This is a good place to register a philosophical objection to this principle. First, let us note that there are abundant counter-examples to the above principle in many fields of science and engineering. Consider, for instance, problems in multi-body dynamics. There is a reasonably simple solution to the two-body (such as sun and earth) problem. The three-body problem (sun, earth, and moon) is harder to solve and may have "chaotic" solutions. The general n-body problem can be solved only in limited cases. But if one considers *a very large number* of bodies, such as 10^{23} gaz molecules, a very simple stochastic solution can be found using the methods of statistical mechanics. This leads to the equation of state of a gaz (like $PV = nRT$ where P is the pressure, etc...). Here the effects of complexity (measured as the number of bodies) do not accumulate, but rather, tend to cancel. In analog electronics, one differential amplifier stage (containing few transistors) has a less "interpretable" behavior than an operational amplifier with multiple stages, 100 transistors, and extremely high gain. Indeed, it is the more "complicated" operational amplifier that is considered an "ideal" amplifier.

Another counter-example comes from the VLSI industry. Clearly a microprocessor is vastly more complicated (with complexity measured by the number of transistors) than the NAND gate, yet it remains highly "interpretable" and "accurate". The common wisdom is that the astronomically high complexities achieved in the VLSI industry are obtained through *designer self-discipline* in which the designer adheres strictly to concepts of modularity, functionality, and scalability. Without this additional *structure* imposed on the VLSI chip, the chip would very soon become "uninterpretable" and hence undesignable.

Hence a modification of the principle of incompatibility that relates four concepts (complexity, interpretability, accuracy, and structure) must be proposed:

Proposition 2. *(Qualified Principle of Incompatibility) As the complexity of a system increases, our ability to make precise and yet significant statements about its behavior diminishes* unless *additional* structure *is imposed on the system, especially with view to modularity, functionality, and scalability.*

By **modularity** we mean that the system is easily decomposable into **modules** that are interfaceable according to known and well-understood standards. By **functionality** we mean that every module performs a small number of well-understood functions. By **scalability** we mean that the cost of the overall system scales well with the number of modules used. These three concepts lead to *a structure that offers a way out of the dilemma posed by the Principle of Incompatibility.*

Therefore, to design large, interpretable, and accurate fuzzy rule-based systems, additional structural constraints must be imposed. This was attempted in [3]. Briefly, the MFs are required to be (a) strongly local, (b) well-ordered, and (c) sufficiently differentiable. Moreover, (d) the consequent polynomials are required to be written in a rule–centered form $Y_l(x) = P_l\left(x - r^l\right)$, where $x \in \Re^n$, $P_l\left(\cdots\right)$ is a polynomial, and r^l is the "center" of the lth rule. This additional structure leads to the general application-independent interpretation that the consequent polynomials are Taylor series expansions of the underlying map about the rule centers with respect to the inputs.

Whether the proposed structure defies Zadeh's Principle of Incompatibility requires more careful consideration, and can be ascertained only after a thorough analysis on the impact of the proposed structure on modularity, functionality, and scalability of fuzzy rule-based systems.

In the following sections, we will present (in Sections 2–4) the structure proposed in [3] and discuss its impact on interpretability (in Section 4), modularity (Section 5.4–5.6), functionality (Section 4.3), and scalability (Sections 5.1-5.3).

2 The Sugeno Approximator

The output $\hat{y}$ of the Sugeno approximator is obtained by a weighted average of the y^k in Eq. (3) and representing the crisp output values recommended by the different rules:

$$\hat{y} = \frac{\sum_{k=1}^{K} \rho_k(x) y^k}{\sum_{j=1}^{K} \rho_j(x)} = \sum_{k}^{K} \beta_k(x) y^k, \tag{6}$$

where $\rho_k(x)$ is the truth value of the antecedent of the kth rule evaluated with input x or $\rho_k(x) = A_1^k(x_1) \cdot \ldots \cdot A_n^k(x_n)$ where $A_i^k(x_i)$ is the MF of x_i in its attribute tested by the kth rule. Hence $\beta_k(x)$ are the normalized truth values of the kth rule,

$$\beta_k(x) = \frac{\rho_k(x)}{\sum_j \rho_j(x)}, \qquad k = 1, \cdots, K, \tag{7}$$

and the β_k's possess the addition-to-unity properties $0 \leq \beta_k \leq 1$, and $\sum_k \beta_k = 1$. It would be very beneficial if the unnormalized truth values satisfy the addition-unity-property:

$$0 \leq \rho_k \leq 1, \text{ and } \sum_k \rho_k = 1 \quad \Rightarrow \quad \beta_k(x) = \rho_k(x) \tag{8}$$

in which case no normalization is needed. Because of the above two properties, $\hat{y}$ is called [7] a convex combination of the y^k.

Often a complex search algorithm is used to optimize the parameters of the FRBS. For example, Karr [6] examined the feasibility of using a genetic algorithm to find MFs for a high-performance controller for the pole-cart system. The coefficients of the consequent polynomials can be found using a least squares approach due to Takagi and Sugeno [14]. Defining the vectors

$$X = [1, x_1, x_2, \cdots, x_n] \in \Re^{n+1}, \qquad \mathbf{c}^k = [c_0^k, c_1^k, \cdots, c_n^k]^T \in \Re^{n+1}, \tag{9}$$

the output can be written as $\hat{y} = \sum_{k=1} \beta_k X \mathbf{c}^k$ or

$$\hat{y} = [\beta_1(x)X, \beta_2(x)X, \cdots, \beta_K(x)X]\mathbf{C} = [\beta(x) \otimes X]\mathbf{C}, \tag{10}$$

where $\otimes$ represents the Kroenecker product,

$$\beta(x) = [\beta_1(x), \beta_2(x), \cdots, \beta_K(x)]$$

is a column vector of the coefficients of the kth rule, and

$$\mathbf{C} = \begin{bmatrix} \mathbf{c}^1 \\ \vdots \\ \mathbf{c}^K \end{bmatrix}$$

is the vector of the unknown coefficients of the consequent polynomials of the K rules.

Let (x^j, y^j) be the jth training input/output pair out of a total of J pairs, with $x^j \in \Re^n$ and $y^j \in \Re$. The training data can be summarized in the J equations

$$\begin{bmatrix} \beta\left(x^1\right) \otimes X^1 \\ \beta\left(x^2\right) \otimes X^2 \\ \cdots \\ \beta\left(x^J\right) \otimes X^J \end{bmatrix} \mathbf{C} = \mathbf{ZC} = \mathbf{Y} = \begin{bmatrix} y^1 \\ \vdots \\ y^J \end{bmatrix} \tag{11}$$

where $\mathbf{Z}$ is a $J \times K(n+1)$ matrix, X^j is as defined in Eq. (9) but for the jth input vector, and $\beta\left(x^j\right)$ represents the truth values of the rules evaluated at the vector x^j.

Equation (11) is a usually an overdetermined system of linear equations. A least-squares solution can be found by minimizing the error index $\mathbf{J} = \sum_{j=1}^{J}(y^j - \hat{y}^j)^2$ where $\hat{y}^j$ is the output of the Sugeno controller corresponding to the jth input sample x^j. The least-squares solution is given by $\mathbf{C} = \mathbf{Z}^{\dagger}\mathbf{Y}$ where $\mathbf{Z}^{\dagger} = \left(\mathbf{Z}^T\mathbf{Z}\right)^{-1}\mathbf{Z}^T$ is the pseudo-inverse of $\mathbf{Z}$. This solution is called the **batch** or **global** solution. This solution can be implemented recursively as

$$\mathbf{C}^{(j+1)} = \mathbf{C}^{(j)} + \mathbf{S}^{(j+1)} \cdot z_{j+1}^T \cdot \left(y^{j+1} - z_{j+1} \cdot \mathbf{C}^{(j)}\right), \tag{12}$$

$$\mathbf{S}^{(j+1)} = \mathbf{S}^{(j)} - \frac{\mathbf{S}^{(j)} \cdot z_{j+1}^T \cdot z_{j+1} \cdot \mathbf{S}^{(j)}}{1 + z_{j+1} \cdot \mathbf{S}^{(j)} \cdot z_{j+1}^T}, \tag{13}$$

where z_j is the jth row vector of matrix $\mathbf{Z}$ defined in (11), $\mathbf{S}^{(j)}$ is a square $n(k+1) \times n(k+1)$ covariance matrix at the jth iteration (i.e., after the jth training pair has been acquired and used), and $\mathbf{C}^{(j)}$ the corresponding coefficient vector. Then the global solution is $\mathbf{C}^{(J)}$ at the final iteration. The initial estimates, $\mathbf{C}^{(0)}$ and $\mathbf{S}^{(0)}$, are usually chosen as $\mathbf{C}^{(0)} = 0$ and $\mathbf{S}^{(0)} = \alpha I$ where α is a large number and I is the identity matrix [14].

3 Local and Differentiable MFs

In Section 1, we concluded that local MFs with little overlap exhibit high interpretability. The triangular MFs with only two functions overlapping at a time seems to be a good choice. Unfortunately, the derivatives of a triangular function are discontinuous at its boundaries and center.

In [3], MFs which are both local in support and sufficiently differentiable were advocated, and they were based on spline functions. The use of spline functions for the description of MFs is not new [18, 2,12, 9], but not in the same way or for the purpose of improving interpretability. For example, Shimojima et al. [12] used a training algorithm to come up with "optimal"

MFs which are expressed as a combination of splines. Lane et al. [9] used cubic splines as squashing functions in neural networks, but the issue of interpretability was not discussed there. Often, when cubic splines are used, three of them are allowed to overlap at a time for every input variable, which is usual in the literature of cubic splines, but is detrimental to the possibility of high interpretability.

3.1 Local Differentiable MFs

Consider a scalar variable $x \in D \subset \Re$ having M attributes $a^1, a^2, \cdots, a^M$. The MF of the mth attribute is denoted $\mu^m(x) = \mu(x \text{ is } a^m)$. The triangular MF is an example of a local MF. Its locality properties can be generalized as follows:

Definition 1. A MF $q(x; L, g, U)$ is a local MF of order ν, centered at g, and of support $[L, U]$, where $L \leq g \leq U$, if (a) $q(x; L, g, U) > 0$ for $L < x < U$ and is zero otherwise; (b) $q(x; L, g, U)$ is monotone increasing over $[L, g]$ and monotone decreasing over $[g, U]$; (c) $q(g; L, g, U) = 1$; and (d) the νth derivative of $q(x; L, g, U)$ with respect to x is continuous everywhere; that is, $\forall x \in D \subset \Re$.

Note that the νth derivative and lower order derivatives of these MFs must vanish at the boundaries L and U. Moreover, it is often desired that these derivatives vanish at the centers as well.

Order-1 Local MFs. An order-1 local MF centered at 0 and of support $[-h_l, h_r]$ is given by a nonsymmetric spline:

$$Q_3(x; -h_l, 0, h_r) = \begin{cases} 1 - 3x^2/h_r^2 + 2x^3/h_r^3, & \text{for } 0 \leq x \leq h_r, \\ 1 - 3x^2/h_l^2 - 2x^3/h_l^3, & \text{for } -h_l \leq x \leq 0, \\ 0 & \text{otherwise,} \end{cases} \tag{14}$$

$$Q_3\left(x; L, g, U\right) = Q_3\left(x - g; L - g, 0, U - g\right),$$

shown in Figure 2. This function is differentiable everywhere and is twice differentiable everywhere except at its boundaries. The first derivative

$$\frac{d}{dx} Q_3(x; -h_l, 0, h_r) = \begin{cases} -6x/h_r^2 + 6x^2/h_r^3, & \text{for } 0 \leq x \leq h_r, \\ -6x/h_l^2 - 6x^2/h_l^3, & \text{for } -h_l \leq x \leq 0, \\ 0 & \text{otherwise,} \end{cases} \tag{15}$$

$$\frac{d}{dx} Q_3\left(x; L, g, U\right) = \frac{d}{dx} Q_3\left(x - g; L - g, 0, U - g\right),$$

is continuous everywhere including at the boundaries.

Higher-order Local MFs. To construct a local MF $q(x; -1, 0, 1)$ which is ν-times differentiable, one can proceed as follows. The function is first

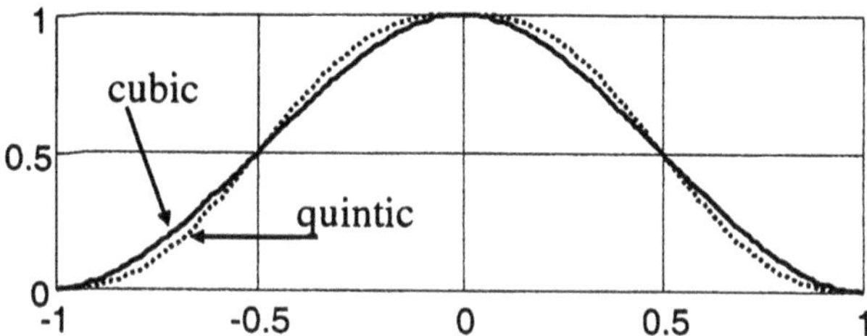

Fig. 2. (a) Cubic and quintic splines over [-1,1] are differentiable everywhere including the points $x = \pm 1$. The quintic spline is twice differentiable everywhere. It is "flatter" at the center and boundaries.

constructed for $0 \leq x \leq 1$. Denote this function by $p(x)$, $0 \leq x \leq 1$, and let it be a polynomial satisfying the boundary conditions

$$\begin{aligned} p(0) &= 1, \quad p(1) = 0, \\ p'(0) &= p''(0) = \cdots p^{(\nu)}(0) = 0, \\ p'(1) &= p''(1) = \cdots p^{(\nu)}(1) = 0. \end{aligned} \tag{16}$$

Since there are $2\nu + 2$ conditions, the polynomial must be of order $2\nu + 1$. To construct $q(x; -1, 0, 1)$ over the interval $-1 \leq x \leq 0$, we only need to use $1 - p(x+1)$. Applying the above procedure for $\nu = 2$ yields the polynomial $p(x) = 1 - 10x^3 + 15x^4 - 6x^5$ and the quintic "mother" function

$$Q_5(x; -1, 0, 1) = \begin{cases} 1 - 10x^3 + 15x^4 - 6x^5, \text{ for } 0 \leq x \leq 1, \\ 1 + 10x^3 + 15x^4 + 6x^5 \quad \text{for } -1 \leq x \leq 0, \\ 0 \qquad\qquad\qquad\qquad \text{otherwise;} \end{cases} \tag{17}$$

shown in Figure 2. Nonsymmetrical quintic splines can be constructed by substituting x with x/h_l for $-h_l \leq x \leq 0$ and with x/h_r for $0 \leq x \leq h_r$.

3.2 Orderly Bases for Scalar Inputs

A set of local MFs can be assembled to effectively cover an input x over the interval $\left[g^0, g^{M+1}\right]$ using as parameters only the points $g^0, g^1, \cdots, g^M, g^{M+1}$, not necessarily equally spaced, ordered in ascending order, and henceforth called the grid of x. We define the grid

$$G[x] = [g^0, g^1, \cdots, g^M, g^{M+1}]. \tag{18}$$

Definition 2. The functions $\mu^m(x) : \Re \to [0,1]$ for $m = 1, 2, \cdots, M$ form an orderly local MF (OLMF) basis of order ν for $x \in D \subset \Re$ over the grid $G[x]$ if every MF $\mu^m(x)$ can be generated from a "mother" MF $q(x;\ L,\ g,\ U)$, which is local and of order-ν, according to the formula

$$\mu^m(x) = q(x; g^{m-1}, g^m, g^{m+1}), \qquad m = 1, \cdots, M. \tag{19}$$

Note that the adjacent MFs $\mu^{m-1}(x)$ and $\mu^m(x)$ totally overlap in the sense that the center of μ^m is the upper bound of the support of μ^{m-1}. Moreover

$$\mu^m\left(g^i\right) = \delta^i_m, \qquad \text{and} \qquad \left.\frac{\partial \mu^m(x)}{\partial x}\right|_{g^i} = 0$$

at any g^i, where δ^i_m is the Kroenecker delta function ($\delta^i_m = 1$ when $i = m$ and zero otherwise).

The spline functions in Eqs. (14) and (17) can be used as "mother" functions to generate OLMF bases that possess the addition-to-unity property [3].

Consider now an arbitrary value η of x for which we would like to determine the relevant rules (those with nonzero truth value.) Introduce the set of indices

$$\eta \Cap G[x] = \{m \text{ such that } \mu^m(\eta) \neq 0\}, \qquad \text{for} \qquad \eta \in D \subset \Re \tag{21}$$

which has at most 2 elements: $\eta \Cap G[x] = \{m, m+1\}$ when $g^m < \eta < g^{m+1}$, and $\eta \Cap G[x] = \{m\}$ when $\eta = g^m$. We will assume that the elements of this set are ordered in ascending order.

3.3 For Many Inputs

When the input to the inference engine is multi-dimensional, $x \in D \subset \Re^n$, one can simply define an OLMF basis for each component separately as before, and then combine them using an outer product operation. For instance, assume that x_i has M_i attributes a_i^1, a_i^2, $\cdots$, $a_i^{M_i}$ and the OLMF basis is defined using the list of centers $G[x_i] = \left[g_i^0, g_i^1, \cdots, g_i^{M_i}, g_i^{M_i+1}\right]$ where g_i^j represents the jth center for x_i. The MF of x_i in its mth attribute is denoted $\mu_i^m(x) = \mu(x_i \text{ is } a_i^m)$. The $n-$dimensional grid is the outer product of the individual grids

$$G[x] = G[x_1] \times G[x_2] \times \cdots \times G[x_n], \tag{22}$$

and can be thought of as an $n-$dimensional array.

We can generalize the notation in Eq. (21) to define the set of indices of all rules adjacent to $x = \eta$. Namely,

$$\eta \Cap G[x] = \eta \Cap G[x_n] \times \eta \Cap G[x_{n-1}] \times \cdots \times \eta \Cap G[x_1], \tag{23}$$

which will contain 2^n n–tuples. When η does not fall on any grid line (hyperplane in general), we will have

$$\eta \Cap G[x] = \{m_n, m_n + 1\} \times \{m_{n-1}, m_{n-1} + 1\} \times \cdots \times \{m_1, m_1 + 1\}. \tag{24}$$

Note that $\times$ is not commutative. Then a given rule is represented by the $n-$tuple $(m_n, m_{n-1}, \cdots, m_2, m_1)$. The order shown indicates our preference

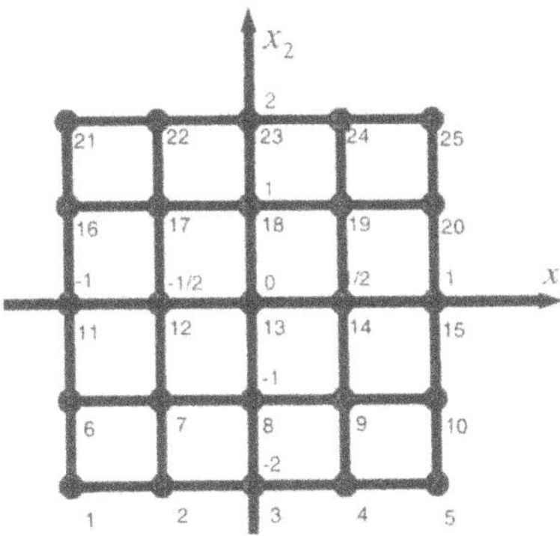

Fig. 3. The centers of 25 rules arranged in a 5 × 5 grid.

to make the index of the attribute of x_1 the "least significant" one. This means that once the attributes of x_2, x_3, etc... have been "fixed", we "cycle" through all the M^1 attributes of x_1 thus resulting in M^1 adjacent rules. The corresponding index k of the rule is then given by the formula

$$k\left(m_n, m_{n-1}, \cdots, m_2, m_1\right) = 1 + \sum_{i=1}^{n}\left(m_i - 1\right)\prod_{j=0}^{i-1} M^j \tag{25}$$

with $M^0 = 1$ and M^i is the number of attributes or linguistic values of x_i. With a slight abuse of notation we can rewrite Eq. (24) using Eq. (25) as $\eta \Cap G[x] = \{k_1, k_2, \cdots\}$ where the k's refer to the rule indices.

Example 1. If $G[x_2] = [-2,-1,0,1,2]$ and $G[x_1] = \left[-1,-\frac{1}{2},0,\frac{1}{2},1\right]$, as shown in Figure 3, then for $\eta = \left[0.4,\ -0.5\right]^T$, we have $-0.5 \Cap G\left[x_2\right] = \{2,3\}$ and $0.4 \Cap G\left[x_1\right] = \{3,4\}$ and hence

$$\eta \Cap G[x] = \{2,3\} \times \{3,4\} = \{(2,3),(2,4),(3,3),(3,4)\} \tag{26}$$

With a slight abuse of notation we can write $\eta \Cap G[x] = \{8,9,13,14\}$ referring to the rule indices. ■

Definition 3. The center of the kth rule is a vector of the centers of the attributes of all x_i tested by the kth rule, or

$$r^k = [g_1^{A_1^k}, g_2^{A_2^k}, \cdots, g_n^{A_n^k}] \tag{27}$$

Notation 1 *We denote $\beta_k\left(x\right)$, the normalized truth of the kth rule, as $\beta\left(x; r^k\right)$ where r^k is the center of that rule.*

Theorem 1. *The OLMF bases generated using cubic and quintic splines satisfy the addition-to-unity property in Eq. (8). Hence* $\beta_k(x) = \rho_k(x)$.

Proof. It can be checked by direct substitution that $Q_3(x; -1, 0, 1) + Q_3(x; 0, 1, 2) = 1$ for $0 \le x \le 1$. This can be used repeatedly to show that Q_3 generates an OLMF basis having the addition-to-unity property for $x \in \Re$. Using mathematical induction over the number of inputs, the proof for $x \in \Re^n$ can be established. The proof for the quintic spline is very similar. ■

From this point forward, we will assume an OLMF basis of order ν generated from splines and hence possessing the addition-to-unity property.

Using mathematical induction, we can show that

$$\beta_k(x)|_{r^l} = \delta_l^k \quad \text{and} \quad \left.\frac{\partial \beta_k(x)}{\partial x}\right|_{r^l} = 0 \tag{28}$$

for all l and k. Moreover $\beta_k(x) = 0$ outside the polytope connecting the adjacent centers.

4 The Interpretable Sugeno Approximator

4.1 Rule-Centered Consequent Polynomials

The FRBS introduced in [3] is a Sugeno approximator that has two main characteristics: (1) On the antecedent side, the MFs form an OLMF basis of degree ν; and (2) on the consequent side, rule-centered consequent polynomials of degree ν are used. This FRBS shall be called the Interpretable Sugeno Approximator (ISA). A typical rule has the form

$$R^k: \text{IF } x_1 \text{ is } A_1^k \text{ AND } x_2 \text{ is } A_2^k \text{ AND } \cdots \text{ AND } x_n \text{ is } A_n^k \tag{29}$$

$$\text{THEN } y^k = b_0^k + b_1^k(x_1 - r_1^k) + \cdots + b_n^k(x_n - r_n^k), \tag{30}$$

where the consequent polynomial is clearly centered at r^k. Comparing Eq. (??) with Eq. (3), we must have $c_i^k = b_i^k$ for $i > 0$ and

$$c_0^k = b_0^k - \left(b_1^k r_1^k + \cdots + b_n^k r_n^k\right). \tag{31}$$

Quadratic and higher-order consequent polynomials can also be written in the rule-centered form:

$$Y_k(x) = b^{k0} + b^k(x - r^k) + \frac{1}{2}\left(x - r^k\right)^T B^k \left(x - r^k\right) + \cdots. \tag{32}$$

4.2 Interpretation of the Consequent Polynomials

The main result related to the ISA is the following [3]:

Theorem 2. *Suppose that $\hat{y}(x)$ is a Sugeno approximator such that*

1. *the antecedent-side of the Sugeno-type fuzzy inference system uses an OLMF basis of order ν_a (the ν_a derivative is everywhere continuous);*
2. *the consequent-side is written in the rule-centered form shown in Eq. (32) and the polynomials $Y_k(x)$ are of degree ν_c;*

Then for $\nu_c \leq \nu_a$, every $Y_k(x)$ can be interpreted as a truncated Taylor series expansion of order ν_c of $\hat{y}(x)$ about the center $x = r^k$ of the kth rule.

Proof. Let us consider the lth rule centered at $x = r^l$, whose consequent polynomial is $Y_l(x)$. Clearly $Y_l(r^l) = b_0^l$ and the truth value of that rule is $\beta_l(r^l) = 1$ from Eq. (28). From Eq. (28), the truth value of every other rule $\beta_l(r^k) = 0$. Hence, the output of the ISA is simply $\hat{y}(r^l) = b_0^l$, the constant term in $Y_l(x)$. Similarly, the partial derivative of $\hat{y}$ with respect to x_p as evaluated at $x = r^l$ is the coefficient of $(x_i - r_i^l)$ in $Y_l(x)$. To show this, we note that

$$\frac{\partial \hat{y}}{\partial x_p} = \sum_k \left(\frac{\partial \beta_k}{\partial x_p} Y_k(x) + \beta_k(x) \frac{\partial Y_k}{\partial x_p} \right) \tag{33}$$

where

$$\frac{\partial}{\partial x_p} Y_k(x) = b_p^k + \sum_j B_{pj}^k (x_j - r_j^k) + \ldots \tag{34}$$

Using Eqs. (28) in Eq. (33) at $x = r^l$ yields

$$\left. \frac{\partial \hat{y}}{\partial x_p} \right|_{x=r^l} = \left. \frac{\partial Y_l}{\partial x_p} \right|_{x=r^l} = b_p^l = p^{th} \text{ component of } b^l \text{ in Eq. (32)} \tag{35}$$

The coefficients of higher-order terms can be similarly shown to be higher-order derivatives. It is important to note that if $\nu_c > \nu_a$, then the polynomial terms of order $\nu > \nu_a$ cannot be interpreted as Taylor series terms, but those of order $\nu \leq \nu_a$ maintain the interpretability in terms of Taylor series. ■

If the MFs were not differentiable at their boundary (as when triangular or trapezoidal functions are used), the term $\frac{\partial \beta_k}{\partial x_p}$ would not vanish when needed, and besides, the derivative $\frac{\partial \hat{y}}{\partial x_p}$ does not exist there. This is why the differentiability condition on the OLMF basis is needed.

Example 2. Approximate the function

$$y(x) = x_1^2 + 3x_2^2 + \cos \frac{\pi x_1 x_2}{2} \tag{36}$$

by an ISA $\hat{y}(x)$ defined over the grid in Figure 3. The general form of the consequent polynomial for the rule centered at $x = \eta$ is

$$Y(x;\eta) = y(\eta) + \left(2\eta_1 - \frac{\pi\eta_2}{2}\sin\frac{\pi\eta_1\eta_2}{2}\right)(x_1 - \eta_1) + \left(6\eta_2 - \frac{\pi\eta_1}{2}\sin\frac{\pi\eta_1\eta_2}{2}\right)(x_2 - \eta_2) \quad (37)$$

for every η in the grid $G[x_1] \times G[x_2]$. ■

Theorem 3. *The ISA of the function $y(x) + \alpha\beta_L(x)$, where $\beta_L(x)$ is the normalized truth value of the Lth rule, is the same as the ISA of $y(x)$ except that the constant term coefficient of the Lth rule is incremented by α.*

Theorem 4. *The ISA of the function $\alpha + \beta y(x)$ can be obtained from that of $y(x)$ by multiplying every consequent coefficient by β then adding α to the constant term coefficient in every rule.*

Proofs. The above two theorems follow by examining the Taylor series of $y(x) + \alpha\beta_L(x)$ and $\alpha + \beta y(x)$ about the rule centers. ■

4.3 Advantages of the ISA

Unlike neural networks and similar FRBS, the ISA form maintains its interpretability even when the input and output spaces are of high dimensions.

This *interpretability derives from that of derivatives* as was illustrated in [3]. Derivatives often have a strong physical interpretation (such as sensitivity derivatives of the aerodynamics of airfoils), and are often easier to compute than the function itself, both numerically and analytically. The interpretability of the ISA offers *meaningful initialization of the consequent coefficients* when using learning or other computationally intensive algorithms. For example: If the function values are given at a grid, finite differences can be used to approximate the derivatives and can be used as initial values to be input to the training algorithms.

Conversely, if only noisy samples of the function-to-be-approximated are available, the coefficients of a trained ISA may represent a robust approximation of the Taylor series expansion at the rule centers. This is therefore a *novel approach to computing Taylor series expansions from noisy data.*

Another major advantage is the resulting *simplification of training and testing.* For example, if values of the map and its partial derivatives are available at the centers of the rules, no training would be necessary. Another advantage is that *training and testing can be conducted locally,* meaning with fewer rules at a time. Hence a training example will only affect the rules which are immediately relevant.

There are also *computational advantages.* For example, *the evaluation of the rule base is sparse.* This really follows from the locality property of the OLMF basis and does not depend on the form of the consequent, and therefore, it is true for many usual fuzzy systems with triangular and trapezoidal MFs. Essentially for n inputs each with m attributes, there are in general m^n rules to be evaluated. But if an OLMF basis is used, only 2^n rules have nonzero truth values and therefore require evaluation. The rules can be arranged systematically and which rules are to be applied can be determined very efficiently thus reducing the computation time. For $n = 4$ and $m = 10$, this results in a savings factor of 625 which is quite significant.

Another computational advantage is possibly that *fewer rules are needed* when higher-order Taylor series expansions are used. This may complicate the design and learning phases but may significantly reduce the evaluation computations.

Experience has shown [7] that the accuracy of the approximation is not sensitive to the shape of the MFs as long as the MFs are qualitatively similar. The accuracy is much more sensitive to the density of the MFs. Hence the accuracy of the ISA should be generally similar to other TSK approximators with local and unimodal MFs of similar density. Similarly, the difference between the cubic and quintic membership functions is not likely to impact the accuracy much. The simulations conducted in the course of this work also point to the same conclusion.

In principle, the constraints imposed on the shapes of the MFs are likely to reduce the accuracy of the approximation but rather marginally. The decrease in accuracy is easily offset by the advantages of the ISA, such as its ease of training. These advantages will encourage the use of ISAs with denser grids, hence leading to significantly better accuracy. *Hence, for the same computational effort, one can in practice obtain ISAs having a superior accuracy as well as a superior interpretability.* This assertion runs counter to the Principle of Incompatability, but agrees with the Modified Principle of Incompatability. See Section 5.3 for some numerical evidence.

A major computational and conceptual advantage, which we will explore in the next section, is that *basic operations on fuzzy inference systems can now be conducted in a well understood fashion without excessive computation or training.*

5 Basic Operations on the ISA

The ultimate test of interpretability is the ability to use the interpretation meaningfully and efficiently. In the context of Sugeno approximators, this is best illustrated by the ability to obtain new approximators from existing ones *without training.* The interpretation of the consequent polynomials as Taylor series allows that very well, which has an important impact on the modularity, functionality, and scalability of FRBSs.

This is true, for instance, for the operations shown in Figure 4. Illustrated in Figure 4a is the problem of initializing the coefficients of an ISA on a fine

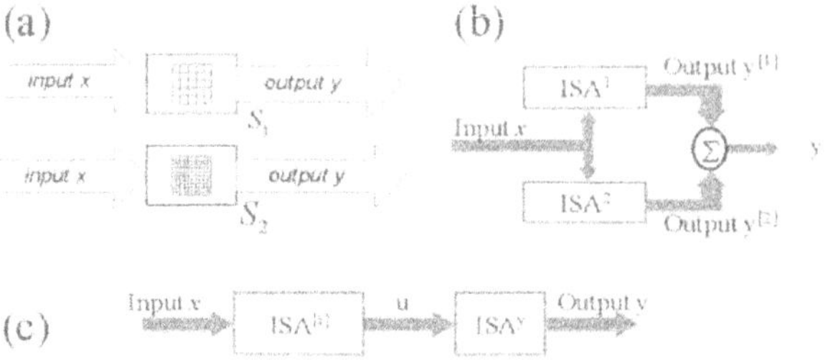

Fig. 4. Some operations on ISAs. (a) $\mathcal{S}_1$ is an ISA defined on a coarse grid. $\mathcal{S}_2$ is an equivalent ISA defined on a finer grid. (b) A parallel combination of ISAs. (c) A series combination.

input grid when an ISA defined on a coarser grid has already been found; or vice versa. This problem is solved efficiently using the Taylor series interpretation in Section 5.1. Another problem is to replace a linear (parallel) combination of ISAs as shown in Figure 4b by a single ISA; again without training. This problem is solved in Section 5.4. The problem of replacing a series combination shown in Figure 4c by a single ISA without training is solved in Section 5.6.

In short, effective design, manipulation, testing, training, initialization, and combination of ISAs (without training) is now enabled by the Taylor series interpretation. This will be amply illustrated in the sections that follow. The foundation of a calculus of fuzzy inference systems therefore emerges.

5.1 Resampling the ISA

Let $\mathcal{S}_1$ be a multi-input single-output ISA given on a nominal grid $G_{[1]}[x] = [r^1, r^2, \cdots, r^K]$. In the following, we discuss a general procedure to produce directly and without learning another ISA $\mathcal{S}_2$, defined on a different but given grid $G_{[2]}[x] = [q^1, q^2, \cdots]$, that approximates $\mathcal{S}_1$. To distinguish quantities belonging to $\mathcal{S}_i$ we will use subscripts or superscripts of the form of $[i]$.

The problem is solved once the consequent polynomial of a generic rule centered at $\eta \in G_{[2]}[x]$ is found. If $\eta \in G_{[1]}[x]$ as well, then the corresponding consequent polynomial is inherited. In general however, $\eta \notin G_{[1]}[x]$ but only few centers in $G_{[1]}[x]$ will be adjacent to η. Their indices are denoted by $\eta \Cap G_{[1]}[x] = \{k_1, k_2, \cdots\}$.

First, we match the value of $\mathcal{S}_2$ to that of $\mathcal{S}_1$ at η, a rule center in $\mathcal{S}_2$:

$$y^{[2]}(\eta) = y^{[1]}(\eta) = \sum_k \beta_k^{[1]}(\eta) Y_k^{[1]}(\eta) = \sum_{k \in \eta \Cap G_{[1]}} \beta_k(\eta) Y_k(\eta) \Big|^{[1]} \tag{38}$$

which then gives the constant term of the consequent polynomial of $\mathcal{S}_2$ centered at η. Next we match the first derivatives of $\mathcal{S}_1$ and $\mathcal{S}_2$ at the same rule center to obtain the coefficient of $(x_p - \eta_p)$ in the same rule

$$\left.\frac{\partial y^{[2]}}{\partial x_p}\right|_{x=\eta} = \left.\frac{\partial y^{[1]}}{\partial x_p}\right|_{x=\eta} = \sum_{k \in \eta \Cap G_{[1]}} \left.\left(\frac{\partial \beta_k}{\partial x_p} Y_k + \beta_k \frac{\partial Y_k}{\partial x_p} \right)\right|_{x=\eta}^{[1]} \tag{39}$$

Clearly we need to evaluate $\frac{\partial \beta_k}{\partial x_p}$ and $\frac{\partial Y_k}{\partial x_p}$ at $x = \eta$. (The first needs Eq. (15). The second is straightforward). Coefficients of higher-order terms can be determined through a similar development.

5.2 Local Learning

When the number of inputs or the number of MFs is increased, the learning time increases exponentially. Often the learning algorithm terminates due to a memory overload. This is because the learning algorithm outlined in Section 2 *updates all the Sugeno coefficients for every sample.*

One can modify the learning procedure to take advantage of the local support of the MFs. Essentially, the iteration is conducted over the rule centers instead of iterating over the samples. In other words, at the kth iteration, one isolates all the samples $\{j_1, \cdots, j_k\}$ falling within the polytope centered at the rule center r^k of the kth rule. Hence for the kth iteration, only the consequent polynomial coefficients c^k of the kth rule constitute the unknowns. Clearly, c^k must satisfy the training conditions

$$\begin{bmatrix} \beta\left(x^{j_1}\right) \otimes X^{j_1} \\ \cdots \\ \beta\left(x^{j_k}\right) \otimes X^{j_k} \end{bmatrix} c^k = \begin{bmatrix} y^{j_1} \\ \vdots \\ y^{j_k} \end{bmatrix} \tag{40}$$

which can be written as $\tilde{Z}^k c^k = \tilde{y}^k$. The system is usually overdetermined and can be solved in a least-squares sense using the pseudo-inverse of $\tilde{Z}^k$. We will call this "local learning" since only local samples are used thus resulting in reduced matrix dimensions.

If the recursive least-squares iteration is used on this local system, a local covariance matrix relating the coefficients of one rule results. Gathering the covariance matrices for all the rules *will not* yield the global covariance matrix of Section 2. This is because the blocks off the diagonal of the global covariance matrix cannot be constructed from the local covariance matrices. Hence, local learning is less accurate than global learning.

Table 1 shows the training time obtained as a result of applying local and global learning algorithms to learn simple functions. For 4 inputs the global learning algorithm could no longer be used whereas the local learning performed well.

Number of inputs	MFs per Input	Global Learning Time (sec)	Local Learning Time (sec)
1	5	7.05	1.6
2	5	49.5	9.1
3	5	1215	51.0
4	5	Memory Error	268.0

Table 1. A comparison between global and local learning.

Because of the dramatic improvement in learning time, local learning can use more training epochs for the same amount of computations. As a result, the accuracy of the local learning algorithms is often quite close to that of global learning, and may sometimes be even superior. More research in this direction is needed, and the next section provides some numerical evidence.

5.3 Learning with Resampling

A dramatic improvement in learning speed can be obtained using the resampling procedure in Section 5.1. A typical learning mechanism requires performing a number of learning epochs in order to improve the approximation quality. This could be computationally expensive if the grid is fine. The alternative is to start with a coarse grid, learn, then resample to a finer grid at the next learning epoch. This is applicable whether local or global learning is used.

This approach was tested on a simple function, $y(x) = \sqrt{2}\cos(x + \pi/4)$, and the results are summarized in Figure 5. First, learning was conducted on a fixed grid consisting of 35 MFs passing through 10 epochs of learning. After each epoch there is a slight improvement in the quality of the approximation (the indicator was the RMS error) until the final epochs where the improvement slows down and the error converges.

Next, the same function is learned using a coarse grid consisting of 5 MFs. As theoretically predicted, the RMS error is much higher than that of the fine grid. Resampling to 15 MFs was applied which increased the RMS error even more. Another epoch of learning was applied through which there was a dramatic improvement of the RMS error. Alternating between learning and resampling was continued until the fine grid of 35 MFs was reached. Figure 5 reveals that after each resampling there is a slight increase in the RMS error but the learning that follows is characterized by a dramatic decrease in the error. At the fourth learning epoch the RMS error reached a value that learning on a fine grid could not decay to even after its 10 epochs of learning.

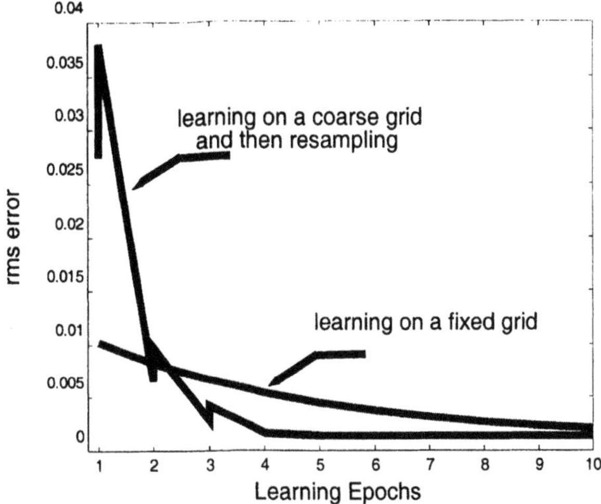

Fig. 5. Plot of approximation error vs. training epoch. When one alternates learning with refining the grid, a better approximation is obtained much faster than when learning on a fixed fine grid.

5.4 Linear (Parallel) Combination of ISAs

Let us assume that there are two functions $y^{[1]}(x)$ and $y^{[2]}(x)$ whose linear combination $y(x) = \alpha_1 y^{[1]}(x) + \alpha_2 y^{[2]}(x)$ is to be approximated by a single ISA. Without loss of generality, we assume that all ISAs are defined on the same grid. (If not, the grids can be resampled.) In this case, the consequent polynomial $Z(x;\eta) = \alpha_1 Y^{[1]}(x;\eta) + \alpha_2 Y^{[2]}(x;\eta)$ determines the ISA approximation of $z(x)$. This is true even if $Y^{[i]}(x;\eta)$ contains quadratic and higher-order terms.

Example 3. Find the ISA approximating the linear combination $z(x) = y^{[1]}(x) + \alpha y^{[2]}(x)$ where $y^{[1]}(x)$ is the same as $y(x)$ shown in Eq. (36), and

$$y^{[2]}(x) = \frac{3.5}{1 + x_1^2 + x_2^2} \tag{41}$$

Solution. Assume differentiable approximations based on the same x-grid and an OLMF basis of order 1. Then $Y^{[1]}(x;\eta)$ is shown in Eq. (37). Moreover,

$$Y^{[2]}(x;\eta) = \frac{3.5}{1 + \eta_1^2 + \eta_2^2} - \frac{7\eta_1(x_1 - \eta_1) + 7\eta_2(x_2 - \eta_2)}{(1 + \eta_1^2 + \eta_2^2)^2} \tag{42}$$

For $\eta = [1, 1]^T$, the consequent polynomial of $z(x)$ is:

$$Z(x;\eta) = \left(4 + \frac{3.5\alpha}{3}\right) + \left(2 - \frac{\pi}{2} - \frac{7}{9}\alpha\right)(x_1 - 1) + \left(6 - \frac{\pi}{2} - \frac{7}{9}\alpha\right)(x_2 - 1) \tag{43}$$

■

5.5 Differentiation of ISAs

Let $y(x)$ be approximated by ISA^y where $x \in \Re^n$ and $y \in \Re$. We seek to find a multi-output ISA to approximate $u(x) = \frac{\partial y}{\partial x}$, the matrix of partial derivatives or the Jacobian of the function $y(x)$. Clearly, we need ISA^y to be order 1 or higher; otherwise, ISA^y would not be differentiable. For simplicity and without loss of generality, we assume that all ISAs are based on the same $x-$grid and hence all ISAs have the same antecedent forms. Let ISA$^{[i]}$ be the ISA mapping $x \in \Re^n$ to $u_i = \frac{\partial y}{\partial x_i} \in \Re$, whose consequent polynomial has the form

$$U_k^{[i]}(x) = b_i^k + \sum_j \left(x_j - r_j^k\right) B_{ij}^k + \cdots \tag{44}$$

where B_{ij}^k is the $(i,j)^{th}$ element of the symmetric matrix B^k. The constant term of every rule of ISA^u will be dropped in the differentiation. The coefficient of the linear term $\left(x_i - r_i^k\right)$ of the kth rule of ISA^u becomes the constant term of the kth rule of ISA$^{[i]}$. The coefficient of the $\left(x_i - r_i^k\right)\left(x_j - r_j^k\right)$ term of the kth rule of ISA^u becomes the coefficient of the $\left(x_j - r_j^k\right)$ term of the kth rule of ISA$^{[i]}$ (and by symmetry, the coefficient of the $\left(x_i - r_i^k\right)$ term of the kth rule of ISA$^{[j]}$).

Example 4. Let $y(x)$ be that given in Example 2. Using the same $x-$grid, the consequent polynomials of the derivatives $u_1 = \frac{\partial y}{\partial x_1} \in \Re$ and $u_2 = \frac{\partial y}{\partial x_2} \in \Re$ are

$$U^{[1]}(x;\eta) = 2\eta_1 - \frac{\pi\eta_2}{2}\sin\frac{\pi\eta_1\eta_2}{2} \tag{45}$$

$$U^{[2]}(x;\eta) = 6\eta_2 - \frac{\pi\eta_1}{2}\sin\frac{\pi\eta_1\eta_2}{2} \tag{46}$$

and they are simply constant. This is because $Y(x;\eta)$ was determined up to linear terms only. ■

Integration of ISAs can be performed by assigning the coefficients of the constant terms of $U^{[i]}(x;\eta)$ to the coefficients of the linear terms of $Y(x;\eta)$. The constant terms in $Y(x;\eta)$ represent "constants of integration". Theorem 3 can be used to assign the coefficient of the constant terms by matching the desired values of $Y(x;\eta)$ at every grid point.

Example 5. If $U^{[i]}(x;\eta)$ in Eqs. (45,46) approximate the derivatives $u_i = \frac{\partial y}{\partial x_i}$, find an ISA approximating $y(x)$.

Solution. Clearly ISA^y described in Eq. (37), which approximates Eq. (36), would be a valid integral or anti-derivative. This particular anti-derivative will have the values of $y(\eta)$ in Eq. (36) for η being one of the 25 grid points of $G[x]$ shown in Figure 3. For example, at $\eta = r^{13} = [0,0]^T$, we have $y\left(r^{13}\right) = y\left(0\right) = 2$. If $y^{new}(r^{13}) = 6$ is desired, the constant term of the 13th rule must be 6 instead of 2 because $\beta_{13}(r^{13}) = 1$. The resulting ISA would then approximate the function $y^{new}(x) = y(x) + 4\beta_{13}(x)$. See Theorem 3. ■

5.6 Composition (Series Combination) of ISAs

The composition of two mappings, $y(u)$ and $u(x)$, represented by ISAs is next considered. Let $x \in \Re^n$, $u \in \Re^m$, and, for simplicity, $y \in \Re$. Can we by simple operations form the ISA of $z\left(x\right) = y(u\left(x\right))$ from the two given ISAs?

Let ISA$^{[i]}$ represent the ISA mapping $x \in \Re^n$ to u_1, and ISA^y be that mapping $u \in \Re^m$ to $y \in \Re$. The sought ISA is denoted ISA^z and it maps $x \in \Re^n$ to $z \in \Re$. Without loss of generality, we assume that ISA$^{[1]}$, ... ISA$^{[m]}$ as well as ISA^z are all defined on the same x−grid.

The problem is solved once a procedure to express the consequent polynomial $Z\left(x;\eta\right)$ at a typical rule center $\eta \in G\left[x\right]$ is obtained. The constant term of $Z\left(x;\eta\right)$ is simply $y\left(u\left(\eta\right)\right)$, and is determined using Eq. (38) m times (for ISA$^{[1]}$, ..., ISA$^{[m]}$) to find $u = \left[u_1, \cdots, u_m\right]^T$ and once to find y (using ISA^y) from u.

To compute the coefficients of the linear terms in $Z\left(x;\eta\right)$, we use the chain rule of differentiation. For instance, the coefficient of the $(x_j - \eta_j)$ term in the rule centered at $x = \eta$ is given by

$$\left.\frac{\partial z}{\partial x_j}\right|_{x=\eta} = \sum_l \left(\left.\frac{\partial y}{\partial u_l}\right|_{u=u(\eta)} \cdot \left.\frac{\partial u_l}{\partial x_j}\right|_{x=\eta} \right). \tag{47}$$

The term $\left.\frac{\partial u_l}{\partial x_j}\right|_{x=\eta}$ is simply a consequent coefficient in ISA$^{[l]}$ and needs no computation. The term $\left.\frac{\partial y}{\partial u_l}\right|_{u=u(\eta)}$ requires (once $u\left(\eta\right)$ is computed) the computation of derivatives of ISA^y from Eq. (39) which is not very difficult.

For higher–order coefficients, the chain rule is used repeatedly, which is straightforward but tedious.

Example 6. Find ISA^z representing $z(x) = u \circ y = y\left(u(x)\right)$ where $y(u) = \sin\frac{\pi u}{2}$ and $u(x) = y^{[2]}(x)$ in Eq. (41) from the ISAs representing $y\left(u\right)$ and $u\left(x\right)$. Assume that $G\left[u\right] = \left[\cdots, -1, -\frac{1}{2}, 0, \frac{1}{2}, 1, \cdots\right]$ and $G\left[x_i\right] = \left[\cdots, -1, 0, 1, \cdots\right]$

Solution. Let us construct the consequent polynomial of $z(x)$ at the rule center $x = \eta = [1,1]^T$. Clearly, $u(\eta) = \frac{7}{6}$. For this value, there are two relevant rules of ISA^y; namely those centered at $u = \gamma_1 = 1$ and $u = \gamma_2 = \frac{3}{2}$. The consequent polynomial of ISA^y of the rule centered at $u = \gamma$ is given by

$$Y(u;\gamma) = \sin\frac{\pi\gamma}{2} + \frac{\pi}{2}\cos\frac{\pi\gamma}{2}(u-\gamma),$$

and hence $Y\left(\frac{7}{6};1\right) = 2.0$, and $Y\left(\frac{7}{6};\frac{3}{2}\right) = 2.15$. The corresponding truth values of these 2 rules are $\beta\left(u = \frac{7}{6};\gamma_1 = 1\right) = 0.7407$ and $\beta\left(u = \frac{7}{6};\gamma_2 = \frac{3}{2}\right) = 0.2593$ respectively. Hence the constant term of the rule of the desired ISA^z centered at $\eta = [1,1]$ is the convex combination of $Y\left(\frac{7}{6};1\right)$ and $Y\left(\frac{7}{6};\frac{3}{2}\right)$:

$$z(\eta) = 0.7407 \times 2.0 + 0.2593 \times 2.1547 = 2.04. \tag{48}$$

This is the constant term of $Z(x;\eta)$.

The coefficients of the linear terms are found from Eq. (47). First we need the derivative $\frac{\partial y}{\partial u}$ at $u = \frac{7}{6}$ which is not a grid point of ISA^y, and hence must be determined using the resampling formula in Eq. (39), written using Notation 1,

$$\begin{aligned}\frac{\partial y}{\partial u}\left(u=\frac{7}{6}\right) &= \frac{\partial\beta\left(\frac{7}{6};1\right)}{\partial u}Y\left(\frac{7}{6};1\right) + \beta\left(\frac{7}{6};1\right)\frac{\partial Y\left(\frac{7}{6};1\right)}{\partial u} \\ &+ \frac{\partial\beta\left(\frac{7}{6};\frac{3}{2}\right)}{\partial u}Y\left(\frac{7}{6};\frac{3}{2}\right) + \beta_k\left(\frac{7}{6};\frac{3}{2}\right)\frac{\partial Y\left(\frac{7}{6};\frac{3}{2}\right)}{\partial u} \\ &= \left(-\frac{5}{6}\right)\left(\sin\frac{\pi}{2} + \frac{\pi}{2}\cos\frac{\pi}{12}\right) + 0.7407\left(-\frac{\pi^2}{2^2}\sin\frac{\pi}{12}\right) \\ &+ \left(\frac{8}{6}\right)\left(\sin\frac{3\pi}{4} + \frac{\pi}{2}\cos\frac{-\pi}{4}\right) + 0.2593\left(-\frac{\pi^2}{2^2}\sin\frac{-\pi}{4}\right) \\ &= 0.3054.\end{aligned}$$

The derivatives $\left.\frac{\partial u}{\partial x_j}\right|_{x=\eta}$ can be simply read from Eq. (42) because $u(x)$ and $z(x)$ are defined on the same grid. They are simply $\frac{7}{6}$. Hence, $\left.\frac{\partial y}{\partial x_1}\right|_{x=\eta} = 0.3054 \times \frac{7}{6} = 0.3563$. Similarly, $\left.\frac{\partial y}{\partial x_2}\right|_{x=\eta} = 0.3563$. In short, the rule of ISA^z centered at $\eta = [1,1]^T$ has the consequent polynomial

$$Z(x;[1,1]^T) = 2.04 + 0.3563(x_1 - 1) + 0.3563(x_2 - 1). \tag{49}$$

■

6 Conclusions

The Principle of Incompatibility poses a severe dilemma to the interpretability of large FRBS. We argued that imposing additional structure on the

FRBS may offer a way out of that dilemma, and a Qualified Principle of Incompatibility was hereby proposed. It was also argued that the Interpretable Sugeno Approximator (ISA) has enough additional structure to allow a size-independent interpretation that can be adapted to any application. The additional structure consists of (a) requiring the membership functions to be local, well ordered, and sufficiently differentiable everywhere, and (b) requiring the consequent polynomials to be rule centered. This structure leads to the interpretation of every consequent polynomial as a Taylor series expansion about the rule center.

The Taylor series interpretation seems to enable an algebra and even a calculus of ISAs. This impacts positively the modularity, functionality, and scalability of the ISA. According to the Qualified Principle of Incompatibility, these properties are good indicators that the structure imposed may offer a way out of the interpretability vs. accuracy dilemma.

Much work remains to be done to fully demonstrate the modularity, functionality, and scalability of the ISA. It is not currently clear whether additional structure needs to be imposed on the ISA, before it can escape the restrictions of the Principle of Incompatibility, or whether there are structures that are more advantageous. For instance, the locality of the MFs used in the ISA seems to impose sparsity on the rulebase, but whether this satisfactorily alleviates the curse of dimensionality (a scalability issue) remains to be addressed.

Acknowledgments: This research is partly supported by the NASA Center of Aerospace Research (CAR) at North Carolina A&T State University, Contract Number NAGW-2924, and by the US Office of Naval Research through the contract N00014-96-1123.

7 References

1. J. J. Buckley, Sugeno-type controllers are universal controllers. *Fuzzy Sets & Systems*, 53:299-303, 1993.
2. P. Bauer, E. P. Klement, A. Leikermoser, and B. Moser, Interpolation and approximation of real input–output functions using fuzzy rule bases. In Kruse, R., Gebhardt, J., and palm, R. (eds.) *Fuzzy Systems in Computer Science,* pages 245-254, Vieweg, Braunschweig, 1994.
3. M. Bikdash, A highly interpretable form of the Sugeno inference system, *IEEE Transactions on Fuzzy Systems,* 7(6):686-696, 1999.
4. J.-S. Jang, ANFIS: Adaptive network-based fuzzy inference systems, *IEEE Transactions on Systems, Man, and Cybernetics,* 23 (3):665–685, May 1993.
5. J. Y. R. Jang, C.-T. Sun, and E. Mizutani, *Neuro-fuzzy and soft computing,* Prentice Hall, New Jersey, 1997.
6. C. L. Karr, Design of an adaptive fuzzy logic controller using a genetic algorithm, *Proceedings of the Fourth International Conference on Genetic Algorithms*, pages 450-457, 1991.

7. B. Kosko, *Fuzzy Engineering,* Prentice Hall, 1997.
8. K. Hornick, M. Stinchcombe, and H. White, Multi-layer feedforward networks are universal approximators, *Neural Networks,* 2:359-366, 1989.
9. S. H. Lane, M. G. Flax, D. A. Handelmanand, and J. J. Gelfand, Multilayer perceptrons with b-spline receptive field functions, In J. E. Moody and D. Touretzky, Editors, *Advances in Neural Information Processing Systems, 3:*684-692, San Mateo, CA, 1991, Morgan Kaufman.
10. E. H., Mamdani, and Assilian, An experiment in linguistic synthesis with a fuzzy logic controller, *International Journal of Man and Machine Studies*, 7:1-13, 1975.
11. H. Nomura, I. Hayashi, and N. Wakami, A self-tuning method of fuzzy reasoning by genetic algorithm, *Proceedings of the International Fuzzy Systems and Intelligent Control Conference IFSICC '92),* Louisville, KY, pages 236-245, 1992.
12. K. Shimojima, T. Fukuda, F. Arai, Self–tuning fuzzy inference based on spline function, In *Proceedings of the IEEE International Conference on Fuzzy Systems*, pages 690-695, June 1994.
13. M. Sugeno, and G. T. Kang, Structure identification of fuzzy model. *Fuzzy Sets and Systems,* 28:15-33, North-Holland Publishing Company, 1988.
14. T. Takagi, and M. Sugeno, Fuzzy identification of systems and its applications to modeling and control, *IEEE Transactions on Systems, Man, Cybernetics,* SMC-15(1):116-132, 1985.
15. L. X. Wang, Fuzzy systems are universal approximators, in *Proceedings of the IEEE International Conference on Fuzzy Systems,* San Diego, March 1992.
16. L. X. Wang, and J. M. Mendel, Generating fuzzy rules by learning from examples, *IEEE Transactions on Systems, Man, Cybernetics,* 22(6):1414-1427, 1992.
17. L. X. Wang, and J. M. Mendel, Fuzzy basis function, universal approximation, and orthogonal least squares, *IEEE Transactions on Neural Networks,* 3(5):1414-1427, September 1992.
18. C.-H. Wang, T.-T. Wang, Lee, and P.-S. Tseng, Fuzzy B-spline MFs (BMF) and its applications in fuzzy neural control, *IEEE Transactions on Systems, Man and Cybernetics,* 25(5):841-851, May 1995.
19. L. A. Zadeh, Fuzzy sets, *Information Control,* 8:338-353, 1965.
20. L. A. Zadeh, Outline of a new approach to the analysis of complex systems and decision processes, *IEEE Transactions on Systems, Man, and Cybernetics*, SMC-3:28-44, 1973.

Hierarchical Genetic Fuzzy Systems: Accuracy, Interpretability and Design Autonomy

Myriam Regattieri Delgado, Fernando Von Zuben, and Fernando Gomide

Department of Computer Engineering and Industrial Automation,
State University of Campinas, 13083-970, Campinas, SP, Brazil
e-mails: {myriam,vonzuben,gomide}@dca.fee.unicamp.br

Abstract. This chapter addresses hierarchical evolutionary rule-based fuzzy modeling, focusing on accuracy, interpretability and design autonomy issues. Special attention is given to interpretability in terms of visibility, simplicity, compactness, and consistency. As a consequence, fuzzy modeling is viewed as a decision making problem where accuracy, interpretability and autonomy are goals. The approach assumes that goals can be handled via corresponding single-objective ϵ-constrained decision making problems whose solution is produced by a hierarchical evolutionary process based on genetic algorithms, namely, a hierarchical genetic fuzzy system. In addition to performance improvement and interpretability constraints fulfillment, the hierarchical approach allows automatic tuning of a number of critical parameters and increases autonomy by minimizing user intervention. The fitting, generalization, and interpretation characteristics of the resulting fuzzy models are discussed using function approximation and classification problems.

1 Introduction

Models are essential in the human behavior since they enable human beings to predict the impact of their actions. Rule-based models are considered a linguistic representation of the mental model created inside the brain about a certain system by means of experience [9]. Fuzzy rule-based systems and models use fuzzy set theory to express expert knowledge in various domains. However, in many applications the knowledge required may not be easily available, and humans may be unable to extract knowledge out of massive amounts of numerical data. This situation motivates the development of computer techniques to extract and represent knowledge in a fuzzy rule-based system form. Data-driven fuzzy modeling, or fuzzy modeling (FM) for short, has attracted interest of many researches [20,22,23,25].

Recently, numerous works and applications combining fuzzy set theory and evolutionary computation have appeared, and there is an increasing concern about the integration of these two areas. In particular, several works explore the use of genetic algorithms (GAs) to design fuzzy systems. These are hybrid approaches named Genetic Fuzzy Systems (GFS) [16]. A GFS is basically a fuzzy system augmented by a learning process based on GAs [6,4]. GFS brought considerable attention of researchers from many areas [15,17,18].

Genetic algorithms enrich optimization tools for fuzzy systems, particularly when the most significant design decisions can be encoded into a genetic-type representation - a chromosome. However, when complex design decisions must be made, one level representation, i.e. the representation of the complete solution by only one chromosome can be inadequate. It would certainly be more appropriate to optimize a larger set of parameters encoded at different levels, producing a hierarchical genetic fuzzy system (HGFS) [8].

In FM, a fundamental aspect is the trade-off between two important criteria: accuracy and interpretability. The issue of accuracy is critical when models are used, for example, in dynamic systems where the predicted value is fed back and small errors will be propagated and reflected as errors in the long term behavior [9]. This kind of situation requires accurate fuzzy models and precise fuzzy modeling is the only goal. Although important, accuracy is not the only critical aspect in FM. In the last years, model interpretability has gained special attention [9,19,21,25,27,30]. One of the aspects that distinguish FM (considered *grey-box* modeling) from black-box techniques, like early versions of neural nets, is that fuzzy models are, to a certain degree, suitable to interpretation analysis [27]. When the focus is interpretability, linguistic fuzzy modeling generates systems for which the language is easily interpretable by human beings. Despite this natural human interpretation, there is no well-established definition of interpretability for fuzzy models of systems [21]. Here, interpretability is identified by four characteristics: visibility, simplicity, consistency, and compactness. Visibility is the absence of gaps and full overlapping in domains granularities. Simplicity can be measured by the length of rules, i.e. the total of relevant features present in the antecedent part of each individual rule. Consistency and compactness are associated with the rule base. Compactness is measured by the total of fuzzy rules in the rule base. A consistent rule-base presumes the absence of conflicting rules, i.e., rules with the similar antecedents, but with very different consequent parameters.

Besides accuracy and interpretability, another important aspect in FM is design autonomy. Here autonomy means the degree of involvement of the system designer with the design procedure adopted for model development. More precisely, the fewer the parameters defined *a priori* by the designer, the higher the degree of design autonomy.

This chapter views fuzzy modeling as a multi-objective decision making problem considering accuracy, interpretability and autonomy as goals. The solution assumes that the multiple goals can be treated by translating the multi-objective problem into single-objective ϵ-constrained problems [3]. The solution of these constrained decision making problems is produced by a hierarchical evolutionary process based on genetic algorithms. This evolutionary approach uses the same strategy to evolve fuzzy systems as the one suggested by Delgado *et al.* [8], based on HGFS. In this case, the evolutionary parameters are adjusted not only to improve the models performance (accuracy), but also to guarantee the interpretability of the resulting fuzzy models. The hierarchical approach induces a minimum designer intervention by means of automatic tuning of many critical parameters of e.g., Mamdani or Takagi-Sugeno (TS) fuzzy models. In this chapter, we emphasize TS models only because the resulting fuzzy model can be easily identified using the input-output data. Moreover, TS models generally represents the behavior of complex nonlinear systems with a small number of rules, although they do not have a clear semantics due to the functional nature of its consequent.

The chapter is organized as follows. Section 2 addresses important issues of fuzzy modeling. In Sect. 3, FM is introduced as a multi-objective decision making problem transformed into single-objective ϵ-constrained problems. Section 4 addresses the hierarchical evolutionary process and its encoding scheme, evolutionary algorithm and operators. Section 5 details local and global approaches to find rules consequent parameters. Section 6 presents simulation results for function approximation problems with noiseless and noisy data, and a pattern classification problem. The remarks of Sect. 7 concludes the chapter.

2 Fuzzy Modeling

Three fundamental issues in fuzzy modeling concerns accuracy, interpretability, and design autonomy, respectively. In what follows, we discuss each of these important issues.

- **accuracy**: Accuracy is usually measured by an error criterion. In function approximation problems, the most common measurement is the mean squared error defined as:

$$\text{MSE} = \frac{1}{N} \sum_{p=1}^{N} (y_p - \hat{y}_p)^2 \ , \tag{1}$$

  where y_p is the target output for the p-th input pattern, $p = 1, \cdots, N$ and $\hat{y}_p$ is the predicted output provided by the fuzzy system.
- **interpretability**: A fundamental issue in FM is the interpretability or the transparency of the resulting fuzzy system, i.e. it must be easy to be understood by human beings [21]. However, as pointed out by Setnes *et al.* [27], many approaches in the literature claim fuzzy systems to be interpretable, while they are actually used as completely black-box systems. Here, the interpretability is identified and analyzed based on four characteristics:
  - **visibility**: FM can be used to obtain not only accurate, but also visible fuzzy rules acquired from data. Here, visibility is associated with two different characteristics:
    * γ-completeness: This criterion, known as coverage property [9], fixes a minimum (γ) of overlapping degree in the universe partition to guarantee a granulation without gaps [14,20].

    * α-overlapping: This criterion, also called distinguishability property [9], fixes a maximum (α) of overlapping degree for the universe granulation [8].
  - **simplicity**: Long linguistic rules with many antecedent conditions are not easy to understand [19]. As discussed in Delgado *et al.* [8], the inclusion of *don't care* conditions [18], simplifies the fuzzy rules and improves the generalization capabilities of the resulting fuzzy system. The simplicity of each rule is evaluated by its length, measured by the number of features or variables minus the number of irrelevant features identified by *don't care* conditions.
  - **compactness**: The size of the rule set is inversely proportional to the interpretability. The compactness of a rule-base is evaluated by the number of fuzzy rules [19].
  - k**-consistency**: The k-consistency condition defined in a classification context [14] is here adapted to measure the total of conflicting rules in the rule-base. Then, the 0-consistency condition (called consistency condition) means the absence of rules with the same antecedent and different consequent parameters.
- **design autonomy**: Automatic systems are self-regulating devices provided with the laws and strategies according to which they control their behavior. To be autonomous the system must first be automatic, and then be capable of developing its own control strategies[28]. Here, by design autonomy we mean the degree with which the design methodology is independent of the user intervention. When many fuzzy model parameters must be decided by the user, we assume that FM is non-automatic. Therefore, it is non-autonomous. Furthermore, when the system has time to study a large number of examples, or to speculate about how it could cope with unforeseen circumstances, the autonomy characteristic of the system may be scaled down. In this sense, data-driven approaches to design fuzzy systems, with no *a priori* information about the final structure of the system, and with minimal user intervention, can be considered semi-autonomous fuzzy modeling approaches.

So far, no definitive FM methodology that, at the same time, maximizes accuracy, interpretability and design autonomy has been developed, and demand for mechanisms to automatic design and tune accurate and interpretable fuzzy systems still exists. Many data-driven approaches have been proposed. Initially, neuro-fuzzy algorithms [20,23] have been oriented to minimize numerical errors, but parameters such shape and number of membership functions, or inference mechanism and operators, must be provided by the designer. These algorithms are often optimized using gradient descent techniques. The use of these optimization techniques with no restriction associated with the final configuration, sometimes generates fuzzy sets with *too much* or *absolutely no* overlap and thereby making the interpretation of the model unwieldy.

In the last years, besides accuracy, model interpretability has gained special attention [9,19,21,27,30]. For example, Setnes *et al.* [27] discuss the compromise between precision and transparency. Product-space clustering generates the rules. The consequent parameters of the rules, expressed by singletons, are transformed into linguistic labels. Jin [21] proposes an efficient approach for FM of high dimensional systems. The structure and parameters of the fuzzy system are optimized using GA and gradient-based methods. The resulting fuzzy system is interpretable, simplified, and the dependencies between the inputs and outputs are given by constant consequent functions. Espinosa & Vandewalle [9] present an interesting discussion about the trade-off between accuracy and interpretability of fuzzy models. The authors in [9] describe a fuzzy rule extractor with linguistic integrity. The algorithm is a two-step approach. The first step uses clustering and projection techniques to find good initial positions for the fuzzy sets in the input domains. The second step reduces the complexity of the model using the concept of semantic integrity. Singletons in the consequent part, obtained by a least squared approach, are transformed into linguistic labels.

Here in this chapter, the hierarchical approach is introduced to find an accurate-feasible solution for the FM problem, with minimum designer intervention, and with all the interpretability constraints being observed.

3 Multi-objective Decision Making and Genetic Fuzzy Systems

Fuzzy modeling requires the consideration of multiple criteria in the design process. Recently, multiple goals in fuzzy system design has gained more interest as discussed in [5,11,13,17,18,21].

Generally speaking, a multi-objective optimization problem can be formulated as:

$$\min \boldsymbol{f}(\boldsymbol{x}) \ , \\ \text{s.t. } \boldsymbol{x} \in \Omega$$

where $\boldsymbol{f}(\boldsymbol{x}) = [f_1(\boldsymbol{x}), f_2(\boldsymbol{x}), \cdots, f_r(\boldsymbol{x})]^T$ is the vector of objectives, $f_i(.) : \Omega \rightarrow \Re, i = 1, \cdots, r$, and $\Omega \subseteq \Re^n$ is the subset of feasible solutions for which $\boldsymbol{f}$ is defined. In multi-objective optimization, the set of efficient solutions, also called non-dominated or Pareto-optimal, is composed of all those elements of the input space for which the corresponding vector of objectives cannot be improved in any of its components without degradation in another component.

Multi-objective optimization problems can be solved by different techniques [3]. Some techniques adopt the transformation of the original problem into a set of single-objective problems. The two most common ways to obtain the non-dominated solutions for the original problem are: 1) weighted method; 2) ϵ-constrained method. The former assumes an association to aggregate all objectives settled as a (often linear) combination of all criteria to define a single function to be optimized. The later comprises the solution of single-objective problems subject to the set of ϵ-constraints associated with the other objectives, except the one taken as reference.

In FM, more complex multi-objective decision making problems emerge. For these problems, objective functions are heterogeneous in the sense that they are not a function of the same set of variables. Nevertheless multi-objective decision theory still is a source of inspiration for many approaches. Genetic algorithms [24] are useful alternatives to carry out search for feasible solutions and Pareto-optimal solutions, resulting in multi-objective genetic fuzzy systems. Many methods adopt an aggregation formula (weighted sum) of the different criteria that translates the result into a fitness measure of a solution of the original problem [17,18,21]. However, it is often hard to choose the appropriate weights. Experiments show that small changes in weights may guide to completely different results. Alternatively, Pareto-optimal solutions can be generated and the decision-maker may choose the preferred solution [5,11,13].

In this chapter, FM will be viewed as a multi-objective decision making problem for which accuracy, interpretability and design autonomy are the goals, and feasible solutions are achieved from ϵ-constrained optimization problems. The aim here is to minimize the error criterion to improve accuracy, but subject to visibility, simplicity, compactness and consistency constraints. Therefore, the corresponding ϵ-constrained problem is expressed as:

$$\min_{v \in \Omega(\epsilon)} \quad \text{MSE} = \frac{1}{N} \sum_{p=1}^{N} (y_p - \hat{y}_p)^2 \qquad (Accuracy)$$

$$s.t. \begin{cases} k\text{-consistency} = \epsilon_1 = 0\text{-consistency} & (Consistency) \\ \text{TotRules} \leq \epsilon_2 = \text{MaxRules} & (Compacteness) \\ \text{TotMF} \leq \epsilon_3 = \text{MaxGranularity} & (Simplicity) \\ \alpha\text{-overlapping} \leq \epsilon_4 = \alpha_{\max} & (Visibility) \\ \gamma\text{-completeness} \geq \epsilon_5 = \gamma_{\min} & (Visibility) \end{cases}$$

where the vector $\epsilon = \{\epsilon_1, \epsilon_2, \epsilon_3, \epsilon_4, \epsilon_5\}$, defined in $\varepsilon = \{\epsilon : \Omega(\epsilon) \neq \emptyset\}$ is associated with interpretability requirements, and $\boldsymbol{v}$ is the decision variable.

The formulation above is a single objective ϵ-constrained version of the original multi-objective problem. If a solution provided by genetic algorithms is a unique solution of the ϵ-constrained problem or if it solves all the ϵ-

constrained problems when we consider each constrain as a reference objective function, than this solution is Pareto-optimal.

For genetic operators, the violation of visibility conditions needs repairing procedures to relocate the membership functions. Repairing algorithms represent a major advantage of evolutionary techniques to adjust membership function parameters when contrasted with classical optimization techniques. Next section details the hierarchical evolutionary process pointing out how the constraints are satisfied during evolution. Because interpretability in the TS consequent functions is not considered as a restriction, this topic will be separately treated in Sect. 6.

4 The Hierarchical Evolutionary Approach

The hierarchical evolutionary process is structured in modules that consider the membership functions (or partition set) at the first level, the population of individual rules at the second level, the population of sets of rules at the third level and the population of fuzzy systems at the fourth level [8]. This FM approach may be applied to evolve Takagi-Sugeno fuzzy models or Mamdani fuzzy models. The encoding scheme uses real and integer encoding (depending on the level) and four populations (each one associated with one level) evolve interactively, as detailed below.

4.1 Encoding Scheme

A partition set individual (level I) contains all the membership functions defined in the universes of the variables involved. The chromosome is formed by the concatenation of all the partition sets associated with each variable. Figure 1 shows the details of the relation between the genotype and phenotype at level I.

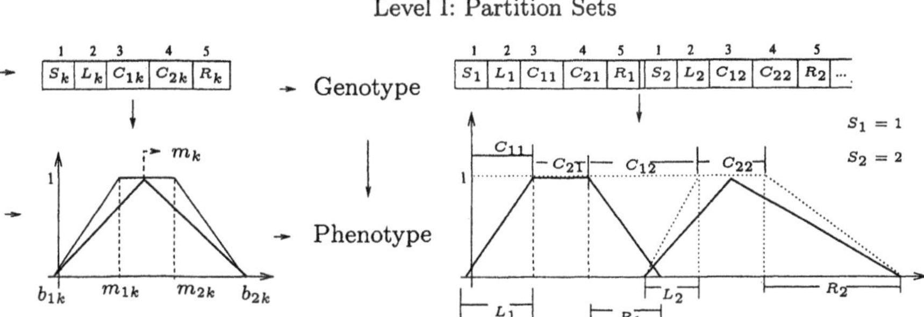

Fig. 1. Membership function encoding

Each membership function is represented by a sequence of 5 alleles

$$\{S_k, L_k, C_{1k}, C_{2k}, R_k\} ,$$

where k identify the position of the function in the corresponding universe. The membership functions can assume three different shapes or types: (1) trapezoidal, (2) triangular, and (3) Gaussian. The type or shape of the function is defined by the allele (S_k) at site 1 of the sequence of 5 alleles. Null values for the first alleles ($S_k = 0$) mean *don't care* conditions. The first allele assumes integer values, and all remaining alleles in the sequence assume real values relative to the reference position (C_{1k}) that occurs at site 3. This reference position measures the distance of the membership function k to the membership function $k-1$. For the first membership function, $k = 1$, the reference measures the distance from the lower limit of the universe. The absolute position of a specific term in the universe depends on the position of the previous one. The absolute values for the kth membership function are expressed as: $m_{1k} = m_{2k-1} + C_{1k}$; $m_{2k} = m_{1k} + C_{2k}$; $b_{1k} = m_{1k} - L_k$; $b_{2k} = m_{2k} + R_k$. For instance, to obtain the phenotype of trapezoidal membership functions, alleles at site 2 to 5 are converted to their absolute values $\{b_{1k}, m_{1k}, m_{2k}, b_{2k}\}$. In the case of triangular functions, after finding all the absolute values, the center m_k is found by the average between m_{1k} and m_{2k}. For Gaussian functions, the modal value is calculated in the same way as in the triangular case, but a modification is required to calculate the dispersion: the maximum dispersion is found by $\Delta_k = (L_k + R_k)/2$, whereas the dispersion is set as $\sigma_k = \Delta_k/3$. Thus, the Gaussian membership function limits will not exceed $3\sigma_k$.

The chromosomes representing individual rules and rule-base individuals (level II and III, respectively) use integer encoding. Fuzzy systems individuals (level IV) use real values to encode t-norm parameters p_t and integer codes for the remaining alleles. The chromosomes at all levels have hierarchical relationships, as summarized in Fig. 2.

Each member of the population of individual rules (level II in Fig. 2) represents a fuzzy proposition. This population accepts different combinations of membership functions as identified by their indexes (the order in the partition set). Null values indicate *don't care* conditions. The genetic operators prioritizes simpler individual rules (rules with more *don't care* conditions).

The population of set of rules (level III) has individuals formed by indexes to identify the corresponding rules. The length of the chromosome determines the maximum number of fuzzy rules, but smaller rule-bases are always aimed at first.

Each individual at level IV represents a fuzzy system. At this level, the code of each chromosome associates a specific set of rules (allele at site 8) and a partition set (allele at site 9), with a subset of operators used to define the inference mechanism (alleles at site 1 to 7). When Takagi-Sugeno fuzzy models are chosen, alleles at site 4 to 7 (which are associated with Mamdani

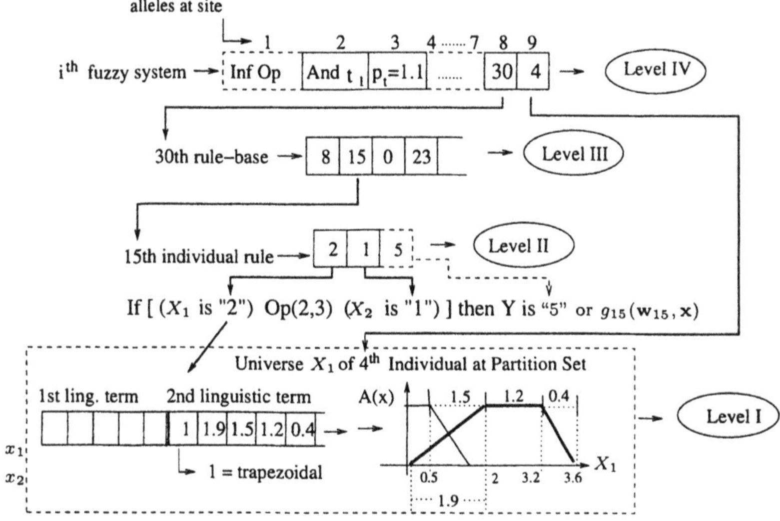

Fig. 2. Encoding and hierarchical relations among individuals at the different levels (rules with two antecedent variables are assumed)

fuzzy models and identify the rule-semantic, rule aggregation operator and defuzzification method) are not considered. In this case, only alleles at sites 2,3,8 and 9 are evolved.

The *i-th* fuzzy system depicted at level IV in Fig. 2, uses the *30-th* set of fuzzy rules at level III and the *4-th* partition set individual at level I. Each individual rule that composes its rule-base aggregates the antecedent part by the t-norm $\boldsymbol{t}_1$ given by:

$$a\,\boldsymbol{t}_1\,b = \frac{1}{1+\sqrt[p_t]{(\frac{1-a}{a})^{p_t}+(\frac{1-b}{b})^{p_t}}}, \tag{2}$$

with an associated parameter $p_t = 1.1$ (see Pedrycz & Gomide [26] for more details). For example, the *15-th* individual rule of the *i-th* fuzzy system could be given by:

If (X_1 is "2") $\boldsymbol{t}_1$ (X_2 is "1") then Y is $g_{15}(\mathbf{w}_{15}, \mathbf{x})$, for TS fuzzy models, with $\mathbf{w}_{15}$ given by (4) or (5); and

If (X_1 is "2") $\boldsymbol{t}_1$ (X_2 is "1") then Y is "5", for Mamdani fuzzy models.

In Fig. 2, for the *i-th* fuzzy system, the linguistic terms indexed by "2" and "1" (and eventually "5" for Mamdani models at the $15-th$ fuzzy rule) are defined in the *4-th* chromosome of level I.

4.2 Hierarchical Evolutionary Algorithm

The design process uses a GA strategy to produce improved fuzzy model parameters along generations. The main steps of the hierarchical evolutionary algorithm are summarized as follows:

1. Start with Generation = 1;
2. Initialize populations for each module;
3. If Takagi-Sugeno fuzzy modeling
   (a) Compute the optimal parameter for the consequent of each individual at level IV (see (4) and (5) in Sect. 5);
4. Calculate the fitness of each individual at all population levels as follows:
   (a) Fuzzy System (level IV): $fit_{\mathrm{FS}}(i)$ is based on the fuzzy system performance;
   (b) Rule-Base (level III): $fit_{\mathrm{RB}(k)} = max\left(fit_{\mathrm{FS}(b)}, \cdots, fit_{\mathrm{FS}(d)}\right)$, where $b, \cdots, d$ are the fuzzy systems of which the rule-base (k) is part.
   (c) Individual Rule (level II): $fit_{\mathrm{IR}(j)} = mean\left(fit_{\mathrm{RB}(m)}, \cdots, fit_{\mathrm{RB}(p)}\right)$, where $m, \cdots, p$ are the rule-bases of which the individual rule (j) is part.
   (d) Partition Set (level I): $fit_{\mathrm{PS}(q)} = max\left(fit_{\mathrm{FS}(x)}, \cdots, fit_{\mathrm{FS}(z)}\right)$, where $x, \cdots, z$ are the fuzzy systems of which the partition set (q) is part.
5. If the stop condition does not hold, do:
   (a) From level IV to level I apply the evolutionary operations (selection, crossover and mutation) to form a new population;
   (b) Generation = Generation + 1;
   (c) Return to step 3;

The initialization phase (step 1) comprises random and deterministic generations of population at each level. At the partition set (level I), chromosomes are generated to uniformly distribute the membership functions over the associated universes. At level II, different fuzzy propositions encoded by integer chromosomes are randomly generated. *Don't care* conditions represented by null values are introduced to attend the simplicity criterion. At the third level, each integer chromosome that represents a set of fuzzy rules is randomly generated. Rule exclusions are admitted to attend the compactness criterion. At level IV, all the alleles are randomly initialized (except the alleles at sites associated with parameter p_t). Alleles at sites associated with antecedent aggregation and rule semantics (this later is specific to Mamdani models) are generated to cover all the possible norm operators. Alleles at site associated with t-norm parameters p_t are initialized with the value $p_t = 2.0$. Alleles at sites 8 and 9 are randomly generated and define which rule-base and partition set will be used by the current fuzzy system.

After all the population levels have been initialized, the next step involves the optimization (only for TS models) of the consequent parameters of each fuzzy system of level IV. This procedure will be detailed in Sect. 5.

Fitness calculation (step 4) is performed after all fuzzy system parameters definition. This process starts at level IV and finishes at level I. At level IV, fitness is evaluated by decoding the chromosome representation and measuring the fitness function that depends on the performance of each fuzzy system when applied to the problem under consideration. An individual that does not participate in any other higher level individual receives fitness = 0.

At step 5, either the stop condition (maximum number of generations or error criterion) is verified or the algorithm continues. The evolutionary operators work downward, that is, selection, crossover and mutation are applied from the top level (IV) to the bottom level (I).

4.3 Evolutionary Operators

Selection is the first evolutionary operator applied and uses the tournament technique [24] to select 80% of the individuals. In this technique, we choose 10% of randomly selected chromosomes to participate in each tournament, and the best (individual with the highest fitness) wins. The remaining 20% of the population is chosen by a technique that favors diversity in the population. This diversity criterion focus the choice of the "most diverse" chromosomes when compared with the one with the highest fitness. The measure of diversity is either the Euclidean distance for real-coded chromosomes or the Hamming distance for integer codes. The selection process uses an elitist strategy to ensure that the chromosome with the highest fitness survives to compose the next generation. The second evolutionary operator is the 1-point crossover. The crossover point is randomly chosen. Mutation, the last operator, is applied to real and integer encoding parameters (see [24] for more details of genetic operators in real and integer codes).

At the partition level, two visibility conditions must be fulfilled by the set of membership functions to achieve the interpretability requirements: 1) the γ-*completeness* [20], and 2) the maximum degree α of overlapping. These conditions state that, given a value x of one of the inputs within the operation range, we can always find a linguistic term A such that $\mu_A(x) \geq \gamma$, and no more than one linguistic term B such that $\mu_B(x) \geq \alpha$. This assumption means a minimum (γ) and maximum (α) overlapping degrees among the membership functions during evolution. Figure 3 shows examples of interpretability analysis in the partition set.

Because each allele value is relative to each other, any point can be chosen for mutation or crossover. The genetic operators change the shape and location of membership functions but the minimum and maximum overlapping degrees are always fulfilled. In the case of visibility restriction violation, a repairing procedure is used to attend the γ-completeness or α-overlapping criteria, relocating the membership functions when it is necessary.

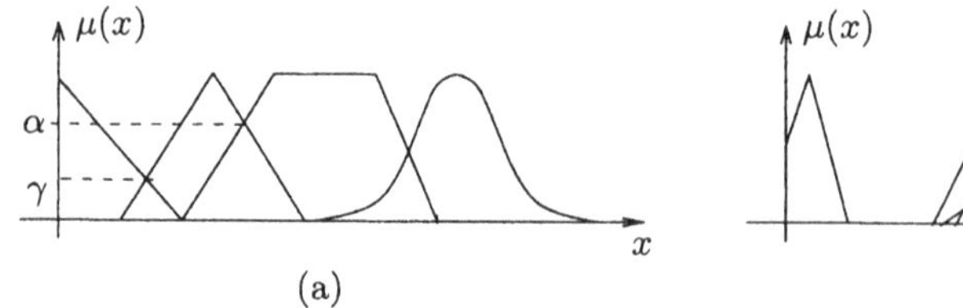

Fig. 3. Interpretability analysis: (**a**) an interpretable partition set; (**b**) a non-interpretable partition set

At the second level, crossover and mutation are applied to produce different combination of linguistic terms of each proposition. Mutation changes the current value by a new one chosen from the set $\{0, 1, ..., L_i\}$, where $L_i \leq \epsilon_3$ means the maximum number of linguistic terms for the i-th variable.

At the rule-base level, crossover and mutation change the integer indexes associated with individual rules. At this level, alleles (representing individual rules) with lower fitness have higher probability of being changed by mutation. The new values are chosen among the values of the set $\{0, 1, ..., S_{II}\}$, where S_{II} means the size of the population at level II. Null values are possible (indicating a rule elimination). The length of the chromosome is defined by the constraint ϵ_2.

At the fuzzy system level, crossover produces combination of two parents. The mutation of integer alleles changes a value by a new one chosen among all the possibilities. For example, the new value of alleles at site 2 (see Fig. 2) is chosen from $\{1, \cdots, 9\}$, where the indexes represent the t-norms accepted for antecedent aggregation [26]; for alleles at site 8, the new values are chosen from $\{0, 1, ..., S_{III}\}$; and for alleles at site 9 are chosen from $\{0, 1, ..., S_I\}$, where S_{III} and S_I are the population sizes at levels III and I, respectively. Uniform mutation operator is applied to alleles at sites associated with parameter p_t because it achieved better results when compared with non-uniform mutation. The possibility of adopting different inference operators gives flexibility to the resulting fuzzy system and improves the degree of design autonomy.

5 Optimization of Consequent Parameters of TS Models

The evolutionary approach described in the previous section assumes the choice of one among two of the most common fuzzy models, Mamdani and Takagi-Sugeno fuzzy models, respectively. Here, we focus on TS models [29] because they are useful to handle complex systems with a reduced number of fuzzy rules (improving compactness), and allow the use of numerically efficient estimation algorithms to find the consequent function (improving accuracy) parameters.

In TS models, the consequent of each rule is a parametric functional relationship of the input variables. Suppose that we have a TS fuzzy model composed of m fuzzy rules $R_j, j = 1, \cdots, m$, of the form:
R_j: If X_1 is A_1^j ... and X_n is A_n^j then Y is $g_j(\boldsymbol{w}_j, \boldsymbol{x})$,
where $\boldsymbol{x} = [x_1 \cdots x_n]$ is the input vector, and the Q-dimensional vector $\boldsymbol{w}_j$ contains the parameters of function $g_j(.)$. The most common types of function found in the literature are constant ($Q = 1$) and linear ($Q = n + 1$). Here, nonlinear TS functions ($Q = \frac{n(n-1)}{2} + 2n + 2$) can also be considered, as shown below:

$$g_j(\boldsymbol{w}_j, \boldsymbol{x}) = w_{j0} + w_{j1}x_1 + \cdots + w_{jn}x_n + w_{j(n+1)}x_1x_1 + \cdots + w_{j(2n)}x_1x_n + \\ + w_{j(2n+1)}x_2x_2 + \cdots + w_{j(\frac{n(n-1)}{2}+2n)}x_nx_n + w_{j(\frac{n(n-1)}{2}+2n+1)}x_1x_2 ... x_n \ .$$

The structure and parameters of TS models are determined by the evolutionary algorithm described in Sect. 4. After the specification of all these parameters, the elements of the vector $\boldsymbol{w}_j \in \Re^Q$ can be computed using optimization techniques, where $\boldsymbol{w}_j = [w_{j0} \cdots w_{jQ-1}]^T$, $j = 1, ..., m$, and T means transpose. Two optimization approaches will be discussed in the next subsections.

5.1 Global Estimation

In global estimation, all the rules are considered simultaneously and the least squares method is used to find the optimal set $\boldsymbol{w}^* = [(\boldsymbol{w}_1^*)^T \cdots (\boldsymbol{w}_m^*)^T]^T$, where $\boldsymbol{w}^* \in \Re^{mQ}$ (m is the number of rules).

Let $\mu_j(\boldsymbol{x}_p)$ be the result of antecedent aggregation of the jth fuzzy rule for the pth input $\boldsymbol{x}_p = [x_{1p} \cdots x_{np}]$, $p = 1, \cdots, N$. Then, the pth output of a fuzzy system is given by:

$$y(\boldsymbol{x}_p) = \frac{\sum_{j=1}^{m} \mu_j(\boldsymbol{x}_p) g_j(\boldsymbol{w}_j, \boldsymbol{x}_p)}{\sum_{j=1}^{m} \mu_j(\boldsymbol{x}_p)} \ . \quad (3)$$

Define $\beta_j(\boldsymbol{x}_p) = \beta_j^p = \frac{\mu_j(\boldsymbol{x}_p)}{\sum_{j=1}^{m} \mu_j(\boldsymbol{x}_p)}$ and let the vector $\boldsymbol{\lambda}_j(\boldsymbol{x}_p) \in \Re^Q$ be given by:

$$\boldsymbol{\lambda}_j(\boldsymbol{x}_p) = [\beta_j^p \quad \beta_j^p x_{1p} \quad \cdots \quad \beta_j^p x_{np} \quad \beta_j^p x_{1p}x_{1p} \\ \beta_j^p x_{2p}x_{2p} \quad \cdots \quad \beta_j^p x_{np}x_{np} \quad \beta_j^p x_{1p} ... x_{np}]^T \ .$$

Then, (3) becomes $y(\boldsymbol{x}_p) = (\boldsymbol{\lambda}(\boldsymbol{x}_p))^T \boldsymbol{w}$, where $\boldsymbol{w} = [\boldsymbol{w}_1^T \cdots \boldsymbol{w}_m^T]^T$, and $\boldsymbol{\lambda}(\boldsymbol{x}_p) \in \Re^{mQ}$ is defined as $\boldsymbol{\lambda}(\boldsymbol{x}_p) = [\boldsymbol{\lambda}_1^T(\boldsymbol{x}_p) \cdots \boldsymbol{\lambda}_m^T(\boldsymbol{x}_p)]^T$. Now, let matrix $\Lambda \in \Re^{N \times mQ}$ be as follows

$$\Lambda = [\boldsymbol{\lambda}(\boldsymbol{x}_1) \quad \boldsymbol{\lambda}(\boldsymbol{x}_2) \quad \cdots \quad \boldsymbol{\lambda}(\boldsymbol{x}_N)]^T \ .$$

Then, the predicted output $\hat{\boldsymbol{y}} = [y(\boldsymbol{x}_1)\cdots y(\boldsymbol{x}_N)]^T \in \Re^N$ is given by the matrix equation

$$\hat{\boldsymbol{y}} = \Lambda \boldsymbol{w} \, .$$

Considering $\boldsymbol{y}_d$ as the desired output, we can find $\boldsymbol{w}^*$, solving the least squares optimization problem $\min_{\boldsymbol{w}} \frac{1}{2} \parallel \hat{\boldsymbol{y}} - \boldsymbol{y}_d \parallel_2^2$ ($\parallel \cdot \parallel_2$ is the Euclidean norm), that produces:

$$\boldsymbol{w}^* = (\Lambda^T \Lambda)^{-1} \Lambda^T \boldsymbol{y}_d \, . \tag{4}$$

As pointed by Golub and Van Loan [12], $\boldsymbol{w}^*$ produces the solution $\hat{\boldsymbol{y}}^* = \Lambda \boldsymbol{w}^*$, but as $\text{rank}(\Lambda) = mQ < N$, $\boldsymbol{w}^* = \arg \min_{\boldsymbol{w}} \left[\frac{1}{2} \parallel \hat{\boldsymbol{y}} - \boldsymbol{y}_d \parallel_2^2\right]$ may not imply $\hat{\boldsymbol{y}}^* = \boldsymbol{y}_d$. As another aspect of the solution, the choice of more complex consequent functions introduces a higher number Q of parameters to be optimized. If Q is higher than necessary, some parameters have to be pruned to avoid redundant terms in the consequent of any fuzzy rule. Then, to guarantee that $\Lambda^T \Lambda$ is not poorly conditioned, the pruning procedure described in Delgado *et al.* [7] is applied. This procedure improves the condition number of matrix $\Lambda^T \Lambda$ by eliminating the columns of Λ that most contribute to the degradation of the condition number. The elimination of all the consequent parameters associated with a specific rule indicates the redundancy of this rule in the rule-base. Then, together with the genetic operators, the pruning procedure improves the interpretability of the fuzzy models through the generation of compact and consistent rule bases.

5.2 Local Estimation

The local estimation optimizes the consequent of each rule independently of the remaining ones [17,30]. Thus, for N input patterns, the fuzzy system output is $\hat{\boldsymbol{y}} = \hat{\boldsymbol{y}}_j = \Lambda_j \boldsymbol{w}_j$ where $\Lambda_j \in \Re^{N \times Q}$ is given by

$$\Lambda_j = [\boldsymbol{\lambda}_j(\boldsymbol{x}_1) \; \boldsymbol{\lambda}_j(\boldsymbol{x}_2) \; \cdots \; \boldsymbol{\lambda}_j(\boldsymbol{x}_N)]^T \, .$$

In this case, the optimal consequent $\boldsymbol{w}_j^* = [w_{j0}^* \cdots w_{j(\frac{n(n-1)}{2}+2n+1)}^*]^T$ can be obtained by the locally weighted optimization approach:

$$\min_{\mathbf{w}_j} \frac{1}{2} \parallel \hat{\mathbf{y}} - \mathbf{y}_d \parallel_{\Psi_j}^2 = \min_{\mathbf{w}_j} \frac{1}{2} \left[(\hat{\mathbf{y}} - \mathbf{y}_d)^T \Psi_j \, (\hat{\mathbf{y}} - \mathbf{y}_d) \right] ,$$

where Ψ_j is the diagonal matrix:

$$\Psi_j = \begin{bmatrix} \beta_j^1 & 0 & 0 \cdots & 0 \\ 0 & \beta_j^2 & 0 \cdots & 0 \\ \vdots & \vdots & \vdots \ddots & \vdots \\ 0 & 0 & 0 \cdots & \beta_j^N \end{bmatrix}$$

Then, the optimal set of parameters for rule R_j is given by:

$$\boldsymbol{w}_j^* = (\Lambda_j^T \Psi_j \Lambda_j)^{-1} \Lambda_j^T \Psi_j \boldsymbol{y}_d \,. \qquad (5)$$

If $\Lambda_j^T \Psi_j \Lambda_j$ matrix is poorly conditioned, the pruning algorithm should be applied to improve its condition number.

6 Experiments and Results

This section considers function approximation and classification problems to illustrate the performance of HGFS and to discuss model interpretability and accuracy issues.

Section 6.1 refers to function approximation with 2 inputs and 1 output system and noiseless training data. Simulations are performed to compare the results provided by HGFS, ANFIS, and the classic feed-forward backpropagation neural network (MLP). Different trade-offs between accuracy and compactness of the rule base are provided by the simulations of HGFS and model interpretability resulting from the choice of alternative modeling tools is discussed.

In Sect. 6.2 noisy 1 input and 1 output system training data is used to verify and to compare the performance of ANFIS with HGFS using different consequent functions. Trade-offs between interpretability and accuracy for TS consequent part are addressed in Sect. 6.2 as follows:

- interpretability improvement and accuracy reduction: this case considers constant consequent functions and local estimation techniques to optimize these parameters. To further improve interpretability, the real numbers at the consequent could be translated into linguistic labels, e.g., by the methods of [9,25,27].
- accuracy improvement and interpretability reduction: this case assumes global estimation and nonlinear consequent.

Section 6.3 addresses the well-known Iris pattern classification data to investigate the performance of HGFS in higher dimensional spaces, and to compare HGFS with ANFIS and the approach of Castellano and Fanelli [2]. The approach based on the ϵ-constrained method is compared with the weighted method to solve the multi-objective decision making problem of fuzzy modeling.

In all simulations presented, the population of individual rules (level II) did not change during evolution because it represents the set S of all possible combinations of linguistic terms (specific rules (SR)) plus the rules containing *don't care* conditions (general rules (GR)), i. e., $S = \text{SR} \cup \text{GR}$. For applications with higher numbers of inputs and higher numbers of linguistic terms, it is not possible to cover all the possibilities. In this case, evolution at this level becomes necessary, and occurs as discussed in Sect. 4.

6.1 Function Approximation with Noiseless Data

A function approximation problem is solved to compare the performance of HGFS with ANFIS (a partially interpretable neuro-fuzzy system) and with multilayer feed-forward backpropagation network MLP (a black-box approach). The HGFS computes the optimal parameters of the nonlinear consequent function by (4).

The problem is to approximate the function

$$F_1 : [-10, 10]^2 \rightarrow \Re, \text{ where } f_1(x_1, x_2) = \frac{\sin(x_1)}{x_1}\frac{\sin(x_2)}{x_2},$$

as depicted in Fig. 5(a). Approximation uses N equally spaced training samples (x_{1p}, x_{2p}, f_p), $p = 1, \cdots, N$.

Both, ANFIS and MLP have been trained during 6000 epochs. HGFS evolved during 1000 generations. HGFS uses fixed population sizes in each module: 100 individuals in the population of fuzzy systems (level IV), 80 individuals in the population of set of fuzzy rules (level III) and 20 individuals in the population of partition set (level I). Since the population of individual rules (level II) contains all possible combinations of the linguistic terms, no evolution is necessary at this level. The evolutionary operators (selection, crossover and mutation) are those detailed in Sect. 4. Crossover and mutation genetic operators are performed with probability $P_C = 0.2$ and $P_M = 0.08$, respectively, for all the three levels (I, III and IV) that evolve. All these evolutionary parameters have been chosen after many tests using different values. The fitness of the fuzzy system is computed using the root mean squared error (RMSE): fitness $= 1/\sqrt{\text{MSE}}$, where the MSE is given by (1).

The constraint ϵ_3 is fixed at 5 which means a maximum granularity of $[5, 5]$ for each variable. The constraints $\{\epsilon_1, \epsilon_4, \epsilon_5\}$ are chosen as follows: $\epsilon_1 = 0$, $\epsilon_4 = 0.9$, and $\epsilon_5 = 0.15$ to define the k-consistency, overlapping and visibility criteria, respectively. These constraints aim at visibility at the partition sets, and consistent rule-bases with an acceptable level of accuracy. Constraint ϵ_2 is tested with different values to find a good trade-off between accuracy/compactness in 6 different simulations, i.e. $\epsilon_2 = m$, where $m \in \{5, 8, 10, 15, 20, 25\}$. This means that each individual (rule-base) in the population of level III contains at most m fuzzy rules. At this level, the effective number of fuzzy rules is allowed to change during evolution since mutation or crossover may exclude or reintroduce a rule in the rule-base. Figure 4 illustrates non-dominated solutions obtained from the 6 simulations performed.

Table 1 summarized the results for the fuzzy system evolved by HGFS (with $\epsilon_2 = 10$) and for fuzzy models produced by ANFIS and MLP.

ANFIS uses a granularity of $[5, 5]$ which means 25 fuzzy rules. Different membership functions have been tested and the best fitting abilities were achieved using Gaussians. The results of Table 1 show the RMSE for the training data set (RMSE tr), the testing data set (RMSE ts), and the total

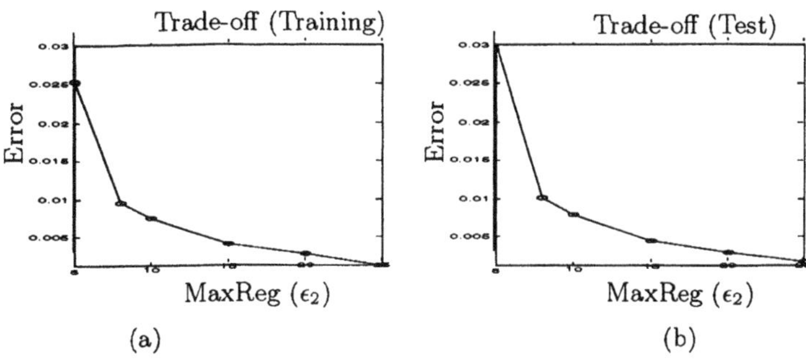

Fig. 4. Trade-offs between error and compactness of the rule base: (a) for training data; (b) for test data

Table 1. Approximation results for noiseless data

Approach	RMSE tr	RMSE ts	Size
HGFS	0.008	0.009	10
ANFIS	0.003	0.003	25
MLP	0.013	0.014	25

of neurons or fuzzy rules of the models (Size). The training error is computed for $N = 225$, whereas testing error is found using 2500 samples equally spaced within $\Omega = [-10, 10]^2$. The $\boldsymbol{t}$-norm of antecedent aggregation (AND) resulted from evolution was the algebraic product.

Table 1 shows that HGFS has a satisfactory performance because it achieves a good trade-off between accuracy and interpretability, and has a high degree of design autonomy.

Figure 5 depicts the generalization capabilities of the models of Table 1.

The results also emphasize the compact nature of the model derived by HGFS, because with only 10 fuzzy rules the HGFS achieved an accuracy level comparable with the ANFIS and MLP approaches.

6.2 Function Approximation with Noisy Data

When noisy data is used to construct system models, trade-offs between bias and variance of the modeling error becomes a major concern because this is intrinsically related with the accuracy and generalization capabilities of the model. To gain an insight on such an important issue, next we consider a set k independent sets of training samples

$$\Phi = \{T_1 \cdots, T_k\} ,$$

$T_i = \{(\boldsymbol{x}_{1i}, y_{1i}), (\boldsymbol{x}_{2i}, y_{2i}), \cdots, (\boldsymbol{x}_{Ni}, y_{Ni})\}$, to estimate the parameters of the consequent function of TS fuzzy models.

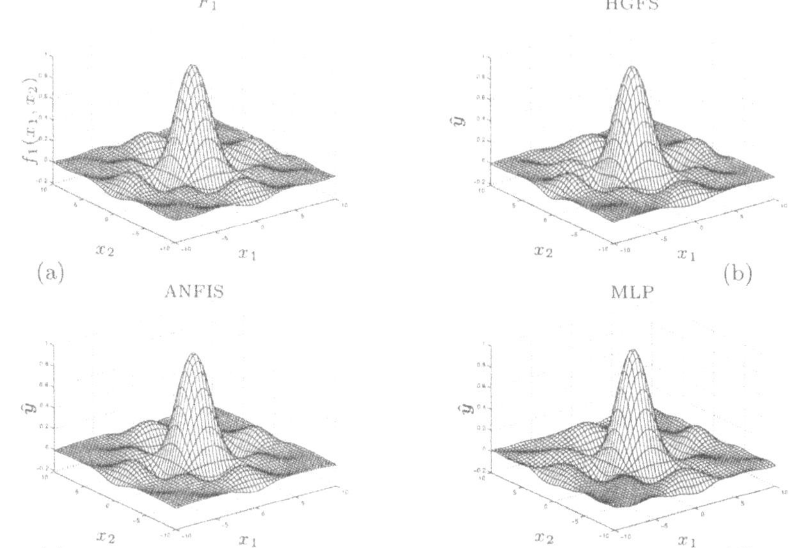

Fig. 5. (**a**) Original function F_1; approximation produced by (**b**) HGFS; (**c**) ANFIS; (**d**) Neural Network (MLP)

After estimation of the consequent parameters $\boldsymbol{w}$ from a training sample $T = T_i$, the fuzzy model can be used to predict $\hat{f}(\boldsymbol{x})$. The difference between the prediction $\hat{f}(\boldsymbol{x})$ and the realization y of the random variable Y is the prediction error [1]. The quality of $\hat{f}$ as a predictor of Y, for a specific realization T of the training samples, can be given as the expectation with respect to $p(Y|\boldsymbol{x})$. Assuming $\mathrm{E}[Y|\boldsymbol{x}] = f(x)$, the squared error for a fixed T and $\boldsymbol{x}$ is:

$$\mathrm{E}\left[\left(Y - \hat{f}(\boldsymbol{x}|T)\right)^2\right] = \underset{\substack{\uparrow \\ \text{Reducible part}}}{\left(f(\boldsymbol{x}) - \hat{f}(\boldsymbol{x}|T)\right)^2} + \underset{\substack{\uparrow \\ \text{Irreducible part}}}{\mathrm{E}\left[(y - f(\boldsymbol{x}))^2\right]} .$$

The reducible part of the error measures the quality of the estimate $\hat{f}(\boldsymbol{x}|T)$ for a particular T. The quality of the estimator $\hat{f}$ is given by the expectation over all possible sets of training samples $\Phi = \{T_1 \cdots, T_\infty\}$, and can be decomposed into bias and variance components:

$$\mathrm{E}_T\left[\left(f(\boldsymbol{x}) - \hat{f}(\boldsymbol{x}|T)\right)^2\right] = \underset{\substack{\uparrow \\ \text{bias}}}{\left(f(\boldsymbol{x}) - \mathrm{E}_T[\hat{f}(\boldsymbol{x}|T)]\right)^2} + \underset{\substack{\uparrow \\ \text{variance}}}{\mathrm{E}_T\left[\left(\hat{f}(\boldsymbol{x}|T) - \mathrm{E}_T[\hat{f}(\boldsymbol{x}|T)]\right)^2\right]}$$

Assuming $T_i = (\boldsymbol{x}_{pi}, y_{pi})$, $p = 1, \cdots, N$ we have:

$$\text{MSE Tr} = \frac{1}{N} \sum_{p=1}^{N} (\mathrm{E}_{bias}(\boldsymbol{x}_p) + \mathrm{E}_{var}(\boldsymbol{x}_p)) . \quad (6)$$

Next we assume that data available to approximate function f_2 is disturbed by a random, normally distributed noise $\mathcal{N}[\mu, \sigma]$, as follows:

$$F_2 : [0,1] \rightarrow \Re, \ F_2 = f_2 + \mathcal{N}[\mu, \sigma]$$

where $f_2(x) = \sin(6\pi x) + 2x + \exp(-(1/0.03^2)(x-0.59)^2)$, $\mu = 0$ and $\sigma = 0.3$. See Fig. 7(a). This function is interesting because it is smooth, but has a dominant nonlinearity given by $\sin(6\pi x)$, and a local $\exp(-(1/0.03^2)(x - 0.59)^2)$.

We consider 1000 sets of training data samples $\Phi_{Tr} = \{T_1 \cdots, T_{1000}\}$. For every T_i, each of which with $N = 300$ pairs of noisy data, the consequent parameters are estimated by the HGFS and the predicted values $\hat{f}(\boldsymbol{x})$ obtained. Table 2 shows the mean squared error given by (6), as well as its bias and variance components in the approximation of function f_2. The average test error $\overline{\text{MSE}}$ Ts is obtained over the mean of MSE for each estimation using $\Phi_{Ts} = \{Ts_1, .., Ts_{1000}\}$, where each Ts_i is formed by 3500 data points. For these simulations we assume 6 trapezoidal membership functions uniformly distributed along the range of the input, with a small overlapping degree ($\gamma = \alpha = 0.1$).

Table 2. Bias and variance components for a small overlapping degree

Approach	Bias	Variance	MSE Tr	$\overline{\text{MSE}}$ Ts
HGFS Global quad	0.008	0.005	0.01	0.16
HGFS Global const	0.21	0.002	0.2	2.48
HGFS Local quad	0.068	0.004	0.07	0.85
HGFS Local const	0.22	0.002	0.2	2.52

Table 2 shows four different combinations for the consequent part of the rules, ranging from the most accurate and less interpretable models (HGFS with global estimation of nonlinear consequent functions as given by (4)) to the less accurate and most interpretable ones (HGFS with local estimation of constant consequent functions as given by (5), with $Q = 1$). The results show that when we are only interest in accuracy, the use of global approaches to optimize nonlinear consequent functions is the best choice. However, when the focus is the interpretability, the combination of local optimization and constant consequent functions can be more appropriate. This can be justified by the following advantages of constant consequent functions: the transformation of real numbers into linguistic labels, and the view of local description of TS models.

Figure 6 illustrates the results shown in Table 2. Here, the bias and variance components of the mean square error (for Φ_{T_r} and Φ_{T_s}) can be visually compared. It can be noticed that function f_2 would require a higher granularity of the membership functions around $x = 0.6$, and this can not be devised a priori. In Fig. 6, the universe has a uniform partition leading to a high degree of interpretability. However, all the approaches were unable to map the local nonlinearities. This suggests that membership functions tuning seems to be necessary to achieve a desired level of accuracy.

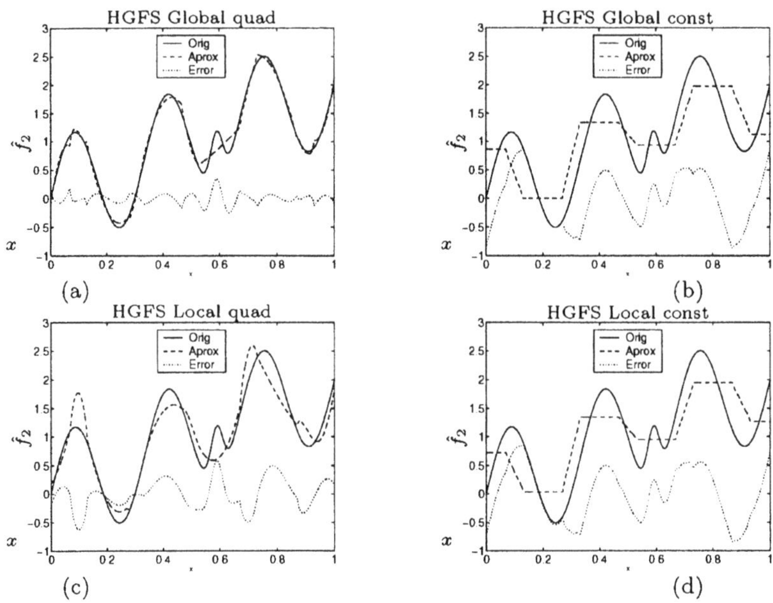

Fig. 6. Consequent parameter optimization with noisy data (F_2): (**a**) global optimization of nonlinear consequent functions; (**b**) global optimization of constants; (**c**) local optimization of nonlinear consequent functions; (**d**) local optimization of constants

Although the previous analysis is important from theoretical point of view, in practice only one realization may be available, instead of many. Henceforth, to test the approaches within a framework often encountered in practice, the evolutionary design process is performed using a single realization T_i whose samples are split into training and test sets.

Table 3 shows the mean squared error for training (MSE tr) and testing (MSE ts) of best fuzzy systems for all the evolutionary approaches (at generation 100), compared with ANFIS trained for 100 epochs. We assume

N equally spaced samples (x_p, y_p), $p = 1, \cdots, N = 300$ during training as illustrated in Fig. 7(a), and 3500 testing points. The constraint associated with simplicity is defined by $\epsilon_3 = 6$ to indicate a maximum of 6 membership functions for variable x. Since we have only one variable this also means the maximum of rules in each rule-base ($\epsilon_2 = 6$). The remaining constraints are the same as defined in the case of noiseless data.

Table 3. Function f_2 (results after evolution or training process)

Approach	MSE tr	MSE ts	Rules
ANFIS	0.0732475	0.103288	6
HGFS Global quad	0.0706621	0.0982288	6
HGFS Global const	0.139271	0.145969	6
HGFS Local quad	0.0955811	0.127486	5
HGFS Local const	0.171676	0.174976	6

Figure 7 illustrates the generalization capabilities of the approaches shown in Table 3. The higher flexibility associated with nonlinear consequent functions and global estimation approaches does not imply over-fitting because any excessive flexibility is pruned every time the corresponding least square procedure is ill conditioned [8]. For the ϵ constraints chosen, the only approach that was able to map the local nonlinearity was the one that combines global optimization and nonlinear consequent functions (HGFS *Global quad*). For the remaining case this particularity has not been observed because the restriction in the number of rules forces the approximation process to pay attention to the dominant nonlinearities only. Fuzzy systems with constant consequent functions require more than 6 rules to simultaneously describe the dominant and local nonlinearities. The number of rules required is inversely proportional to the smoothness of the function to be approximated, and this is a strong disadvantage of rules with linear consequent functions.

Figure 8 illustrates the universe partition and rule-base parameters of the best fuzzy system of HGFS *Global quad* approach at generation 100. In Fig. 8, each fuzzy rule has the form:

$$R_j : \text{ If } X \text{ is } A^j \text{ then } Y \text{ is } g_j = w_{j0} + w_{j1}x + w_{j2}x^2 \; .$$

As shown by Fig. 7(c) and 8, for HGFS *Global quad* approach, even under hard constraints, the required higher granularity around $x = 0.6$ has been achieved and attends the interpretability requirement for the universe partition.

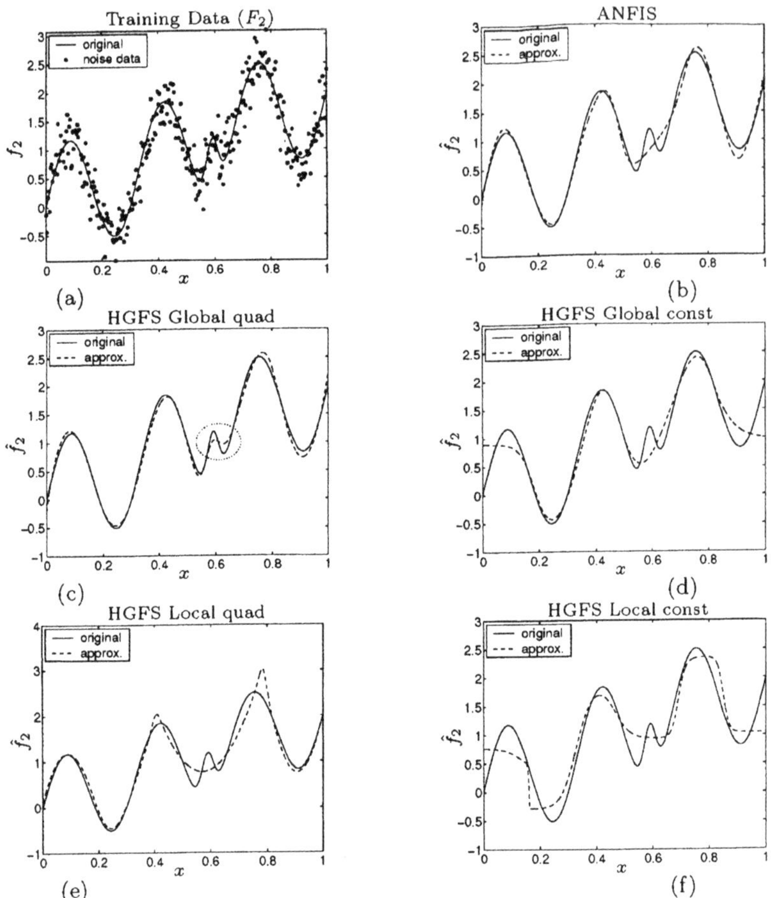

Fig. 7. Approximation with noise data (generation 100): (**a**) Original Function; Approximation by: (**b**) ANFIS (global optimization of linear consequent functions); (**c**) HGFS with global optimization of nonlinear consequent functions (good performance at the region in evidence); (**d**) HGFS with global optimization of constant consequent functions; (**e**) HGFS with local optimization of nonlinear consequent functions; (**f**) HGFS with local optimization of constant consequent functions

6.3 Pattern Classification

The performance of the HGFS has also been evaluated with the Iris classification data set. Popularized by Fisher [10], this three-class classification problem has 150 four-dimensional vectors representing 50 samples of each species: *Iris setosa* (Cl_1), *Iris versicolor* (Cl_2), and *Iris virginica* (Cl_3). The four dimensional pattern vector $x_p = (x_{1p}, x_{2p}, x_{3p}, x_{4p}), p = 1, \cdots, 150$, has been normalized within the range [0,1]. The attributes are as follows: x_{1p} is

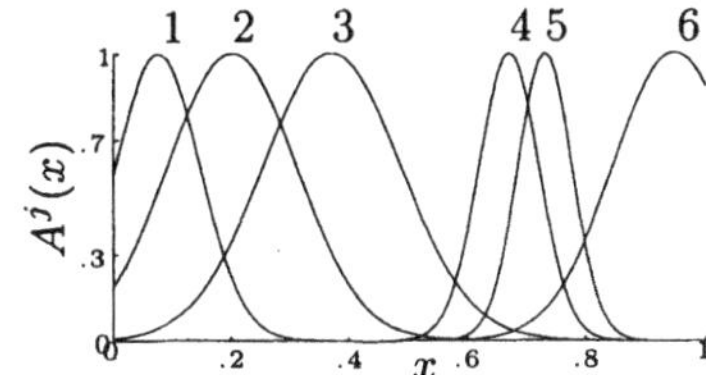

Rule	A^j	w_{j0}	w_{j1}	w_{j2}
1	1	6.0	4.2	0
2	2	-20.9	167.9	-598.8
3	3	45.8	-95.2	26.7
4	4	445.3	-1451.5	1193.0
5	5	-398.4	969.2	- 594.5
6	6	155.0	-339.2	186.4

Fig. 8. The partition set and rule base for the best system of the HGFS *Global quad* approach at generation 100

the sepal length, x_{2p} is the sepal width, x_{3p} is the petal length, and x_{4p} is the petal width.

The training (Tr) and testing (Ts) sets have 75 patterns each (25 patterns randomly chosen from each class). The purpose is twofold: to evaluate the generalization capabilities in a classification context, and to compare with the results provided by Castellano and Fanelli [2], because their approach (Compact minFuzzy [2]) has outperformed (in terms of trade-off between accuracy/compactness) many other approaches that solved the Iris classification problem.

Here, HGFS with non-linear consequents optimized by the global method is used to solve the multi-objective problem of fuzzy modeling based on ϵ-constrained method proposed in Sect. 3. To test the benefits of this methodology, the proposed approach is compared with HGFS based on a weighted method, following the one proposed in [19] which assumes that the fitness of each fuzzy system is given by:

$$\text{fitness} = W_{error} \frac{1}{\text{Error}_{tr}} - W_{comp}\, m \,,$$

where W_{error} and W_{comp} are positive weights, and m define the total of fuzzy rules in the rule base.

If we denote by f the fuzzy system evolved by HGFS, then the following assignment is devised to relate a value of $f(\boldsymbol{x}_p)$ with each class Cl_i: $Cl_1 \rightarrow f(\boldsymbol{x}_p) = 1$; $Cl_2 \rightarrow f(\boldsymbol{x}_p) = 2$; $Cl_3 \rightarrow f(\boldsymbol{x}_p) = 3$.

For the HGFS approaches, the population sizes were fixed as 10, 70 and 30 for the partition set, rule-base and fuzzy system levels, respectively. The selection for reproduction was, again, based on the tournament strategy and the elitist selection, with the same crossover and mutation rates adopted in Sect. 6.1. For comparison purposes, in the ϵ-constrained method, the constraint ϵ_2 is defined as 4 resulting in a maximum of 4 fuzzy rules in each rule-base; the constraint $\epsilon_3 = 2$ fixed a maximum of 2 linguistic terms for each variable (equally spaced in the range $[0, 1]$). For the weighted method, different weights have been tested, but the best results were achieved with $W_{erro} = 10$ and $W_{comp} = 0.1$. The ANFIS system assumed the same partition for all variables, that is, each with 2 Gaussian membership functions. This means a total of 16 fuzzy rules.

Table 4 shows the classification errors for the test set obtained by Castellano and Fanelli [2], ANFIS, and HGFS.

Table 4. IRIS data simulation results

Approach	Cycles	Rules	Misclassification	
			Training	Test
ANFIS	100	16	0	6
Compact minFuzzy [2]	-	5	0	4
HGFS (weighted method)	100	7	0	7
HGFS (ϵ-constrained method)	100	4	0	3

As Table 4 shows, the fuzzy classifier provided by proposed approach (HGFS based on the ϵ-constrained method) outperforms the other approaches because it achieved the best compromise between classification performance (classification rate of 96% in the test set) and system complexity (a total of 4 fuzzy rules). The results emphasize the benefits of translating the multi-objective problem into a single-objective ϵ-constrained problem, for which the solution may be produced by a hierarchical evolutionary process based on genetic algorithms.

It is important to point out that some approaches presented in the literature achieve 100% of correct classification in Iris data when all the 150 available patterns were used in the training process. This means that the cardinalities of training set Tr and test set Ts are given by $S_{Tr} = 150$ and $S_{Ts} = 0$, respectively. When we consider all the available patterns in the training process, the HGFS was also able to correctly classify all the patterns. Although, since the aim here is to test the generalization capability of the proposed approach, the attention has been given to the simulations considering patterns distributed among training and test sets.

At the end of the evolutionary process, for the fuzzy system considering HGFS (ϵ-constrained method), the resulting rule-base produced is detailed bellow:

R_1 : If x_3 is *low* and x_4 is *low* then $y = 0.86 - 0.3x_1 + 0.19x_2 + 0.31x_3 + 0.09x_4 - 0.14x_1^2 - 0.23x_2^2 - 2.86x_3^2$

R_2 : If x_1 is *low* and x_2 is *low* and x_3 is *high* then $y = 1.15 - 1.97x_1 - 6.06x_2 + 1.82x_3 + 3.41x_4 + 5.18x_2^2$

R_3 : If x_1 is *high* and x_3 is *low* and x_4 is *high* then $y = 9.7 - 0.17x_1 + 1.97x_2 + 2.65x_3 - 33.7x_4 + 0.36x_1^2 - 3.44x_2^2 - 0.8x_3^2 + 31x_4^2$

R_4 : If x_1 is *high* and x_2 is *high* and x_4 is *low* then $y = 1.52 - 0.49x_1 + 0.2x_2 + 3.83x_3 + 0.32x_4 + 0.29x_1^2 - 0.42x_2^2 - 2.42x_3^2 - 0.24x_4^2 + 0.33x_1x_2x_3x_4$

The antecedent aggregation '**and**' evolved is given by (2) with $p_t = 1.66$. The linguistic terms *low* and *high* are shown in Fig. 9. As it can be noted, a good level of accuracy is achieved and all constraints, needed to ensure the interpretability criterion, are fulfilled: visibility, simplicity (*don't care* conditions in all the fuzzy rules), compactness (only 4 fuzzy rules), and consistency.

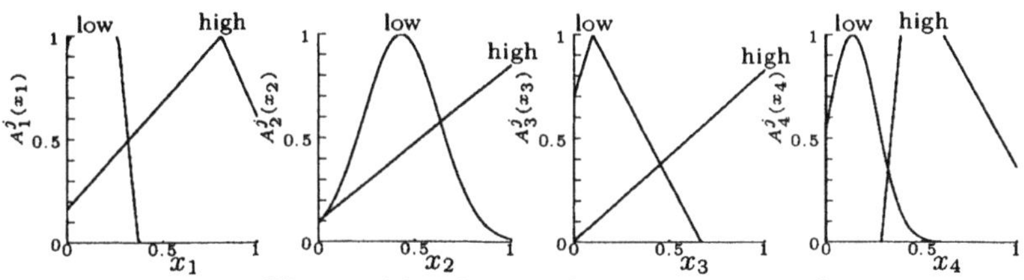

Fig. 9. The partition for variables x_1, x_2, x_3, and x_4

7 Conclusions

This chapter has discussed accuracy, interpretability and autonomy criteria in fuzzy rule-based modeling within the hierarchical evolutionary approach. Accuracy, interpretability and autonomy were envisioned from the multi-objective decision making framework and associated single-objective ϵ-constrained problems. Therefore, in addition to precision, interpretability of the rule-based model is achieved through a set of constraints imposed for the search process. Interpretability is based on visibility, simplicity, compactness and consistency goals. The use of HGFS improves performance in terms of accuracy, satisfies interpretability constraints, and provides an automatic adjustment mechanism for a number of critical parameters which increases autonomy by minimizing user intervention.

The choice among different classes of TS consequent functions and two alternatives to find their parameters produce models with different tradeoffs between accuracy and interpretability: less-accurate-more-interpretable-models that adopt constant consequent functions optimized by local methods, and less-interpretable-more-accurate-models with nonlinear consequent functions optimized by global methods.

Acknowledgments

Myriam Regattieri Delgado acknowledges CEFET/PR and CAPES–PICDT. Fernando Von Zuben the CNPq grant 300910/96-7, and Fernando Gomide the CNPq grant 300729/86-3 for their support.

References

1. Berthold M., Hand D.J. (1999) Intelligent Data Analysis: an Introduction. Springer-Verlag
2. Castellano G., Fanelli A.M. (2000) A Staged Approach for Generation and Compression of Fuzzy Classification rules. In: The 9th IEEE Int. Conf. on Fuzzy Systems, USA, May, 2000. 42-47
3. Chankong, V. and Haimes Y.Y. (1983) Multiobjective Decision Making - Theory and Methodology. Elsevier
4. Cordón O., Herrera F., Hoffmann F. and Magdalena L. (2001) Genetic Fuzzy Systems. Evolutionary Tuning and Learning of Fuzzy Knowledge Bases. Advances in Fuzzy Systems - Applications and Theory, World Scientific.
5. Cordón O., del Jesus M.J., Herrera F. and Villar P. (2001) A Multiobjective Genetic Algorithm for Feature Selection and Granularity Learning in Fuzzy-rule-based Classification Systems. In: 9thIFSA World Cong. and 20thNAFIPS Int. Conf., Canada, 2001. 1253-1258
6. Cordón O., Gomide F., Herrera F., Hoffmann F and Magdalena L. (2001) Ten Years of Genetic Fuzzy Systems: Current Framework and New Trends. In: 9thIFSA World Cong. and 20thNAFIPS Int. Conf., Canada, 1241-1246
7. Delgado M.R., Von Zuben F. and Gomide F. (2000) Optimal Parameterization of Evolutionary Takagi-Sugeno Fuzzy Systems. In: 8th Conf. on Inf. Processing and Management of Uncert. in Knowledge-Based Systems, Spain, July, 2000. 650-657
8. Delgado M.R., Von Zuben F. and Gomide F. (2001) Hierarchical Genetic Fuzzy Systems. Information Sciences. **136**, 29-52
9. Espinosa J., Vandewalle J. (2000) Constructing Fuzzy Models with Linguistic Integrity from Numerical Data - AFRELI Algorithm. IEEE Trans. on Fuzzy Systems. **8**, 591-600
10. Fisher R.A. (1936) The Use of Multiple Measurements in Taxonomic Problems. Ann. Eugen. 179-188
11. Gacôgne .L (1997) Research of Pareto Set by Genetic Algorithm, Application to Multicriteria Optimization of Fuzzy Controller. In: 5th Eur. Cong. on Intelligent Techniques and Soft Computing, Germany, Sep., 1997. 837-845
12. Golub G.H., Van Loan C.F. (1996) Matrix Computations. Johns Hopkins Series in the Mathematical Sciences. Johns Hopkins University Press
13. Gómez-Skarmeta A.F., Jiménez F. and Ibáñez J. (1998) Pareto-optimality in Fuzzy Modeling. In: 6th Eur. Cong. on Intelligent Techniques and Soft Computing, Germany, Sep., 1998. 694-700
14. Gonzales A., Perez R. (1998) Completeness and Consistency Conditions for Learning Fuzzy Rules. Fuzzy Sets and Systems. **96**, 37-51
15. Heider H., Drabe T. (1997) A Cascaded Genetic Algorithm for Improving Fuzzy Systems Design. Int. Journal of Approximate Reasoning. **17**, 351-368
16. Herrera F., Magdalena L. (1998) Introduction: Genetic Fuzzy Systems. Int. Journal of Intelligent Systems. **13**, 887-890
17. Hoffmann F., Nelles O. (2000) Structure Identification of TSK-fuzzy Systems Using Genetic Programming. In: 8th Conf. on Inf. Processing and Management of Uncert. in Knowledge-Based Systems, Spain, July, 2000. 438-445
18. Ishibuchi H., Nakashima T. (1999) Genetic-algorithm-based Approach to Linguistic Approximation of Nonlinear Functions with Many Input Variables. In: 8th IEEE Int. Conf. on Fuzzy Systems, Korea, Aug., 1999. 779-784

19. Ishibuchi H., Takeuchi D. and Nakashima T. (2001) GA-based Approaches to Linguistic Modeling of Nonlinear Functions. In: 9thIFSA World Cong. and 20thNAFIPS Int. Conf., Canada, July, 2001. 1229-1234
20. Jang J.-S. (1993) ANFIS: Adaptive-network-based Fuzzy Inference Systems. IEEE Trans. on Syst., Man, and Cybern. **23**, 665-685
21. Jin Y. (2000) Fuzzy Modeling of High-Dimensional Systems: Complexity Reduction and Interpretability Improvement. IEEE Trans. on Fuzzy Systems. **8**, 212-221
22. Karr C.L. (1991) Genetic Algorithms for Fuzzy Controllers. AI Expert. **6**, 26-33
23. Lin C.-T., George Lee C.S. (1996) Neural Fuzzy Systems. Prentice Hall
24. Michalewicz Z. (1996) Genetic Algorithms + Data Structures = Evolution Programs. Springer-Verlag
25. Nozaki K., Ishibuchi H. and Tanaka H. (1997) A Simple but Powerful Heuristic Method for Generating Fuzzy Rules from Numerical Data. Fuzzy Sets and Systems. **86**, 251-270
26. Pedrycz W., Gomide F. (1998) An Introduction to Fuzzy Sets: Analysis and Design. MIT Press, Cambridge
27. Setnes M., Babuška R. and Verbruggen H.B. (1998) Rule-Based Modeling: Precision and Transparency. IEEE Trans. on Syst., Man, and Cybern. - Part C. **28**, 165-169
28. Steels L. (1995) When Are Robots Intelligent Autonomous Agents? Journal of Robotics and Autonomous Systems. **15**, 3-9
29. Takagi T., Sugeno M. (1983) Derivation of Fuzzy Control Rules from Human Operator's Control Actions. In: IFAC Symp. on Fuzzy Inf., Knowledge Repres. and Decis. Analysis, France, July, 1983. 55-60
30. Yen J., Wang L. and Gillespie C.W. (1999) Improving the Interpretability of TSK Fuzzy Models by Combining Global Learning and Local Learning. IEEE Trans. on Fuzzy Systems. **6**, 530-537

About the trade-off between accuracy and interpretability of Takagi-Sugeno models in the context of nonlinear time series forecasting

Antonio Fiordaliso

Faculté Polytechnique de Mons,
Service de Mathématique et de Recherche Opérationnelle
Rue de Houdain, 9
7000, Mons, Belgique

Abstract. The focus of this chapter is related to the question of how to find appropriate Takagi-Sugeno (TS) rules in the framework of (chaotic) time series forecasting. We propose a generalization of the conventional TS system (GTS) allowing to evaluate the importance of any rule in the inference process. The added value of this feature has been put in light on two well-known chaotic time series.

Local (clustering-based) and global (gradient-based) learning strategies for GTS systems are compared in terms of interpretability and accuracy. It appears that there is an unavoidable compromise between these two objectives. Global learning seems to be superior in term of accuracy but cannot achieve the same level of interpretability as the local approach.

We also present an application of the GTS model in the field of forecasts combination with the target of generating interpretable final rules. This has led us to put additional constraints in the model. More precisely, we force the local linear models (rules conclusions) to achieve convex combination of the forecasts (input patterns). The proposed system may be useful for a wide range of applications where a consensus is required, including forecasts synthesis, controllers and patterns classifiers aggregation.

1 Introduction

In recent years, fuzzy systems have attracted an enormous interest. These methods have proven to be a powerful tool for modelling and control of processes not easily tractable with conventional approaches [14]. Their intrinsic characteristics have been massively exploited in a wide range of applications related to decision making, process control, reasoning under uncertainty, functions approximation, patterns recognition, nonlinear filtering and image processing (e.g., see [14,32,41]). Applications in the field of time series analysis and forecasting can also be found (e.g., see [3,21,59]).

Compared to "black-box" modelling techniques like neural networks, fuzzy rule-based models are *to a certain degree* more transparent to interpretation and analysis. But in many practical situations where the identification of fuzzy models is data driven, interpretability is not automatically achieved.

One reason of this comes from the fact that such models may exhibit some redundancy and hence, unnecessary complexity.

These last years, an increasing attention has been focused on the development of formal procedures for fuzzy models analysis and design. Several methods have been proposed to automatically generate a *concise* set of fuzzy rules for a given problem. Among the proposed techniques, let us cite for example, incremental and pruning strategies (e.g., see [3,18]), rules simplification techniques (e.g., see [31,51]) or linguistic approximation methods (e.g., see [49]). However, the identification of a concise set of *interpretable* rules is still an extremely complex process with no guarantee of success.

The process of identifying a fuzzy inference system generally requires two types of tuning. The first one (structural tuning) concerns the structure of the rules and deals with problems such as the number of fuzzy if-then rules and the number of membership functions for each input. Though several methods have been proposed to automate this task (see for example [9,22,30,38,51]), the identification of rules structure is still an extremely difficult process where human intervention is generally required. Once a satisfactory structure is available, it is relatively easy to adjust the membership functions (parametric tuning). Several techniques have been developed including gradient-based algorithms [22,43], Karr's genetic algorithm [30] or Lin's reinforcement learning method [38].

Section 3 introduces some learning strategies commonly used for Takagi-Sugeno systems (TS, in the sequel). We distinguish two families: the local learning methods and the global ones. We illustrates the fact that a local learning strategy is recommended when the building of interpretable rules is a main objective. As a contrary, when accuracy is the main objective, a global learning approach is recommended.

Another question of crucial importance for fuzzy systems identification is related to the adequate number of rules to use. We know a large class of fuzzy sytems (including TS systems) are universal approximators [6], but how many rules are really needed? From a learning-generalization point of view, it is well known that overparametrized structures often fail to generalize on new data (overfitting). Hence, the structure of the fuzzy model to discover must be *rich enough* to learn the inputs-outputs associations of the training set, but *not too rich* (parcimony principle) in order to avoid the system to model the noise inherent to the data set. This preoccupation also exists in other disciplines such as statistics or neural networks: the use of information criteria [48], complexity penalization techniques (weight decay [27], weight elimination [57]) or regularization methods [45] are some illustrations of this trade-off.

In section 4, we propose a decremental algorithm which is clearly in the spirit of the parcimony principle. The key idea is to associate a parameter with each rule corresponding to the rule importance in the inference process. Then, the algorithm attempts to find redundant or least sensitive rules in order to remove them until a stable structure is found. Applications of that autostructuration algorithm will be considered in the field of highly complex nonlinar time series forecasting.

This chapter is organized as follows. TS systems are briefly described in section 2 and some classical parametric adjustment strategies are discussed and compared in section 3. Section 4 presents the decremental approach and illustrates the ability of the algorithm to detect redundancies in a TS system. This feature is successfully exploited for chaotic time series forecasting in section 5. In section 6, the proposed model is used to combine several forecasting models, focusing our attention on the interpretability of the final combining rules. The last section provides the conclusions. In order to facilitate the reading of this chapter, an appendix contains a list of abbreviations that will be used in the sequel along with their meaning.

2 Takagi-Sugeno fuzzy systems

2.1 Design

Takagi-Sugeno fuzzy systems [52] form a very special class of fuzzy systems because the conclusion of each rule is crisp (not a fuzzy set). A typical single antecedent fuzzy rule in a Takagi-Sugeno model of order d (TS) has the following form:

$$R_k: \quad \textbf{If } \boldsymbol{x}_t \text{ is } A_k \textbf{ then } \hat{y}_{t,k} = P_k^{(d)}(\boldsymbol{x}_t), \qquad k = 1, 2, \ldots, c$$

where $\boldsymbol{x}_t$ is the input variable ($\boldsymbol{x}_t \in \boldsymbol{R}^n$), A_k is a fuzzy set of $\boldsymbol{R}^n$ and $P_k^{(d)}(\boldsymbol{x}_t)$ is a polynomial of order d in the components $\boldsymbol{x}_{t,i}$ of $\boldsymbol{x}_t$. In the sequel, we will suppose $d = 1$. For convenience, we will write the conclusion $\hat{y}_{t,k}$ as

$$\hat{y}_{t,k} = \boldsymbol{x}_t' \boldsymbol{\beta}_k \tag{1}$$

where $\boldsymbol{\beta}_k = (\beta_1, \ldots, \beta_n)'$. An intercept is allowed in the conclusion $\hat{y}_{t,k}$ if we suppose $x_{t,1} = 1$ (bias term). Output $\hat{y}_t$ relative to input $\boldsymbol{x}_t$ obtained after aggregating a set of c TS-rules can be written as a weighted sum of the individual conclusions:

$$\hat{y}_t = \sum_{k=1}^{c} \pi_k(\boldsymbol{x}_t) \hat{y}_{t,k} \tag{2}$$

with

$$\pi_k(\boldsymbol{x}_t) = \frac{\mu_{A_k}(\boldsymbol{x}_t)}{\sum_{j=1}^{c} \mu_{A_j}(\boldsymbol{x}_t)} \tag{3}$$

where μ_{A_k} is the membership function related to the fuzzy set A_k.

Links can be drawn between TS systems and other models such as Radial Basis Functions Neural Networks [45], Switching Regression Models [46], Mixture of Distributions [40], Fuzzy c-Regression Models [26] or Mixture of Experts [29].

2.2 The universal approximation property

It has been shown that TS systems (along with several other fuzzy models) are universal approximators (see [25] for a review). This mainly motivates their use in fields related to functions approximation such as time series forecasting or control. Several proofs of the universal approximation property are based on the Stone-Weierstrass theorem (see e.g. [55,56]). Other papers show the uniform approximation capability of specific families of fuzzy systems not fulfilling the hypotheses of the Stone-Weierstrass theorem. This mainly concerns fuzzy systems that do not form an algebra (see e.g. [33,42,60]). Other existing proofs are adaptations of some universal approximation results known in the field of neural networks (see e.g. [5,6]).

A universal approximation theorem is an existence theorem. As such, it does not help in specifying the satisfactory model. For this reason, the results on the universal approximation property of various models (including fuzzy systems) are rather academic. Note also that condition for arbitrary accuracy is the exponential growth of the size of the system (and consequently, the computational time) (see e.g. [34–36,42]). There must, therefore, be a trade-off between accuracy (resolution) and complexity.

3 System parameters identification

Let us see how a TS system can be tuned to approximate a mapping $\mathcal{F}$: $\boldsymbol{R}^p \rightarrow \boldsymbol{R}$ when a collection of "examples" $(\boldsymbol{x}_t, y_t)$ $(t = 1, \dots, N)$ is available to estimate the unknown parameters $\boldsymbol{\theta}$ ($\boldsymbol{\theta}$ is a generic notation representing any parameter to adjust, t is a temporal index taking values in $\{1, \dots, N, N+1, \dots, T\}$). The N "examples" $(t = 1, \dots, N)$ form a set which will be referred to as the learning set or in-sample data, while the remaining examples $(t = N+1, \dots, T)$ form the test set (out-of-sample data). The learning set is used to estimate the parameters (identification phase), while the test set will be used to test the system in the presence of inputs not previously seen or learned (generalization phase).

3.1 Global learning

There are basically two kind of strategies to train TS systems. The first one is a global learning where tuning is achieved by minimizing a global cost function E on the learning set. Let us denote by e_t the approximation error related to pattern $(\boldsymbol{x}_t, y_t)$:

$$e_t = y_t - \hat{y}_t. \tag{4}$$

It is usual to take a quadratic error function E of the following form

$$E = \sum_{t=1}^{N} E_t$$

where $E_t = \frac{1}{2}e_t^2$ is the instantaneous quadratic error.

Output $\hat{y}_t$ given in equation (2) is derivable with respect to any parameter to adjust, so that the global error E is a derivable (nonlinear) function of $\boldsymbol{\theta}$. Gradient descent techniques can be used to minimize the error metric E online or off-line. More sophisticated nonlinear optimization methods, such as quasi-Newton or Levenberg-Marquardt [39] methods can also be applied.

Also, it is worthwile to notice that the cost E is linear in $\boldsymbol{\beta}_k$. As a result, these parameters could be identified by solving an (overdetermined) linear set of equations in the least squares sense. In view of this, pseudo-inverse methods and orthogonal decomposition techniques such as SVD (Singular Values Decomposition) or QR decomposition are useful.

3.2 Local learning

A second class of methods is known as local learning techniques (see for example [1,26,29]). As opposed to global learning, local learning strategies encourage the linear experts (parametrized by $\boldsymbol{\beta}_k$) to compete for the minimization of E. Product space clustering [1] is one practical method to achieve this competition. This method simply consists in producing clusters in the product space (cartesian product of input and output spaces) and then, to project these clusters in the input space. Local linear models can then be adjusted by weighted least squares methods. If we use fuzzy clustering techniques with adaptive metrics [19,24], one can obtain input membership functions with individual metric by projecting the variance-covariance matrixes of the product space clusters onto the input space. This allows for asymmetrical membership functions.

3.3 Trade-off between accuracy and interpretability

To illustrate the differences between local and global learning, let us consider the $N = 200$ data points (x_t, y_t) generated by the following process:

$$y_t = 10^{-4}x_t^3 \sin(10^{-3}x_t^2) + \epsilon_t, \quad x_t \in [0, 100] \tag{5}$$

where ϵ_t is a Gaussian noise with mean $\mu = 0$ and variance $\sigma^2 = 25$. These data points are represented in Figs. 1a and 2a ('+' symbols). A TS system with $c = 4$ rules has been trained respectively globally and locally to approximate this data set.

Global learning: LM+SVD In this example, we have used hyperbasis functions (generalized Gaussians [45]) for the coding of the input linguistic terms:

$$\mu_{A_k}(\boldsymbol{x}_t) = G(\|\boldsymbol{x}_t - \boldsymbol{m}_k\|_{\boldsymbol{S}_k}^2) \tag{6}$$

$$G(y) = e^{-\frac{y}{2}} \tag{7}$$

$$\|\boldsymbol{x}_t - \boldsymbol{m}_k\|_{\boldsymbol{S}_k}^2 = (\boldsymbol{x}_t - \boldsymbol{m}_k)'\boldsymbol{S}_k'\boldsymbol{S}_k(\boldsymbol{x}_t - \boldsymbol{m}_k) \tag{8}$$

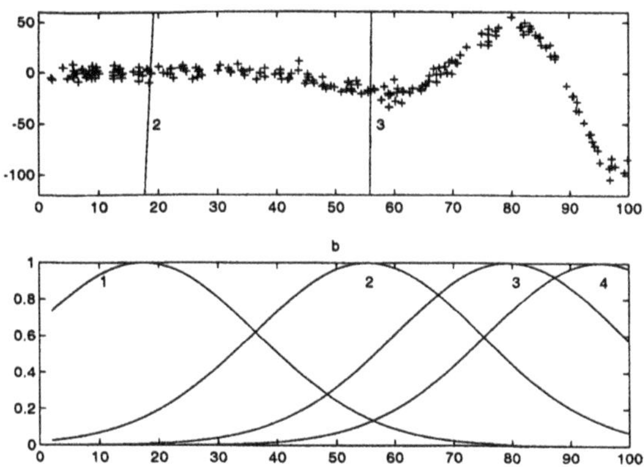

Fig. 1. Example 3.3, learning method LM+SVD. (**a**) Data points (x_t, y_t) (+) and the linear models adjusted by SVD. (**b**) Input membership functions

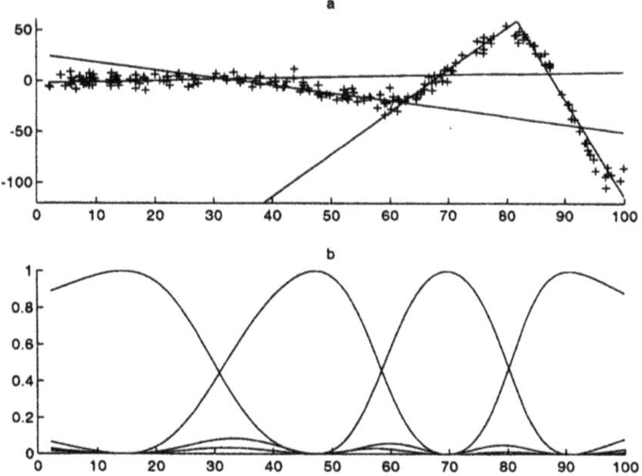

Fig. 2. Example 3.3, learning method GK+WLS. (**a**) Points (x_t, y_t) (+) and the linear models adjusted by WLS. (**b**) Input membership functions

where $\boldsymbol{S}_k$ are symmetrical matrixes, meaning that the resulting Gaussian can rotate around its center. Initialization of the centers $\boldsymbol{m}_k$ and metrics $\boldsymbol{S}_k$ is explained in [18]. The membership functions are adjusted by the Levenberg-Marquardt (LM) method and are shown in Fig. 1b. Fig. 1a also represents the linear models identified by SVD decomposition.

Local learning: GK+WLS The fuzzy clustering method of Gustafson and Kessel (GK) [24] is used to cluster the product space $\boldsymbol{R}^2$ in four overlapping regions (the overlapping parameter is set to its standard value, i.e. 2). GK is initialized with FCM (Fuzzy c-means algorithm [4]). Around each cluster center, a hyperplane is fitted by WLS (Weighted Least Squares) (see Fig. 2a). The membership functions coding the input linguistic terms are obtained by projection and are depicted in Fig. 2b.

Remarks

- The global learning strategy gives a more accurate approximation: we obtain RMS=6.5199 for LM+SVD and RMS=8.1474 for GK+WLS (RMS is the Root Mean Square error). This constatation will be true in general, since global learning minimize a global cost, while local learning minimizes a sum of local costs.
- The global method generates Gaussian membership functions with possible high spreads; this is because no constraint has been added in the cost function E to be minimized. In contrast, the domains of expertise μ_{A_k} of the local linear models are well localized.
- Only two of the four linear models are visible in Fig. 1a, while the four linear models obtained in Fig. 2a by WLS could be used to locally approximate (explain) the noisy function. Actually, the identifcation of the hyperplanes in Fig. 1b results from the minimization of four local costs, while a global cost is considerd in Fig. 1a.

The experiments of this section illustrate the fact that there is an unavoidable compromise between interpretability and accuracy. Global (gradient-based) learning seems to be superior in term of accuracy (see also section 5.1) but cannot achieve the same level of interpretability as the local (clustering-based) approach (see also section 6).

4 Generalized Takagi-Sugeno systems

This section introduces a generalized version of TS models that incorporates an intensity parameter useful for pruning operations.

4.1 The GTS model

We slightly modify the classical first-order Takagi-Sugeno output computation by introducing a supplementary real parameter ρ_k controlling the importance of rule R_k in the inference process. The final output $\hat{y}_t$ corresponding to input $\boldsymbol{x}_t$ is now computed as in equation (2), but with

$$\pi_k(\boldsymbol{x}_t) = \frac{\gamma_k \mu_{A_k}(\boldsymbol{x}_t)}{\sum_{j=1}^{c} \gamma_j \mu_{A_j}(\boldsymbol{x}_t)} \tag{9}$$

$$\gamma_k = g(\rho_k) \tag{10}$$

$$g(x) = \frac{1}{1+e^{-x}}. \tag{11}$$

Function g enables the weights ρ_k to be normalized in the sense that γ_k will always verify: $0 \leq \gamma_k \leq 1$, $(k = 1, \ldots, c)$. This enables to interpret γ_k as an intensity parameter: a small γ_k value means that the corresponding rule can be deleted from the system without altering too much the system output value. This sensitivity parameter will be used in the sequel to detect and remove redundant rules.

Remark that if we set $g(\rho_k) = \gamma$ (with $\gamma \neq 0$; $k = 1, \ldots, c$), then the resulting system is a conventional TS system. For this reason, we will call *Generalized* Takagi-Sugeno model, the approximation system described by relations (2), (9), (10) and (11).

System output $\hat{y}_t$ is a nonlinear function of the new parameter ρ_k and one can use any nonlinear optimization procedure for the ρ_k adjustment. Note that the interpretation of $g(\rho_k)$ as an intensity parameter requires to adjust the ρ_k by minimizing a global cost since we want γ_k to measure the contribution of R_k in the (global) approximation process. The LM algorithm can be used to tune γ_k (see [18] for details).

4.2 Detecting and removing redundancies

Let us consider $N = 200$ points $z_t = (x_t, y_t)'$ from the model

$$y_t = |x_t| + \epsilon_t \tag{12}$$

where ϵ_t is a Gaussian noise with mean $\mu = 0$ and variance $\sigma^2 = 0.01$.

Fig. 3a shows the approximation obtained when $c = 3$ GTS-rules are used. The input membership functions and the local linear models are adjusted as previously (GK+WLS). The ρ_k parameters are adjusted by the LM algorithm. Fig. 3b represents the $\gamma_k \mu_{A_k}$ functions.

Let us denote by γ_k^* the normalized intensities: $\gamma_k^* = \gamma_k / \sum \gamma_j$. In this example, we have: $\gamma_1^* = 0.4607$, $\gamma_2^* = 0.1135$ and $\gamma_3^* = 0.4258$. The low value for γ_2^* leads us to suspect R_2 as redundant.

4.3 A decremental algorithm for autostructuration

The basic idea of the algorithm (DEC) is to start with an oversized bank of rules and then, gradually detect and remove redundant rules from the system. The resulting structure must be as small as possible in order to avoid overfitting.

A portion of the learning set is used to estimate the accuracy of the approximation models rather than to adjust the system parameters. This cross-validation technique helps to prevent the system to model details of the learning set because its evaluation is made on a different set. DEC is meant to end up with a structure producing the lowest error on the validation set. The search of that minimal structure is mainly guided by the analysis of the γ_k^* values (see [18] for implementation details).

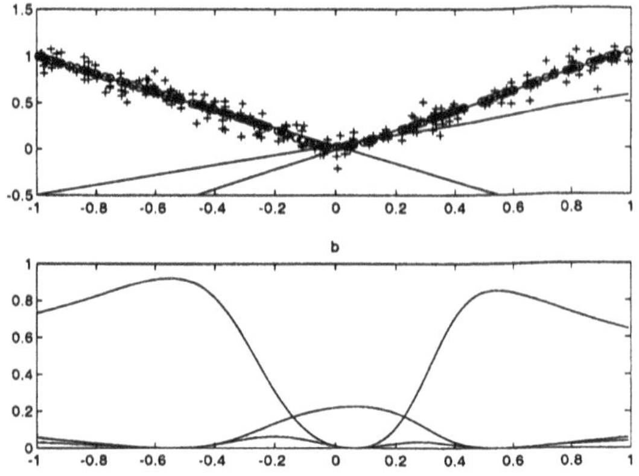

Fig. 3. Three GTS rules trained by method GK+WLS. (**a**) Points (x_t, y_t) (+), approximations $(x_t, \hat{y}_t)$ (o) and the local linear models. (**b**) $\gamma_k \mu_{A_k}(x)$

The originality of DEC is that the rules sensitivity analysis is directly linked to the parametric adjustment phase since the observation of the (self-adjusted) γ_k parameters guides the pruning operations (other rules simplification techniques proposed for example by Sun [51] or Kaymak and Babuška [31] rely on the computation of a similarity measure for merging purposes, so that structural and parametric operations are dissociated).

5 Applications to chaotic time series forecasting

5.1 The Mackey-Glass chaotic time series

Data The Mackey-Glass equation (MG_τ) [8] is a differential equation with delay τ that can be written as follows:

$$\frac{du}{dt} = \frac{a u_{t-\tau}}{1 + u_{t-\tau}^k} - b u_t \tag{13}$$

where in our case, $a = 0.2$, $b = 0.1$, $k = 10$ and $\tau = 17$.

The kind of trajectory obtained by integrating relation (13) mainly depends on the τ values. When $\tau > 16.8$, the dynamical system (13) has a strange attractor and the observed trajectories are chaotic [8].

The data points are the benchmark used by Platt [44]. The $N = 500$ points $(\boldsymbol{x}_t, y_t)$ forming the learning set are built from a portion of the $\{u_t\}$ trajectory ($200 \leq t \leq 700$) as follows (see Fig. 4):

$$\begin{aligned} \boldsymbol{x}_t &= (u_t, u_{t-6}, u_{t-12}, u_{t-18})' \\ y_t &= u_{t+h} \end{aligned}$$

where h is the forecast horizon ($h = 85$). 500 points of the same trajectory ($5000 \leq t \leq 5500$) are used for the test (see Fig. 4).

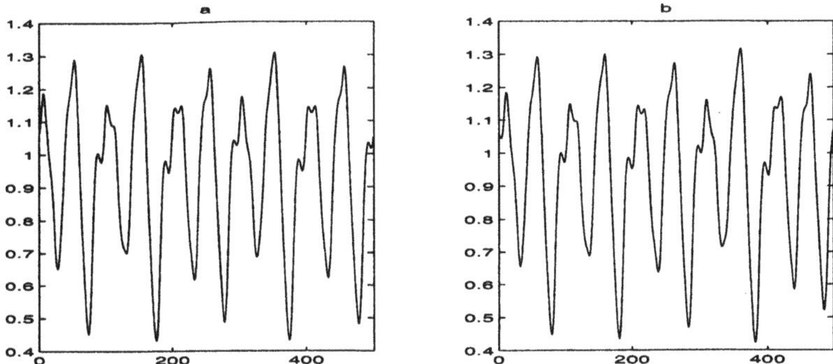

Fig. 4. MG_{17} data. (**a**) Learning time series. (**b**) Testing time series

Initialization and learning The GTS-DEC system is initialized with $c_{init} = 20$ rules. The 100 last patterns of the learning set forms the validation set. In the sequel, NER will denote the Normalized Error [15]. The NER is simply the root mean square error RMS divided by the standard deviation of the data. A NER value inferior to 1 indicates that the forecasts are better than the constant mean predictor.

Some preliminary tests have shown that local learning strategy GK+WLS can't achieve a good approximation on the training set, even with a great number of rules. To give an idea of this, we have computed the NER criterion corresponding to $c_{init} = 50$ locally adjusted rules (GK+WLS) and to $c_{init} = 20$ globally adjusted rules (LM+SVD). We obtain NER=0.0770 with GK+WLS, while the global learning LM+SVD enables to reach 0.0420. This constatation motivates the use of a global learning strategy in the sequel (we will use method LM+SVD described in section 3.3). The membership functions are generalized Gaussians (see equations (6), (7) and (8)).

Results Table 1 reports the results of our system GTS-DEC compared with 5 other forecasting methods. INC is an incremental method for fuzzy models adjustment introduced by Bersini and al. [3] based on the use of a hierarchical tree obtained with the compatible merging technique of Babuška [1]. INC has been trained and tested on the same data sets used by DEC. The INC (and DEC) system parameters have been trained by global learning. RAN (Resource Allocating Network [44]) is a neural network which can be seen as a variant of the Radial Basis Functions (RBF) neural network. BP is a direct Multilayer Perceptron trained with the backpropagation algorithm. The results relative to methods RAN, BP, RBF, B-Spline and INC used for comparison come from the papers of Platt [44] and Bersini et al. [3].

Table 1. Results on test set (series MG_{17}, $h = 85$)

Methods	NER
GTS-DEC-14	0.044
INC-14	0.090
RAN	0.066
BP	0.050
Hashing B-Spline	0.050
RBF	> 0.1

DEC progressively removes 6 of the 20 initial rules. The results of table 1 reveal the high quality at the GTS-DEC-14 forecasts. One possible explanation concerning the better performance of GTS-DEC-14 versus INC-14 is perhaps that the policy consisting in progressively adding one rule where it is necessary in the input space supposes that INC is able to correctly detect these zones. But one may think that the greater the dimensionality of the problem, the more difficult these zones will be to localize. In the worst case, the system can even omit to cover some important regions. As a contrary, the decremental approach avoids this problem from the beginning, since the initial overparametrized structure ensures covering to excess.

5.2 The Laser NH_3 chaotic time series

This time series comes from [28] and is related to the fluctuations in a far-infrared NH_3 Laser (see Fig. 5). This time series exhibits a low-dimensional chaotic behaviour: the estimations of the fractal dimension of the underlying attractor vary in the interval $[2.0, 2.2]$ (see [20]). This time series is one of the well-known benchmarks proposed in the Santa Fe competition [20].

The Laser NH_3 benchmark is a time series of length 1000. We use the first 500 points for the learning phase and the next 500 points for the test. The input patterns $\boldsymbol{x}_t$ are the lag vectors $\boldsymbol{x}_t = (1, u_t, u_{t-1}, \ldots, u_{t-(\nu-1)})'$, where ν is the size of the window (state space reconstruction technique [53]). Takens' theorem [53] gives us an indication about the appropriate size ν to consider: $d \leq \nu \leq 2d + 1$, where d is the fractal dimension of the underlying attractor. In the sequel, we will choose $\nu = 5$.

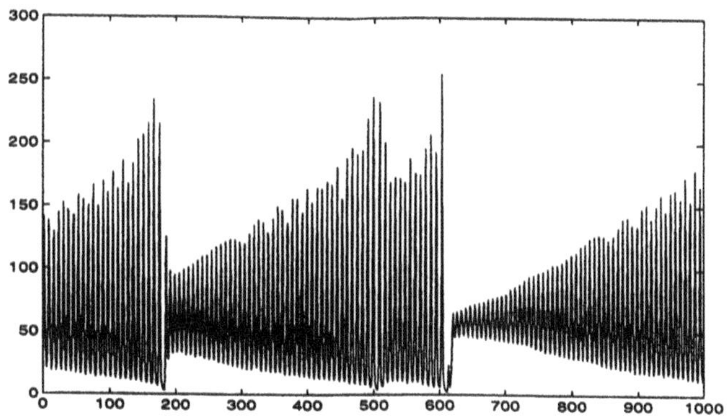

Fig. 5. The Laser NH_3 time series (1000 data points)

The system parameters adjustment is achieved as previously explained for the Mackey-Glass time series (global learning LM+SVD with $c_{init} = 20$ rules). We compare the GTS-DEC forecasts at horizons $h = 1$ and $h = 6$ with those obtained with 3 other forecasting methods described below.

- KNN-FS is a k nearest neighbours method introduced by Farmer and Sidorowich [15]. We also consider KNN-SM, a simplified version of KNN-FS proposed by Sugihara and May [50].
- DRNN (Dynamic Recurrent Neural Network) is a totally interconnected recurrent neural network [54]. The DRNN system has been trained to forecast the Laser NH_3 time series at horizon $h = 6$ with the same partitioning of the data (500-500) [54]. Hence, a comparison with KNN and GTS is possible. Two architectures have been tested: 5 neurons (DRNN-5) and 10 neurons (DRNN-10).

The final number of rules generated by DEC is $c_{end} = 13$, when $h = 1$ and $c_{end} = 19$, when $h = 6$. This is coherent with the fact that the complexity of the associations to learn increases with the forecast horizon h.

Table 2 reports the results on the portion of the time series dedicated to the test when $h = 1$ and $h = 6$. MAE is the Mean Absolute Error. COR is the correlation coefficient between true values and forecasts.

Figs. 6 shows the plots of the forecasts versus the true values when $h = 1$. The ideal forecasts are on the solid straight line. The best forecasts are those of the GTS-DEC-13 and KNN-FS models; the fuzzy system gives the best performances relatively to the NER and COR criteria, while the MAE is slightly better for KNN-FS. Fig. 6 shows a tendency of the DRNN model to systematically overestimate (resp. underestimate) the true values when $y_t \leq 100$ (resp. $y_t \geq 100$). We suspect some kind of saturation effect linked to

Table 2. Results on tests sets (Laser NH_3 time series)

Methods	COR	NER	MAE
h=1			
KNN-FS	0.9738	0.2305	3.4946
KNN-SM	0.9549	0.2959	5.3094
DRNN-5	0.9339	0.3576	10.2107
DRNN-10	0.9086	0.4183	12.3117
GTS-DEC-13	0.9785	0.2080	3.7903
h=6			
KNN-FS	0.9016	0.3883	6.6395
KNN-SM	0.9227	0.4383	7.1747
DRNN-5	0.8928	0.4544	11.1919
DRNN-10	0.8890	0.4653	12.2272
GTS-DEC-19	0.8693	0.5262	7.3204

the use of sigmodal transfer functions in this network. As a result, DRNN-5 and DRNN-10 are the less accurate models at $h = 1$.

When $h = 6$, a degradation of the performances can be observed for the 5 models. Observation of table 2 reveals that the best forecasts are those of the KNN models. DRNN forecasts are slightly better than those produced by GTS-DEC-19 for criteria NER and COR, but considerably worse relatively to the MAE. Actually, for this criterion, GTS-DEC-19 compares to KNN methods. The systematic bias of the DRNN forecasts are still observed.

Fig. 7 shows the individual forecasting errors for the test time series ($h = 6$). Actually, one can see the magnitude of the GTS-DEC-19 errors is high near $t = 0$ and around $t = 100$ (collapsing cycles). Such errors contribute more in the computation of the NER than for the MAE. This explains why the NER performance of GTS-DEC-19 is bad. Outside the collapsing cycles, the fuzzy system errors are considerably smaller (in absolute value) than those of the other models. Actually, this is confirmed by observation of table 3 reporting the errors criteria computed on time period $[150, 500]$ (this portion represents 70 percents of the whole test time series). It clearly appears that GTS-DEC-19 achieves the best performances: our system yields a MAE (resp. NER) reduction of about 56% (resp. 40%) over the best competitive forecasting model (KNN-FS).

Now, one has to note that contrary to KNN, GTS-DEC (and DRNN) does not exploit the topology of the underlying fractal attractor. This explains why the collapsing cycles are more accurately predicted by KNN. Also, the window size ν has been optimized for KNN, but this has not been done for

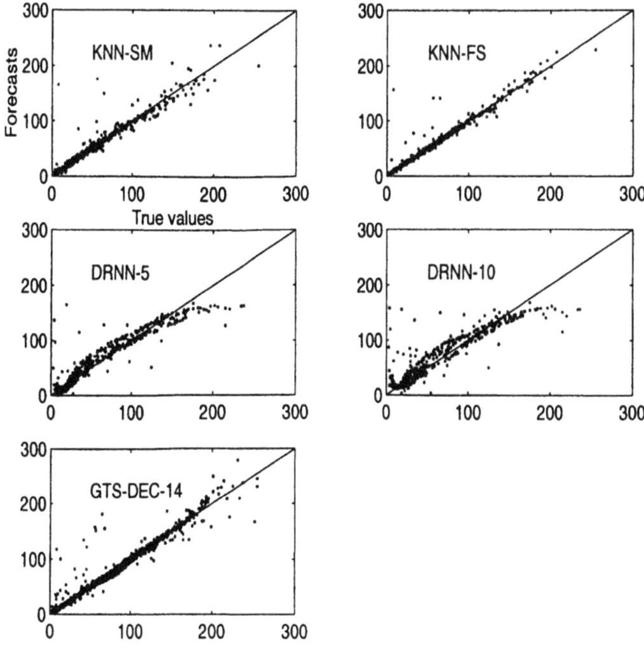

Fig. 6. Points $(y_t, \hat{y}_t)$ relative to the testing Laser time series $(h = 1)$

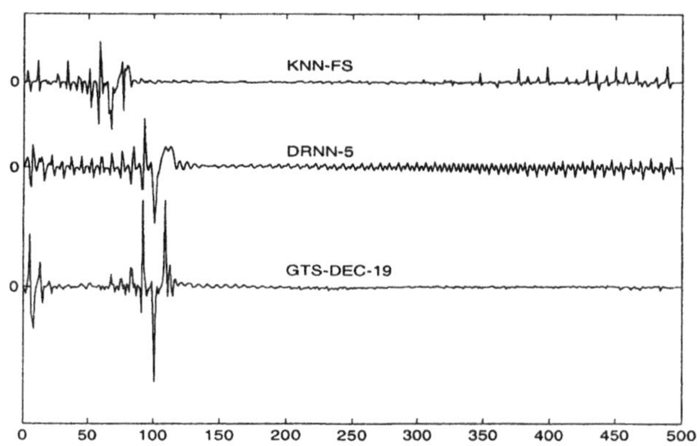

Fig. 7. Forecasting errors on test set (Laser time series, h=6)

DRNN and GTS since it would have been too much time consuming. In view of this, the performance of the fuzzy system is remarkable.

Table 3. Results on test set (NH_3 series, $h = 6$, time period $[150, 500]$)

Methods	COR	NER	MAE
KNN-FS	0.9837	0.1827	3.5399
KNN-SM	0.9737	0.2287	4.0001
DRNN-5	0.9648	0.2638	7.5177
DRNN-10	0.9584	0.2920	8.6876
GTS-DEC-19	0.9976	0.0737	1.9967

6 Application to the case of combining forecasts

In this section, we consider the GTS forecasts combining methodology previously developed in [16]. One of the conclusions of [16] was that GTS forecasts combinations were at least as accurate as standard combining methods, but the combining rules we obtained via global learning techniques were not interpretable. The motivation of this section is to build GTS-based combining models focusing our attention on the question related to the analysis and interpretation of the final combining rules.

6.1 Some basics about forecasts aggregation

Many studies and empirical tests have shown the advantage of aggregating forecasts (see [10]). After the seminal papers of Reid [47] and Bates and Granger [2], many combining methods have been proposed such as the historical weighting of forecasts [13], Bayesian methods (e.g., see [7,58]), the minimum-variance method [12] or regression based methods [23].

The usual approach to combining forecasts consists in taking a vector $\boldsymbol{F}_t(h)$ of forecasts at time t with horizon h:

$$\boldsymbol{F}_t(h) = (F_{t1}(h), F_{t2}(h), \ldots, F_{tp}(h))'$$

(in the sequel, we will omit h) and building a new forecast C_t as

$$C_t = w_{t0} + \boldsymbol{F}'_t \boldsymbol{w}_t \tag{14}$$

where $\boldsymbol{w}_t = (w_{t1}, \ldots, w_{tp})'$ is the combining vector computed at time t. In other words, combination C_t given in equation (14) is a linear mixture of the individual forecasts F_{ti}. Linear mixture does not mean C_t is necessarily a linear function of F_{ti}: this will depend on the way weights w_{ti} are computed at time t. A purely linear combination scheme will result if the combining weights are constants or do not depend on F_{ti}.

6.2 Combining forecasts with GTS models

It is clear that output $\hat{y}_t$ of a GTS fuzzy system performs the combination given in relation (14) when $\boldsymbol{x}_t = (1, F_{t1}, F_{t2}, \ldots, F_{tp})'$. Rewriting vector $\boldsymbol{\beta}_k$ like this: $\boldsymbol{\beta}_k = (\beta_0, \beta_1, \ldots, \beta_p)'$, the combination weights are:

$$w_{tj} = \sum_{k=1}^{c} \pi_k(\boldsymbol{x}_t)\beta_{kj}, \quad j = 0, \ldots, p. \tag{15}$$

This decomposition property results from the linearity of the local output models $\hat{y}_{t,k}$. Hence, combination (14) could not be achieved with other fuzzy systems because they realize *nonlinear* mixture operations.

Actually, each rule of a GTS system performs a special linear combination of forecasts which is encoded in the corresponding rule conclusion $\hat{y}_{t,k}$. These combinations can be modulated via the γ_k parameters and are aggregated using the weighted mean given in equation (2). Moreover, the contribution of each rule relatively to one given forecasts vector $\boldsymbol{x}_t$ depends on its *localization* ($\mu_{A_k}(\boldsymbol{x}_t)$) in the input space. The idea of exploiting these spatial correlations is new in the field of combining forecasts and has been applied with success in [16] to aggregate linear and nonlinear forecasts.

6.3 Improving interpretability with constrained experts

We have previously mentioned that the type of learning has a direct impact on the interpretability of the resulting fuzzy model. The behaviour of the experts $\hat{y}_{t,k}$ is completely different according to the kind of learning method used. When local learning is considered, these experts compete for the determination of their expertise domains μ_{A_k}, while they cooperate in the other case. As a result, more interpretable fuzzy rules could be obtained by local learning since the competition allows to dissociate their actions.

We can go one step further by forcing the fuzzy rules to encode a special meaning in term of combining strategy. More precisely, we shall now consider the forecasting models as a group of individuals having its own opinion F_{ti} ($i = 1, \ldots, p$) about the value to be predicted at time t. The actions $\hat{y}_{t,k}$ of each of the c fuzzy combining rules will consist in pooling these opinions. The consensus will be obtained by constraining the local experts to achieve convex combinations of the forecasts:

$$\beta_{k0} = 0; \quad \sum_{j=1}^{p} \beta_{kj} = 1; \quad \beta_{kj} \geq 0 \quad (j = 1, \ldots, p). \tag{16}$$

This has the immediate advantage of interpreting β_{kj} as the confidence level of expert k relatively to forecast F_{tj}. In the sequel, we will denote by 01+ the set of constraints given in equation (16). A GTS-01+ rule R_k can thus be interpreted as a mixer whose action can be modulated by γ_k. This particular interpretation of a fuzzy rule may help in giving more intuition about TS fuzzy systems processing operations.

β_{kj} identification can be achieved using the LSIE method (Least Squares with Inequality and Equality constraints, see [37]). Remark that a constraint solution $\boldsymbol{\beta}_k$ always exists for each rule.

6.4 A case study

The time series considered here is the laser NH_3 time series (see section 5.2). The individual forecasting models are the previously described KNN-FS, DRNN and GTS-DEC. The out-of-sample forecast errors are shown in Fig. 7 (500 points). Actually, one can see that the magnitude of the GTS-DEC errors is high near $t = 0$ and around $t = 100$ (these instants correspond to collapsing cycles in the time series). Elsewhere, the fuzzy system forecast errors are smaller (in absolute value) than those of the other models.

The first 400 data points (out-of-sample forecasts) are used for the (combining forecasts) learning phase and the last 100 patterns are dedicated to the test. We've trained a GTS-DEC system with $c_{init} = 20$ initial rules with local FMLE+WLS (l), constrained local (l01+) and global LM+SVD (g) strategies. Method (l01+) is FMLE+WLS where the local linear models are adjusted with the LSIE algorithm.

The results are reported on table 4. We compare our forecasts combination system with other traditional combining methods including the popular simple average of individual forecasts (SAV), variant of regression based methods and the bayesian method OUTPERF (outperformance method) proposed by Bunn [7] (see [10] for details about these benchmarks). The classical linear regression model is denoted by REG; REG0 is a regression model with no constant term; REG+ is a regression model with non negative parameters; REG01 is a linear regression without independent term and with coefficients summing up to one; REG01+ adds to REG01 the additional constraint that the coefficients must be non negative. The combining weights generated by REG01+ and OUTPERF are constrained in the 01+ sense and thus, a comparison with the constrained GTS model is possible.

NER and MAE are respectively the Normalized ERror and the Mean Absolute Error.

Respectively 5, 10 and 3 final rules are obtained by DEC for the cases (l), (g) and (l01+). We remark the fuzzy combinations are better (in terms of NER and MAE) than those produced by the 7 other combining systems. Notice however that only GTS-DEC-3-l01+ competes with the best forecasting model (GTS-DEC-19). The fact that one forecasting model is more accurate than the others leads us to suspect that the best combining strategy is in fact the models *selection*. It appears that the selection operated by GTS-3-l01+ is better than those produced by the 01+ constrained methods REG-01+ and OUTPERF.

Table 4. Forecasting and combining models errors (NH_3 series, testing part)

Models	NER	MAE
DRNN-5	0.2683	11.1687
KNN-FS	0.2174	6.7038
GTS-DEC-19	0.0866	2.7328
SAV	0.1473	5.5069
REG	0.2168	7.1174
REG0	0.2070	6.4590
REG+	0.2168	7.1174
REG01	0.1975	6.4388
REG01+	0.1975	6.4388
OUTPERF	0.1149	4.0743
GTS-DEC-5-l	0.1104	4.0397
GTS-DEC-10-g	0.0989	3.8086
GTS-DEC-3-l01+	0.0884	3.1936

The mixtures achieved by GTS-DEC-3-l01+ are reported on table 5. The GTS-3-l01+ mixing operations consist in considering 3 combinations of 2 forecasting models; rule R_1 deals with models GTS and KNN with a preference for model GTS-DEC-19, while rules R_2 and R_3 combine DRNN and KNN-FS giving more weight to the latter model. These 3 strategies are next mixed via the (defuzzification) process (see equation (2)).

Table 5. Mixtures achieved by system GTS-DEC-3-l01+ (NH_3 time series)

	β_{1j}	β_{2j}	β_{3j}
DRNN	0.0000	0.2212	0.3474
KNN-FS	0.1725	0.7788	0.6526
GTS-19	0.8275	0.0000	0.0000

Let us consider the following quantities Δ_t defined as follows:

$$\Delta_t = \max_{i \neq j} |F_{ti} - F_{tj}|, \quad 1 \leq i, j \leq p. \tag{17}$$

The evolution of Δ_t over time enables to detect zones where the forecasts values are similar, or as a contrary, dissimilar. Observation of Figs. 8 and 9 gives more details about the combining process. Subplots (b), (c) and (d) report the time evolution of $\pi_k(\boldsymbol{x}_t)$ (proportions expressing the rules contributions in the final output computation). The graph of Fig. 8e shows three distinct zones relative to the time evolution of Δ_t (zones A, B and C). Actually, when the 3 forecasts values are quite similar (zones B), rule R_2 is dominant and the GTS-DEC-3-l01+ selects model KNN-FS (see subplots b, c, d and e). When the dissimilarities Δ_t are more important (zones A), confidence goes to model GTS-DEC-19 (rule R_1). One can see rule R_3 is never activated, except when large dissensions exist between the forecasting models (zones C). These zones are relative to sudden transitions observed in the time series between high and low amplitude oscillations (see Fig. 8a). These structural changes are difficult to forecast for the 3 models. In such cases, the fuzzy combining system simultaneously activates rules R_1 and R_3 and proceeds to a weighted average of the three forecasting models.

In order to show the importance of the input patterns localization for the combining weight assignment procedure, we show the $\boldsymbol{F}_t$ scatter plot in Figs. 10 and 11 where a bar of length $100 \times \pi_k(\boldsymbol{x}_t)$ has been placed on each data point $\boldsymbol{F}_t$. Observation of Fig. 10 reveals that rule R_2 is mainly dominant in the central region of the cloud - where the forecasts are similar - while rule R_1 is influent on a larger region spreading from north to south. Activation of rule R_3 is linked to the presence of points far from the cloud (outliers). These input patterns could be seen as anomalies, or atypic points describing brutal transitions localized in zones C (cf. Fig. 8e). Such zones do not appear in test phase (see Fig. 9e) and so rule R_3 does not cooperate in the forecasts consensus (see Fig. 9d).

The main objective of this experiment was to show that a GTS-DEC-l01+ model can be transparent to interpretation and analysis. The action of the 3 final rules - automatically obtained by DEC - are well understood. This is possible because of the 01+ constraints and the local learning.

7 Conclusions

This chapter is motivated by the building of Takagi-Sugeno (TS) based inference systems allowing for accuracy and interpretation in the field of time series forecasting. Unfortunately, there is an unavoidable compromise between interpretability and accuracy. Global (gradient-based) learning seems to be superior in term of accuracy but cannot achieve the same level of interpretability as the local (clustering-based) approach. When the forecasting accuracy is the most important target, it appears that the appropriate identification method is global. As a contrary, when the objective is the emergence of transparent final rules, than the local approach is preferred. We have pointed out that local learning implies a competition between the rules, while global learning favours rules cooperation and creates coupling effects between free parameters that mask the system operations.

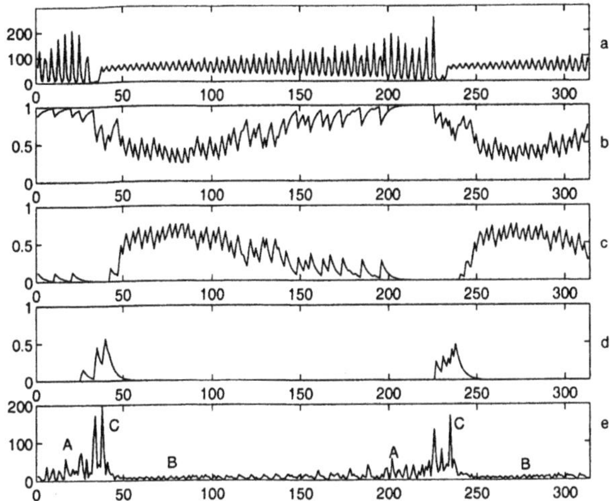

Fig. 8. (**a**) NH_3 series (learning part). (**b**) $\pi_1(x_t)$ versus time t (rule 1). (**c**) $\pi_2(x_t)$ versus t (rule 2). (**d**) $\pi_3(x_t)$ versus t (rule 3). (**e**) Plot of Δ_t

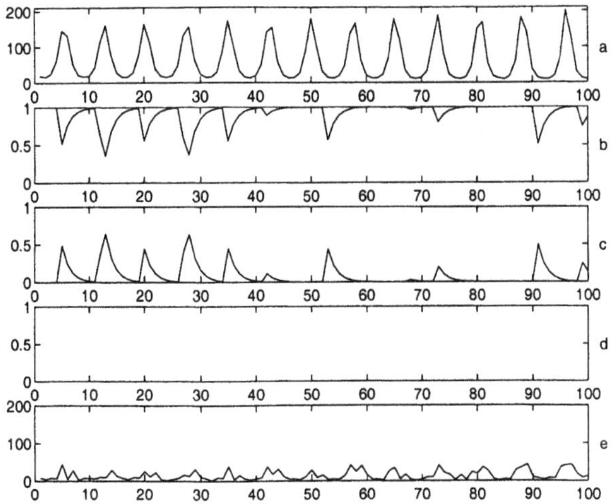

Fig. 9. (**a**) NH_3 series (testing part). (**b**) $\pi_1(x_t)$ versus time t (rule 1). (**c**) $\pi_2(x_t)$ versus t (rule 2). (**d**) $\pi_3(x_t)$ versus t (rule 3). (**e**) Plot of Δ_t

In view of building self-structuring systems, we have presented a pruning procedure that performs a sensitivity analysis of the inference rules and progressively eliminates redundancies. This algorithm has been successfully applied to forecast highly nonlinear (chaotic) time series. The comparison with other forecasting methods has shown the high quality of our forecasts.

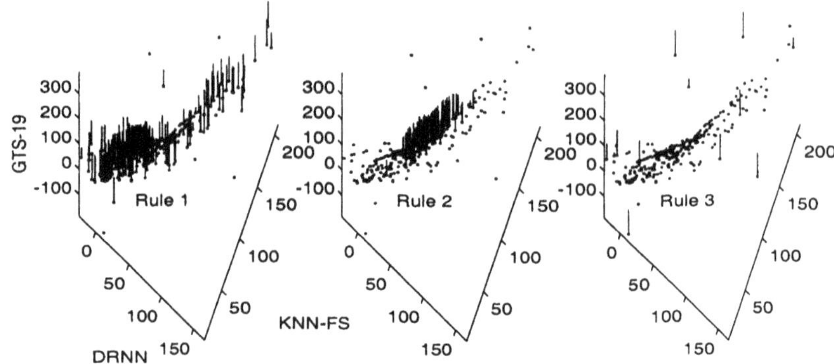

Fig. 10. Laser NH_3 time series (learning phase). The scatter plot shows the input data points F_t (the length of the bars is $100 \times \pi_k(x_t)$)

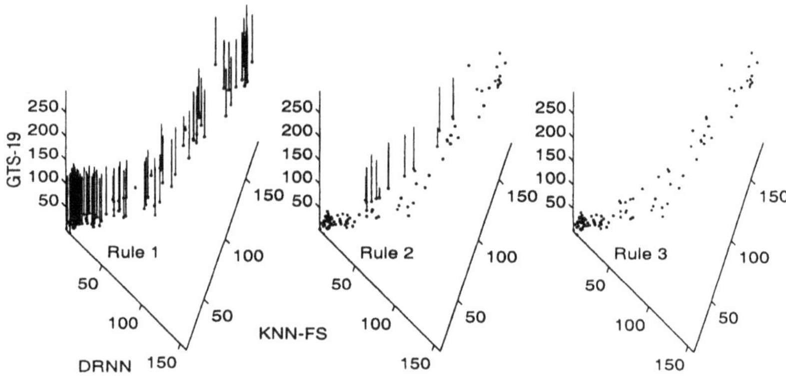

Fig. 11. Laser NH_3 time series (testing phase). The scatter plot shows the data F_t. The length of the bars is $100 \times \pi_k(x_t)$

A second application of our system deals with the forecasts combination technique focusing on the interpretability of the final rules. This has led us to put additional constraints in the model. Focusing on forecasts combination, it appears that the local learning of constrained rules allows soft selections of the individual models guided by the localization of the forecasts vector in the input space. This feature has been exploited in a practical case from which intuitive combining strategies have emerged. The proposed combining system may be useful for a wide range of applications where a consensus is needed. That constrained model has been applied in control for combining linear observers estimates [17].

Acknowledgment: This research work is part of the CELOFA project (Centre d'Etude de la Logique Floue et ses Applications) supported by Région Wallonne in Belgium. See http://mathro.fpms.ac.be/CELOFA/.

Appendix

In order to facilitate the reading, we have put in table 6 a list of the most important abbreviations used in this chapter along with their meaning.

Table 6. List of the most important abbreviations used in this chapter

Abbreviation	Meaning
TS	**T**akagi-**S**ugeno fuzzy system
GTS	**G**eneralized **T**akagi-**S**ugeno model
DEC-n	Decremental aproach **DEC** leading to **n** rules
INC-n	Incremental aproach **INC** leading to **n** rules
GTS-DEC-n-m	**m** is the learning mode: **g** (global), **l** (local),
	l01+ (local and constrained)
FMLE	**F**uzzy **M**aximum **L**ikelihood **E**stimates
GK	**G**ustafson and **K**essel clustering method
WLS	**W**eighted **L**east **S**quares method
LSIE	**L**east **S**quares with **I**nequality and **E**quality constraints
SVD	**S**ingular **V**alues **D**ecomposition method
LM	**L**envenberg-**M**arquardt optimization method
RBF	**R**adial **B**asis **F**unctions neural network
KNN-m	**K** **N**earest **N**eighbours. **m** is the implementation mode:
	FS (Farmer and Sidorowich), **SM** (Sugihara and May)
DRNN-n	**D**ynamic **R**ecurrent **N**eural **N**etwork with **n** neurons
OUTPERF	**OUTPERF**ormance combining method of Bunn
REG	Linear **REG**ression model
REGm	Constrained **REG**. **m** denotes the type of constraint.
	Possible modes: **0**, **+**, **01**, **01+** (see text)
SAV	**S**imple **AV**erage combining method
RMS	**R**oot **M**ean **S**quare error
NER	**N**ormalized **ER**ror
MAE	**M**ean **A**bsolute **E**rror

References

1. Babuška, R. (1996) Fuzzy Modeling and Identification. Thesis, Delft University of Technology
2. Bates, J.M., Granger, C.W.J. (1969) The combination of forecasts. Operational Research Quarterly **20**, 451–468
3. Bersini, H. Duchateau, A., Bradshaw, N. (1997) Using incremental learning algorithms in the search for minimal and effective fuzzy models. Proceedings of the FUZZ-IEEE'97 Conference, 1417–1422
4. Bezdek, J. (1981) Pattern recognition with fuzzy objective function algorithms. Plenum Press, New York
5. Buckley, J.J. (1992) Universal fuzzy controllers. Automatica Journal **28**, 1245–1248
6. Buckley, J.J., Hayashi, Y. (1993) Numerical relationships between neural networks, continuous functions and fuzzy systems. Fuzzy Sets and Systems **60**, 1–8
7. Bunn, D.W. (1975) A Bayesian approach to the linear combination of forecasts. Operational Research Quarterly **26**, 325–329
8. Castagli, M. (1989) Nonlinear prediction of chaotic time series. Physica D **35**, 335–356
9. Chiu, S.L. (1994) Fuzzy model identification based on cluster estimation. Journal of Intelligent and Fuzzy Systems **2**, 267–278
10. Clemen, R.T. (1989) Combining forecasts: a review and annotated bibliography. International Journal of Forecasting **5**, 559–583
11. Combs, W.E., Andrews, J.E. (1998) Combinatorial rule explosion eliminated by a fuzzy rule configuration. IEEE Transactions on Fuzzy Systems **6**, 1–11
12. Dickinson, J.P. (1975) Some comments on the combination of forecasts. Operational Research Quarterly **26**, 205–210
13. Doyle, P., Fenwick, I.A. (1976) Sales forecasting using a combination of approaches. Long-Range Planning **9**, 60–69
14. Dubois, D., Prade, H., Yager, R.R. (1997) Fuzzy Information Engineering: A Guided Tour of Applications. Wiley and Sons, New York
15. Farmer, J.D., Sidorowich, J.J. (1987) Predicting chaotic time series. Physical Review Letters **59**, 845–848
16. Fiordaliso, A. (1998) A nonlinear forecasts combination method based on Takagi-Sugeno fuzzy systems. International Journal of Forecasting **14**, 367–379
17. Fiordaliso, A. (1999) Systèmes flous et prévision de séries temporelles. Hermès, Paris
18. Fiordaliso, A. (2000) Autostructuration of fuzzy systems by rules sensitivity analysis. International Journal for Fuzzy Sets and Systems **118**, 281–296
19. Gath, I., Geva, A.B. (1989) Unsupervised optimal fuzzy clustering. IEEE Transactions on Pattern Analysis and Machine Intelligence **11**, 773–781
20. Gershenfeld, N.A., Weigend, A.S. (1993) The future of time series: learning and understanding. In: Gershenfeld, N.A., Weigend, S.A. (Eds.): Time series prediction: forecasting the future and understanding the past. Addison Wesley, Palo Alto, 1–70.
21. Geyer, A., Taudes, A. (1990) A fuzzy time series analyzer. In: Janko, W.H., Roubens, M., Zimmermann, H.-J. (Eds.): Progress in Fuzzy Sets and Systems. Kluwer Academic Publishers, London, 63–74

22. Gorrini, V., Salom, T., Bersini, H. (1995) Self-structuring systems for function approximation. Proceedings of the FUZZ-IEEE/IFES'95 Conference, 531–536
23. Granger, C.W.J., Ramanathan, R. (1984) Improved methods of combining forecasts. Journal of Forecasting **3**, 197–204
24. Gustafson, E.E., Kessel, W.C. (1979) Fuzzy clustering with a fuzzy covariance matrix Proceedings of the IEEE CDC, 761–766
25. Hartani, R., Nguyen, H.T., Bouchon-Meunier, B. (1996) Sur l'approximation universelle des systèmes flous. RAIRO-APII-JESA **30**, 645–663
26. Hataway, R., Bezdek, J. (1993) Switching regression models and fuzzy clustering. IEEE Transactions on Fuzzy Systems **1**, 195–204
27. Hinton, G., (1989) Machine Learning: Paradigms and Methods. Artificial Intelligence **40**, 185–234
28. Hübner, U., Abraham, N.B., Weiss, C.O. (1989) Dimensions and entropies of chaotic intensity pulsations in a single-mode far-infrared NH3 laser. Physical Review A **40**, 6354–6365
29. Jacobs, R.A., Jordan, M.I., Nowlan, S.J., Hinton, G.E. (1991) Adaptive Mixture of Local Experts. Neural Computation **3**, 79–87
30. Karr, C.L. (1991) Design of an adaptive fuzzy logic controller using a genetic algorithm. Proc. 4th Int. Conf. Genetic Algorithms, 450–457
31. Kaymak, U., Babuška, R. (1995) Compatible cluster merging for fuzzy modeling. Proceedings of the FUZZ-IEEE/IFES'95 Conference, 897–904
32. Klir, G.J., Yuan, B. (1995) Fuzzy Sets and Fuzzy Logic - Theory and Applications. Prentice-Hall, London
33. Kosko, B. (1994) Fuzzy systems as universal approximators. IEEE Trans. Comput. **43**, 1329–1333
34. Kreinovich, V.Y. (1991) Arbitrary nonlinearity is sufficient to represent all functions by neural networks: a theorem. Neural Networks **4**, 381–383
35. Kurkova, V. (1991) Kolmogorov's theorem is relevant. Neural Computation **3**, 617–622
36. Kurkova, V. (1992) Kolmogorov's theorem and multilayer neural networks. Neural Networks **5**, 501–506
37. Lawson, C.L., Hanson, R.J. (1974) Solving Least Squares Problems. Prentice Hall, New York
38. Lin, C.T., Lee, C.S.G. (1994) Reinforcement structure/parameter learning for neural-network-based fuzzy logic control systems. IEEE Transactions on Fuzzy Systems **2**, 46–63
39. Marquardt, D.W. (1963) An algorithm for least squares estimation of nonlinear parameters. Journal of the Society for Industrial and Applied Mathematics **11**, 431–441
40. McLachlan, B., Basford, K. (1988) Mixture Models: Inference and Applications to Clustering. Dekker, Basel
41. Morabito, F.C. (1997) Advances in Intelligent Systems. IOS Press, Amsterdam
42. Nguyen, H.T., Kreinovich, V. (1993) On approximation of controls by fuzzy systems. Proc. Fifth IFSA, 1414–1417
43. Nomura, H., Hayashi, I., Wakami, N. (1992) A learning method of fuzzy inference rules by descent method. Proc. IEEE Int. Conf. Fuzzy Syst., 203–210
44. Platt, J. (1991) Resource-allocating network for function interpolation. Neural Computation **3**, 213–225
45. Poggio, T., Girosi, F. (1990) Networks for approximation and learning. Proceedings of the IEEE **78**, 1481–1497

46. Quandt, R.E., Ramsey, J.B. (1972) A new approach to estimating switching regressions. Journal of the American Statistical Society **67**, 306–310
47. Reid, D.J. (1968) Combining three estimates of gross domestic products. Economica **35**, 431–444
48. Schwarz, G. (1978) Estimating the Dimension of a Model. Annals of Statistics **6**, 461–464
49. Sugeno, M., Yasukawa, T. (1993) A fuzzy-logic-based approach to qualitative modeling. IEEE Transactions on Fuzzy Systems **1**, 7–31
50. Sugihara, G., May, R.M. (1990) Nonlinear forecasting as a way of distinguishing chaos from measurement error in time series. Nature **344**, 734–741
51. Sun, C.T. (1994) Rulebase structure identification in an adaptive network based fuzzy inference system. IEEE Transactions on Fuzzy Systems **2**, 64–73
52. Takagi, T., Sugeno, M. (1985) Fuzzy identification of systems and its application to modelling and control. IEEE Transactions on Systems, Man and Cybernetics **15**, 116–132
53. Takens, F. (1981) Detecting strange attractors in turbulence. Lecture Notes in Mathematics **898**, 336–381
54. Teran, R., Draye, J.P. et al. (1996) Predicting a Chaotic Time Series Using a Dynamical Recurrent Neural Network. Proceedings IWISP '96, 115–118
55. Wang, L.-X. (1992) Fuzzy systems are universal controllers. Proc. IEEE International Conference on Fuzzy Systems, 1163–1170
56. Wang, L.-X, Mendel, J.M. (1992) Fuzzy basis functions, universal approximation and orthogonal least-squares learning. IEEE Trans. Neural Networks **3**, 807–814
57. Weigend, A., Huberman, B., Rumelhart, D. (1992) Predicting sunspots and exchange rates with connectionnist methods. In: M. Castagli, S. Eubank (Eds.): Nonlinear Modelling and Forecasting. Addison-Wesley, New York, 395–432
58. Winkler, R.L. (1981) Combining probability distributions from dependent information sources. Management Science **27**, 479–488
59. Wong, F.S., Wang, P.Z. (1991) A fuzzy neural network for Forex rate forecasting. In: Terano, T., Sugeno, M., Mukaidono, M., Shigemasu, K. (Eds.): Fuzzy Engineering Toward Human Friendly Systems. IOS Press, Amsterdam, 535–545
60. Zeng, X., Singh, M.J. (1994) Approximation theory of fuzzy systems: SISO case. IEEE Trans. on Fuzzy Systems **2**, 162–176

Accurate, transparent and compact fuzzy models by multi-objective evolutionary algorithms

Fernando Jiménez[1], Antonio F. Gómez-Skarmeta[1], Gracia Sánchez[1], Hans Roubos[2], and Robert Babuška[2]

[1] Dept. Ingeniería de Información y las Comunicaciones, University of Murcia, Spain
[2] Control Engineering Laboratory, Delft University of Technology, the Netherlands

Abstract. Interpretability aspects of fuzzy models have received quite some attention in recent years and may be obtained by using transparent rule-structures and well characterized fuzzy membership functions. Moreover, model compactness is important for the interpretability and is related to the number of rules and fuzzy sets. Besides these two criteria, the model accuracy should always be taken into account. In this way, several criteria appear in fuzzy modeling and then multi-objective evolutionary algorithms are a suitable, because these are able to capture several non-dominated solutions in a single run of the algorithm. For fuzzy modeling, we describe two multi-objective evolutionary algorithms that consider all three objectives. Differences between both algorithms arise in the fuzzy sets considered, trapezoidal and gaussian respectively. The algorithms apply an accuracy criterium and a transparency criterium, based on fuzzy set similarity, while compactness is achieved by a specific technique, incorporated ad hoc within the evolutionary algorithms. Finally, we propose a decision process to find the most satisfactory non-dominated solution. Results are shown for three approximation problems that were studied before by others authors.

1 Introduction

In recent years, fuzzy modeling, as a complement to conventional modeling techniques, has become an active research topic and has found successful applications in many areas. Most fuzzy models are built based on operator's experience and knowledge, However, for complex or new processes there may not be enough experience available [27]. In such a situation, unsupervised data-based learning techniques may appear very useful. The problem can be stated as follows. Given a set of data for which we presume some functional dependency, the question arises whether there is a suitable methodology to derive (fuzzy) rules from the data that characterize the unknown function as precisely as possible. The last decade, already many approaches were proposed for the automatic generation of fuzzy if-then rules from numerical data without the use of domain experts [15]. However, in most cases the accuracy aspect prevails and interpretability aspects are partly ignored. We will introduce a method that takes both aspects into account.

Transparency and model interpretability for data-based fuzzy models received quite some interest in recent literature [16,13,1,14]. Based on this literature, we introduce three important criteria to be optimized in fuzzy model identification: compactness, transparency and accuracy. Different measures for these criteria are proposed here. Compactness is related to the size of the model, i.e. the number of rules, the number of fuzzy sets and the number of inputs for each rule, while transparency is related to linguistic interpretability of the fuzzy sets and locality of the rules [2,22]. Often one is interested in the local behavior of the global nonlinear model. Such information may be obtained by constraining the model-structure during identification or by using these criteria by multi-objective optimization techniques, like multi-objective Evolutionary Algorithms.

The effectiveness of Evolutionary Algorithms (EAs) in solving multi-objective optimization problems has been widely recognized in recent years. Proof of this can be seen in the growing number of special sessions and workshops on multi-objective evolutionary optimization incorporated into the framework of prestigious, international congresses and in the recent appearance of the First International Conference on Multi- Criteria Evolutionary Optimization, held in Zurich in March 2001.

Most evolutionary approaches to multi-objective fuzzy modeling consist of multiple EAs, usually designed to achieve a single task each, which are then applied sequentially. In these cases, each EA optimizes the model attending to a single criterion separately, which is an impediment for global search. Simultaneous optimization of all criteria is more appropriate. Therefore, others used approaches based on classical multi-objective techniques in which multiple objectives are aggregated into a single function to be optimized [6,20]. In this way, the EA obtains a compromise solution that consist of the weighted criteria. Another promising method to handle multi-criteria optimization problems is the multi-objective EA based on the Pareto optimality notion, in which all objectives are optimized simultaneously to find multiple non-dominated solutions in a single run of the EA [4,9,24]. The solution for a such a multi-objective optimization problem is a set of so-called Pareto optimal solutions.

A multi-objective EA incorporates the Pareto concept to identify multiple solutions through a single run of the algorithm. This practically leaves obsolete the classical tendency to aggregate the different objectives using a weight vector or similar approach to obtain a single function which is then optimized. In that case, the method often requires several executions of the EA with different weights, in order to identify one Pareto solution in each run. In this aspect, multi-objective optimization, based on the Pareto-optimality concept, distinguishes itself from related optimization methods, like gradient techniques, simulated annealing or neural networks.

Pareto-based multi-objective evolutionary approaches can also be considered from the fuzzy modeling perspective [7,10]. The advantage of the the classical optimization approach with aggregated objectives, a single solution is obtained without further interaction with the decision maker. However, it may often be difficult to define a good aggregation function. Then, if the solution cannot be accepted, new runs of the EA are required until a satisfying solution is found. The advantages of the Pareto approach are that no aggregation function need to be defined, and that the decision maker can choose the most appropriate solution according to the current decision environment at the end of the EA run. Moreover, if the decision environment changes, it is not always necessary to run the EA again. Another solution may be chosen out of the family of non-dominated solutions that has already been obtained.

In this chapter, we propose two multi-objective Evolutionary Algorithms to search for multiple non-dominated solutions for fuzzy modeling problems. In section 2, the fuzzy model structure is given and a model reduction method is discussed. Section 3 then introduces the criteria that are taken into account to achieve accurate and interpretable models. Followed by section 4, where the main components of applied multi-objective EAs are described. Section 5 proposes an optimization model for fuzzy modeling and a decision making strategy. In section 6, experiments with the EA for three test problems are shown and compared with results from literature. Section 6, concludes the paper and indicates lines for future research.

2 Fuzzy Model Identification

2.1 Fuzzy model structure

We consider rule-based models of the Takagi-Sugeno (TS) type [26] which are especially suitable for the approximation of dynamic systems. The rule consequents are often taken to be linear functions of the inputs:

$$R_i : \textbf{If } x_1 \text{ is } A_{i1} \textbf{ and } \ldots x_n \text{ is } A_{in} \textbf{ then} \tag{1}$$
$$\hat{y}_i = \zeta_{i1} x_1 + \ldots, \zeta_{in} x_n + \zeta_{i(n+1)}, \; i = 1, \ldots, M$$

Here $\mathbf{x} = [x_1, x_2, \ldots, x_n]^T$ is the input vector, $\hat{y}_i$ is the output of the ith rule, A_{ij} ($j = 1, \ldots, n$) are fuzzy sets defined in the antecedent space by membership functions $\mu_{A_{ij}} : \mathrm{R} \rightarrow [0, 1]$, $\zeta_{ij} \in \mathrm{R}$ ($j = 1, \ldots, n+1$) are the consequent parameters, and M is the number of rules. The total output of the model is computed by aggregating the individual contributions of the rules:

$$\hat{y} = \sum_{i=1}^{M} p_i(\mathbf{x}) \hat{y}_i \tag{2}$$

where $p_i(\mathbf{x})$ is the normalized firing strength of the ith rule:

$$p_i(\mathbf{x}) = \frac{\prod_{j=1}^{n} \mu_{A_{ij}}(x_j)}{\sum_{i=1}^{M} \prod_{j=1}^{n} \mu_{A_{ij}}(x_j)} \tag{3}$$

We apply two different membership functions to describe the fuzzy sets A_{ij} in the rule antecedents: trapezoidal and gaussian.

Membership function for trapezoidal fuzzy sets can be expressed as:

$$\mu_{A_{ij}}(x) = \max\left(0, \min\left(\frac{x - a_{ij}}{b_{ij} - a_{ij}}, 1, \frac{c_{ij} - x}{c_{ij} - d_{ij}}\right)\right) \tag{4}$$

where $a_{ij}, b_{ij}, c_{ij}, d_{ij} \in [l_j, u_j]$, with $a_{ij} \leq b_{ij} \leq c_{ij} \leq d_{ij}$.

For gaussian fuzzy sets we consider the following membership function:

$$\mu_{A_{ij}}(x) = \begin{cases} \exp\left(-\frac{(s_{ij}-x)^2}{2{\sigma^l_{ij}}^2}\right) & if\ x < c_{ij} \\ \exp\left(-\frac{(x-s_{ij})^2}{2{\sigma^r_{ij}}^2}\right) & if\ x \geq c_{ij} \end{cases} \tag{5}$$

In this case the fuzzy model is defined by a Radial Basis Function (RBF) Neural Network [11]. The number of neurons in the hidden layer of an RBF neural network is equal to the number of rules in the fuzzy model. The firing strength of the ith neuron in the hidden layer matches the firing strength of the ith rule in the fuzzy model. We apply an asymmetric gaussian membership function defined by three parameters, the center s_{ij}, left variance σ^l_{ij} and rigth variance σ^r_{ij}. Therefore, each neuron in the hidden layer has these three parameters that define its firing strength value. The neurons in the output layer perform the computations for the first order linear function described in the consequents of the fuzzy model, therefore, the ith neuron of the output layer has the parameters ζ_{ij} that correspond to the linear function defined in the ith rule of the fuzzy model.

2.2 Rule set simplification techniques

Automated approached to fuzzy modeling often introduce redundancy in terms of several similar fuzzy sets that describe almost the same region in the domain of some variable. According to some similarity measure, two or more similar fuzzy sets can be merged to create a new fuzzy set representative for the merged sets [21]. This new fuzzy set substitutes the ones merged in the rule base. The merging process is repeated until fuzzy sets for each model variable cannot be merged, i.e., they are not similar. This simplification may results in several identical rules, which are removed from the rule set.

We consider the following similarity measure between two fuzzy sets A and B:

$$S(A,B) = \frac{|A \cap B|}{|A \cup B|} \tag{6}$$

If $S(A,B) > \theta_S$ (we use $\theta_S = 0.6$) then fuzzy sets A and B are merged in a new fuzzy set C as follows:

Trapezoidal fuzzy sets:

$$\begin{array}{l} a_C = min\{a_A, a_B\} \\ b_C = \alpha b_A + (1-\alpha) b_B \\ c_C = \alpha c_A + (1-\alpha) c_B \\ d_C = \max\{d_A, d_B\} \end{array} \tag{7}$$

Gaussian fuzzy sets:

$$\begin{array}{l} \sigma_C^l = \frac{s_C - min\{s_A - 3\sigma_A^l, s_B - 3\sigma_B^l\}}{3} \\ s_C = \alpha s_A + (1-\alpha) s_B \\ \sigma_C^r = \frac{s_C - min\{s_A - 3\sigma_A^r, s_B - 3\sigma_B^r\}}{3} \end{array} \tag{8}$$

where $\alpha \in [0,1]$ determines the influence of A and B on the new fuzzy set C.

3 Criteria for Fuzzy Modeling

3.1 Multi-objective Identification

Identification of fuzzy models from data requires the presence of multiple criteria in the search process. In multi-objective optimization, the set of solutions is composed of all those elements of the search space for which the corresponding objective vector cannot be improved in any dimension without degradation in another dimension. These solutions are called *non-dominated* or *Pareto-optimal.* Given two decision vectors $\mathbf{x}$ and $\mathbf{y}$ in a universe U, $\mathbf{x}$ is said to *dominate* $\mathbf{y}$ if $f_i(\mathbf{x}) \leq f_i(\mathbf{y})$, for all objective functions f_i, and $f_j(\mathbf{x}) < f_j(\mathbf{y})$, for at least one objective function f_j, for minimization. A decision vector $\mathbf{x} \in U$ is said to be *Pareto-optimal* if no other decision vector dominates $\mathbf{x}$.

The Pareto-optimality concept should be integrated within a decision process in order to select a suitable compromise solution from all non-dominated alternatives. In a decision process, the decision maker expresses preferences which should be taken into account to identify preferable non-domination solutions. Approaches based on weights, goals and priorities have been used more often.

3.2 Three criteria for fuzzy modeling

We consider three main criteria to search for an acceptable fuzzy model: (i) accuracy, (ii) transparency, and (iii) compactness. It is necessary to define quantitative measures for these criteria by means of appropriate objective functions which define the complete fuzzy model identification.

The accuracy of a model can be measured with the *mean squared error:*

$$MSE = \frac{1}{K} \sum_{k=1}^{K} (y_k - \hat{y}_k)^2 \tag{9}$$

where y_k is the true output and $\hat{y}_k$ is the model output for the kth input vector, respectively, and K is the number of data samples.

Many measures are possible for the second criterion, transparency. Nevertheless, in this paper we only consider one of most significant, *similarity*, as a first starting point. The similarity S among distinct fuzzy sets in each variable of the fuzzy model can be expressed as follows:

$$S = \max_{\substack{i,j,k \\ A_{ij} \neq B_{kj}}} S(A_{ij}, B_{kj}),$$
$$i = 1, \ldots, M, \; j = 1, \ldots, n, \; k = 1, \ldots, M \tag{10}$$

This is an aggregated similarity measure for the fuzzy rule-based model with the objective to minimize the maximum similarity between the fuzzy sets in each input domain.

Finally, measures for the third criterion, the compactness, are the number of rules M and the number of different fuzzy sets L of the fuzzy model. We assume that models with a small number of rules and fuzzy sets are compact. In summary, we have considered three criteria for fuzzy modeling, and we have defined the following measures for these criteria:

Criteria	Measures
Accuracy	MSE
Transparency	S
Compactness	M, L

4 Multi-Objective Evolutionary Algorithms

A highly important aspect of multi-objective evolutionary optimization is that of bestowing a good diversity mechanism on the algorithm. Diversity techniques in multi- objective Evolutionary Computation were originally put forward by Goldberg [5] at the end of the eighties and their importance lies fundamentally in two facts. Firstly, multiple solutions captured in a simple EA execution should cover all the Pareto-optimal fronts which make up the solution. This means that the algorithm has to search for non dominated solutions in a diversified way. Secondly, all the non dominated individuals of the population should have an equal probability of being selected, since they are all equally good. This fact may lead to the genetic drift phenomenon which causes the EA population to converge to just a small region of the solution space. Diversity techniques are, therefore, paramount in multi-objective and multi-modal optimization, and have usually been referred to as *niche formation techniques*. According to Goldberg [5], niche formation techniques can be classified into two categories: *implicit* and *explicit*. With implicit techniques, diversity is achieved through the selfsame generational substitution used by the EA, with the *pre-selection scheme* the *crowding factor model* being the most usual. In the case of explicit techniques, *sharing function* is typically defined to determine the degree of participation in each individual of the population and this is used to degrade, as a penalty, the fitness of each individual.

4.1 Multi-objective EAs characteristics

We propose two multi-objective EAs which consider trapezoidal and gaussian fuzzy sets respectively. The common characteristics of both multi-objective EAs are the following:

1. The proposed algorithms are a Pareto-based multi-objective EAs for fuzzy modeling, i.e., it has been designed to find, in a single run, multiple non-dominated solutions according to the Pareto decision strategy. There is no dependence between the objective functions and the design of the EAs, thus, any objective function can easily be incorporated. Without loss of generality, the EA minimizes all objective functions.
2. Constraints with respect to the fuzzy model structure are satisfied by incorporating specific knowledge about the problem. The initialization procedure and variation operators always generate individuals that satisfy these constraints.
3. The EAs have a variable-length, real-coded representation. Each individual of a population contains a variable number of rules between 1 and *max*, where *max* is defined by a decision maker. Fuzzy numbers in the antecedents and the parameters in the consequent are coded by floating-point numbers.
4. The initial population is generated randomly with a uniform distribution within the boundaries of the search space, defined by the learning data and model constraints.
5. The EAs search for among simplified rule sets, i.e, all individuals in the population has been previously simplified (after initialization and variation), which is an added ad hoc technique for transparency and compactness. So, all individuals in the population have a similarity S between 0 and 0.6.
6. Chromosome selection and replacement are achieved by means of a variant of the preselection scheme. This technique is, implicitly, a niche formation technique and an elitist strategy. Moreover, an explicit niche formation technique has been added to maintain diversity respect to the number of rules of the individuals. Survival of individuals is always based on the Pareto concept.
7. The EAs variation operators affect at the individuals at different levels: (i) the rule set level, (ii) the rule level, and (iii) the parameter level.

4.2 Representation of solutions and constraint satisfaction

An individual I for this problem is a rule set of M rules as follows:

$$R_1: \ A_{11} \ \dots \ A_{1n} \quad \zeta_{11} \ \dots \ \zeta_{1n} \ \zeta_{1(n+1)}$$
$$\dots$$
$$R_M: A_{M1} \ \dots \ A_{Mn} \ \zeta_{M1} \ \dots \ \zeta_{Mn} \ \zeta_{M(n+1)}$$

The constraints on the domain of the variables for a fuzzy model are given by the semantic of a fuzzy number. Thus, a trapezoidal fuzzy number A_{ij} $(i = 1, \dots, M, \ j = 1, \dots, n)$ can be represented by means of four real values $a_{ij}, b_{ij}, c_{ij}, d_{ij} \in [l_j, u_j]$, with $a_{ij} \leq b_{ij} \leq c_{ij} \leq d_{ij}$. A gaussian fuzzy number A_{ij} $(i = 1, \dots, M, \ j = 1, \dots, n)$ can be represented by means of three real values $s^l_{ij}, s_{ij}, s^r_{ij} \in [l_j, u_j]$, with $s^l_{ij} = s_{ij} - 3\sigma^l_{ij} \leq s_{ij} \leq s^r_{ij} = s_{ij} + 3\sigma^r_{ij}$. The consequent parameters are also real values constrained by a domain, i.e. $\zeta_{ij} \in [l, u]$ $(i = 1, \dots, M, \ j = 1, \dots, n+1)$. Other constraint are related with the number of rules M of the model, which can be defined between a lower number 1 and a upper number max fixed by the decision maker.

In the following sections we describe easy initialization and variation procedures to generate random individuals which satisfy these constraints.

4.3 Initial population

Initial population is completely random, except that the number of individuals with M rules, for all $M \in [1, max]$, should be between $minNS$ and $maxNS$ to ensure diversity respect to the number of rules, where $minNS$ and $maxNS$, with $0 \leq minNS \leq \frac{PS}{max} \leq maxNS \leq PS$ (PS is the population size), are the minimum and maximum niche size respectively (see next subsection).

To generate an individual with M rules, the procedure is as follows: for each trapezoidal fuzzy number A_{ij} $(i = 1, \dots, M, \ j = 1, \dots, n)$, four random real values from $[l_j, u_j]$ are generated and sorted to satisfy the constraints $a_{ij} \leq b_{ij} \leq c_{ij} \leq d_{ij}$ (three values $s^l_{ij}, s_{ij}, s^r_{ij} \in [l_j, u_j]$, with $s^l_{ij} \leq s_{ij} \leq s^r_{ij}$ in the gaussian case). Parameters ζ_{ij} $(i = 1, \dots, M, \ j = 1, \dots, n+1)$ are real values generated at random from $[l, u]$. After, the individual is simplified according to the procedure described in a previous section.

4.4 Selection and generational replacement

We use a variant of the preselection scheme [5] which has been one of the results of previous works for general constrained multi-objective optimization problems by EA [12].

In each iteration of the EA, two individuals are picked at random from the population. These individuals are crossed $nChildren$ times and children mutated resulting in $nChildren$ pairs of candidates (in addition, the offspring are trained by a gradient method [11] in the gaussian case). Next, the offspring are reduced by using the previous rule set simplification technique. Finally, the best of the first offspring replaces the first parent, and the best of the second offspring replaces to the second parent only if:

- the offspring is better than the parent, and
- the number of rules of the offspring is equal to the number of rules of the parent, or the niche count of the parent is greater than $minNS$ and the niche count of the offspring is smaller than $maxNS$

An individual I is better than another individual J if I dominates J. The best individual of a collection is any individual I such that there is no other individual J which dominates I. The niche count of an individual I is the number of individuals in the population with the same number of rules as I.

Note that the preselection scheme is an implicit niche formation technique to maintain diversity in the populations because an offspring replaces an individual similar to itself (one of their parents). Implicit niche formation techniques are more appropriate for fuzzy modeling than explicit techniques, such as sharing function, which can provoke an excessive computational time. However, we need an additional mechanism for diversity with respect to the number of rules of the individuals in the population. One of the reasons is that the number of rules is an integer parameter and the variation operators can generate individuals with quite different numbers of rules of the parents. The preselection scheme is not effective in such a case. The added explicit niche formation technique ensures that the number of individuals with M rules, for all $M \in [1, max]$, is greater or equal to $minNS$ and smaller or equal to $maxNS$. Moreover, the preselection scheme is also an elitist strategy because the best individual in the population is replaced only by a better one.

4.5 Variation operators

As already said, an individual is a set of M rules. A rule is a collection of n fuzzy numbers (antecedent) plus $n+1$ real parameters (consequent), and a fuzzy number is composed of four real numbers. In order to achieve an appropriate exploitation and exploration of the potential solutions in the search space, variation operators working in the different levels of the individuals are necessary. In this way, we consider three levels of variation operators: rule set level, rule level, and parameter level. After a sequence of crossovers and mutations, the offspring are simplified according to the rule set simplication procedure as described previously.

Five crossover operators, four mutation operators and a training operator are used in the EAs. In the following, $\alpha \in [0, 1]$ is a random number from a uniform distribution.

Rule set level variation operators

- **Crossover1**: Given two parents $I_1 = (R_1^1 \ldots R_{M_1}^1)$ and $I_2 = (R_1^2 \ldots R_{M_2}^2)$, this operator exchanges information about the number of rules of the parents and information about the rules of the parents, but no rule is internally crossed. Two children are produced: $I_3 = (R_1^1 \ldots R_a^1 R_1^2 \ldots R_b^2)$

and $I_4 = (R^1_{a+1} \cdots R^1_{M_1} R^2_{b+1} \cdots R^2_{M_2})$, where $a = round(\alpha \cdot M_1)$ and $b = round((1-\alpha) \cdot M_2)$. The number of rules of the children is between M_1 and M_2.

- **Crossover2**: This operator increases the number of rules of the two children as follows: the first child contains all M_1 rules of the first parent and $\min\{\max - M_1, M_2\}$ rules of the second parent; the second child contains all M_2 rules of the second parent and $\min\{\max - M_2, M_1\}$ rules of the first parent.
- **Mutation1**: This operator deletes or adds, both with equal probability, one rule in the rule set. For deletion, one rule is randomly deleted from the rule set. For rule-addition, one rule is randomly generated, according to the initialization procedure described, and added to the rule set.
- **Training (only applied in the gaussian case)**: This operator modifies the rule set by training the RBF neural network with a gradient method (100 iterations, $\mu = 0.01$). The purpose of this operator is to improve the accuracy of the individuals.

Rule level variation operators

- **Crossover3**: Given two parents $I_1 = (R^1_1 \ldots R^1_i \ldots R^1_{M_1})$ and $I_2 = (R^2_1 \ldots R^2_j \ldots R^2_{M_2})$, this operator produces two children $I_3 = (R^1_1 \ldots R^3_i \ldots R^1_{M_1})$ and $I_4 = (R^2_1 \ldots R^4_j \ldots R^2_{M_2})$, with $R^3_i = \alpha R^1_i + (1-\alpha) R^2_j$ and $R^4_j = \alpha R^2_j + (1-\alpha) R^1_i$, where i, j are random indexes from $[1, M_1]$ and $[1, M_2]$ respectively.
- **Crossover4**: Given two parents $I_1 = (R^1_1 \ldots R^1_i \ldots R^1_{M_1})$ and $I_2 = (R^2_1 \ldots R^2_j \ldots R^2_{M_2})$, this operator produce two children $I_3 = (R^1_1 \ldots R^3_i \ldots R^1_{M_1})$ and $I_4 = (R^2_1 \ldots R^4_j \ldots R^2_{M_2})$, where R^3_i and R^4_j are obtained with the *uniform* crossover.
- **Mutation2**: This operator removes a randomly chosen rule and inserts a new one which is randomly generated by the rule-initialization procedure.

Parameter level variation operators

- **Crossover5**: Given two parents, and one rule of each parent randomly chosen, this operator crosses the fuzzy numbers corresponding to a random input variable or the consequent parameters. The crossover is arithmetic.
- **Mutation3**: This operator mutates a random fuzzy number or the consequent of a random rule. The new fuzzy number or consequent is generated at random.
- **Mutation4**: This operator changes the value of one of the antecedent fuzzy sets a, b, c or d of a random fuzzy number, or a parameter of the consequent ζ, of a randomly chosen rule. The new value of the parameter is generated at random within the constraints by a non-uniform mutation.

5 Optimization Model and Decision Making

After preliminary experiments in which we have checked different optimization models, the following remarks can be maded:

1. Instead of minimizing of the number of rules M we have decided to search for rules sets with a number of rules within an interval $[1, max]$ where a decision maker can feel comfortable. The explicit niche formation technique ensures the EA always contains a minimum of representative rule sets for each number of rules in the populations. Then, we do not minimize the number of rules during the optimization, but we will take it into account at the end of the run, in a posteriori decision process applied to the last population.
2. It is very important to note that a very transparent model will be not accepted by a decision maker if the model is not accurate. In most fuzzy modeling problems, excessively low values for similarity hamper accuracy, for which these models are normally rejected. Alternative decision strategies, as *goal programming*, enable us to reduce the domain of the objective functions according to the preferences of a decision maker. Then, we can impose a goal g_S for similarity, which stop minimization of the similarity in solutions for which goal g_S has been reached.
3. The measure L (number of different fuzzy sets) is considerably reduced by the rule set simplification technique. So, we do not define an explicit objective function to minimize L.

According to the previous remarks, we finally consider the following optimization model:

$$\begin{array}{l} Minimize\ f_1 = MSE \\ Minimize\ f_2 = \max\{g_S, S\} \end{array} \tag{11}$$

At the end of the run, we consider the following a posteriori decision process applied to the last population to obtain the final compromise solution:

1. Identify the set $X^* = \{x_1^*, \ldots, x_p^*\}$ of non-dominated solutions according to:

$$\begin{array}{l} Minimize\ f_1 = MSE \\ Minimize\ f_2 = \max\{g_S, S\} \\ Minimize\ f_3 = M \end{array} \tag{12}$$

2. Choose from X^* the most accurate solution x_i^*; remove x_i^* from X^*;
3. If solution x_i^* is not accurate enough or there is no solution in the set X^* then STOP (no solution satisfies);
4. If solution x_i^* is not transparent or compact enough then go to step 2;
5. Show the solution x_i^* as output.

Computer aided inspection shown in Figures 3, 4 and 6 can help in decisions for steps 2 and 3.

6 Experiments and results

In this section, the multi-objective EAs are applied to the identification of fuzzy models for three well-known data sets studied by other authors. The following values for the parameters of the EA were used in the simulations: population size 100, crossover probability 0.8, mutation probability 0.4, training probability 0.1 (gaussian case), number of children for the preselection scheme 10, minimum number of individuals for each number of rules 5, and maximum number of individuals for each number of rules 20. All crossover and mutation operators are applied with the same probability. The EAs stops when the solutions satisfy the decisor maker.

6.1 Example 1

Consider the 2nd order nonlinear plant studied by Wang and Yen in [28,29]:

$$y(k) = g(y(k-1), y(k-2)) + u(k) \tag{13}$$

with

$$g(y(k-1), y(k-2)) = \frac{y(k-1)y(k-2)(y(k-1)-0.5)}{1+y^2(k-1)y^2(k-2)} \tag{14}$$

The goal is to approximate the nonlinear component $g(y(k-1), y(k-2))$ of the plant with a fuzzy model. As in [28], 400 simulated data points were generated from the plant model (13). Starting from the equilibrium state $(0,0)$, 200 samples of identification data were obtained with a random input signal $u(k)$ uniformly distributed in $[-1.5, 1.5]$, followed by 200 samples of evaluation data obtained using a sinusoidal input signal $u(k) = \sin(2\pi k/25)$. The resulting signals and the real surface are shown in Figure 1.

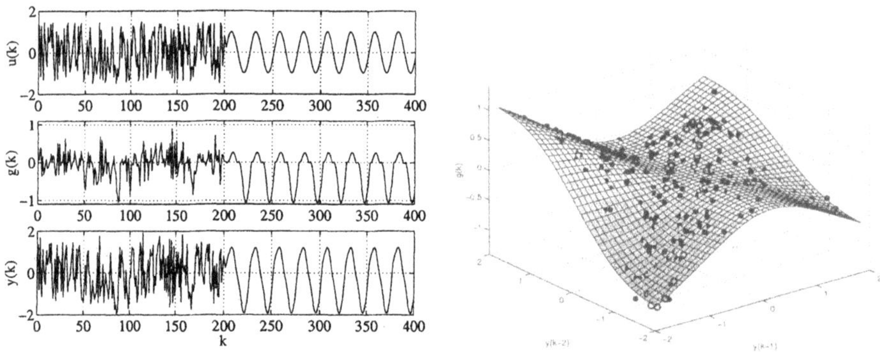

Fig. 1. *Left*: Input $u(k)$, unforced system $g(k)$, and output $y(k)$ of the plant in (13). *Right*: Real surface.

We show results obtained with the EA (trapezoidal case) by using the optimization model (11) ($max = 5$, $g_S = 0.25$). Figure 2 shows the non-dominated solutions in the last population according to (12). One can appreciate the effectiveness of the preselection technique and the added explicit niche formation technique to maintain diversity in the populations. Goal-based model have the disadvantage that it is necessary to choose, a priori, a good goal for the problem, although this value is representative of the maximum degree of overlapping of the fuzzy sets allowed by a decisor.

According to the described decision process, we finally choose a compromise solution showed in Figure 3 by means of different graphics for the obtained model. Figure 3(a) shows the local model. The surface generated by the model is shown in Figure 3(b), fuzzy sets for each variable are showed in Figure 3(c), and finally, the identification and validation results as well as the prediction error are shown in Figure 3(d).

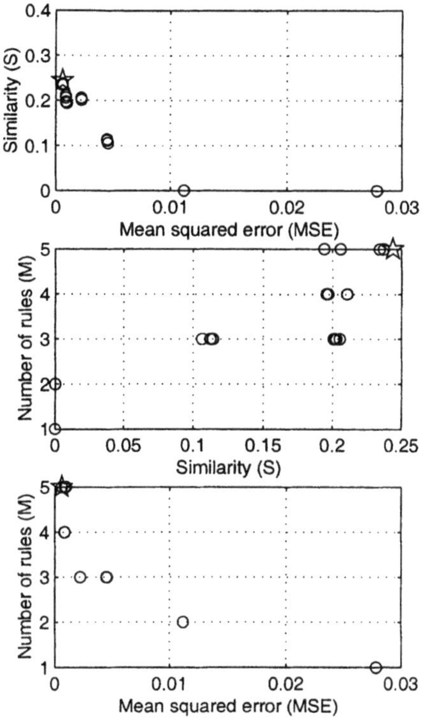

Fig. 2. Non-dominated solutions according to (12) obtained with the Pareto-based multi-objective EA (trapezoidal case) by using the optimization model (11). Solution marked with * is the final compromise solution.

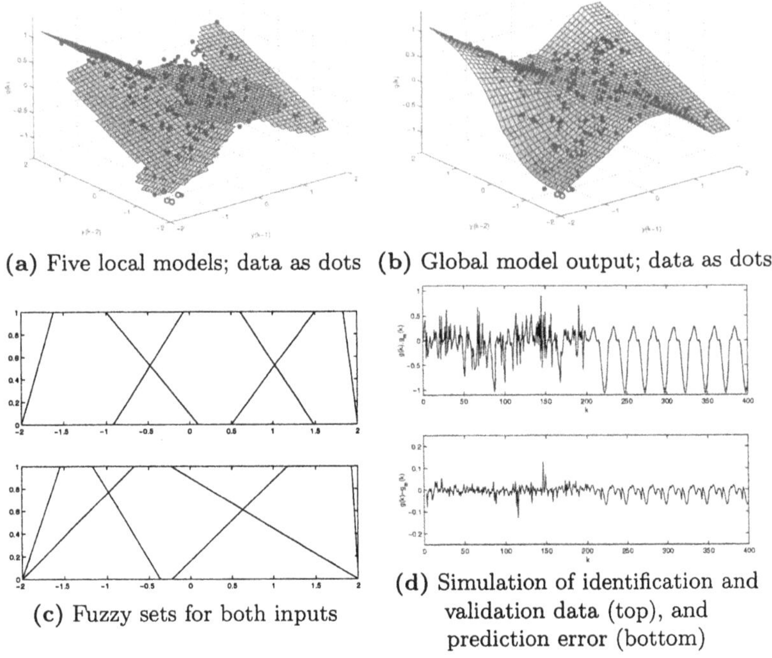

(a) Five local models; data as dots (b) Global model output; data as dots

(c) Fuzzy sets for both inputs (d) Simulation of identification and validation data (top), and prediction error (bottom)

Fig. 3. Accurate, transparent and compact fuzzy model for the plant model (13).

We compared our results, with those obtained by the four different approaches proposed in [29] and [19]. The best results obtained for in each case are summarized in Table 1, with an indication of the number of rules, number of different fuzzy sets, consequent type, and obtained MSE for training and evaluation data. In [29], the low MSE on the training data is in contrast with the MSE for the evaluation data which indicates overtraining. The solution in [19] is similar to the solutions in this paper with respect to the accuracy, transparency and compactness, but hybrid techniques (initial fuzzy clustering and a sequence of specific genetic algorithms) were required in [19]. Solutions in this paper are obtained with a single EA and they have been chosen among different alternatives, which is an advantage for an appropriate decision process.

6.2 Example 2

With a constant input air flow-rate, the pressure in the fermentor tank (variable y) is controlled by the opening of the outlet valve (variable u). Based on simplified assumptions, a first-principle physical model can be derived for this

Table 1. Fuzzy models for the dynamic plant. All models are of the Takagi-Sugeno type.

Ref.	No. of rules	No. of sets	Consequent	MSE train	MSE eval
[29]	36 rules (initial)	12 (B-splines)	Linear	$1.9 \cdot 10^{-6}$	$2.9 \cdot 10^{-3}$
	24 rules (optimized)	-	Linear	$2.0 \cdot 10^{-6}$	$6.4 \cdot 10^{-4}$
[19]	7 rules (initial)	14 (triangular)	Linear	$1.8 \cdot 10^{-3}$	$1.0 \cdot 10^{-3}$
	5 rules (optimized)	5 (triangular)	Linear	$5.0 \cdot 10^{-4}$	$4.2 \cdot 10^{-4}$
This paper[1]	5 rules	6 (trapezoidal)	Linear	$5.9 \cdot 10^{-4}$	$8.8 \cdot 10^{-4}$

[1] Solution corresponds to the solution marked with * in Figure 2, and Figure 3

process. The setting of the outlet valve results in a certain transient behavior of the pressure, which can be described by a first-order nonlinear differential equation. As shown in [18], a fuzzy linear model provides a simple and elegant solution to the problem. In this way, the process can be represented by means of a TS rule set of the following form:

$$\textbf{If } y(k) \text{ is } A_i \textbf{ and } u(k) \text{ is } B_i \textbf{ then}$$
$$y(k+1) = a_i y(k) + b_i u(k) + c_i, \quad i = 1, \ldots, M.$$

This rule base represents a nonlinear first-order regression model $y(k+1) = f(y(k), u(k))$, where $y(k)$ and $u(k)$ are the pressure and the valve position at time k, respectively. The membership functions of the antecedent A_i and B_i, as well as the consequent parameters a_i, b_i and c_i are estimated from the data by the proposed multi-objective EA (trapezoidal case), by using the optimization model (11) ($g_S = 0.3$, $max = 5$) and the proposed a posteriori decision strategy.

We compared our results, with those obtained by the two different approaches proposed in [17] and [3]. Solution in [17] is obtained by means of hyperplanar fuzzy clustering (Gustafson-Kessel's algorithm [8]) with projections of the fuzzy clusters in each domain and making the extensional hull of the fuzzy sets obtained to approximate them by trapezoidal fuzzy sets. The obtained solution is transparent and compact, containing 3 rules and 6 fuzzy sets with mean squared error $4.516 \cdot 10^{-4}$. In [3], in order to obtain the coeficient of the linear consequent using the recursive least-squares algorithm (or a stationary Kalman Filter), the grade of membership of the data to the antecedent of the fuzzy rules is considered using directly the grade of membership of the data to the fuzzy clusters found in product space of input-output variables. Moreover as in this kind of model we are searching for local linear models presents in the data, a modification we have adopt is not to use only a Fuzzy C-Means algorithm, but as an inicialization to a Gustafson-Kessel fuzzy clustering algorithm. In this way, this algorithm is adequate to detect the fuzzy partitions that better fulfill the assumption of fuzzy linear models. The obtained solution is compact but non transparent, containing 4 rules and 4 n-dimensional fuzzy sets, with mean squared error $6.4 \cdot 10^{-5}$. Solution in this paper is obtained with a single multi-objective EA and it has been chosen among different alternatives, which is an advantage for an appropriate decision process. Non-dominated solutions according to (12) are summarized in Table 2, with an indication of the number of rules, number of different fuzzy sets, obtained MSE for training data and similarity S of the fuzzy sets. The accurate, transparent and compact solution with 2 rules, 4 different fuzzy sets, mean squared error $4.845 \cdot 10^{-5}$ and similarity 0.235 is finally chosen with the proposed a posteriori decision strategy. This solution is showed in Figure 3 by means of different graphics for the obtained model. Figure 4(a) shows the local model, the surface generated by the model is shown in Figure 4(b), fuzzy sets for each variable are showed in Figure 4(c), and finally, the prediction error is showed in Figure 4(d).

No. rules	No. fuzzy sets	MSE	S
1	1	$9.989 \cdot 10^{-4}$	0.0
2	4	$4.845 \cdot 10^{-5}$	0.235
3	5	$2.778 \cdot 10^{-5}$	0.248
4	6	$2.470 \cdot 10^{-5}$	0.232
5	7	$2.306 \cdot 10^{-5}$	0.232

Table 2. Non-dominated solutions according to (12) obtained with the multi-objective EA (trapezoidal case) for the pressure dynamics of a laboratory fermentor.

6.3 Example 3

We consider the modeling of the rule base given in [25]:

R_1: If x_1 is A_1 (3, 9) then $y = 1.0x_1 + 0.5x_2 + 1.0$

R_2: If x_1 is A_2 (3, 9) and x_2 is B_1 (4, 13) then $y = -0.1x_1 + 4.0x_2 + 1.2$

R_3: If x_1 is A_3 (3, 9, 11, 18) and x_2 is B_2 (4, 13) then $y = 0.9x_1 + 0.7x_2 + 9.0$

R_4: If x_1 is A_4 (11, 18) and x_2 is B_2 (4, 13) then $y = 0.2x_1 + 0.1x_2 + 0.2$

The corresponding surface is shown in Figure 5. In [23] a model with four rules was identified from sampled data ($N = 546$) by the supervised clustering algorithm, which was initialized with 12 clusters. This model was optimized using a Genetic Algorithm to result in a MSE of 1.6. Tables 3 and 6.3 show results obtained with the Pareto-based multi-objective EAs for trapezoidal and gaussian cases respectively ($g_S = 0.25$, $max = 5$).

We finally choose a compromise solution (4-rules fuzzy model) according to the described decision process. Figures 6 and 7 show, for both trapezoidal and gaussian cases respectively, the local model, the surface generated by the model, fuzzy sets for each variable and the prediction error.

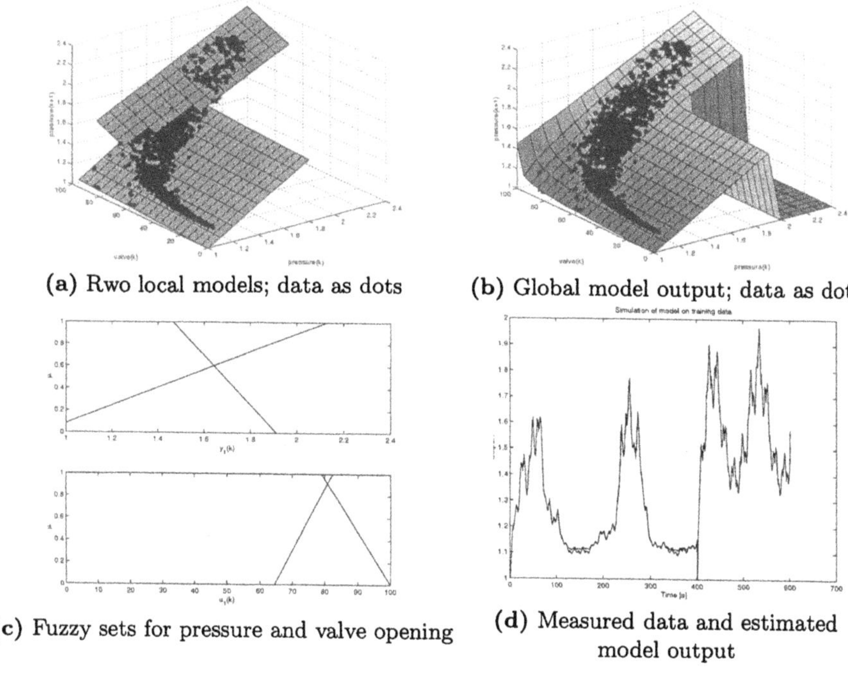

(a) Rwo local models; data as dots

(b) Global model output; data as dots

(c) Fuzzy sets for pressure and valve opening

(d) Measured data and estimated model output

Fig. 4. Accurate, transparent and compact fuzzy model for the pressure dynamics of a laboratory fermentor.

No. rules	No. fuzzy sets	MSE	S
1	1	56.696	0.0
2	3	14.226	0.042
3	4	2.945	0.134
4	5	1.233	0.249
5	5	0.993	0.247

Table 3. Non-dominated solutions according to (12) obtained with the multi-objective EA for the example in [25] (trapezoidal case).

7 Conclusions and future research

In this paper we present Pareto-based multi-objective evolutionary algorithms considering trapezoidal and gaussian fuzzy sets to obtain interpretable fuzzy models. Criteria such as accuracy, transparency and compactness have been propose and are taken into account in the optimization process. Some of these criteria have been partially incorporated into the EA by means of ad hoc techniques. Advantages of the gaussian case arise with the possibility of training the RBF neural networks associated with the fuzzy models in order

No. rules	No. fuzzy sets	MSE	S
1	2	125.391	0.0
2	4	25.606	0.348
3	5	2.187	0.349
4	5	1.017	0.349
5	5	0.910	0.350

Table 4. Non-dominated solutions according to (12) obtained with the multi-objective EA for the example in [25] (guassian case).

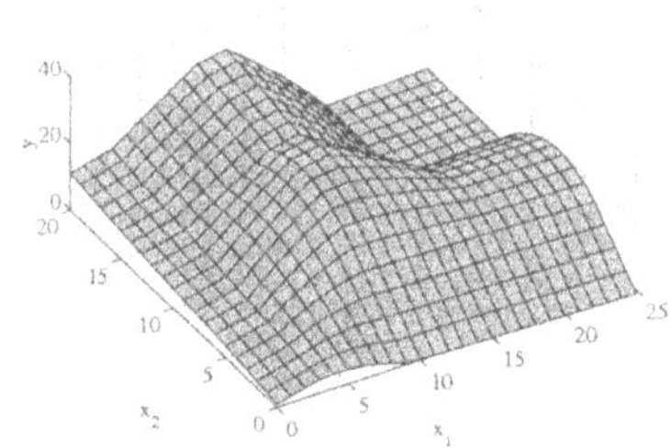

Fig. 5. Real surface for the example in [25].

to obtain more accuracy. In addition, several new ideas to reduce computational load and improve the global search capabilities, have been incorporated in the evolutionary algorithms. An implicit niche formation technique (preselection) in combination with other explicit techniques with low computational costs was introduced to maintain diversity. These niche formation techniques are appropriate in fuzzy modeling if cases where excessive amount of data are required. Excessive computational times would result if sharing functions were used. Elitism is also implemented by means of the preselection technique. A goal-based approach has been proposed to obtain more accurate fuzzy models.

The main difference between the proposed EAs and other approaches for fuzzy modeling is the reduced complexity because we use a single EA for generating, tuning and simplification of the fuzzy model. Moreover, human intervention is only required at the end of the run in choosing one of the multiple non-dominated solutions. Results obtained are good in comparison with other iterative techniques reported in literature, with the advantage that the proposed techniques identifies a set of alternative solutions. We also proposed an easy decision process with a posteriori decision process to choose finally a compromise solution.

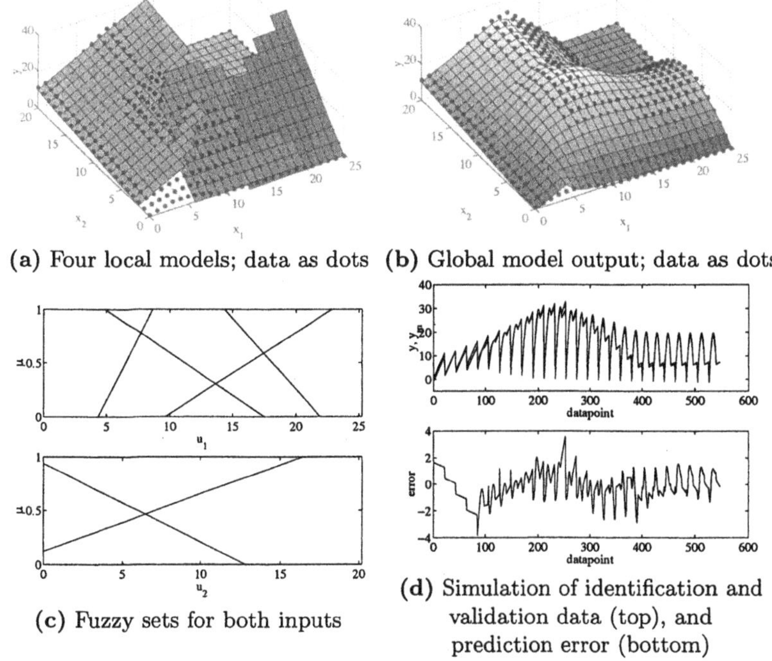

(a) Four local models; data as dots (b) Global model output; data as dots

(c) Fuzzy sets for both inputs

(d) Simulation of identification and validation data (top), and prediction error (bottom)

Fig. 6. Accurate, transparent and compact fuzzy model for the example in [25] (trapezoidal fuzzy sets).

References

1. O. Cordón and F. Herrera. A proposal for improving the accuracy of linguistic modeling. *IEEE Transactions on Fuzzy Systems*, 8(3):335–344, 2000.
2. J. Valente de Oliveira. Semantic constraints for membership function optimization. *IEEE Transactions on Fuzzy Systems*, 19(1):128–138, 1999.
3. M. Delgado, A.F. Gomez Skarmeta, and F. Martín. Generating fuzzy rules using clustering based approach. In *Third European Congress on Fuzzy and Intelligent Technologies and Soft Computing*, pages 810–814. Aachen, Germany, Agosto, 1995.
4. C.M. Fonseca and P.J. Fleming. An overview of evolutionary algorithms in multiobjective optimization. *Evolutionary Computation*, 3(1):1–16, 1995.
5. D.E. Goldberg. *Genetic Algorithms in Search, Optimization, and Machine Learning*. Addison-Wesley, 1989.
6. A.F. Gómez-Skarmeta and F. Jiménez. Fuzzy modeling with hibrid systems. *Fuzzy Sets and Systems*, 104:199–208, 1999.
7. A.F. Gómez-Skarmeta, F. Jiménez, and J. Ibánez. Pareto-optimality in fuzzy modelling. In *EUFIT'98*, pages 694–700, Aachen, Alemania, 1998.

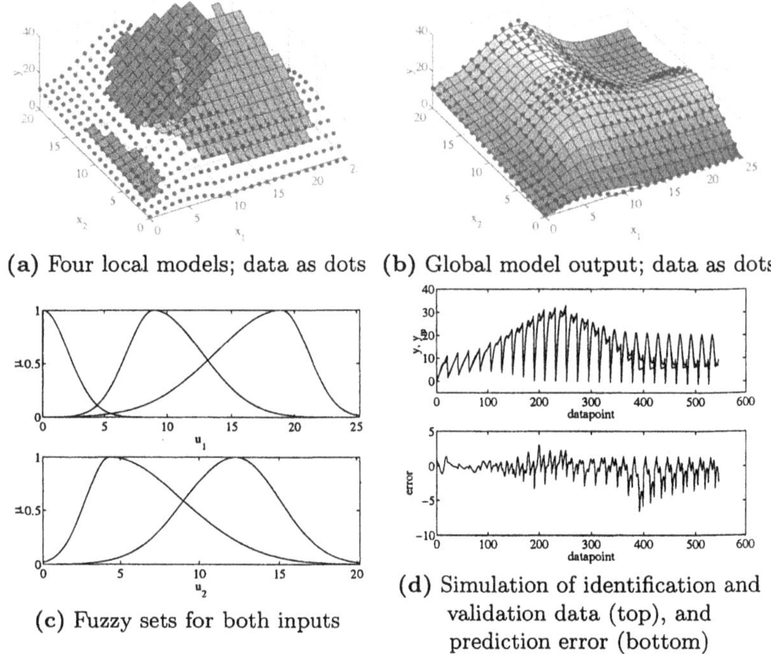

(a) Four local models; data as dots (b) Global model output; data as dots

(c) Fuzzy sets for both inputs

(d) Simulation of identification and validation data (top), and prediction error (bottom)

Fig. 7. Accurate, transparent and compact fuzzy model for the example in [25] (gaussian fuzzy sets).

8. D. E. Gustafson and W. C. Kessel. Fuzzy clustering with a fuzzy covariance matrix. In *IEEE Int. Conf. on Fuzzy Systems*, pages 761–766, 1979. San Diego.
9. J. Horn and N. Nafpliotis. Multiobjective optimization using the niched pareto genetic algorithm. IlliEAL Report No. 93005, July 1993.
10. H. Ishibuchi, T. Murata, and I.B. Türksen. Single-objective and two-objective genetic algorithms for selecting linguistic rules for pattern classification problems. *Fuzzy Sets and Systems*, 89:135–150, 1997.
11. J.S.R. Jang, C.T. Sun, and E. Mizutani. *Neuro-Fuzzy and Soft Computing.* Prentice Hall, 1997.
12. F. Jiménez, J.L. Verdegay, and A.F. Gómez-Skarmeta. Evolutionary techniques for constrained multiobjective optimization problems. In *Genetic and Evolutionary Computation Conference (GECCO-99), Workshop on Multi-Criterion Optimization Using Evolutionary Methods*, pages 115–116, Orlando, Florida, USA, 1999.
13. Y. Jin. Fuzzy modeling of high-dimensional systems. *IEEE Transactions on Fuzzy Systems: Complexity Reduction and Interpretability Improvement.*, 8:212–221, 2000.
14. T.A. Johansen, R. Shorten, and R. Murray-Smith. On the interpretation and identification of dynamic Takagi-Sugeno fuzzy models. *IEEE Transactions on Fuzzy Systems*, 8(3):297–313, 2000.

15. F. Klawonn and R. Kruse. Constructing a fuzzy controller from data. *Fuzzy Sets and Systems*, 85:177–193, 1997.
16. H. Pomares, I. Rojas, J. Ortega, J. Gonzalez, and A. Prieto. A systematic approach to a self-generating fuzzy rule-table for function approximation. *IEEE Transactions on Systems, Man, and Cybernetics - Part B: Cybernetics*, 30(3):431–447, 2000.
17. H.B. Verbruggen R. Babuska. A new identification method for linguistic fuzzy models. In *Open Meeting of the FALCON working group*, pages 1–8. Aachen, Germany, September, 1995.
18. H.B. Verbruggen R. Babuska. applied fuzzy modeling. In *IFAC Symposium on Artificial Intelligence in Real time Control*, 1994. Valencia, Spain.
19. H. Roubos and M. Setnes. Compact fuzzy models through complexity reduction and evolutionary optimization. In *proceedings 9th IEEE Conference on Fuzzy System*, pages 762–767, San Antonio, USA, May 7-10, 2000.
20. J.A. Roubos and M. Setnes. Compact and transparent fuzzy models and classifiers through iterative complexity reduction. *IEEE Transactions on Fuzzy Systems*, 9(4):516–524, 2001.
21. M. Setnes, R. Babuška, U. Kaymak, and H.R. van Nauta Lemke. Simlilarity measures in fuzzy rule simplification. *IEEE Transaction on Systems, Man and Cybernetics, Part B: Cybernetics*, 28(3):376–386, 1999.
22. M. Setnes, R. Babuška, and H. B. Verbruggen. Rule-based modeling: Precision and transparency. *IEEE Transactions on Systems, Man and Cybernetics, Part C: Applications & Reviews*, 28:165–169, 1998.
23. M. Setnes and J.A. Roubos. Transparent fuzzy modeling using fuzzy clustering and GA's. In *18th International Conference of the North American Fuzzy Information Processing Society*, pages 198–202, New York, USA, June 10-12, 1999. NAFIPS.
24. N. Srinivas and K. Deb. Multiobjective optimization using nondominated sorting in genetic algorithms. *Evolutionary Computation*, 2(3):221–248, 1995.
25. M. Sugeno and G.T. Kang. Structure identification of fuzzy model. *Fuzzy Sets and Systems*, 28:15–23, 1988.
26. T. Takagi and M. Sugeno. Fuzzy identification of systems and its application to modeling and control. *IEEE Transactions on Systems, Man and Cybernetics*, 15:116–132, 1985.
27. L. Wang and R. Langari. Complex systems modeling via fuzzy logic. *IEEE Trans. on System Man and Cybernetics*, 26:100–106, 1996.
28. L. Wang and J. Yen. Extracting fuzzy rules for system modeling using a hybrid of genetic algorithms and Kalman filter. *Fuzzy Sets and Systems*, 101:353–362, 1999.
29. J. Yen and L. Wang. Application of statistical information criteria for optimal fuzzy model construction. *IEEE Transactions on Fuzzy Systems*, 6(3):362–371, 1998.

Transparent Fuzzy Systems in Modelling and Control

Andri Riid and Ennu Rüstern

Department of Computer Control, Tallinn Technical University, Ehitajate tee 5, 19086, Tallinn, Estonia. E-mail: andri@dcc.ttu.ee.

Abstract. This chapter deals with low-level transparency of fuzzy systems that is necessary to ensure reliable interpretation of linguistic information provided by fuzzy systems. It is shown that for different types of fuzzy systems different definitions of transparency apply. Particular attention is paid to transparency protection mechanisms for data-driven optimisation algorithms such as gradient descent and genetic algorithms that otherwise would destroy the semantics of fuzzy systems in the course of optimisation. The need for transparency in fuzzy control is discussed and further illustrated by a control application of truck backer-upper.

1 Introduction

First fuzzy controllers based on Zadeh's studies [1] of human-machine interaction, were simple expert systems designed on the basis of human operator experience as, for example, the steam engine controller of Mamdani and Assilian [2]. Although this approach has produced many successful applications e.g. [3], the common complaint is that the design procedure relies heavily on human judgement and is therefore more heuristics than exact science.

The important step toward the automation of fuzzy system design was taken in [4] where Takagi-Sugeno (TS) rules and least squares procedure for the identification of fuzzy system parameters from data where simultaneously introduced. Thereafter, a gradient descent backpropagation technique adopted from neural network research was proposed for parameter adaptation [5], creating the concept of "neuro-fuzzy" systems. Jang combined these techniques into ANFIS [6] that is presently one of the most effective approximation algorithms. Recently, application of clustering methods and genetic algorithms as well as further combinations of different algorithms for fuzzy system design have become popular [7,8,9].

Data-driven optimisation of fuzzy systems and consequent reduction of human role in the design process has resulted in greatly improved accuracy. One, however, cannot ignore the fact that on linguistic level these numerically improved fuzzy models and controllers are usually completely meaningless to an human observer.

In fact, surprisingly little attention (see section 3 for details) is devoted to the issue that perhaps the most attractive property of fuzzy systems that lies in the capacity to process information in linguistic terms is ignored or sacrificed to numerical accuracy. The aim of the present chapter is to establish the mechanisms that would preserve the semantics of fuzzy systems even with the application of data-driven optimisation techniques instead of letting them to be destroyed.

The chapter is organised as follows: in next section the definitions of main types of fuzzy systems are given. 3rd section presents the definition of transparency for standard and 0th order TS systems and transparency conditions/constraints based on this definition are derived. Transparency paradox of 1st order TS systems is discussed in section 3.2. In section 4 the issue of transparency protection in fuzzy modelling is considered and in the 5th section relevance of transparency in fuzzy control is discussed. In section 6 several of previously described techniques are applied for the control of truck backer-upper demonstrating the advantages of transparent control for this particular problem.

2 Fuzzy systems

Presently, two main types of fuzzy systems are distinguished:

- Standard (linguistic, Mamdani) fuzzy systems [1,2].

Standard fuzzy systems consist of a number of rules that specify linguistic relation between the linguistic labels of input and output variables of the system. A fuzzy rule (1) is a statement where the premise and the consequent consist of fuzzy propositions with A_{ir} and B_{jr} denoting the linguistic labels of ith input variable x_i and output variable y ($i = 1 \dots N$), respectively, associated with the rth rule ($r = 1 \dots R$).

$$\text{IF } X_1 \text{ is } A_{1r} \text{ AND} \dots \text{ AND } X_i \text{ is } A_{ir} \dots \text{ AND } X_N \text{ is } A_{Nr} \text{ THEN } Y \text{ is } B_{1r} \tag{1}$$

Note that (1) expresses the input-output relationship in linguistic terms. To give the relationship in numerical terms, a special inference function (2) is used

$$y = Y\left(\bigcup_{r=1}^{R}\left(\bigcap_{i=1}^{N}\mu_{ir}(x_i)\right)\cap\gamma_r\right), \tag{2}$$

where μ_{ir} and γ_r, denote normal and convex fuzzy subsets or membership functions (MFs) having one-to-one correspondence with the respective linguistic labels in (1); x_i denotes the numerical value of the ith input variable; $\cap$ and $\cup$ denote the t- and s-norms that act as inference operators, respectively, and $Y(*)$ denotes the defuzzification function (centre-of-gravity (CoG), mean-of-maximum (MoM), etc.).

- (First-order) Takagi-Sugeno systems [4]

TS rules (3) are interpreted in terms of local linear models (y_r) as the consequent fuzzy proposition is replaced by a linear combination of inputs.

$$\text{IF } X_1 \text{ is } A_{1r} \text{ AND } X_2 \text{ is } A_{2r} \ldots \text{ AND } X_i \text{ is } A_{ir} \ldots \text{ AND } X_N \text{ is } A_{Nr} \text{ THEN} \quad y_r = p_{0r} + p_{1r}x_1 + \ldots + p_{ir}x_i + \ldots + p_{Nr}x_N \tag{3}$$

p_{ir} ($i = 0\ldots N$) in (3) denote the consequent coefficients. Because t-norms and s-norms in first-order TS systems (3) are commonly product and sum, inference function (2) reduces to

$$y = \sum_{r=1}^{R}\left(\prod_{i=1}^{N}\mu_{ir}(x_i)\right)\left(p_{0r} + \sum_{i=1}^{N} p_{ir}x_i\right) \Big/ \sum_{r=1}^{R}\prod_{i=1}^{N}\mu_{ir}(x_i) \tag{4}$$

A special case of TS systems called 0th order TS systems is obtained if the consequent function is a constant ($\forall p_{ir} = 0$, $i = 1\ldots N$, $r = 1\ldots R$):

$$\text{IF } U_1 \text{ is } A_{1r} \ldots \text{ AND } U_i \text{ is } A_{ir} \ldots \text{ AND } U_N \text{ is } A_{Nr} \text{ THEN } y_r = p_{0r} \tag{5}$$

$$y = \sum_{r=1}^{R}\prod_{i=1}^{N}\mu_{ir}(x_i)p_{0r} \Big/ \sum_{r=1}^{R}\prod_{i=1}^{N}\mu_{ir}(x_i) \tag{6}$$

0th order TS systems can also be regarded as a special case of standard fuzzy systems (with consequent fuzzy sets defined as fuzzy singletons) and semantically their interpretation is closer to standard fuzzy systems than to 1st order TS systems (3-4).

3 Fuzzy system transparency

The use of the term (transparency) in present chapter is based on [10] where transparency is defined as a property that enables us to understand the influence of each system parameter on the system output as well as on [11] where fuzzy systems are characterised as being transparent to interpretation.

Fuzzy system transparency is closely related to the concept of linguistic interpretability but these are not matching terms and, in our opinion, it is very important to see the distinction. Interpretability is a property of fuzzy systems (1-6) that exists by default, being established with linguistic rules and fuzzy sets associated with these rules, even the rules of 1st order TS systems can be interpreted. Transparency, on the other hand, is not a default property of fuzzy systems and should be regarded as a measure of how valid or how reliable is the linguistic interpretation of the system. It will be shown shortly that for standard fuzzy systems and 0th order TS systems, transparency has binary character, for 1st order TS systems it is a continuous variable.

Most authors, however, do not make this distinction; some of them do not pay attention to transparency at all and consequently assume that transparency like interpretability is a default property of fuzzy systems (sometimes regarded characteristic to standard and 0th order TS systems only as in [12]); others do emphasise that transparency of fuzzy systems is not guaranteed by default [13, 14] but use the terms in parallel.

There are two aspects of transparency in fuzzy systems. First one is related to the readability of rules that basically boils down to the overall complexity of the system. Improvement of readability through the use of moderate number of variables, rules and fuzzy subsets or by avoiding the inconsistency of the rule base, however, does not provide the solution to the problem of destroyed semantics. To solve the problem, one should concentrate on low-level transparency that grows out from conformity between the linguistic layer and the inference function of a fuzzy system.

In fact, very few authors [13, 14, 15, 16] have investigated the latter issue in any detail. The most important of these works is perhaps [16] that lists a set of properties (moderate number of MFs; natural zero positioning, normality, coverage and distinguishability of MFs) that fuzzy systems should meet and proposes mathematically formulated constraints for preserving the last two, incorporated into the cost function of the gradient descent algorithm. These works dealing with low-level transparency, however, aim for certain balance between transparency and accuracy and the results can be generally applied only to a limited class of systems/algorithms.

It is claimed that "currently there exists no well-established definition of transparency of a fuzzy system" and "there are no definite criteria for the distinguishability of a fuzzy partition" [13]. Hopefully, solutions proposed to these problems in [17], further developed in [18] and [19] being summarised here, help to fill the void.

3.1 Transparency of standard fuzzy systems

Let us consider the properties listed in [16]. It is arguable if coverage and natural zero positioning have anything to do with transparency [14]. Normality, on the other hand, is the standard assumption in fuzzy systems. Only distinguishability of input MFs (directly related to the overlap of input MFs) is vital to transparency as shown in the following.

The effect of overlap of input MFs to system output can be most conveniently observed in two-dimensional space that we do by constructing five otherwise equivalent SISO fuzzy systems, made up of 6 rules with 0%, 25%, 50%, 75% and 100% overlap degree, respectively. Although other system parameters (including minimum t-norm, maximum s-norm and CoG defuzzification) are fixed, in each case quite a different result is obtained (Fig. 1). With 0% overlap, no interpolation occurs, the system behaves as a multi-level relay and its output abruptly switches from one rule centroid to another. With 25% overlap the input intervals where the output has a constant value, are still present with some interpolation between the neighbouring rules.

With 50% overlap, the interval where system output is the explicit contribution of a given rule is reduced to a single point. With larger overlap, however, at least two rules contribute simultaneously for any given input, thus system output is always the result of interpolation. This makes the contribution of the observed rule invisible in system output. We suggest that such feature is undesirable. The phenomenon is driven to extreme with 100% overlap where all rules are fully activated simultaneously and system output has constant value, equalling to the centroid of the union of output fuzzy sets.

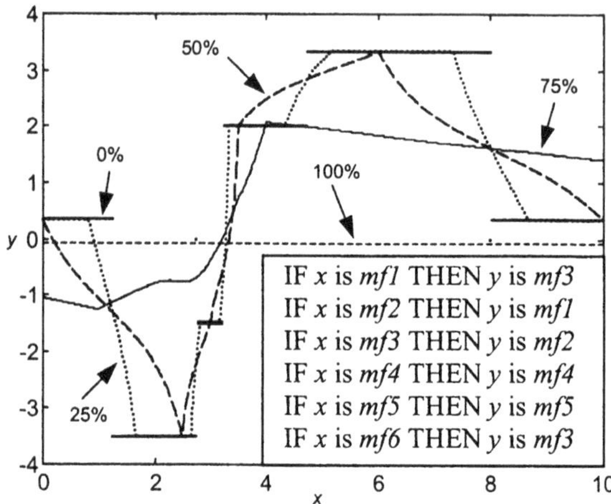

Fig. 1. Numerical input-output mapping of five fuzzy systems.

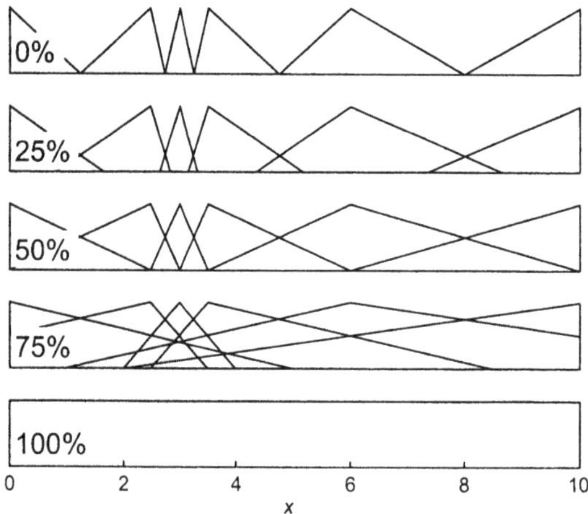

Fig. 2. Input MFs of observed systems.

Let us consider again the case of 50% overlap and let us refer to the point in input-output space where the explicit contribution of a given rule takes place and the rule under observation is fully activated as transparency checkpoint. When overlap of input MFs is equal or smaller than 50%, transparency checkpoints do exist. Closer inspection reveals that the input co-ordinate of the transparency checkpoint is equal to the centre of the fired MF (where $\mu(x) = 1$). Building up on the analogy, the desired output y for the transparency checkpoint should be the centre of the respective output MF, where $\gamma(y) = 1$. This ensures that the interpretation of the rule that we are able to obtain by combining the information from the rule base and MF definition base has good correspondence with the inferred numerical values (conformity!). This is exactly what we call transparency. The ideology of transparency checkpoints extends to MISO (and MIMO) systems and is covered by the following definition.

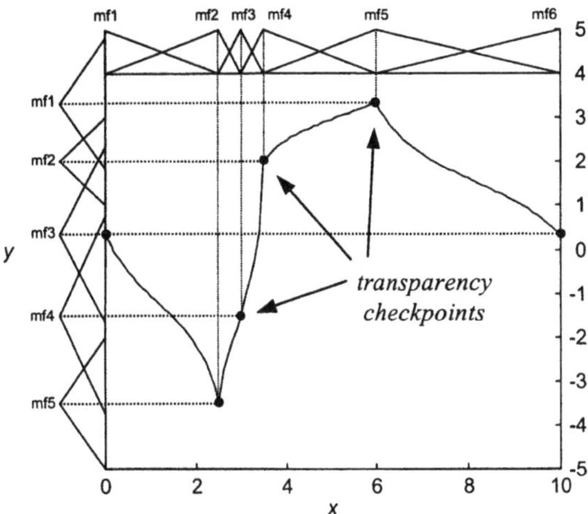

Fig. 3. Transparent fuzzy system.

Definition 1: r^{th} rule of the standard MISO fuzzy system (1) is transparent if it's activation degree

$$\tau_r = \bigcap_{i=1}^{N} \mu_{ir}(x_i) = 1 \tag{7}$$

results in the system output

$$y = \text{core}(\gamma_r), \tag{8}$$

where $\text{core}(\gamma_r) = \{y \in Y | \gamma_r(y) = 1\}$

A standard fuzzy system (1-2) can thus be regarded transparent only if all its rules are transparent (Fig.3).

In order to satisfy (7-8), certain conditions concerning input and output MFs of the system must be satisfied. The condition for input MFs is given:

$$\forall x_i \in X_i : \sum_{s=1}^{S_i} \mu_i^s(x_i) \le 1 , \tag{9}$$

where S_i denotes the number of fuzzy subsets defined for x_i. (9) implies that overlap of input MFs should not exceed 50%. Note that if (9) is strictly equal to 1, a fuzzy partition (Ruspini partition) is established.

For output MFs the following condition applies:

$$\mathrm{Y}_{cog}(\gamma_r(y)) = \frac{\int_{y_{\min}}^{y_{\max}} y\gamma_r(y)dy}{\int_{y_{\min}}^{y_{\max}} \gamma_r(y)dy} = \mathrm{core}(\gamma_r(y)) \tag{10}$$

It must be taken into account that with several MF types (e.g. Gaussian), (9) cannot be easily satisfied because of non-compact support of the MFs. Input MFs must therefore be "local" according to the following definition.

Definition 2: A MF $\mu_A(x)$, defined by three (or four) parameters a, b, c (or d), $(x, a, b, c, d \in X)$, is said to be local if the following conditions hold:

$$\begin{cases} a \le b \le c \\ a = \min(\mathrm{supp}(A)) \\ b = \mathrm{core}(A) \\ c = \max(\mathrm{supp}(A)) \end{cases} \quad \text{or} \quad \begin{cases} a \le b \le c \le d \\ a = \min(\mathrm{supp}(A)) \\ b = \min(\mathrm{core}(A)) \\ c = \max(\mathrm{core}(A)) \\ d = \max(\mathrm{supp}(A)) \end{cases}, \tag{11}$$

where $\mathrm{supp}(A) = \{x \in X | \mu_A(x) > 0\}$.

It is easy to see that commonly used MFs such as triangular or trapezoid satisfy the respective conditions. Other examples of local MFs that can be found from literature are squared-cosine and cubic spline MFs [20].

The conditions (9,10) also apply for 0th order TS systems. Note that output MFs of 0th order TS systems are symmetrical by definition.

For transparent fuzzy systems, we are able to predict the output at transparency checkpoints. Between these points the output is the result of interpolation that takes place between individual rules. The nature of interpolation is determined by fuzzy system parameters - defuzzification method, inference operators, shape of membership functions [17].

3.2 Transparency of 1st order TS systems

1st order TS systems are interpreted in terms of local linear models [11]. Overall system output, however, is interpolated from individual rules and quite often local models cannot be recognised in system output because of large degree of interpolation. Interpolation issues of 1st order TS systems are considered in detail in [7] and two kinds of interpolation are distinguished: (i) S-type interpolation that produces intuitively expected results; (ii) V-type interpolation that has some undesirable properties but on the other hand is better suited for continuous, smooth function approximation. For these two types of interpolation clear order of preference cannot be given.

Based on the interpretation of TS rules, a measure of transparency error (12), can be constructed that estimates the difference between the global output y and locally fired model y_r.

$$\varepsilon_{tr} = \sqrt{\frac{\sum_{k=1}^{K}((y(k) - \underset{\max \tau_r(k)}{y_r}(k)))^2}{K}} \tag{12}$$

where $\underset{\max \tau_r(k)}{y_r}$ denotes the output of the rth rule with the highest activation degree for the kth input-output pair and $y(k)$ is the corresponding global output.

Let us consider another example where five otherwise equivalent 1st order TS systems are obtained by varying the overlap and the magnitude of the cores of input MFs (Fig. 4). We construct separate examples for V-type and S-type interpolation (Figs. 5 and 6, respectively).

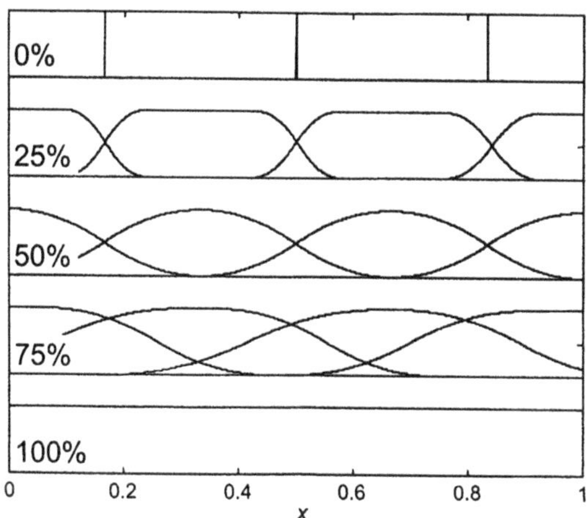

Fig. 4. Input MFs of five TS systems.

Note that in the case of 100% overlap system rule base is replaced by single "average rule". If the overlap of input MFs equals 50%, and the core of the rth rule is a single point, the existence of the transparency checkpoint in output space where $y = y_r$ (like in case of 0th order TS systems) is guaranteed. With smaller overlap and larger cores, the region where system output is the contribution of a single rule increases and vice versa; consequently, the relationship between interpolation and transparency error is rather straightforward. With 0% overlap, transparency error is reduced to zero; the resulting perfect piecewise linear system, however, is interpolation-free and can be hardly considered a fuzzy system anymore. This is the transparency paradox of 1st order TS systems.

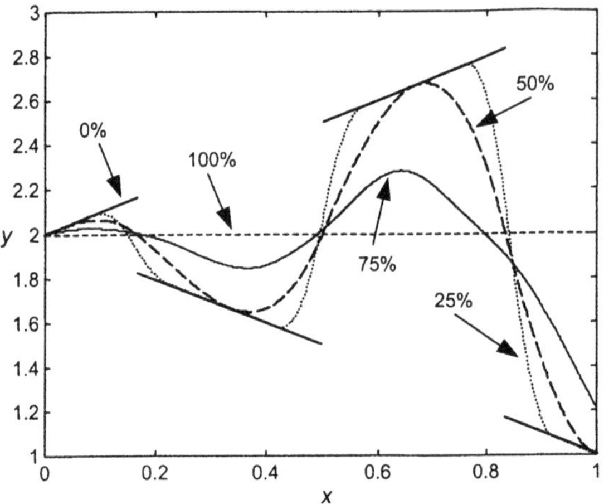

Fig. 5. Input-output relation of TS systems with S-type interpolation and varying overlap degree of input MFs.

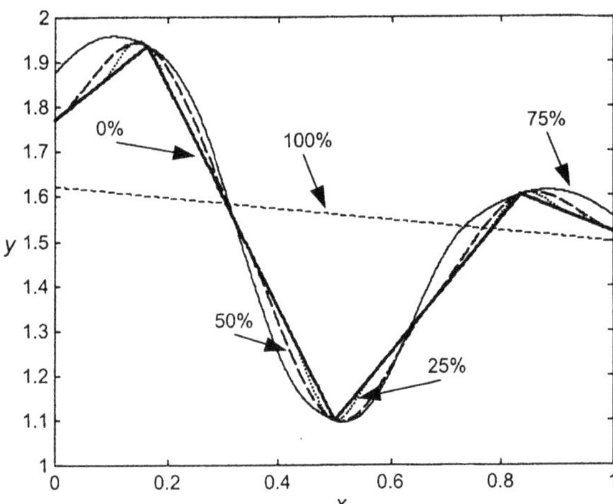

Fig. 6. Input-output relation of TS systems with V-type interpolation and varying overlap degree of input MFs.

Although V-interpolation provides lower transparency error than S-type interpolation, it is shown in [19] that generally we have no reliable means for controlling the type of interpolation. Input MF constraint (9) remains relevant, by using MFs with multi-point cores, ε_{tr} can further reduced but this, however, does not always ensure low transparency error (particularly in case of S-interpolation). Moreover, we are not able to derive explicit transparency constraints for consequent parameters of (4) that would fulfil the purpose [1]. One possibility is to redefine the consequent function of 1st order TS systems to obtain interpolation properties that favour transparency as attempted in [21, 22]. Application of transparency-sensitive modelling algorithms is another possibility, discussed in the next section.

4 Modelling with transparency protection

With some identification algorithms transparency constraints of Mamdani and 0th order TS systems (9-10) can be satisfied by a suitable a priori selection of MF parameters. This is for example true for Wang-Mendel method [23] or Babuska's combined approach of Gustafson-Kessel (GK) clustering and least square estimation (LSE) for 0th order TS systems [7] where input fuzzy sets extracted from GK clusters form a fuzzy partition and output MFs identified through LSE are symmetrical by definition.

The issue of transparency protection, however, specifically arises with iterative learning algorithms, which have become very popular recently, such as gradient descent or genetic algorithms where MFs undergo many modifications and in unconstrained mode this generally leads to a non-transparent model.

The problem can be solved by (i) imposing constraints on membership functions that prevent the system from becoming non-transparent (ii) employing special membership functions that make transparency a default property of a fuzzy system (iii) multi-objective optimisation [16]. First two methods are directly applicable to standard fuzzy systems where the transparency constraints have binary nature as is shown in next two sections.

For 1st order TS systems the possibilities are limited. Local least squares method is considered for transparency enhancement in [24]. Multi-objective optimisation where a certain balance between accuracy and transparency is sought may also be applicable to 1st order TS systems. The problem is that transparency protection restricts the approximation capabilities of adaptation algorithms (not unexpected as trade-off between accuracy and interpretability is a long known fact) and seems to influence 1st order TS systems more seriously.

---

[1] Interpretable Sugeno Approximator (ISA) proposed by Bikdash [20] uses local input MFs and rewrites the consequent functions in "centered" form that allows to interpret the consequent polynomials as Taylor series coefficients. It is, however, easy to see that ISA only provides transparency checkpoints and does not otherwise influence interpolation properties of TS systems. Consequently, ISA does not reduce transparency error. Besides, consequent coefficients are not exactly meaningless which is what they are called in [20].

4.1 Gradient descent

One of the most popular optimisation algorithms for fuzzy systems is gradient descent that is based on the minimisation of the cost function

$$\varepsilon = \frac{1}{2}[y - \tilde{y}]^2, \tag{13}$$

where y denotes the output of the fuzzy model (e.g. 2,4 or 6) and $\tilde{y}$ is the reference output.

To minimise the cost function (13) through the modifications of MF parameters, differential calculus is used that computes the necessary updates of optimised parameters c

$$\Delta c = -\alpha \frac{\partial \varepsilon}{\partial c}, \tag{14}$$

where α is the learning rate. Typically, a fuzzy model becomes less transparent as the approximation error decreases.

(14) implies that inference function y must be differentiable. This restricts gradient descent to TS systems (3-6).

One possibility to protect transparency of a fuzzy system trained with gradient descent is to verify the fulfilment of transparency conditions before every parameter update. If the new parameter value violates transparency conditions, the update cannot be applied to the given parameter.

More efficient way to protect system transparency is to use such definition of MFs that make transparency a default property of the system.

Note that output MFs of 0th order TS systems satisfy (10) by definition, hence transparency preservation problem reduces to (9). To protect input transparency in the similar manner one may use Jager (neighbour-oriented) definition of triangular fuzzy sets [25], where each fuzzy subset is defined so that its edge parameters b_i^j, c_i^j equal the centres of the neighbouring sets, a_i^{j-1}, a_i^{j+1}, respectively (Fig. 10). Thus, a fuzzy partition is permanently maintained. Another advantage of Jager partition (15) is that the number of adjustable antecedent parameters is reduced.

$$\mu_i^j(x_i) = \begin{cases} \dfrac{x_i - a_i^{j-1}}{a_i^j - a_i^{j-1}}, & a_i^{j-1} < x_i < a_i^j \\ \dfrac{a_i^{j+1} - x_i}{a_i^{j+1} - a_i^j}, & a_i^j < x_i < a_i^{j+1} \\ 0, \quad a_i^{j+1} < x_i < a_i^{j-1} & \end{cases} \tag{15}$$

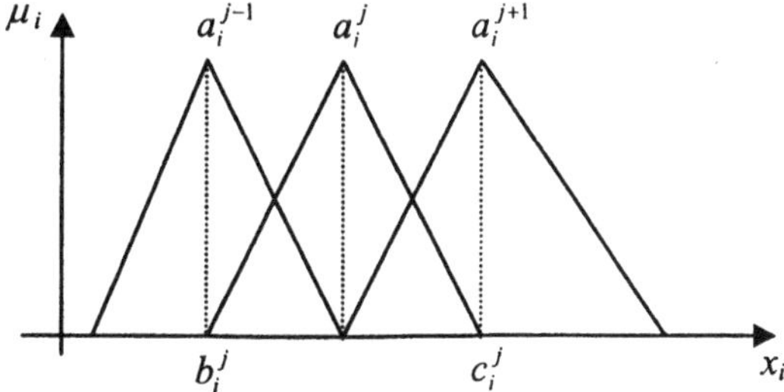

Fig. 7. Neighbour-oriented definition of triangular membership functions.

Usage of (15) means that traditional gradient descent based update formulas given in [5,26,27] are no longer valid and one must use the update rules derived by Jager [25]. Jager designed the algorithm originally for triangular MFs but expressions similar to (15) for other types of transparent MFs and respective update rules for their parameters can easily be derived.

4.2 Genetic algorithms

Genetic algorithms (GAs), which are search techniques based on the mechanisms of natural genetics have been applied to many different problems like function optimisation, routing problem, scheduling, design of neural networks, system identification, digital signal processing, computer vision, control and machine learning. They search large spaces efficiently without the need for derivative information.

GAs have also become very popular in fuzzy modelling [8] and control [28] because of robustness that allows the training of fuzzy systems of arbitrary configuration and arbitrary sets of parameters and flexibility in the construction of the cost (fitness) function.

The general principle of GA is as follows: fuzzy system parameters to be trained (usually encoded into binary alphabet) are joined together to form a chromosome. The initial population consisting of N chromosomes evolves to the next generation through the genetic operations of crossover and mutation. After each step, chromosomes are decoded, fuzzy systems corresponding to those chromosomes are reconstructed and the fitness of these systems is evaluated. Chromosomes with better fitness have higher probability to survive to the next generation. If the algorithm is properly designed it will converge to an optimal solution although it usually requires many cycles (generations). Because of high computational cost, GAs are often combined with other optimisation algorithms such as clustering [9] and gradient descent [8].

As with gradient descent one possibility to rule out non-transparency is to constrain the parameters of MFs, according to conditions (9-10). Note that in [9] a conservative form of this approach is proposed. The validation takes place when system is being reconstructed from a chromosome.

Alternatively, we can make use of Jager partition for input MFs and some definition of symmetrical output MFs, e.g triangular MFs.

$$\gamma^{t}(y)=\begin{cases}-(s^{t}/2)(y_{j}-a^{t})+1, & a^{t}-s^{t}/2<y_{j}<a^{t}\\ (s^{t}/2)(y_{j}-a^{t})+1, & a^{t}<y_{j}<a^{t}+s^{t}/2,\\ 0, & a^{t}+s^{t}/2<y_{j}<a^{t}-s^{t}/2\end{cases} \quad (16)$$

where s^t and a^t are the centre and the spread of (16), respectively.

The part of the chromosome that corresponds to MF parameters (15) and (17) is depicted in Fig. 8.

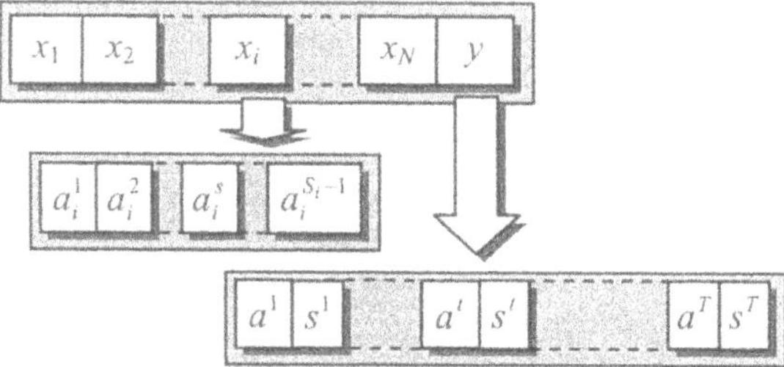

Fig. 8. Chromosome configuration.

5 Transparent fuzzy systems in control

Traditional classification of fuzzy control [29] divides fuzzy controllers into four main categories. We briefly describe these four categories and discuss the relevance of transparency in each case.

- Controllers designed on the basis of expert experience and control engineering knowledge

Knowledge based controllers are typically open loop controllers (there is no feedback involved), state feedback controllers or set point controllers with and without additional inputs. The design methodology is ill-defined and problem dependent. Two more problems associated with this approach can be pointed out. One is the possible inadequacy of the expert as the controller cannot be better than expert's knowledge. Another and even more serious issue is expert's possible inability to express his control experience or general knowledge effectively with the tools of fuzzy logic, either because he/she does not understand the properties of fuzzy systems very well or cannot formulate the control rules verbally because it is only his/her body that knows how to control the process/system, not the mind.

By definition, transparency is vital to this type of controllers, otherwise the expert knowledge appears in distorted form and consequently the controller performance is sub-optimal.

- Controllers modelled on the existing controllers

Fuzzy controllers of this type try to mimic some other working controller. During the training it is connected so that it has access to the inputs and outputs of the working controller. After it is found to have learnt the expected task, it is put online and replaces the original controller. The approach is very useful if the controller to be emulated is a human being (who is unable to express his control knowledge verbally) or if the original control algorithm is very expensive to implement. The possible disadvantage of the controller is that it cannot be better than the original controller and often may be worse because there always exists certain modelling error.

Transparency of the controller is not the necessary requirement, as we are primarily concerned with the numerical performance of the controller. Transparency, however, may be useful as it allows the validation of the controller by an expert.

- Model-based fuzzy control

Model-based fuzzy control uses a given (typically fuzzy) open loop model of the plant under control to derive the set of fuzzy rules for the fuzzy controller and is therefore principally different from previous two approaches where it is implicitly assumed that no model exists. Examples of model-based fuzzy control are model-based predictive control [30,31], inverse fuzzy process model based control [32] and fuzzy gain-scheduling methods [33, 34]. As stated, this type of control involves the generation of a fuzzy model of the controlled process as the preliminary step and typically numerical accuracy of the model is the primary concern with the exception of [35] where the linguistic inversion of the controlled process is proposed.

With this type of control we are again typically more interested in the numerical properties of the controller (and of the model) that makes transparency unnecessary. The exception is the linguistic inversion of the process model [35] where transparency of the model plays the key role and it is interesting to note that exact model inversion technique proposed in [7] assumes certain properties of a model (including transparency).

- Self-learning fuzzy controllers (adaptive fuzzy control)

In adaptive fuzzy control, the focus is on the automatic on-line synthesis and tuning of fuzzy controller parameters, which will ensure that the performance objectives are met even if the plant parameters change in time. Generally, these techniques can be split into two categories: direct and indirect adaptive fuzzy control. In indirect adaptive fuzzy control, there is an identifier mechanism that produces a model of the plant, which is then used to specify the controller e.g. [36]. Thus the distinction between model-based and adaptive fuzzy control is sometimes imaginary. In direct adaptive control, a model of the plant is not estimated; instead, we tune the controller parameters directly using plant data [37, 38].

Presently, the role of transparency in adaptive fuzzy control is unexplored (but very interesting) research topic.

6 Application of transparent fuzzy modelling and control

Truck backer-upper problem was first presented in [39] where the learning potential of neural networks was utilised to obtain the self-learning controller through temporal backpropagation. In more recent works, several authors have replaced or complemented neural networks with genetic algorithms, e.g [40]. The basic shortcoming of all these data-driven techniques is the computational cost.

On the other hand, the problem is an ideal test bed for fuzzy control systems because nearly anyone is able to drive the truck to the desired position given some time to adjust himself to the controls and the potential of fuzzy logic for implementing expert knowledge is well-known. Several applications can be found from literature [23,41]. We use this control example to illustrate the presented ideas about transparent fuzzy systems in control using the results of [42].

6.1 Truck backer-upper system

The system used in the simulations is supplied with MATLAB as a demo. The truck as in [23,41] corresponds to the cab part of the Nguyen-Widrow's truck and trailer, referred to as simplified Nguyen-Widrow problem. The truck position is determined by three state variables $x = [-20, 20]$, $y = [0, 25]$, and, $\Phi = [-90°, 270°]$ - the angle between truck's onward direction and the x-axis (Fig. 9). The width and length of the truck are 4 and 2 meters, respectively.

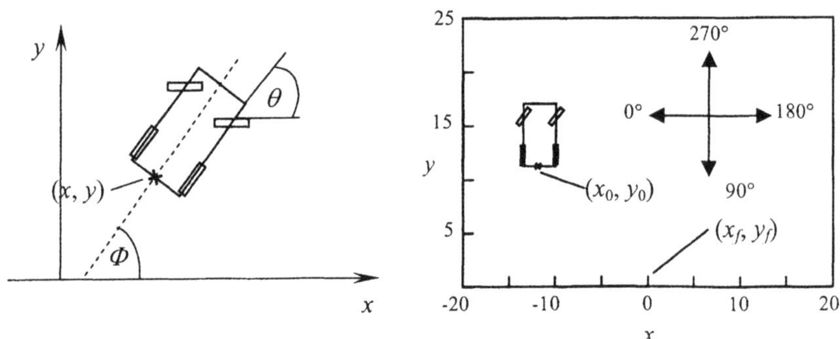

Fig. 9. Truck backer-upper system.

Truck must arrive from the initial position (x_0, y_0, Φ_0) to the loading dock ($x_f = 0$, $y_f = 0$) at a right angle ($\Phi_f = 90°$). Truck only moves backward with the fixed speed 2m/s. To control the truck at every stage appropriate steering angle θ = [-45°, 45°] must be provided. Thus controller is a function of state variables

$$\theta = f(x, y, \Phi) \tag{17}$$

For the evaluation of results we use the error measure (18) that combines angle and position errors into one expression (see [42] for details).

$$\varepsilon_c = \text{abs}(x_f - x(T_f)) + 0.0267 \cdot \text{abs}(\Phi_f - \Phi(T_f)), \tag{18}$$

where T_f is the duration of the experiment.

Additionally, for each backing experiment, a secondary criterion (19) is produced that measures the length and smoothness of backing trajectory.

$$\eta = \frac{\int_0^{T_f} \sqrt{(dx/dt)^2 + (dy/dt)^2}\, dt}{\sqrt{(x_0 - x_f)^2 + (y_0 - y_f)^2}} \tag{19}$$

6.2 Truck backer-upper controllers

We concentrate on two modes of fuzzy control described in the section 5: knowledge-based control and mimicking control.

6.2.1 Knowledge-based controllers

With knowledge-based controller design, it is obvious that transparency is the necessary and fundamental property of the designed controller to guarantee that there is no information distortion in the process where expert knowledge is translated into fuzzy rules. The case of truck-backer upper system, however, presents a knowledge acquisition problem, as will be promptly shown.

Selection of the controller function (19) implies that the following rule base format should be used

$$\text{IF } x \text{ is } A \text{ and } y \text{ is } B \text{ and } \Phi \text{ is } C \text{ THEN } \theta \text{ is } D, \tag{22}$$

where A, B, C and D are the linguistic labels of fuzzy sets into which the system variables are partitioned.

Knowledge-based design of the fuzzy controller is based on our (human) understanding of the driving process. The problem with (22) is so-called curse of dimensionality. Employing the input partition {5 3 7}, for example, would result in 105 rules that all must be specified individually.

One possibility to reduce the design load is to assume that enough clearance between the truck and the loading dock exists so that the truck y-position co-ordinate can be neglected from (22) as is done in [41].

The second problem is that although we can drive the car to the loading dock manually from almost any position, the knowledge how to do that does not translate easily into fuzzy rules. This is because the skill is in our hands. As a result the design procedure becomes time-consuming and frustrating especially when the number of tuned parameters is large (even with only two input variables) and may result in poorly performing controller as the final result.

To come out from this situation hierarchical control system was proposed in [39]. It is observed that in interaction with the real world we are engaged in a continuous process of constructing representations of that environment and our experience of it that may operate at different levels of consciousness. The driving process (how we handle the car controls) is generally carried out on subconscious level, which is the reason why the information cannot be successfully extracted.

Conscious models, on the other hand that are based on "common sense", can be easily expressed in linguistic terms. It is easy to see that truck backing is actually a mixture of conscious and subconscious actions. We choose the backing trajectory consciously but the control actions are the result of subconscious reasoning.

This two-level approach can be implemented with the control block that consists of a fuzzy supervisor and a PD controller (Fig. 10). The supervisor provides the setpoint Φ_r (desired truck angle) for the given state; based on the difference between the desired and actual angle, appropriate steering angle θ is then determined by PD controller.

Hierarchical control system serves two purposes simultaneously – it facilitates focused knowledge extraction and allows us to cope with the curse of dimensionality – y-coordinate can be included at small additional cost.

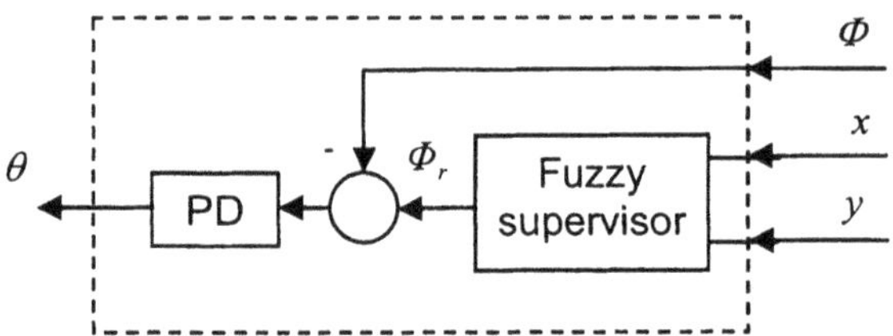

Fig. 10. Block diagram of fuzzy hierarchical control system.

The rule base of the supervisor is easily configured, e.g. grey region in Fig. 11 reads as

$$\text{IF } x \text{ is } mf4 \text{ AND } y \text{ is } mf3 \text{ THEN } \Phi_r \text{ is } 90°, \tag{20}$$

(20) is in good accordance with the general idea what angle the truck in this particular area should maintain. The rest of the rules are based on the same principle. Two parameters of PD controller are also easy to determine.

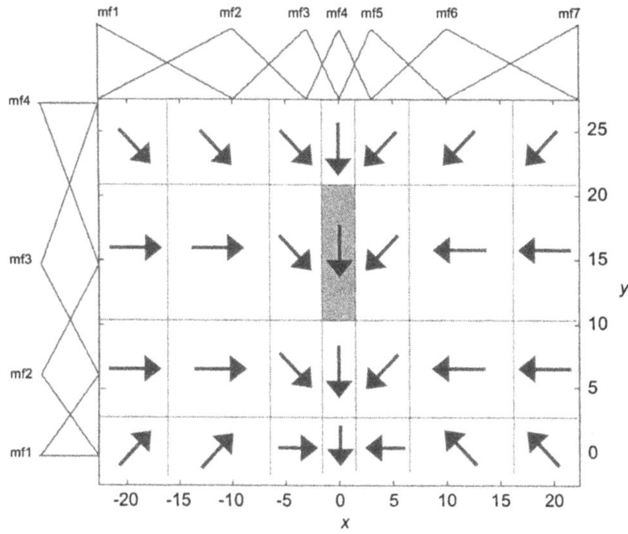

Fig. 11. Rule base of the two-input fuzzy supervisor.

6.2.2 Controllers modelled on human operator

Even if the information on how to drive the truck cannot be extracted from the driver it is still possible to record his control actions and to use the collected data to train the controller. If the data is carefully selected, good results can be expected. Two problems, however, arise. The modelling algorithms available are not perfect; there always exists modelling error. Secondly, with non-transparent techniques, the controller cannot be validated otherwise than by its control performance and we do not know what it has learned. That presents a potential danger of unexpected behaviour.

Training data was collected from 31 truck backing experiments with 8 upward, 6 leftward, 6 rightward and 11 downward initial angles. Starting positions were chosen so that different backing trajectories would be present (Fig. 12). To reduce the computational load, most of data was filtered out so that the final data set, consisting of 642 input-output pairings, corresponds to the situation as if information had been available every third second only (normal sampling interval is 0.1s).

For modelling we used ANFIS [5] and Gustafson-Kessel clustering in combination with least squares procedure [7]. The ANFIS algorithm lacks transparency protection but has excellent approximation properties as shown in [5]; application of GK/LSE, on the other hand, produces the transparent model of human controller.

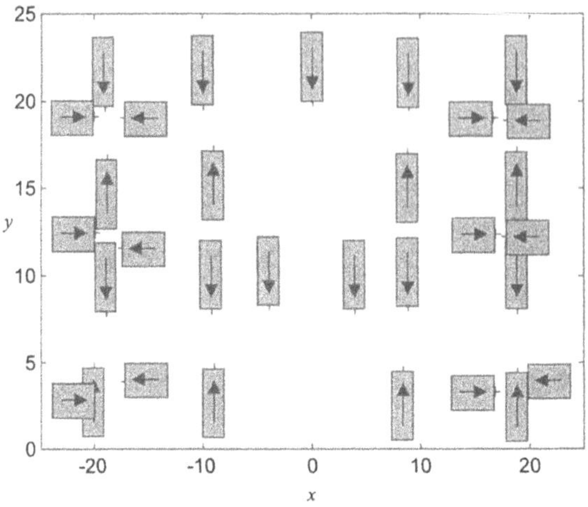

Fig. 12. The initial positions of backing trajectories for training data set.

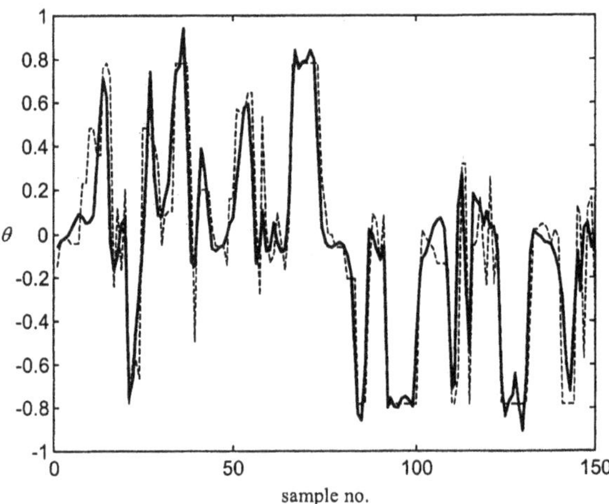

Fig. 13. Modelling results (excerpt) - target steering angle (dashed line), modelled steering angle (bold line).

The determination of parameters such as number of input MFs and fuzzy rules was based on trial and error, to establish the configuration by what "reasonably low" approximation error could be achieved. ANFIS was applied to 1st order Takagi-Sugeno system with input partition of {7 3 9} and GK/LS model was identified as a 0th order Takagi-Sugeno system with the same partition. Final modelling root mean square errors for ANFIS (2500 epochs, RMSE = 0.2129) and GK/LS (RMSE = 0.2048) are very similar, though.

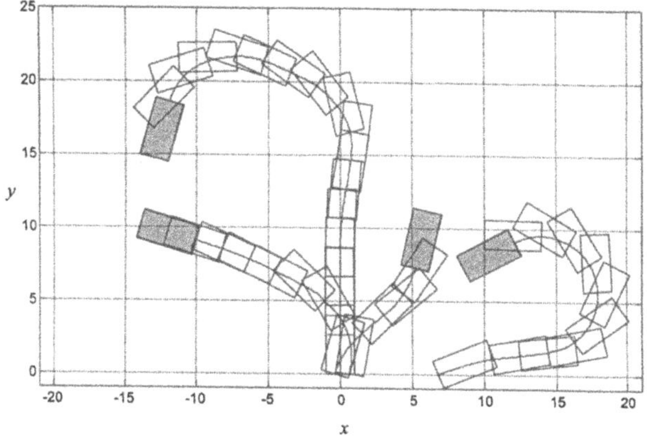

Fig. 14. Backing up with knowledge-based controller (initial positions of the truck are indicated with grey colour).

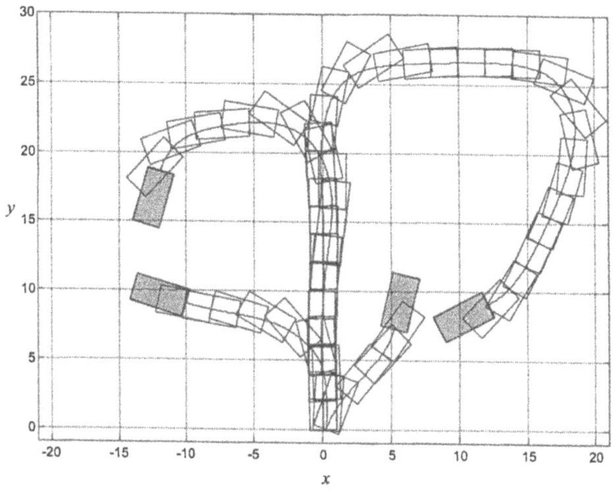

Fig. 15. Backing up with non-transparent human operator model (ANFIS).

6.3 Control results

The comparison of designed controllers is based on backing the truck from randomly chosen initial positions (Table 1). The control performance measures of knowledge-based controller (KBC), hierarchical knowledge based controller (HKBC), non-transparent data-driven controller (NTDDC) and transparent data-driven controller (TDDC) are given in Table 2 and selected backing trajectories (experiments no. 2, 4, 6, 10) are depicted in Figs. 14-17.

Table 1. Initial conditions.

N_e	1	2	3	4	5	6	7	8	9	10
x_0	-3.7	-10.1	18.4	5.2	-14.9	12.1	-14.3	-0.9	-4.4	-11.9
y_0	18.0	8.5	17.7	7.3	5.0	9.1	7.0	9.1	16.2	18.6
Φ_0	-12.5	154.7	92.1	76.4	252.5	206.9	158.0	162.8	265.4	253.6

Expert defined controller shows the weakest performance - partly because it does not make account of y co-ordinate (consequently it cannot guarantee success for initial positions close to the loading dock), partly because tuning of the controller is based more on trial-and-error than on focused knowledge translation.

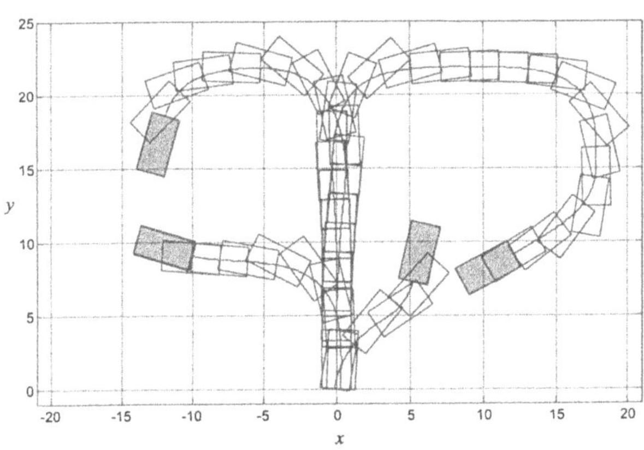

Fig. 16. Backing up with transparent human operator model (GK/LSE).

Table 2. Control results.

	KBC		HKBC		NTDDC		TDDC	
N_e	ε_c	η	ε_c	η	ε_c	η	ε_c	η
1	0.5779	1.4083	0.1039	1.3485	0.1008	3.7809	0.1201	2.8395
2	0.7674	1.2732	0.1911	1.1289	0.4641	1.2311	0.3734	1.2612
3	0.1537	1.3544	0.0645	1.1899	0.3314	1.5973	0.2869	1.2447
4	0.9941	1.0687	0.2803	1.0463	0.8084	0.9905	0.3977	1.0463
5	0.4259	1.6403	0.1745	1.3285	0.2427	1.5894	0.2683	1.7103
6	1.1177	1.9538	0.0165	2.2642	0.0669	3.9683	0.0378	3.3606
7	1.0205	1.1295	0.2342	1.1132	0.5431	1.2841	0.2866	1.2272
8	1.1797	1.1351	0.4236	1.1679	0.8096	1.1570	0.4827	1.1679
9	0.1434	1.7561	0.0372	1.7204	0.1809	1.7263	0.0895	2.0004
10	0.0192	1.5208	0.0039	1.3940	0.0873	1.5389	0.0341	1.5027

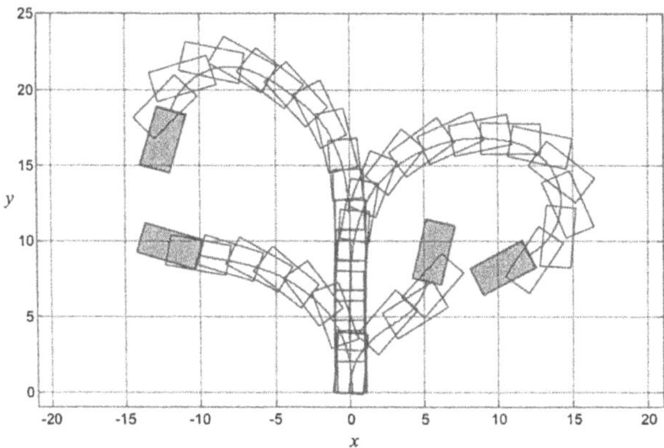

Fig. 17. Backing up with hierarchical control system.

Controllers modelled on human operator show better performance (Figs. 15-16). Due to approximation error, however, their performance is sub-optimal. It is interesting to note that ANFIS-approximated controller shows erratic behaviour on some occasions (Fig. 15) although the approximation errors of both algorithms were in the same range. Due to non-transparency of the controller we are not able to validate the rules and on the other hand, this non-transparency may be the reason why truck occasionally goes "berserk".

Finally, hierarchical control system allows more efficient design and provides smooth and economic truck trajectories with superior control accuracy compared to other approaches (Fig. 17).

7 Concluding remarks

In this chapter the systematic approach to transparency problem of fuzzy systems was presented. Transparency that is distinguished from linguistic interpretability (the latter is considered a default property of the observed classes of fuzzy systems) measures validity or reliability of the linguistic interpretation. Transparency as defined for standard and 0th order TS systems (8-9) implies that fuzzy system transparency is of binary character for these types of systems. Taking the transparency definition as the basis, transparency constraints for standard and 0th order TS systems were derived and mechanisms for preserving transparency in iterative modelling were discussed.

For 1st order TS systems the situation is a bit different. Although transparency checkpoints can be similarly defined, this does not guarantee low transparency error (12) because interpolation in 1st order TS systems has undesirable properties from transparency viewpoint. Additional means for improving transparency of 1st order TS systems were discussed including the use of MFs with multi-point cores (e.g. trapezoid MFs) and transparency-sensitive identification algorithms for consequent parameters (local least squares, iterative algorithms with multi-objective optimisation criterion).

Transparency is of primary importance in linguistic analysis and synthesis of control systems [35]. The application of transparent control presented in this chapter clearly demonstrates that transparency is vital to this branch of intelligent control that seeks solutions by emulating the mechanisms of reasoning and decision processes of human beings. It must be stressed that best results are obtained if besides transparency preservation other aspects of modelling such as complexity reduction and focused knowledge acquisition are taken into account. Possible implication to presently black-box techniques of fuzzy control where numerical accuracy is the primary concern is presently unclear, however, this line of research will be our first concern in near future.

References

1. L.A. Zadeh. Outline of a New Approach to the Analysis of Complex Systems and Decision Processes. *IEEE Trans. Systems, Man and Cybernetics*, 3:28-44, 1973.
2. E.H. Mamdani and S. Assilian. An experiment in linguistic synthesis with a fuzzy logic controller. *Int. J. Man-Machine Studies*, 7:1-13, 1975.
3. L.P. Holmblad and J.J. Ostergaard. Control of Cement Kiln by Fuzzy Logic. In M.M-Gupta and E. Sanchez, editors, *Approximate Reasoning in Decision Analysis*, pages 389-400. Amsterdam, Holland, 1983.
4. T. Takagi and M. Sugeno. Fuzzy identification of systems and its applications to modeling and control. *IEEE Trans. Systems, Man and Cybernetics*, SMC-15(1):116-132, 1985.
5. L.-X. Wang and J.M. Mendel. Back-propagation fuzzy system as nonlinear dynamic system identifiers. *Proc. 1st IEEE Int. Conf. on Fuzzy Systems*. pages 1409-1416, San Diego, CA, USA, 1992.
6. J.-S.R. Jang. ANFIS: Adaptive-network-based fuzzy inference system. *IEEE Trans. Systems, Man and Cybernetics*, 23(3):665-685, 1993.
7. R. Babuska. *Fuzzy Modeling and Identification*, Ph.D. dissertation. Technical University of Delft, Delft, Holland, 1997.
8. J. Liska and S.S. Melsheimer. Complete design of fuzzy logic systems using genetic algorithms. *Proc. 3rd IEEE Int. Conf. on Fuzzy Systems*, pages 1377 – 1382, Orlando, FL, USA, 1994.
9. J.A. Roubos and M. Setnes. Compact fuzzy models through complexity reduction and evolutionary optimization. *Proc IEEE Int. Conf. Fuzzy Systems*, pages 762-767. San Antonio, TX, USA, 2000.
10. M. Brown and C.J. Harris. *Neurofuzzy Adaptive Modelling and C*ontrol, Prentice Hall, Englewood Cliffs. 1994.
11. M. Setnes, R. Babuska and H.B. Verbruggen. Rule-Based Modeling: Precision and Transparency. *IEEE Trans. Systems, Man and Cybernetics*. 28(1):165-169, 1998.
12. D. Nauck, F. Klawonn and R. Kruse. *Foundations of Neuro-Fuzzy Systems*. Wiley, Chichester, UK, 1997.
13. Y. Jin. Fuzzy modeling of high-dimensional systems: Complexity reduction and interpretability improvement. *IEEE Trans. Fuzzy Systems*, 8(2):212-220, 2000.

14. R. Babuska. Construction of fuzzy systems - Interplay between precision and transparency. *Proc. European Symp. Intelligent Techniques (ESIT)*, pages 445-452, Aachen, Germany, 2000.
15. A. Lotfi, H.C. Andersen and A.C. Tsoi. Interpretation preservation of adaptive fuzzy inference systems. *Int. J. Approximate Reasoning*, 15(4):379-394, 1996.
16. J.V. de Oliveira. Semantic constraints for membership function optimization. *IEEE Trans. Systems, Man and Cybernetics*, 29(1):128-138, 1999.
17. A. Riid and E. Rüstern. Transparent fuzzy systems and modeling with transparency protection. *Proc. IFAC Symp. on Artificial Intelligence in Real Time Control*, pages 229-234. Budapest, Hungary, 2000.
18. A. Riid, P. Jartsev and E. Rüstern. Genetic algorithms in transparent fuzzy modeling. *Proc. 7th Biennal Baltic Electronic Conf. 2000*, pages 91-94. Tallinn, Estonia, 2000.
19. A. Riid, R. Isotamm and E. Rüstern. Transparency analysis of 1st order Takagi-Sugeno systems. *Proc 10th Int. Conf. System-Modeling-Control*, 2:165-170. Zakopane, Poland, 2001.
20. M. Bikdash. A highly interpretable form of Sugeno inference systems. *IEEE Trans. Fuzzy Systems*, 7(6):686-696, 1999.
21. R. Babuska, C. Fantuzzi, U. Kaymak and H.B. Verbruggen. Improved inference for Takagi-Sugeno models. *Proc. 5th IEEE Int. Conf. Fuzzy Systems*, pages 701–706, New Orleans, LA, USA, 1996.
22. A. Fiordaliso. A constrained Takagi-Sugeno fuzzy system that allows for better interpretation and analysis. *Fuzzy Sets and Systems*, 118(2):207-318, 2001.
23. L.-X. Wang and J.M. Mendel. Generating fuzzy rules by learning from examples. *IEEE Trans. Systems, Man, and Cybernetics*, 22(6): 1414-1427, 1992.
24. J. Yen, L. Wang, and W. Gillespie. Improving the Interpretability of TSK Fuzzy Models by Combining Global Learning and Local Learning. *IEEE Transactions on Fuzzy Systems*, 6(4): 530-537, 1998.
25. R. Jager. *Fuzzy Logic in Control*, Ph.D. dissertation. Technical University of Delft, Delft, Holland, 1995.
26. F. Guely and P. Siarry. Gradient descent method for optimizing various fuzzy rule bases. *Proc. 2nd IEEE Int. Conf. Fuzzy Systems*, pages 1241-1246. San Francisco, CA, USA, 1993.
27. H. Nomura, I. Hayashi and N. Wakami. A learning method of fuzzy inference by descent method. *Proc. 1st IEEE Int. Conf. on Fuzzy Systems*, pages 485-491. San Diego, CA, USA, 1992.
28. C.L. Karr and E.J. Gentry. Fuzzy control of pH using genetic algorithms. *IEEE Transactions on Fuzzy Systems*, 1(1): 46-53, 1993.
29. C.C. Lee. Fuzzy logic in control systems: Part I and Part II. *IEEE Trans. Systems, Man and Cybernetics*, 20(2):404-435, 1990.
30. J. J. Espinosa, M. L. Hadjili, V. Wertz and J. Vandewalle. Predictive control using fuzzy models - Comparative study. *Proc. European Control Conference.* Karlsruhe Germany, 1999.
31. R. Babuska, M. A. Botto, J.S. da Costa, H.B. Verbruggen. Neural and fuzzy modeling in nonlinear predictive control: a comparative study. *Proc. Computational Engineering in Systems Applications: Symposium on Control, Optimization and Supervision*, pages 1049-1054. Lille, France, 1996.

32. J. Abonyi, L. Nagy and F. Szeifert. "Indirect Model Based Control Using Fuzzy Model Inversion. *Proc. IEEE Int. Conf. Intelligent Systems*, pages 951-956. Vienna, Austria, 1998.
33. K.J. Hunt and T.A. Johansen. Design and analysis of gain-scheduled local controller networks. *International Journal of Control*, 66(5):619–651, 1997.
34. P. Korba and P.M. Frank. An applied optimization-based gain-scheduled fuzzy control. *Proc. American Control Conference*, pages 3383–3387, Chicago, CH, USA, 2000.
35. M. Braae and D.A. Rutherford. Theoretical and Linguistic Aspects of the Fuzzy Logic Controller. *Automatica*, 15(5):553-577, 1979.
36. J-S. R. Jang. Self-learning fuzzy controllers based on temporal back propagation. *IEEE Trans. Neural Networks*, 3(5):714-723, 1992.
37. T.J. Procyk and E.H. Mamdani. A linguistic self-organizing process controller. *Automatica*, 15:15-30, 1979.
38. H.R. Berenji and P. Khedkar. Learning and tuning fuzzy logic controllers through reinforcements. *IEEE Trans. Neural Networks*, 3(5):724-740, 1992.
39. D. Nguyen and B. Widrow. Neural networks for self learning control systems. *IEEE Contr. Syst. Mag.*, 10(2): 18-23, 1990.
40. M. Schoenauer and E. Ronald. Neuro-genetic truck backer-upper controller. *Proc. First Int. Conf. Evolutionary Comp.*, pages 720-723. Orlando, FL, USA, 1994.
41. S.-G. Kong and B. Kosko. Adaptive fuzzy systems for backing up a truck-and-trailer. *IEEE Trans. Neural Networks*, 3(5):211-223, 1992.
42. A. Riid and E, Rüstern. Fuzzy logic in control: Truck backer-upper problem revisited. *Proc. IEEE 10th Int. Conf. Fuzzy Systems*, 1. Melbourne, Australia, 2001.

Uniform Fuzzy Partitions with Cardinal Splines and Wavelets: Getting Interpretable Linguistic Fuzzy Models

Adolfo R. de Soto**

E. I. Industrial e Informática
University of León
24071 Spain
e-mail: ddears@unileon.es

Abstract. In an abridged form, two main steps are habitual to build a Fuzzy Rule Model from a data set:

1. To select the shape and distribution of the linguistic labels set for each variable,
2. To tune the set of rules to the training data set.

Most of methods begin with a uniform fuzzy partition on the universe of each variable and, then, they use different learning mechanism to tune the model to data set, operation that, habitually, change the parameters and position of the linguistic labels. These changes on the labels can produce a final fuzzy sets with a difficult interpretation, having, for example, an excessive overlapping between labels, but getting a good performance of the fuzzy system with respect to the data set.

When the fuzzy system is used as a tool of data mining and the final system must be presented to a human expert, who want to evaluate the behavior of the data, the fuzzy sets should be easily interpreted and hence they should constitute, or at least to be near of, a fuzzy partition. Perhaps these requirements are incompatible with a good model accuracy and a trade-off between both requisites is necessary.

It is clear that with a strict uniform fuzzy partition a lost in flexibility of the model is obtained, but also it is clear that by means of a refinement of the partition it is possible to get a better accuracy, at the expense of having more labels.

The present work has as objective to show the utility of the Wavelet Theory as a tool to obtain a good initial uniform partition. The theory of spline functions gives a collection of functions very suitable to model linguistic labels and, by means of multiresolution analysis, it is possible to apply the wavelet theory using spline functions. This way, a mechanism to evaluate the change of resolution is obtained. Moreover, the wavelet theory gives a method to localize changes in the data, and so to allow a better way to distribute linguistic labels.

** This work has been partially supported by project TIC2000-1420 of Spanish Plan R+D+I and project 2002/29 of the regional Spanish government Junta de Castilla y León.

1 Introduction

After more of thirty-six years of the seminal paper by L. A. Zadeh [32] about fuzzy set theory, and twenty-seven years of the Mandani' paper [18] with the first application to control theory, the fuzzy rule base systems have reached a very important role in many application areas, and the research work goes on with a high activity. At the present time, it would be very difficult to make a complete list of the different applications of this systems in fields as control theory, decision making theory, pattern recognition, robotics, database systems, information systems, among others. Still more difficult would be to give a list of methods to build fuzzy rule systems.

Basically, there are two ways of building a fuzzy rule system: from the expert knowledge or from a data set. The former exploit the fuzzy sets capability to represent linguistic terms, while the latter utilize the universal approximation property of this systems. Of course, this methods can be mixed and a fuzzy rule system is usually tunned by means of a data set and, on the other hand, the automatic extraction methods habitually starts with an adequate fuzzy partition. In both cases, the main property of fuzzy rule systems would be preserved, namely, its high interpretable capacity.

A fuzzy rule system is an easy way to build a non-linear system. When a human expert writes down their rules, he is indicating the zones where the system behavior is different; from simple local rules a complex global behavior is obtained. This characteristic is the strength of these systems, they are an excellent tool for modeling complex systems from simple rules. In some methods, when the tunning process is applied, this quality is seriously damaged. Perhaps it is not a critical problem when the fuzzy rule system is going to be used in a control system because of a high accuracy must be reached in these cases, specially if the system do not required a human design maintenance which can depend on the previously expert knowledge. But when a high interaction between the fuzzy rule system and the expert human is needed, either for maintenance or for knowledge extraction for example, the interpretable capacity must be a main goal. If we wish to conserve a high interpretable capacity, we must preserve a adequate local behavior. For this reason, a global measure of accuracy perhaps is not a good enough criteria to build a fuzzy rule system, at least as unique measure.

In natural language, the graduate linguistic terms form a multiresolution system. There are many situations where a great detail is not necessary and general terms as "tall" or "short" are enough, but in other cases a finer resolution is needed and terms as "very tall" or "quite short" must be used. In the same way, it is possible to build a multiresolution family of fuzzy rule systems [11]. A solution to get high interpretability and good accuracy could consist in to use some kind of multiresolution. We can fix the shape of the linguistic labels and to introduce more labels if it were necessary.

The wavelets theory [8,6,14] with the multiresolution point of view [17] has many similarities with fuzzy rule system theory. Both theories exploit their high locality to get an adequate data modelling, both theories can be considered functional theories, in spite of the functional approach is not very developed in fuzzy theory, both theories are full of heuristics to select either the appropriate wavelet or the appropriate fuzzy rule system, and finally both theories present a multiresolution approach to their own problems.

Several works use wavelet theory in the soft computing field [24]. With respect to fuzzy rule systems an important link appears with B-splines functions. The theory of spline functions has a strong mathematical background since seventies [20,10]. They have been used in interpolation theory, graphic computing or signal and image processing. The use of B-splines in wavelet theory starts very soon, in the Mallat works [17], and it was developed in later works as [7,30,4,3]. The main reason was the capacity of B-splines to constitute a multiresolution system on the space of square integrable functions.

In this work, the foundations of several algorithms to take advantage of B-spline and wavelet theory and its application in a fuzzy rule system context are given. In the section 2, the so called TSK-fuzzy rule systems (see chapter 1 in this volume) are viewed as an interpolation problem. Thanks of that, it is possible to make a good use of the method presented in section 3 to calculate, in an efficient form, the B-spline and cardinal spline approximation to a sample. The sections 5 and 6 are dedicate to show the link between splines and wavelets and to show the fast discrete wavelet transform. Finally, the section 7 put all together and summarize the process to take advantage from the previous results.

2 Equivalence between TSK-Fuzzy Rule Systems and Function Approximation

Let S be a Takagi-Sugeno-Tang fuzzy rule system of degree 0 with sum-product connectives. With n input variables and one output variable, the rules of these systems take the expression:

$$r_{\boldsymbol{k}}: \quad \text{if } x_1 \text{ is } \phi_{k_1} \text{ and } \dots \text{ and } x_n \text{ is } \phi_{k_n} \text{ then } c_{\boldsymbol{k}}. \tag{1}$$

The calculated function f_S of this system is

$$f_S(\boldsymbol{x}) = \frac{\sum_{\boldsymbol{k}} c_{\boldsymbol{k}} \phi_{k_1}(x_1) \dots \phi_{k_n}(x_n)}{\sum_{\boldsymbol{k}} \phi_{k_1}(x_1) \dots \phi_{k_n}(x_n)}. \tag{2}$$

where $\boldsymbol{x} = (x_1, \dots, x_n)$ is a n-dimensional point. Let us suppose that $\boldsymbol{k} = (k_1, \dots, k_n)$ belongs to a regular grid G in the $\mathbb{R}^n$ space and that the system S has one rule by each different value of $\boldsymbol{k}$, with some values $c_{\boldsymbol{k}}$ possibly zero. Moreover, let us suppose that the value of the sum $\sum_{\boldsymbol{k}} \phi_{k_1}(x_1) \dots \phi_{k_n}(x_n)$ do not depends of the point $\boldsymbol{x}$, and without loose of generality, we can suppose that is equal to 1 for any $\boldsymbol{x} \in \mathbb{R}^n$

Taking the tensor product $\phi_{\boldsymbol{k}}(\boldsymbol{x}) = \phi_{k_1}(x_1)\ldots\phi_{k_n}(x_n)$ is evident that the function 3 can be rewritten as

$$f_S(\boldsymbol{x}) = \sum_{\boldsymbol{k}\in G} c_k \phi_{\boldsymbol{k}}(\boldsymbol{x}). \tag{3}$$

This expression is similar to the general expression of the interpolation problem: given a function f in a set of points G, the objective is to extend the function to its full domain D_f through the expression

$$f(x) = \sum_{\boldsymbol{k}\in G} W(x, \boldsymbol{k}) f(\boldsymbol{k}). \tag{4}$$

Starting on this point we work in the 1-dimensional space to simplify the notation. The n-dimensional case will be made by means of tensor product.

Habitually the function W must be verified two basic properties:

1. $W(k, k) = 1$ for all $k \in G$
2. $\sum_{k\in G} W(x, k) = 1$

which guarantee the equality between the extended function values in the points of the grid with the sample values $f(k)$. Indeed, this is the main difference between equations (3) and (4), in the former one, the values c_k do not have to be equal to sample values. The general interpolation formula has a special case when all function $W(x, k)$ are given by a family of functions $\{\phi_k(x)\}_{k\in G}$; then, the expression (4) must be written exactly as (3).

The usual definition of the continuous dot product (for real value functions) is

$$\langle f_1, f_2 \rangle = \int_{-\infty}^{\infty} f_1(x) f_2(x) dx \ .$$

If both sides of equation (3) are multiplied by a function ϕ_j of the family, and then

$$\langle \phi_j(x), f_S(x) \rangle = \sum_{k\in G} c_k \langle \phi_j(x), \phi_k(x) \rangle \ ,$$

and by denoting with Φ the matrix of all products $\langle \phi_k(x), \phi_j(x) \rangle$, an explicit expression for the coefficients c_k is obtained:

$$c_k = \sum_{j\in G} \Phi_{kj}^{-1} \langle \phi_j(x), f_S(x) \rangle. \tag{5}$$

where Φ_{kj}^{-1} is the (k,j)-element of the inverse matrix of Φ .

The matrix Φ is invertible as long as the basis set of functions is linearly independent. In the special case of an orthonormal basis, Φ reduces to the identity matrix:

$$\Phi_{jk} = \Phi_{jk}^{-1} = \delta_{jk} \ .$$

The solution given by (5) is equivalent to minimize the least-square norm of the difference between $f_S(x)$ and $\sum_{k \in G} c_k \phi_k(x)$. Approximating the integral by a finite sum on the regular grid G, we arrive at the approximate equality:

$$\langle \phi_j(x), f_S(x) \rangle = \int_{-\infty}^{\infty} \phi_j(x) f_S(x) dx \approx \sum_{k \in G} \phi_j(x) f_S(x) \; .$$

Substituting last equality into equations (5) and (3) we obtain the expression of the general interpolation function when a family $\{\phi_k\}_{k \in G}$ is used:

$$W(x,k) = \sum_{m \in G} \sum_{l \in G} \Phi^{-1}_{ml} \phi_m(x) \phi_l(k) \; .$$

Fixing the grid and the sample data in their points, last expression gives a method to obtain a TSK-fuzzy system as long as the function $W(x,k)$ have an compatible shape with a fuzzy set, but it is not habitually the case. Moreover, the calculation of the functions $W(x,k)$ can be very expensive, it is necessary to get an inverse matrix and it is possible that we do not have an explicit available expression for this functions.

There is some family of functions $\{\phi_k\}$ which resolve many of last problems. We consider now the case where the functions ϕ_k are translations of a unique function $\beta(x)$:

$$\phi_k(x) = \beta(x-k)$$

and the equation (3) is written as

$$f_S(x) = \sum_{k \in G} c_k \beta(x-k) = c * \beta(x) \; .$$

The operation $*$ represents the discrete convolution operator. In the continuous case, the convolution operator is defined as

$$f * g(x) = \int_{-\infty}^{\infty} f(t) g(x-t) dt \; .$$

If we want to calculate c_k it is necessary to make an inverse recursive filtering (an deconvolution). Next section shows a efficient method to achieve this objective in the particular case of B-splines.

3 Using interpolating B-splines

The B-splines [1] are the basic building blocks for splines. A spline is a continuous piecewise polynomial functions of degree n with derivatives up to order $n-1$ that are continuous everywhere on the real line [20,10]. The joining points of the polynomials are called knots. Splines have been used in several

[1] The B may stand for basis or basic

works in the context of soft computing [15,21,31,2]. Habitually, when a set of equally spaced simple knots is chosen the splines are called cardinal splines. We reduce our exposition to cardinal splines.

Any cardinal spline ϕ^n can be represented as

$$\phi^n(x) = \sum_{k\in\mathbb{Z}} c_k \beta^n(x-k) \ .$$

where $\beta^n(x)$ denotes the normalized B-spline of order n define below. The function $\phi^n(x)$ is uniquely determined by its B-spline coefficients c_k .

The zeroth-order B-spline is defined as

$$\beta^0(x) = \begin{cases} 1 & 0 \le x < 1 \\ 0 & \text{otherwise} \end{cases}$$

and the B-splines of a higher order can be define by a repetitive convolution of the zeroth-order spline β^0:

$$\beta^n(x) = \beta^0(x) * \overset{n+1}{\cdots} * \beta^0(x) \ . \tag{6}$$

There is an explicit expression of the B-Splines [20]:

$$\beta^n(x) = \frac{1}{n!} \sum_{k=0}^{n+1} (-1)^k \binom{n}{k} (x-k)_+^n \tag{7}$$

where $x_+ = max(0,x)$. The B-spline of nth-order reaches the maximum value in $x_0 = \frac{n+1}{2}$. It is possible to define it as a symmetrical function centered in the origin shifting the definition (6) by x_0 .

The family $\{\beta^n(x-k)\}_{k\in\mathbb{Z}}$ has many interesting properties:

1. They are compactly supported. In fact they are the shortest possible polynomial splines.
2. $\sum_{k\in\mathbb{Z}} \beta^n(x-k) = 1$ for all x .
3. B-spline converge to a Gaussian function when its degree increases.
4. As it was said above, it is a basis of all splines of degree n .
5. Its range is always included in the interval $[0,1]$.
6. In any interval, only a finite number of functions $\beta^n(x-k)$ are non-null. In fact, $\beta^n(x)$ is not null in the interval $(0, n+1)$.

The B-splines are extensively used in image and signal processing, computer graphics and interpolation theory [26,25,23]. It is possible to consider uniform knots distribution different of integer numbers or non-uniform distributions, but we only consider integer valued knot distributions.

It is the interpolation theory approach what we want to use here. We need to calculate the coefficients c_k of the equation (3) when the functions ϕ_k are shifted B-splines of degree n . In the case of B-splines of degree 0 or 1, the problem is trivial because it reduces to an interpolating general problem and taking $c_k = f_S(k)$ is enough. The case $n > 1$ is more complicated. Traditionally, the B-spline interpolation problem has been approached in a matrix framework using a band diagonal system of equations and applying standard numerical techniques as forward/backward substitution or LU decomposition [10,19]. Following the works [27,29,28] this coefficients can be calculated by means of a simpler recursive procedure (in terms of signal processing by means of a digital filtering technique). The process can be summarized as follows.

Let b_m^n be the discrete B-spline kernel which is obtained by sampling the B-spline of degree n expanded by a factor of m:

$$b_m^n \stackrel{def}{=} \beta^n(k/m) .$$

We want

$$\sum_{l\in\mathbb{Z}} c(l)\beta^n(x-l)\bigg|_{x=k} = f_S(k) .$$

where we have changed the notation c_l by $c(l)$.

Using the discrete B-spline, this can be rewritten as a convolution

$$f_S(k) = (b_1^n * c)(k) \tag{8}$$

The solution is found by inverse filtering

$$c_k = (b_1^n)^{-1} * f_S(k) .$$

The method to obtain the expression $(b_1^n)^{-1}$ is shown in [27]. It requires to take the z-transform[2] of (8) and resolving in the z-space. In the cited paper [27] a table of all solutions until degree 5 is given. In the general case, the method gives rise to next recursive filter equations [28]:

$$\begin{aligned} c^+(k) &= f_S(k) + z_i c^+(k-1),\ (k = 2,\ldots,K) \\ c(K) &= d_i(2y^+(K) - f_S(K)) \\ c(k) &= z_i(c(k+1) - c^+(k)),\ (k = K-1,\ldots,1) ; \end{aligned}$$

where K is the sample size, and where $d_i = -z_i/(1-z_i^2)$ is a scaling constant which depends of the B-spline degree considered. The method need to fix the sample behavior out of boundary. Habitually a mirror image approach will be enough, which can be managed by extending the sample with the relations

$$\begin{aligned} f_S(-k) &= f_S(k+1),\ k = 0,\ldots,K-1 \\ f_S(k) &= f_S(2K-k),\ k = K,\ldots,2K-1 . \end{aligned}$$

[2] The z-transform of a sequence b_k is given by $B(z) = \sum_k b_k z^{-1}$, with z in the complex space

In this case, the recursion begins with

$$c^+(1) = \sum_{k=1}^{k_0} z_i^{|k-1|} f_S(k)$$

where k_0 is chosen to ensure that $z_i^{|k_0|}$ is smaller than some prescribed level of precision [28]. Last approach to find the coefficients c_k is about twice faster than traditional matrix methods.

A different method to calculate the coefficients c_k is proposed by Chui in [5]. The author defines a family of quasi-interpolation operators Q_r which allows to reach a good approximation for coefficients c_k. For any function f, the operator mapping $Q_r f$ gives a projection into the cardinal splines space of degree n. The operator Q_r preserves the polynomials of degree $n-1$ when $r > \frac{n-2}{2}$ and it is local, in the sense of $(Q_k f)(x)$ depends only of a shifted compact $J + x$ of a fix compact set J. The sequence converge to the cardinal spline interpolation operator Q_∞ which is uniquely determined by the interpolation property:

$$(Q_\infty f - f)(l) = 0 \ \forall l \in \mathbb{Z} .$$

For example for cubic B-splines, Q_1 is defined by

$$(Q_1 f)(x) = \sum_{k \in \mathbb{Z}} \frac{1}{6}(-f(k+1) + 8f(k) - f(k-1))\tilde{\beta}_3(x + 2 - l) .$$

where $\tilde{\beta}$ is a centered B-spline defined as

$$\tilde{\beta}^n = \beta^n(x + \frac{n+1}{2}) .$$

We have work in the real line but using the tensor product everything can be applied to a n-dimensional space. Hence we have an efficient method to extract a first TSK-fuzzy system for B-splines of degree n when we fix a grid and we have the sample values in the grid points. This can be generalized to cardinal splines.

4 Cardinal Splines

Let ϕ be a cardinal spline of degree n defined by

$$\phi^n(x) = \sum_{k \in \mathbb{Z}} p(k)\beta^n(x-k) = (p * \beta^n)(x)$$

and let S^n be the space of polynomial spline functions

$$S^n = \left\{ g : g(x) = \sum_{k \in \mathbb{Z}} \hat{c}_k \phi^n(x-k) \right\} .$$

Now, from a sample $\{f_S(k)\}_{k\in\mathbb{Z}}$ we want to obtain the coefficients $\hat{c}_k$ that interpolates f_S. In terms of convolution operators and, thanks to previous section, we know how to approximate f_S with B-splines

$$f_S(x) = \sum_{k\in\mathbb{Z}} c_k \beta^n(x-k) = (c * \beta^n)(x) ,$$

where $c_k = (b^n)^{-1} * f_S(k)$ and now we want to obtain

$$f_S(x) = \sum_{k\in\mathbb{Z}} \hat{c}_k \phi^n(x-k) = (\hat{c} * \phi^n)(x) .$$

Putting all together, we have

$$f_S = c * \beta^n = \hat{c} * \phi^n = \hat{c} * p * \beta^n$$

and the equality $c = \hat{c} * p$ holds. Using the expression for c_k obtained in last section, it results

$$\hat{c}_k = (p)^{-1} * c(k) = (p * b^n)^{-1} * f_S(k) .$$

All of this is valid if the family of functions $\{\phi^n(x-k)\}_{k\in\mathbb{Z}}$ is a basis of S^n, which is true provided that p is an invertible convolution operator from the space of all square finite sequences into itself [30].

Again, the coefficients $\hat{c}_k$ can be calculated by digital filtering, and it can result in a more efficient algorithm that the traditional approach based on matrix formulation.

Until now, we have shown efficient methods to obtain a TSK fuzzy system if we have the sample in a regular grid. In this case, we can use any adequate cardinal spline as fuzzy sets and to build the fuzzy rule system in an efficient way. It is evident that it will not always be the case. In spite of this, the methods can serve if a kind of summarize can be applied on data and so, to get the sample values in the grid. This approach would allows to make a first approximation to get a fuzzy rule system and to substitute the traditional methods of fuzzy clustering. The digital filtering point of view gives the possibility of to design efficient recursive algorithms which can operate on line.

Another possibility consists in to think in multiresolution terms. Let us consider two grids G and G'. If we had efficient methods to calculate the fuzzy rule system in G from the fuzzy rule system in G' perhaps we could work with both grids, one of them with a higher resolution and the other one with a lower resolution. The system could give a first low resolution answer in a very fast manner and to give a high resolution answer on demand. A feasible approach is given by wavelet theory.

5 Splines and Wavelets

The wavelet theory use the dilations and translations of a function, called the mother wavelet, to get the projections of any square integrable real function, the elements of $L^2(\mathbb{R})$, over these small wavelets and so to obtain its wavelet transform. Let ψ be a function in $L^2(\mathbb{R})$, its dilations and translations can be given by the general expression

$$\psi_{a,b}(x) = \frac{1}{\sqrt{|a|}}\psi(\frac{x-b}{a})$$

with $a, b \in \mathbb{R}$ and $a \neq 0$. The constant $1/\sqrt{|a|}$ is a normalizing factor. It is very habitual to work with the discrete dyadic case: considering shifts (translations by integers) and dyadic dilation by taking $a = 2^{-j}$ and $b = k2^{-j}$, with $j, k \in \mathbb{Z}$ and it is the case that we study in this work. However another dilations and irregular grid are perfectly possible.

The discrete wavelet transform of a function $f \in L^2(\mathbb{R})$ with respect to the wavelet generated by ψ is given by the expression

$$\mathcal{W}(j,k) = \langle f, \psi_{j,k}\rangle = \int_{-\infty}^{+\infty} f(x)\psi_{j,k}(x)dx$$

when the work space is the real numbers.

Interesting cases of wavelets are obtained when the set of functions $\{\psi_{j,k}\}_{j,k\in\mathbb{Z}}$ constitute an orthonormal basis of the space of functions $L^2(\mathbb{R})$. Then any function $f \in L^2(\mathbb{R})$ can be represented as an infinite sum of its projections over each $\psi_{j,k}$:

$$f = \sum_{k,j\in\mathbb{Z}} \langle f, \psi_{j,k}\rangle\psi_{j,k}.$$

If the wavelet functions were adequate to use as fuzzy sets, they could be used to build TSK fuzzy rule systems straightly thanks to last expression, but it is not the habitual case. The functional expressions of the wavelet are in many cases unknown, in the majority of the cases out of the $[0,1]$ range and with a very irregular behavior.

Bellow we are giving the ordinary method to build wavelets functions, the multiresolution analysis. In this method appear a class of functions, the scale functions, which allows have a bigger capacity to choose adequate functions in wavelet theory to use as fuzzy sets, and, so, to make a good use of the wavelet theory in this field.

6 Multiresolution analysis

One of the most habitual definition of multiresolution analysis is the following.

Definition 1. A multiresolution analysis of $L^2(\mathbb{R})$ is a sequence of closed subspaces $\{V_j\}_{j\in\mathbb{Z}}$ of $L^2(\mathbb{R})$ such that the following conditions hold for all $j \in \mathbb{Z}$:

1. $V_j \subset V_{j+1}$,
2. $\cup_{j\in\mathbb{Z}} V_j$ is dense in $L^2(\mathbb{R})$ and $\cap_{j\in\mathbb{Z}} V_j = \{0\}$,
3. $f(x) \in V_j \iff f(2x) \in V_{j+1}$,
4. $f(x) \in V_0 \iff f(x+1) \in V_0$,
5. A scaling function $\varphi \in V_0$, with a non-vanishing integral, exists such that the collection $\{\varphi(\cdot - k) : k \in \mathbb{Z}\}$[3] is a Riesz basis of V_0.

The scaling function φ generate the multiresolution analysis because of the functions

$$\varphi_{j,k}(x) = 2^{j/2}\varphi(2^j x - k)$$

are a Riesz basis for V_j. In spite of this condition, the collection $\{\varphi_{j,k}\}_{k\in\mathbb{Z}}$ is not a Riesz basis for $L^2(\mathbb{R})$

The projections $P_j f$ of any function $f \in L^2(\mathbb{R})$ on V_j is a approximation to f in the 2^{-j} scale. This approximation have the expression

$$P_j f(x) = \sum_k c_{j,k}\varphi_{j,k}(x) \ ,$$

with $c_{j,k} = \langle f, \varphi_{j,k}\rangle = \int_{\mathbb{R}} f(x)\varphi_{j,k}(x)dx$. Since the union of the V_j is dense in $L^2(\mathbb{R})$, any function f can be arbitrarily approximate by this projections.

In many cases no explicitly expression for φ is available, but it can be evaluate through the refinement equation in dyadic points. The refinement equation, also called dilation equation or two-scale difference equation, play a crucial role in the fast wavelet algorithm and is given by the expression:

$$\varphi(x) = 2\sum_k h_k \varphi(2x - k), \tag{9}$$

since $\varphi \in V_0 \subset V_1$ and $\varphi(2x-k)$ is a Riesz basis for V_1.

By integrating both sides of (9), and dividing by the (non-vanishing) integral of φ, we see that

$$\sum_k h_k = 1 \ .$$

Moreover, under very general conditions, the equation (9) and the normalization

$$\int_{-\infty}^{\infty} \varphi(x) = 1$$

determinate φ completely.

The B-spline functions can be serve as scaling functions and generated the nth order spline multiresolution analysis $\{V_j^n\}_{j\in\mathbb{Z}}$ where each V_j^n is defined

[3] Habitually, it is forced that $\sum_{k\in\mathbb{Z}} \varphi(x-k) = 1$.

as the closed span of β^n, i.e. the space of all functions $f \in L^2(\mathbb{R})$ generate by finite linear combinations of $\beta^n(2^j x - k)$. The hat functions are the B-spline of order one. The refinement equation for B-splines take the expression:

$$\beta^n(x) = 2^{-n} \sum_{k=0}^{(n+1)} \binom{n+1}{k} \beta^n(2x - k + 1).$$

In the context of multiresolution analysis the wavelet function is introduced as follow. Let W_j be a complementary space of V_j in V_{j+1}. Then

$$V_{j+1} = V_j \oplus W_j,$$

and any element of V_{j+1} can be written, in a unique way, as the sum of an element of W_j and an element of V_j. The space W_j contains the "detail" information needed to go from an approximation at resolution j to an approximation at resolution $j+1$. It holds that $\oplus_j W_j = L^2(\mathbb{R})$, because of $W_j = V_{j+1} \ominus V_j$ and $\oplus_{j=-n}^{n} W_j = V_{n+1} \ominus V_{-n}$, with V_n going to $L^2(\mathbb{R})$ and V_{-n} to the constant function zero when n go to infinity. A function ψ is a wavelet if the collection of functions $\{\psi(x-k) : k \in \mathbb{Z}\}$ is a Riesz basis of W_0. In this case the set of functions $\{\psi_{j,k} : j, k \in \mathbb{Z}\}$ is a Riesz basis of $L^2(\mathbb{R})$.

The refinement equation has its counterpart in terms of the wavelet function. As $\psi \in W_0 \subset V_1$ then a set of constants g_k exist such that

$$\psi(x) = 2 \sum_k g_k \varphi(2x - k). \tag{10}$$

The subspace W_j is not uniquely determined from V_j and V_{j+1}. Many types of wavelets can be defined [8]. For example, when W_j es an orthogonal complement of V_j in V_{j+1} the orthogonal wavelets are obtained. In [16] the conditions to obtain orthonormal wavelets are summarize. It is possible to begin with a scale function and by applying an orthonormalization process to get a orthonormal wavelet. In particular, the B-splines functions can be taken as scale functions. However, the orthonormalization process applied to B-spline functions does not produce adequate properties to be used as fuzzy sets.

Another wavelet type more interesting for our proposes are the bi orthogonal wavelets. The orthogonality property puts a strong limitation on the construction of wavelets. The generalization to bi-orthogonal wavelets has been considered to gain more flexibility. Here, a dual scaling function $\tilde{\varphi}$ and a dual wavelet $\tilde{\psi}$ exist that generate a dual multiresolution analysis with subspaces $\tilde{V}_j$ and $\tilde{W}_j$ such that

$$\tilde{V}_j \perp W_j \qquad \text{and} \qquad V_j \perp \tilde{W}_j$$

and consequently

$$\tilde{W}_j \perp W_j \qquad \text{for} \quad j \neq j'$$

or equivalently

$$\langle \tilde{\varphi}, \psi(\cdot - l)\rangle = \langle \tilde{\psi}, \psi(\cdot - l)\rangle = 0 .$$

Moreover, the dual functions also have to satisfy

$$\langle \tilde{\varphi}, \varphi(\cdot - l)\rangle = \delta_l \qquad \text{and} \qquad \langle \tilde{\psi}, \psi(\cdot - l)\rangle = \delta_l .$$

The role of the basis (φ and ψ) and the dual functions $\tilde{\varphi}$ and $\tilde{\psi}$ can be interchanged.

The dual functions $\tilde{\varphi}$ and $\tilde{\psi}$ define a multiresolution analysis, and hence they must satisfy

$$\tilde{\varphi} = 2\sum_k \tilde{h}_k \tilde{\varphi}(2x - k) \tag{11}$$

$$\tilde{\psi} = 2\sum_k \tilde{g}_k \tilde{\varphi}(2x - k) . \tag{12}$$

The refinement equation (9) and the general expression of the wavelet function obtained by resolution given in (10) permit to obtain a fast algorithm to get the dyadic discrete wavelet transform.

The projection operators take the form

$$P_j f(x) = \sum_l \langle f, \tilde{\varphi}_{j,l}\rangle \varphi_{j,l}(x) Q_j f(x) = \sum_l \langle f, \tilde{\psi}_{j,l}\rangle \varphi_{j,l}(x)$$

and

$$f = \sum_{k,l} \langle f, \tilde{\psi}_{j,l}\rangle \psi_{j,l} .$$

It is satisfied that

$$\begin{aligned} \tilde{h}_{k-2l} &= \langle \tilde{\varphi}(x - l), \varphi(2x - k)\rangle \\ \tilde{g}_{k-2l} &= \langle \tilde{\psi}(x - l), \varphi(2x - k)\rangle \end{aligned}$$

In particular, by writing $\phi(2x - k) \in V_1$ in the basis of V_0 W_0 we obtain that

$$\varphi(2x - k) = \sum_l \tilde{h}_{k-2l}\varphi(x - l) + \sum_l \tilde{g}_{k-2l}\psi(x - l) .$$

6.1 Fast Dyadic Wavelet Transform

Let $P_{j+1}f$ be a approximation to the function f at the scale $j + 1$, i.e.

$$P_{j+1}f = \sum_k c_{j+1,k}\varphi_{j+1,k}(x).$$

As $V_{j+1} = V_j \oplus W_j$, we have that

$$P_{j+1}f = P_j f + d_j = \sum_k c_{j,k}\varphi_{j,k}(x) + \sum_k d_{j,k}\psi_{j,k}(x).$$

The fast wavelet transform is an algorithm to compute the coefficients $c_{j,k}$ and $d_{j,k}$ from the coefficients $c_{j+1,k}$ and vice versa. The algorithm does not use the functions φ and ψ, only the direct refinement equation coefficients: h_k, g_k and dual refinement equation coefficients: $\tilde{h}_k$ and $\tilde{g}_k$. The algorithm can be applied in two directions: from a high resolution to a low resolution, and vice versa, from a low resolution to a finer one.

First direction is called decomposition, or analytic, process and it involves the dual refinement equation coefficients. It can be proved that

$$c_{j,l} = \langle P_{j+1}f, \tilde{\varphi}_{j,l}\rangle = \sqrt{2}\langle P_{j+1}f, \sum_k \tilde{h}_{k-2l}c_{j+1,k}\rangle$$
$$= \sum_k h_{k-2l}c_{j+1,k}$$

by bi-orthogonality, and similarly

$$d_{j,l} = \sum_k \tilde{g}_{k-2l}c_{j+1,k}.$$

This transform can be inverted by the reconstruction, or synthesis, process which is given by the expression:

$$c_{j+1,k} = \sum_l h_{k-2l}c_{j,l} + \sum_l g_{k-2l}d_{j,l}.$$

In [16], an pseudo code implementation of the periodic fast wavelet transform algorithm is given. For one level, the algorithm complexity is multiplicative in sample size and refinement equation coefficients length. This algorithm requires a sample with a power of two number of data. Using the lifting scheme [22] this limitation can be avoided. The lifting scheme has several advantages with respect to the traditional fast discrete wavelet transform algorithm: it does not need a regularly spaced set of data, its behavior with the boundary conditions is better and it does not extra memory positions to calculate the detail coefficients. The reference [9] gives a method to factoring any bi-orthogonal wavelet transform with finite filters into lifting steps. Finite filters mean that the set of non-null coefficients $\{h_k, g_k\tilde{h}_k, \tilde{g}_k\}_{k\in\mathbb{Z}}$ is finite.

6.2 Spline Wavelets

It is well known that the B-spline β^n generates a multiresolution analysis of L_2. In [7] the wavelet ψ_n corresponding to the nth order cardinal B-spline-wavelet was calculated:

$$\psi_n(x) = \sum_{j=0}^{3n+1} q_{n,j}\beta^n(2x-j) \qquad \text{for all} \quad n \in \mathbb{N} \cup 0$$

where

$$q_{n,j} = \frac{(-1)^j}{2^n}\sum_{l=0}^{n+1}\binom{n+1}{l}\beta_{2n+1}(j-l+1)\,.$$

Last equation and the refinement equation for B-splines:

$$\beta^n = \sum_{j=0}^{n+1} p_{n,j}\beta^n(2x-j)\;;$$

with

$$p_{n,j} = \begin{cases} 2^{-n}\binom{n+1}{l} & 0 \le l \le n+1 \\ 0 & \text{otherwise}\,, \end{cases}$$

give the primal filters for B-splines of degree n by taking $h_k = p_{n,k}$ and $g_k = q_{n,k}$. The dual filters are can be calculated and are given, for example, in [5]. These filters are not finite and an approximation is needed to apply them.

7 Putting all together

The fast wavelet transform allows to obtain the approximation of a function in a coarser scale given another approximation in another finer scale. Thanks to the properties of the scale and wavelet functions this process can be done in both directions: from a finer resolution to a coarser one and vice versa.

Given a sample of points regularity spaced it is possible to apply the fast wavelet transform to obtain two subgroups of data, each of them with the half of size with respect the first one. A subgroup corresponds to the sample in a coarse resolution and the another to the detail information; moreover last subgroup allows revert the process. This process can be done in each variable when we are working in the n-dimensional space [11]. To close the circle is necessary to determine, with a certain accuracy and in a computationally favorable way, the coefficients to begin the fast wavelet transform. By means of the algorithm of section 3 it is possible to obtain a first TSK fuzzy rule system as a approximation problem with B-splines in a efficient manner. This first fuzzy rule system has as rules as sample points. When the fast wavelet is applied for first time, we obtain the detail coefficients which give the error of going from the finer sample, to a coarser sample, so it is possible to make an analysis of the accuracy of the new system. We can apply the habitual compression process with wavelets: analyzing the absolute value of the wavelet coefficients $d_{j-1,k}$ in the localizations k, it can be detected

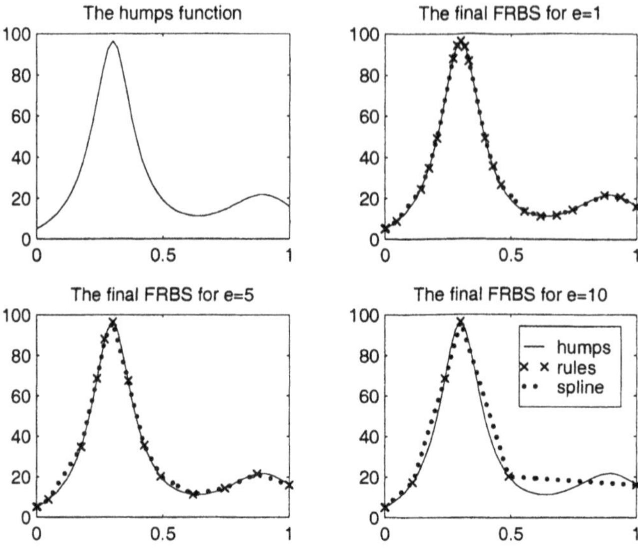

Fig. 1. The humps example

the zones where the local error is smaller and discard these values by making them equal to zero. Undone the wavelet transformation we get a better local approximation in the place where it is really needed. And all of this is made with B-splines functions, which have a easy interpretation as fuzzy sets and a set of adequate mathematical properties such as differentiability and locality. For example, these properties in [2] allows to give a mathematically interpretation of a lightly modified TSK model in terms of Taylor series.

Lastly, works as [3] and [1] allows to use last process in a non-uniform approach.

7.1 A simple example

As an example, we apply the process previously described to a unidimensional function and with lineal b-splines. Let us consider the function

$$humps(x) = \frac{1}{(x-0.3)^2+0.01} + \frac{1}{(x-0.9)^2+0.04} - 6 \qquad x \in [0,1]$$

which appears in the Matlab 5.1 packet. In figure 7.1 a graphical representation of this function is showed.

The process to obtain a approximated fuzzy rule system takes following steps:

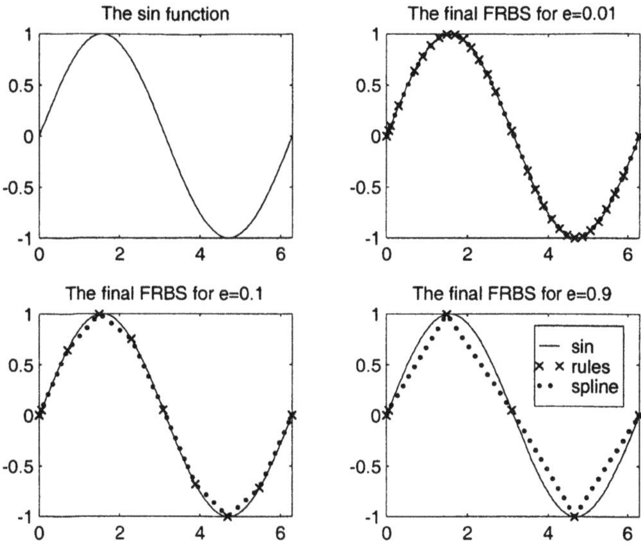

Fig. 2. The sin function

1. Take an initial partition of the interval $[0, 1]$. For example, take a regular partition $X = [x_1, \ldots, x_n]$ and let $Y = humps(X)$ be the mapping values on X.
2. Fix a positive number ϵ as the threshold error.
3. Apply to Y the filter $(-\frac{1}{2}, 1, -\frac{1}{2}$ to calculate the detail coefficients d_i. This filter is the corresponding decomposition detail filter of Cohen-Daubechies-Feauveau bi-orthogonal wavelet transform.
4. Make zero all details with $|d_i| < \epsilon$.
5. Take the points (x_i, y_i) such as the associated detail coefficient d_i is no zero.
6. Back to the point 3 with the new X and new Y calculated, until no variation on X and Y is produced.

Some requirements are to apply correctly the algorithm. In first place, the cardinality of X must always be an even number, and, hence, it can be necessary to insert a new point; this is made by lineal interpolation. Also the boundaries require a special processing: the point set (X, Y) is extended in both edges by a lineal extrapolation. Due to this, it is necessary to force the algorithm to maintain the lower interval boundary, if not, it will always disappeared.

In the figure 7.1 appears the result of applying the algorithm with different thresholds. Each cross-shaped point matchs with a fuzzy rule defined by a b-spline, i.e. a triangular fuzzy set. The TSK-fuzzy rule system calculates a spline which is given by the dotted line in the figure. As can be viewed, the initial objective is reached. We obtain different approximations to the function, when we fix different threshold values. With more rules, a better approximation is reached, but this impose a penalty on the fuzzy rule database complexity and linguistic interpretability. With less rules, a worst approximation but a easier fuzzy rule database. And in all cases, using only fuzzy partitions with triangular fuzzy sets. In the figure 7.1 can be the same for the sin function. [4].

References

1. Aldroubi, A.; Gröchening, K. . Non-uniform sampling and reconstruction in shift-invariant spaces. *To appear in SIAM reviews*, pages 1–43, 2001.
2. M. Bikdash. A highly interpretable form of Sugeno inference systems. *IEEE Transactions On Systems, Man And Cybernetics*, 7(6):686–696, 1999.
3. M.D. Buhmann and C.A. Micchelli. Spline prewavelets for non-uniform knots. *Numer. Math.*, 61:455–474, 1992.
4. Chen, D. Characterization of Bi-orthogonal Cardinal Spline Wavelet Bases *Joint Mathematics Meetings*, Cincinnati, 1994.
5. C.K. Chui. *An Introduction to Wavelets.* Academic Press, 1992.
6. C.K. Chui, editor. *Wavelets: A Tutorial in Theory and Applications.* Academic Press, 1992.
7. C.K. Chui and J-Z. Wang. On compactly supported spline wavelets and a duality principle. *Trans. Amer. Math. Society*, 330(2):903–914, April 1992.
8. Ingrid Daubechies. *Ten Lectures on Wavelets.* CBMS-NSF Regional Conference Series in Applied Mathematics. SIAM, 1992.
9. W. Daubechies, I.; Sweldens. Factoring wavelet transforms into lifting steps. *J. Fourier Anal. Appl.*, 4(3):245–267, 1998.
10. C. de Boor. *A practical guide to Splines.* Springer Verlag, 1978.
11. A.R. de Soto. Building a hierarchical set of fuzzy rules using wavelet theory. In *Inter. Conference on Information Processing and Management of Uncertainty Knowledge-Based Systems*, volume 3, pages 1764–1769, Madrid, July 2000.
12. A.R. de Soto and Recasens, J. Modelling a linguistic variable as hierarchical family of partitions induced by an indistinguishability operator *Fuzzy Sets and Systems*, 121(3), pages 57–67, Madrid, July 2000.
13. A.R. de Soto and Trillas, E. Second Thoughts on Linguistic Variables. In *Proc. 18th Inter. Conference of the North American Fuzzy Information processing Society (NAFIPS'99*, pages 37–41, New York, 1999.
14. R.A. DeVore and B.J. Lucier. Wavelets. *Acta Numerica*, 1:1–56, 1992.
15. H. Ichihashi. Efficient algorithms for acquiring fuzzy rules from examples. In H.T. Nguyen, M. Sugeno, R. Tong, and R.R. Yager, editors, *Theoretical Aspects of Fuzzy Control*, pages 261–281. John Wiley & Sons, 1995.
16. B. Jawerth and W. Sweldens. An overview of wavelet based multiresolution analysis. *SIAM Review*, 36(3):377–412, 1994.
17. S.G. Mallat. Multiresolution approximations and wavelet orthogonal bases of $l^2(\mathrm{R})$. *Trans. Amer. Math. Society*, 315:69–87, 1989.

[4] These examples were done with Matlab packet. The routines to reproduce the examples are available and they can be solicitated to the author

18. E. A. Mamdani. Application of fuzzy algorithms for control of simple dynamic plant. In *Proceedings of IEEE 121*, volume 12, pages 1585–1588, 1974.
19. W.H. Press, S. A. Teukolsky, W.T. Vetterling, and B.P. Flannery. *Numerical Recipes in C. The Art of Scientific Computing.* Cambridge University Press, Second edition, 1992.
20. I.J. Schoenberg. *Cardinal spline interpolation.* SIAM, 1 edition, 1973.
21. K. Shimojima, T. Fukuda, and F. Arai. Self-tunning fuzzy inference based on spline functions. In *Proc. Of 3rd IEEE Int. Conf. On Fuzzy Systems. FUZZ-IEEE'94.*, volume 1, pages 690–695, Orlando, 1994.
22. W. Sweldens and P. Schrder. *Building your own Wavelets at home.* ACM SIGGRAPH Course, 1996.
23. P. Thévenaz, T. Blu, and M. Unser. Interpolation revisited. *IEEE Transactions on Medical Imaging*, 19(7):739–758, 2000.
24. Thuillard, M. Fuzzy-Wavelets: Theory and Applications. *Proc. ELITE'98*, pages 1149–1158, Aachen, September 1998,
25. A.B. Tucker, editor. *The Computer Science and Engineering Handbook.* CRC Press, 1996.
26. M. Unser. Splines. a perfect fit for signal and image processing. *IEEE Transactions on Signal Processing*, 16:22–38, November 1999.
27. M. Unser, A. Aldroubi, and M. Eden. Fast b-spline transforms for continuous image representation and interpolation. *IEEE Transactions on Pattern Analysis and Machine Intelligence*, 13(3):277–285, March 1991.
28. M. Unser, A. Aldroubi, and M. Eden. B-spline signal processing: Part i - efficient design and applications. *IEEE Transactions on Signal Processing*, 41(2):834–846, February 1993.
29. M. Unser, A. Aldroubi, and M. Eden. B-spline signal processing: Part i - theory. *IEEE Transactions on Signal Processing*, 41(2):821–832, February 1993.
30. M. Unser, A. Aldroubi, and M. Eden. The l2 polynomial spline pyramid. *IEEE Transactions on Pattern Analysis and Machine Intelligence*, 15(4):364–379, April 1993.
31. C-H. Wang, W-Y. Wnag, T-T. Lee, and P-S. Tseng. Fuzzy b-spline membership function (bmf) and its applications in fuzzy-neural control. *IEEE Transactions On Systems, Man And Cybernetics*, 25(5):841–851, 1995.
32. L.A. Zadeh. Fuzzy sets. *Information and Control*, 8:338–353, 1965.

SECTION 6

ASSESSMENTS ON THE INTERPRETABILITY LOSS

Relating the theory of partitions in MV-logic to the design of interpretable fuzzy systems

Paolo Amato[1] and Corrado Manara[2]

[1] Soft Computing and Nano-Organics Operations
STMicroelectronics, Via C. Olivetti 2, 20041 Agrate Brianza (MI), Italy
e-mail: paolo.amato@st.com

[2] Dipartimento di Scienze dell'Informazione
Università degli Studi di Milano, Via Comelico 39, 20135 Milano, Italy
e-mail: manara@dsi.unimi.it

Abstract. The problem of interpretability of a set of membership functions asks whether each fuzzy set can be associated with a linguistic value and how coherent the association is with respect to any other possible association.

In the same way a component of a classical crisp partition is immediately interpreted, the interpretability of a set of membership functions is guaranteed by the act of partitioning the domain. Although many authors have proposed constraints on membership functions in order to guarantee the semantic interpretability of a fuzzy model, formal definitions of these constraints are usually not well connected with the theory of partitions in many-valued and fuzzy logic.

Our aim here is to relate the theory of partitions in MV-logic to the design of interpretable fuzzy systems, and to analyze the concepts of refinement and joint refinement of partitions.

1 Introduction

Whenever a human expert is able to describe the variables of a system in terms of linguistic values and if-then rules, the problem arises of attaching suitable fuzzy sets to the values of each variable, thus obtaining a fuzzy model of the system. The main problem is to find the most appropriate fuzzy sets for the given description. In the opposite direction, in most data driven methodologies (such as those currently used in Soft Computing [39]), the problem is to find suitable linguistic values for given fuzzy sets. The latter may typically occur as the output of a learning algorithm, or of a general automated modeling process.

Soft Computing techniques are increasingly used for approximation tasks (e.g., in system modeling and control problems). Here the only guiding principle leading to an optimal fuzzy model is the minimization of the gap between system output and training data. Accordingly, the familiar parameters of membership functions (e.g., center and width) are treated just like any other parameter of the system. Consequently, one can neither ensure that the resulting membership functions can be interpreted as linguistic values, nor that the linguistic values have a meaningful association with a set of membership functions.

When the approximation process is unconstrained, the resulting model may have several drawbacks. Firstly, it is a black-box model giving us no insight into the original system. Moreover, this model, as well as the classical model, is often trapped in local minima. For this reason several authors introduced constraints on membership functions. Among the first papers dealing with such constraints let us quote [16] and [31]. In [34], the author considers the tradeoff between precision and "transparency". In [8], some constraints are presented, together with a discussion on their efficient implementability. These constraints include coverage, distinction, normality, and the request that membership functions should be as few and distinguishable as possible. The paper [9] presents an algorithm whose output is a model that respects similar constraints.

Our aim is to investigate, from a formal point of view, the problem of interpreting a fuzzy model obtained by applying a data-driven methodology. We present a conceptual framework (Section 3 and 4) from which the (common-sense) desiderata described in Section 2 can be derived. Moreover, we analyze the relationships among different descriptions of the same linguistic variable: how new descriptions can be generated from the old ones (which is referred to as *refinement* in the partition terminology) and how it is possible to obtain a common description (Section 5). In Section 6 we analyze the role of partition theory in the formalization of fuzzy modeling. Then in Section 7 we give an example of constrained learning, incorporating the notions of the preceding sections. In Section 8 we present some concluding remarks.

2 How to guarantee interpretability?

A *linguistic variable* [38] is a variable whose values are words, represented quantitatively by membership functions. To guarantee the semantic interpretability of a fuzzy model, one usually assumes that the set of membership functions $\{\mu_1, \mu_2, \ldots, \mu_n\}$, associated to a certain linguistic variable defined on a universe U, satisfies the following desiderata:

Coverage. Since the basic idea behind a linguistic variable is to granulate the universe U into a set of linguistic values, each point in U should belong to at least one fuzzy set (intended to represent a linguistic value) with a degree greater than zero. This means that (the supports of) the membership functions should cover the entire universe U:

$$\forall\, x \in U\ \exists\, i \in \{1, 2, \ldots, n\}\ :\ \mu_i(x) > 0.$$

Distinction. A linguistic variable is completely defined by its syntax (its name, the universe U and the set of membership functions) and its semantics (the association of a linguistic value to each membership function). To have soundness, the same linguistic value should not be associated to two distinct fuzzy sets, and the same fuzzy set should not share two distinct linguistic values. In data driven methodology any two membership functions should be distinct:

$$\forall\, i,j \in \{1,2,\ldots,n\}, i \neq j \;:\; \mu_i \neq \mu_j.$$

Normality. In fuzzy set theory we are free to define membership functions which never attain the value 1. But, when a fuzzy set represents a linguistic value, it is natural to think that there exists at least one point in U which satisfies the crisp definition of that linguistic value (otherwise, it does not make any sense to introduce that value in the chosen universe). Then each membership function should be normal:

$$\forall\, \mu_i\; \exists\, x \in U \;:\; \mu_i(x) = 1.$$

Distinguishability. Relative to an appropriately chosen *similarity measure* [37,33] $S(\mu_i, \mu_j)$, the degree to which two membership functions are equal should be low:

$$\forall\, i,j \in \{1,2,\ldots,n\}, i \neq j \;:\; S(\mu_i, \mu_j) \leq \varepsilon,$$

where $0 \leq \varepsilon < 1$ is a threshold degree of similarity. Note that the constraint of distinction is included here because $\mu_i = \mu_j$ if and only if $S(\mu_i, \mu_j) = 1$. Thus, distinguishability implies distinction, but the reasons behind these two desiderata are different. On the one side, distinction relates to the internal coherence of the set of membership functions. On the other side, distinguishability relates to an efficient granulation of the universe U, and the threshold value ε can be chosen as low as desired.

Parsimony. The number of membership functions should be moderate. In fact the association of labels (linguistic values) to fuzzy sets is more easily accomplished when the number of fuzzy sets is low. However, having a large number of fuzzy sets does not always conflict with a meaningful and sound interpretation. Thus, while it is a good rule of thumb to require a moderate number of fuzzy sets, there is no reason to fix a specific threshold.

From a formal point of view, the problem of *interpretability* of a set of membership functions asks (i) whether each fuzzy set can be associated with a linguistic value and (ii) how coherent the association is with respect to any other possible association. Here, the actual number of fuzzy sets, as well as their distinguishability degree, has no relevance. However, the smaller is the number of membership functions, the simpler is their association with a set of labels.

An answer to questions (i) and (ii) can be found in Zadeh's words [40]: "The finite ability of sensory organs to resolve detail necessitates a *partitioning* of objects (points) into granules... For example, a perception of age may be described as very young, young, middle-aged, old and very old, with very young, young, etc., constituting the granules of the variable Age". It is the act of partitioning the domain that guarantees the interpretability of the fuzzy sets of a linguistic variable.

This is straightforward for classical boolean partitions. In fact each element of a boolean partition is a crisp set disjoint from the other elements, and all the elements together cover the whole universe. Then, each element of a boolean partition can be associated to a certain label in a sound way (even though this association will be very restrictive and of little use for applications). Thus, since the first, second and third of the above conditions characterize the set of membership functions as a special kind of partition of the domain U, from a formal point of view they do guarantee interpretability.

The last two conditions pertain to the main idea behind Soft Computing, that is, "to exploit the tolerance for imprecision and uncertainty" [39]. On the one hand, as soon as the number of membership functions increases, the precision of the model may increase, too. However, if the membership functions do not form a partition, the precision may well decrease. In fact, for some hybrid neuro-fuzzy systems with an *a priori* fixed number of membership functions on each input variable, it happens that, given a set of training patterns, the performance of the trained system degenerates when the chosen number of fuzzy sets on each input is greater than a given threshold.

On the other hand, very large numbers of membership functions rather witness our inability to exploit uncertainty. In the limit case, when the length support of each membership function becomes comparable with the precision of our measurements, or even, when the number of membership functions approaches the number of data, a fuzzy system boils down to a numerical system. To avoid this problem, we can impose an upper bound on the number of fuzzy sets of each linguistic variable, or we can use a similarity measure [37] to minimize the number of similar fuzzy sets [33]. The latter solution is also useful when the fuzzy partition under consideration is obtained by a refinement process that generates new membership functions from old ones.

As stated above, from a formal point of view interpretability means the possibility of coherently associating a linguistic value to each fuzzy set on a given input domain. However in less formal contexts, interpretability asks also (iii) how much this association process is feasible for a human being. In this last question distinguishability plays a role which is more relevant than that played by parsimony. In fact, on one side a person may be able to give a linguistic label to each member in a family of well distinguished fuzzy sets even when their number is high. On the other side, he (or she) can have serious difficulties in associating two different linguistic values to two distinct but poorly distinguishable fuzzy sets. This aspect is hard to manage in a formal way. The distinguishability concept is not completely taken care of by similarity measures. Also other factors, more related to psychology than to mathematics, should be taken into consideration.

In the following we focus our attention on the first two questions about interpretability and, consequently, present a conceptual framework from which the first, second and third desiderata can be derived.

3 Basic notions

In this section we recall the definition of boolean partition and we give some definitions and results about Łukasiewicz logic and MV-algebras. These notions will be used in the next section, where the generalization of boolean partitions to the many-valued case is discussed.

By a universe U we mean an arbitrary nonempty set.

3.1 From boolean to nonboolean partitions

Given a universe U, a *boolean partition* of U is a finite set of boolean functions $\{h_1, h_2, \ldots, h_n\}$ defined over U, satisfying the following conditions:

- *Exhaustiveness*: $\bigvee_{i=1}^{n} h_i = 1$. Thus the pointwise maximum of the h_i's is constantly equal to 1.
- *Incompatibility*: $\forall\ i, j \in \{1, 2, \ldots, n\}, i \neq j\ :\ h_i \wedge h_j = 0$. Thus the pointwise minimum of h_i and h_j is 0.

Sometimes one also assumes that no h_i coincides with the zero constant function (*nontriviality*).[1]

Several definitions of nonboolean partition have been proposed in the literature. According to Ruspini [32], a finite set of $[0, 1]$-valued functions $\{h_1, h_2, \ldots, h_n\}$ over U is a partition if and only if

$$h_1 + h_2 + \ldots + h_n = 1. \tag{1}$$

In the special case when the functions h_i are $\{0, 1\}$-valued, we recover boolean partitions (possibly with zero functions).

[1] From the set-theoretical point of view, we can equivalently state the following conditions, for the sets $u_1, u_2, \ldots, u_n \subseteq U$ corresponding to the given boolean functions:

- $\bigcup_{i=1}^{n} u_i = U$;
- $\forall\ i, j \in \{1, 2, \ldots, n\}, i \neq j\ :\ u_i \cap u_j = \emptyset$;
- $\forall\ i \in \{1, 2, \ldots, n\}\ :\ u_i \neq \emptyset$.

Identifying boolean formulas φ_i and their associated boolean functions h_i, Ruspini's condition can be expressed in the classical propositional calculus by the familiar formula

$$(\varphi_1 \vee \varphi_2 \vee \ldots \vee \varphi_n) \wedge \bigwedge_{i,j,i \leq j} \neg(\varphi_i \wedge \varphi_j), \tag{2}$$

where $\vee$ denotes boolean disjunction and $\wedge$ ($\bigwedge$) denotes boolean conjunction.

In the infinite-valued calculus of Łukasiewicz and in its algebraic counterparts, known as Chang's MV-algebras, the notion of partition is well developed. We refer to [5] for a comprehensive account on Łukasiewicz logic and MV-algebras. Here we give a brief review of the main definitions and results to be used in the sequel.

3.2 Łukasiewicz logic and MV-algebras

The infinite-valued calculus of Łukasiewicz is the generalization of the boolean propositional calculus where truth values range over the unit real interval $[0,1]$, and negation $\neg$ and disjunction $\oplus$ are interpreted as follows:

$$\neg x \stackrel{\text{def}}{=} 1 - x, \tag{3}$$

$$x \oplus y \stackrel{\text{def}}{=} \min(1, x + y), \tag{4}$$

for all $x, y \in [0,1]$. Restricting to the two-element set $\{0,1\}$, one recovers the familiar boolean calculus and algebra.

Writing $x \Rightarrow y$ instead of $\neg x \oplus y$, the axioms for this logic are the following, where φ, ρ and τ denote arbitrary formulas:

1. $\varphi \Rightarrow (\rho \Rightarrow \varphi)$;
2. $(\varphi \Rightarrow \rho) \Rightarrow ((\rho \Rightarrow \tau) \Rightarrow (\varphi \Rightarrow \tau))$;
3. $((\varphi \Rightarrow \rho) \Rightarrow \rho) \Rightarrow ((\rho \Rightarrow \varphi) \Rightarrow \varphi)$;
4. $(\neg\varphi \Rightarrow \neg\rho) \Rightarrow (\rho \Rightarrow \varphi)$.

Given a set Φ of formulas, a *proof* from Φ is a finite string of formulas $\varphi_1, \ldots, \varphi_n$ such that, for each $i \in \{1, \ldots, n\}$, either φ_i is an axiom, or $\varphi_i \in \Phi$, or φ_i follows by *modus ponens* from φ_j and φ_k, i.e. φ_k coincides with the formula $\varphi_j \Rightarrow \varphi_i$, with $j, k \in \{1, \ldots, i-1\}$. An *evaluation* is a function α from the set of formulas to $[0,1]$, interpreting operation symbols as in (3) and (4), and letting the variables represent identity functions in the usual way. The problem of characterizing all functions admitting a representation by Łukasiewicz formulas was solved by McNaughton [20].

Definition 1 (McNaughton function) *A function* $f\colon [0,1]^n \to [0,1]$ *is a* McNaughton function *if and only if it is continuous, piecewise linear and each piece has integer coefficients.*

Let $\mathcal{M}_n$ be the set of McNaughton functions of n variables. The constant functions $\mathbf{1}$ and $\mathbf{0}$ belong to $\mathcal{M}_n$. Given $f, g \in \mathcal{M}_n$, define $f \oplus g = \min(f+g, \mathbf{1})$ (pointwise minimum and addition) and $\neg f = \mathbf{1} - f$ (pointwise subtraction). It is easy to see that $f \oplus g, \neg f \in \mathcal{M}_n$. Let Φ_n be the set of Łukasiewicz formulas over n variables and define a function $\Xi: \Phi_n \to \mathcal{M}_n$, associating to any such formula τ the McNaughton function $\Xi(\tau)$ obtained by evaluating variables over the unit interval $[0,1]$ and interpreting operation symbols $\oplus$ and $\neg$ as above. McNaughton proved the following result:

Theorem 1 (McNaughton's theorem) *For each function $f \in \mathcal{M}_n$ there exists a formula $\tau \in \Phi_n$ such that $\Xi(\tau) = f$.*

Introduced by Chang [3], MV-algebras stand to Łukasiewicz propositional logic as boolean algebras stand to the classical propositional calculus. An *MV-algebra* is a structure $A = (A, 0, \oplus, \neg)$ where $\oplus$ is an associative and commutative operation with neutral element 0, and $\neg$ is an operation such that:

- $\neg\neg x = x$,
- $x \oplus \neg 0 = \neg 0$,
- $\neg(\neg x \oplus y) \oplus y = \neg(\neg y \oplus x) \oplus x$.

Chang's completeness theorem [4], [5] states:

Theorem 2 (Completeness theorem) *An equation holds in every MV-algebra if and only if it holds in the MV-algebra $[0,1]$ equipped with negation $\neg x = 1 - x$ and disjunction $x \oplus y = \min(1, x+y)$.*

Definition 2 (Homomorphism and semisimple MV-algebra) *Let A and B be MV-algebras. A function $g: A \to B$ is a* homomorphism *if and only if it satisfies the following conditions, for each $x, y \in A$:*

- $g(0) = 0$,
- $g(x \oplus y) = g(x) \oplus g(y)$,
- $g(\neg x) = \neg g(x)$.

When $B = [0,1]$ the homomorphism g is said to be real. *An MV-algebra A is* semisimple *if and only if for every $x \in A$, with $x \neq 0$, there is a real homomorphism g such that $g(x) \neq 0$.*

The class of semisimple MV-algebras is important for applications. Chang proved the following theorem ([4]):

Theorem 3 (Representation theorem) *Up to isomorphism, every semisimple MV-algebra A is an algebra of continuous $[0,1]$-valued functions over some compact Hausdorff space X, the unit element $1 \in A$ coinciding with the constant function 1 over X.*

Since boolean algebras are a particular case of semisimple MV-algebras, this theorem generalizes *Stone's representation theorem* stating that each boolean algebra is isomorphic to an algebra of continuous $\{0,1\}$-valued functions over a totally disconnected compact Hausdorff space.

Definition 3 (Free MV-algebra) *An MV-algebra A with a distinguished subset Y of elements is said to be* free over (the generating set) Y *if and only if for every MV-algebra B and every function $g: Y \to B$, g can be uniquely extended to a homomorphism $\tilde{g}$ of A into B.*

As a consequence of Chang's completeness theorem, a free MV-algebra with free generators can be described as an MV-algebra of McNaughton functions. Hence, McNaughton functions stand to MV-algebras as boolean functions stand to boolean algebras.

4 MV-partitions

By the main theorem of [22], up to categorical equivalence, MV-algebras are the same as lattice-ordered abelian groups with strong unit. As a consequence, every MV-algebra has a unique, genuine group-theoretical addition $+$, and one can unambiguously express the fact that in an MV-algebra A "linearly independent elements sum up to one". This condition simultaneously generalizes all three conditions (exhaustiveness, incompatibility, nontriviality) of boolean partitions [25]. As is well known, the Łukasiewicz disjunction '$\oplus$' precisely expresses truncated addition and, again identifying functions and formulas, whenever $h_1 + h_2 + \ldots + h_n = 1$, then Łukasiewicz disjunctions actually coincide with sums [27]. By [22, 3.2-3] the following formula $\psi(\varphi_1, \varphi_2, \ldots, \varphi_n)$ expresses Ruspini's condition using only the Łukasiewicz connectives:

$$\varphi_1 \oplus \ldots \oplus \varphi_{i-1} \oplus \varphi_{i+1} \oplus \ldots \oplus \varphi_n = \neg\varphi_i \ (i \in \{1, 2, \ldots, n\}). \qquad (5)$$

This formula generalizes the above boolean formula (2). As well as formula (2) cannot express that each element is nonzero, this formula doesn't capture the linear independence of the elements.

While the paper [25] gives an abstract MV-algebraic definition of nonboolean partition, we are only interested in the special case when the algebra A is semisimple. In our present context, when only MV-algebras of $[0,1]$-valued functions are considered, Mundici's definition [25] can be specialized as follows:

Definition 4 (Concrete MV-partition) *Let U be a universe. A* concrete MV-partition[2] *of U is a finite set $\Pi = \{h_1, h_2, \ldots, h_n\}$ of $[0,1]$-valued functions over U, together with a function assigning to each h_i an integer $m_i \geq 1$ ($i \in \{1, 2, \ldots, n\}$), satisfying the following conditions:*

[2] We shall henceforth omit the adjective "concrete". There will not be any danger of confusion with the more general definition of MV-partitions given by Mundici.

1. $m_1 h_1 + m_2 h_2 + \ldots + m_n h_n = \sum_{i=1}^{n} m_i h_i = 1$,
2. *the set* $\{h_1, h_2, \ldots, h_n\}$ *(equivalently, the set* $\{m_1 h_1, m_2 h_2, \ldots, m_n h_n\}$*) is linearly independent in the* rational *vector space of all real-valued functions over* U*: in other words, whenever* $0 = \lambda_1 h_1 + \lambda_2 h_2 + \ldots + \lambda_n h_n$ *with integer coefficients* λ_i *then* $\lambda_i = 0$ *for all* $i = 1, \ldots, n$.

Each element h_i is called a *miniblock*, while $m_i h_i$ is called a *block* of the MV-partition. The integers m_i are the *multiplicities*; by linear independence, they are uniquely determined. In the particular case when A is a boolean algebra, we have the usual definition of partition with all the multiplicities equal to 1 and no void component. Membership functions are tacitly understood as blocks.

We shall first prove that partitions satisfying Definition 4 automatically satisfy the properties of coverage and distinction:

Proposition 4 *Let* Π *be an MV-partition* $\{h_1, h_2, \ldots, h_n\}$ *of* U*. Then*

1. $\forall\, x \in U\ \exists\, i \in \{1, 2, \ldots, n\}\ :\ h_i(x) > 0$.
2. $\forall\, i, j \in \{1, 2, \ldots, n\}, i \neq j\ :\ h_i \neq h_j$.

PROOF. Immediate by the assumed linear independence of the miniblocks. □

4.1 Irredundant MV-partitions

To relate MV-partitions with the normality condition, we must strengthen Definition 4 by adding an *irredundancy* condition. As observed in [18], while different definitions of nonboolean partitions are many-valued generalizations of the boolean conditions, "it may happen that one element of a fuzzy partition is (fully or partly) a subset of the union of the other elements, which means that the information contained in this element is somewhat redundant, as it is (fully or partly) available from the other elements".

Intuitively, the following definition is to prevent any miniblock from being contained in a positive linear combination of the others:

Definition 5 (Irredundant MV-partition) *Given a universe* U*, an* irredundant MV-partition *of* U *is an MV-partition* $\Pi = \{h_1, h_2, \ldots, h_n\}$ *of* U*, with multiplicity* m_i *associated to each* h_i*, satisfying the following additional condition:*

3. *For every* $i \in I = \{1, 2, \ldots, n\}$ *there do not exist integer coefficients* $\lambda_j \geq 0$ *with* $j \in I \backslash \{i\}$ *such that* $h_i \leq \sum_{j \in I \backslash \{i\}} \lambda_j h_j$.

A further analysis of MV-partitions can be made under increasingly restrictive topological conditions on the common domain, domΠ, of the miniblocks in Π. We start with the assumption that domΠ is compact Hausdorff. Recall that a compact Hausdorff space is a space in which any two distinct points have disjoint open neighbourhoods, and every open cover has a finite subcover. By Stone's theorem, our present analysis also applies to the special case of boolean partitions.

Proposition 5 *Let the universe U be compact and Hausdorff and let Π be a set of continuous $[0,1]$-valued functions $\{h_1, h_2, \ldots, h_n\}$ over U, with multiplicities $\{m_1, m_2, \ldots, m_n\}$, such that $\sum_{i=1}^{n} m_i h_i = 1$. Then the following conditions are equivalent:*

(i) Π is an irredundant MV-partition;
(ii) Π is normal.

PROOF. $(ii) \Rightarrow (i)$ We shall prove that condition (ii) implies linear independence and irredundancy of Π. Let $x_i \in U$ be a point such that $h_i(x_i) = 1$. As a matter of fact, if $0 = \lambda_1 h_1 + \lambda_2 h_2 + \ldots + \lambda_n h_n$, then only h_i is nonzero at x_i $(i = 1, 2, \ldots, n)$. Then $\lambda_i = 0$ and Π is linear independent.

Fix $j \in \{1, 2, \ldots, n\}$. Since $\sum_{i=1}^{n} m_i h_i = 1$, $\sum_{i=1, i \neq j}^{n} m_i h_i$ is equal to 0 at x_j. Then $h_j(x_j) > \sum_{i=1, i \neq j}^{n} m_i h_i(x_j)$ and no positive linear combination of the h_i $(i \neq j)$ can dominate h_j at x_j, whence irredundancy immediately follows.

$(i) \Rightarrow (ii)$ By way of contradiction, assume $\exists\, h_i \in \Pi\, \forall\, x \in U \ : \ m_i h_i(x) < 1$. Fix $\overline{x} \in U$.
Then $\exists\, k \ : \ m_i h_i(\overline{x}) < k \sum_{j=1, j \neq i}^{n} m_j h_j(\overline{x}) = \sum_{j=1, j \neq i}^{n} k m_j h_j(\overline{x})$. Then there exist coefficients $\lambda_j^{\overline{x}}$ such that $h_i(\overline{x}) < \sum_{j=1, j \neq i}^{n} \lambda_j^{\overline{x}} h_j(\overline{x})$.

Since U is a compact Hausdorff space, for every $\overline{x}$ there is an open set $B(\overline{x})$ such that $\forall\, x \in B(\overline{x}) \ : \ h_i(x) < \sum_{j=1, j \neq i}^{n} \lambda_j^{\overline{x}} h_j(x)$ and $\bigcup B(\overline{x})$ covers U. Let λ_j equal to $\max_{\overline{x}} \lambda_j^{\overline{x}}$. Then $h_i < \sum_{j=1, j \neq i}^{n} \lambda_j h_j$, which is a contradiction. □

4.2 Connectedness vs. unimodal MV-partitions

In most non-boolean applications, domΠ satisfies additional connectedness properties. To avoid pedantic definitional details, we shall simply assume that domΠ is homeomorphic to some n-cube $[0,1]^n$.

Definition 6 (Unimodal MV-partition) *Let Π be a set of continuous $[0,1]$-valued functions $\{h_1, h_2, \ldots, h_n\}$, defined on a space homeomorphic to $[0,1]^n$, and forming an MV-partition. We say that Π is* unimodal *if and only if for every $i \in \{1, 2, \ldots, n\}$ and for every $\theta \in [0,1]$ the* upper level set[3]

$$A(h_i, \theta) = \{x \in [0,1]^n \ : \ h_i(x) \geq \theta\}$$

is (nonempty and) homeomorphic to some closed m-cube $(0 \leq m \leq n)$.

[3] also called *θ-cut* in fuzzy sets theory

Since by assumption $A(h_i, 1) \neq \emptyset$ then Π is normal, hence, by Proposition 5, it is irredundant.

In the particular case when $n = 1$, unimodality means that each $A(h_i, \theta)$ is either a closed interval or a single point. Thus our present definition, in this case, generalizes Khinchine's definition [17].

In many fuzzy models the membership functions of a single variable defined over a closed interval of $\mathbb{R}$ turn out to have an overlap of $\frac{1}{2}$ (see below). See [30] for some motivation.

Proposition 6 *Let Π be a unimodal MV-partition $\{h_1, h_2, \ldots, h_n\}$ formed by continuous $[0,1]$-valued functions over a closed interval $U = [a, b]$ of $\mathbb{R}$, with multiplicity m_i associated to each h_i. Then the overlap of the set $\{m_1 h_1, m_2 h_2, \ldots, m_n h_n\}$ is $\frac{1}{2}$. In other words, the maximum value of the intersection of any two overlapping functions $m_i h_i$ and $m_j h_j$ (i.e., $m_i h_i \wedge m_j h_j \neq 0$) is equal to $\frac{1}{2}$.*

PROOF. For every $i \in \{1, \ldots, n\}$ let $[a_i, b_i]$ the closed interval such that $\forall x \in [a_i, b_i] : m_i h_i(x) = 1$. Without loss of generality, we suppose $\forall i \in \{1, \ldots, n-1\} : b_i < a_{i+1}$. These intervals are disjoint and $\forall x \in [a_i, b_i] \forall j \neq i : m_j h_j = 0$. Then $\forall x \in (b_i, a_{i+1}) : m_{i+1} h_{i+1}(x) = 1 - m_i h_i(x)$. Hence, any two consecutive functions overlap and the maximum value of their intersection is $\frac{1}{2}$. □

4.3 Schauder bases

For his constructive proof of McNaughton's theorem, Mundici in [23] introduced a significant class of MV-partitions in free MV-algebras, known as Schauder bases. Some terminology from rational polyhedral geometry is needed to define this class. See [10] for more details.

A finite set of rational points $\{v_0, \ldots, v_m\} \subseteq \mathbb{Q}^n$ is *affinely independent* if and only if the set of rational vectors $\{v_1 - v_0, \ldots, v_m - v_0\}$ is linearly independent over $\mathbb{R}$. A *rational simplex* is the convex hull of a finite set of affinely independent points in $\mathbb{Q}^n$, its *vertices.* A *face* of a simplex is the convex hull of a subset of its vertices. A finite set $\mathcal{S}$ of rational simplices is a *rational simplicial complex* if and only if together with each simplex it contains all its faces, and any two simplices intersect in a common face. The *support* of $\mathcal{S}$ is the union of all its simplices, denoted $|\mathcal{S}|$.

Let $\mathcal{S}$ be a rational simplicial complex in $[0,1]^n$. If $|\mathcal{S}| = [0,1]^n$, we call $\mathcal{S}$ *complete.* Let v be a vertex of $\mathcal{S}$. Then $v = (r_1/s_1, \ldots, r_n/s_n)$ for uniquely determined positive integers r_i, s_i such that $s_i \neq 0$ and r_i and s_i are relatively prime. The least common multiple of the set $\{s_1, \ldots, s_n\}$ is said to be the *denominator* of v, written $\mathrm{den}(v)$. The *Schauder hat* at v in $\mathcal{S}$ is the unique continuous piecewise-linear function $h_v \colon [0,1]^n \to [0,1]$ which attains rational value $1/\mathrm{den}(v)$ at v, is equal to 0 in any other vertex of $\mathcal{S}$, and is linear on each simplex of $\mathcal{S}$. In general, h_v has linear pieces with non-integer coefficients. This is unfortunate, because it implies that h_v is not a

McNaughton function. To overcome this problem one can consider the following important notion. Let S be an n-dimensional simplex in $\mathcal{S}$ with vertices $v_0, \ldots, v_n$. Let us write $v_j = (r_{j1}/s_{j1}, \ldots, r_{jn}/s_{jn})$, r_{ji} and s_{ji} relatively prime positive integers, $s_{ji} \neq 0$. Passing to homogeneous coordinates, we obtain $v_j^{hom} = (\bar{r}_{j1}, \ldots, \bar{r}_{jn}, \mathrm{den}(v_j)) \in \mathbb{Z}^{n+1}$, where $\bar{r}_{ji} = \mathrm{den}(v_j) r_{ji}/s_{ji}$. We then say that S is *unimodular* if and only if the $(n+1) \times (n+1)$ matrix whose jth row coincides with v_j^{hom} has its determinant equal to ± 1. A complete rational simplicial complex $\mathcal{S}$ is *unimodular* if and only if all its n-simplices are unimodular. In this case, we call $\mathcal{S}$ a *unimodular triangulation* of $[0,1]^n$.

Unimodularity of $\mathcal{S}$ is necessary and sufficient for each Schauder hat of $\mathcal{S}$ to be an element of $\mathcal{M}_n$. As the reader will recall, the latter denotes the set of McNaughton functions of n variables. We call the set of Schauder hats at the vertices of $\mathcal{S}$ a *Schauder basis* of $\mathcal{M}_n$, and we denote it by $\mathbf{H}_{\mathcal{S}}$. Direct inspection, using Proposition 5, gives the following:

Proposition 7 *For every unimodular triangulation $\mathcal{S}$ of $[0,1]^n$, the Schauder basis $\mathbf{H}_{\mathcal{S}}$ is an irredundant MV-partition of $[0,1]^n$.*

A tedious but routine argument shows:

Proposition 8 *For every unimodular triangulation $\mathcal{S}$ of $[0,1]^n$, the Schauder basis $\mathbf{H}_{\mathcal{S}}$ is a unimodal MV-partition of $[0,1]^n$.*

4.4 Examples

In this subsection we give some examples of sets of membership functions on a universe. We show particular cases of partitions to stress the importance to give precise definitions. In every partition shown, all the multiplicities are equal to 1. The set of membership functions in Figure 1 is not a partition, while that in Figure 2 satisfies the Ruspini's Condition 1. Figure 3 shows an MV-partition. This partition is not irredundant because one of the membership functions is contained in another one. Also in Figure 2 one block is completely contained in another. But the partition doesn't satisfy the linear independence property. The additional condition of irredundancy (normality) is shown in Figure 4 and 5. The difference between the two figures is that the latter is composed by unimodal blocks and, then, it has the properties stated in Proposition 6. We are aware that a partition as shown in Figure 4 is not very common or easy to obtain even when the set of membership functions is the result of some optimization process. But, in general, it can happen that membership functions have different local maxima and it may be the case that they satisfy Definition 5.

Although the examples given so far refer to a one-dimensional universe, the definitions presented in the previous sections are general; they can be applied to every multidimensional domain U (the same fact is true for all the results presented except for Proposition 6). Figure 6 shows a unimodal MV-partition on a two-dimensional universe.

Note that the partitions shown in Figure 5 and 6 are two *normalized* Schauder bases. An algorithm to get them and their multiplicities can be found in Section 5.

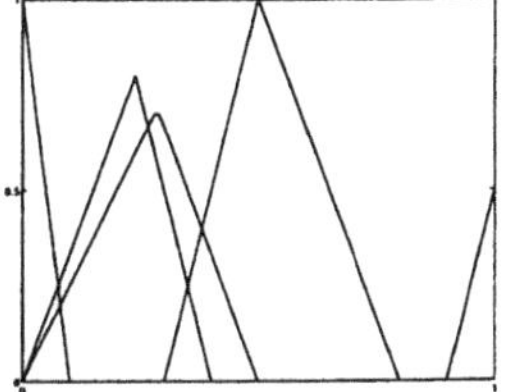

Fig. 1. A set of membership functions that is not a partition

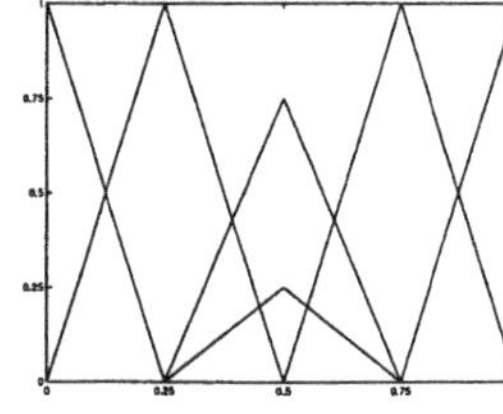

Fig. 2. A Ruspini partition

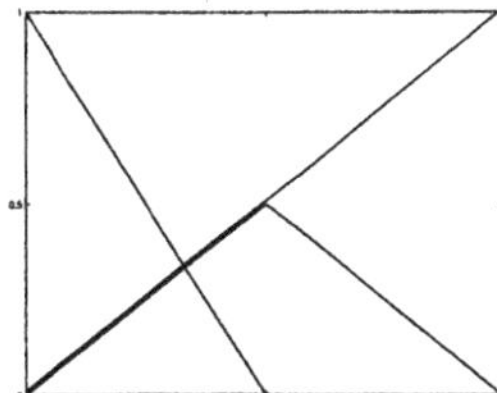

Fig. 3. An MV-partition

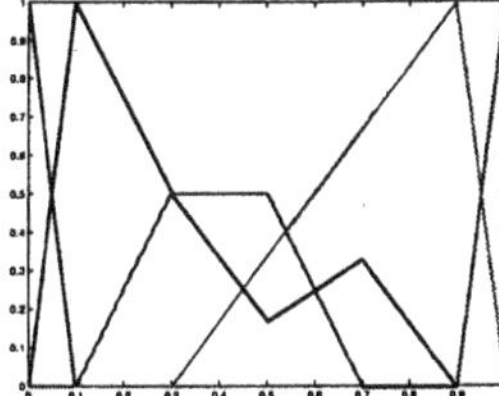

Fig. 4. An irredundant MV-partition

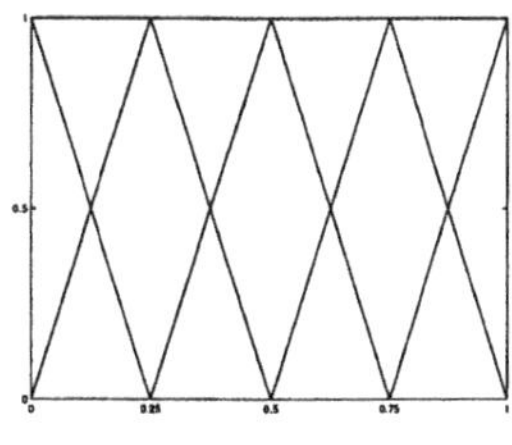

Fig. 5. A unimodal MV-partition

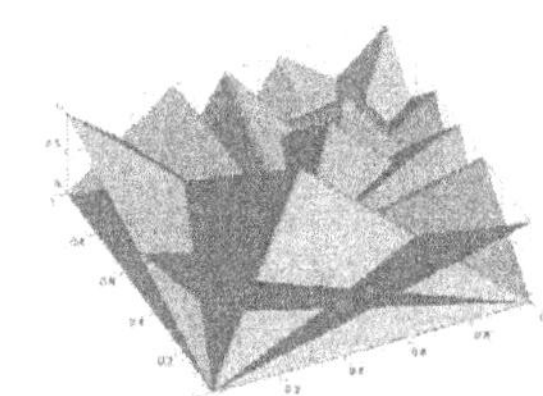

Fig. 6. A unimodal MV-partition (two-dimensional)

5 Refinement of partitions

A partition P is a *refinement* of a partition Q if each miniblock of Q is a sum of miniblocks of P. This generalizes the classical boolean notion. In this section we show that any two MV-partitions on the same universe have a joint refinement, thus generalizing another property of boolean partitions. As a consequence, for any refinement process eventually leading to "optimal" partitions, if we use MV-partitions, then our starting point is immaterial. This fact may be relevant for certain Soft Computing techniques where dependence on the initial conditions may be undesirable. Needless to say, joint refinements are necessary for the fusion of different fuzzy representations of the same knowledge base (for example, when several different imprecise descriptions of the same control function are given by various human experts).

5.1 Refinement of MV-partitions

The study of MV-partitions (and then irredundant MV-partitions) is only at the beginning. Refinement and joint refinement are key requisites for any good generalization of the boolean definition. In this sense, some motivations for the use of the MV-partition theory can be given starting from the boolean theory and its algebraic properties.

Definition 7 (Boolean Refinement) *Given a universe U, for any two boolean partitions of U, Π and Π', we say that Π' is a* refinement *of Π if and only if each block of Π is a join of blocks of Π'.*

While the poset of boolean partitions equipped with refinement forms a lattice, this is no longer the case for Ruspini's partitions. In fact, in this case refinement is not even a partial order but merely a preorder. This is because the multiplicities are not uniquely determined (for instance, if in $\mathbb{Q}$ we select 1 as a strong order unit, then 1/2 and 1/4 are blocks of a Ruspini's partition with multiplicites 1 and 2, respectively. But 1/4 can be the single block of a partition with multiplicity 4.) However, the set of MV-partitions equipped with refinement actually is a poset. Moreover, it is possible to prove that this poset is lower directed, that is, any two elements have a lower bound — MV-partitions admit joint refinement. Formally ([25]):

Definition 8 (Refinement) *Given a universe U, for any two MV-partitions of U, $\Pi = \{h_1, h_2, \ldots, h_n\}$ with multiplicities m_i associated to each h_i ($i = 1, \ldots, n$), and $\Pi' = \{g_1, g_2, \ldots, g_p\}$ with multiplicities e_k associated to each g_i ($k = 1, \ldots, p$), we say that Π' is a* refinement *of Π if and only if each h_i is obtainable as a (necessarily unique) linear combination of $\{g_1, g_2, \ldots, g_p\}$ with integer coefficients ≥ 0.*

Definition 9 (Joint Refinement) *Given a universe U, for any two MV-partitions of U, Π_1 and Π_2 we say that Π' is a* joint refinement *of Π_1 and Π_2 if and only if Π' is simultaneously a refinement of Π_1 and of Π_2.*

As shown in [24], the joint refinability of any two MV-partitions is a corollary of the *ultrasimplicial property* of lattice-ordered abelian groups. The latter property was established by Marra [19], and is a deep algebraic result. Summing up,

Theorem 9 *Given a universe U, for any two MV-partitions Π_1 and Π_2 of U, a joint refinement Π' of Π_1 and Π_2 always exists.*

5.2 Refinement of Schauder bases

We now consider refinement when MV-partitions are Schauder bases. In this special case we observe that the definition of refinement can be reformulated giving a constructive technique, and, moreover, particular properties can be proved.

Definition 10 (Refinement of a Schauder basis) *Let $\mathbf{H}$ be a Schauder basis for some (necessarily unique) unimodular triangulation $\mathcal{S}$ of $[0,1]^n$. We say that $\mathbf{H}'$ is* one-step star refinement *of $\mathbf{H}$ if and only if it is obtained from $\mathbf{H}$ as follows:*

1. *Pick a subset of Schauder hats* $\mathcal{H} = \{h_1, \ldots, h_q\} \subseteq \mathbf{H}$ *and let* $h_{\mathcal{H}} = h_1 \wedge \ldots \wedge h_q$;
2. *For each* $j = 1, \ldots, q$ *replace* h_j *by* $h_j \odot \neg h_{\mathcal{H}} = \neg(\neg h_j \oplus h_{\mathcal{H}})$;
3. *if* $h_{\mathcal{H}} \neq 0$ *put* $h_{\mathcal{H}}$ *in* $\mathbf{H}'$.

We say that $\mathbf{H}^*$ *is a* star refinement *of* $\mathbf{H}$ *if and only if it is obtained from* $\mathbf{H}$ *via a path* $\mathbf{H}^0, \mathbf{H}^1, \ldots, \mathbf{H}^t = \mathbf{H}^*$ *(for arbitrary* $t \in \mathbb{N}$*), where each* $\mathbf{H}^i$ *is a one-step star refinement of* $\mathbf{H}^{i-1}$. *If* $q = 2$ *and* $h_{\mathcal{H}} \neq 0$, *we speak of* binary starring.

It is possible to prove the following theorem:

Theorem 10 *For any two Schauder bases* $\mathbf{H}$ *and* $\mathbf{L}$ *there is a star refinement* $\mathbf{H}^*$ *such that every element of* $\mathbf{L}$ *is a (truncated) sum of elements of* $\mathbf{H}^*$. *Moreover, all one-step star refinements leading from* $\mathbf{H}$ *to* $\mathbf{H}^*$ *may be assumed to be binary.*

This is a consequence of this crucial theorem [6] in toric algebraic geometry:

Theorem 11 (De Concini-Procesi Lemma) *Let* $\mathcal{S}_1$ *and* $\mathcal{S}_2$ *be unimodular triangulations of* $[0,1]^n$. *Then there exists a third unimodular triangulation* $\mathcal{T}$ *such that*

- $\mathcal{T}$ *is obtained from* $\mathcal{S}_1$ *via a finite sequence of binary starrings;*
- $\mathcal{T}$ *refines* $\mathcal{S}_2$.

As an example we now consider the one-dimensional case. Here Schauder hats are continuous piecewise linear functions $h: [0,1] \to [0,1]$. Each linear piece of h is a straight line with integer coefficients. These coefficients are determined by the *Farey partition* only. For every $n = 0, 1, 2, \ldots$, the nth Farey partition $\mathcal{F}_n$ of $[0,1]$ is defined as:

$$\mathcal{F}_0 \stackrel{\text{def}}{=} \{0,1\}, \mathcal{F}_1 \stackrel{\text{def}}{=} \left\{0, \frac{1}{2}, 1\right\}, \mathcal{F}_2 \stackrel{\text{def}}{=} \left\{0, \frac{1}{3}, \frac{1}{2}, \frac{2}{3}, 1\right\}, \ldots$$

Thus, $\mathcal{F}_{n+1}$ is obtained by inserting between any two consecutive elements $\frac{a}{b}$ and $\frac{c}{d}$ of $\mathcal{F}_n$ their *mediant* $\frac{a+c}{b+d}$, with $0 = \frac{0}{1}$ and $1 = \frac{1}{1}$. In most texts in number theory, one includes in $\mathcal{F}_n$ only those rationals whose denominator is less than or equal to n, but this restriction is immaterial for our purpose here. As proved by Cauchy in 1816, following Farey's observations, all fractions in $\mathcal{F}_n$ are automatically in irreducible form, every irreducible fraction $\frac{p}{q} \in [0,1]$ occurs in $\mathcal{F}_{q-1}$, and $\frac{a}{b} \leq \frac{a+c}{b+d} \leq \frac{c}{d}$. Moreover, every interval $[\frac{a}{b}, \frac{c}{d}]$, determined by any two consecutive fraction $\frac{a}{b}$ and $\frac{c}{d}$ in $\mathcal{F}_n$ has the unimodularity property (see Section 4).

The elements of $\mathcal{F}_n$ can be displayed in an increasing order as follows: $0 < \alpha < \ldots < \gamma < \delta = \frac{c}{d} < \varepsilon < \ldots < \omega < 1$. The graph of a Schauder hat of $\mathcal{F}_n$

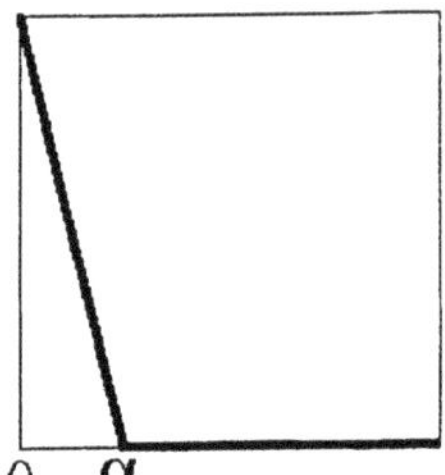

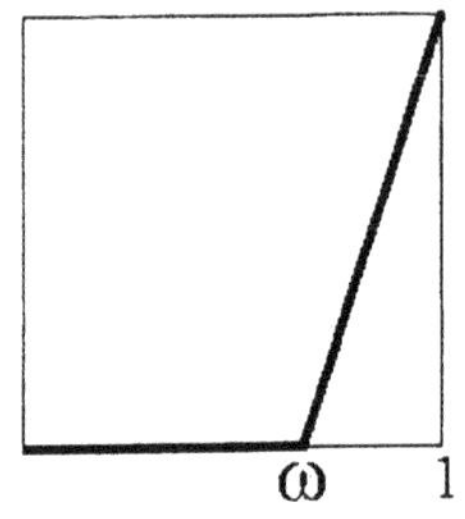

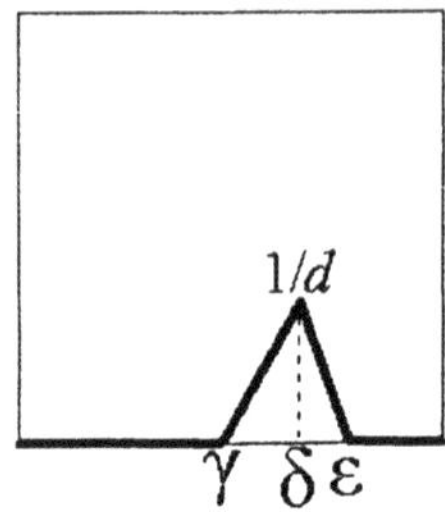

Fig. 7. One-dimensional Schauder hats

at δ consists of the four segments joining the points $(0,0), (\gamma,0), (\delta,\frac{1}{d}), (\varepsilon,0)$ and $(1,0)$ (see figure 7). The point $(\delta,\frac{1}{d})$ is the *vertex* of the hat and d is its *multiplicity* l_δ. The function $l_\delta h_\delta$ is called the normalized hat of $\mathcal{F}_n$ at δ. For every $n = 0,1,2,\ldots$, we shall denote by $\mathcal{S}_n$ the naturally ordered sequence $\{h_1,\ldots,h_u\}$ of all Schauder hats of $\mathcal{F}_n$, with $u = 2^n + 1$. The operation of inserting between any two consecutive elements h_i, h_{i+1} of $\mathcal{S}_n$ the mediant hat is the starring. The following propositions formalize the construction of $\mathcal{S}_{n+1}$ from $\mathcal{S}_n$:

Proposition 12 *Let $h_1,\ldots,h_u$ be the hats of $\mathcal{S}_n$, in their natural order, and with their respective multiplicities $\mu_1,\ldots,\mu_u$. Let $k_1,\ldots,k_{2u-1}$ be the hats of $\mathcal{S}_{n+1}$, with their respective multiplicities $\xi_1,\ldots,\xi_{2u-1}$. Then we have:*

$$\begin{aligned}
k_1 &= h_1 - (h_1 \wedge h_2)\\
&\quad \text{with } \xi_1 = \mu_1 = 1;\\
k_{2u-1} &= h_u - (h_{u-1} \wedge h_u)\\
&\quad \text{with } \xi_{2u-1} = \mu_u = 1;\\
k_{2i} &= h_i \wedge h_{i+1}\\
&\quad \text{with } \xi_{2i} = \mu_i + \mu_{i+1}, \text{ for every } i = 1,2,\ldots,u-1;\\
k_{2i-1} &= h_i - (h_i \wedge (h_{i-1} \vee h_{i+1}))\\
&\quad \text{with } \xi_{2i-1} = \mu_i, \text{ for every } i = 2,3,\ldots,u-1.
\end{aligned}$$

Proposition 13 *Each of the extremal hats h_1 e h_u of $\mathcal{S}_n$ is the sum of two hats of $\mathcal{S}_{n+1}$, while all the remaining hats of $\mathcal{S}_n$ are obtainable as a sum of three hats of $\mathcal{S}_{n+1}$.*

Figure 8 shows the sequences of Schauder hats $\mathcal{S}_0$,$\mathcal{S}_1$,$\mathcal{S}_2$,$\mathcal{S}_3$ obtained by starring.

Note that we have considered *complete* sequences of Schauder hats, in the sense that all the mediant hats are inserted getting to $\mathcal{S}_{n+1}$ from $\mathcal{S}_n$. For practical purpose, one only inserts Farey mediants when needed.

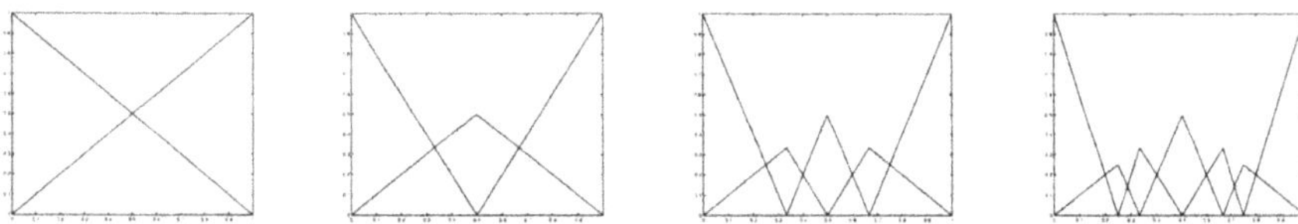

Fig. 8. The sequences of Schauder hats $\mathcal{S}_0, \mathcal{S}_1, \mathcal{S}_2, \mathcal{S}_3$ (from left to right) obtained by starring

5.3 Refinement of irredundant MV-partitions

One can now naturally ask whether the joint refinement property also holds for MV-partitions satisfying stronger conditions. For irredundant MV-partitions over general (disconnected) universes, joint refinement is not ensured in general. As a matter of fact, the following counterexample shows that, on the three-element universe $U = \{a, b, c\}$, there exist two irredundant MV-partitions having no irredundant refinement (but having, by Theorem 9, some joint refinement).

Counterexample Let $U = \{a, b, c\}$ be a universe. By the notation (u, v, w) we mean a miniblock attaining the value u at a, v at b and w at c. Let $\Pi_1 = \{(1, 0, \varepsilon), (0, 1, 1-\varepsilon)\}$ with unit multiplicities and $\Pi_2 = \{(1, 0, \delta), (0, 1, 1-\delta)\}$ with unit multiplicities (Figures 9 and 10). Suppose ε, δ irrationals and

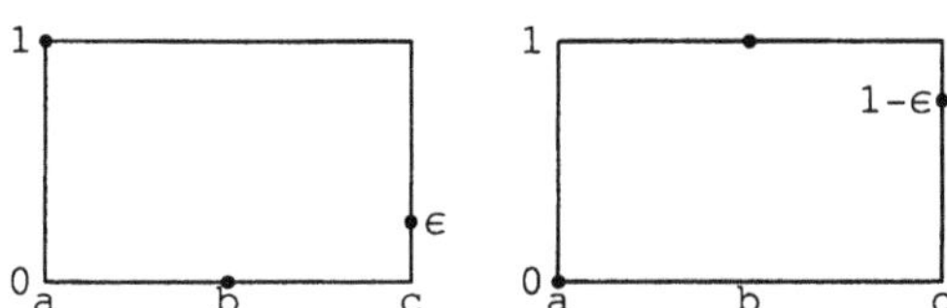

Fig. 9. Irredundant MV-Partition Π_1

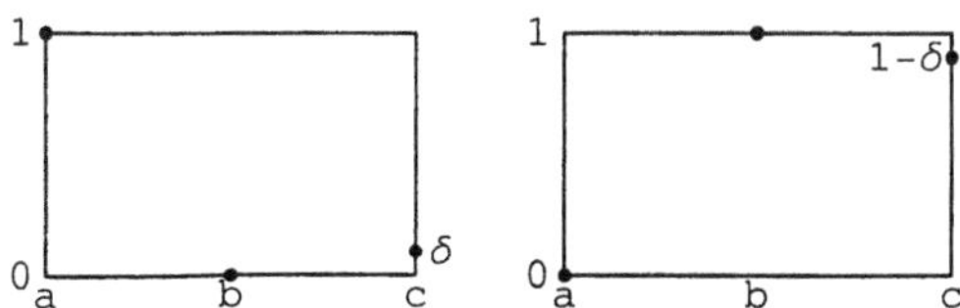

Fig. 10. Irredundant MV-Partition Π_2

$\varepsilon > \delta$. Π_1 and Π_2 are normal MV-partitions but their joint refinement $\Pi' = \{(1, 0, \delta), (0, 1, 1-\varepsilon), (0, 0, \varepsilon - \delta)\}$ with multiplicities $\{1, 1, 1\}$ is not normal (Figure 11). Π' is a joint refinement of Π_1 and Π_2 because $(1, 0, \varepsilon) = (1, 0, \delta)+$

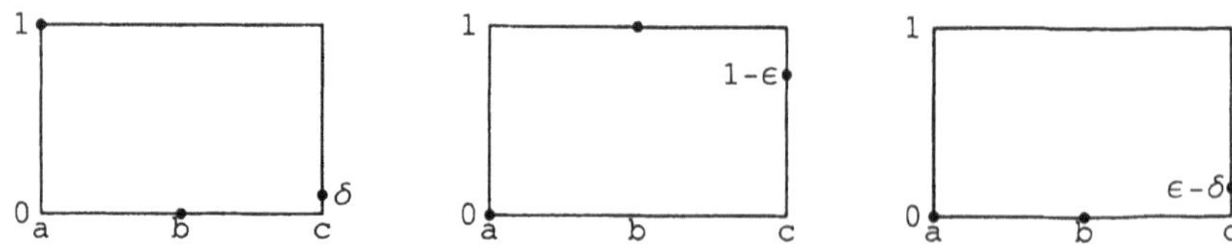

Fig. 11. Joint refinement of Π_1 and Π_2

$(0,0,\varepsilon-\delta)$, $(0,1,1-\delta)=(0,1,1-\varepsilon)+(0,0,\varepsilon-\delta)$ and the remaining blocks are the same as those of Π'. Thus Π' is an MV-partition (the blocks sum up to one, satisfying Ruspini's condition and they are linearly independent), but it is not irredundant: the block $(0,0,\varepsilon-\delta)$ is less than the block $(0,1,1-\varepsilon)$. Evidently, there cannot exist a normal joint refinement.

Since the joint refinement property is important in partition theory, one can naturally ask under which conditions this property holds for irredundant MV-partitions. This investigation is still in progress and we conjecture that these conditions involve a deeper analysis of the unimodality property and its strenghtenings.

Further developments concern the connection of the concept of refinement with some recent contributions about the use of *hierarchical* fuzzy partitions (see for example [14] and [7]).

6 The role of partitions in IF-THEN descriptions

In this section we analyze the role of partition theory in the logical analysis of human experts' descriptions of a system.

Such description usually has the form of a *conjunction* of implications:

$$\begin{aligned} &\text{If } H_1 \text{ then } A_1\\ \text{and }&\text{if } H_2 \text{ then } A_2\\ &\vdots\\ \text{and }&\text{if } H_n \text{ then } A_n. \end{aligned} \tag{6}$$

In boolean logic, under the assumption that the premises $H_1, H_2, \ldots, H_n$ form a partition, expression (6) is equivalent to a *disjunction* of conjunctions as follows:

$$\begin{aligned} \text{Either }& H_1 \text{ and (then) } A_1\\ \text{or }& H_2 \text{ and (then) } A_2\\ &\vdots\\ \text{or }& H_n \text{ and (then) } A_n. \end{aligned} \tag{7}$$

Having to choose between (6) and (7) to represent the expert's description by means of formulas of a predetermined many-valued logic, it is natural to choose a logic where a nonboolean generalization of the notion of partition is well developed. As shown in Section 3, the infinite-valued calculus of Łukasiewicz affords a satisfactory nonboolean extension of the notion of partition. Here, however, formulations (6) and (7) are no longer equivalent. Our preference goes to the 'granular' disjunctive representation (7) because the latter only uses apparently associative and commutative connectives (also satisfying the usual axioms of t-norms and t-conorms [12]). There is little doubt whether the 'Either ... or' disjunction in (7) is to be interpreted as a sum. To see this, suppose for the moment that all conclusions A_i were tautologies (whose effect is the same as "do nothing"). Then, for any reasonable meaning of the conjunction connective "and then", (7) boils down to a formula which is a reformulation of Ruspini's condition:

Either H_1 occurs or H_2 occurs or ... or H_n occurs.

In the previous section, we have seen that the Łukasiewicz disjunction '$\oplus$' is a good choice for the disjunction connective. There remains to choose the conjunction connective 'and (then)' in (7). Some considerations about the most plausible choice of this t-norm can be found in [27] and [1], where the role of distributivity is stressed. A possible and often used t-norm is here multiplication '$\cdot$'. This is so because in the additive group of real numbers with the natural order, multiplication is the only order-preserving commutative associative operation that distributes over addition and that has 1 as the neutral element [11] (see also [26, Proof of Theorem 2.4]). With this choice, the above disjunction of conjunctions acquires the form $\varphi = H_1 \cdot A_1 \oplus H_2 \cdot A_2 \oplus \ldots \oplus H_n \cdot A_n$. The use of MV-logic (and the tensor product) for the formalization of fuzzy modeling has been shown, for example, in [29] and [1].

7 A simple example of constrained learning

In this section we give a simple example of a learning algorithm in which the defining parameters for membership functions must obey MV-partitioning, irredundancy and unimodality constraints. In particular we consider a hybrid neuro-fuzzy system (HNF) [28] with one input variable, where the neural network is a feedforward multilayer perceptron [13] and the associated fuzzy system is Tagaki-Sugeno-Kang[4] [36,35]. We use this system to learn (the fuzzy description of) a function starting from a set of input-output data. Although we restrict our attention to a simple-minded one-dimensional system,

[4] In this paper we do not address the question of how interpretability constraints can affect the accuracy of a fuzzy system. In future works we will analyze this point taking into account less potentially accurate but more interpretable fuzzy systems like the Mamdani one.

the relevance of this section lies in showing how the highly theoretical reasoning of the previous sections can be easily incorporated in common learning algorithms, and in describing the refinement process.

Usually in the literature (see for example [2] and [15]) the synaptic weights of a HNF are the parameters of the input fuzzy sets (in the case of a triangular membership function there are three parameters), and the parameters for the rule consequents. Figure 12 depicts an example of an HNF with two input variables and one output.

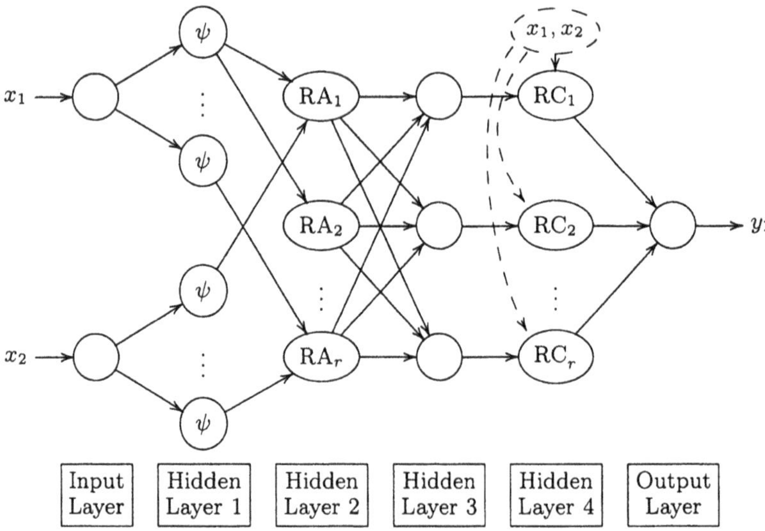

Fig. 12. An example of hybrid neuro-fuzzy system with two input variables

In the example we are going to consider the chosen t-norm and t-conorm are product and sum respectively, and the learning algorithm takes as input two distinct objects: a set of training data and a partition of the input domain composed by a given number of triangular membership functions (Figure 13, right). The chosen learning algorithm is the common backpropagation one (see for example [13]). The only modification consists in taking care of nondifferentiable points of the triangular membership functions (standard backpropagation requires differentiable activation function for the neurons of the network).

In virtue of what has been shown in the previous sections, in this example the conditions guaranteeing that the input fuzzy sets form an irredundant unimodal partition consist in (i) taking as only free parameter of each fuzzy set its center, and (ii) univocally determining the two other parameters from the previous and consequent fuzzy sets for the same input variable. Thus, in this case, the constrained learning algorithm is simpler than the corresponding unconstrained one, because the number of free parameters is reduced to one third. Moreover, since MV-partitions satisfy Ruspini's condition, the HNF needs no more a normalization layer (hidden layer 3 in Figure 12) and its formal representation becomes the following:

$$N(x) = \sum_{i=1}^{r} (a_{i0} + a_{i1}x) \cdot \psi(x, m_{i-1}, m_i, m_{i+1}),$$

where r is the number of the rules (in this trivial case each input fuzzy set is associated to only one rule), ψ is a one-dimensional triangular membership function, for $i = 2, \ldots, r$ m_i is the vertex of the ith input fuzzy set, m_0 is any negative value , $m_1 = 0, m_r = 1$ and m_{r+1} is any value greater than 1, and a_{i0}, a_{i1} are the parameters for the ith rule consequent. The exact value of both m_0 and m_{r+1} are immaterial, because the domain considered for input domain is the interval $[0, 1]$ (consequently the first and last membership functions are 'fake' triangular functions).

We used also a simple a simple pruning algorithm which, at each step of the learning algorithm, cuts down similar fuzzy sets. Since the fuzzy sets are normal and the maximum overlap between two of them is $\frac{1}{2}$, the most part of similarity measures (support overlap, intersection area,...) has little relevance. Then, as a similarity criteria we have taken the distance between the vertices of consecutive membership functions. Let ϵ be a threshold value. If two consecutive fuzzy sets $\psi(x, m_{i-1}, m_i, m_{i+1})$ and $\psi(x, m_i, m_{i+1}, m_{i+2})$ are such that $|m_i - m_{i+1}| \leq \epsilon$, then the pruning algorithm join them together in the fuzzy set $\psi(x, m_{i-1}, \frac{m_i + m_{i+1}}{2}, m_{i+2})$. Consequently even the rules associated to the two original fuzzy sets are joint together in a new one, whose consequent parameters are $\frac{a_{i,0} + a_{i+1,0}}{2}$ and $\frac{a_{i,1} + a_{i+1,1}}{2}$.

Figures 13, 14, 15 show two simple applications of this constrained learning process. The starting point is the random sampling of the function $\sqrt{x}$ with 50 samples (case 1) and 200 samples (case 2). The input partitions

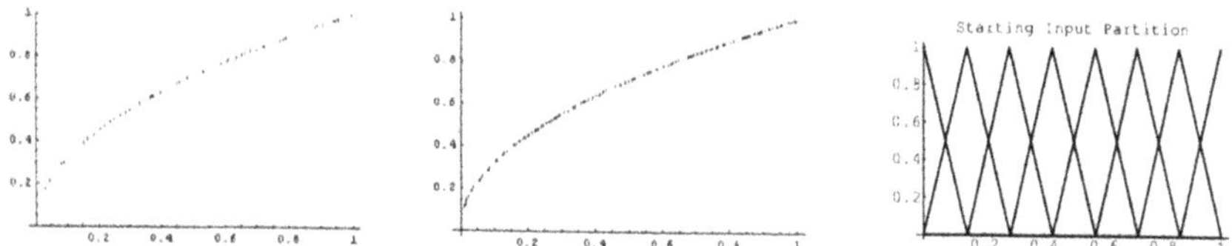

Fig. 13. Sampling of the function to be approximated in case 1 (left) and case 2 (center), and the input partition at the beginning of the learning process (right).

shown in figure 14 can be jointly refined. Using Marra's constructive proof in [19], we produce the joint refinement shown in Figure 14 (on the right). Indeed, every block of the input partitions is a combination of the blocks

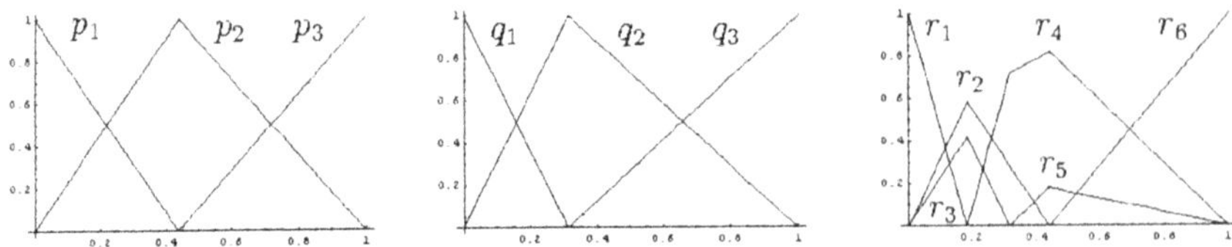

Fig. 14. The input partition at the end of the learning process in case 1 (left) and case 2 (center), and their joint refinement (right)

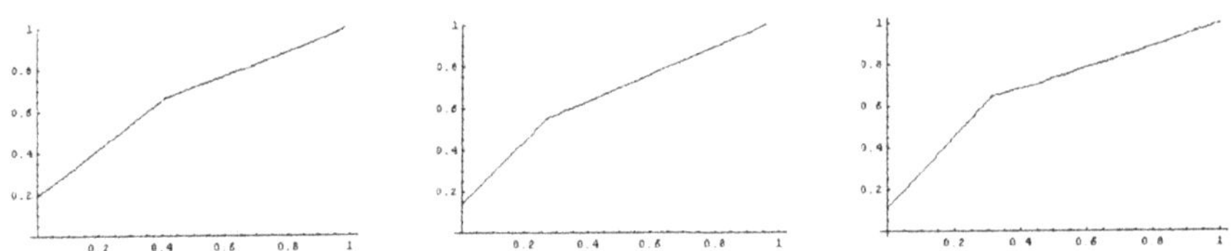

Fig. 15. The output of the trained neural network in case 1 (left) and case 2 (center), and and the output of the neural network with the elements of joint refined partition as input fuzzy sets (right)

of this new set of functions: $p_1 = r_1 + r_2, p_2 = r_3 + r_4 + r_5, p_3 = r_6, q_1 = r_1 + r_3, q_2 = r_2 + r_4, q_3 = r_5 + r_6$ It is then clear that the functions r_i positively span the constant function 1, because our original partitions did it. However, we still need to check that the r_i's are linearly independent, thus yielding an MV-partition. To do this, it is sufficient to proceed in the following way: for every block r_i, let $X_i = \{x_{i_1}, \ldots, x_{i_n}\}$ be the set of *nodes* of the block, i.e. the nondifferentiable points in the domain of the block, and let $X = \bigcup X_i$. Then we consider the matrix in which the ith column is the vector of the evaluations of the block r_i at every $x \in X$. Since the rank of this matrix equals the number of its columns, these vectors are linearly independent. Hence, so are the blocks r_i. As we have shown in Section 5 there is no guarantee to obtain a joint irredundant MV-partition starting from two MV-partitions satisfying this additional condition: the example confirms this, as well as the need of a deeper analysis.

Even though the elementary examples given here are one-dimensional, the only pay off in considering n-dimensional examples is the complication of the algorithmic machinery. Moreover, in our opinion, it is just in dealing with multidimensional systems that the theory gets its full exploitation. For example, in fuzzy modeling there is the problem of exponential growth of rule bases, covering all the possible cases, as the number of input increases. The reason of this exponential growth is that n-dimensional fuzzy sets are usually obtained as cartesian product of one-dimensional ones (taken as building block for their supposed interpretability). On the contrary, once interpretability (or, at least, a logical notion of interpretability) has been defined in terms of partition theory, it is possible to directly manage n-dimensional fuzzy sets and, in particular, sets which are not cartesian product of less dimensional sets. In this way more efficient way of covering all the possible cases can be considered, and in doing so the size of rule bases can be reduced.

8 Conclusions

While fuzzy systems are often claimed to be transparent and physically interpretable, they are mostly used as mere black boxes. In this paper we have stressed that interpretability is not granted by fuzzy modeling *per se*. A necessary prerequisite is to give constraints on membership functions at the beginning of the modeling process.

Although many authors have proposed constraints on membership functions for automatic optimization processes, formal definitions of these constraints are largely missing. This holds in particular for partitions in many-valued and fuzzy logic, for which several desiderata have been introduced in the literature. However, when one carefully considers the significance of such properties, one hardly gets more insight than that is given by common practice using "rules of thumb". Consider for instance, the parsimony property. Here a somewhat 'magical' number 7 ± 2 turns out to be, for several authors, an upper bound on the number of membership functions, its justification being the supposed inability of the human mind to manage more than 7 ± 2 labels [21].

In this paper we have stressed that only a part of the proposed constraints pertains to the (formal) interpretation of the fuzzy sets of a linguistic variable; we have focused attention on the fundamental notion of partition, for its role in if-then descriptions. We have shown that, once one decides to work with partitions having (joint) refinement, then deep mathematical results, stemming from the well established theory of MV-algebras and their corresponding logic, may give some insight on what should be assumed.

Acknowledgements

The authors are grateful to Daniele Mundici and Mirko Navara for their invaluable comments and suggestions.

The second author was supported by the European Union under Project ICA 1-CT-2000-70002 MIRACLE and by MURST Project on logic, algebraic and algorithmic methods for the treatment of uncertainty.

References

1. P. Amato and M. Porto. An algorithm for the automatic generation of a logical formula representing a control law. *Neural Nerwork World*, 10(5):777–786, 2000.

2. C.K. Chak, G. Feng, and J. Ma. An adaptive fuzzy neural network for MIMO system model approximation in high-dimensional spaces. *IEEE Transactions on Systems, Man, and Cybernetics*, 28(3):436–446, 1998.
3. C. C. Chang. Algebraic analysis of many valued logics. *Transactions of the American Mathematical Society*, 88:74–80, 1958.
4. C. C. Chang. A new proof of the completeness of the Łukasiewicz axioms. *Transactions of the American Mathematical Society*, 93:74–90, 1959.
5. R. Cignoli, I. D'Ottaviano, and D. Mundici. *Algebraic Foundations of Many-valued Reasoning.* Kluwer, Dordrecht, 2000.
6. C. De Concini and C. Procesi. Complete symmetric varieties. II. Intersection theory. In *Algebraic groups and related topics (Kyoto/Nagoya, 1983)*, pages 481–513. North-Holland, Amsterdam, 1985.
7. O. Cordón, F. Herrera, and I. Zwir. Linguistic modeling by hierarchical systems of linguistic rules. *IEEE Transactions on Fuzzy Systems*, to appear.
8. J. Valente de Oliveira. Semantic constraints for membership function optimization. *IEEE Transactions on Systems, Man, and Cybernetics*, 29(1):128–138, 1999.
9. J. Espinosa and J. Vandewalle. Constructing fuzzy models with linguistic integrity from numerical data-AFRELI algorithm. *IEEE Transactions on Fuzzy Systems*, 8(5):591–600, 2000.
10. G. Ewald. *Combinatorial Convexity and Algebraic Geometry.* Springer-Verlag, Berlin, 1996.
11. L. Fuchs. *Partially Ordered Algebraic Systems.* Pergamon Press, Oxford, 1963.
12. P. Hájek. *Metamathematics of fuzzy logic.* Kluwer, Dordrecht, 1998.
13. S. Haykin. *Neural Networks: A comprehensive Foundation.* IEEE, Piscataway, 1999.
14. F. Herrera and L. Martínez. A model based on linguistic 2-tuples for dealing with multigranularity hierarchical linguistic contexts in multiexpert decision-making. *IEEE Transactions on Systems, Man and Cybernetics. Part B: Cybernetics*, 31(2):227–234, 2001.
15. C. Juang and C. Lin. An on-line self constructing neural fuzzy inference network and its applications. *IEEE Transactions on Fuzzy Systems*, 6(1):12–32, 1998.
16. Y. Kajitani, K. Kuwata, R. Katayama, and Y. Nishida. An automatic fuzzy modeling with constraints of membership functions and a model determination for neuro and fuzzy model by plural performance indices. In *Proceedings IFES'91*, pages 586–597, 1991.
17. A. Y. Khinchine. On unimodal distributions. *Izv. Nauch. -Issl. Inst. Mat. Mech., Tomsk.*, 2:1–7, 1938.
18. E. P. Klement and B. Moser. On the redundancy of fuzzy partitions. *Fuzzy Sets and Systems*, 85:195–201, 1997.
19. V. Marra. Every abelian ℓ-group is ultrasimplicial. *Journal of Algebra*, 225:872–884, 2000.
20. R. McNaughton. A theorem about infinite-valued sentential logic. *The Journal of Symbolic Logic*, 16(1):1–13, 1951.
21. G. A. Miller. The magical number seven, plus or minus two: some limits on our capacity for processing information. *The Psychological Review*, 63(2):81–97, 1951.
22. D. Mundici. Interpretation of AF C^*-algebras in Łukasiewicz sentential calculus. *Journal of Functional Analysis*, 65:15–63, 1986.

23. D. Mundici. A constructive proof of McNaughton's theorem in infinite-valued logics. *Journal of Symbolic logic*, 59:596–602, 1994.
24. D. Mundici. Uncertainty measures in MV algebras, and states of AF C^*-algebras. *Notas de la Sociedad de Matematica de Chile*, 15(1):43–54, 1996.
25. D. Mundici. Nonboolean partitions and their logic. *Soft Computing*, 2(1):18–22, 1998.
26. D. Mundici. Tensor product and the Loomis-Sikorski theorem for MV-algebras. *Advances in Applied Mathematics*, 22:227–248, 1999.
27. D. Mundici. Reasoning on imprecisely defined functions. In V. Novák and I. Perfilieva, editors, *Discovering the world with fuzzy logic*, pages 331–366. Springer-Verlag, Berlin, 2000.
28. D. Nauck, F. Klawonn, and R. Kruse. *Neuro-Fuzzy Systems*. John Wiley & Sons, Chichester, 1997.
29. V. Novák. Fuzzy control from the logical point of view. *Fuzzy Sets and Systems*, 66:159–173, 1994.
30. W. Pedrycz. Why triangular membership functions? *Fuzzy Sets and Systems*, 64:21–30, 1994.
31. W. Pedrycz and J. Valente De Oliveira. Optimization of fuzzy models. *IEEE Transactions on Systems, Man, and Cybernetics*, 26(4):627–636, 1996.
32. E. Ruspini. A new approach to clustering. *Information and Control*, 15:22–32, 1969.
33. M. Setnes, R. Babuška, U. Kaimak, and H.R. van Nauta Lemke. Similarity measures in fuzzy rule base simplification. *IEEE Transactions on Systems, Man, and Cybernetics*, 28(3):376–386, 1998.
34. M. Setnes, R. Babuška, and H.B. Verbruggen. Rule-based modeling: Precision and transparency. *IEEE Transactions on Systems, Man, and Cybernetics*, 28(1):165–169, 1998.
35. M. Sugeno and G.T. Kang. Structure identification of fuzzy model. *Fuzzy Sets and Systems*, 28:15–33, 1988.
36. T. Takagi and M. Sugeno. Fuzzy identification of systems and its applications to modeling and control. *IEEE Transactions on Systems, Man, and Cybernetics*, 15(1):116–132, 1985.
37. L. Xuecheng. Entropy, distance measure and similarity measure on fuzzy sets and their relations. *Fuzzy Sets and Systems*, 52:305–318, 1992.
38. L. Zadeh. The concept of a linguistic variable and its applications to approximate reasoning I. *Information Science*, 8:199–249, 1975.
39. L. Zadeh. Soft computing and fuzzy logic. *IEEE Software*, 11(6):48–56, 1994.
40. L. Zadeh. Toward a theory of fuzzy information granulation and its centrality in human reasoning and fuzzy logic. *Fuzzy Sets and Systems*, 90:111–127, 1997.

A Formal Model of Interpretability of Linguistic Variables

Ulrich Bodenhofer[1] and Peter Bauer[2]

[1] Software Competence Center Hagenberg
A-4232 Hagenberg, Austria
e-mail: ulrich.bodenhofer@scch.at
[2] COMNEON Software
A-4040 Linz, Austria
e-mail: peter.bauer@comneon.com

Abstract. The present contribution is concerned with the interpretability of fuzzy rule-based systems. While this property is widely considered to be a crucial one in fuzzy rule-based modeling, a more detailed investigation of what "interpretability" actually means is still missing. So far, interpretability has often been associated with heuristic assumptions about shape and mutual overlapping of fuzzy membership functions. In this chapter, we attempt to approach this problem from a more general and formal point of view. First, we clarify what, in our opinion, the different aspects of interpretability are. Following that, we propose an axiomatic framework for the interpretability of linguistic variables (in Zadeh's sense) which is underlined by examples and application perspectives.

1 Introduction

The epoch-making idea of L. A. Zadeh's early work was to utilize what he called "fuzzy sets" as mathematical models of linguistic expressions which cannot be represented in the framework of classical binary logic and set theory in a natural way. The introduction of his seminal article on fuzzy sets [37] contains the following remarkable words:

> *"More often than not, the classes of objects encountered in the real physical world do not have precisely defined criteria of membership. [...] Yet, the fact remains that such imprecisely defined "classes" play an important role in human thinking, particularly in the domains of pattern recognition, communication of information, and abstraction."*

Fuzzy systems became a tremendously successful paradigm – a remarkable triumph which started with well-selling applications in consumer goods implemented by Japanese engineers. The reasons for this development are manifold; however, we are often confronted with the following arguments:

1. The main difference between fuzzy systems and other control or decision support systems is that they are parameterized in an interpretable

way – by means of rules consisting of linguistic expressions. Fuzzy systems, therefore, allow rapid prototyping as well as easy maintenance and adaptation.
2. Fuzzy systems offer completely new opportunities to deal with processes for which only a linguistic description is available. Thereby, they allow to achieve a robust, secure, and reproducible automation of such tasks.
3. Even if conventional control or decision support strategies can be employed, re-formulating a system's actions by means of linguistic rules can lead to a deeper qualitative understanding of its behavior.

We would like to raise the question whether fuzzy systems, as they appear in daily practice, really reflect these undoubtedly nice advantages. One may observe that the *possibility to estimate the system's behavior by reading and understanding the rule base only* is a basic requirement for the validity of the above points. If we adopt the usual wide understanding of fuzzy systems (rule-based systems incorporating vague linguistic expressions), we can see, however, that this property – let us call it *interpretability* – is *not guaranteed by definition.*

In our opinion, interpretability should be *the* key property of fuzzy systems. If it is neglected, one ends up in nothing else than black-box descriptions of input-output relationships and any advantage over neural networks or conventional interpolation methods is lost completely.

The more fuzzy systems became standard tools for engineering applications, the more Zadeh's initial mission became forgotten. In recent years, however, after a relatively long period of ignorance, an increasing awareness of the crucial property of interpretability has emerged [1,2,6,9,17,34–36], where the present book is intended to bundle these forces by presenting a comprehensive overview of recent research on this topic. So far, the following questions have been identified to have a close connection to interpretability:

1. Does the inference mechanism produce results that are technically and intuitively correct?
2. Is the number of rules still small enough to be comprehensible by a human expert?
3. Do the fuzzy sets associated to the linguistic expressions really correspond to the human understanding of these expressions?

The first question has to be approached from the side of approximate reasoning [14–16] and relational equations [10,21,26]. There is no standard way to tackle the problems associated with the second question (see the introductory chapter of this book for an overview of different ideas).

The given chapter is solely devoted to the third question. So far, there is a shallow understanding that the third question is related to shape, ordering, and mutual overlapping of fuzzy membership functions. We intend to approach this question more formally. This is accomplished making the inherent relationships between the linguistic labels explicit by formulating them as (fuzzy) relations. In order to provide a framework that is as general as possible, we consider linguistic variables in their most general form.

2 Preliminaries

Throughout the whole chapter, we do not explicitly distinguish between fuzzy sets and their corresponding membership functions. Consequently, uppercase letters are used for both synonymously. For a given non-empty set X, we denote the set of fuzzy sets on X with $\mathcal{F}(X)$. As usual, a fuzzy set $A \in \mathcal{F}(X)$ is called *normalized* if there exists an $x \in X$ such that $A(x) = 1$.

Triangular norms and conorms [24] are common standard models for fuzzy conjunctions and disjunctions, respectively. In this chapter, we will mainly need these two concepts for intersections and unions of fuzzy sets. It is known that couples consisting of a *nilpotent t-norm* and its dual t-conorm [18,24] are most appropriate choices as soon as fuzzy partitions are concerned [11,26]. The most important representatives of such operations are the so-called Łukasiewicz operations:

$$T_{\mathbf{L}}(x,y) = \max(x+y-1,0)$$
$$S_{\mathbf{L}}(x,y) = \min(x+y,1)$$

The intersection and union of two arbitrary fuzzy sets $A, B \in \mathcal{F}(X)$ with respect to the Łukasiewicz operations can then be defined as

$$(A \cap_{\mathbf{L}} B)(x) = T_{\mathbf{L}}\big(A(x), B(x)\big) \quad \text{and}$$
$$(A \cup_{\mathbf{L}} B)(x) = S_{\mathbf{L}}\big(A(x), B(x)\big),$$

respectively. We restrict to these two standard operations in the following – for the reason of simplicity and the fact that they perfectly fit to the concept of fuzzy partitions due to Ruspini [33]; recall that a family of fuzzy sets $(A_i)_{i\in I} \subseteq \mathcal{F}(X)$ is called *Ruspini partition* if the following equality holds for all $x \in X$:

$$\sum_{i\in I} A_i(x) = 1$$

Furthermore, recall that a fuzzy set $A \in \mathcal{F}(X)$ is called *convex* if the property

$$x \leq y \leq z \implies A(y) \geq \min\big(A(x), A(z)\big)$$

holds for all $x, y, z \in X$ (given a crisp linear ordering $\leq$ on the domain X) [4,27,37].

Lemma 1. [4] *Let X be linearly ordered. Then an arbitrary fuzzy set $A \in \mathcal{F}(X)$ is convex if and only if there exists a partition of X into two connected subsets X_1 and X_2 such that, for all $x_1 \in X_1$ and all $x_2 \in X_2$, $x_1 \leq x_2$ holds and such that the membership function of A is non-decreasing over X_1 and non-increasing over X_2.*

As a trivial consequence of the previous lemma, a fuzzy set whose membership function is either non-decreasing or non-increasing is convex.

3 Formal Definition

Since it has more or less become standard and offers much freedom, in particular with respect to integration of linguistic modifiers and connectives, we closely follow Zadeh's original definition of linguistic variables [38–40].

Definition 1. A *linguistic variable* V is a quintuple of the form

$$V = (N, G, T, X, S),$$

where N, T, X, G, and S are defined as follows:

1. N is the name of the linguistic variable V
2. G is a grammar
3. T is the so-called *term set*, i.e. the set linguistic expressions resulting from G
4. X is the universe of discourse
5. S is a $T \to \mathcal{F}(X)$ mapping which defines the semantics – a fuzzy set on X – of each linguistic expression in T

In this chapter, let us assume that the grammar G is always given in *Backus-Naur Form (BNF)* [32].

In our point of view, the ability to interpret the meaning of a rule base qualitatively relies deeply upon an intuitive understanding of the linguistic expressions. Of course, this requires knowledge about inherent relationships between these expressions. Therefore, if qualitative estimations are desired, these relationships need to transfer to the underlying semantics, i.e. the fuzzy sets modeling the labels. In other words, interpretability is strongly connected to the preservation of inherent relationships by the mapping S (according to Def. 1).

The following definition gives an exact mathematical formulation of this property.

Definition 2. Consider a linguistic variable $V = (N, T, X, G, S)$ and an index set I. Let $R = (R_i)_{i \in I}$ be a family of relations on the set of verbal values T, where each relation R_i has a finite arity a_i. Assume that, for every relation R_i, there exists a relation Q_i on the fuzzy power set $\mathcal{F}(X)$ with the same arity.[1] Correspondingly, we abbreviate the family $(Q_i)_{i \in I}$ with Q. Then the linguistic variable V is called *R-Q-interpretable* if and only if the following holds for all $i \in I$ and all $x_1, \ldots, x_{a_i} \in T$:

$$R_i(x_1, \ldots, x_{a_i}) \implies Q_i(S(x_1), \ldots, S(x_{a_i})) \tag{1}$$

[1] Q_i is associated with the "semantic counterpart" of R_i, i.e. the relation that models R_i on the semantical level.

Remark 1. The generalization of Def. 2 to fuzzy relations is straightforward. If we admit fuzziness of the relations R_i and Q_i, the implication in (1) has to be replaced by the inequality

$$R_i(x_1,\ldots,x_{a_i}) \leq Q_i(S(x_1),\ldots,S(x_{a_i})).$$

4 A Detailed Study by Means of Practical Examples

In almost all fuzzy control applications, the domains of the system variables are divided into a certain number of fuzzy sets by means of the underlying ordering – a fact which is typically reflected in expressions like "small", "medium", or "large". We will now discuss a simple example involving orderings to illustrate the concrete meaning of Def. 2.

Let us consider the following linguistic variable:

$$V = (\text{"v1"}, G, T, X, S)$$

The grammatical definition G is given as follows:

$$\begin{array}{ll} \perp & := \langle\text{atomic}\rangle \; ; \\ \langle\text{atomic}\rangle & := \langle\text{adjective}\rangle \mid \langle\text{adverb}\rangle \; \langle\text{adjective}\rangle \; ; \\ \langle\text{adjective}\rangle & := \text{"small"} \mid \text{"medium"} \mid \text{"large"} \; ; \\ \langle\text{adverb}\rangle & := \text{"at least"} \mid \text{"at most"} \; ; \end{array}$$

Obviously, the following nine-element term set can be derived from G:

$$\begin{aligned} T = \{ & \text{"small"}, \text{"medium"}, \text{"large"}, \\ & \text{"at least small"}, \text{"at least medium"}, \\ & \text{"at least large"}, \text{"at most small"}, \\ & \text{"at most medium"}, \text{"at most large"} \} \end{aligned}$$

The universe of discourse is the real interval $X = [0, 100]$.

Taking the "background" or "context" of the variable into account, almost every human has an intuitive understanding of the qualitative meaning of each of the above linguistic expressions, even if absolutely nothing about the quantitative meaning, i.e. the corresponding fuzzy sets, is known. This understanding, to a major part, can be attributed to elementary relationships between the linguistic values. According to Def. 2, let us assume that these inherent relationships are modeled by a family of relations $R = (R_i)_{i \in I}$.

In our opinion, the most obvious relationships in the example term set T are orderings and inclusions. Therefore, we consider the following two binary relations (for convenience, we switch to infix notations here):

$$R = (\preceq, \sqsubseteq) \tag{2}$$

The first relation $\preceq$ stands for the ordering of the labels, while the second one corresponds to an inclusion relation, e.g. $u \sqsubseteq v$ means that v is a more general term than u.

First of all, one would intuitively expect a proper ordering of the adjectives, i.e.

$$\text{“small”} \preceq \text{“medium”} \preceq \text{“large”}. \tag{3}$$

Moreover, the following monotonicities seem reasonable for all adjectives u, v (atomic expressions from the set {“small”, “medium”, “large”}):

$$u \sqsubseteq \text{“at least” } u$$
$$v \sqsubseteq \text{“at most” } v$$

$$u \preceq v \Longrightarrow \text{“at least” } u \preceq \text{“at least” } v$$
$$u \preceq v \Longrightarrow \text{“at most” } u \preceq \text{“at most” } v$$

$$u \preceq v \Longrightarrow \text{“at least” } v \sqsubseteq \text{“at least” } u$$
$$u \preceq v \Longrightarrow \text{“at most” } u \sqsubseteq \text{“at most” } v$$

Figures 1 and 2 show Hasse diagrams which fully describe the two relations $\preceq$ and $\sqsubseteq$ (note that both relations are supposed to be reflexive, a fact which, for the sake of simplicity, is not made explicit in the diagrams).

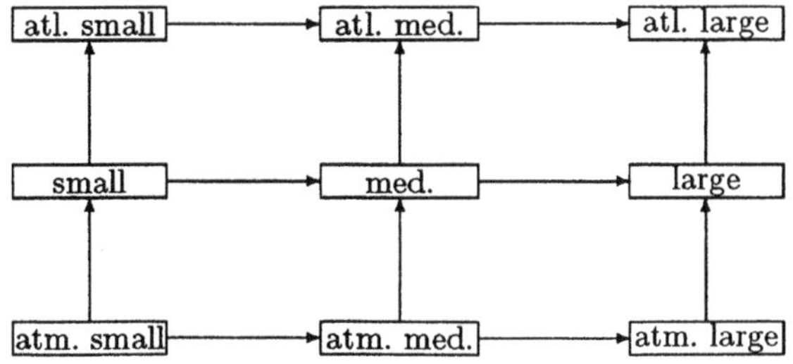

Fig. 1. Hasse diagram of ordering relation $\preceq$

Now we have to define meaningful counterparts of the relations in R on the semantical level, i.e. on $\mathcal{F}(X)$. We start with the usual inclusion of fuzzy sets according to Zadeh [37].

Definition 3. Consider two fuzzy sets of $A, B \in \mathcal{F}(X)$. A is called a *subset* of B, short $A \subseteq B$, if and only if, for all $x \in X$, $A(x) \leq B(x)$. Consequently, in this case, B is called a *superset* of A.

For defining a meaningful counterpart of the ordering relation $\preceq$, we adopt a simple variant of the general framework for ordering fuzzy sets proposed in [4,5], which includes well-known orderings of fuzzy numbers based on the extension principle [23,25].

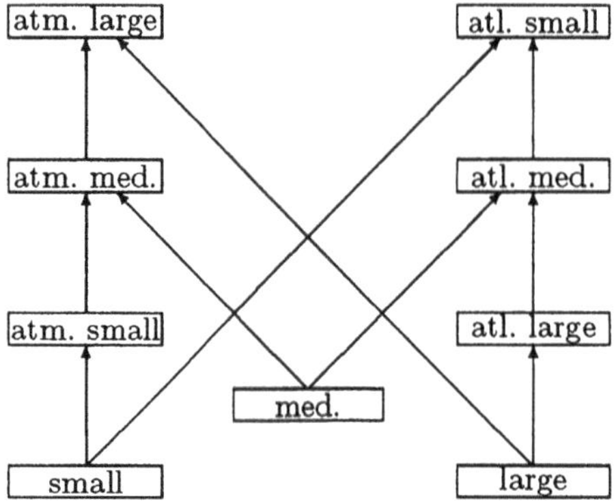

Fig. 2. Hasse diagram of inclusion relation $\sqsubseteq$

Definition 4. Suppose that a universe X is equipped with a crisp linear ordering $\leq$. Then a preordering $\lesssim$ of fuzzy sets can be defined by

$$A \lesssim B \iff \big(\mathrm{ATL}(B) \subseteq \mathrm{ATL}(A) \text{ and } \mathrm{ATM}(A) \subseteq \mathrm{ATM}(B)\big),$$

where the operators ATL and ATM are defined as follows:

$$\mathrm{ATL}(A)(x) = \sup\{A(y) \mid y \leq x\}$$
$$\mathrm{ATM}(A)(x) = \sup\{A(y) \mid y \geq x\}$$

Figure 3 shows an example what the operators ATL and ATM give for a non-trivial fuzzy set. It is easy to see that ATL always yields the smallest superset with non-decreasing membership function, while ATM yields the smallest superset with non-increasing membership function. For more details about the particular properties of the ordering relation $\lesssim$ and the two operators ATL and ATM, see [4,5].

Summarizing, the set of counterpart relations Q looks as follows (with the relations from Defs. 3 and 4):

$$Q = (\lesssim, \subseteq) \tag{4}$$

Now R-Q-interpretability of linguistic variable V (with definitions of R and Q according to (2) and (4), respectively) specifically means that the following two implications hold for all $u, v \in T$:

$$u \preceq v \implies S(u) \lesssim S(v) \tag{5}$$
$$u \sqsubseteq v \implies S(u) \subseteq S(v) \tag{6}$$

This means that the mapping S plays the crucial role in terms of interpretability. In this particular case, R-Q-interpretability is the property that

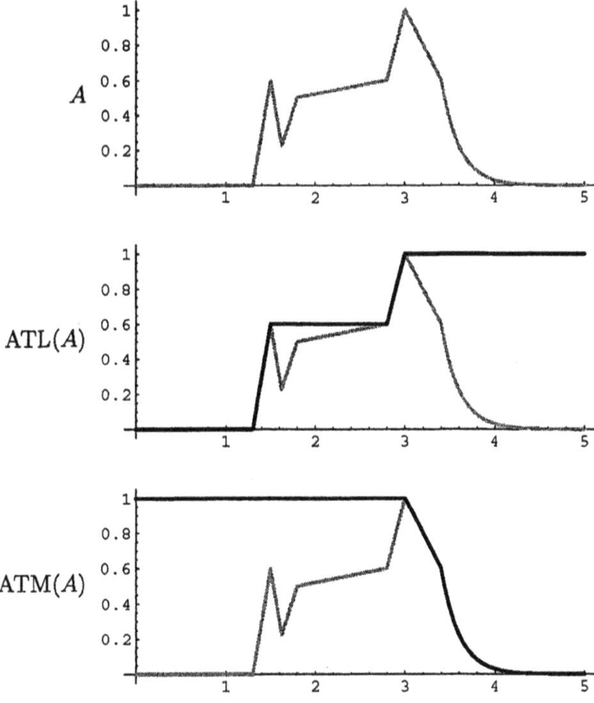

Fig. 3. A fuzzy set $A \in \mathcal{F}(\mathbb{R})$ and the results which are obtained when applying the operators ATL and ATM

an ordering or inclusion relationship between two linguistic terms is never violated by the two corresponding fuzzy sets. From (3) and (5), we can deduce the first basic necessary condition for the fulfillment of R-Q-interpretability – that the fuzzy sets associated with the three adjectives must be in proper order:

$$S(\text{“small”}) \lesssim S(\text{“medium”}) \lesssim S(\text{“large”}) \tag{7}$$

It is easy to observe that this basic ordering requirement is violated by the example shown in Fig. 4, while it is fulfilled by the fuzzy sets in Fig. 5.

In order to fully check R-Q-interpretability of V, the semantics of linguistic expressions containing an adverb (“at least” or “at most”) have to be considered as well. The definition of linguistic variables does not explicitly contain any hint how to deal with the semantics of such expressions. From a pragmatic viewpoint, two different ways are possible: one simple variant is to define a separate fuzzy set for each expression, regardless whether they contain an adverb or not. As a second traditional variant, we could use fuzzy modifiers – $\mathcal{F}(X) \longrightarrow \mathcal{F}(X)$ functions – for modeling the semantics of adverbs. In this example, it is straightforward to use the fuzzy modifiers introduced

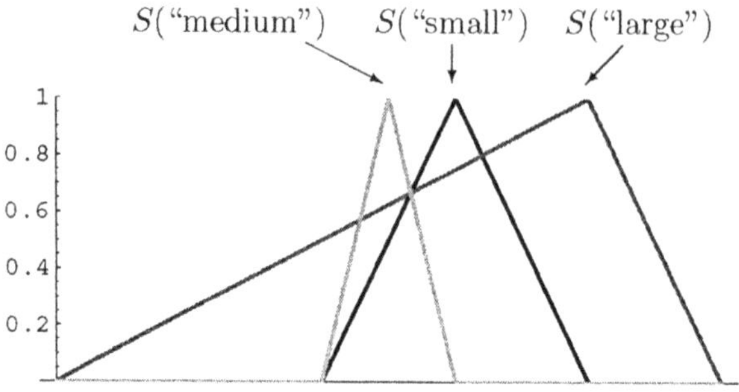

Fig. 4. A non-interpretable setting

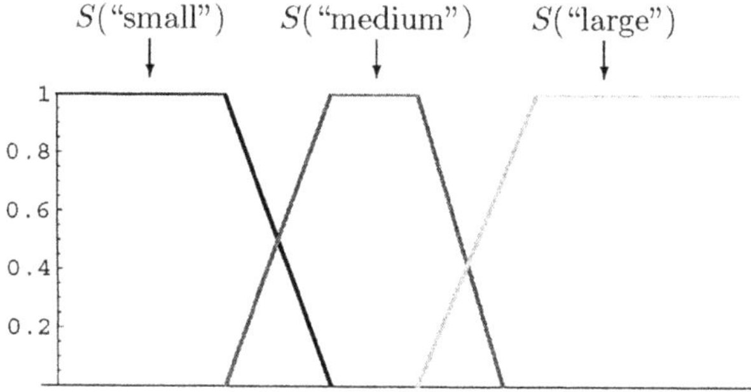

Fig. 5. An example of an interpretable setting

in Def. 4 (see [3,4,12] for a detailed justification):

$$S(\text{"at least"}A) = \mathrm{ATL}(S(A))$$
$$S(\text{"at most"}A) = \mathrm{ATM}(S(A))$$

Since it is by far simpler and easier to handle with respect to interpretability, we strongly suggest the second variant.

In case that we use the above fuzzy modifiers for modeling the two adverbs "at least" and "at most", we are now able to formulate a necessary condition for the fulfillment of R-Q-interpretability in our example.

Theorem 1. *Consider the linguistic variable V and the two relation families R and Q as defined above. Provided that the mapping S always yields a normalized fuzzy set, the following two statements are equivalent:*

(i) V is R-Q-interpretable
(ii) $S(\text{"small"}) \lesssim S(\text{"medium"}) \lesssim S(\text{"large"})$

Proof. (i)⇒(ii): Trivial (see above).
(ii)⇒(i): The following basic properties hold for all normalized fuzzy sets $A, B \in \mathcal{F}(X)$ [4]:

$$A \subseteq \mathrm{ATL}(A) \tag{8}$$
$$A \subseteq \mathrm{ATM}(A) \tag{9}$$

$$\mathrm{ATL}(\mathrm{ATL}(A)) = \mathrm{ATL}(A) \tag{10}$$
$$\mathrm{ATM}(\mathrm{ATM}(A)) = \mathrm{ATM}(A) \tag{11}$$

$$\mathrm{ATL}(\mathrm{ATM}(A)) = \mathrm{ATM}(\mathrm{ATL}(A)) = X \tag{12}$$

$$A \subseteq B \implies \mathrm{ATL}(A) \subseteq \mathrm{ATL}(B) \tag{13}$$
$$A \subseteq B \implies \mathrm{ATM}(A) \subseteq \mathrm{ATM}(B) \tag{14}$$

Since the relations $\subseteq$ and $\lesssim$ are reflexive and transitive [4,5], it is sufficient to prove the relations indicated by arrows in the two Hasse diagrams (see Figs. 1 and 2).
Let us start with the ordering relation. The validity of the relations in the middle row is exactly assumption (ii). The relations in the two other rows follow directly from the following two relationships which can be proved easily using (10), (11), and (12):

$$A \lesssim B \implies \mathrm{ATL}(A) \lesssim \mathrm{ATL}(B)$$
$$A \lesssim B \implies \mathrm{ATM}(A) \lesssim \mathrm{ATM}(B)$$

The three vertical relationships in Fig. 1 follow directly from

$$\mathrm{ATM}(A) \lesssim A \lesssim \mathrm{ATL}(A)$$

which can be shown using (8), (9), (13), and (14).
The relations in the Hasse diagram in Fig. 2 follow from (8), (9), and the definition of the preordering $\lesssim$ (cf. Def. 4). □

Obviously, interpretability of V in this example (with respect to the families R and Q) does not fully correspond to an intuitive human understanding of interpretability. For instance, all three expressions "small", "medium", and "large" could be mapped to the same fuzzy set without violating R-Q-interpretability. The intention was to give an example which is just expressive enough to illustrate the concrete meaning and practical relevance of Def. 2.

In order to formulate an example in which R-Q-interpretability is much closer to a human-like understanding of interpretability (e.g. including separation constraints), we have to consider an extended linguistic variable

$$V' = (\text{"v2"}, G', T', X', S').$$

The extended grammar G' is given as follows:

$$
\begin{array}{ll}
\bot & := \langle\text{exp}\rangle \mid \langle\text{bounds}\rangle \ ; \\
\langle\text{exp}\rangle & := \langle\text{atomic}\rangle \mid \langle\text{atomic}\rangle\ \langle\text{binary}\rangle\ \langle\text{atomic}\rangle \ ; \\
\langle\text{atomic}\rangle & := \langle\text{adjective}\rangle \mid \langle\text{adverb}\rangle\ \langle\text{adjective}\rangle \ ; \\
\langle\text{adjective}\rangle & := \text{“small”} \mid \text{“medium”} \mid \text{“large”} \ ; \\
\langle\text{adverb}\rangle & := \text{“at least”} \mid \text{“at most”} \ ; \\
\langle\text{binary}\rangle & := \text{“and”} \mid \text{“or”} \ ; \\
\langle\text{bounds}\rangle & := \text{“empty”} \mid \text{“anything”} \ ;
\end{array}
$$

It is easy to see that the corresponding term set T' has the following elements: the grammar admits nine atomic expressions (three adjectives plus two adverbs times three adjectives; note that this subset coincides with T from the previous example). Hence, there are $9 + 2 \cdot 9^2 = 171$ expressions of type $\langle\text{exp}\rangle$. Finally adding the two expressions of type $\langle\text{bounds}\rangle$, the term set T' has a total number of 173 elements.

As in the previous example, we would like to use an inclusion and an ordering relation. Since the two relations $\lesssim$ and $\subseteq$ are defined for arbitrary fuzzy sets, we can keep the relation family Q as it is. If we took R as defined above, R-Q-interpretability would be satisfied under the same conditions as in Th. 1. This example, however, is intended to demonstrate that partition and convexity constraints can be formulated level of linguistic expressions, too. Therefore, we extend the inclusion relation $\sqsubseteq$ as follows. Let us consider a binary relation $\dot{\sqsubseteq}$ on T'. First of all, we require that $\dot{\sqsubseteq}$ coincides with $\sqsubseteq$ on the set of atomic expressions ($u, v \in T$):

$$(u \sqsubseteq v) \Longrightarrow (u \,\dot{\sqsubseteq}\, v) \tag{15}$$

Of course, we assume that the two binary connectives are non-decreasing with respect to inclusion, commutative, and that the "and" connective yields subsets and the "or" connective yields supersets (for all $u, v, w \in T$):

$$(v \,\dot{\sqsubseteq}\, w) \Longrightarrow (u \text{ “and” } v) \,\dot{\sqsubseteq}\, (u \text{ “and” } w) \tag{16}$$

$$(v \,\dot{\sqsubseteq}\, w) \Longrightarrow (u \text{ “or” } v) \,\dot{\sqsubseteq}\, (u \text{ “or” } w) \tag{17}$$

$$(u \text{ “and” } v) \,\dot{\sqsubseteq}\, (v \text{ “and” } u) \tag{18}$$

$$(u \text{ “or” } v) \,\dot{\sqsubseteq}\, (v \text{ “or” } u) \tag{19}$$

$$(u \text{ “and” } v) \,\dot{\sqsubseteq}\, u \tag{20}$$

$$u \,\dot{\sqsubseteq}\, (u \text{ “or” } v) \tag{21}$$

Next, let us suppose that "anything" is the most general and that "empty" is the least general expression, i.e., for all $u \in T'$,

$$u \,\dot{\sqsubseteq}\, \text{“anything”} \quad \text{and} \quad \text{“empty”} \,\dot{\sqsubseteq}\, u. \tag{22}$$

Now we can impose reasonable disjointness constraints like

$$\text{“small and at least medium”} \,\dot{\sqsubseteq}\, \text{“empty”} \tag{23}$$

$$\text{“at most medium and large”} \,\dot{\sqsubseteq}\, \text{“empty”} \tag{24}$$

and coverage properties:

$$\text{"small or medium"} \sqsubseteq \text{"at most medium"} \tag{25}$$
$$\text{"at most medium"} \sqsubseteq \text{"small or medium"} \tag{26}$$
$$\text{"anything"} \sqsubseteq \text{"at most medium or large"} \tag{27}$$

$$\text{"medium or large"} \sqsubseteq \text{"at least medium"} \tag{28}$$
$$\text{"at least medium"} \sqsubseteq \text{"medium or large"} \tag{29}$$
$$\text{"small or at least medium"} \sqsubseteq \text{"anything"} \tag{30}$$

Finally, let us assume that "small" and "large" are the two boundaries with respect to the ordering of the labels:

$$\text{"at most small"} \sqsubseteq \text{"small"} \tag{31}$$
$$\text{"anything"} \sqsubseteq \text{"at least small"} \tag{32}$$

$$\text{"at least large"} \sqsubseteq \text{"large"} \tag{33}$$
$$\text{"anything"} \sqsubseteq \text{"at most large"} \tag{34}$$

If we denote the reflexive and transitive closure of $\sqsubseteq$ with $\sqsubseteq'$, we can finally write down the desired family of relations:

$$R' = (\preceq, \sqsubseteq') \tag{35}$$

In order to study the R'-Q-interpretability of V', we need to define the semantics of those expressions that have not been contained in T. Of course, for the expressions in T, we use the same semantics as in the previous example, i.e., for all $u \in T$, $S'(u) = S(u)$. Further, let us make the convention that the two expressions of type $\langle$bounds$\rangle$ are always mapped to the empty set and the whole universe, respectively:

$$S'(\text{"empty"}) = \emptyset \qquad S'(\text{"anything"}) = X$$

The "and" and the "or" connective are supposed to be "implemented" by the intersection and union with respect to the Łukasiewicz t-norm and its dual t-conorm (for all $u, v \in T$):

$$S'(u \text{ "and" } v) = S'(u) \cap_{\mathbf{L}} S'(v)$$
$$S'(u \text{ "or" } v) = S'(u) \cup_{\mathbf{L}} S'(v)$$

Now we are able to fully characterize R'-Q-interpretability for the given example (the linguistic variable V').

Theorem 2. *Provided that S' yields a normalized fuzzy set for each adjective, V' is R'-Q-interpretable if and only if the following three properties hold together:*

1. $S'(\text{"small"}) \lesssim S'(\text{"medium"}) \lesssim S'(\text{"large"})$
2. $S'(\text{"small"})$, $S'(\text{"medium"})$, *and* $S'(\text{"large"})$ *are convex*
3. $S'(\text{"small"})$, $S'(\text{"medium"})$, *and* $S'(\text{"large"})$ *form a Ruspini partition*

Proof. First of all, let us assume that V' is R'-Q-interpretable. The first property follows trivially as in the proof of Th. 1. Now, taking (23) into account, we obtain from R'-Q-interpretability that

$$\begin{aligned} 0 &\geq T_{\mathbf{L}}\big(S'(\text{"small"})(x), S'(\text{"at least medium"})(x)\big) \\ &= T_{\mathbf{L}}\big(S'(\text{"small"})(x), \mathrm{ATL}(S'(\text{"medium"}))(x)\big) \\ &\geq T_{\mathbf{L}}\big(S'(\text{"small"})(x), S'(\text{"medium"})(x)\big), \end{aligned}$$

i.e. that the $T_{\mathbf{L}}$-intersection of $S'(\text{"small"})$ and $S'(\text{"medium"})$ is empty. Analogously, we are able to show that the $T_{\mathbf{L}}$-intersection of $S'(\text{"medium"})$ and $S'(\text{"large"})$ is empty, too. Since $S'(\text{"medium"}) \lesssim S'(\text{"large"})$ implies that

$$\mathrm{ATL}\big(S'(\text{"large"})\big) \subseteq \mathrm{ATL}\big(S'(\text{"medium"})\big),$$

it follows that, by the same argument as above, that the $T_{\mathbf{L}}$-intersection of $S'(\text{"small"})$ and $S'(\text{"large"})$ is empty as well. Now consider (25) and (26). R'-Q-interpretability then implies the following (for all $x \in X$):

$$\begin{aligned} S_{\mathbf{L}}\big(S'(\text{"small"})(x), S'(\text{"medium"})(x)\big) &= S'(\text{"at most medium"})(x) \\ &= \mathrm{ATL}\big(S'(\text{"medium"})\big)(x) \end{aligned}$$

Taking (27) into account as well, we finally obtain

$$\begin{aligned} 1 &\leq S_{\mathbf{L}}\big(S'(\text{"at most medium"})(x), S'(\text{"large"})(x)\big) \\ &= S_{\mathbf{L}}\big(S'(\text{"small"})(x), S'(\text{"medium"})(x), S'(\text{"large"})(x)\big) \end{aligned}$$

which proves that the $S_{\mathbf{L}}$-union of all fuzzy sets associated to the three adjectives yields the whole universe X. Since all three fuzzy sets are normalized and properly ordered, not more than two can have a membership degree greater than zero at a given point $x \in X$. This implies that the three fuzzy sets form a Ruspini partition [26].

From (31) and (33) and R'-Q-interpretability, we can infer that

$$\begin{aligned} \mathrm{ATM}\big(S'(\text{"small"})\big) &= S'(\text{"small"}), \\ \mathrm{ATL}\big(S'(\text{"large"})\big) &= S'(\text{"large"}), \end{aligned}$$

hence, $S'(\text{"small"})$ has a non-increasing membership function and $S'(\text{"large"})$ has a non-decreasing membership function. Both fuzzy sets, therefore, are convex. Since the three fuzzy sets S'("small"), S'("medium"), and S'("large") form a Ruspini partition, while only two can overlap to a positive degree, we have that S'("medium") is non-decreasing to the left of any value x for which $S'(\text{"medium"})(x) = 1$ and non-increasing to the right. Therefore, by Le. 1, S'("medium") is convex, too.

Now let us prove the reverse direction, i.e. we assume all three properties and show that R'-Q-interpretability must hold. By (15) and the first property from Th. 2, we can rely on the fact that all correspondences remain preserved for cases that are already covered by Th. 1. Therefore, it is sufficient to show that S' preserves all relationships (16)–(34). Clearly, the preservation of (16)–(21) follows directly from elementary properties of t-norms and t-conorms [24]. The inclusions (22) are trivially maintained, since S' is supposed to map "empty" to the empty set and "anything" to the universe X. The preservation of the two disjointness conditions (23) and (24) follows from the fact that S'("small"), S'("medium"), and S'("large") form a Ruspini partition and that the first property holds. The same is true for the six coverage properties (25)–(30). Since all three fuzzy sets are convex and form a Ruspini partition, S'("small") must have a non-increasing membership function and S'("large") must have a non-decreasing membership function. Therefore, the following needs to hold:

$$\text{ATM}(S'(\text{"small"})) = S'(\text{"small"})$$
$$\text{ATL}(S'(\text{"large"})) = S'(\text{"large"})$$

This is a sufficient condition for the preservation of inclusions (31) and (33). Then the preservation of (32) and (34) follows from the fact, that $\text{ATL}(\text{ATM}(A)) = X$ for any normalized fuzzy set A (cf. (12)). Since $\sqsubseteq'$ and $\subseteq$ are both supposed to be transitive ($\sqsubseteq'$ being the transitive closure of the intermediate relation $\sqsubseteq$), the preservation of all other relationships follows instantly. □

At first glance, this example might seem unnecessarily complicated, since the final result is nothing else than exactly those common sense assumptions – proper ordering, convexity, partition constraints – that have been identified as crucial for interpretability before in several recent publications (see [8] for an overview). However, we must take into account that they are not just heuristic assumptions here, but necessary conditions that are enforced by intuitive requirements on the level of linguistic expressions. From this point of view, this example provides a sound justification for exactly those three crucial assumptions.

Now the question arises how the three properties can be satisfied in practice. In particular, it is desirable to have a constructive characterization of the constraints implied by requiring R'-Q-interpretability. The following theorem provides a unique characterization of R'-Q-interpretability under the assumption that we are considering real numbers and fuzzy sets with continuous membership functions – both are no serious restrictions from the practical point of view. Fortunately, we obtain a parameterized representation of all mappings S' that maintain R'-Q-interpretability.

Theorem 3. *Assume that X is a connected subset of the real line and that S', for each adjective, yields a normalized fuzzy set with continuous membership function. Then the three properties from Th. 2 are fulfilled if and only if there exist four values $a, b, c, d \in X$ satisfying $a < b \leq c < d$ and two continuous non-decreasing $[0,1] \to [0,1]$ functions f_1, f_2 fulfilling $f_1(0) = f_2(0) = 0$ and $f_1(1) = f_2(1) = 1$ such that the semantics of the three adjectives are defined as follows:*

$$S'(\text{“small”})(x) = \begin{cases} 1 & \textit{if } x \leq a \\ 1 - f_1\left(\frac{x-a}{b-a}\right) & \textit{if } a < x < b \\ 0 & \textit{if } x \geq b \end{cases} \tag{36}$$

$$S'(\text{“medium”})(x) = \begin{cases} 0 & \textit{if } x \leq a \\ f_1\left(\frac{x-a}{b-a}\right) & \textit{if } a < x < b \\ 1 & \textit{if } b \leq x \leq c \\ 1 - f_2\left(\frac{x-c}{d-c}\right) & \textit{if } c < x < d \\ 0 & \textit{if } x \geq d \end{cases} \tag{37}$$

$$S'(\text{“large”})(x) = \begin{cases} 0 & \textit{if } x \leq c \\ f_2\left(\frac{x-c}{d-c}\right) & \textit{if } c < x < d \\ 1 & \textit{if } x \geq d \end{cases} \tag{38}$$

Proof. It is a straightforward, yet tedious, task to show that the fuzzy sets defined as above fulfill the three properties from Th. 2. Under the assumption that these three properties are satisfied, we make the following definitions:

$$\begin{aligned} a &= \sup\{x \mid S'(\text{“small”})(x) = 1\} \\ b &= \inf\{x \mid S'(\text{“medium”})(x) = 1\} \\ c &= \sup\{x \mid S'(\text{“medium”})(x) = 1\} \\ d &= \inf\{x \mid S'(\text{“large'})(x) = 1\} \end{aligned}$$

The two functions f_1, f_2 can be defined as follows:

$$\begin{aligned} f_1(x) &= S'(\text{“medium”})\big(a + x \cdot (b-a)\big) \\ f_2(x) &= S'(\text{“large”})\big(c + x \cdot (d-c)\big) \end{aligned}$$

Since all membership functions associated with adjectives are continuous, the functions f_1 and f_2 are continuous. Taking the continuity and the fact that the three fuzzy sets associated to the adjectives form a Ruspini partition into account, it is clear that the following holds:

$$S'(\text{"small"})(a) = 1$$
$$S'(\text{"small"})(b) = S'(\text{"small"})(c) = S'(\text{"small"})(d) = 0$$

$$S'(\text{"medium"})(b) = S'(\text{"medium"})(c) = 1$$
$$S'(\text{"medium"})(a) = S'(\text{"medium"})(d) = 0$$

$$S'(\text{"large"})(d) = 1$$
$$S'(\text{"large"})(a) = S'(\text{"large"})(b) = S'(\text{"large"})(c) = 0$$

These equalities particularly imply that $f_1(0) = f_2(0) = 0$ and $f_1(1) = f_2(1) = 1$ holds. As a consequence of convexity and Le. 1, we know that the membership function of S'("medium") to the left of b is non-decreasing. Analogously, we can infer that the same is true for S'("large") to the left of d. Therefore, f_1 and f_2 are non-decreasing. To show that the three representations (36)–(38) hold is a routine matter. □

5 Applications

5.1 Design Aid

As long as the top-down construction of small fuzzy systems (e.g. two-input single-output fuzzy controllers) is concerned, interpretability is usually not such an important issue, since the system is simple enough that a conscious user will refrain from making settings which contradict his/her intuition.

In the design of complex fuzzy systems with a large number of variables and rules, however, interpretability is a most crucial point. Integrating tools which guide the user through the design of a large fuzzy system by preventing him/her from making non-interpretable settings accidentally are extremely helpful. As a matter of fact, debugging of large fuzzy systems becomes a tedious task if it is not guaranteed that the intuitive meanings of the labels used in the rule base are reflected in their corresponding semantics.

To be more precise, our goal is not to bother the user with additional theoretical aspects. Instead, the idea is to integrate these aspects into software tools for fuzzy systems design, but not necessarily transparent for the user, with the aim that he/she can build interpretable fuzzy systems in an even easier way than with today's software tools. Theorem 3 gives a clue how this could be accomplished. This result, for one particular example, clearly identifies how much freedom one has in choosing interpretable settings. The example is not quite representative, since three linguistic expressions are a quite restrictive assumption. However, the extension to an arbitrary finite number of such expressions is straightforward, no matter whether we consider such a typical "small"-"medium"-"large" example or a kind of symmetric setting (e.g. "neg. large", "neg. medium", "neg. small", "approx. zero", "pos. small", "pos. medium", "pos. large") as it is common in many fuzzy control applications. In all these cases, the requirements for interpretability are similar and, by Th. 3, the resulting set of degrees of freedom is an increasing chain of values that mark the beginning/ending of the kernels of the fuzzy sets and a set of continuous non-decreasing functions that control the shape of the transitions between two neighboring fuzzy sets. While linear transitions are common and easy to handle, smooth transitions by means of polynomial functions with higher degree may be beneficial in some applications as well.

As simple examples, the following three polynomial $[0,1] \rightarrow [0,1]$ functions of degrees 1, 3, and 5 perfectly serve as transition functions in the sense of Th. 3. They produce membership functions that are continuous (p_1), differentiable (p_3), and twice differentiable (p_5), respectively:

$$p_1(x) = x$$
$$p_3(x) = -2x^3 + 3x^2$$
$$p_5(x) = 6x^5 - 15x^4 + 10x^3$$

Figure 6 shows examples of interpretable fuzzy partitions with three fuzzy sets using the transition functions p_1, p_3, and p_5.

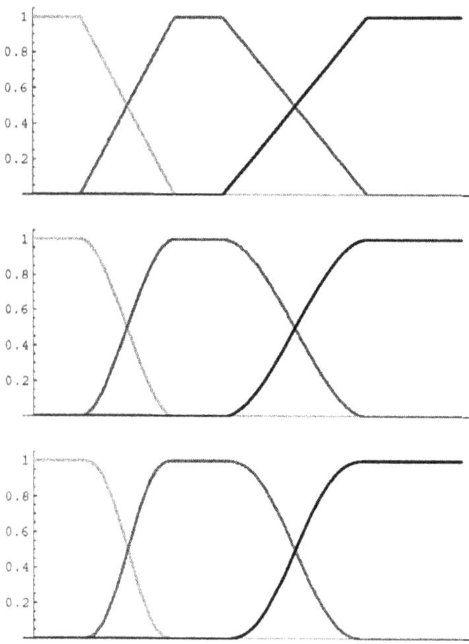

Fig. 6. Three interpretable fuzzy partitions with polynomial transitions of degree 1 (top), 3 (middle), and 5 (bottom)

5.2 Tuning

Automatic design and tuning of fuzzy systems has become a central issue in machine learning, data analysis, and the identification of functional dependencies in the analysis of complex systems. In the last years, a vast number of scientific publications dealt with this problem. Most of them, however, disregarded the importance of interpretability – leading to results which are actually black-box functions that do not provide any meaningful linguistic information (typical pictures like in Fig. 4 can be found in an enormous number of papers).

One may argue that proper input-output behavior is the central goal of automatic tuning. To some extent, this is true; however, as stated already in Sect. 1, this is not the primary mission of fuzzy systems.

Again, Th. 3 gives a clear indication how the space of possible solutions among interpretable settings may be parameterized – by an ascending chain of transition points (given a set of transition functions). Note that this kind of parameterization even leads to a reduction of the search space. Parameterizing three trapezoidal fuzzy sets independently requires a total number of twelve parameters and most probably leads to difficulty interpretable results. Requiring interpretability (as described in the previous section) leads to a set of only four parameters. It is true that such a setting is much more restrictive. However, in our opinion, it is not necessarily the case that requiring interpretability automatically leads to a painful loss of accuracy. The requirement of interpretability implies more constraints that have to be taken into account and, therefore, is more difficult to handle for many tuning algorithms, no matter whether we consider genetic algorithms, fuzzy-neuro methods, or numerical optimization. As a recent investigation has shown, it is indeed possible to require interpretability while, at the same time, maintaining high accuracy and robustness [22]. Many other studies have also come up with tuning algorithms that produce interpretable *and* accurate results [2,7,13,17,35].

5.3 Rule Base Simplification

The examples in Sect. 4 used an inclusion relation $\sqsubseteq$ and its counterpart on the semantical side – the inclusion relation $\subseteq$. Both relations are preorderings which particularly implies that their symmetric kernels are equivalence relations. As easy to see, the relation $\subseteq$ is even an ordering on $\mathcal{F}(X)$, which implies that the symmetric kernel is the crisp equality relation, i.e. $A \subseteq B$ and $B \subseteq A$ hold together if and only if the two membership functions coincide exactly.

Now let us assume that we are given a linguistic variable V and two relation families R and Q, where R contains an inclusion relation $\sqsubseteq$ and Q contains its counterpart $\subseteq$. If we have two linguistic labels u and v for which $u \sqsubseteq v$ and $v \sqsubseteq v$ hold, R-Q-interpretability guarantees that the inclusions $S(u) \subseteq S(v)$ and $S(v) \subseteq S(u)$ hold, i.e. $u \sqsubseteq v$ and $v \sqsubseteq v$ are sufficient conditions that the membership functions of $S(u)$ and $S(v)$ are equal. This means that the equivalence relation defined as (for two $u, v \in T$)

$$(u \equiv v) \iff (u \sqsubseteq v \text{ and } v \sqsubseteq u)$$

may be considered as a set of *simplification rules*, while R-Q-interpretability corresponds to the validity of these rules on the semantical side.

Let us recall the second example (linguistic variable V' defined as in Sect. 4). The two inequalities (25) and (26) together imply

$$\text{“small or medium”} \equiv \text{“at most medium”}.$$

This could be read as a replacement rule

$$\text{“small or medium”} \longrightarrow \text{“at most medium”},$$

with the meaning that, in any linguistic expression, “small or medium” can be replaced by “at most medium”. If we assume interpretability, we can be sure that this replacement is also semantically correct. If we incorporate as many reasonable relationships in the relation $\sqsubseteq$ as possible such that interpretability can still be fulfilled, we are able to provide a powerful set of simplification rules.

Of course, very simple grammars do not necessitate any simplification. However, if we have to consider very complex rule bases like they appear in grammar-based rule base optimization methods (e.g. inductive learning [28–31] or fuzzy genetic programming [19,20]), simplification is a highly important concern. The methodology presented here allows to deal with simplification in a symbolic fashion – assuming interpretability – without the need to consider the concrete semantics of the expressions anymore.

6 Conclusion

This chapter has been devoted to the interpretability of linguistic variables. In order to approach this key property in a systematic and mathematically exact way, we have proposed to make implicit relationships between the linguistic labels explicit by formulating them as (fuzzy) relations. Then interpretability corresponds to the preservation of this relationships by the associated meaning. This idea has been illustrated by means of two extensive examples. These case studies have demonstrated that well-known common sense assumptions about the membership functions, such as, ordering, convexity, or partition constraints, have a sound justification also from a formal linguistic point of view. In contrast to other investigations, the model proposed in this chapter cannot just be applied to simple fuzzy sets, but also allows smooth integration of connectives and ordering-based modifiers.

By characterizing parameterizations which ensure interpretability, we have been able to provide hints for the design and tuning of fuzzy systems with interpretable linguistic variables. Finally, we have seen that interpretability even corresponds to the fact that symbolic simplification rules on the side of linguistic expressions still remain valid on the semantical side.

Acknowledgements

Ulrich Bodenhofer is working in the framework of the K*plus* Competence Center Program which is funded by the Austrian Government, the Province of Upper Austria, and the Chamber of Commerce of Upper Austria.

References

1. R. Babuška. Construction of fuzzy systems – interplay between precision and transparency. In *Proc. European Symposium on Intelligent Techniques (ESIT 2000)*, pages 445–452, Aachen, September 2000.
2. M. Bikdash. A highly interpretable form of Sugeno inference systems. *IEEE Trans. Fuzzy Systems*, 7(6):686–696, December 1999.
3. U. Bodenhofer. The construction of ordering-based modifiers. In G. Brewka, R. Der, S. Gottwald, and A. Schierwagen, editors, *Fuzzy-Neuro Systems '99*, pages 55–62. Leipziger Universitätsverlag, 1999.
4. U. Bodenhofer. *A Similarity-Based Generalization of Fuzzy Orderings*, volume C 26 of *Schriftenreihe der Johannes-Kepler-Universität Linz*. Universitätsverlag Rudolf Trauner, 1999.
5. U. Bodenhofer. A general framework for ordering fuzzy sets. In B. Bouchon-Meunier, J. Guitiérrez-Ríoz, L. Magdalena, and R. R. Yager, editors, *Technologies for Constructing Intelligent Systems 1: Tasks*, pages 213–224. Springer, 2002. (to appear).
6. U. Bodenhofer and P. Bauer. Towards an axiomatic treatment of "interpretability". In *Proc. 6th Int. Conf. on Soft Computing (IIZUKA2000)*, pages 334–339, Iizuka, October 2000.
7. U. Bodenhofer and E. P. Klement. Genetic optimization of fuzzy classification systems – a case study. In B. Reusch and K.-H. Temme, editors, *Computational Intelligence in Theory and Practice*, Advances in Soft Computing, pages 183–200. Physica-Verlag, Heidelberg, 2001.
8. J. Casillas, O. Cordón, F. Herrera, and L. Magdalena. Finding a balance between interpretability and accuracy in fuzzy rule-based modelling: An overview. In J. Casillas, O. Cordón, F. Herrera, and L. Magdalena, editors, *Trade-off between Accuracy and Interpretability in Fuzzy Rule-Based Modelling*, Studies in Fuzziness and Soft Computing. Physica-Verlag, Heidelberg, 2002.
9. O. Cordón and F. Herrera. A proposal for improving the accuracy of linguistic modeling. *IEEE Trans. Fuzzy Systems*, 8(3):335–344, June 2000.
10. B. De Baets. Analytical solution methods for fuzzy relational equations. In D. Dubois and H. Prade, editors, *Fundamentals of Fuzzy Sets*, volume 7 of *The Handbooks of Fuzzy Sets*, pages 291–340. Kluwer Academic Publishers, Boston, 2000.
11. B. De Baets and R. Mesiar. T-partitions. *Fuzzy Sets and Systems*, 97:211–223, 1998.

12. M. De Cock, U. Bodenhofer, and E. E. Kerre. Modelling linguistic expressions using fuzzy relations. In *Proc. 6th Int. Conf. on Soft Computing (IIZUKA2000)*, pages 353–360, Iizuka, October 2000.
13. M. Drobics, U. Bodenhofer, W. Winiwarter, and E. P. Klement. Data mining using synergies between self-organizing maps and inductive learning of fuzzy rules. In *Proc. Joint 9th IFSA World Congress and 20th NAFIPS Int. Conf.*, pages 1780–1785, Vancouver, July 2001.
14. D. Dubois and H. Prade. What are fuzzy rules and how to use them. *Fuzzy Sets and Systems*, 84:169–185, 1996.
15. D. Dubois, H. Prade, and L. Ughetto. Checking the coherence and redundancy of fuzzy knowledge bases. *IEEE Trans. Fuzzy Systems*, 5(6):398–417, August 1997.
16. D. Dubois, H. Prade, and L. Ughetto. Fuzzy logic, control engineering and artificial intelligence. In H. B. Verbruggen, H.-J. Zimmermann, and R. Babuška, editors, *Fuzzy Algorithms for Control*, International Series in Intelligent Technologies, pages 17–57. Kluwer Academic Publishers, Boston, 1999.
17. J. Espinosa and J. Vandewalle. Constructing fuzzy models with linguistic integrity from numerical data - AFRELI algorithm. *IEEE Trans. Fuzzy Systems*, 8(5):591–600, October 2000.
18. J. Fodor and M. Roubens. *Fuzzy Preference Modelling and Multicriteria Decision Support*. Kluwer Academic Publishers, Dordrecht, 1994.
19. A. Geyer-Schulz. *Fuzzy Rule-Based Expert Systems and Genetic Machine Learning*, volume 3 of *Studies in Fuzziness*. Physica-Verlag, Heidelberg, 1995.
20. A. Geyer-Schulz. The MIT beer distribution game revisited: Genetic machine learning and managerial behavior in a dynamic decision making experiment. In F. Herrera and J. L. Verdegay, editors, *Genetic Algorithms and Soft Computing*, volume 8 of *Studies in Fuzziness and Soft Computing*, pages 658–682. Physica-Verlag, Heidelberg, 1996.
21. S. Gottwald. *Fuzzy Sets and Fuzzy Logic*. Vieweg, Braunschweig, 1993.
22. J. Haslinger, U. Bodenhofer, and M. Burger. Data-driven construction of Sugeno controllers: Analytical aspects and new numerical methods. In *Proc. Joint 9th IFSA World Congress and 20th NAFIPS Int. Conf.*, pages 239–244, Vancouver, July 2001.
23. E. E. Kerre, M. Mareš, and R. Mesiar. On the orderings of generated fuzzy quantities. In *Proc. 7th Int. Conf. on Information Processing and Management of Uncertainty in Knowledge-Based Systems (IPMU '98)*, volume 1, pages 250–253, 1998.
24. E. P. Klement, R. Mesiar, and E. Pap. *Triangular Norms*, volume 8 of *Trends in Logic*. Kluwer Academic Publishers, Dordrecht, 2000.
25. L. T. Kóczy and K. Hirota. Ordering, distance and closeness of fuzzy sets. *Fuzzy Sets and Systems*, 59(3):281–293, 1993.
26. R. Kruse, J. Gebhardt, and F. Klawonn. *Foundations of Fuzzy Systems*. John Wiley & Sons, New York, 1994.
27. R. Lowen. Convex fuzzy sets. *Fuzzy Sets and Systems*, 3:291–310, 1980.
28. R. S. Michalski, I. Bratko, and M. Kubat. *Machine Learning and Data Mining*. John Wiley & Sons, Chichester, 1998.
29. S. Muggleton and L. De Raedt. Inductive logic programming: Theory and methods. *J. Logic Program.*, 19 & 20:629–680, 1994.
30. J. R. Quinlan. Induction of decision trees. *Machine Learning*, 1(1):81–106, 1986.

31. J. R. Quinlan. Learning logical definitions from relations. *Machine Learning*, 5(3):239–266, 1990.
32. A. Ralston, E. D. Reilly, and D. Hemmendinger, editors. *Encyclopedia of Computer Science*. Groves Dictionaries, Williston, VT, 4th edition, 2000.
33. E. H. Ruspini. A new approach to clustering. *Inf. Control*, 15:22–32, 1969.
34. M. Setnes, R. Babuška, and H. B. Verbruggen. Rule-based modeling: Precision and transparency. *IEEE Trans. Syst. Man Cybern., Part C: Applications and Reviews*, 28:165–169, 1998.
35. M. Setnes and H. Roubos. GA-fuzzy modeling and classification: Complexity and performance. *IEEE Trans. Fuzzy Systems*, 8(5):509–522, October 2000.
36. J. Yen, L. Wang, and C. W. Gillespie. Improving the interpretability of TSK fuzzy models by combining global learning and local learning. *IEEE Trans. Fuzzy Systems*, 6(4):530–537, November 1998.
37. L. A. Zadeh. Fuzzy sets. *Inf. Control*, 8:338–353, 1965.
38. L. A. Zadeh. The concept of a linguistic variable and its application to approximate reasoning I. *Inform. Sci.*, 8:199–250, 1975.
39. L. A. Zadeh. The concept of a linguistic variable and its application to approximate reasoning II. *Inform. Sci.*, 8:301–357, 1975.
40. L. A. Zadeh. The concept of a linguistic variable and its application to approximate reasoning III. *Inform. Sci.*, 9:43–80, 1975.

Expressing Relevance Interpretability and Accuracy of Rule-Based Systems

Witold Pedrycz

Department of Electrical & Computer Engineering
University of Alberta, Edmonton, Canada
(pedrycz@ee.ualberta.ca)

and

Systems Research Institute, Polish Academy of Sciences
01-447 Warsaw, Poland

Summary: We discuss a problem of synthesis and analysis of rules based on experimental numeric data and study their interpretation capabilities. Two descriptors of the rules being viewed individually and en block are introduced. The relevance of the rules is quantified in terms of the data being covered by the antecedents and conclusions standing in the rule. While this index describes each rule individually, the consistency of the rule deals with the quality of the rule viewed vis-à-vis other rules. It expresses how much the rule "interacts" with others in the sense that its conclusion is distorted by the conclusion parts coming from other rules. We show how the rules are formed by means of fuzzy clustering and their quality evaluated by means of the above indexes. We also discuss a construction of a granular mapping (that is a mapping between fuzzy clusters in the input and output spaces) and quantify its performance (approximation capabilities at the numeric level). Global characteristics of a set of rules are also discussed and related to the number of information granules formed in the space of antecedents and conclusions.

Keywords rule-based systems, information granulation, fuzzy clustering, relevance and consistency of rules

1 Introduction

Rule-based systems are granular and highly modular models [1,4-9,11-14]. The granularity of rules becomes fully reflected in the form of the antecedents and conclusions and is quantified in the language of sets, fuzzy sets, rough sets, probability, to name the main formal vehicles. The evident transparency of these systems is their genuine asset. Several design issues remain still open and tend to become even more profound as we move toward developing larger systems

- ✓ The origin of the rules and their eventual quantification (e.g., confidence or relevance of relationships where the confidence measure is expressed in relation to experimental data)
- ✓ The dimensionality problem becomes of concern when the number of variables in the rules increases (then the number of rules tend to explode at the exponential rate)

Fuzzy clustering gives rise to fuzzy sets and fuzzy relations being thus constructed in a fully algorithmic fashion. They are examples of information granules regarded as a basic building canvass of numerous fuzzy models. For given collections of such information granules, studying and quantifying relationships between them leads to the emergence of the rule-based models. The preliminary development step of this nature is referred to as a link analysis – a phase in which we reveal and quantify dependencies between information granules before proceeding with further detailed construction and refinement of the rule-based architecture.

More descriptively, one can consider the outcome of the link analysis to arise as a web of connections between the granules. As fundamentally we distinguish between input and output variable(s), there are two essential facets of the analysis that deal with the following aspects

(a) expressing strength between the granules in the input and output space (relevance of the rule), and

(b) completing a consistency (crosstalk) analysis in which we quantify an interaction between the given link and the links that are invoked owing to the interaction (overlap) between the information granules. In the language of the calculus of rule-based systems, the notion of relevance is linked with the notion of strength of the rule whereas the second aspect of consistency is concerned with an interaction between the information granules.

In the sequel, the mapping between the information granules constructed in the input and output space is quantified in terms of its approximation capabilities being regarded at the level of numeric data.

The material is arranged into 6 sections. First, in Section 2 we discuss clustering as a basic means of information granulation and pose the problem of constructing rules. Section 3 covers a discussion of two descriptors of the rules, namely rule relevance and rule consistency. Experimental studies are covered in Section 4 while conclusions are included in Section 5.

2 Clustering as a vehicle of information granulation and a quest for a web of information granules

Quite often, the Fuzzy C-Means (FCM) algorithm arises as a basic vehicle of data granulation. As the method is well-known in the literature, cf. [2], we will not discuss it here but rather clarify the notation and cast the method in the setting of the problem at hand.

The input and output data spaces are denoted by X and Y, respectively. The numeric data set under discussion assumes a form of input – output pairs (x(k), y(k)), k=1,2,…,N where xk $\in$ Rn and yk $\in$ Rm. The results of clustering carried out for X and Y separately are discrete fuzzy sets (more precisely, fuzzy relations) defined over these data sets. In general, we end up with "c[1]" fuzzy relations in X, say A1, A2, …A c[1] and "c[2]" fuzzy relations in Y, namely B1, B2, …, B c[2]. Technically, they are organized in a form of partition matrices so we have

$$U[1] = \begin{bmatrix} A_1 \\ A_2 \\ \dots \\ A_{c[1]} \end{bmatrix} \qquad U[2] = \begin{bmatrix} B_1 \\ B_2 \\ \dots \\ B_{c[2]} \end{bmatrix}$$

Our ultimate goal is a formation of a thorough and constructive description of a web of directed links between Ai and Bj. In a nutshell, such task gives rise to a collection of rules

- if Ai then Bj

A schematic view of this construct is portrayed in Figure 1.

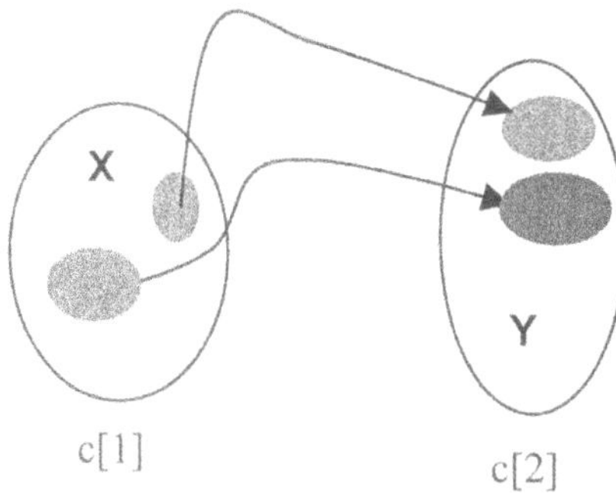

Fig. 1. Rule-based system as a web of links between information granules (shadowed regions) formed in the space of antecedents and conclusions through fuzzy clustering

The techniques of fuzzy clustering, no matter what type of objective function we employ, share several main features whose understanding is of importance in the framework of this problem. Clusters are direction-free (relational) constructs: in the clustering process there are no provisions as to a possible direction between the variables. This implies that the resulting granules do not accommodate any request that relates to the mapping itself and are not formed in a way they reflect a directionality component. In contrast, any mapping (function) is a directional construct. If we establish a link between Ai and Bj, its nature needs to be inspected with respect to the directionality of the rule obtained in this manner. More formally, if there is no directionality component (or it is ignored), we are essentially assessing properties of a Cartesian product of the two information granules, namely $A_1 \times B_j$

Fuzzy clusters (information granules) are sound building blocks of rule-based systems as they provide an answer to the two challenges outlined at the beginning of the previous section. First, they reflect the structure of the data so we anticipate that any cluster comes with enough experimental evidence behind it (otherwise it would not have been formed in the first place). Second, as we are concerned with fuzzy relations rather than fuzzy sets, the dimensionality problem does not arise. For instance if c[1]=c[2]=c then we have (potentially) "c" meaningful rules in spite of the potentially high dimensionality of the input and output space. Otherwise, as it happens quite often when dealing with the individual fuzzy sets, we end up with a combinatorial explosion of the collection of the rules produced at this level.

In what follows, by proceeding with information granules formed by clustering we synthesize a collection of rules and analyze their properties. The crux of this development is as follows. With each A_i i=1,2,...,c[1] we associate a single granule in the conclusion space, that is we form a mapping

$$A_i \rightarrow B_j \tag{1}$$

(the way of determining the associations will be discussed later on). Following this procedure, we build a collection of rules, R, where its cardinality is equal to c[1]. In a concise way, we can summarize the rules as a list of pairs of indexes of the respective information granules

$$(1, j_1) \quad (2, j_2) \dots (c[1], j\, c[1])$$

where j_k is in the range of integers from 1 to c[2].
The characteristics of this form of assignment between the information granules are quantified in the next Section.

3 Characteristics of rules

There are two main descriptors of the rules. The first one reflects the experimental evidence behind the rule (association). The second captures the relationships between the rules discussed as a collection of entities.

3.1 Relevance of the rule

Let us view a certain rule as a Cartesian product) meaning that we do not consider the "direction" of the rule but look at it as an entity linking two information granules defined in the two different spaces (input-output). The rule (1) comes with the (experimental) relevance equal to

$$\text{rel}(Ai \times Bj) = \sum_{k=1}^{N} A_i(\mathbf{x}_k) t B_j(\mathbf{y}_k) \tag{2}$$

where "t" is a t-norm viewed here as a model of an and logical connective. If rel( ) attains higher values, we say that the rule comes with more experimental relevance (in other words, it is more justifiable from the experimental point of view). For fixed "i", we order all associations (rules) Ai →B1, Ai → B2.... Ai → B c[2] according to the associated relevance level. The highest value of the relevance identifies a rule of the form
Ai → B j(i).

In light of the above realization (properties of t-norms), we come up with a straightforward monotonicity property, namely: If Ai $\subseteq$ A'i and Bj $\subseteq$ B'j then rel(Ai $\times$Bj) $\leq$ rel(A'i $\times$B'j). Intuitively, note that if we increase the size of the information granules, this change contributes to the increasing level of relevance of the particular rule as in this way we tend to "cover" more data and thus elevate experimental evidence of this rule.

The relevance defined above exhibits a close analogy to the notion of rule support encountered in data mining that is articulated in the language of probability theory and reads in the form

$$\text{Support}(A \rightarrow B) = \text{Prob}(A \times B)$$

3.2 Consistency of the rule

The associations we have constructed so far were totally isolated. We have not expressed and quantified possible interactions between the rules. Nevertheless,

this interaction does exist owing to the nature of the overlapping fuzzy sets (relations). Considering two rules, we note that their antecedents and conclusions overlap to a certain degree. The overlap at the condition end could be very different than the one encountered at the conclusion end. The differences between the overlap levels shed light on an important issue of consistency of the rules. In turn, this leads to a detection of conflicting rules (where the term of conflict itself is rather continuous than binary, so we talk about a degree of conflict, or equivalently, a degree of consistency).
The problem of conflicting rules is well known in the literature and has a long path in the research in rule – based systems, especially those in the realm of fuzzy controllers, cf [10]. Bearing in mind the origin of the control knowledge, this effect was attributed mainly to some deficiencies of knowledge acquisition when working with a human expert. As in this study, we are concerned with an automated vehicle of information granulation, the effect of conflict is a result of incompatibility of information granules in the spaces of conditions and conclusions. Before we move on with a detailed quantification of this effect, let us concentrate on Table 1. It summarizes four different scenarios of interaction occurring between two rules A1→ B1 and A2 → B2.

Table 1. Four possible scenarios of interaction between two rules and an evaluation of their of interaction

	B1 and B2 similar	B1 and B2 different
A1 and A2 similar	rules are redundant	rules are conflicting
A1 and A2 different	rules exhibit similar conclusion	rules are different

Noticeably, there is a single entry in the table that is of significant concern, namely the rules exhibiting very similar conditions (A1 and A2 are similar) with different conclusions (B1 and B2). What it really means is that almost the same condition triggers two different (if not opposite) actions. In all other cases in Table 1, there are no worries from the conceptual standpoint (obviously one may question some of them as to the computational efficiency and structural redundancy of the rulebase).

We develop an index that captures the effect of consistency of two rules. In its construction, we follow the observations coming from Table 1. The consistency measure is developed in two steps by

- ✓ expressing a measure of consistency of the rules for a single data point (x(k), y(k))
- ✓ constructing a global performance measure over all data (X, Y)

Note that the term of similarity is invoked at the level of the information granules rather than original data, so in essence we are looking at A1(x(k)) and A2(x(k)) along with B1(y(k)) and B2(y(k)). To express a degree of similarity, we use the formula that is deeply rooted in the language of logic and set theory: we say that two sets are equal if the first is included in the another and vice versa. The continuous version of this statement being realized in the framework of membership grades of the respective fuzzy sets reads as

$$A_1(\mathbf{x}(k)) \equiv A_2(\mathbf{x}(k)) = (A_1(\mathbf{x}(k)) \rightarrow A_2(\mathbf{x}(k))t(A_2(\mathbf{x}(k)) \rightarrow A_1(\mathbf{x}(k)) \tag{3}$$

(note that the implication in logic corresponds to the inclusion operation in set theory). The implication is implemented in the form of the pseudoresiduation operation implied by a certain t-norm, namely

$$a \rightarrow b = \sup\{c \in [0,1] \mid atc \le b\}, a, b \in [0,1]$$

The consistency is low only if A1 and A2 are similar and B1 and B2 are different. This naturally leads to the following expression as a measure of consistemcy

$$A_1(\mathbf{x}(k)) \equiv A_2(\mathbf{x}(k)) \rightarrow (B_1(\mathbf{y}(k)) \equiv B_2(\mathbf{y}(k)) \tag{4}$$

To gain a better insight into the character of this expression, we plot it for the pseudoresiduation generated by a product and minimum, refer to Figure 2.

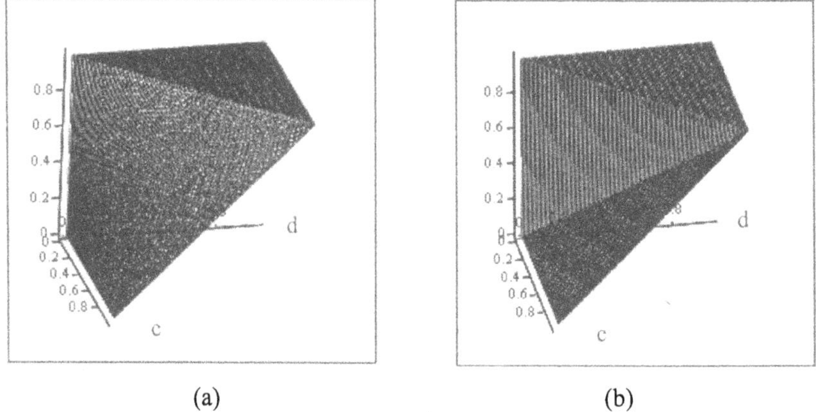

(a) (b)

Fig, 2. Plot of the pseudoresiduation c $\rightarrow$ d induced by the product operation, $c \rightarrow d = \min(1, d/c)$ (a) and minimum, $c \rightarrow d = \min(1, d)$ (b) ("c" and "d" denote a level of similarity in the space of antecedents and conclusions).

The above expression concerns a single data point. Naturally, a sum over the entire data is a legitimate global measure of consistency of the two rules (rule-1 and rule-2),

$$\text{cons}(1,2) = \frac{1}{N}\sum_{k=1}^{N} A_1(\mathbf{x}(k)) \equiv A_2(\mathbf{x}(k)) \rightarrow (B_1(\mathbf{y}(k)) \equiv B_2(\mathbf{y}(k)) \tag{5}$$

Likewise we may like to express a consistency of a given rule versus all other rules in the ruleset R. This leads to the expression

$$\text{Cons}(i, R) = \sum_{\substack{j=1 \\ j \neq i}} \text{cons}(i, j) \tag{6}$$

where "i" and "j" are indexes of the rules in R.

3.3 Accuracy of rule-based mappings between information granules

So far we have discussed the structural aspects of the granules and the ensuing rules. An interesting alternative is to quantify the performance of the resulting numeric mapping (that is a fuzzy rule-based model). Subsequently the dependency between a certain performance index of the rule-based model and the number of information granules becomes of interest. Based on the information granules we constructed in the input and output space and the relationship between them we form a numeric input – output mapping. Note that a granular level we have already designed a correspondence between the granules in the form of the correspondence between their labels, say j = F(i) where "i" and 'j" denote the indexes of the linguistic terms (information granules). This construct is realized as follows. Given a numeric input x we compute its membership to each of c1 information granules defined in the input space. This produces a c1-dimensional vector

$$u = [u1, u2, \ldots, uc1]$$

whose entries are computed as follows

$$u_i(\mathbf{x}) = \frac{1}{\sum_{j=1}^{c_1} \frac{\|\mathbf{x} - \mathbf{v}_i\|^2}{\|\mathbf{x} - \mathbf{v}_j\|^2}} \tag{7}$$

(as a matter of fact this is the same expression that we use to compute the elements of the partition matrix in the FCM algorithm). The vector of the output is determined by taking a weighted sum of the prototypes in the output space v1, v2, ..., vc2 as follows

$$y^{\sim} = \sum_{i=1}^{c_1} u_i \mathbf{v}_{F(i)} \tag{8}$$

The performance of such numeric mapping is evaluated by a sum of squared errors between the output of the model (xx) and the data,

$$Q = \sum_{k=1}^{N} (\mathbf{y}(k) - \mathbf{y}^{\sim}(k))^{T} (\mathbf{y}(k) - \mathbf{y}^{\sim}(k)) \tag{9}$$

where the above summation (k) is taken over all data, k=1, 2, ...,N. One may stress that this type of model is very simple and we did not strive for its further refinement (which could be easily done by considering a more sophisticated type of relationship between the information granules). We were essentially interested in envisioning the model that is data "noninvasive" meaning that it does not impose any constraints on the data.

4 Experiments

In this section, we are concerned with a synthesis and analysis of fuzzy rules for two selected data sets available on the WWW (ftp://ftp.ics.uci.edu/pub/machine-learning-databases/), namely fuel consumption and Boston housing data. In both cases, we use the FCM method set up in the same way across all experiments. The fuzzification factor (m) standing in the standard objective function is equal to 2,

$$Q = \sum_{i=1}^{c} \sum_{k=1}^{N} u_{ik}^{m} d_{ik}^{2}$$

where c = c[1] or c[2].

The dissimilarity between the patterns is expressed in terms of a weighted Euclidean distance. A partition matrix is initialized randomly. Once the information granules (clusters) have been generated, the analysis of the rules is completed in terms of their relevance and consistency. Furthermore we carry out a global analysis as to the number of rules and quantify them as an overall collection and derive characteristics related to their suitability in the description of data. As to the implementation details of t-norms, we use a product operation. The implication is also induced by the same t-norm.

Auto data

This dataset comes from the StatLib library that is maintained at Carnegie Mellon University. The data concerns a city-cycle fuel consumption that is expressed in miles per gallon and has to be predicted on a basis of 7 attributes (number of cylinders, displacement, horsepower, weight, acceleration, model year, and origin). All but the fuel consumption are treated as inputs. City-cycle fuel consumption is an output variable (conclusion).

We choose the number of clusters (information granules) to be equal to c[1] = c[2] = 7 (this is done for illustrative reasons; we discuss this selection in more detail later on). The relevance of the rules are shown in Figure 3.

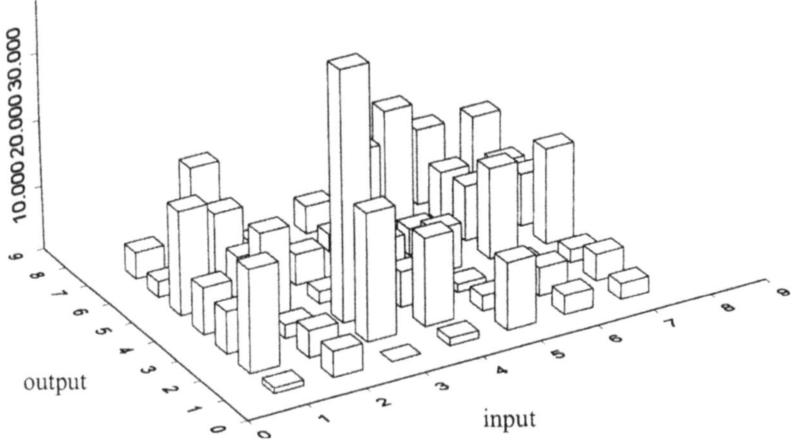

Fig. 3. Relevance of the rules

In each row of the matrix, Figure 3, we have at least one fairly dominant association. By selecting the dominant links, we end up with the following seven rules

1 → 5	2 → 7	3 → 3	4 → 5
5 → 6	6 → 4	7 → 4	

where the above is a schematic summary of the links (associations) between the information granules in the input and output space. Noticeably all rules are quite similar in terms of their relevance. One may anticipate this to be a result of using clusters as generic building blocks of the rules that comes with a similar experimental evidence behind the fuzzy relations).

The consistency levels of these rules are as follows

rule no.	consistency
1	2.707643
2	2.962191
3	3.455391
4	2.634168
5	2.604724
6	2.667142
7	2.687571

Again, these levels of consistency are fairly similar across all the rules with a single exception (rule 3→ 3 is the most consistent with the corresponding value of the consistency level of 3.45).

Now we investigate a situation where there is a significantly different number of the information granules in the input and output spaces. More specifically, we analyze (a) c[1] =3 and c[2]=10 and (b) c[1] =10 and c[2]=3. The results are visualized in Figure 4 and Tables 2 and 3.

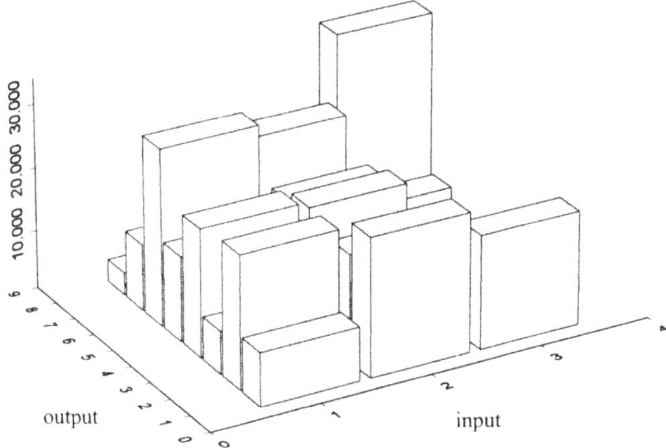

Fig. 4. Relevance of the rules for c[1]=3 and c[2]=10

Table 2. Rules (associations) formed by the information granules for c[1]=3 and c[2]=10

Rule	relevance	consistency
1 → 6	28.63	0.76
2 → 7	23.85	1.02
3 → 8	33.98	1.20

Table 3. Rules (associations) formed by the information granules for c[1]=3 and c[2]=10

rule	relevance	consistency
1→1	39.32	3.26
2→3	18.98	2.23
3→1	33.16	3.43
4→3	18.62	2.27
5→1	23.54	2.35
6→3	19.16	2.25
7→2	26.36	3.54
8→3	18.20	2.25
9→1	22.02	2.30
10→3	22.23	2.31

The most striking is a fact of increasing consistency of the rules with the increasing number of the information granules in the condition space. The consistency goes down significantly when the values of c[2] get lower. The change in c[2] from 10 to 3 decreases the level of consistency by a factor of 2.

In the above series of experiments, the number of the rules is fixed in advance (as we picked up some specific values of c[1] and c[2]). Evidently, the number of information granules (both in the input and output space) and a way in which they are distributed across the spaces induce a certain quality of the rules. The performance of the rules can be quantified in the two terms we have already used, namely the relevance and consistency of the rules. These characteristics were computed for some specific configurations of the values of c[1] and c[2]. Importantly, these two are crucial to the overall collection of the rules. We complete calculations of the relevance and the consistency of the entire system of rules treating them as a function of c[1] and c[2]. The plots of the overall summary of the performance of the rules, rel(R) and cons(R) are shown in Figure 5 and 6, respectively,

$$\mathrm{rel}(R) = \sum_{k:\text{all rules}} \mathrm{rel}(\mathrm{rule}_k) \tag{7}$$

$$\mathrm{cons}(R) = \sum_{k:\text{all rules}} \mathrm{cons}(\mathrm{rule}_k) \tag{8}$$

They reveal important an important feature: the total relevance of the rules depends very much on the values of c[2] and it drops down radically once we move beyond a certain number of the information granules. The confidence level of all rules changes radically with respect to c[1].

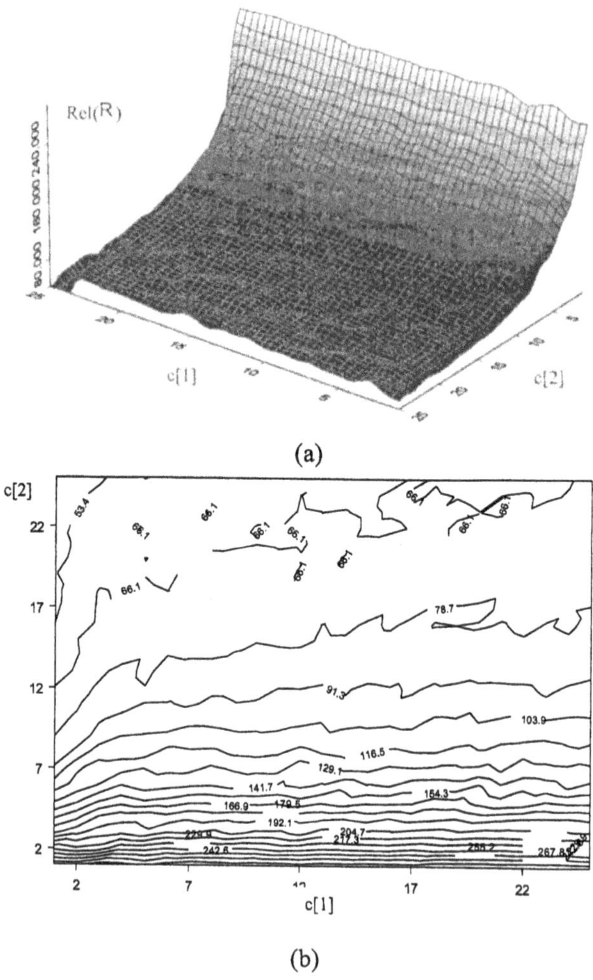

Fig. 5. Relevance rel(R) as a function of c[1] and c[2]: 3D plot (a) and contour plot (b)

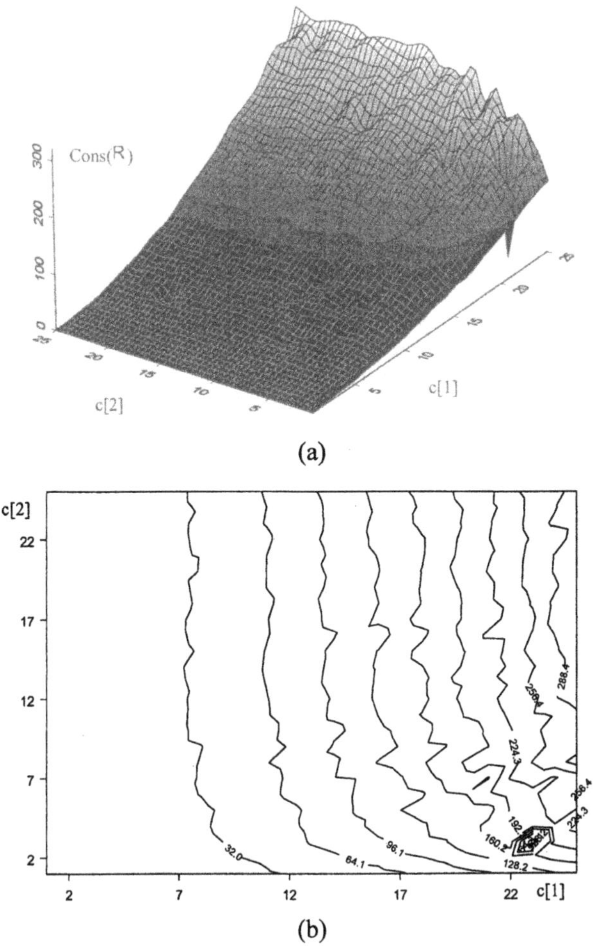

Fig. 6. Consistency cons(R) as a function of c[1] and c[2]: 3D plot (a) and contour plot (b)

If c[2] is too high, we note a significant drop in the relevance level of the rules. The second number, c[1] does not impact the relevance to the same extent. The situation is different for the consistency of the rules: here c[1] plays a dominant role and too low values have an evident impact on the reduced consistency of the ruleset. The changes in c[2] affect the consistency but the pattern of changes is not too evident. In some region we can see that lower values of c[2] contribute to higher consistency values.

Boston housing data. The data describes real estate in the Boston area and is concerned with a number of characteristics such as price, square footage, distance from main employment centers, student-teacher ratio. We consider 8 variables to be inputs and the 6 remaining to be the outputs. The clustering is completed in the same way as for the first experiment.Table 4 summarizes the findings for some selected values of c[1] and c[2].

Table 4. Relevance and confidence of rules for selected values of c[1] and c[2]

Rules c[1]=, c[2]=7	relevance	consistency
1 → 3	16.300497	3.689469
2 → 5	23.272026	3.831919
3 → 7	13.974018	3.400926
4 →6	27.478477	3.719316
5 → 4	15.116308	4.849354
6 → 6	23.940590	3.834462
7 → 5	13.321856	3.334186

Rules c[1]=10, c[2]=3	relevance	consistency
1 → 1	33.544365	4.607216
2 → 3	23.007704	2.892645
3 → 3	22.627285	2.893813
4 → 3	26.637455	2.919845
5 → 2	21.804045	6.648260
6 → 3	28.802927	2.835659
7 → 2	23.232628	6.176850
8 → 1	24.259182	4.682113
9 → 3	24.261139	2.979397
10 → 3	21.016140	2.311069

Rules c[1]=3, c[2]=10	relevance	consistency
1→9	24.538391	0.607769
2→10	33.162212	1.178861
3→9	31.443815	0.571092

The comparison of the results reveals interesting regularities. First, the consistency of the rules is affected by the number of the fuzzy sets defined in the conclusion space (this is the same observation as encountered in the previous data set). Secondly, with the increasing number of the information granules in the input space we note a tendency of increasing values of the relevance of the rules.

The relevance and consistency of the rules, Figure 7 and 8, treated as a function of c[1] and c[2] exhibit the pattern of changes as for the previous data set. Again, the relevance is affected by the values of c[2] while c[1] impacts the consistency of the rules.

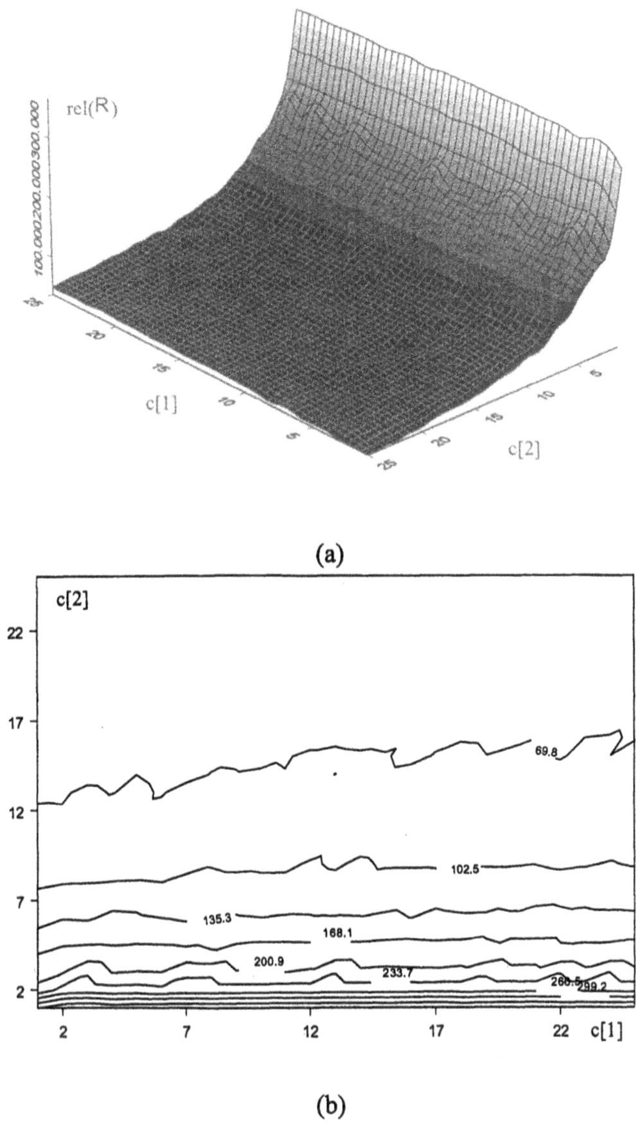

Fig. 7. Relevance of the rules as a function of c[1] and c[2]: 3D plot (a) and contour plot (b)

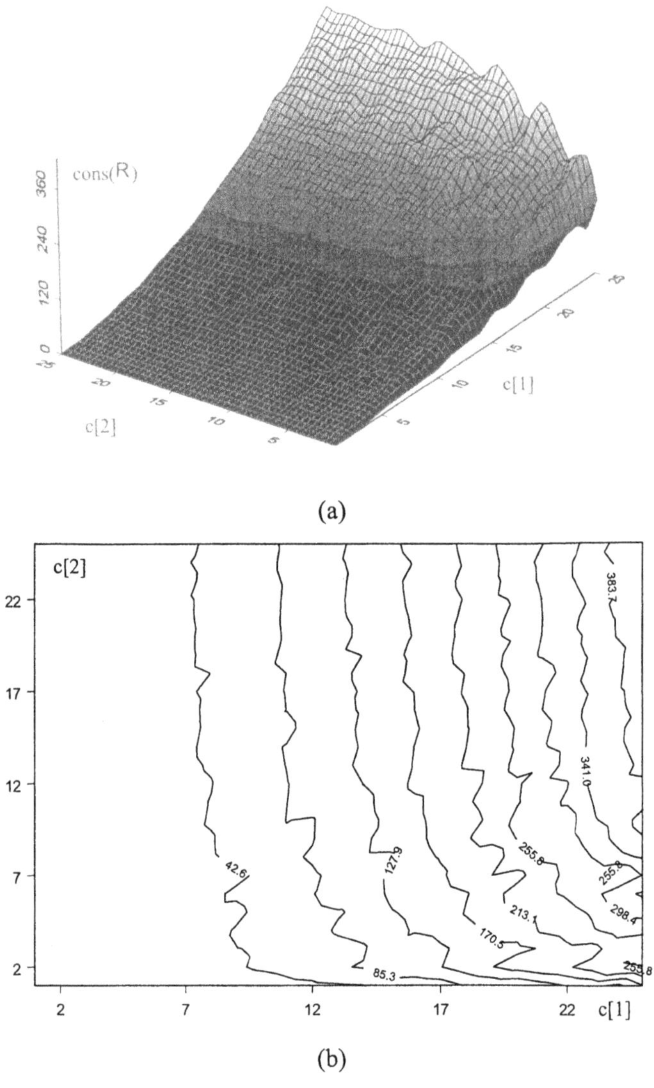

Fig. 8. Consistency of the Boston housing rules as a function of c[1] and c[2]: 3D plot (a) and their contour plot (b)

The experiments lead to some interesting design guidelines of rule-based systems as to the selection of the size of the vocabulary of information granules in the input and output spaces. To assure sufficient relevance and consistency, a feasible region in the c[1] – c[2] plane is delineated in Figure 9. Obviously, some detailed cutoff values of c[1] and c[2] are data-specific yet a general tendency as to the behavior of these two characteristics still holds.

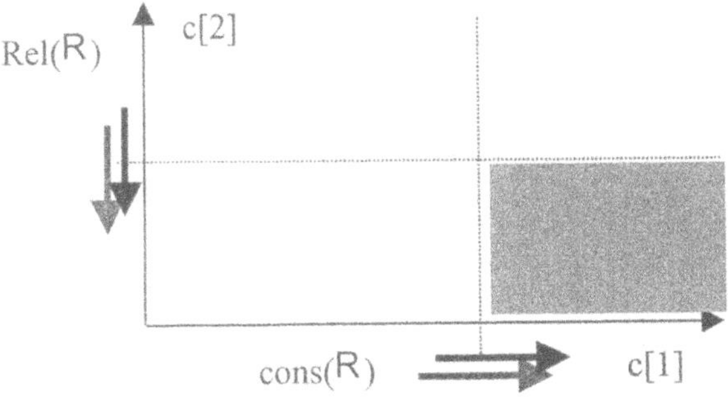

Fig. 9. A feasible region of granularity in the c[1] and c[2] plane

Proceeding with the Boston housing data as we started in the previous section, we compute Q for a variable number of c1 and c2. This helps us understand the form of relationship that occurs between the size of the vocabularies of information granules and the performance of the model.

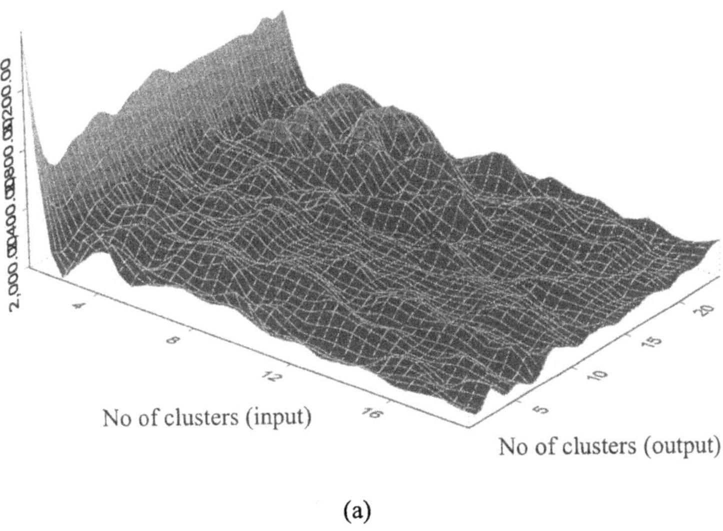

(a)

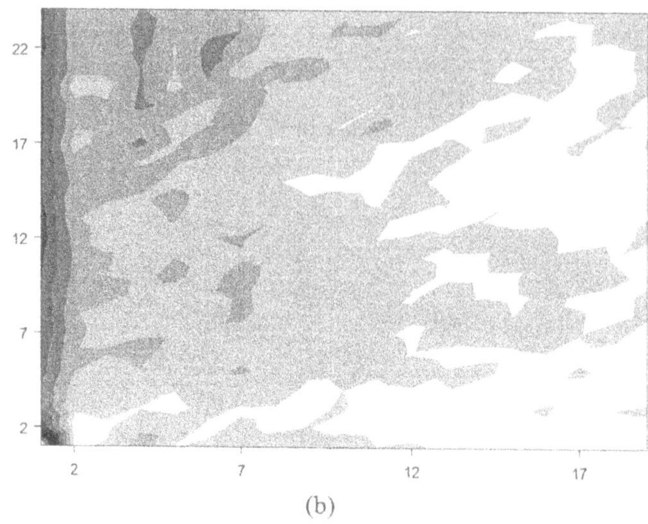

Fig. 10. Performance index as a function of clusters (c1 and c2) constructed in the input and output space: 3D plot (a) and contour plot (b)

The performance index, Figure 11, reveals interesting properties. First, the performance exhibits a monotonic behavior as a function of the information granules constructed in the input space; where this number increases the values of the performance index get lower. The most visible drop in the values of Q is encountered at the lower values of c1. Second, for the fixed number of the clusters in the input space the increase of the number of the clusters in the output may result in different local patterns of behavior. No straight and strong dependency can be discovered. From the modeling perspective we can conclude that there could be a pair of (c1, c2) leading to a minimal value of the performance index. More specifically, we may fix the value of c1 and determine a value of c2 leading to the minimal value of Q. The granular mapping (viz. the mapping between the information granules) can be refined further through standard techniques of gradient-based optimization.

Figure 11 illustrates the performance index for the auto data. Noticeably, there is a strong dependency of the values of Q on the number of the information granules. For low values of c1 the performance index assumes high values. For c1 greater than 4 the approximation error is reduced quite substantially. The landscape of Q is quite rugged resulting in a number of local maxima. Some of them are quite visible. The design guideline concerning the number of the information granules should take into consideration the relevance, consistency and approximation capabilities of the corresponding rule-based model.

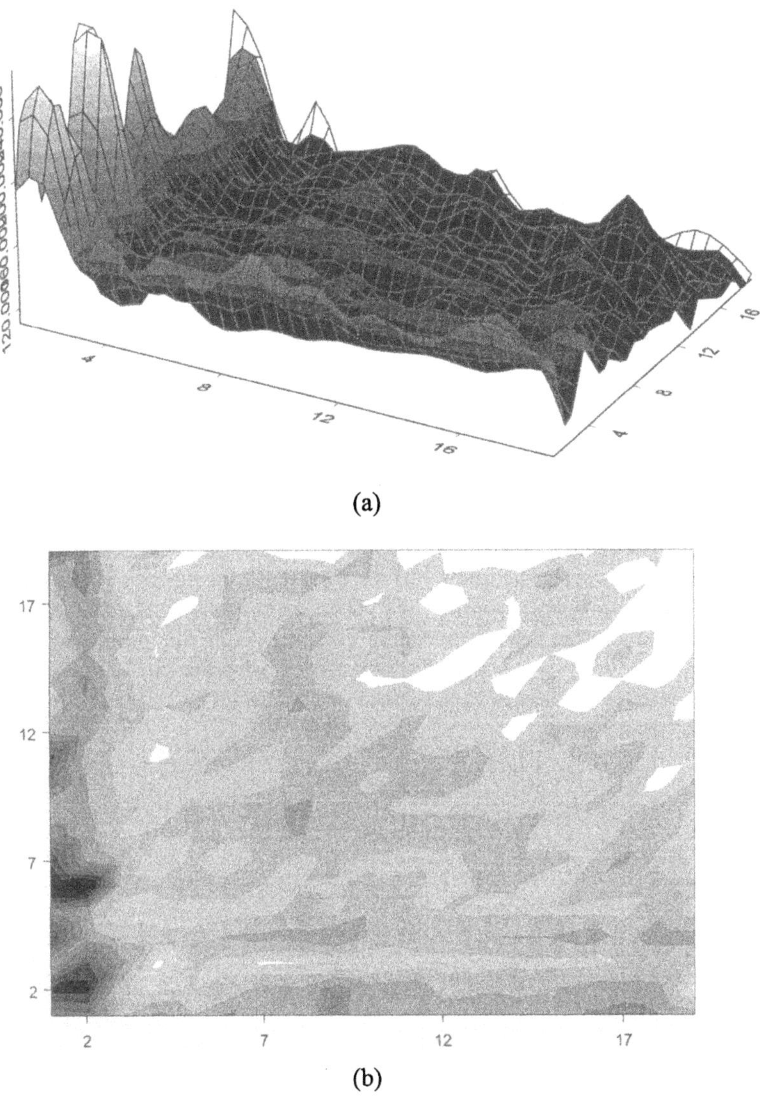

Fig. 11. Auto-data: performance index as a function of clusters (c1 and c2) constructed in the input and output space: 3D plot (a) and contour plot (b)

5 Conclusions

The study focused on the synthesis of information granules (fuzzy sets and fuzzy relations) and generation of rules (associations) composed of them. Such associations are characterized by two indexes. The first one is about the relevance of the rule and expresses how much experimental evidence is behind the association. The second one is about a directional aspect of the construct and describes much a given rule interacts with all others and produces a crosstalk (relational) effect. We also looked into an interesting numeric quantification of the rules with respect to the size of the vocabulary of the information granules both at the antecedent and conclusion part of the rules. It was revealed that there are some cutoff values of these granules beyond which the quality of the rules drops off significantly.

The above analysis imposes a minimal level of structuralization of the rules. The rules developed here are the direct product of data summarization as we use fuzzy clustering to reveal and capture the structure of the data. It should be stressed that the cluster-based rules help avoid combinatorial explosion in cases of high dimensional spaces. The language of relations (rather than fuzzy sets) becomes instrumental in this setting.

The issue of the numeric quality of the rules has not been discussed at all (namely, a problem of expressing the quality of the rules vis-à-vis the original experimental numeric data). In other words, we have not studied the features of the transformation of the inference results (coming from the rules) into numeric representations. This phase is definitely related with the clustering mechanism itself, the number of clusters in the space of conclusion and a way of aggregation of the conclusions.

References

1. A. Bárdossy, L. Duckstein, *Fuzzy Rule-Based Modeling with Application toGeophysical, Biological and Engineering Systems*, CRC Press, Boca Raton, 1995.
2. J.C. Bezdek, *Pattern Recognition with Fuzzy Objective Function Algorithms*, Plenum Press, N. York, 1981.
3. B. Bouchon-Meunier, M. Rifqi, S. Bothorel, Towards general measures of comparison of objects, *Fuzzy Sets and Systems*, 84, 2, 1996, 143-153.
4. 0. Cordón, M.J. del Jesus, F. Herrera, A proposal on reasoning methods in fuzzy rule based classification systems, *Int. J. of Approximate Reasoning*, 20, 1999, 21-45.

5. 0. Cordón, F. Herrera, Villar P., Analysis and guidelines to obtain a good uniform fuzzy partition granularity for fuzzy rule-based systems using simulated annealing, *Int. J of Approximate Reasoning*, Vol. 25, 3, 2000, 187-215.
6. M. Delgado, F. Gomez-Skarmeta, and F. Martin, A fuzzy clustering-based prototyping for fuzzy rule-based modeling, *IEEE Transactions on Fuzzy Systems*, 5(2), 1997, 223-233.
7. M. Delgado, A.F. Gomez-Skarmeta, F. Martin, A methodology to model fuzzy systems using fuzzy clustering in a rapid-prototyping approach, *Fuzzy Sets and Systems*, vol. 97, no.3, 1998, 287-302.
8. D. Dubois, H. Prade, What are fuzzy rules and how to use them, *Fuzzy Sets and Systems*, 84, 1996, 169-185.
9. H. Ishibuchi, K. Nozaki, N. Yamamoto, H. Tanaka, "Selecting fuzzy if-then rules for classification problems using genetic algorithms", *IEEE Transactions on Fuzzy Systems*, Vol. 3: 3, 1995, 260-270.
10. W. Pedrycz, *Fuzzy Control and Fuzzy Systems*, 2nd edition, Research Studies Press, Chichester, 1993.
11. E. H. Ruspini, On the semantics of fuzzy logic, *International Journal of Approximate Reasoning* 5, 1991, 45-88.
12. T. Sudkamp, Similarity, interpolation, and fuzzy rule construction, *Fuzzy Sets and Systems*, vol. 58, no. 1, 1993, 73-86.
13. T. A. Sudkamp, R. J. Hammell II, Granularity and specificity in fuzzy function approximation, in Proc. NAFIPS-98, 105-109, 1998.
14. R. R. Yager, D. P. Filev, *Essentials of Fuzzy Modeling and Control*. J. Wiley, New York, 1994.

Conciseness of Fuzzy Models

Toshihiro Suzuki[1] and Takeshi Furuhashi[2]

[1] Dept. of Information Electronics, Nagoya University, Furo-cho, Chikusa-ku, Nagoya 464-8603, Japan
[2] Dept. of Information Engineering, Mie University, 1515 Kamihama-cho, Tsu 514-8507, Japan

Abstract. Fuzzy models are used to describe input-output relationships of unknown nonlinear systems in an interpretable manner for humans. Interpretability is one of the indispensable features of fuzzy models, which is closely related to their conciseness. The authors introduce the conciseness of fuzzy models, based on observations that humans grasp the input-output relationships by granules. The conciseness measure is then formulated by introducing De Luca and Termini's fuzzy entropy and a new measure is derived from the analogy of relative entropy. This chapter also discusses the conflicting relationships between the conciseness and the accuracy of fuzzy models. A fuzzy modeling with Pareto optimal solutions is presented. Numerical experiments are done to demonstrate the effects of the conciseness measure.

1 Introduction

Fuzzy models [15] have been constructed by acquiring knowledge from experts, but knowledge acquisition through interviews has often been difficult because they seldom have explicit knowledge that can be extracted as fuzzy if-then type rules.

Many methods that automatically derive if-then type fuzzy rules from numerical data have been proposed to overcome the problem of knowledge acquisition. Since tuning of both the antecedent and consequent part of fuzzy rules can be formulated as an optimization problem, evolutionary algorithms have been applied to solve this problem. However, automatically derived fuzzy models are not often linguistically interpretable, as recognized in the literature [12,13,16].

The interpretability of fuzzy models has been evaluated by the number of fuzzy rules, the number of membership functions [6], or the degree of freedom term of AIC (Akaike's Information Criterion) [1,9], but it also depends on other factors such as the shapes and allocation of membership functions, and more on interpreter's prior knowledge.

In this chapter, fuzzy modeling using the genetic algorithm (GA) and a new conciseness measure is presented. Conciseness could be a criterion for the interpretability of fuzzy models, and it is defined by referring to shapes and allocation of membership functions of a fuzzy model.

Discussion on how prior knowledge of the target system affects interpretability of fuzzy models is also presented in this chapter. It is shown that, when the prior knowledge is unavailable, the input-output relationships of a *concise* fuzzy model are easy for us to understand.

This chapter introduces De Luca and Termini's fuzzy entropy [2] as a conciseness measure for evaluation of the shape of a membership function. De Luca and Termini proposed fuzzy entropy as a measure of fuzziness. They used Shannon's function, and defined a measure that became largest at the grade of membership of 0.5. Several authors have attempted to quantify fuzziness and proposed fuzzy entropy [10,11]. This measure has been applicable to various industrial applications, e.g. image processing [17], fuzzy clustering [7], etc.

De Luca and Termini's fuzzy entropy, however, cannot evaluate deviation of a membership function. This chapter also presents a new measure derived on the analogy of relative entropy. This new measure is also a conciseness measure that evaluates the deviation of allocation of membership functions on the universe of discourse. A combination of these two measures is shown to be another good conciseness measure.

A concise fuzzy model is not always interpretable. In the case where a human has prior knowledge about the target system, an interpretable model could be the one that explicitly explains his/her knowledge. Experimental results show that the obtained concise model is interpretable. The results also show that human's knowledge changes the most interpretable model from the most concise model.

Since conciseness is in conflict with the accuracy of fuzzy models, fuzzy modeling using GA with these two criteria is formulated as a multi-objective optimization problem [3–5,14].

The rest of this chapter is organized as follows: Section 2 describes fuzzy modeling using GA with the average measure, Sect. 3 discusses the conciseness of fuzzy models in an illustrative way, Sect. 4 defines the average measure, Sect. 5 shows some numerical results about the proposed measure and, finally, Sect. 6 concludes this chapter.

2 Fuzzy modeling using GA with conciseness measure

This section describes the details of fuzzy modeling using GA with a new conciseness measure. This measure is defined in Sect. 4 as the average measure that follows discussion on the conciseness of fuzzy models in Sect. 3. In this chapter, a very simple procedure of fuzzy modeling using GA is described to clarify feasibility of the average measure. The goal of the fuzzy modeling is to obtain fuzzy models which are interpretable for human beings, not to mention that they should have accurate model output. Thus the fuzzy modeling is considered as a multi-objective optimization problem.

2.1 Fuzzy model

A single-input single-output fuzzy model with simplified fuzzy inference [8] is used in this chapter. The output y of a fuzzy model with the input x is given by

$$y = \sum_{i=1}^{N_{\mathrm{m}}} \mu_i(x) \cdot c_i, \tag{1}$$

where $\mu_i(x)$ and c_i ($i = 1, \cdots, N_{\mathrm{m}}$) are grades of membership in the antecedent part and singletons in the consequent part, respectively. N_{m} is the number of membership functions. With the use of singletons in the consequent part, we can easily see the effect of changes of membership functions in the antecedent part on the actual output.

The rules of such models are, for example, described as follows:

IF x is S **THEN** y is c_1
IF x is M **THEN** y is c_2
IF x is B **THEN** y is c_3.

This is the case where 3 antecedent membership functions are allocated on the universe of discourse, i.e. $N_{\mathrm{m}} = 3$. We assigned linguistic terms S, M and B to antecedent membership functions $\mu_1(x)$, $\mu_2(x)$ and $\mu_3(x)$, respectively.

Given a set of data D from the target system, the singleton in the consequent part of a rule is calculated by

$$c_i = \sum_{k=1}^{N_{\mathrm{d}}} y_k \cdot \mu_i(x_k), \tag{2}$$

where N_{d} is the number of input-output pairs (x_k, y_k) $(k = 1, \ldots, N_{\mathrm{d}})$.

2.2 Membership functions

The following conditions for allocation of membership functions are used in this chapter:

(a) For all $x \in X$, membership functions $\mu_i(x)$ $(i = 1, \cdots, N_{\mathrm{m}})$ satisfy

$$\sum_{i=1}^{N_{\mathrm{m}}} \mu_i(x) = 1. \tag{3}$$

(b) For all x that are not crest points, i.e. x s.t. $\mu_i(x) = 1$, only two membership functions have the grades larger than zero, i.e. $\mu_i(x) > 0$ and for all the others $\mu_i(x) = 0$.

(c) Each membership function is similar with respect to the crest point $x = a$, in the sense that

$$\mu_{l_i}(x) = \mu_{r_i}\left(1 - \frac{1-a}{a}x\right), \tag{4}$$

where

$$\mu_{l_i}(x) = \{\,\mu_i(x)\,|\,x \le a, \mu_i(x) > 0\,\} \tag{5a}$$
$$\mu_{r_i}(x) = \{\,\mu_i(x)\,|\,a \le x, \mu_i(x) > 0\,\}. \tag{5b}$$

(d) All the fuzzy sets are convex.

Fig. 1 shows an example allocation of membership functions that satisfies these conditions. These conditions are introduced for good interpolation

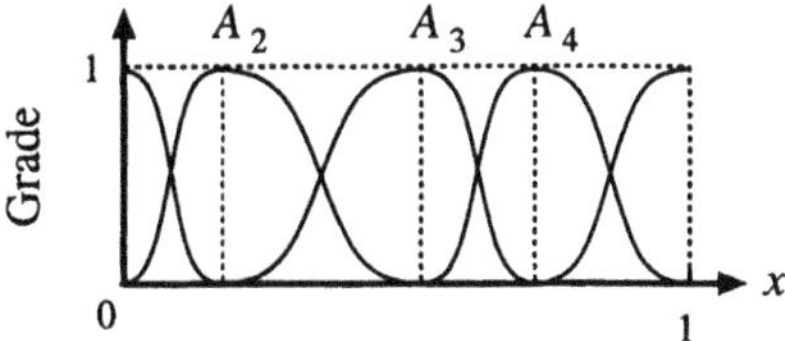

Fig. 1. Example allocation of membership functions that satisfies the conditions

between the fuzzy rules. Good interpolation here means that for any input a model has a correspondent output with some accuracy, given a collection of data. The membership functions satisfying the conditions (a)–(d) that cover the universe of discourse with two properly overlapping convex fuzzy sets at any $x \in X$ ensure the good interpolation. These membership functions are distinct from each other. The conditions (a)–(d), which have been employed for obtaining practical fuzzy models, could also contribute to interpretability of fuzzy models. Under these conditions, freedom lies in number, allocation, and shape of membership functions. The chapter discusses evaluation function for fuzzy models related to these parameters, especially allocation and shape of membership functions. We are to call this performance index as conciseness of fuzzy models in this paper.

2.3 Rank-based evaluation

A rank-based evaluation is used for finding Pareto-optimal solutions with the two criteria: conciseness and accuracy. Conciseness is measured by the average measure defined in eq. (25) in Subsection 4.4, and accuracy is measured by mean squared error given by

$$\frac{1}{N_t}\sum_{k=1}^{N_t}(y_k - \hat{y}_k)^2, \tag{6}$$

where N_t is the number of the test data, y is the output from the target system, and $\hat{y}$ is the model output.

The rank of each chromosome i in the population R_i is given by

$$R_i = 1 + q_i, \tag{7}$$

where q_i is integer, and chromosome i is inferior to q_i chromosomes.

2.4 Chromosome encoding

Since simplified fuzzy inference is employed and allocation of membership functions is restricted with the conditions (a)–(d) in Sect. 2.2, the parameters of a fuzzy model are the following three parameters: the number, the positions of the crest points and the shape of the membership functions.

We assume that the number of membership functions is already determined. This number is also a good measure for the interpretability of fuzzy models, and has been employed for the fuzzy modeling. The chapter concentrates on the remaining two parameters, the allocation and the shape of membership functions. The positions of the crest points and the shape of the membership functions of a fuzzy model are encoded into a chromosome. The length of all the chromosomes in the population is fixed, and each chromosome has $N_m + 1$ genes: N_m for storing the positions of the crest points and one for the shape. An example is shown in Fig. 2, where $N_m = 6$. Fig. 2 (a) and (b) show a chromosome and the allocation of membership functions represented by the chromosome, respectively.

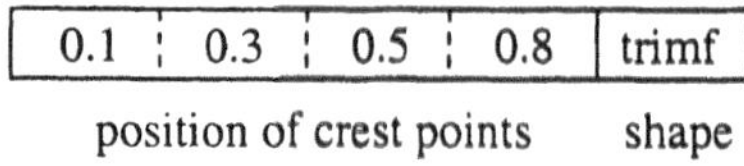

(a) chromosome

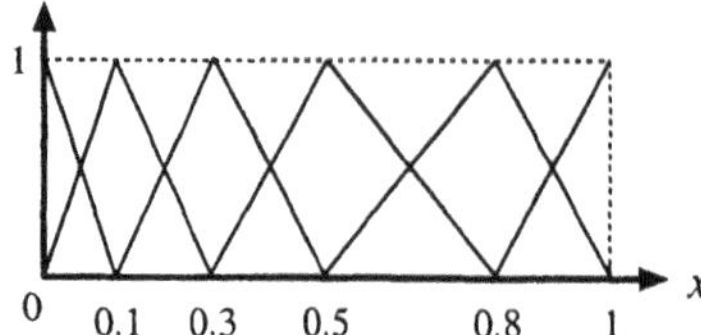

(b) Allocation of membership functions represented by the above chromosome

Fig. 2. Example of a chromosome

The positions of the crest points of the membership functions at both of the ends are always $x = 0$ and 1, respectively. Only the positions of the intermediate crest points appear in a chromosome. The shape parameter is set simple in this chapter as either "trimf", "sigmf1" or "sigmf2", and the shapes of them are shown in Fig. 3.

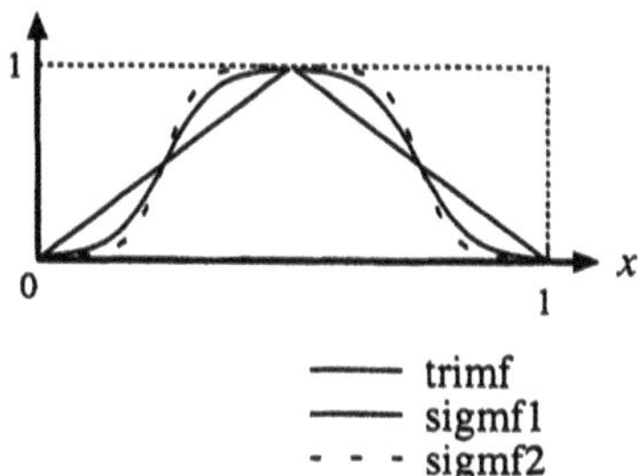

Fig. 3. Shapes of membership functions

The number of the chromosomes in the population was fixed at N_c.

2.5 Genetic operators

Genetic operators are applied in the usual order: selection, crossover and mutation. The selection operation is based on rank which is assigned using the average measure and the accuracy. The roulette wheel selection method is used to select N_c chromosomes. The probability for the roulette wheel selection is given by

$$\frac{P_i}{\sum_{i'} P_{i'}}, \tag{8}$$

where $P_i = 1/R_i$. For the crossover operation, the chromosomes are selected in pairs, and for each pair the two chromosomes are crossed over at a random position. Some chromosomes among the lowest ranked chromosomes are randomly selected and mutated.

3 Conciseness of fuzzy models

Interpretability of fuzzy models heavily depends on human's prior knowledge. If we have profound knowledge about the target system, a model that makes our knowledge explicit could be considered interpretable. Even though fuzzy models have many parameters and the input-output relationships are highly non-linear, our knowledge helps us interpret the relationships. Even a concise model could not be interpretable if it does not fit into our prior knowledge.

In the case where we have no prior knowledge, a concise fuzzy model could be easy for a human to interpret. Assuming that four types of single-input single-output fuzzy models are given as shown in Fig. 4, the rules of which are the same among the four models as the following:

IF x is S **THEN** y is c_1
IF x is M **THEN** y is c_2
IF x is B **THEN** y is c_3.

Even if the membership functions overlap with each other from the conditions in Sect. 2.2, we can see the exact output from the above rules where the input has full membership to either S, M or B. It may be, however, difficult to grasp the output, when the input has a certain membership to one of the membership functions, and at the same time it has a certain membership to another. This is due to the fact that the output of a fuzzy model are dependent on multiple membership functions. When the number of fuzzy rules and the number of membership functions are fixed, the allocation and shape of membership functions greatly affect the output. Conciseness discussed in this chapter comes from this point of view.

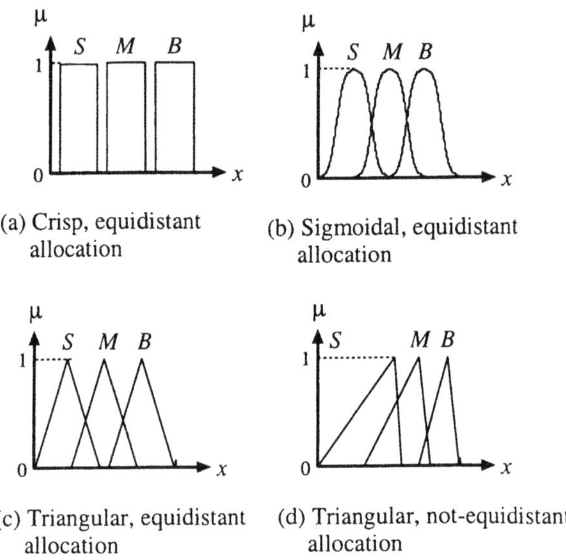

Fig. 4. Membership functions

Fig. 4 (a) shows a case where crisp membership functions S, M and B are equidistantly allocated on the universe of discourse in the antecedent. The output is depicted with granules O_B, O_M and O_S in Fig. 5 (a). This model can be described with the following rules:

IF x is S **THEN** y is O_B
IF x is M **THEN** y is O_M
IF x is B **THEN** y is O_S.

The granules S, M, B, O_B, O_M and O_S help us grasp the input-output relationships. We interpret the models in Fig. 4 in the form of above rules with the granules. The model in Fig. 4 (a) with crisp and equidistantly allocated granules is the most concise, and could be the most interpretable.

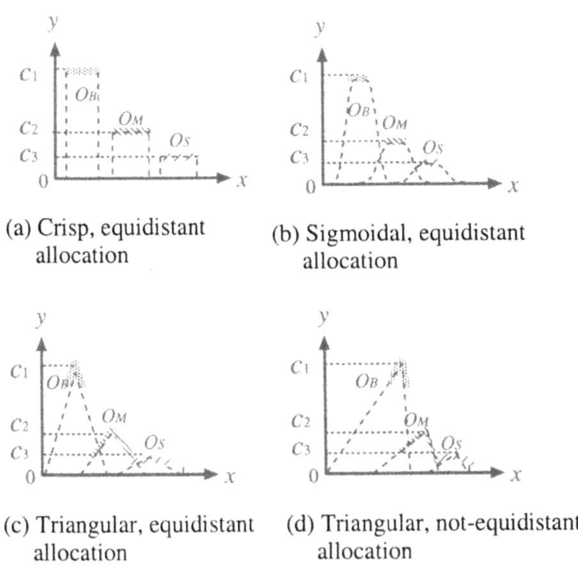

Fig. 5. Output granules

Fig. 4 (b) and (c) show cases with Gaussian and triangular membership functions, respectively. Fig. 4 (d) is the case where triangular membership functions are allocated unevenly. Every model can be described with the above three rules. But it becomes less and less concise, and more and more difficult to grasp the input-output relationships. Thus the model in Fig. 4 (d) could be the most difficult to interpret among the four models in Fig. 4.

Assume that we have the following knowledge about the target system: "*the non-linearity of the system becomes stronger with larger* x." In this case, the interpretability of the models in Fig. 4 changes drastically from the above observation. We may think that the model in Fig. 4 (d) is the most interpretable, because this model fits our knowledge most.

Interpretability of fuzzy models could depend on prior knowledge. For quantitative analysis of interpretability, this chapter limits the discussions in the following sections to the case where prior knowledge is unavailable.

In the above discussion, we have used the phrase "*conciseness of fuzzy models*," which means that "*a fuzzy model is more concise, if the membership functions are more equidistantly allocated on the universe of discourse, and the shapes of membership functions are less fuzzy.*

4 Fuzzy entropy

A quantitative measure of the conciseness of fuzzy models is examined in the following two subsections:

4.1 De Luca and Termini's fuzzy entropy

De Luca and Termini defined fuzzy entropy of fuzzy set A as

$$d(A) = \int_{x_1}^{x_2} \{ -\mu_A(x) \ln \mu_A(x) - (1 - \mu_A(x)) \ln(1 - \mu_A(x)) \} \, dx, \tag{9}$$

where $\mu_A(x)$ is the membership function of fuzzy set A. If $\mu_A(x) = 0.5$ for all x on the support of A, then the fuzzy entropy of fuzzy set A is the maximum.

The fuzzy entropy of the membership functions with various shapes in Fig. 6 has the inequality:

$$d(D) > d(C) > d(B) > d(A) = 0. \tag{10}$$

Thus this fuzzy entropy can distinguish the shapes of membership functions and coincides with the meaning of the conciseness in the previous section. This entropy can be a candidate for the quantitative measure of the conciseness of fuzzy models.

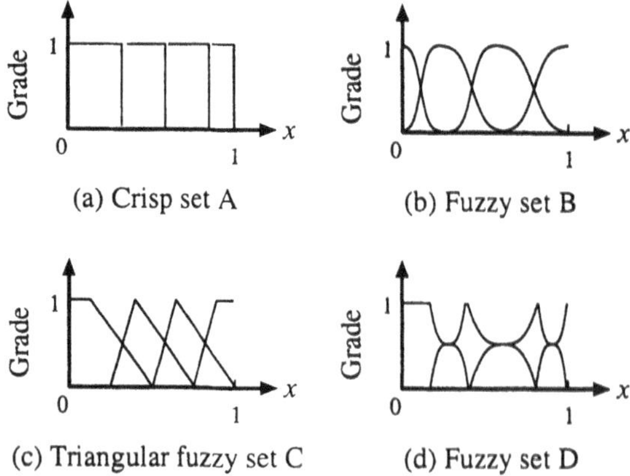

Fig. 6. Membership functions with various shapes

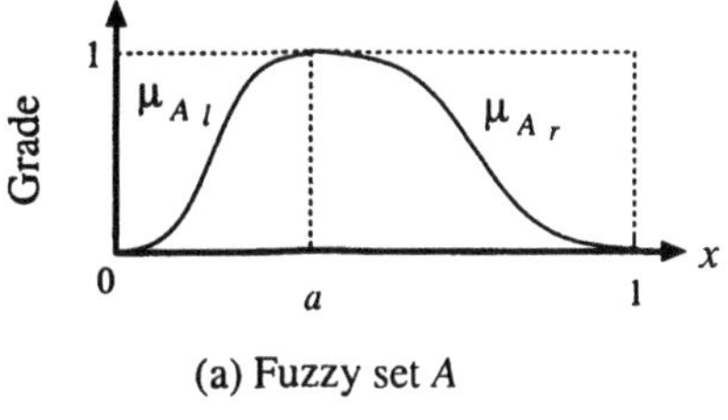

(a) Fuzzy set A

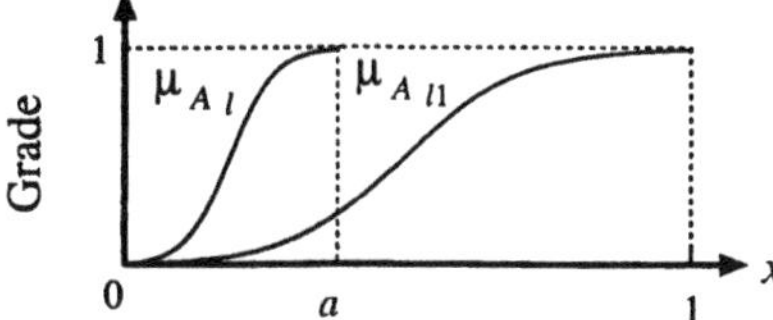

(b) Left-hand side membership functions

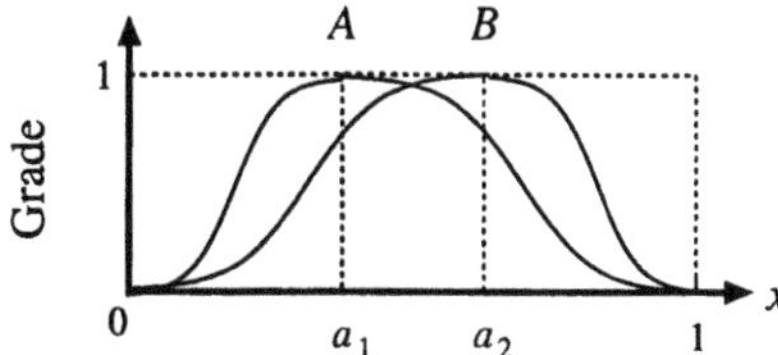

(c) Fuzzy sets A, B, that are similar with respect to vertical line through the crests

Fig. 7. Fuzzy sets

The fuzzy entropy of a fuzzy set in Fig. 7 (a) is studied further. μ_{A_l} denotes the left-hand side membership function from the crest of fuzzy set A, and μ_{A_r} the right-hand side one. In the case in Fig. 7 (b), these one-hand side membership functions are assumed to have the following similarity relation:

$$\mu_{A_l}(at) = \mu_{A_{l1}}(t) \ \ (0 \leq t \leq 1), \tag{11}$$

where a is the position of the crest point of fuzzy set A on x-axis. The fuzzy entropy of fuzzy set A_l is given by

$$\begin{aligned} d(A_l) &= \int_0^a \{-\mu_{A_l}(x) \ln \mu_{A_l}(x) - (1 - \mu_{A_l}(x)) \ln(1 - \mu_{A_l}(x))\}\, dx \\ &= \int_0^1 \{-\mu_{A_l}(at) \ln \mu_{A_l}(at) - (1 - \mu_{A_l}(at)) \ln(1 - \mu_{A_l}(at))\} \cdot a\, dt. \end{aligned} \tag{12}$$

Thus,

$$d(A_l) = a \cdot \int_0^1 \{-\mu_{A_{l1}}(t) \ln(\mu_{A_{l1}}(t)) - (1 - \mu_{A_{l1}}(t)) \ln(1 - \mu_{A_{l1}}(t))\}\, dt$$
$$= a \cdot d(A_{l1}). \tag{13}$$

The same results can be easily obtained for the right-hand side membership functions.

If fuzzy set A in Fig. 7 (c) is similar with respect to $x = a_1$, i.e.

$$\mu_{A_l}(x) = \mu_{A_r}\left(1 - \frac{1 - a_1}{a_1} x\right), \tag{14}$$

then, from eq. (13)

$$d(A_r) = (1 - a_1) d(A_{l1}). \tag{15}$$

Thus the fuzzy entropy of fuzzy set A is given by

$$d(A) = d(A_l) + d(A_r)$$
$$= d(A_{l1}). \tag{16}$$

If fuzzy set B in Fig. 7 (c) is also similar with respect to $x = a_2$, then

$$d(B) = d(B_{l1}), \tag{17}$$

and if the left-hand sides of fuzzy set A and B, μ_{A_l} and μ_{B_l}, are similar with each other,

$$d(A) = d(B). \tag{18}$$

This result means that De Luca and Termini's fuzzy entropy in eq. (9) can be a measure for the shape of a membership function, but it cannot be a measure for evaluation of the deviation of the allocation of membership functions. An example is shown in Fig. 8. The membership functions satisfy eq. (14). The fuzzy entropy of each membership function does not depend on the position of its crest. The fuzzy entropy in Fig. 8 (a) and (b) are the same.

When the conditions (a) and (b) in Sect. 2.2 are given, De Luca and Termini's entropy can be simplified. Assuming that two membership functions $\mu_A(x)$ and $\mu_B(x)$ overlap and for all $x \in [x_1, x_2]$ $\mu_A(x) + \mu_B(x) = 1$, then

$$-\int_{x_1}^{x_2} (1 - \mu_B(x)) \ln(1 - \mu_B(x))\, dx = -\int_{x_1}^{x_2} \mu_A(x) \ln \mu_A(x)\, dx. \tag{19}$$

Thus the entropy of fuzzy sets A and B is reformulated as

$$d(A) + d(B) = -2 \int_{x_1}^{x_2} \{\mu_A(x) \ln \mu_A(x) + \mu_B(x) \ln \mu_B(x)\}\, dx. \tag{20}$$

Under the conditions (a) and (b) in Sect. 2.2, we can use the following measure for evaluation of the shape of membership functions instead of eq. (9).

$$d(A) = -\int_{x_1}^{x_2} \mu_A(x) \ln \mu_A(x)\, dx. \tag{21}$$

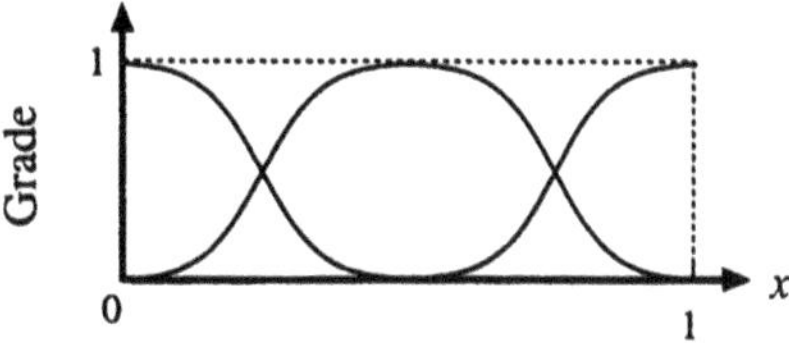

(a) Equidistantly allocated membership functions

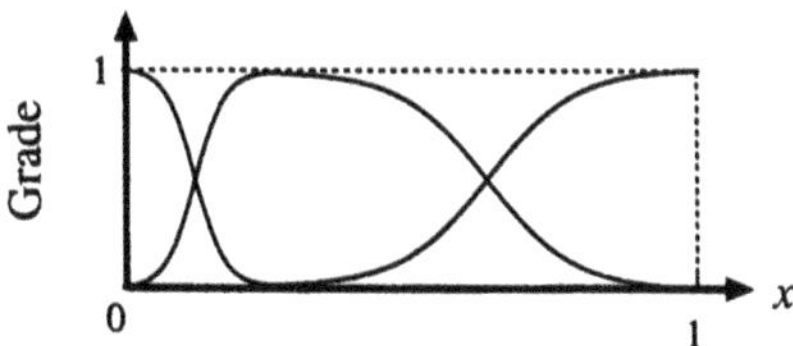

(b) Not-equidistantly allocated membership functions

Fig. 8. Example of allocation of membership functions

4.2 Measure for deviation of membership function

The authors define a quantitative measure of the deviation of a membership function from symmetry on the analogy of relative entropy. The membership function is assumed to satisfy the conditions (a)–(d) in Sect. 2.2. This measure is defined by considering eq. (21).

Definition 1 (Measure for deviation of membership function) *The measure for deviation of fuzzy set A from symmetry is given by*

$$r(A) = \int_{x_1}^{x_2} \mu_C(x) \ln \frac{\mu_C(x)}{\mu_A(x)} \, dx, \tag{22}$$

where x_1 and x_2 are the left and right points of the support of fuzzy set A, respectively; $\mu_A(x)$ is the membership function of fuzzy set A; $\mu_C(x)$ is the symmetrical membership function of fuzzy set C, which has the same support as that of fuzzy set A.

□

Fig. 9 illustrates an example of fuzzy sets A and C.

When the shape of the membership functions in Fig. 9 is triangular, the measure $r_{\mathrm{tri}}(A)$ is expressed as

$$r_{\mathrm{tri}}(A) = s \left[\frac{|d|}{s} - \frac{1}{2} \ln \left\{ 2 \left(\frac{1}{2} + \frac{|d|}{s} \right) \right\} \right] \quad \left(0 < a \leq 1, \; -\frac{1}{2} \leq d \leq \frac{1}{2} \right) \tag{23}$$

where s is the width of support, and d is the deviation of the crest point of fuzzy set A from that of the isosceles triangular fuzzy set C. The value

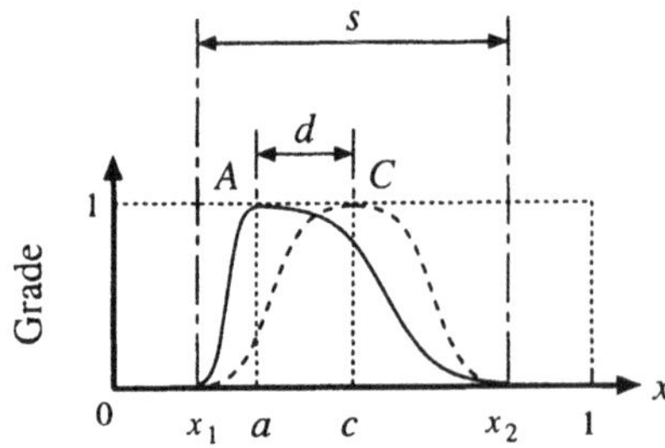

Fig. 9. Membership function A and symmetrical membership function C

of this measure $r_{\mathrm{tri}}(A)$ is monotonically increasing with the absolute value of d. Numerical calculation gives that $r(A)$ is also monotonically increasing with the absolute value of d with any shapes of membership functions that satisfy the conditions (a)–(d) in Sect. 2.2. This measure, which evaluates the deviation of a membership function, is another good candidate for the conciseness measure of fuzzy models. A combination of De Luca and Termini's fuzzy entropy in eq. (21) and the deviation measure in eq. (22) can evaluate the conciseness in Sect. 3.

4.3 Combined measure

One way of combining the two measures is summation. By summing the fuzzy entropy $d(A)$ in eq. (21) and the measure for deviation of a membership function $r(A)$ in eq. (22), a new measure $dr(A)$ is obtained:

$$\begin{aligned} dr(A) &= d(A) + r(A) \\ &= -\int_{x_1}^{x_2} \mu_A(x) \ln \mu_A(x) + \int_{x_1}^{x_2} \mu_C(x) \ln \frac{\mu_C(x)}{\mu_A(x)} \, dx \\ &= -\int_{x_1}^{x_2} \mu_C(x) \ln \mu_A(x) \, dx. \end{aligned} \tag{24}$$

The fuzzy entropy $d(A)$ can evaluate the shape of a membership function, and if the shape is fixed, the measure $r(A)$ can evaluate the deviation of a membership function.

4.4 Average measure

Average measure dr_{avr} is introduced to evaluate the shape and allocation of N_{m} fuzzy sets A_i $(i = 1, \cdots, N_{\mathrm{m}})$ on the universe of discourse X on x-axis. The authors define the average measure dr_{avr} as

$$dr_{\mathrm{avr}} = \frac{1}{N_{\mathrm{m}} - 2} \sum_{i=2}^{N_{\mathrm{m}}-1} dr(A_i), \tag{25}$$

where $dr(A)$ is the combined measure in eq. (24), which evaluates the shape and deviation of a membership function, N_{m} is the number of fuzzy sets $A_i(i = 1, \cdots, N_{\mathrm{m}})$ on the universe of discourse X on x-axis.

5 Numerical results

This section describes numerical results to show usefulness of the average measure for fuzzy modeling. The following single-input/single-output function is used as a target function throughout this section:

$$f(x) = \begin{cases} 1 - 2x & (0 \leq x \leq 0.5) \\ -4x^2 + 8x - 3 & (0.5 < x \leq 1) \end{cases} \tag{26}$$

Fig. 10 depicts this function. Input-output pairs of data were generated using this function. The conditions (a)–(d) in Sect. 2.2 were imposed.

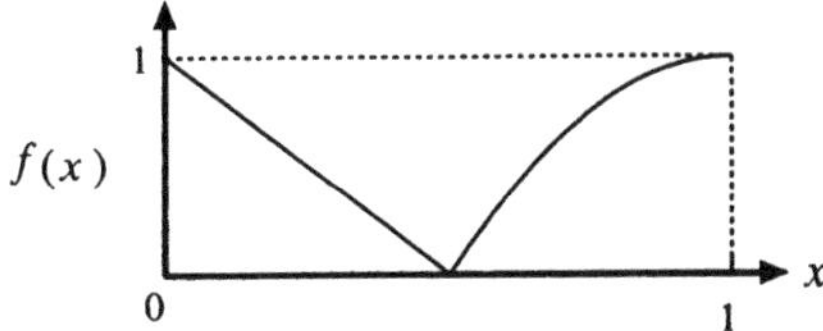

Fig. 10. Modeled function for numerical experiments ($f(x)$ in eq. (26))

5.1 Average measure vs. accuracy of fuzzy models

To examine the relationships between the average measure (dr_{avr}) and the accuracy, 1000 fuzzy models were randomly generated and their average measure and accuracy were calculated. Among them, the fuzzy models along the Pareto front are shown in Fig. 11. In this case, the shape of membership functions was fixed to triangular and the number of membership functions of a fuzzy model was set at 6. Each dot in the figure corresponds to a fuzzy model that has a unique combination of the crest points of membership functions. From Fig. 11, it is observed that the average measure and the accuracy are in conflict as indicated with the broken line.

Fig. 12 (a), (b) and (c) show the allocations of the membership functions of the fuzzy models (a), (b) and (c) on the Pareto front in Fig. 11, respectively. These fuzzy models have the following fuzzy rules:

IF x_1 is A_1 **THEN** $y = c_1$
IF x_2 is A_2 **THEN** $y = c_2$
IF x_3 is A_3 **THEN** $y = c_3$

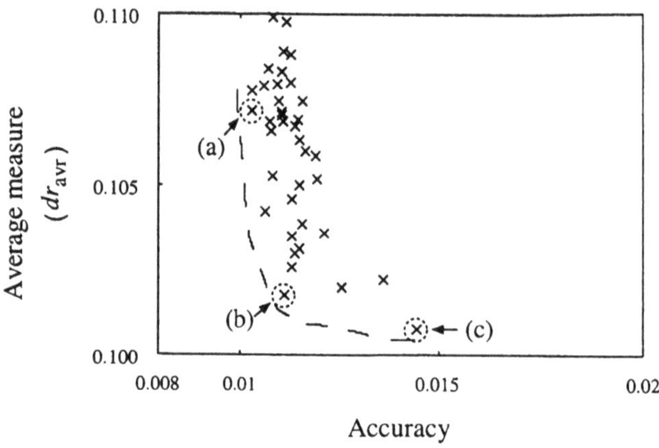

Fig. 11. Average measure vs. accuracy of fuzzy models with triangular membership functions

IF x_4 is A_4 **THEN** $y = c_4$
IF x_5 is A_5 **THEN** $y = c_5$
IF x_6 is A_6 **THEN** $y = c_6$

It was observed that the less the average measure was, the more equidistant the allocation of membership functions was.

Assume that we have no prior knowledge about the target system represented in eq. (10). From the collected input-output pairs of data, we have obtained the models in Fig. 12. The question here is which model is the most interpretable. The model in Fig. 12 (c) has equidistantly allocated membership functions. Although this model is less accurate, it is easier to have a rough idea about the input-output relationships from this model than from other models in Fig. 12. The average measure of the model in Fig. 12 (c) is the smallest, and this measure coincides with the observation of conciseness.

Next, assume that we have prior knowledge about the target system. This knowledge is expressed, for example, as "*it is linear and decreasing on the left half of the universe of discourse, and is sharply rising up from the central point with increasing x.*" In this case, we may think the model in Fig. 12 (a) is the most interpretable among the models in this figure, because this model fits our prior knowledge most. On the other hand, the model in Fig. 12 (c) is the least interpretable now. This case implies that the interpretability of models may depend on our knowledge.

5.2 Fuzzy modeling using GA with average measure

Fuzzy modeling using GA with the average measure and the accuacy was done. The parameters for the numerical experiments were the following: the

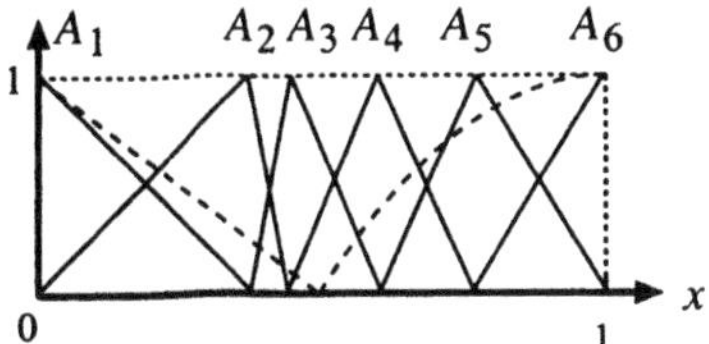

(a) Accuracy: 0.0103
Average measure: 0.107

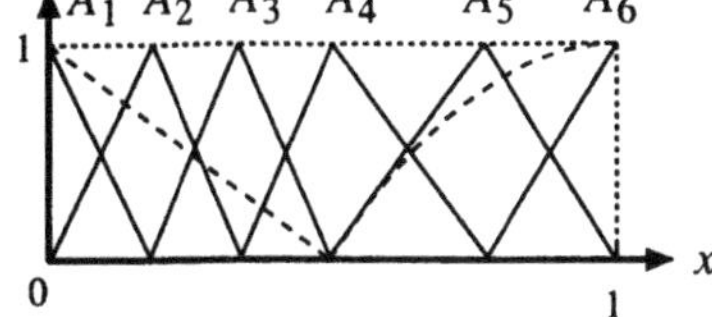

(b) Accuracy: 0.0111
Average measure: 0.102

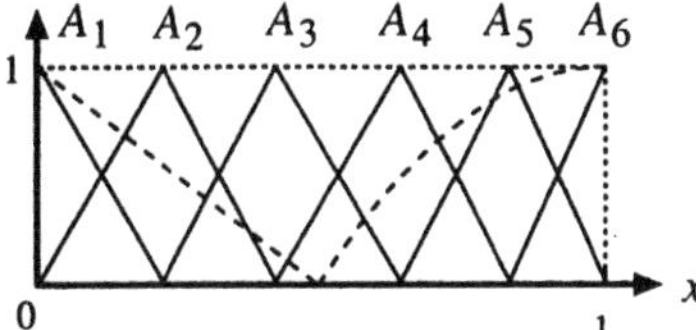

(c) Accuracy: 0.0144
Average measure: 0.101

Fig. 12. Allocation of the membership functions of the fuzzy models (a), (b) and (c) in Fig. 11

number of membership functions was six; the number of chromosomes N_c was set at 50; the shape parameter in a chromosome was either "trimf", "sigmf1" or "sigmf2" in Fig 3; the crossover rate and mutation rate were set at 0.5, 0.05, respectively.

Fig. 13 and Fig. 14 show the results. The dots labeled "trimf", "sigmf1" and "sigmf2" are the initial chromosomes, which were randomly generated, and the dots labeled "trimf_g" and "sigmf1_g" are the chromosomes after 10 generations of genetic operations.

From Fig. 13, it is observed that the fuzzy models are distributed on the Pareto front at the 10th generation as indicated with the broken line. Fig. 14 (a) and (b) show the allocations of the membership functions of the fuzzy models (a) and (b), which were on the Pareto front in Fig. 13, respectively. A conflict between the average measure and the accuracy enabled the successful search for variety of concise fuzzy models with good accuracy.

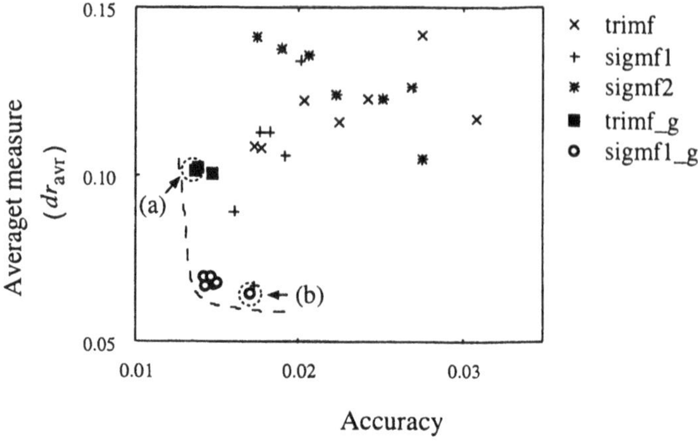

Number of membership functions: $N_m = 6$

Fig. 13. Fuzzy models acquired after 10 generations of genetic operations

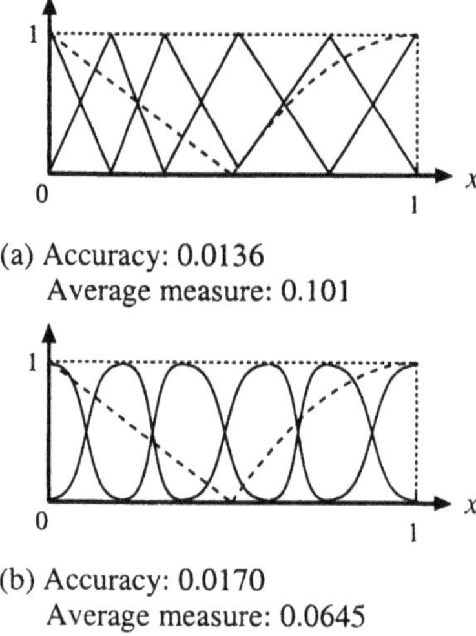

Fig. 14. Allocation of membership functions of acquired fuzzy models

6 Conclusion

This chapter presented a fuzzy modeling method using the new measure for the interpretability of fuzzy models. The authors quantified the conciseness of fuzzy models, which could be one of the measure of interpretability, by introducing fuzzy entropy. De Luca and Termini's fuzzy entropy could evaluate the shapes of membership functions, but their entropy could not distinguish similar shaped membership functions. The authors defined a measure for deviation of a membership function from symmetry. This is another measure for the conciseness of fuzzy models. With De Luca and Termini's measure and the measure for deviation, a combined measure was derived. Based on the combined measure, an average measure was defined to evaluate the shape and allocation of membership functions of a fuzzy model. The average measure was in conflict with the accuracy of fuzzy models when the membership functions were near triangular. Numerical results showed that the average measure was effective for fuzzy modeling formulated as a multi-objective optimization problem.

References

1. Akaike, H. (1973) Information Theory and an Extension of the Maximum Likelihood Principle. 2nd International Symposium on Information Theory. 267–281
2. De Luca, A., Termini, S. (1972) A Definition of a Nonprobabilistic Entropy in the Setting of Fuzzy Sets Theory. Information and Control. **20**, 301–312
3. Fonseca, C. M., Fleming, P. J. (1993) Genetic Algorithms for Multiobjective Optimization: Formulation, Discussion and Generalization. Proc. of the Fifth International Conference on Genetic Algorithms. 416–423
4. Goldberg, D. E. (1989) Genetic Algorithms in Search, Optimization, and Machine Learning. Addison-Wesley.
5. Louis, S. J., Rawlins, G. J. E. (1993) Pareto Optimality, GA-Easiness and Deception. Proc. of the Fifth International Conference on Genetic Algorithms. 118–123
6. Matsushita, S., Furuhashi, T., et al. (1996) Selection of Input Variables Using Genetic Algorithm for Hierarchical Fuzzy Modeling. Proc. of 1996 The First Asia-Pacific Conference on Simulated Evolution and Learning. 106–113
7. Miyamoto, S., Mukaidono, M. (1997) Fuzzy *c*-means as a regularization and maximum entropy approach. Proc. of the 7th Int'l Fuzzy Systems Association World Congress (IFSA'97). 86–92
8. Mizumoto, M. (1987) Fuzzy Control Under Various Approximate Reasoning Methods. Proc. of Second IFSA Congress. 143–146
9. Nomura, H., Araki, S., Hayashi, I., Wakami, N. (1992) A Learning Method of Fuzzy Reasoning by Delta Rule. Proc. of Intelligent System Symposium. 25–30
10. Pal, N. R. (1999) On Quantification of Different Facets of Uncertainty. Fuzzy Sets and Systems. **107**, 81–91
11. Pal, N. R., Bezdek, J. C. (1994) Measuring Fuzzy Uncertainty. IEEE Trans. on Fuzzy Systems. **2**, 107–118
12. Setnes, M., Babuška, R., Verbruggen, H.B. (1998) Rule-Based Modeling: Precision and Transparency. IEEE Trans. Syst., Man, Cybern., pt.C. **28**, 165–169
13. Setnes, M. Roubos, H. (2000) GA-Fuzzy Modeling and Classification: Complexity and Performance. IEEE Trans. Fuzzy Syst. **8**, 509–522
14. Schaffer, J. D. (1985) Multiple Objective Optimization with Vector Evaluated Genetic Algorithms. Proc. of the First International Conference on Genetic Algorithms and Their Applications. 93–100
15. Takagi, T. and Sugeno, M. (1985) Fuzzy Identification of Systems and its Applications to Modeling and Control. IEEE Trans. on Systems, Man, and Cybernetics. **15**, 116–132

16. Valente de Oliveira, J. (1999) Semantic Constraints for Membership Function Optimization. IEEE Trans. Syst., Man, Cybern., pt.A. **29**, 128-138
17. Zenzo, S. D., Cinque, L. (1998) Image Thresholding Using Fuzzy Entropies. IEEE Trans. on Systems, Man, and Cybernetics-Part B. **28**, 15–23

Exact trade-off between approximation accuracy and interpretability: solving the saturation problem for certain FRBSs*

Domonkos Tikk[1,2] and Péter Baranyi[1,2]

[1] Dept. of Telecommunications & Telematics,
Budapest University of Technology and Economics,
1117 Budapest, Magyar Tudósok Körútja 2., Hungary,
e-mails: tikk@ttt.bme.hu, baranyi@ttt-202.ttt.bme.hu
[2] Intelligent Integrated Systems Japanese–Hungarian Laboratory
1111 Budapest, Műegyetem rakpart 3., Hungary

Abstract. Although, in literature various results can be found claiming that fuzzy rule-based systems (FRBSs) possess the universal approximation property, to reach arbitrary accuracy the necessary number of rules are unbounded. Therefore, the inherent property of FRBSs in the original sense of Zadeh, namely that they can be characterized by a semantic relying on linguistic terms is lost. If we restrict the number of rules, universal approximation is not valid anymore as it was shown for, including others, Sugeno and TSK type models [10,19]. Due to this theoretic bound there is recently a great demand among researchers on finding trade-off techniques between a required accuracy and the number of rules, and as such, they attempt to determine the (optimal) number of rules as a function of accuracy. Naturally, to obtain such results one has to restrict somehow the set of continuous functions, usually requiring some smoothness conditions on the approximated function. In terms of approximation theory this is the so-called saturation problem, the determination of optimal order and class of approximation. Hitherto, saturation classes and orders have not been determined for FRBSs and neural networks. In this paper we solve the saturation problem for a special type of fuzzy controller, for the generalized KH-interpolator, being a suitable inference method in sparse rule bases.

1 Introduction

Considering the success of fuzzy controllers with nonlinear plants, several authors raised the question, what is the reason of this versatility of fuzzy controllers. In the beginning of the 90's an answer was given from the mathematical point of view [2,9,24]: FRBSs are universal approximators in the sense that it is possible to construct such rule based fuzzy controllers that approximate any (nonlinear) continuous functions with arbitrary accuracy,

* This research was supported by the Hungarian Scientific Research Fund (OTKA) Grants No. D034614, F30056, and T34212 and by the Hungarian Ministry of Education Grant No. FKFP 0180/2001. Péter Baranyi was supported by the Zoltán Magyary Scholarship.

i.e., the fuzzy control system can implement any necessary control function. It was a start of an universal approximation theorem "boom", which ended up with the conclusion that almost all types of fuzzy systems independently from their design parameters (such as shape of the rules, inference function, defuzzification technique, form of the consequents, but also density and structure of the rule base), i.e. including methods for hierarchical rule bases [25] and fuzzy rule interpolation algorithms [21,22] are universal approximator.

On the one hand, the early universal approximation results have been usually criticized due to their solely existential nature [5,23]. In the recent past, there arose an effort to give constructive proofs, or to determine the number of rules as a function of the accuracy (see e.g. [8,26,27]). On the other hand, it was pointed out that if the number of rules are very large, the inherent property of FRBSs in the original sense of Zadeh, namely that they can be characterized by a semantic relying on linguistic terms is lost. Moreover, if the number of rules is bounded as a practical limitation, the universal approximation property does not hold anymore [10,19]. More precisely, it was proven for T-controllers (this class of fuzzy controllers has tensor product form consequents in the rules and includes Sugeno [14], and TSK controllers [15]) are nowhere dense in the space of continuous functions. This statement is recalled and analyzed in details in Section 2.

Due to this theoretic bound there is recently a great demand among researchers on finding trade-off techniques between a required accuracy and the number of rules, and as such, they attempt to determine the number of rules as a function of accuracy. Naturally, to obtain such results one has to restrict somehow the set of continuous functions, usually requiring some smoothness conditions on the approximated function. In terms of approximation theory this is the so-called saturation problem, the determination of optimal order and class of approximation. Hitherto, saturation classes and orders have not been determined for FRBSs and neural networks. In this paper we solve the saturation problem for a special type of fuzzy controller, for the generalized KH interpolator, being a suitable inference method in sparse rule bases [6]; for a brief survey of KH and other interpolation methods see Section 3.1. The saturation problem of generalized KH interpolator is solved via its analogy with Shepard function [12], an interpolatory operator thoroughly studied by approximation theorists (see e.g. [3,4,13,16,17]), and is presented in Section 3.

2 Nowhere denseness of T-controllers

In this section we show that if practical limitation for boundedness of rule base is taken into consideration, the universal approximation property of is lost [10,19].

The results are valid for T-controllers, being defined below. We will show that under the assumption of bounded number of rules T-controllers are nowhere dense in the L^p space (space of continuous function equipped with the norm $\| \cdot \|_p$), which means that the set of T-controllers lies "almost" discretely in the L^p space.

2.1 Preliminaries

Let us turn to define the class of T-controllers. This class contains fuzzy controllers with consequents that can be expressed as tensor product of univariate functions. For convenience, we call them T-controllers, where T stands for the abbreviation of "tensor" (product consequent).

Definition 1. Let us consider a fuzzy rule base with rules of the structure

$$R_{i,j} : \text{ If } x \text{ is } A_i \text{ and } y \text{ is } B_j \text{ then } \sum_{k=1}^{m_{i,j}} f_{k,(i,j)}(x) g_{k,(i,j)}(y) \tag{1}$$

where the number of the rules is bounded by n_1 in the first and by n_2 in the second variable. The functions $f_{k,(i,j)}(x)$ and $g_{k,(i,j)}(y)$ are arbitrary univariate functions, furthermore, there is an overall upper bound m that for all indices $m_{i,j} \leq m$ $(1 \leq i \leq n_1,\ 1 \leq j \leq n_2)$ holds. For brevity, when context makes clear that the denotation concerns the rule $R_{i,j}$ we write f_k and g_k instead of $f_{k,(i,j)}(x)$ and $g_{k,(i,j)}(x)$, respectively. We refer to FRBSs containing rules of form (1) as T-controllers.

Let us denote the set of FRBSs containing rules of the form (1) by

$$\mathbf{T}_{(n_1,n_2),m}(X \times Y) \tag{2}$$

where the pair (n_1, n_2) is an upper bound for the number of rules in the input spaces X and Y, and m is the upper bound for the number of functions summed in the consequent part of the rules. For $p \in [1, \infty]$ we introduce the set

$$\mathbf{T}^{(p)}_{(n_1,n_2),m}(X \times Y) \tag{3}$$

which is the subset of (2) and $L^p(X \times Y)$, as well. Therefore, the set (3) is equipped with the L^p-norm $\| \cdot \|_p$. The elements of sets (3) and (2) are called T-controllers of order $((n_1, n_2), m)$.

Observe that the set of Sugeno controllers [14] is a subset of T-controllers where indices $m_{i,j}$ equal to 1, $f_1 = c_{i,j}$ and $g_1 = 1$ for all i and j. Further, let us consider the generalization of Sugeno controller, the so-called Takagi–Sugeno–Kang (TSK) controller [15]. The set of TSK controllers is also a subset of the set of T-controllers. The TSK controller models the controlled system by describing its nature by means of fuzzy rules with bounded polynomial consequent as

$$\text{If } x \text{ is } A_i \text{ and } y \text{ is } B_j \text{ then } P^{ij}_m(x, y) \tag{4}$$

where $P_m^{ij}(x,y)$ is a polynomial of maximum order m having the form

$$P_m^{ij}(x,y) = \sum_{p=0}^{m}\sum_{q=0}^{m} a_{pq}^{ij} x^p y^q. \tag{5}$$

If indices $m_{i,j} = (m+1)^2$, and functions $f_{pq} = a_{pq}^{ij} x^p$ and $g_{pq} = y^q$ $(1 \leq p,q \leq m)$ then (1) and (4) are equivalent.

The input-output function of T-controller $t \in \mathbf{T}_{(n_1,n_2),m}(\Omega)$, $t : \Omega \to \mathbf{R}$ is given by

$$t(x,y) = \frac{\sum_{i=1}^{n_1}\sum_{j=1}^{n_2} \tilde{\mu}_i(x)\tilde{\upsilon}_j(y)\sum_{k=1}^{m_{i,j}} f_{k,(i,j)}(x) g_{k,(i,j)}(y)}{\sum_{i=1}^{n_1}\sum_{j=1}^{n_2} \tilde{\mu}_i(x)\tilde{\upsilon}_j(y)} \tag{6}$$

where the families of membership functions provide complete cover of the input space Ω. We can get

$$t(x,y) = \sum_{i=1}^{n_1}\sum_{j=1}^{n_2} \tilde{\mu}_i(x)\tilde{\upsilon}_j(y)\sum_{k=1}^{m_{i,j}} f_{k,(i,j)}(x) g_{k,(i,j)}(y)$$

from (6), requiring the families of membership functions to be fuzzy partitions in the Ruspini sense [11]. Without loss of generality, from now we consider Ω to be $[0,1]^2$.

2.2 Results and discussion

Moser has proven the nowhere denseness of Sugeno controllers in two steps [10]. This technique was adopted also for T-controllers [19]. First, it is shown in Lemma 1 that a special function cannot be approximated arbitrary well by T-controllers. Then it is presented in Theorem 1 that for every T-controller and for arbitrary $\varepsilon > 0$ there exists a function ω in the ε-environment of the T-controller, which cannot be approximated with arbitrary accuracy by T-controllers. This immediately implies that T-controllers are nowhere dense in the space of continuous functions. The construction of ω in Theorem 1 is based on the special function of the lemma.

Lemma 1. *Let $\omega : [0,1]^2 \to \mathbf{R}$ be a function of the form $\omega(x,y) = \alpha$ if $y \geq x$, $\omega(x,y) \neq \alpha$ else, where $\alpha \in \mathbf{R}$. Then for each $p \in [1,\infty]$ and $n_1, n_2 \in \mathbf{N}$ there holds*

$$\inf\left\{\|\omega - t\|_p \,\middle|\, t \in \mathbf{T}_{(n_1,n_2),m}^{(p)}\right\} > 0. \tag{7}$$

Theorem 1. *To each $p \in [1,\infty]$, $\varepsilon > 0$ and $t \in \mathbf{T}_{(n_1,n_2),m}^{(p)}(\Omega)$, $n_1, n_2, m \in \mathbf{N}$ there is a continuous function*

$$\omega \in L^p(\Omega)\backslash \operatorname{cl}\left(\mathbf{T}_{(n_1,n_2),m}^{(p)}(\Omega)\right)$$

fulfilling $\|\omega - t\|_p < \varepsilon$.

Here $\mathrm{cl}\left(\mathbf{T}^{(p)}_{(n_1,n_2),m}(\Omega)\right)$ denotes the closure of the subset $\mathbf{T}^{(p)}_{(n_1,n_2),m}(\Omega)$ of $L^p(\Omega)$ with respect to the topology induces by the L^p-norm $\|\cdot\|_p$.

As Theorem 1 guarantees that to each $\varepsilon > 0$ and $t \in T^{(p)}_{(I_1,I_2),M}\left([0,1]^2\right)$ there is a function $\omega \in L^p([0,1]^2)\backslash \mathrm{cl}T^{(p)}_{(I_1,I_2),M}\left([0,1]^2\right)$ with $\|\omega - t\|_p < \varepsilon$, or equivalently, there is no inner point of the set $T^{(p)}_{(I_1,I_2),M}\left([0,1]^2\right)$, because every ε-environment of an arbitrary element of the set contains functions not included in the closure of the set. Thus, we immediately obtain that $T^{(p)}_{(I_1,I_2),M}\left([0,1]^2\right)$ is nowhere dense in $L^p([0,1]^2)$.

Theorem 1 points out if we restrict the number of rules the corresponding set of T-controllers are nowhere dense in the space of continuous functions. This contributes to the opinion that not the great variety of design parameters but rather the number of rules used in the controller is the reason of the universal approximator property of various fuzzy controllers. Comparing this result with theorems stating the universal approximator property of various classes of fuzzy controllers, the presented result reveals the practical limitation of fuzzy controllers for function approximation.

3 The saturation of KH controllers

Recently, it was pointed out [22] that a special version of the interpolative fuzzy KH interpolator (described below; Section 3.1) is universal approximator in the sense that it is able to approximate any continuous function on a compact domain with arbitrary accuracy with respect to the L_p norm $p \in [1,\infty]$. This result is purely existential, no trade-off between the number of the rules and the accuracy is given.

However, in the same paper it was remarked without any proof that the KH interpolator can be considered as a fuzzy generalization of the Shepard interpolator (Section 3.2). Now we show their exact connection in Section 3.3. Based on the link of between Shepard and KH interpolator we can make up the deficiency of missing "trade-off" type characterization of approximation by KH interpolator, and derive theorems for its approximation rate in Subsection 3.4.

Here we remark again, that the price to be paid for obtaining better approximation characteristics for subset of continuous functions is certain conditions on the subset itself. This feature is an immediate consequence of Theorem 1. Usually these conditions require some smoothness property on the approximated function, such as the modulus of continuity, the Lipschitz constant or some other smoothness property.

3.1 The KH interpolator

The fuzzy rule based interpolation technique (or briefly: fuzzy interpolation) was proposed to provide an inference mechanism suitable for rule bases containing gaps. The "classical", so-called tomato, example of fuzzy interpolation is the following. Let us given two linguistic rules in the form:

If the tomato is **green**, then it is **unripe**

If the tomato is **red**, then it is **ripe**

What happens if the observation is that tomato is **yellow**? In such a case the rule selecting mechanism of classical inference algorithms (Mamdani, TSK), which choose active rules for an observation based on matching antecedents, cannot create a conclusion. However, knowing that the ripeness and the color of tomatoes are gradual notions, and on the basis of the yellowness" of the actual tomato (its distance between the linguistic terms red and green), we can conclude intuitively that answer: the tomato is **half-ripe**.

This intuitive reasoning mechanism was the model of the first fuzzy interpolation method introduced in [6] (termed KH interpolation). Technically, it creates the conclusion by means of its α-cuts based on the Extension Principle, and the Resolution Principle. The conclusion is calculated in a such way that the ratio of distances between neighboring rule antecedents and the observation, should be identical with the one between the corresponding rule consequents and the conclusion (in the simplest one dimensional input case). For every α value of the important level set (e.g., for triangular and trapezoidal shaped membership function that is 0 and 1), it determines the conclusion as

$$B^*_{\alpha C} = \frac{\sum_{i=1}^{n} B_{i\alpha C} \frac{1}{d_C(A^*_{\alpha C}, A_{i\alpha C})}}{\sum_{k=1}^{n} \frac{1}{d_C(A^*_{\alpha C}, A_{k\alpha C})}}, \tag{8}$$

where A_i and B_i $(i = 1, \ldots, n)$ denote the antecedents neighboring the observation A^*, and the corresponding consequents fuzzy sets, respectively. $C \in \{L, U\}$, where L and U refer to the lower and upper extreme of the α-cut. Finally, the function $d_C : \mathbf{R} \times \mathbf{R} \to [0, +\infty)$ is an appropriate lower/upper distance function (cf. [7]). If $n = 2$, the approach is termed linear interpolation.

For the applicability of the method the involved fuzzy sets should be ordered as:

$$A_i \prec_X A^* \prec_X A_{i+1}, \quad B_i \prec_Y B_{i+1} \tag{9}$$

where $\prec_X$ and $\prec_Y$ are proper partial orderings on the multidimensional input and the one dimensional output space, respectively. This expressions mean that the observation should be flanked by antecedents from both sides (no extrapolation is allowed), and a similar ordering should exist for the corresponding consequents.

In [22], the stabilized version of the KH interpolation is introduced, when we take λth power of the distance, where the exponent λ is not smaller then N, the dimension of the antecedents:

$$B^*_{\alpha C} = \frac{\sum_{i=1}^{n} B_{i\alpha C} \frac{1}{d_C^{\lambda}(A^*_{\alpha C}, A_{i\alpha C})}}{\sum_{k=1}^{n} \frac{1}{d_C^{\lambda}(A^*_{\alpha C}, A_{k\alpha C})}}, \quad (\lambda \geq N) \tag{10}$$

Now, we recall the universal approximation theorem of the stabilized KH approach [22].

Definition 2. Let $\mathbf{R}^N \supset \Omega = [a_1, b_1] \times \cdots \times [a_N, b_N]$, further let $\{\Gamma_n\}_{n=1}^{\infty}$ be a sequence of finite subsets of Ω with $\#\Gamma_n = n$. If

$$\forall \varepsilon > 0 \; \exists n_0 \; \forall \omega \in \Omega \; \forall n \geq n_0 : \left| \frac{\#(\Gamma_n \cap \omega)}{\#\Gamma_n} - \frac{|\omega|}{|\Omega|} \right| < \varepsilon \tag{11}$$

then the set Γ_n are uniformly distributed on the domain Ω. Here $\#(\Gamma_n \cap \omega)$ denotes the cardinality of the finite set $(\Gamma_n \cap \omega)$ and $|\omega|$ is the Lebesque measure of ω.

Theorem 2. *Consider the L_p norm $\|\cdot\|_p$ with $p \in [1, \infty]$, the domain $\mathbf{R}^N \supset \Omega = [a_1, b_1] \times \cdots \times [a_N, b_N]$ and a continuous function $f : \Omega \to \mathbf{R}$, then for all $x \in \Omega$ the expression*

$$\lim_{n \to \infty} K_n^{\lambda}(f, x) := \lim_{n \to \infty} \sum_{k=1}^{n} f\left(x_k^{(n)}\right) \frac{\dfrac{1}{\left\|x - x_k^{(n)}\right\|_p^{\lambda}}}{\sum\limits_{j=1}^{n} \dfrac{1}{\left\|x - x_j^{(n)}\right\|_p^{\lambda}}} \tag{12}$$

is equal to $f(x)$, where measurement points $x_k^{(n)}$ are uniformly distributed on Ω in the sense of (11), *and $\lambda \geq N$.*

Without the loss of generality, in the next we suppose that $\overline{\Omega} = [0, 1]$, because with a proper linear transformation every compact domain can be mapped into each other. From now, we mean the stabilized KH interpolator (10) on the term "KH interpolator".

3.2 The Shepard operator

The Shepard interpolation method was first introduced in [12] for arbitrarily placed bivariate data as

$$S_0(f, x, y) = \begin{cases} f(x_i, y_i), & \text{if } (x, y) = (x_i, y_i) \\ & \text{for some } i \\ \left(\sum\limits_{i=0}^{n} f(x_i, y_i)/d_i^{\lambda}\right) \Big/ \left(\sum\limits_{i=0}^{n} 1/d_i^{\lambda}\right), & \text{otherwise} \end{cases} \tag{13}$$

where measurement points x_i, y_i $(i = 0, \ldots, n)$ are irregularly spaced on the domain of $f \in \mathbf{R}^2 \to \mathbf{R}$, $\lambda > 0$, and $d_i = [(x - x_i)^2 + (y - y_i)^2]^{1/2}$. This function can be used typically when a surface model is required to interpolate scattered spatial measurements (e.g. pattern recognition, geology, cartography, earth sciences, fluid dynamics and many others).

Beside the application oriented investigation of Shepard's method such as [1], an increasing interest has arisen from mathematical researchers to examine the approximation property of formula (13). For a more general analysis, formula (13) was generalized to

$$S_n^\lambda(f,x) = \frac{\sum_{i=0}^n f(x_i)(x-x_i)^{-\lambda}}{\sum_{i=0}^n (x-x_i)^{-\lambda}}, \qquad \lambda > 0,\ n = 1,2\ldots \tag{14}$$

for an arbitrary $f \in C[0,1]$, where x_i $(i = 0,\ldots,n)$, in general, denotes the nodes of the equidistant distribution of the domain $[0,1]$. We recall that fixing the domain to the interval $[0,1]$ does not mean any restriction.

The possible use of rational functions of type (14) as approximating means was first discovered by J. Balázs. (After his name this operator is often termed Balázs–Shepard operator in the literature of approximation theory.) The properties of the (14) operator was widely investigated by mostly Hungarian and Italian mathematicians; see e.g. [3,4,13,16,17].

3.3 The fuzziness of the approximation

As it was shown in [22], for a fixed $\alpha \in [0,1]$ and $C \in \{L,U\}$ the input-output function of the (stabilized) KH interpolator, $K_n^\lambda(f,x)$ coincides with the Shepard operator. Therefore, we can consider the family of $K_n^\lambda(f,x)$ functions as a generalization of the $S_n^\lambda(f,x)$. Here, first we aim to clarify, what is nature of this generalization, how the approximated functions for various α and/or C values differ. It is obvious, that the family of KH functions tailor the same approximated function, if all the involved fuzzy sets are crisp. In the next, we investigate how the conclusion depends on the fuzziness of the antecedents and consequents.

First, let us only assume that the modulus of continuity of the approximated function $f : [0,1] \to \mathbf{R}$ is known:

$$\omega(f,n^{-1}) = \max_{\substack{x,y\in[0,1]\\ \|x-y\|\le n^{-1}}} |f(x)-f(y)|. \tag{15}$$

Due to the uniform distribution (11) of the knot points we can estimate the difference of the adjacent knot points by n^{-1}.

Therefore, we can estimate the support of a consequent fuzzy set, or in other words, its fuzziness by this quantity:

$$\max_{i=1}^{n}(\text{supp}(B_i)) \le \omega(f,n^{-1}). \tag{16}$$

Note, that this is a very rough estimation. We can sharpen it easily under certain circumstances to be discussed later.

Let us estimate now the difference of two $K_n^\lambda(f,x)$ operators. The largest difference appears when the minimum and the maximum of the support are calculated. That is when $\alpha = 0$ is fixed, and $C = L$ and $C = U$, respectively.

Let

$$K_1^{(n)} = \frac{\sum_{i=1}^{n} (\inf B_{i0})^{(n)} \frac{1}{d^\lambda((\inf A_0^*)^{(n)}, (\inf A_{i0})^{(n)})}}{\sum_{j=1}^{n} \frac{1}{d^\lambda((\inf A_0^*)^{(n)}, (\inf A_{j0})^{(n)})}}$$

$$K_2^{(n)} = \frac{\sum_{i=1}^{n} (\sup B_{i0})^{(n)} \frac{1}{d^\lambda((\sup A_0^*)^{(n)}, (\sup A_{i0})^{(n)})}}{\sum_{j=1}^{n} \frac{1}{d^\lambda((\sup A_0^*)^{(n)}, (\sup A_{j0})^{(n)})}}$$

the two farthest point of the interpolated conclusion. The superscript $^{(n)}$ refers to the knot point system consisting of n points. Then

$$\begin{aligned}
\lim_{n\to\infty} |K_1^{(n)} - K_2^{(n)}| &\le \left| \frac{\sum_{i=1}^{n} (\inf B_{i0})^{(n)} \frac{1}{d^\lambda((\inf A_0^*)^{(n)}, (\inf A_{i0})^{(n)})}}{\sum_{j=1}^{n} \frac{1}{d^\lambda((\inf A_0^*)^{(n)}, (\inf A_{j0})^{(n)})}} \right. \\
&\quad \left. - \frac{\sum_{i=1}^{n} \left((\inf B_{i0})^{(n)} + \omega(f, n^{-1})\right) \frac{1}{d^\lambda((\sup A_0^*)^{(n)}, (\sup A_{i0})^{(n)})}}{\sum_{j=1}^{n} \frac{1}{d^\lambda((\sup A_0^*)^{(n)}, (\sup A_{j0})^{(n)})}} \right| \\
&\le \left| \frac{\sum_{i=1}^{n} (\inf B_{i0})^{(n)} \frac{1}{d^\lambda((\inf A_0^*)^{(n)}, (\inf A_{i0})^{(n)})}}{\sum_{j=1}^{n} \frac{1}{d^\lambda((\inf A_0^*)^{(n)}, (\inf A_{j0})^{(n)})}} \right. \\
&\quad \left. - \frac{\sum_{i=1}^{n} (\inf B_{i0})^{(n)} \frac{1}{d^\lambda((\sup A_0^*)^{(n)}, (\sup A_{i0})^{(n)})}}{\sum_{j=1}^{n} \frac{1}{d^\lambda((\sup A_0^*)^{(n)}, (\sup A_{j0})^{(n)})}} \right| + \omega(f, n^{-1}) \\
&\le \omega(f, n^{-1})
\end{aligned} \tag{17}$$

As n converges to zero, and the knot points are uniformly distributed, the support of antecedents and observation fuzzy sets should also vanish due to the condition of their ordering (cf. (9)). Therefore, the two expressions of distance become identical, hence the absolute value of the difference of the fractions vanishes. Note, that the support of the consequents does not necessarily vanishes as n tends to zero.

The obtained result means that the maximal difference between two operators in a family of Shepard type approximations is bounded by the modulus of continuity of the approximated functions, and this value can be theoretically arbitrary large. Observe that the result only depends on the estimated support size (16). However, it is not very reasonable to model a significant change in the approximated function with only one consequent fuzzy set. Therefore, the rough estimation of (16) and (17) can be improved, if we suppose that antecedents and consequents are fuzzy numbers modelling the measured input-output samples of the approximated function. In this case we can set the consequents' maximum support length to δ, being e.g. the margin of error of the measuring tool:

$$\max_{i=1}^{n}(\mathrm{supp}(B_i)) \leq \delta. \tag{18}$$

With analog reasoning as in (17), we can get

$$\lim_{n\to\infty} |K_1^{(n)} - K_2^{(n)}| \leq \delta.$$

Based on the previous thoughts we can state the following

Proposition 1. *Under the conditions of Theorem 2, the fuzziness of the conclusion of the stabilized KH interpolator is bounded by the maximum fuzziness of the consequent fuzzy sets, if n, the number of knot points tends to ∞.*

Because of Proposition 1, the theorems for Shepard operators are convertible for the KH interpolators, bearing in mind that all derived results have an uncertainty factor (or fuzziness) due to the fuzziness of the consequents.

3.4 Main results

The approximation property and the saturation problem of the Shepard operator were investigated for various λ values in [3,13,16,17]. Here we recall the proof of Szabados [17], which gives a complete analysis for all $\lambda \geq 1$. When $\lambda < 1$ the operator does not converge for all $f(x) \in C[0,1]$, so it is of no interest for our investigations.

Theorem 3. [17] *The approximation order of the operator $S_n^\lambda(f,x)$ is*

$$\|f(x) - S_n^\lambda(f,x)\| = \begin{cases} O\left(\omega(f,n^{-1})\right), & \text{if } \lambda > 2 \\ O(n^{1-\lambda}) \int_{1/n}^{1} t^{-\lambda}\omega(f,t)\mathrm{d}t, & \text{if } 1 < \lambda \leq 2 \\ O(\log^{-1} n) \int_{1/n}^{1} t^{-\lambda}\omega(f,t)\mathrm{d}t, & \text{if } \lambda = 1 \end{cases} \tag{19}$$

for any $f \in C[0,1]$.

Proof. Let us suppose equispaced knot points systems on the unit interval, i.e.

$$x_i = \frac{i}{n}; \qquad \{x_i\}^{(n)} = \{x_i | i = 0,\ldots,n\}. \tag{20}$$

Further, let $x_j \in \{x_i\}^{(n)}$, $j = 0,\ldots,n$ be the closest knot point to an arbitrary $x \in [0,1]$, $x \notin \{x_i\}^{(n)}$:

$$\left|x - \frac{j}{n}\right| = \min_{0\leq i\leq n} \left|x - \frac{i}{n}\right|,$$

(if this definition is not unique take any of the two possibilities). Then

$$|f(x) - S_n^\lambda(f,x)| \le \frac{\sum_{i=0}^n \left|f(x) - f\left(\frac{i}{n}\right)\right| \cdot \left|x - \frac{i}{n}\right|^{-\lambda}}{\sum_{i=0}^n \left|x - \frac{i}{n}\right|^{-\lambda}}.$$

The order of the denominator is evidently

$$\left(\sum_{i=0}^n \left|x - \frac{i}{n}\right|^{-\lambda}\right)^{-1} = O\left(n^{-\lambda} \log^{-\lceil \frac{1}{\lambda} \rceil} n\right), \qquad (\lambda \ge 1), \tag{21}$$

so

$$\begin{aligned}
|f(x) - S_n^\lambda(f,x)| &\le \left|f(x) - f\left(\frac{j}{n}\right)\right| \\
&\quad + O\left(n^{-\lambda} \log^{-\lceil \frac{1}{\lambda} \rceil} n\right) \sum_{i \ne j} \left|f(x) - f\left(\frac{i}{n}\right)\right| \cdot \left|x - \frac{i}{n}\right|^{-\lambda} \\
&\le \omega\left(f, \frac{1}{2n}\right) \\
&\quad + O\left(n^{-\lambda} \log^{-\lceil \frac{1}{\lambda} \rceil} n\right) \sum_{i \ne j} \left(\frac{|j-i| - 1/2}{n}\right)^{-\lambda} \omega\left(f, \frac{|j-i| - 1/2}{n}\right) \\
&\le \omega\left(f, n^{-1}\right) + O\left(n^{-\lambda} \log^{-\lceil \frac{1}{\lambda} \rceil} n\right) \sum_{i=1}^n \left(\frac{i}{n}\right)^{-\lambda} \omega\left(\frac{i}{n}\right)
\end{aligned} \tag{22}$$

Because $\omega(f, k/n) \le k\omega(f, 1/n)$, for $\lambda > 2$ we have

$$\begin{aligned}
|f(x) - S_n^\lambda(f,x)| &\le \omega\left(f, n^{-1}\right) \left\{1 + O\left(n^{-\lambda}\right) \sum_{i=1}^n \frac{i^{1-\lambda}}{n^{-\lambda}}\right\} \\
&= O\left(\omega\left(f, n^{-1}\right)\right).
\end{aligned}$$

If $1 \le \lambda \le 2$, then

$$|f(x) - S_n^\lambda(f,x)| \le \omega\left(f, n^{-1}\right) + O\left(n^{1-\lambda} \log^{-\lceil \frac{1}{\lambda} \rceil} n\right) \int_{1/n}^1 t^{-\lambda} \omega(f,t) dt,$$

whence the last two statements of the theorem follows. □

The theorem gives some immediate results on the saturation problem. For its precise characterization we need the following definition.

Definition 3. A function $f : [0,1] \to \mathbf{R}$ is called Lipschitz continuous with Lipschitz coefficient α (notation: $f \in \operatorname{Lip} \alpha$) if

$$|f(x) - f(y)| \le \alpha |x - y| \text{ for all } x, y \in [0,1]. \tag{23}$$

From Theorem 3 it is obvious that if $f(x) \in \text{Lip}\,1$ and $\lambda > 2$, then the saturation order is

$$\|f(x) - S_n^\lambda(f,x)\| = O(n^{-1}). \tag{24}$$

In fact, as it is proved in [13], (24) holds if and only if $f(x) \in \text{Lip}\,1$. Furthermore

$$\|f(x) - S_n^\lambda(f,x)\| = o(n^{-1}), \tag{25}$$

if and only if $f(x) = \text{const}$. Thus the saturation problem is completed for $\lambda > 2$.

If $\lambda = 2$, even with $f(x) \in \text{Lip}\,1$ Theorem 3 gives only

$$\|f(x) - S_n^\lambda(f,x)\| = O\left(\frac{\log n}{n}\right). \tag{26}$$

This result can be improved to $O(n^{-1})$ under stronger restriction on $f(x)$:

Theorem 4. [17] *If* $f'(x) \in [0,1]$ *and*

$$\int_0^1 t^{-1}\omega(f',t)\mathrm{d}t < \infty, \tag{27}$$

further

$$f'(0) = f'(1) = 0 \tag{28}$$

then

$$\|f(x) - S_n^2(f,x)\| = O(n^{-1}) \tag{29}$$

It is also shown that we need both conditions, so expressions (27)–(28) on $f(x)$ cannot be weakened. This is because, e.g. $f(x) = x$ satisfies (27) but not (28), and on the other hand

$$f(x) = \frac{x(1-x)}{\log(x(1-x))}$$

satisfies (28) but not (27), and (29) does not hold for either function. However, the saturation problem is not solved for $\lambda = 2$, because the converse result of Theorem 4, that is (29) implies conditions (27) and (28), has not been proved yet.

Even less is known for the $1 \le \lambda \le 2$ case, when only best error estimates are obtained. From Theorem 3,

$$O(n^{1-\lambda}), \quad 1 < \lambda < 2 \qquad \text{and} \qquad O(\log^{-1} n), \quad \lambda = 1 \tag{30}$$

provided that

$$\int_0^1 t^{-\lambda}\omega(f,t)\mathrm{d}t < \infty \tag{31}$$

holds.

The results of Theorems 3 and 4 can be carried over directly for KH interpolators under the following conditions:

1. The approximated function f is univariate, i.e. the input universe of the interpolator is one dimensional.
2. The knot points are equispaced on the unit interval (20).

The second condition can be weakened, and changed to uniform distribution, since (11) ensures asymptotically the same behavior, which yield that the expression (21) and the estimate (22) remain valid in the proof of Theorem 3, if we substitute the equispaced system (20) by (11).

Corollary 1. *The KH interpolator* (12) *saturates with order* $O(n^{-1})$ *on the class of the functions* $f(x) \in \mathrm{Lip}\,1$, *if* $\lambda > 2$, *the knots point system satisfies* (11), *and the input is one dimensional.*

Proof. Because of $f(x) \in \mathrm{Lip}\,1$

$$\omega(f, n^{-1}) = \max_{\substack{x,y\in[0,1] \\ \|x-y\|\leq n^{-1}}} |f(x) - f(y)| \leq \frac{1}{n}$$

thus (17) can be estimated as

$$|K_1^{(n)} - K_2^{(n)}| = O(n^{-1}).$$

Hence, as each member of the the family of K_n^λ $(\lambda > 2)$ saturate on the class of Lip 1 with order $O(n^{-1})$ and the difference between the family members are also in order $O(n^{-1})$, then the saturation order of the whole family is $O(n^{-1})$ with the class Lip 1. □

The following statement is derived from Theorem 4.

Corollary 2. *The best approximation order of the KH interpolator* $K_n^\lambda(f, x)$ $(1 \leq \lambda \leq 2)$ *is*

$$\|f(x) - K_n^\lambda(f, x)\| = \begin{cases} O(n^{-1}), & \text{if } \lambda = 2, \\ O(n^{1-\lambda}), & \text{if } 1 \leq \lambda < 2, \\ O(\log^{-1} n), & \text{if } \lambda = 1, \end{cases}$$

on the class of the functions $f(x) \in \mathrm{Lip}\,1$, *if* $f(x)$ *further satisfies the conditions* (27) *and* (28) *for* $\lambda = 2$, *and the condition* (31) *for* $1 \leq \lambda < 2$, *and furthermore if the knots point system is in accordance with* (11), *and the input is one dimensional.*

The proof is similar as in Corollary 1.

The KH operator does not always give directly interpretable fuzzy conclusion. This problem was investigated by several researchers, and alternative solutions and methods were proposed. Among those, the so-called MACI (Modified Alpha-Cut based Interpolation) method [20] is the most advantageous, because it eliminates the abnormality problem and maintains the low computational requirements of the KH method. Moreover, its generalized version also possesses the universal approximation property [20,21]. The general MACI method tailors the conclusion as a finite sum of KH interpolators. Based on this property it is easy to show that analog results can be established as Corollaries 1 and 2 for MACI method (see also [18]).

This property of KH interpolation beside its theoretical importance, has also practical relevance. In such cases when a control function which is known only on some section of the input universe (i.e. there is no full α-cover of at least one input universe), but it can be assumed that it behaves relatively nicely on its entire domain (i.e. the conditions of Corollaries 1 and 2 hold) we can approximate the control function by KH interpolator in the given saturation order with linguistic type of rules.

4 Conclusion

This paper dealt with two important issues of approximation and modelling by FRBSs. The first part addresses the problem of universal approximation vs. interpretability. It was recalled that universal approximation property of T-controllers is not valid, if the number of rules are restricted, where the set of T-controllers include a considerably large subset of important fuzzy modelling methods, such as e.g. TSK models. This result revealed the importance of finding exact functional dependence between the accuracy of the modelling and the interpretability of the system in terms of the number of fuzzy rules.

The second part of the paper determined approximation order and class for another class of FRBSs, the fuzzy KH interpolators, by solving the saturation problem. It means that we gave the optimal order and class of approximation for stabilized KH controllers under certain condition.

References

1. R. E. Barnhill, R. P. Dube, and F. F. Little. Properties of Shepard's surfaces. *Rocky Mountain J. Math*, 13:365–382, 1991.
2. J. L. Castro. Fuzzy logic controllers are universal approximators. *IEEE Trans. on SMC*, 25:629–635, 1995.
3. G. Criscuolo and G. Mastroianni. Estimates of Shepard interpolatory procedure. *Acta Math. Hung.*, 61:79–91, 1993.
4. B. Della Vecchia, G. Mastroianni, and V. Totik. Saturation of the Shepard operators. *Appr. Theory and its Appl.*, 6(4):76–84, 1990.
5. E. P. Klement, L. T. Kóczy, and B. Moser. Are fuzzy systems universal approximators? *Int. J. General Systems*, 28(2–3):259–282, 1999.
6. L. T. Kóczy and K. Hirota. Approximate reasoning by linear rule interpolation and general approximation. *Internat. J. Approx. Reason.*, 9:197–225, 1993.
7. L. T. Kóczy and K. Hirota. Ordering, distance and closeness of fuzzy sets. *Fuzzy Sets and Systems*, 60:281–293, 1993.

8. L. T. Kóczy and A. Zorat. Fuzzy systems and approximation. *Fuzzy Sets and Systems*, 85:203–222, 1997.
9. B. Kosko. Fuzzy systems as universal approximators. *IEEE Tr. on Computers*, 43(11):1329–1333, 1994.
10. B. Moser. Sugeno controllers with a bounded number of rules are nowhere dense. *Fuzzy Sets and Systems*, 104(2):269–277, 1999.
11. E. H. Ruspini. A new approach to clustering. *Information Control*, 15:22–32, 1969.
12. D. Shepard. A two dimensional interpolation function for irregularly spaced data. In *Proc. of the 23rd ACM International Conference*, pages 517–524, 1968.
13. G. Somorjai. On a saturation problem. *Acta Math. Acad. Sci. Hungar.*, 32:377–381, 1978.
14. M. Sugeno. An introductory survey of fuzzy control. *Information Science*, 36:59–83, 1985.
15. M. Sugeno and G. T. Kang. Structure identification of fuzzy model. *Fuzzy Sets and Systems*, 28:15–33, 1988.
16. J. Szabados. On a problem of R. DeVore. *Acta Math. Acad. Sci. Hungar.*, 27:219–223, 1976.
17. J. Szabados. Direct and converse approximation theorems for Shepard operator. *J. Approx. Th. and its Appl.*, 7:63–76, 1991.
18. D. Tikk. Notes on the approximation rate of KH controllers. Submitted to *Fuzzy Sets and Systems.*
19. D. Tikk. On nowhere denseness of certain fuzzy controllers containing prerestricted number of rules. *Tatra Mountains Math. Publ.*, 16:369–377, 1999.
20. D. Tikk and P. Baranyi. Comprehensive analysis of a new fuzzy rule interpolation method. *IEEE Trans. on Fuzzy Systems*, 8(3):281–296, 2000.
21. D. Tikk, P. Baranyi, Y. Yam, and L. T. Kóczy. Stability of a new interpolation method. In *Proc. of the IEEE Int. Conf. on System, Man, and Cybernetics (IEEE SMC'99)*, volume III, pages 7–9, Tokyo, Japan, October, 1999.
22. D. Tikk, I. Joó, L. T. Kóczy, P. Várlaki, B. Moser, and T. D. Gedeon. Stability of interpolative fuzzy KH-controllers. *Fuzzy Sets and Systems*, 125(1):105–119, January 2002.
23. D. Tikk, L. T. Kóczy, and T. D. Gedeon. A survey on the universal approximation and its limits in soft computing techniques. Research Working Paper RWP-IT-01-2001, School of Information Technology, Murdoch University, Perth, W.A., 2001. p. 20.
24. L. X. Wang. Fuzzy systems are universal approximators. In *Proc. of the IEEE Int. Conf. on Fuzzy Systems*, pages 1163–1169, San Diego, 1992.
25. L. X. Wang. Analysis and design of hierarchical fuzzy systems. *IEEE Trans. on FS*, 7(5):617–624, 1999.
26. H. Ying. Sufficient conditions on uniform approximation of multivariate functions by general Takagi–Sugeno fuzzy systems with linear rule consequents. *IEEE Trans. on SMC, Part A*, 28(4):515–520, 1998.
27. K. Zeng, N.-Y. Zhang, and W.-L. Xu. A comparative study on sufficient conditions for Takagi–Sugeno fuzzy systems as universal approximators. *IEEE Trans. on FS*, 8(6):773–780, 2000.

SECTION 7

INTERPRETATION OF BLACK-BOX MODELS AS FUZZY RULE-BASED MODELS

Interpretability improvement of RBF-based neurofuzzy systems using regularized learning

Yaochu Jin

Future Technology Research, Honda R&D Europe
63073 Offenbach/Main, GERMANY
email: Yaochu_Jin@de.hrdeu.com

Abstract. Radial-basis-function (RBF) networks are mathematically equivalent to a class of fuzzy systems under mild conditions. Therefore, RBF networks have widely been used in learning of neurofuzzy systems to improve the performance. However, in most cases, the interpretability of fuzzy system will get lost after neural network learning. This chapter proposes a learning method using interpretability based regularization for neurofuzzy systems. This method can either be used in extracting interpretable fuzzy rules from RBF networks or in improving the interpretability of RBF-based neurofuzzy systems. Two simulation examples are presented to show the effectiveness of the proposed method.

1 Introduction

Jang and Sun [5] have shown that radial basis function (RBF) networks and a simplified class of fuzzy systems are functionally equivalent under some mild conditions. This functional equivalence has made it possible to combine the features of these two systems and has initiated a new model called neurofuzzy system, which has a very strong learning capability. On the other hand, however, a fuzzy system that has been trained using learning algorithms may lose its interpretability or transparency, which is one of the most important features of fuzzy systems.

In this chapter, the relationship between RBF networks and fuzzy systems is re-examined. We emphasize the differences rather than the equivalence between these two models. It is argued that the essential difference between RBF networks and fuzzy systems is the interpretability, which enables fuzzy systems to be easily comprehensible. Based on the discussions on their relationships, a method for extracting interpretable fuzzy rules from trained RBF networks or for improving the interpretability of RBF-based neuro-fuzzy systems using regularization techniques is proposed. Simulation studies are carried out on an example from process modeling to show how fuzzy rules with good interpretability can be extracted from RBF networks.

It should be mentioned that the method proposed in this work is quite different from the existing techniques for rule extraction from neural netwoks [17]. For example, a wide class of existing methods extract symbolic rules from multilayer perceptrons [18]. Although fuzzy rule extraction has been studied in [3], the work is mainly based on a special feedforward neural network structure. In our work, fuzzy rules are extracted from RBF networks by investigating the difference between interpretable fuzzy rules and RBF networks. Since fuzzy rules and RBF networks are mathematically equivalent, emphasis of this work has been laid on interpretability, which is most critical for the semantic meanings of fuzzy rules.

2 Relations Between RBF Networks and Fuzzy Systems

In this section, we first briefly review the functional equivalence between RBF networks and a class of fuzzy systems. A definition of interpretability of fuzzy systems is then proposed. Finally, the conditions on converting an RBF network to a fuzzy system are discussed.

2.1 Functional Equivalence Between RBF Networks and Fuzzy Systems

Radial basis function neural networks are one of the most important models of artificial neural networks. They were proposed in [13,12] and [14] among others in the context of different research motivations. Generally, an RBF network with a single output can be expressed as follows:

$$y = \sum_{j=1}^{N} f_j \varphi_j \left(\| \boldsymbol{x} - \boldsymbol{\mu}_j \|, \boldsymbol{\sigma}_j \right), \tag{1}$$

where, $\boldsymbol{x}$ is the input vector, $\varphi_j(\cdot)$ is called the j-th radial-basis function or the j-th receptive field unit, $\boldsymbol{\mu}_j$ and $\boldsymbol{\sigma}_j$ are the center and the variance vectors of the j-th basis function, and f_j is the weight or strength of the j-th receptive field unit. If the basis functions of the RBF network are Gaussian functions and the output is normalized, an RBF network can be described as:

$$y = \frac{\sum_{j=1}^{N} f_j \Pi_{i=1}^{m_j} \exp \left[- \left(\frac{x_i - \mu_{ij}}{\sigma_{ij}} \right)^2 \right]}{\sum_{j=1}^{N} \Pi_{i=1}^{m_j} \exp \left[- \left(\frac{x_i - \mu_{ij}}{\sigma_{ij}} \right)^2 \right]} \tag{2}$$

where $1 \le m_j \le M$ is the dimension of the j-th basis function, M is the dimension of the input space, and N is the number of hidden nodes.

Several supervised and unsupervised learning methods as well as evolutionary computation based optimization algorithms have been developed to find optimal values of the neural network parameters. Almost all of these algorithms can be applied to the neurofuzzy systems.

The theory of fuzzy sets and fuzzy inference systems [20] originated from a completely different research field. Fuzzy inference systems are composed of a set of if-then rules. In the field of modeling and control, there are two types of popular fuzzy models, namely, the Mamdani fuzzy model [11] and the Sugeno-Takagi fuzzy model [16].

A Sugeno-Takagi fuzzy model has the following form of fuzzy rules:

$$R_j : \text{ If } x_1 \text{ is } A_{1j} \text{ and } x_2 \text{ is } A_{2j} \text{ and ... and } x_M \text{ is } A_{Mj}, \quad \text{Then } y = g_j(x_1, x_2, ..., x_M), \tag{3}$$

where $g_j(\cdot)$ is a crisp function of x_i. The overall output of the fuzzy model can be obtained by:

$$y = \frac{\sum_{j=1}^{N} [g_j(\cdot) T_{i=1}^{m_j} \varphi_{ij}(x_i)]}{\sum_{j=1}^{N} T_{i=1}^{m_j} \varphi_{ij}(x_i)} \tag{4}$$

where $1 \leq m_j \leq M$ is the number of input variables that appear in the rule premise, M is the number of inputs, $\varphi_{ij}(x_i)$ is the membership function for fuzzy set A_{ij} and T is a t-norm for fuzzy conjunction. It is noticed that the RBF network expressed in equation (2) and the fuzzy systems described by equation (4) are mathematically equivalent provided that multiplication is used for the t-norm in fuzzy systems and both systems use Gaussian basis functions. Here, we will not re-state the restrictions proposed in [5], however, we will show that although these restrictions do result in the mathematical equivalence between RBF networks and fuzzy systems, they do not guarantee the equivalence of the two models in terms of the semantic meanings.

2.2 Interpretability Conditions for Fuzzy Systems

The main difference between radial-basis-function networks and fuzzy systems is the interpretability. Generally speaking, neural networks are considered to be black-boxes and therefore no interpretability conditions are imposed on conventional neural systems. On the other hand, fuzzy systems are supposed to be inherently comprehensible, especially when the fuzzy rules are obtained from human experts. However, interpretability of fuzzy systems cannot be guaranteed during data based rule generation and adaptation. For this reason, interpretability of fuzzy systems has received increasing attention in the recent years [9,7,8,6,2,15,19,4]. In the following, we propose some major conditions a fuzzy system should fulfill to be interpretable:

1. All fuzzy subsets are convex and normal. In addition, the fuzziness of all fuzzy subsets should be neither too small nor too large.
   Remarks: There are several definitions for the fuzziness measure. Fig. 1 shows three different fuzzy sets for the linguistic term "Middled aged". It is obvious that membership functions in Fig. 1 (a) and (c) my harm the interpretability.

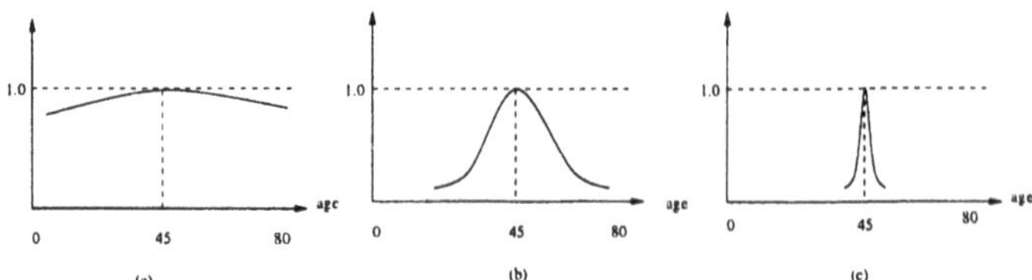

Fig. 1. Different fuzzy membership functions for "Middle aged".

2. The fuzzy partition [1] of all variables in the fuzzy system are both complete and well distinguishable. In addition, the number of fuzzy subsets in a fuzzy partition is limited.
   Remarks: The completeness and distinguishability condition makes it possible to assign a clear physical meaning to each fuzzy subset in a fuzzy partition. Therefore, it is the most important aspect for the interpretability of fuzzy systems. Usually, this also leads to a small number of fuzzy subsets. A quantitative description of the completeness and distinguishability condition can be expressed as follows:

$$\delta_1 \leq S(A_i, A_{i+1}) \leq \delta_2 \tag{5}$$

   where, A_i and A_{i+1} are two arbitrary neighboring fuzzy subsets in a fuzzy partition, $S(A_i, A_{i+1})$ is a fuzzy similarity measure between them, δ_1 and δ_2 are the lower and upper thresholds of the fuzzy similarity measure, where a positive δ_1 guarantees the completeness and a δ_2 that is sufficiently smaller than one maintains good distinguishability [8].
3. Fuzzy rules in the rule base are consistent with each other and consistent with the prior knowledge, if available.
   Remarks: Although the performance of fuzzy systems is believed to be insensitive to the inconsistency of the fuzzy rules to a certain degree, seriously inconsistent fuzzy rules will undoubtedly result in incomprehensible fuzzy systems. We argue that fuzzy rules are inconsistent in the following situations:
   - The condition parts are the same, but the consequent parts are completely different. For example,
     R1: If $x1$ is $A1$ and $x2$ is $A2$, then y is Positive Large;
     R2: If $x1$ is $A1$ and $x2$ is $A2$, then y is Negative Large
   - Although the condition parts are seemingly different, they are physicals the same. However, the consequents of the rules are totally different.
     R1: If $x1$ is $A1$ and $x2$ is $A2$, then y is Positive Large;
     R2: If $x1$ is $A1$ and $x3$ is $A3$, then y is Negative Large

[1] The word "partition" in this chapter is not used in a strictly sense. Therefore, it does not necessarily satisfy the mathematical definition of a fuzzy partition.

Although "$x2$ is $A2$" and "$x3$ is $A3$" appear to be different conditions, they might imply the same situation in some cases. For example, for a chemical reactor, a statement "temperature is high" may imply "conversion rate is high".

- The conditions in a rule premise are contradictory, e.g. "If the sun is bright and the rain is heavy".
- The actions in the rule consequent part are contradictory. For example, in the rule "If x is A then y is B and z is C." However, "y is B and z is C" cannot happen simultaneously.

4. The number of variables that appear in the premise part of the fuzzy rules should be as small as possible. In addition, the number of fuzzy rules in the rule base should also be small. These two aspects deal with the compactness of the rule structure.

The interpretability conditions impose implicit constraints on the parameters of fuzzy systems. While interpretability is one of the most important feature of fuzzy systems, there are generally no interpretability requirements on the parameters of RBF networks. In this sense, RBF networks and fuzzy systems are not the same even if they are functionally equivalent.

2.3 Conversion of An RBFN into Fuzzy Rules

The central point in converting an RBFN into a Sugeno fuzzy model is to ensure that the extracted fuzzy rules are interpretable, i.e. easy to understand. In order to convert an RBF network to an interpretable fuzzy rule system, the following conditions should be satisfied:

1. The basis functions of the RBF network are Gaussian functions.
2. The output of the RBF network is normalized.
3. The basis functions within each receptive field unit of the RBF network are allowed to have different variances.
4. Certain numbers of basis functions for the same input variable but within different receptive field units should share a mutual center and a mutual variance. If Mamdani fuzzy rules are to be extracted, then some of the output weights of the RBF network should also share.

Conditions 3 and 4 are necessary for good interpretability of the extracted fuzzy system. Without condition 3, the first part of condition 4 cannot be realized. As the most important condition, condition 4 requires that some weights in the RBF network should share. For the sake of simplicity, we use 'weights' to refer to both the parameters of basis functions (centers and variances) and the output weights of RBF networks in the following text. The weight sharing condition ensures a good distinguishability for the fuzzy partition, which is the most essential feature for the interpretability of fuzzy systems. Given a fuzzy system in Fig. 2, the RBF network that is directly converted from the fuzzy system is illustrated in Fig. 3. If we take a closer

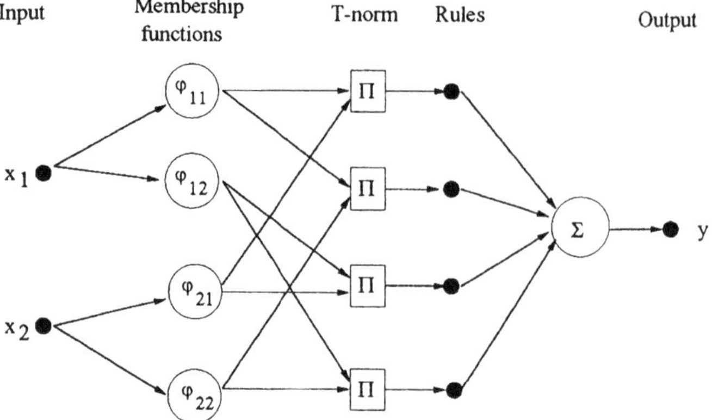

Fig. 2. A fuzzy system.

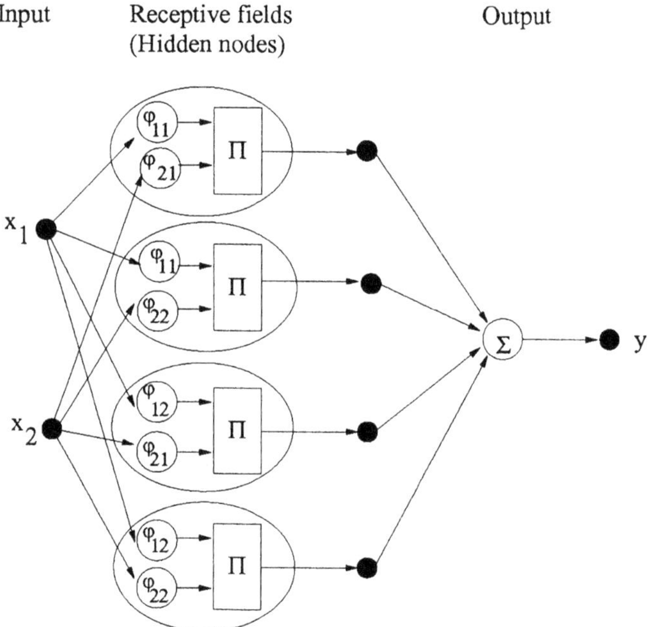

Fig. 3. The RBF network converted from the fuzzy system.

look at the RBF network in Fig. 3, we notice that some of the basis functions are identical. On the other hand, if a fully connected RBF network with N hidden nodes is directly converted into a fuzzy system, each variable of the fuzzy system will have N sub-fuzzy sets. If N is large (e.g., $N > 10$), it will be difficult to understand the fuzzy system. However, we find it difficult to define the weight sharing condition explicitly because we cannot require that the structure of the extracted fuzzy system should be the same as its original structure, therefore, we do not know beforehand which weights should share. Additionally, the completeness of the fuzzy partition should be considered together with the weight sharing condition in the course of rule extraction. These problems will be treated in detail in the next section.

Note that the consistency condition is not considered here, because we suppose that it has been taken into account in generating the initial fuzzy system [8]. Nevertheless, measures can be taken to prevent the rule extraction algorithm from generating seriously inconsistent rules, namely, rules with the same premise but different consequents.

3 Fuzzy Rule Extraction from RBF Networks

Assume now we have a trained RBF network. This RBF network is either directly trained using the common training algorithms for RBF networks [13,12], or it is converted from an optimized fuzzy system [7]. As we have discussed in the last section, to extract interpretable fuzzy rules from an RBF network, some weights should share. Thus, extracting interpretable fuzzy rules from RBF networks can be treated as fine training of the RBF network with regularization [1] such that similar weights in the RBF network share the same value.

Before the regularization can be applied, it is necessary to specify, which weights should be identical before the weight sharing algorithm can be employed. Thus, the first step toward rule extraction from RBF networks is to determine which weights, including the parameters of basis functions and the output weights, should share.

3.1 Specification of Shared Weights

Since the rule structure of the fuzzy system is unknown, we do not know in advance, which weights should share. However, it is straightforward to imagine that the weights that is going to share a same value should be similar before the regularization is applied. Therefore, we have to first identify similar weights using a distance measure. Currently, several distance measures (or similarity measures) are available. Among them, Euclidean distance is very simple and has been widely used. The Euclidean distance between two membership functions (basis functions) $\varphi_i(\mu_i, \sigma_i)$ and $\varphi_j(\mu_j, \sigma_j)$, where μ_i, μ_j are centers and σ_i, σ_j are the variances can be defined as:

$$d(\varphi_i, \varphi_j) = \sqrt{(\mu_i - \mu_j)^2 + (\sigma_i - \sigma_j)^2}. \quad (6)$$

For a Gaussian basis function or membership function, each has two elements, namely, the center and the variance. For the output weights, each vector has only one element. Suppose that a given input variable x_i has M different basis functions $\varphi_{ij} (j = 1, 2, \cdots, M)$ with the center μ_{ij} and the variance σ_{ij}, the procedure to determine similar basis functions can be described as follows:

1. List the basis functions (φ_{ij}) in the order of increasing sequence with regard to their center values. Let U_{ik} be the k-th set for x_i containing similar basis functions. Two basis functions are considered similar if the distance between them is less than d_i, where d_i is a prescribed threshold. The regularization algorithm will drive the similar basis functions in set U_{ik} to share the same parameters. Put φ_{ij} to U_{ik}, let $j, k = 1$, $\varphi_i^0 = \varphi_{i1}$.
2. If $d(\varphi_i^0, \varphi_{ij+1}) < d_i$, put φ_{ij+1} to U_{ik}; else $k = k + 1$, put φ_{ij+1} to U_{ik} and let $\varphi_i^0 = \varphi_{ij+1}$.
3. $j = j + 1$, if $j \leq M$, go to step 2; else stop.

The prescribed distance threshold d_i is very important because it determines both the distinguishability and the completeness of the fuzzy partitions. Suppose $\hat{\mu}_{ik}$ and $\hat{\sigma}_{ik}$ are the averaged center and variance of the basis functions in U_{ik}, then the fuzzy partition constructed by $\hat{\mu}_{ik}$ and $\hat{\sigma}_{ik}$ should satisfy the completeness and distinguishability condition described in equation (5).

In practice, we find that the performance of the extracted fuzzy system is not satisfactory if we simply choose $\hat{\mu}_{ik}$ and $\hat{\sigma}_{ik}$ to be the values to share by the basis functions in U_{ik}. In other words, a direct merge of similar basis functions will degrade the performance seriously. In the following subsection, we will introduce an adaptive weight sharing method to improve the performance of the extracted fuzzy system.

3.2 Adaptive Weight Sharing

We do not directly require that the weights in the same set should be identical. Instead, we realize weight sharing by regularizing the RBF network. In the following, we present the weight sharing algorithm with regard to the RBF model described in equation (2).

Regularization of neural networks is realized by adding an extra term to the conventional cost function:

$$J = E + \lambda \cdot \Omega, \quad (7)$$

where E is the conventional cost function, λ is the regularization coefficient ($0 \leq \lambda < 1$), and Ω is the regularization term for weight sharing. The cost function E is expressed as:

$$E = \frac{1}{2}(y - y^t)^2, \tag{8}$$

where, y is the output of the neural network and y^t is the target value. In the following, we assume that Sugeno fuzzy rules are to be extracted, that is to say, the output weights of the RBF network are not regularized. In this case, the regularization term Ω has the following form:

$$\Omega = \frac{1}{2}\sum_i \sum_k \sum_{\varphi_{ij} \in U_{ik}} (\mu_{ij} - \bar{\mu}_{ik})^2 + \frac{1}{2}\sum_i \sum_k \sum_{\varphi_{ij} \in U_{ik}} (\sigma_{ij} - \bar{\sigma}_{ik})^2 \tag{9}$$

where $\bar{\mu}_{ik}$ and $\bar{\sigma}_{ik}$ are the center and variance to be shared by the basis functions φ_{ij} in set U_{ik}. Empirically, the averaged center $\hat{\mu}_{ik}$ and variance $\hat{\sigma}_{ik}$ of set U_{ik} are used as the initial values for $\bar{\mu}_{ik}$ and $\bar{\sigma}_{ik}$. The gradients of J are:

$$\frac{\partial J}{\partial \mu_{ij}} |_{\varphi_{ij} \in U_{ik}} = \frac{\partial E}{\partial \mu_{ij}} + \lambda(\mu_{ij} - \bar{\mu}_{ik}) \tag{10}$$

$$\frac{\partial J}{\partial \sigma_{ij}} |_{\varphi_{ij} \in U_{ik}} = \frac{\partial E}{\partial \sigma_{ij}} + \lambda(\sigma_{ij} - \bar{\sigma}_{ik}) \tag{11}$$

$$\frac{\partial J}{\partial \bar{\mu}_{ik}} = -\lambda \sum_{\varphi_{ij} \in U_{ik}} (\mu_{ij} - \bar{\mu}_{ik}) \tag{12}$$

$$\frac{\partial J}{\partial \bar{\sigma}_{ik}} = -\lambda \sum_{\varphi_{ij} \in U_{ik}} (\sigma_{ij} - \bar{\sigma}_{ik}), \tag{13}$$

where

$$\frac{\partial E}{\partial \mu_{ij}} = (y - y^t)(f_j - y)(x_i - \mu_{ij}) \prod_{i=1}^{m_j} \left[\exp\left(-\frac{(x_i - \mu_{ij})^2}{\sigma_{ij}^2} \right) \right] / \sigma_{ij}^2 \tag{14}$$

$$\frac{\partial E}{\partial \sigma_{ij}} = (y - y^t)(f_j - y)(x_i - \mu_{ij})^2 \prod_{i=1}^{m_j} \left[\exp\left(-\frac{(x_i - \mu_{ij})^2}{\sigma_{ij}^2} \right) \right] / \sigma_{ij}^3 \tag{15}$$

Strictly speaking, the value of $\bar{\mu}_{ik}$ and $\bar{\sigma}_{ik}$ also depends on the value of μ_{ij} and σ_{ij}. However, $\bar{\mu}_{ik}$ and $\bar{\sigma}_{ik}$ do not change in each learning cycle, but only when the elements in U_{ik} changes. Although the specification algorithm introduced in the last subsection must be applied in each iteration of the neural network learning to improve the performance of the extracted fuzzy system, it does not necessarily happen the elements in U_{ik} change. Recall that the weights of the RBF network are determined not only by the regularization term, but also by the conventional error term. That is to say, it is possible that a basis function φ_{ij} that is originally put in U_{ik} will be classified to another set after an iteration of learning. By specifying the shared weights during network learning, a more optimal rule structure will be obtained. It is also necessary to check the completeness condition on the fuzzy partition constructed by $(\bar{\mu}_{ik}, \bar{\sigma}_{ik})$ during learning. The incompleteness of the fuzzy partition can be avoided by temporarily stopping the adaptation of the shared parameters $(\bar{\mu}_{ik}, \bar{\sigma}_{ik})$. The completeness of a fuzzy partition can be checked using Equation (5).

If Mamdani fuzzy rules are to be extracted, a similar specification of the shared weights should be carried out on the output weights and an extra term should be added to Ω. Then, a regularized learning algorithm for the output weights can also be derived. The resulting weights can be explained as the centers of the membership functions for the output variable.

A flowchart of the rule extraction process is illustrated in Fig. 4.

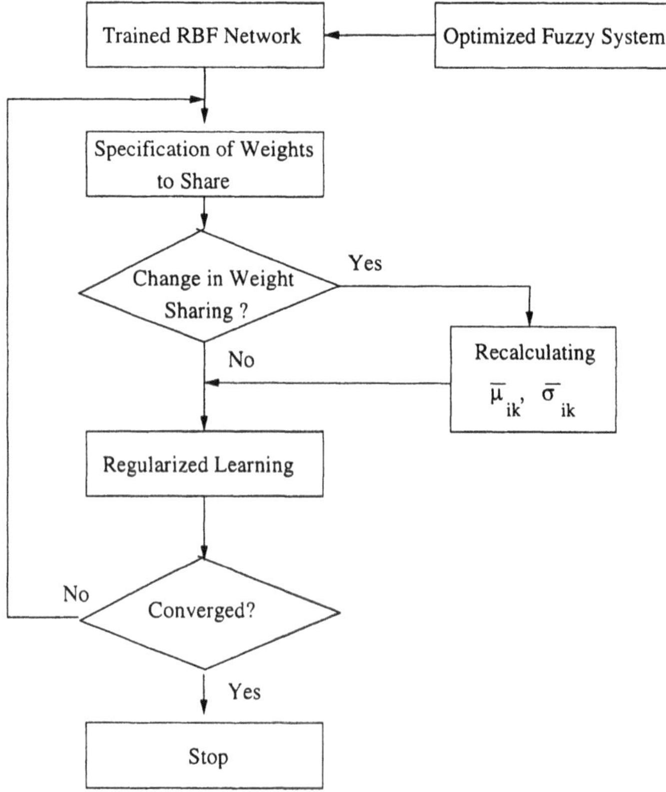

Fig. 4. The flowchart of rule extraction.

4 Numerical Examples

4.1 The Lorenz System

The Lorenz system studied in this paper is described by the following differential equations [10]:

$$\frac{dx}{dt} = -y^2 - z^2 - a(x - F) \tag{16}$$

$$\frac{dy}{dt} = xy - bxz - y + G \tag{17}$$

$$\frac{dz}{dt} = bxy + xz - z \tag{18}$$

where $a = 0.25$, $b = 4.0$, $F = 8.0$ and $G = 1.0$. In our simulation, we predict $x(t)$ from $x(t-1)$, $y(t-1)$ and $z(t-1)$. 2000 data sets are generated using the fourth order Runge-Kutta method with a step length of 0.05, where 1000 pairs of data are used for training and the other 1000 for test. An RBF network is obtained with 5 receptive field units. Fig. 5 shows the approximation results of the RBF network. From Fig. 5, it can be seen that the performance is satisfying. However, most basis functions are hard to distinguish (see Fig. 6 and therefore it is difficult to associate proper semantic meanings to the basis functions. In this sense. the RBF network cannot be considered as an interpretable fuzzy system.

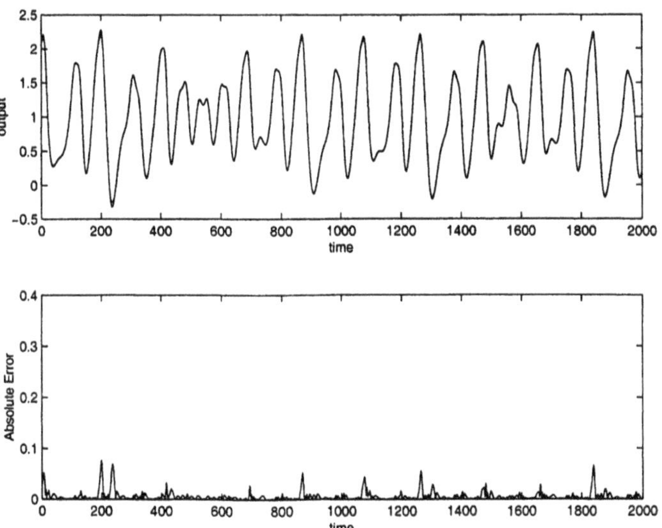

Fig. 5. Output and approximation error of the RBF network.

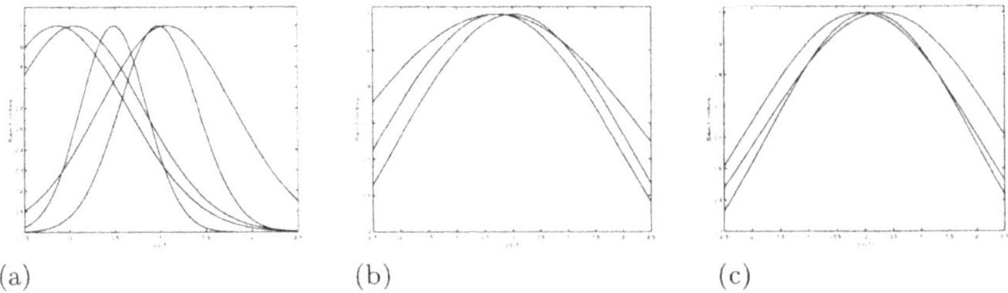

Fig. 6. The basis functions of the RBF network: (a) $x(t-1)$, (b) $y(t-1)$ and (c) $z(t-1)$.

To extract interpretable fuzzy system from the RBF network, the algorithm for extracting fuzzy rules is implemented. The algorithm consists of two parts: specification of shared weights and network training with regularization. The approximation result of the extracted fuzzy systems are shown in Fig. 7. Although the performance of the extracted fuzzy system is a little worse than that of the RBF network, the interpretability of the extracted fuzzy system has been significantly improved (see Fig. 8). According to the distribution of the membership functions, a proper linquistic meaning can be associated to each fuzzy set. For input $x(t-1)$, " **Negative Small (NS)**" can be assigned to the membership function $(-0.49, 1.69)$, " **Positive Small (PS)**" to $(0.37, 1.24)$, " **Positive Middle (PM)**" to $(1.01, 0.80)$ and " **Positive Large (PL)**" to $(2.01, 1.49)$. Similarly, for $y(t-1)$, " **Negative Small (NS)**" is assigned to $(-0.35, 4.11)$, " **Zero (ZO)**" to $(0.03, 3.53)$; for $z(t-1)$, " **Zero (ZO)**" is assigned to $(-0.04, 2.79)$. In this way, some intelligible knowledge about the Lorenz system in terms of interpretable fuzzy rules can be acquired.

4.2 Process Modeling

The data used in the following simulation are generated to simulate an industrial process. We use this example because it is a high-dimensional system with deliberately added biased noises. In this simulated system, there are 11 inputs and one output with 20,000 data for training and 80,000 data for test.

An RBF network with 27 hidden nodes is obtained using the training data. The RMS errors on training and test data are 0.189 and 0.207. Although the performance is satisfying, the RBF model is hard to understand when we take a look at the basis functions of, particularly 6 of the 11 inputs, see Figures 9 and 10.

To extract interpretable fuzzy rules from the RBF network, the proposed algorithm is employed. After training, the RMS errors of the fuzzy system on the training and test data become 0.191 and 0.213 respectively, which have slightly increased as expected. What is very encouraging is that the number of fuzzy subsets in the fuzzy partitions are significantly reduced and the

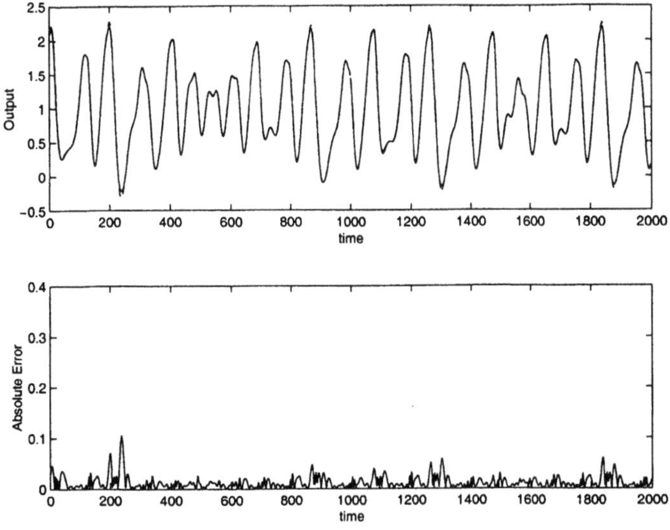

Fig. 7. Output and approximation error of the extracted fuzzy system.

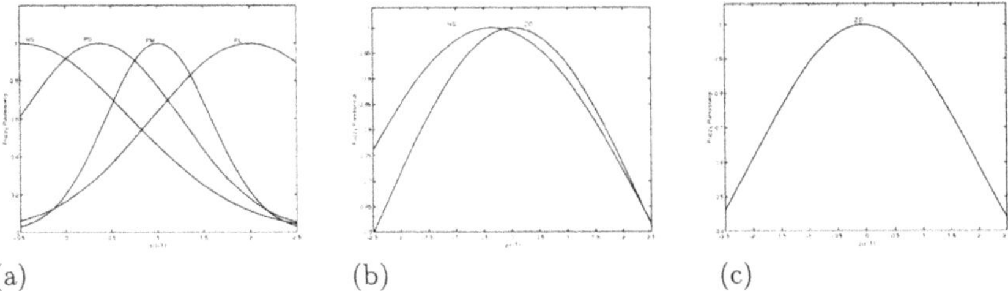

(a) (b) (c)

Fig. 8. The membership functions of the fuzzy system: (a) $x(t-1)$, (b) $y(t-1)$ and (c) $z(t-1)$.

distinguishability is greatly improved, see Figures 11 and 12. With these well distinguishable fuzzy partitions, it is possible to associate a linguistic term to each fuzzy subset and thus interpretable fuzzy rules can be obtained. This shows that the proposed algorithm is effective for high-dimensional systems.

5 Conclusions

The relationships between RBF networks and interpretable fuzzy systems have been discussed. A definition for an interpretable fuzzy system has also been suggested. Conditions for converting RBF networks to fuzzy systems have been proposed. In order to extract interpretable fuzzy rules from an RBF network, an adaptive weight sharing algorithm has been introduced. We have shown that an RBF network and a fuzzy system are not fully equivalent

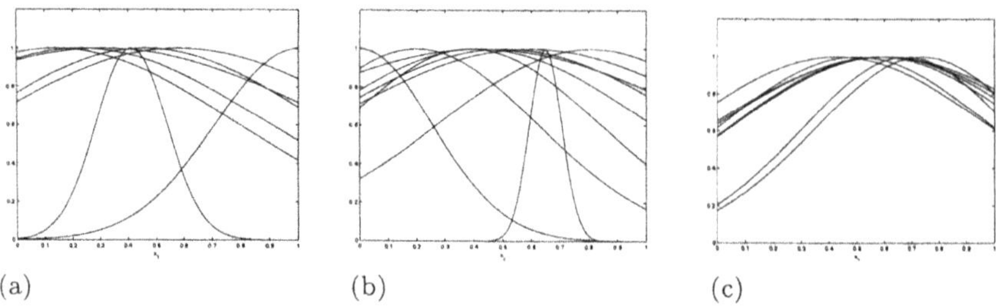

Fig. 9. The basis functions of the RBF network: (a) x_1, (b) x_2 and (c) x_3.

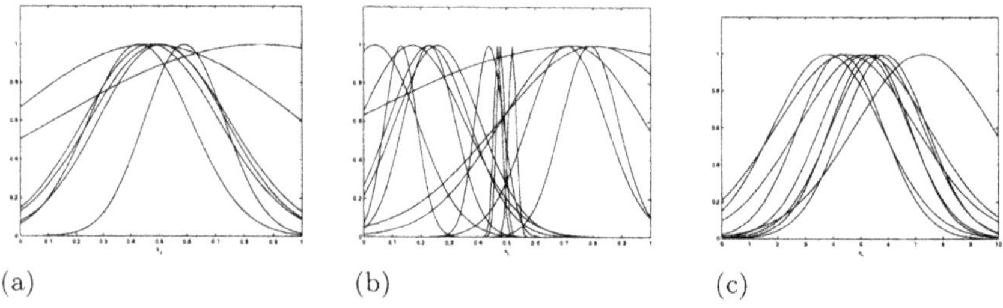

Fig. 10. The basis functions of the RBF network: (a) x_4, (b) x_5 and (c) x_6.

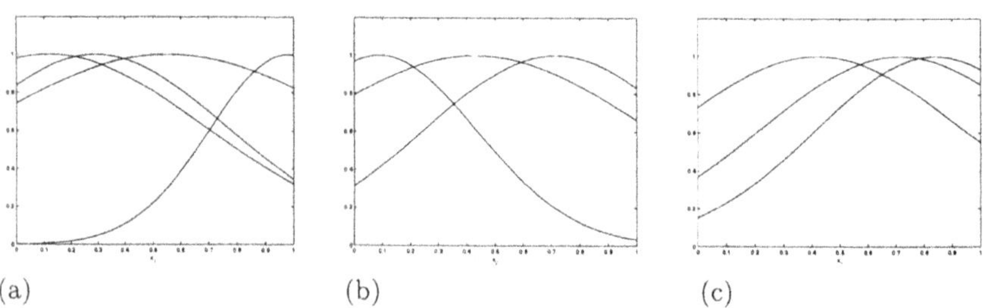

Fig. 11. The membership functions of the fuzzy system: (a) x_1, (b) x_2 and (c) x_3.

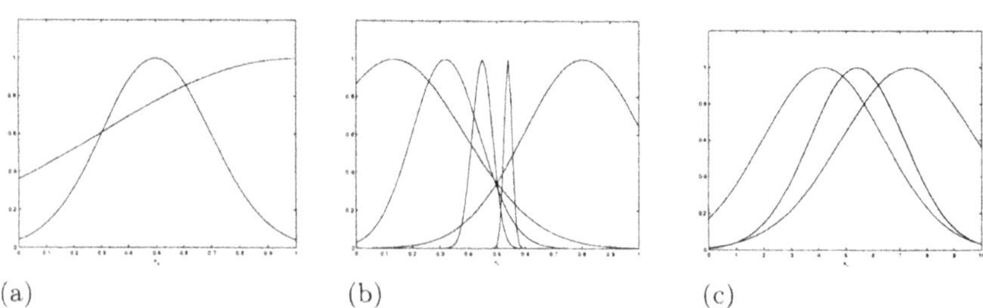

Fig. 12. The membership functions of the fuzzy system: (a) x_4, (b) x_5 and (c) x_6.

in terms of their semantic meanings and that the extraction of interpretable fuzzy rules from RBF networks is both important and feasible for gaining a deeper insight into the logical structure of the system to be approximated. Simulation studies have been carried out on a high-dimensional system to show the effectiveness of the proposed method.

ACKNOWLEDGEMENTS

Part of this work was done when the authors were with the Institut für Neuroinformatik, Ruhr-Universität Bochum. This work was supported in part by German Ministry of the Research under the grant AENEAS.

References

1. C. M. Bishop. *Neural Networks for Pattern Recognition.* Oxford University Press, Oxford, UK, 1995.
2. M.-Y. Chow, S. Altrug, and H.J. Trussell. Heuristic constraints enforcement for training of and knowledge extraction from a fuzzy/neural architeture - Part I: Foundations. *IEEE Transactions on Fuzzy Systems*, 7(2):143–150, 1999.
3. Y. Hayashi. A neural expert system with automated extraction of fuzzy if-then rules and its application to medical diagnosis. *Advances in Neural Information Processing Systems*, 3:578–584, 1990.
4. H. Ishbuchi, T. Nakashima, and T. Murada. Multi-objective optimization in linguistic rule extraction from numerical data. In *Proceedings of 1st International Conferemce on Evolutionary Multi-criterion Optimization*, pages 588–602, 2001.
5. J.-S. R. Jang and C.-T. Sun. Functional equivalence between radial basis functions and fuzzy inference systems. *IEEE Trans. on Neural Networks*, 4:156–158, 1993.
6. Y. Jin. Fuzzy modeling of high-dimensional systems: Complexity reduction and interpretability improvement. *IEEE Transactions on Fuzzy Systems*, 8(2):212–221, 2000.
7. Y. Jin, W. von Seelen, and B. Sendhoff. An approach to rule-based knowledge extraction. In *Proceedings of IEEE Int. Conf. on Fuzzy Systems*, pages 1188–1193, Anchorage, AL, 1998.
8. Y. Jin, W. von Seelen, and B. Sendhoff. On generating FC^3 fuzzy rule systems from data using evolution strategies. *IEEE Trans. on Systems, Man, and Cybernetics*, 29:829–845, 1999.
9. A. Lofti and A.C. Tsoi. Interpretation preservation of adaptive fuzzy inference systems. *Int. Journal of Approximating Reasoning*, 15, 1996.
10. E.N. Lorenz. Irregulatiry: A fundamental property of the atmosphere. *Tellus*, A(36):98, 1984.
11. E. H. Mamdani and S. Assilian. An experiment in linguistic synthesis with a fuzzy logic controller. *Int. Journal of Man-Machine Studies*, 7:1–13, 1975.
12. J. Moody. Fast learning in multi-resolution hierarchies. *Advances in Neural Information Processing Systems*, 1:29–39, 1989.
13. J. Moody and C. Darken. Fast learning in networks of locally-tuned processing units. *Neural Computation*, 1:181–194, 1989.

14. M. Powell. Radial basis functions for multivariable interpolation: A review. In C. Mason and M.G. Cox, editors, *Algorithms for Approximation*, pages 143–167. Oxford, UK: Oxford University Press, 1987.
15. M. Setnes, R. Babuska, and B. Verbruggen. Rule-based modeling: Precision and transparency. *IEEE Transactions on Systems, man, and Cybernetics - Part C: Applications and Reviews*, 28(1):165–169, 1998.
16. T. Takagi and M. Sugeno. Fuzzy identification of systems and its applications to modeling and control. *IEEE Trans. on Systems, Man, and Cybernetics*, 15:116–132, 1985.
17. A. Tickle, R. Andrews, M. Golea, and J. Diederich. The truth will come to light: Directions and challenges in extracting the knowledge embedded within trained artificial neural networks. *IEEE Transactions on Neural Networks*, 9(6):1057–1068, 1998.
18. G. Towell and J. Shavlik. Extracting refined rules from knowledge-based neural networks. *Machine Learning*, 13:71–101, 1993.
19. J. Valente de Oliveira. On the optimization of fuzzy systems using bio-inspired strategies. In *IEEE Proceedings of International Conference on Fuzzy Systems*, pages 1129–1134, Anchorage, Alaska, 1998. IEEE Press.
20. L.A. Zadeh. Outline of a new approach to the analysis of complex systems and decision processes. *IEEE Trans. on Systems, Man, and Cybernetics*, 3:18–44, 1973.

Extracting Fuzzy Classification Rules from Fuzzy Clusters on the Basis of Separating Hyperplanes

Birka von Schmidt[1] and Frank Klawonn[2]

[1] Institute of Flight Guidance
German Aerospace Center
Lilienthalplatz 7
D-38108 Braunschweig, Germany
[2] Department of Computer Science
University of Applied Sciences Braunschweig/Wolfenbuettel
Salzdahlumer Str. 46/48
D-38302 Wolfenbuettel, Germany

Abstract. Fuzzy clustering provides a (fuzzy) classification of data into different classes. From the result of a fuzzy cluster analysis fuzzy classification rules can be derived. The most common techniques for this derivation of rules are based on projections of the clusters. The corresponding rules classify only approximately in the same way as the fuzzy clusters themselves, since a certain loss of information has to be tolerated caused by the projections. In this paper, we propose to compute the class or cluster boundaries induced by the fuzzy clusters explicitly and to build up fuzzy rules that reflect exactly these boundaries.

1 Introduction

Fuzzy clustering (for an overview see for example [8]) is a method for classifying data. Usually fuzzy clustering is applied in the context of unsupervised classification, where the data from the training are not assigned to classes. But there are also methods to use fuzzy clustering in the case of supervised classification. Since a classifier derived from fuzzy clustering uses multi-dimensional membership functions, the corresponding classifier is often transformed into a fuzzy classifier using if-then rules in order to have a better interpretation and understanding of the classifier. Projection is a very common technique to derive rules from fuzzy clusters. However, projection means always a loss of information so that the rule-based classifier does not have the same performance as the original one. This means that a certain degradation of accuracy is tolerated for the sake of interpretability.
The properties of fuzzy classifiers based on if-then rules are well-examined and it is important to note that the structure of the class boundaries for standard fuzzy if-then classifiers and for fuzzy clustering classifiers is identical. We therefore propose not to derive rules by projecting fuzzy clusters, but to construct a fuzzy if-then classifier directly from the class boundaries induced by fuzzy clustering.

For this purpose we just assume that the classification problem is a piecewise linear one as it is in the case of the boundaries given by hyperplanes. We can use the Łukasiewicz-t-norm to construct a fuzzy classification system that draws exactly these boundaries.

2 Fuzzy Classification Systems

This section provides a brief introduction into the background in fuzzy classification and fuzzy clustering that is needed for understanding the connection that we are going to establish in this paper.

2.1 Fuzzy Classifiers

We consider classification problems. Our data are real-valued tuples or vectors with n components. We therefore assume that our data lie within $\mathbb{R}^n$ or - in case of normalised data - within $[0,1]^n$. We usually assume that the data lie within some box $X_1 \times \ldots \times X_n \subseteq \mathbb{R}^n$, where the sets X_i are intervals. To each datum one of the classes $\mathcal{C}_1, \ldots, \mathcal{C}_c$ or *unknown* is assigned. Including the class *unknown* into our classification means that we do not require that each datum must be classified in terms of the meaningful classes $\mathcal{C}_1, \ldots, \mathcal{C}_c$. This means that the subsets of $X_1 \times \ldots \times X_n$ associated with the classes (including *unknown*) induce a partition of $X_1 \times \ldots \times X_n$.
This partition or, equivalently, the assignment of the elements of $\mathbb{R}^n$ or $X_1 \times \ldots \times X_n \subset \mathbb{R}^n$ to the classes must either be specified by some expert or has to be learned from a finite training data set for which the classes are known. A fuzzy max-min classifier is characterised as follows. There is a finite set $\mathcal{R}$ of rules of the form

R: If x_1 is $\mu_R^{(1)}$ and ... and x_n is $\mu_R^{(n)}$ then class is $\mathcal{C}^{(R)}$,

where $\mathcal{C}^{(R)}$ is one of the classes $\mathcal{C}_1, \ldots, \mathcal{C}_c$. The $\mu_R^{(i)}$ are assumed to be fuzzy sets on the intervals X_i, i.e. $\mu_R^{(i)} : X_i \to [0,1]$, where X_i is an interval. In order to keep the notation simple, we denote the fuzzy sets $\mu_R^{(i)}$ directly in the rules. In real systems one would replace them by suitable linguistic values like *positive big*, *approximately zero*, etc. and associate the linguistic value with the corresponding fuzzy set.
A single rule is evaluated by interpreting the conjunction in terms of the minimum, i.e.

$$\mu_R(p_1, \ldots, p_n) = \min_{i \in \{1, \ldots, n\}} \left\{ \mu_R^{(i)}(p_i) \right\} \tag{1}$$

is the degree to which rule R fires.

$$\mu_C^{(\mathcal{R})}(p_1, \ldots, p_n) = \max \left\{ \mu_R(p_1, \ldots, p_n) \mid \mathcal{C}^{(R)} = \mathcal{C} \right\} \tag{2}$$

is the degree to which the point $(p_1, \ldots, p_n)$ is assigned to class $\mathcal{C}$.
Finally, the data point $(p_1, \ldots, p_n)$ must be assigned to a unique class (defuzzification) by

$$\mathcal{R}(p_1, \ldots, p_n) = \begin{cases} \mathcal{C} & \text{if for all } \mathcal{C}' \neq \mathcal{C}: \\ & \mu_{\mathcal{C}}^{(\mathcal{R})}(p_1, \ldots, p_n) > \mu_{\mathcal{C}'}^{(\mathcal{R})}(p_1, \ldots, p_n) \\ \textit{unknown} & \text{otherwise.} \end{cases}$$

This means that we assign the point $(p_1, \ldots, p_n)$ to the class of the rule with the maximum firing degree. If there is more than one class having the maximum firing degree, the class of the datum is *unknown*. Note that $\mathcal{R}$ denotes the set of rules as well as the associated mapping assigning to each point the corresponding class based on these classification rules.
We can easily generalise fuzzy max-min classifiers to fuzzy s-t classifiers where t is a an arbitrary t-norm and s a t-conorm by replacing the minimum in (1) by t and the maximum in (2) by s.
An early overview on fuzzy classification is given in [14].
Although fuzzy classifiers were used in practical application for many years already, studies about their fundamental properties have been published only in recent years. In [10] it was shown that max-min classifiers cannot solve linearly separable classification problems for $n > 2$ exactly, but by the use of other t-norms or t-conorms than min, respectively max, any linearly separable problem can be solved by a fuzzy classifier. In [18] it was proved that max-min classification depends locally on only two variables.
Nürnberger et al. [15,16] investigate the class boundaries of two- and three-dimensional data that can be generated by fuzzy classifiers using different t-norms. Cordón et al. [2] analyse fuzzy classifiers on an experimental basis that do not rely on a classification based on the rule that best fits the input. L. Kuncheva provides a very thorough and detailed analysis of fuzzy classifiers in [13].
It is important to note that in case of max-min classifiers as well as for other types of classifiers based on the Łukasiewicz t-norm or the bounded sum t-conorm the separation boundaries between classes can be described by hyperplanes, if triangular or trapezoidal membership functions are used in the rules. The dimension of these separating hyperplanes is always $n-1$ if n attributes are used for the classification.

2.2 Fuzzy Clustering

We restrict our considerations to the most basic fuzzy clustering techniques: The fuzzy c-means algorithm with its variants.
The purpose of the fuzzy c-means algorithm [1] is to (fuzzy) partition a finite data set $\{x_1, \ldots, x_N\} \subseteq \mathbb{R}^n$ into a fixed number K of clusters. Each cluster is represented by a prototype $v_i \in \mathbb{R}^n$ and for each datum x_j we have to determine a membership degree $u_{ij} \in [0, 1]$ to cluster v_i. The prototypes and

membership degrees should be chosen in such a way that they minimise the sum of the weighted distances of the data to the prototypes, weighted with the corresponding membership degrees, i.e.

$$\sum_{i=1}^{K}\sum_{j=1}^{N} u_{ij}^{m} d_{ij} \tag{3}$$

should be minimised where $d_{ij} = \| v_i - x_j \|^2$ is the squared Euclidean distance between prototype v_i and datum x_j.
The parameter $m > 1$ is called fuzzifier and controls how well-separated fuzzy clusters are. It is not of importance in our context here.
In order to avoid the trivial solution $u_{ij} = 0$ for all i, j – no datum is assigned to any cluster – an additional constraint has to be introduced:

$$\sum_{i=1}^{C} u_{ij} = 1 \qquad \text{for all } j. \tag{4}$$

This means that each datum is required to have an accumulated membership degree of 1 to all clusters. Since in this case the membership degree can be interpreted as the probability that a datum is assigned to a cluster, this kind of clustering is called probabilistic clustering.
In order to have a minimum of the objective function (3) satisfying constraint (4) the following necessary conditions must be satisfied:

$$u_{ij} = \frac{1}{\sum_{k=1}^{C}\left(\frac{d_{ij}}{d_{kj}}\right)^{\frac{1}{m-1}}} \tag{5}$$

$$v_i = \frac{\sum_{j=1}^{N} u_{ij}^{m} x_j}{\sum_{j=1}^{N} u_{ij}^{m}} \tag{6}$$

In order to minimise the objective function (3) these two equations are alternatingly applied until convergence is reached. (Note that in the case $d_{ij} = 0$ the membership degree u_{ij} must be set to 1 for exactly one i for which $d_{ij} = 0$ holds.)
For a new datum $x \in \mathbb{R}^n$ that was not involved in the computation of the prototypes and the membership degrees, we can still determine its membership degree to any cluster v_i, simply by replacing x_j in (5) by x. In this way we can associate a membership function $u_i : \mathbb{R}^n \to [0,1]$ to each cluster v_i.
Due to the probabilistic constraint, the membership functions u_i have some undesired properties. Since the membership degree in (5) depends only on the relative distance of the datum the prototypes, the membership degree tends to $1/C$ for data that are far away from any cluster. Figure 1 illustrates this behaviour by considering one-dimensional data and two prototypes at $v_1 = 0$ and $v_2 = 1$. The membership function shown is the one associated with cluster v_1.

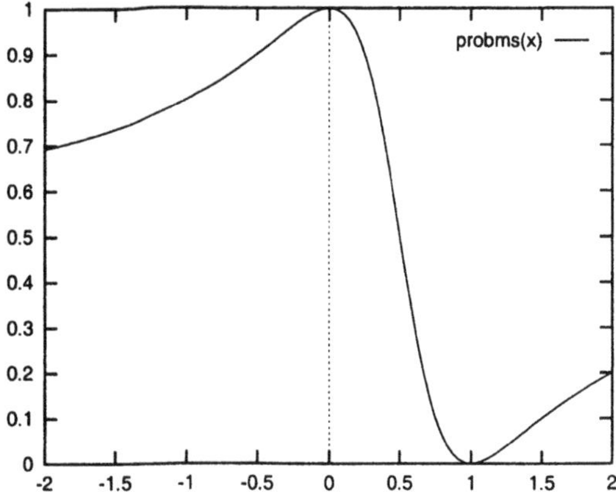

Fig. 1. Probabilistic membership function

In order to reduce this undesired effect, Davé [3] has introduced noise clustering that still uses the probabilistic constraint (4). However, in noise clustering there is one so-called noise cluster that is supposed to take care of outliers. The noise cluster is not represented by a prototype. Instead it is assumed that each datum has a fixed (large) distance to the noise cluster.

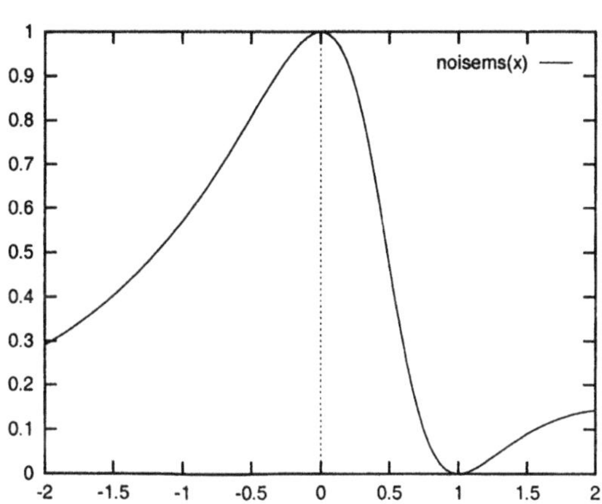

Fig. 2. Noise membership function

Figure 2 shows the same membership function as in figure 1, except that we have introduced a third cluster, the noise cluster to which all data have a constant distance of 2.

Possibilistic clustering drops the probabilistic constraint (4) completely and introduces an additional term into the objective function (3) assigning a penalty to membership degrees near 0. This leads to the formula

$$u_{ij} = \frac{1}{1 + \left(\frac{d_{ij}}{\eta_i}\right)^{\frac{1}{m-1}}}$$

for the membership degrees. η_i is a parameter determining how slow membership decrease with increasing distance from the prototype. The equation for the prototypes remains the same as (3).
Looking at the right ends of the graphs in figures 1,2 and 3 we can see how outliers are treated by the corresponding fuzzy clustering strategies. Although possibilistic clustering seems to be the best choice for handling outliers, there are other disadvantages of possibilistic clustering as they are discussed in [4].

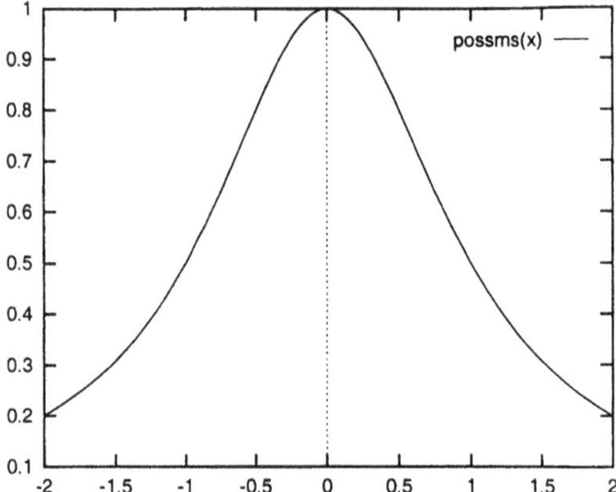

Fig. 3. Possibilistic membership function

Although it is assumed that the number of clusters K is fixed in advance, there are techniques to automatically adapt the number of clusters. For an overview on this topic we refer to [8].
Fuzzy clustering is designed for unsupervised classification, i.e., the assignment of the training data to classes is not known in advance. Nevertheless, fuzzy clustering can also be applied in the case of supervised classification in order to construct a fuzzy classifier. One can for instance cluster the data first ignoring the class information and then assign a class to each cluster, namely the class to which the majority (in terms of the sum of the membership degrees) of the data assigned to this cluster belongs [6,11]. An alternative approach for (partially) classified data is proposed in [17].
A fuzzy classifier based on fuzzy clustering can use the multi-dimensional membership functions u_i directly. However, in order to better understand

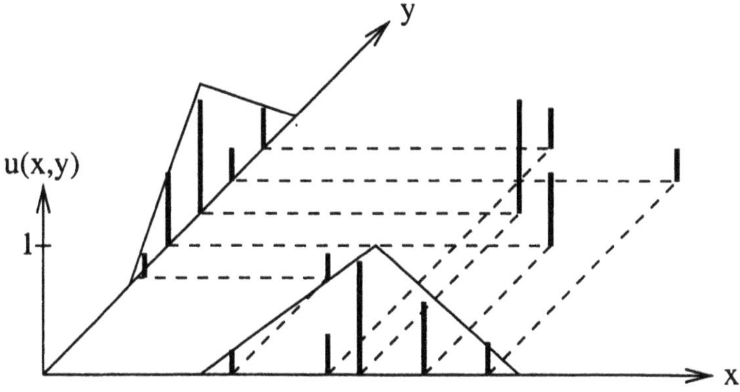

Fig. 4. A projection of a fuzzy cluster

and interpret the classifier, it is desirable to describe the fuzzy classifier in terms of fuzzy rules as they were introduced in the previous subsection. Projecting the fuzzy clusters and approximating the projections by triangular or trapezoidal fuzzy sets is a very common strategy to derive rules from fuzzy clusters [11,20]. In this way each cluster induces a rule using the fuzzy sets derived from the projections and the class associated with the cluster. Figure 4 illustrates the principle idea behind this concept.

Unfortunately, the projection and the approximation of the projections by standard membership functions enforces a certain loss of information, so that the classifier based on the rules does not always classify the data in the same way as it is done by the original multi-dimensional membership functions. When we look at the final classification after defuzzification, the only important parameters for the classification are the distances of the datum to be classified to the prototypes, no matter whether we apply probabilistic, noise or possibilistic clustering. A datum is assigned to the class of the nearest prototype. Therefore, the crisp partition induced by the prototypes has hyperplanes as boundaries between the sets of the partition.

These hyperplanes can be used to construct a fuzzy classifier that produces exactly the same class boundaries. In [19] a rough idea of how a fuzzy classifier can be constructed on the basis of hyperplanes is outlined. Since in [18] it is proved that a max-min classifier decides locally on the basis of two attributes, such a classifier can only model hyperplanes that are parallel to $n-2$ coordinates. To get more flexibility, we use the Łukasiewicz-t-norm instead of the minimum.

Note that the Łukasiewicz-t-norm is nilpotent, i.e. it is not only 0 if at least one of its arguments is 0. Therefore fuzzy sets in a rule base that is valuated by the Łukasiewicz-t-norm tend to be wider than in the case of the minimum. For a comparison of the Łukasiewicz-t-norm with the min-t-norm see figures 6 and 7.

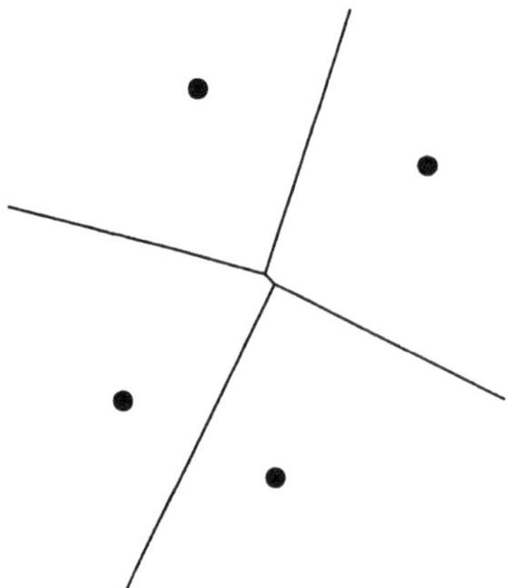

Fig. 5. Cluster boundaries

3 The Łukasiewicz-t-Norm

The Łukasiewicz-t-norm is defined by $\top_{Luk}(a_1, a_2) = \max\{0, a_1 + a_2 - 1\}$, so that the firing degree μ_R of the rules R for the datum $(x_1, \ldots, x_n)$ can be calculated by

$$\mu_R(x_1, \ldots, x_n) = \max\left\{0, \sum_{i=1}^{n} \mu_R^{(i)}(a_i) + 1 - n\right\}$$

with $\mu_R^{(i)}$ being the fuzzy set for the ith dimension for the rules R.

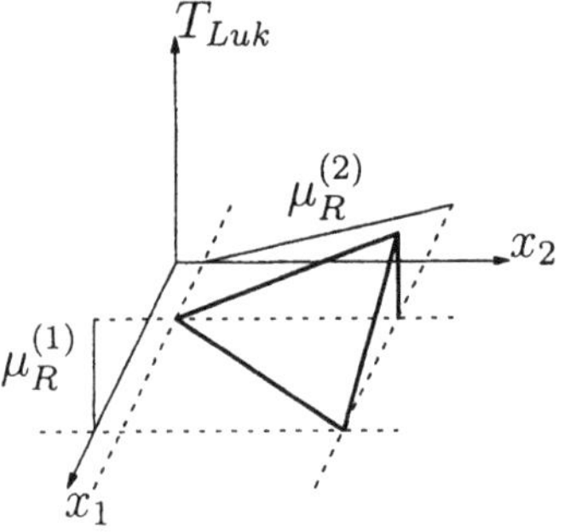

Fig. 6. The Łukasiewicz-t-norm.

Figure 6 illustrates, how the Łukasiewicz t-norm looks like in the two- dimensional case, when we use linear fuzzy sets. The vertical axis shows the firing degree of a rule that consists of the two fuzzy sets $\mu_R^{(1)}$ and $\mu_R^{(2)}$. Because

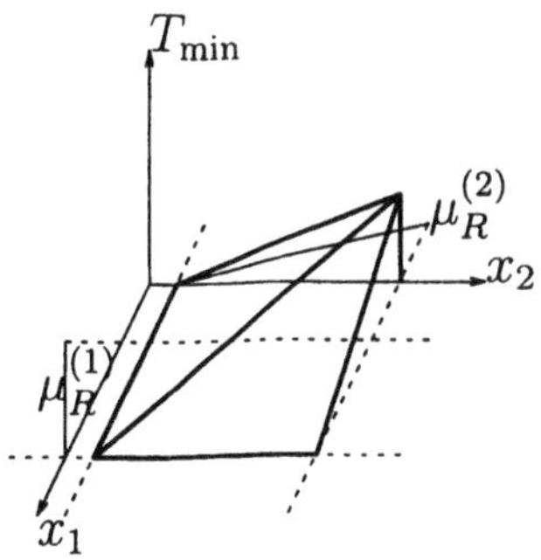

Fig. 7. The Min-t-norm.

of the max in the definition of the Lukasiewicz-t-norm the firing degree is 0 on the left side of the diagonal of the x_1-x_2-rectangle and increases then orthogonally to this line.
In the n-dimensional space such a rule starts with firing degree 0 at a hyperplane and then increases orthogonally to this hyperplane until it reaches the firing degree 1. When we have two different classes on both sides of a hyperplane H, then we just need two rules, one for the first class that starts with 0 at H and increases into the direction of the class that it represents, and the second rule to do the same into the other direction.
In this case at each time we have one rule with a firing degree greater than 0, but when there are several hyperplanes describing the classes, then we have to use several rules for one datum. This is to be described in the following.
We assume that we have only two classes. To get the more general case, we just have to consider one class against all the others and then continue with the second class against the remaining and so on.

4 Łukasiewicz Classification System with *h* Hyperplanes

We have already seen that the classification based on fuzzy clustering is characterised by hyperplanes. We want to construct a classifier that defines exactly the same hyperplanes as class boundaries.
As a first step we do not consider the whole data space but partition it into cuboids of the form $[a_1, b_1] \times \ldots \times [a_n, b_n]$. The cuboids are chosen in such a way that they contain h hyperplanes that intersect in one point and define a convex region. Inside this region we have one class, outside this region another class (compare figure 9). The point of intersection should be placed in one corner of the cuboid, but can also be situated outside.
Figure 8 illustrates the partition of the space for the two- dimensional case. We will treat this point more detailed in section 5.

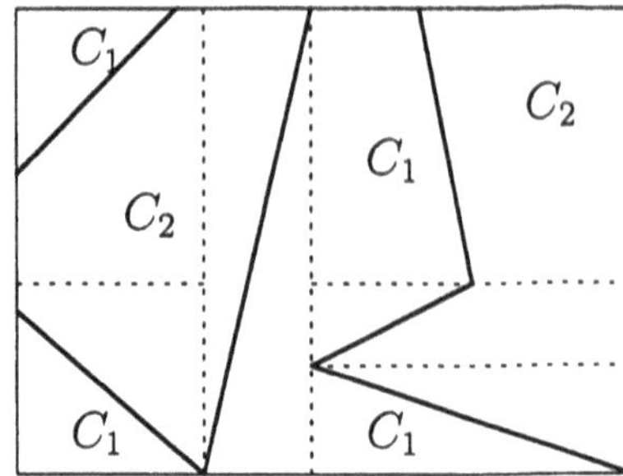

Fig. 8. An example how to partition the space into handy rectangles

In order to be able to illustrate the construction of the rules, we will explain the method for the 3-dimensional case, but the technique easily extends to higher dimensions.
It is possible to partition the space into several n-dimensional cuboids that have hyperplanes for the separation that do not bend. If they bend, we have to divide the cuboid again into several smaller ones as illustrated in figure 8. In order to consider the region as a bounded set, we sometimes have to consider the borders of the cuboid as separating hyperplanes.

4.1 Basic Principles

We have a region that is bounded by the different hyperplanes. The interior of the region belongs to one class, while the outer part belongs to another class. The most interesting points for the construction of the fuzzy classifier are those where in the n-dimensional space n hyperplanes meet. In the 3-dimensional space e.g. planes meet in lines, and these lines meet in points. The very special case that more than n hyperplanes meet in the same point will not be considered here.
We will depict the basic principles of the construction in the three-dimensional case. We have $n = 3$ hyperplanes that meet in one point P_s. They mark a section that belongs to one class, while the surroundings belong to another class (see figure 9). To construct the rules we use a straight line l_M that starts at P_s and is continued inside the section. We can draw it e.g. through the centre of gravity of the section. Figure 10 shows a cut through the section that is orthogonal to this straight line.
Now we construct two auxiliary planes for each plane. These planes mark the points, where the assigned rules adopt the value 0. The first auxiliary plane M_i includes l_M and its image in figure 10 is parallel to H_i, while B_i is situated in the middle between H_i and M_i. The rule R_{B_i} that starts with firing degree 0 at B_i increases twice as fast as R_{M_i}, so that it "overtakes" R_{M_i} exactly at H_i.

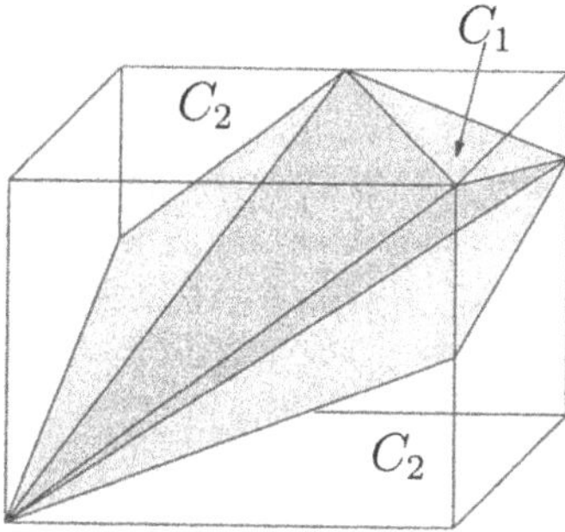

Fig. 9. Three hyperplanes partition the space into two classes- inside and outside the section that is marked by the planes.

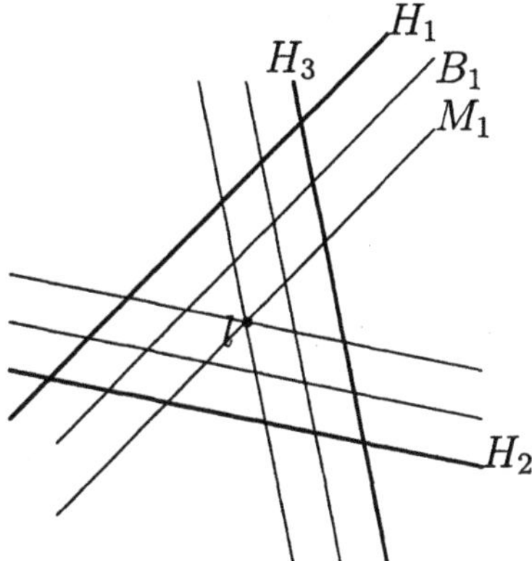

Fig. 10. A cut through the three-dimensional case orthogonally to the straight line l_M.

The resulting system consists of two rules for each hyperplane. One rule R_{M_i} gives a firing degree for the class inside the section. The other rule R_{B_i} has a firing degree lower than that one of R_{M_i} inside the section, but at H_i it adopts the same firing degree, and outside the section it is the winning rule itself.

When we construct these two rules for each hyperplane, we get a fuzzy classification system that solves our classification problem correctly. The next section describes the calculations for the construction of the rules in the n-dimensional case.

4.2 Steps for the Construction of the Classification System

Let P_s be the point where all the hyperplanes $H_1, \ldots, H_h$ meet. We want to distinguish between the section that is bounded by the hyperplanes and the

region outside this section. Two hyperplanes meet in one "line" (hyperplane of dimension $n-2$). These lines of intersection between two hyperplanes are called l_{ij}, $i, j \in \{1, \ldots, n\}$. As we just consider one section marked by the hyperplanes, there are only h lines that are relevant. Each hyperplane is only involved in the definition of two lines, those where it meets its neighbour.
First of all we have to calculate the centre of gravity of the marked section. We need the vectors x_i, $i = 1, \ldots, h$, that are directed from P_s to the point, where l_i meets the border of the rectangle. Then the centre of gravity P_g is calculated by

$$P_g := P_s + \frac{1}{h} \cdot \sum_{i=1}^{h} x_i.$$

By drawing a line between P_s and P_g we get a line

$$l_M : x = P_s + \alpha(P_g - P_s), \qquad (\alpha \in \mathbb{R})$$

that is situated inside the section that we want to describe by the fuzzy classification system.
Now we have to construct the auxiliary planes. The following construction has to be done for each hyperplane H_i separately.

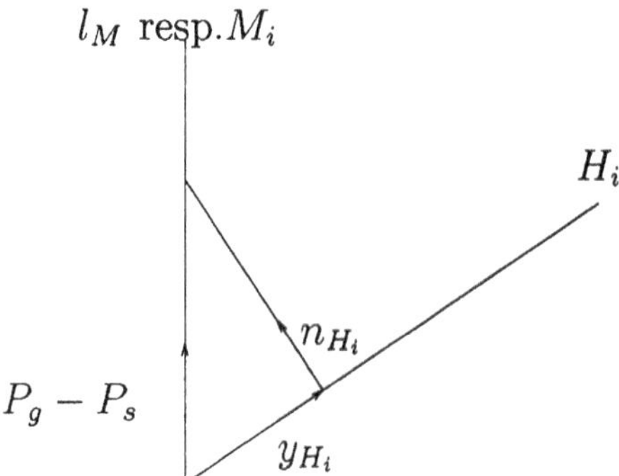

Fig. 11. y_{H_i} is a linear combination of $(P_g - P_s)$ and n_{H_i}.

Construction of the Auxiliary Planes First we need the vector y_{H_i} that belongs to H_i, and that can be written as a linear combination of n_{H_i} and $(P_g - P_s)$ with n_{H_i} being the normal vector of H_i:

$$y_{H_i} = \alpha \cdot n_{H_i} + \beta \cdot (P_g - P_s), \qquad (\alpha, \beta \in \mathbb{R}).$$

The principle can be seen in figure 11, that shows a cut through l_M and H_i. If $\alpha > 0$, then n_{H_i} is pointing into the direction of l_M, otherwise into the other direction.

Now we construct an orthogonal basis for H_i that includes y_{H_i}. Replacing in this basis y_{H_i} by $(P_g - P_s)$, we get a basis for the auxiliary plane M_i. We calculate the normal form $0 = n_{M_i} \cdot x + d_{M_i}$ of M_i with the normal vector n_{M_i}. When choosing a point P_{H_i} of H_i, we calculate $p_{H_i} = n_{M_i} \cdot P_{H_i} + d$. If $p_{H_i} > 0$, then n_{M_i} is pointing into the direction of H_i, otherwise into the other direction. The same can be done for a point P_{M_i} of M_i.
For B_i has to be in the middle between H_i and M_i, we use the normal vectors n_{B_i} and n_{H_i} of the two planes. The normal vector n_{B_i} of B_i is calculated by

$$n_{B_i} := \frac{n'_{M_i} - n'_{H_i}}{| n'_{M_i} - n'_{H_i} |}$$

if n'_{H_i} is pointing towards M_i and n'_{M_i} towards H_i. This can be achieved by using $n'_{M_i} = \mathrm{sgn}(p_{M_i}) \cdot n_{M_i}$ and $n'_{H_i} = \mathrm{sgn}(p_{H_i}) \cdot n_{H_i}$.
As P_s has to belong to B_i, we can calculate $d_{B_i} = -P_s \cdot n_{B_i}$. Then B_i is described by

$$n_{B_i} \cdot x + d_{B_i} = 0.$$

Construction of the Rule R_{M_i} Now we have to determine the rules R_{M_i} and R_{B_i}, that belong to the two planes M_i and B_i. The firing degrees of the rules have to start with $\mu = 0$ at M_i resp. B_i.
As the t-norm is the Łukasiewicz t-norm, we calculate R_{M_i} and R_{B_i} by

$$R_{M_i}(x_1, \ldots, x_n) = \sum_{t=1}^{n} \mu^{(t)}_{R_{M_i}}(x_t) + 1 - n \text{ and } R_{B_i}(x_1, \ldots, x_n) = \sum_{t=1}^{n} \nu^{(t)}_{R_{B_i}}(x_t) + 1 - n.$$

First of all we construct the rule R_{M_i} that has to start with $R_{M_i}(X) = 0$ at any $X \in M_i$ and to increase until it reaches $R_{M_i}(P_i) = 1$ at the corner $P_i = (p_1, \ldots, p_n)$ of the cuboid.
The fuzzy degrees have to be between 0 and 1, therefore all fuzzy degrees have to be 1 in P_i to fulfil $\sum_{t=1}^{n} \mu^{(t)}_{R_{M_i}}(p_t) + 1 - n = 1$. As we want to have linear fuzzy sets, we choose

$$\mu^{(t)}_{R_{M_i}}(x_t) = \begin{cases} 1 - \alpha_t \cdot (x_t - p_t) & \text{if } p_t = a_t \\ 1 - \alpha_t \cdot (p_t - x_t) & \phantom{\text{if }} p_t = b_t \end{cases} \tag{7}$$

with $[a_1; b_i] \times \ldots \times [a_n; b_n]$ being the cuboid. Then the $\alpha_t, t = 1, \ldots, n$, are the unknown values of the fuzzy sets and have to stay between 0 and 1. We can denote this by

$$\mu^{(t)}_{R_{M_i}}(x_t) = 1 + \alpha'_t \cdot (x_t - p_t) \tag{8}$$

with $\alpha'_t = \alpha_t$ if $p_t = a_t$ and $\alpha'_t = -\alpha_t$ if $p_t = b_t$. Now we have to calculate the values α'_t. Let the hyperplane M_i be described by

$$\sum_{t=1}^{n} \gamma_t \cdot x_t + c = 0. \tag{9}$$

The values γ_t are the components of the normal vector n_{M_i} of M_i. We multiply the equation with $\gamma := \frac{-1}{c+\sum_{t=1}^{n}\gamma_t \cdot p_t}$, so that we get

$$\sum_{t=1}^{n} \gamma_t' \cdot x_t - \frac{c}{c+\sum_{t=1}^{n}\gamma_t \cdot p_t} = 0 \tag{10}$$

with $\gamma_t' = \gamma \cdot \gamma_t$ instead of equation (9). As the firing degree of the rule has to be 0 at the hyperplane M_i, this rule has to fulfil the condition

$$\sum_{t=1}^{n} \mu_{R_{M_i}}^{(t)}(x_t)+1-n = \sum_{t=1}^{n}(1-\alpha_t'\cdot(x_t-p_t))+1-n = -\sum_{t=1}^{n}\alpha_t'\cdot(x_t-p_t)+1 = 0 \tag{11}$$

Therefore we define $\alpha_t' := \gamma_t'$. With this construction, the equations (10) and (11) are equivalent:

$$\begin{aligned}
&\mu_{R_{M_i}}(x_1,\dots,x_n) = 0 && \Leftrightarrow \\
&\textstyle\sum_{t=1}^{n}\alpha_t'\cdot x_t - 1 - \sum_{t=1}^{n}\alpha_t \cdot p_t = \gamma\sum_{t=1}^{n}\gamma_t\cdot x_t - 1 - \gamma\sum_{t=1}^{n}\gamma_t\cdot p_t && = \\
&\textstyle\gamma(\sum_{t=1}^{n}\gamma_t\cdot x_t - (-c-\sum_{t=1}^{n}\gamma_t\cdot p_t) - \sum_{t=1}^{n}\gamma_t\cdot p_t) && = \\
&\textstyle\gamma(\sum_{t=1}^{n}\gamma_t\cdot x_t + c) = 0 && \\
&\textstyle\Leftrightarrow \sum_{t=1}^{n}\gamma_t\cdot x_t + c && = 0 \\
&\Leftrightarrow (x_1,\dots,x_n) \in M_1, &&
\end{aligned}$$

so that the $\mu_{R_{M_i}}^{(t)}$ define a rule that starts at the hyperplane M_i with firing degree 0 and increases until it reaches firing degree 1 at the point P_i.
The result is the same, if we have the the normal vector of M_i pointing into the opposite direction, i.e. when we have $-n_{M_i}$ instead of n_{M_i}.
The fuzzy sets that we have constructed also take values that do not belong to $[0,1]$. By scaling and cutting them in the very end we obtain fuzzy sets that range between 0 and 1. This will be described in section 4.2.

Construction of the Rule R_{B_i} Now we turn towards the other rule R_{B_i} that has to start at the hyperplane B_i and increase faster than R_{M_i} until it "overtakes" R_{M_i} at the hyperplane H_i. For the construction we first consider the fuzzy sets ν_i to be linear, i.e. we also allow them to adopt values below 0 or above 1. Later we cut them at 0 and 1 so that we get

$$\nu_{R_{M_i}}^{(t)}(x) = \max\{0, \min\{1, b_t + \beta_t \cdot x\}\}.$$

In figure 12 we illustrate how these rules behave. The horizontal line represents the way from M_i to P_i passing the other two auxiliary planes, while the vertical axis shows the firing degree of the rules (R_{M_i} and R_{B_i})

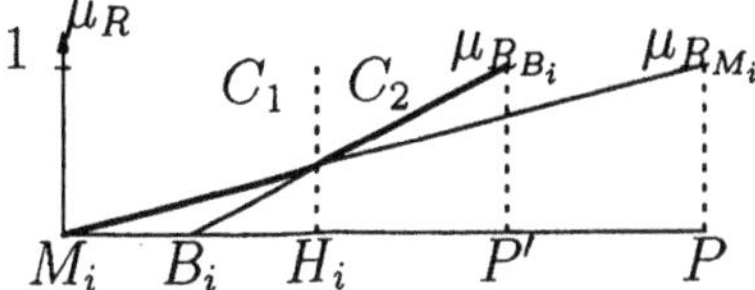

Fig. 12. The rule for the class C_2 'overtakes' the rule for C_2.

We can use the same construction as in the previous section, if we have a point P_i' that fulfils the function that P_i has for R_{M_i}, i.e. that the rule reaches firing degree 1 at P_i'. As B_i is situated in the middle between M_i and H_i, and as R_{B_i} has to reach the same firing degree at H_i as R_{M_i}, the rule R_{B_i} has to increase twice as fast as R_{M_i}.

For the calculation of P_i', we have to consider the construction as shown in figure 13. Let P_i' lie in the plane that is orthogonal to M_i, B_i and H_i and that includes P_i. Let S_{P_i} be the point where H_i and the line from P_i to M_i, that is orthogonal to M_i, meet. The rules R_{M_i} and R_{B_i} are to have the same firing degree on H_i and therefore also in S_{P_i}.

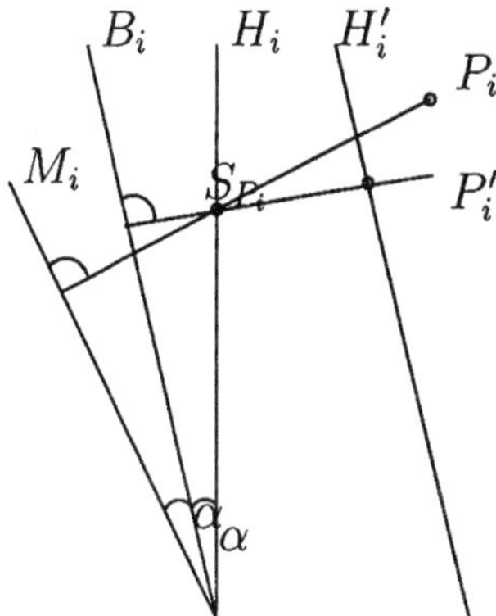

Fig. 13. An orthogonal cut through the planes and P_i.

Now we have to put P_i' that way that

$$\frac{\text{dist}(P_i, M_i)}{\text{dist}(S_{P_i}, M_i)} = \frac{\text{dist}(P_i', B_i)}{\text{dist}(S_{P_i}, B_i)}$$

with the distance $\text{dist}(P_i, H) = P_i \cdot n_{H_i} + d_H$ if a hyperplane H is described by the normal form $P_i \cdot n_{H_i} + d_H = 0$.

$$\frac{\text{dist}(P_i, M_i)}{\text{dist}(S_{P_i}, M_i)} = \frac{\text{dist}(P_i', B_i)}{\text{dist}(S_{P_i}, B_i)}$$

$$\Leftrightarrow (S_{P_i} \cdot n_{B_i} + d_{B_i})(P_i \cdot n_{M_i} + d_{M_i}) = (S_{P_i} \cdot n_{M_i} + d_{M_i})(P_i' \cdot n_{B_i} + d_{B_i})$$

$$\Leftrightarrow \qquad P_i' \cdot n_{B_i}(S_{P_i} \cdot n_{M_i} + d_{M_i}) = \frac{(S_{P_i} \cdot n_{B_i} + d_{B_i})(P_i \cdot n_{M_i} + d_{M_i})}{(S_{P_i} \cdot n_{M_i} + d_{M_i})} - d_{B_i}$$

As the right side of the last equation is a scalar, we get the normal form of a hyperplane H_i' that is parallel to B_i. The firing degree of the rule R_{B_i} is increasing orthogonally to B_i until it reaches 1 at H_i'.
Now we can choose the point $P_i' \in H_i'$. P_i has to belong to H_i' and to be on the line $S'_{P_i} + \alpha \cdot n_{B_i}$, $\alpha \in \mathbb{R}$. When we construct the rule R_{B_i} the same way as we have constructed R_{M_i} in the previous section starting with firing degree 0 at B_i and increasing until it reaches 1 at P_i', then also

$$R_{M_i}(a_1, \ldots, a_n) = R_{B_i}(a_1, \ldots, a_n) \text{ for all } (a_1, \ldots, a_n) \in H_i$$

is fulfilled.

Scaling the Rules Now we have the two rules that we needed to describe the classification performed by the hyperplane H_i. We have to do this for all the hyperplanes $H_1, \ldots, H_h$ separately and then combine them by using the maximum as t-conorm. Now we have to make sure that the rules R_{M_i} and R_{M_j} (resp. R_{B_i} and R_{B_j}) do not disturb each other when they collide at the l_{ij}. Therefore the rules for the two hyperplanes that meet at l_{ij} must have the same firing degree. Anyway the two rules for one hyperplane H_i have the same firing degree $R_{B_i}(a_1, \ldots, a_n) = R_{M_i}(a_1, \ldots, a_n)$ for each point $(a_1, \ldots, a_n)$ of H_i. Now we require the rules for the two hyperplanes H_i and H_j to have the same membership degree $R_{M_i}(a_1, \ldots, a_n) = R_{M_j}(a_1, \ldots, a_n)$ for each point $(a_1, \ldots, a_n)$ of l_{ij}. We can choose any point for our procedure. This can be achieved by scaling the fuzzy sets for the rules. The aim is that a rule R_{M_i} reaches firing degree δ_i in P_i and rule R_{B_i} reaches δ_i in P_i' instead of 1. δ_i has to be in $]0; 1]$ and $\max_{i \in \{1, \ldots, h\}}\{\delta_i\} = 1$.
If we just consider two hyperplanes H_i and H_j meeting at l_{ij} with

$$s_{ij} := \frac{R_{M_i}(a_1, \ldots, a_n)}{R_{M_j}(a_1, \ldots, a_n)}$$

for any $(a_1, \ldots, a_n) \in l_{ij}$, then s_{ij} would be the scaling multiplier. We determine the scaling multiplier for each l_k. By using the equation $s_{il} := s_{ij} \cdot s_{jl}$ we calculate the other scaling multipliers, so that we have one for each pair of hyperplanes. We determine $s := \max_{i,j}\{s_{ij}\} = s_{pq}$, and then the p tells us

the hyperplane H_p that stays the same, while the other H_j, $j \neq p$, are to be scaled with $s_{pj} < 1$.
This means that they are to reach the firing degree s_{pj} in P_i instead of the firing degree 1. As changing the firing degree of a rule in a point results in a complex system of equations, the easiest way to achieve this is to do the same construction as we described it in the previous sections for a point P_i (and resp. P_i') for a new point $\bar{P}_i$ (resp. $\bar{P}_i'$).that is situated more closely to M_i. Let S_{M_i} be the orthogonal projection of P_i on M_i. If τ is defined by $S_{M_i} + \tau \cdot n_{M_i} = P_i$, then we choose a point $\bar{P}_i(S_{M_i} + s_{pj} \cdot \tau \cdot n_{M_i})$ instead of P_i to construct the rules.
If for the point P_i the construction would result in a rule with the fuzzy sets $\mu^{(t)}_{RM_i}(x_t) = 1 - \alpha_t(x_t - p_t)$, then the rule for $\bar{P}_i$ would result in

$$\bar{\mu}^{(t)}_{RM_i}(x_t) = 1 - \alpha_t \cdot \frac{\bar{\gamma}}{\gamma} \cdot (x_t - p_t + \tau \cdot (1 - s_{pj}) \cdot n^{(t)}_{M_i})$$

with $\bar{\gamma} := (c + \sum_{t=1}^{n} \gamma_t(p_t - \tau \cdot (1 - s_{pj}) \cdot n^{(t)}_{M_i}))^{-1}$ and $n^{(t)}_{M_i}$ being the tth coordinate of the normal vector n_{M_i} of the hyperplane M_i. This can easily be shown by calculating

$$\begin{aligned} \mu_{\bar{R}M_i}(x_1, \ldots, x_n) &= \textstyle\sum_{t=1}^{n} \bar{\mu}^{(t)}_{RM_i}(x_t) + 1 - n \\ &= n - \bar{\gamma} \textstyle\sum_{t=1}^{n} \gamma_t(x_t - p_t + b \cdot n^{(t)}_{M_i}) + 1 - n \\ &= 1 + \frac{\sum \gamma_t \cdot x_t - \sum \gamma_t \cdot (p_i - b \cdot n^{(t)}_{M_i})}{c + \sum \gamma_t \cdot (p_i - b \cdot n^{(t)}_{M_i})} \\ &= 1 + \frac{c + \sum \gamma_t \cdot x_t}{c + \sum \gamma_t \cdot (p_i - b \cdot n^{(t)}_{M_i})} \end{aligned}$$

with $b := \tau \cdot (1 - s_{pj})$. $\mu_{\bar{R}M_i}$ is a linear function. For all points $(a_1, \ldots, a_n) \in M_i$ we have $\sum_{t=1}^{n} \gamma_t \cdot a_i + c = 0$ and then we get

$$\mu_{\bar{R}M_i}(a_1, \ldots, a_n) = \frac{0}{c + \sum \gamma_t \cdot (p_i - b \cdot n^{(t)}_{M_i})} = 0,$$

at M_i and for the point $\bar{P}_i = P_i - b \cdot n_{M_i}$ we get

$$\mu_{\bar{R}M_i} = 1 + \frac{c + \sum \gamma_t \cdot (p_t - b \cdot n^{(t)}_{M_i})}{c + \sum \gamma_t \cdot (p_i - b \cdot n^{(t)}_{M_i})} = 1.$$

When having done this with each rule the firing degree of two rules that meet at an l_i is the same everywhere on l_i. This guarantees that those two rules do not disturb each other, when we choose the maximum-t-conorm, so that also the combination of the rules results in the requested classification.
Then inside the section that is bordered by the H_i, $i = 1, \ldots, h$, we have the first class for that all the R_{M_i} are firing and outside the section we have the other class.

The only thing still to be done is do make sure that the membership degrees oft he fuzzy set stay in [0, 1] by calculating

$$\mu^{(t)}_{\tilde{R}_{M_i}}(x_t) = \max\{0, \min\{1, \mu^{(t)}_{\tilde{R}_{M_i}}(x_t)\}\}$$

and resp. $\mu^{(t)}_{\tilde{R}_{B_i}}$. We do not not get a change in the classification: By the time when the firing degree of a rule reaches 1, all the fuzzy sets have already membership degree 1, and from this on, we simply stay at this firing degree. And at the point, where a fuzzy set has membership degree 0, we have already a firing degree 0 for the rule, so that beyond it stays 0 anyway.

5 Segmentation into Cuboids

Until now we just considered one cuboid of the dimension n with hyperplanes of the dimension $n-1$ that go from one side of the rectangle to the other side. The main point is that we have to classify the space by much more complicated separations, that can be combined of many hyperplanes as we have in the case of a piecewise linearly separable problem.
To use the construction that we developed for one cuboid, we have to partition the space into several cuboids that fulfil our conditions.

5.1 Regular Grid

We can cover the space with a regular grid. We mark those points where a hyperplanes does not have to be continued, so that from this point on the separation follows a different hyperplane. Then we construct a grid, that is parallel to the coordinates and that includes all those points. An example can be seen in figure 14. In this example we need fifteen rectangles to cover the space.

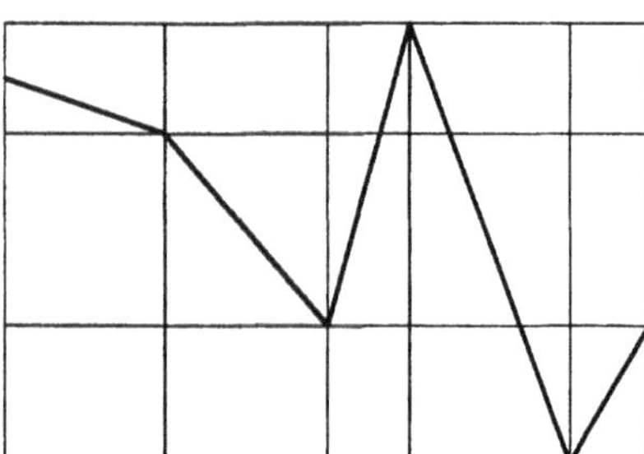

Fig. 14. An example for covering the space with a regular grid.

This method has the advantage, that the resulting fuzzy sets can be transformed into uniformly distributed fuzzy sets by stretching the coordinates, but it can result in an awful lot of cuboids and with a lot of cuboids we get many rules.

5.2 As few cuboids as Possible

There is another possibility. Here we just construct a sort of minimal grid. Again we determine the points, where the directions of the separation planes change, but we just draw a part of the grid until we reach the next part of the grid resp. the next coordinate of one of the determined points. The algorithm for constructing this minimal grid is a separate problem and will not be discussed here.
When we consider the same example as in figure 14, then we just need five rectangles instead of fifteen as we can see in figure 15. The minimal grid is not unique. We could also only use vertical devisions in our example.
This method using only few cuboids has the advantage of constructing only few rules. This makes it faster when calculating the rules themselves, but we need more time to determine the cuboids that are to be considered. Additionally the construction results into fuzzy rules that are not uniformly distributed on the coordinates.

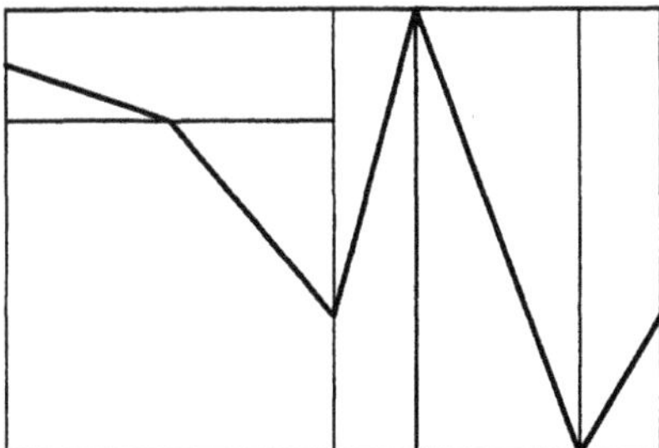

Fig. 15. The same example when using a 'minimal' grid.

6 An Example: The Iris Data Set

To illustrate our procedure, we use the well known iris data [5]. The dataset contains three classes of 50 instances each. The classes are types of iris plants (iris setosa, iris versicolor, iris virginica). There are four input attributes, but clustering only with the last three attributes gives nearly as good results as using all four of them [9]. We refer to the used attributes as x, y and z. The range of these attributes defines the cuboid $[2, 4.4] \times [1, 6.9] \times [0.1, 2.5]$ that is to be considered.

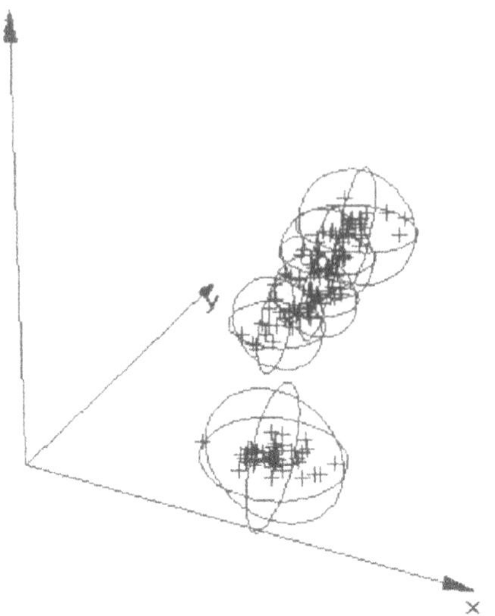

Fig. 16. The Iris Data and its five clusters.

By using the fuzzy c-means algorithm we get five clusters as depicted in figure 16. The prototypes are the following.

$$c_1 = \begin{pmatrix} 3.43 \\ 1.46 \\ 0.25 \end{pmatrix}, c_2 = \begin{pmatrix} 2.52 \\ 3.82 \\ 1.15 \end{pmatrix}, c_3 = \begin{pmatrix} 2.93 \\ 4.53 \\ 1.43 \end{pmatrix}, c_4 = \begin{pmatrix} 2.89 \\ 5.20 \\ 1.93 \end{pmatrix}, c_5 = \begin{pmatrix} 3.13 \\ 5.99 \\ 2.16 \end{pmatrix}.$$

The cluster with the prototype c_1 belongs to class 0, c_2 and c_3 define clusters of class 1 and the clusters of c_4 and c_5 belong to class 3. Therefore we just have to consider the hyperplanes between c_1 and c_2 and between c_3 and c_4. These hyperplanes are

$$H_1 : \begin{pmatrix} -0.34 \\ 0.88 \\ 0.34 \end{pmatrix} \cdot x - 1.55 = 0 \text{ and } H_2 : \begin{pmatrix} -0.05 \\ 0.80 \\ 0.59 \end{pmatrix} \cdot x - 4.75 = 0$$

In figure 17 the prototypes of the clusters are depicted together with the separating hyperplanes between the first and second prototype and between the third and fourth.
We calculate the centre of gravity P_g out of the points, where H_1 and H_2 intersect the surface of the cuboid. H_1 and H_2 do not intersect inside the cuboid, and as we have only two hyperplanes, we use the y-z-boundary of the cuboid as a third hyperplane to determine the intersection point P_s.

$$P_g = \begin{pmatrix} 3.20 \\ 3.83 \\ 1.30 \end{pmatrix} \quad \text{and} \quad P_s = \begin{pmatrix} 2.0 \\ -1.304 \\ 9.938 \end{pmatrix}.$$

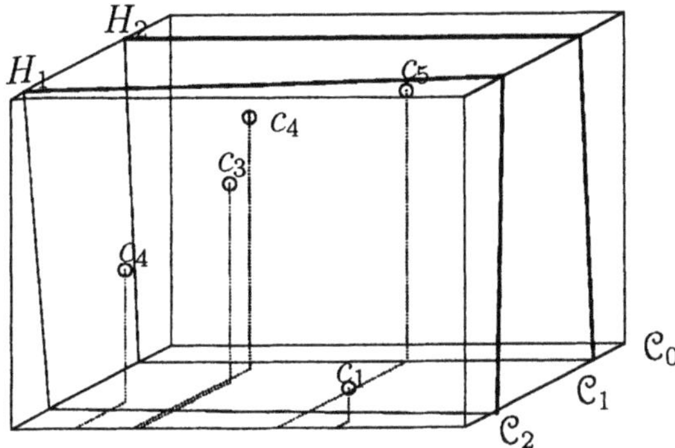

Fig. 17. The prototypes of the clusters and the hyperplanes.

The next step is to calculate the auxiliary planes M_1 and M_2, and with their help B_1 and B_2:

$$M_1 : \begin{pmatrix} -0.39 \\ 0.81 \\ 0.43 \end{pmatrix} \cdot x - 2.42 = 0 \text{ and } M_2 : \begin{pmatrix} -0.07 \\ 0.86 \\ 0.50 \end{pmatrix} \cdot x - 3.72 = 0$$

$$B_1 : \begin{pmatrix} -0.37 \\ 0.85 \\ 0.38 \end{pmatrix} \cdot x - 1.98 = 0 \text{ and } B_2 : \begin{pmatrix} -0.06 \\ 0.83 \\ 0.55 \end{pmatrix} \cdot x - 4.24 = 0$$

It is obvious that the point $P_1 = (4.4,\ 1.0,\ 0.1)$ is chosen to calculate the rules R_{M_1} and $P_2 = (2.0,\ 6.9,\ 2.5)$ for the R_{M_2}. The points $P_1' = (3.95,\ 1.83,\ 0.61)$ and $P_2' = (2.09,\ 5.93,\ 2.00)$ are closer towards the planes, so that the fuzzy sets of the rules R_{B_1} and R_{B_2} turn out to have a greater slope by absolute value than those of R_{M_1} and R_{M_2}. We obtain the following rules:

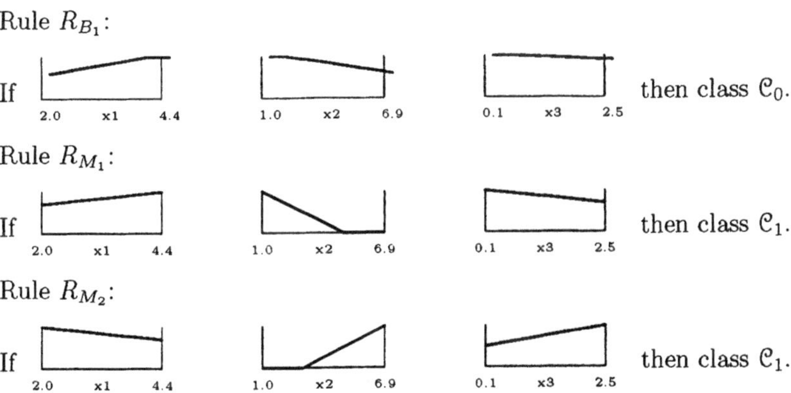

Rule R_{B_2}:

If [2.0 x1 4.4] [1.0 x2 6.9] [0.1 x3 2.5] then class $\mathcal{C}_2$.

Now we have a rule base for a fuzzy classification system that classifies the iris data. The only misclassified data are those that have a greater distance to the prototype of their own class than to another prototype. This fuzzy classification system computes exactly the same classification as the fuzzy clustering does.

7 Conclusions

Fuzzy clustering can be applied to unsupervised and even to supervised classification problems. However the description of classes in terms of cluster prototypes and multidimensional membership functions is not always suitable for interpreting the classification system. Therefore, a number of techniques to derive fuzzy classification rules from the clusters have been proposed. These approaches usually project the clusters onto the single attributes and accept a loss of information leading to a less accurate classifier.
In this paper we have introduced a method that avoids projection and uses the class boundaries directly to derive the classification rules from the clusters. In this way an interpretable rule-based classifier is obtained that maintains the accuracy of the original clusters.
It should be emphasised that our method is restricted to fuzzy c-means clusters and cannot be extended to ellipsoidal clusters as they are computed e.g. in the Gustafson-Kessel algorithm [7], since in this case the cluster boundaries are no longer hyperplanes.

References

1. J.C. Bezdek: Pattern recognition with fuzzy objective function algorithms, Plenum Press, New York (1981)
2. O. Cordón, M. José del Jesus, F. Herrera: Analysing the reasoning mechanism in fuzzy rule based classification systems. Mathware & Soft Computing 5 (1998), 321-332
3. R.N. Davé: Characterisation and detection of noise in clustering. Pattern Recognition Letters 12 (1991), 657-664
4. R.N. Davé, R. Krishnapuram: Robust clustering methods: A unified view, IEEE Transactions on Fuzzy Systems 5 (1997), 270-293
5. R.A. Fisher: The use of multiple measurements in taxonomic problems. Annals of Eugenics 7 (1936), 179-188

6. H. Genther, M. Glesner: Automatic generation of a fuzzy classification system using fuzzy clustering methods. Proc. ACM Symposium on Applied Computing (SAC'94), Phoenix (1994), 180-183
7. D. Gustafson, W. Kessel: Fuzzy clustering with a fuzzy covariance matrix. Proc. IEEE CDC, San Diego (1979), 761-766
8. F. Höppner, F. Klawonn, R. Kruse, T. Runkler: Fuzzy cluster analysis. Wiley, Chichester (1999)
9. A. Keller, F. Klawonn: Fuzzy clustering with weighting of data variables. International Journal of Uncertainty, Fuzziness and Knowledge-Based Systems 8 (2000), 735-746
10. F. Klawonn, E.P. Klement: Mathematical analysis of fuzzy classifiers. In: Mathematical analysis of fuzzy classifiers. In: X. Liu, P. Cohen, M. Berthold (eds.): Advances in intelligent data analysis. Springer, Berlin (1997), 359-370
11. F. Klawonn, R. Kruse: Derivation of fuzzy classification rules from multidimensional data. In: G.E. Lasker, X. Liu (eds.): Advances in intelligent data analysis. The International Institute for Advanced Studies in Systems Research and Cybernetics, Windsor, Ontario (1995), 90-94
12. R. Krishnapuram, J. Keller: A Possibilistic approach to clustering. IEEE Transactions on Fuzzy Systems 1 (1993), 98-110
13. L.I. Kuncheva: How good are fuzzy if-then classifiers? IEEE Transactions on Systems, Man and Cybernetics 30, Part B (2000), 501-509
14. K.D. Meyer Gramann: Fuzzy classification: An overview. In: R. Kruse, J. Gebhardt, R. Palm (eds.): Fuzzy systems in computer science. Vieweg, Braunschweig (1994), 277-294
15. A. Nürnberger, A. Klose, R. Kruse: Discussing cluster shapes of fuzzy classifiers. Proc. 18th Conf. of the North American Fuzzy Information Processing Society (NAFIPS'99), New York (1999), 546-550
16. A. Nürnberger, A. Klose, R. Kruse: Analyzing borders between partially contradicting fuzzy classification rules. Proc. 19th Conf. of the North American Fuzzy Information Processing Society (NAFIPS'00), Atlanta (2000), 59-63
17. W. Pedrycz: Algorithms of fuzzy clustering with partial supervision. Pattern Recognition Letters 23 (1985), 13-20
18. B. von Schmidt, F. Klawonn: Fuzzy max-min classifiers decide locally on the basis of two attributes. Mathware and Soft Computing 6 (1999), 91-108
19. B. von Schmidt, F. Klawonn: Construction of fuzzy classification systems with the Łukasiewicz-t-norm. Proc. 19th Conf. of the North American Fuzzy Information Processing Society (NAFIPS'00), Atlanta (2000), 109-113
20. M. Sugeno, T. Yasukawa: A fuzzy logic-based approach to qualitative modelling. IEEE Transactions on Fuzzy Systems 1 (1993), 7-31

www.ingramcontent.com/pod-product-compliance
Ingram Content Group UK Ltd.
Pitfield, Milton Keynes, MK11 3LW, UK
UKHW021902190726
13853UKWH00003B/1387